T0205753

Lecture Notes in Computer Science 13804

Founding Editors

Gerhard Goos
Karlsruhe Institute of Technology, Karlsruhe, Germany

Juris Hartmanis
Cornell University, Ithaca, NY, USA

Editorial Board Members

Elisa Bertino
Purdue University, West Lafayette, IN, USA

Wen Gao
Peking University, Beijing, China

Bernhard Steffen
TU Dortmund University, Dortmund, Germany

Moti Yung
Columbia University, New York, NY, USA

More information about this series at https://link.springer.com/bookseries/558

Leonid Karlinsky · Tomer Michaeli ·
Ko Nishino (Eds.)

Computer Vision – ECCV 2022 Workshops

Tel Aviv, Israel, October 23–27, 2022
Proceedings, Part IV

 Springer

Editors
Leonid Karlinsky
IBM Research - MIT-IBM Watson AI Lab
Massachusetts, USA

Tomer Michaeli
Technion – Israel Institute of Technology
Haifa, Israel

Ko Nishino
Kyoto University
Kyoto, Japan

ISSN 0302-9743   ISSN 1611-3349 (electronic)
Lecture Notes in Computer Science
ISBN 978-3-031-25068-2   ISBN 978-3-031-25069-9 (eBook)
https://doi.org/10.1007/978-3-031-25069-9

© The Editor(s) (if applicable) and The Author(s), under exclusive license
to Springer Nature Switzerland AG 2023
This work is subject to copyright. All rights are reserved by the Publisher, whether the whole or part of the material is concerned, specifically the rights of translation, reprinting, reuse of illustrations, recitation, broadcasting, reproduction on microfilms or in any other physical way, and transmission or information storage and retrieval, electronic adaptation, computer software, or by similar or dissimilar methodology now known or hereafter developed.
The use of general descriptive names, registered names, trademarks, service marks, etc. in this publication does not imply, even in the absence of a specific statement, that such names are exempt from the relevant protective laws and regulations and therefore free for general use.
The publisher, the authors, and the editors are safe to assume that the advice and information in this book are believed to be true and accurate at the date of publication. Neither the publisher nor the authors or the editors give a warranty, expressed or implied, with respect to the material contained herein or for any errors or omissions that may have been made. The publisher remains neutral with regard to jurisdictional claims in published maps and institutional affiliations.

This Springer imprint is published by the registered company Springer Nature Switzerland AG
The registered company address is: Gewerbestrasse 11, 6330 Cham, Switzerland

Foreword

Organizing the European Conference on Computer Vision (ECCV 2022) in Tel-Aviv during a global pandemic was no easy feat. The uncertainty level was extremely high, and decisions had to be postponed to the last minute. Still, we managed to plan things just in time for ECCV 2022 to be held in person. Participation in physical events is crucial to stimulating collaborations and nurturing the culture of the Computer Vision community.

There were many people who worked hard to ensure attendees enjoyed the best science at the 17th edition of ECCV. We are grateful to the Program Chairs Gabriel Brostow and Tal Hassner, who went above and beyond to ensure the ECCV reviewing process ran smoothly. The scientific program included dozens of workshops and tutorials in addition to the main conference and we would like to thank Leonid Karlinsky and Tomer Michaeli for their hard work. Finally, special thanks to the web chairs Lorenzo Baraldi and Kosta Derpanis, who put in extra hours to transfer information fast and efficiently to the ECCV community.

We would like to express gratitude to our generous sponsors and the Industry Chairs Dimosthenis Karatzas and Chen Sagiv, who oversaw industry relations and proposed new ways for academia-industry collaboration and technology transfer. It's great to see so much industrial interest in what we're doing!

Authors' draft versions of the papers appeared online with open access on both the Computer Vision Foundation (CVF) and the European Computer Vision Association (ECVA) websites as with previous ECCVs. Springer, the publisher of the proceedings, has arranged for archival publication. The final version of the papers is hosted by SpringerLink, with active references and supplementary materials. It benefits all potential readers that we offer both a free and citeable version for all researchers, as well as an authoritative, citeable version for SpringerLink readers. Our thanks go to Ronan Nugent from Springer, who helped us negotiate this agreement. Last but not least, we wish to thank Eric Mortensen, our publication chair, whose expertise made the process smooth.

October 2022

Rita Cucchiara
Jiří Matas
Amnon Shashua
Lihi Zelnik-Manor

Preface

Welcome to the workshop proceedings of the 17th European Conference on Computer Vision (ECCV 2022). This year, the main ECCV event was accompanied by 60 workshops, scheduled between October 23–24, 2022. We received 103 workshop proposals on diverse computer vision topics and unfortunately had to decline many valuable proposals because of space limitations. We strove to achieve a balance between topics, as well as between established and new series. Due to the uncertainty associated with the COVID-19 pandemic around the proposal submission deadline, we allowed two workshop formats: hybrid and purely online. Some proposers switched their preferred format as we drew near the conference dates. The final program included 30 hybrid workshops and 30 purely online workshops. Not all workshops published their papers in the ECCV workshop proceedings, or had papers at all. These volumes collect the edited papers from 38 out of the 60 workshops. We sincerely thank the ECCV general chairs for trusting us with the responsibility for the workshops, the workshop organizers for their hard work in putting together exciting programs, and the workshop presenters and authors for contributing to ECCV.

October 2022

Tomer Michaeli
Leonid Karlinsky
Ko Nishino

Organization

General Chairs

Rita Cucchiara — University of Modena and Reggio Emilia, Italy
Jiří Matas — Czech Technical University in Prague, Czech Republic
Amnon Shashua — Hebrew University of Jerusalem, Israel
Lihi Zelnik-Manor — Technion – Israel Institute of Technology, Israel

Program Chairs

Shai Avidan — Tel-Aviv University, Israel
Gabriel Brostow — University College London, UK
Giovanni Maria Farinella — University of Catania, Italy
Tal Hassner — Facebook AI, USA

Program Technical Chair

Pavel Lifshits — Technion – Israel Institute of Technology, Israel

Workshops Chairs

Leonid Karlinsky — IBM Research - MIT-IBM Watson AI Lab, USA
Tomer Michaeli — Technion – Israel Institute of Technology, Israel
Ko Nishino — Kyoto University, Japan

Tutorial Chairs

Thomas Pock — Graz University of Technology, Austria
Natalia Neverova — Facebook AI Research, UK

Demo Chair

Bohyung Han — Seoul National University, South Korea

Social and Student Activities Chairs

Tatiana Tommasi — Italian Institute of Technology, Italy
Sagie Benaim — University of Copenhagen, Denmark

Diversity and Inclusion Chairs

Xi Yin   Facebook AI Research, USA
Bryan Russell   Adobe, USA

Communications Chairs

Lorenzo Baraldi   University of Modena and Reggio Emilia, Italy
Kosta Derpanis   York University and Samsung AI Centre Toronto,
                                    Canada

Industrial Liaison Chairs

Dimosthenis Karatzas   Universitat Autònoma de Barcelona, Spain
Chen Sagiv   SagivTech, Israel

Finance Chair

Gerard Medioni   University of Southern California and Amazon,
                                    USA

Publication Chair

Eric Mortensen   MiCROTEC, USA

Workshops Organizers

W01 - AI for Space

Tat-Jun Chin   The University of Adelaide, Australia
Luca Carlone   Massachusetts Institute of Technology, USA
Djamila Aouada   University of Luxembourg, Luxembourg
Binfeng Pan   Northwestern Polytechnical University, China
Viorela Ila   The University of Sydney, Australia
Benjamin Morrell   NASA Jet Propulsion Lab, USA
Grzegorz Kakareko   Spire Global, USA

W02 - Vision for Art

Alessio Del Bue   Istituto Italiano di Tecnologia, Italy
Peter Bell   Philipps-Universität Marburg, Germany
Leonardo L. Impett   École Polytechnique Fédérale de Lausanne
                                    (EPFL), Switzerland
Noa Garcia   Osaka University, Japan
Stuart James   Istituto Italiano di Tecnologia, Italy

W03 - Adversarial Robustness in the Real World

Angtian Wang	Johns Hopkins University, USA
Yutong Bai	Johns Hopkins University, USA
Adam Kortylewski	Max Planck Institute for Informatics, Germany
Cihang Xie	University of California, Santa Cruz, USA
Alan Yuille	Johns Hopkins University, USA
Xinyun Chen	University of California, Berkeley, USA
Judy Hoffman	Georgia Institute of Technology, USA
Wieland Brendel	University of Tübingen, Germany
Matthias Hein	University of Tübingen, Germany
Hang Su	Tsinghua University, China
Dawn Song	University of California, Berkeley, USA
Jun Zhu	Tsinghua University, China
Philippe Burlina	Johns Hopkins University, USA
Rama Chellappa	Johns Hopkins University, USA
Yinpeng Dong	Tsinghua University, China
Yingwei Li	Johns Hopkins University, USA
Ju He	Johns Hopkins University, USA
Alexander Robey	University of Pennsylvania, USA

W04 - Autonomous Vehicle Vision

Rui Fan	Tongji University, China
Nemanja Djuric	Aurora Innovation, USA
Wenshuo Wang	McGill University, Canada
Peter Ondruska	Toyota Woven Planet, UK
Jie Li	Toyota Research Institute, USA

W05 - Learning With Limited and Imperfect Data

Noel C. F. Codella	Microsoft, USA
Zsolt Kira	Georgia Institute of Technology, USA
Shuai Zheng	Cruise LLC, USA
Judy Hoffman	Georgia Institute of Technology, USA
Tatiana Tommasi	Politecnico di Torino, Italy
Xiaojuan Qi	The University of Hong Kong, China
Sadeep Jayasumana	University of Oxford, UK
Viraj Prabhu	Georgia Institute of Technology, USA
Yunhui Guo	University of Texas at Dallas, USA
Ming-Ming Cheng	Nankai University, China

W06 - Advances in Image Manipulation

Radu Timofte	University of Würzburg, Germany, and ETH Zurich, Switzerland
Andrey Ignatov	AI Benchmark and ETH Zurich, Switzerland
Ren Yang	ETH Zurich, Switzerland
Marcos V. Conde	University of Würzburg, Germany
Furkan Kınlı	Özyeğin University, Turkey

W07 - Medical Computer Vision

Tal Arbel	McGill University, Canada
Ayelet Akselrod-Ballin	Reichman University, Israel
Vasileios Belagiannis	Otto von Guericke University, Germany
Qi Dou	The Chinese University of Hong Kong, China
Moti Freiman	Technion, Israel
Nicolas Padoy	University of Strasbourg, France
Tammy Riklin Raviv	Ben Gurion University, Israel
Mathias Unberath	Johns Hopkins University, USA
Yuyin Zhou	University of California, Santa Cruz, USA

W08 - Computer Vision for Metaverse

Bichen Wu	Meta Reality Labs, USA
Peizhao Zhang	Facebook, USA
Xiaoliang Dai	Facebook, USA
Tao Xu	Facebook, USA
Hang Zhang	Meta, USA
Péter Vajda	Facebook, USA
Fernando de la Torre	Carnegie Mellon University, USA
Angela Dai	Technical University of Munich, Germany
Bryan Catanzaro	NVIDIA, USA

W09 - Self-Supervised Learning: What Is Next?

Yuki M. Asano	University of Amsterdam, The Netherlands
Christian Rupprecht	University of Oxford, UK
Diane Larlus	Naver Labs Europe, France
Andrew Zisserman	University of Oxford, UK

W10 - Self-Supervised Learning for Next-Generation Industry-Level Autonomous Driving

Xiaodan Liang	Sun Yat-sen University, China
Hang Xu	Huawei Noah's Ark Lab, China

Fisher Yu   ETH Zürich, Switzerland
Wei Zhang   Huawei Noah's Ark Lab, China
Michael C. Kampffmeyer   UiT The Arctic University of Norway, Norway
Ping Luo   The University of Hong Kong, China

W11 - ISIC Skin Image Analysis

M. Emre Celebi   University of Central Arkansas, USA
Catarina Barata   Instituto Superior Técnico, Portugal
Allan Halpern   Memorial Sloan Kettering Cancer Center, USA
Philipp Tschandl   Medical University of Vienna, Austria
Marc Combalia   Hospital Clínic of Barcelona, Spain
Yuan Liu   Google Health, USA

W12 - Cross-Modal Human-Robot Interaction

Fengda Zhu   Monash University, Australia
Yi Zhu   Huawei Noah's Ark Lab, China
Xiaodan Liang   Sun Yat-sen University, China
Liwei Wang   The Chinese University of Hong Kong, China
Xiaojun Chang   University of Technology Sydney, Australia
Nicu Sebe   University of Trento, Italy

W13 - Text in Everything

Ron Litman   Amazon AI Labs, Israel
Aviad Aberdam   Amazon AI Labs, Israel
Shai Mazor   Amazon AI Labs, Israel
Hadar Averbuch-Elor   Cornell University, USA
Dimosthenis Karatzas   Universitat Autònoma de Barcelona, Spain
R. Manmatha   Amazon AI Labs, USA

W14 - BioImage Computing

Jan Funke   HHMI Janelia Research Campus, USA
Alexander Krull   University of Birmingham, UK
Dagmar Kainmueller   Max Delbrück Center, Germany
Florian Jug   Human Technopole, Italy
Anna Kreshuk   EMBL-European Bioinformatics Institute,
                               Germany
Martin Weigert   École Polytechnique Fédérale de Lausanne
                               (EPFL), Switzerland
Virginie Uhlmann   EMBL-European Bioinformatics Institute, UK

| Peter Bajcsy | National Institute of Standards and Technology, USA |
| Erik Meijering | University of New South Wales, Australia |

W15 - Visual Object-Oriented Learning Meets Interaction: Discovery, Representations, and Applications

Kaichun Mo	Stanford University, USA
Yanchao Yang	Stanford University, USA
Jiayuan Gu	University of California, San Diego, USA
Shubham Tulsiani	Carnegie Mellon University, USA
Hongjing Lu	University of California, Los Angeles, USA
Leonidas Guibas	Stanford University, USA

W16 - AI for Creative Video Editing and Understanding

Fabian Caba	Adobe Research, USA
Anyi Rao	The Chinese University of Hong Kong, China
Alejandro Pardo	King Abdullah University of Science and Technology, Saudi Arabia
Linning Xu	The Chinese University of Hong Kong, China
Yu Xiong	The Chinese University of Hong Kong, China
Victor A. Escorcia	Samsung AI Center, UK
Ali Thabet	Reality Labs at Meta, USA
Dong Liu	Netflix Research, USA
Dahua Lin	The Chinese University of Hong Kong, China
Bernard Ghanem	King Abdullah University of Science and Technology, Saudi Arabia

W17 - Visual Inductive Priors for Data-Efficient Deep Learning

Jan C. van Gemert	Delft University of Technology, The Netherlands
Nergis Tömen	Delft University of Technology, The Netherlands
Ekin Dogus Cubuk	Google Brain, USA
Robert-Jan Bruintjes	Delft University of Technology, The Netherlands
Attila Lengyel	Delft University of Technology, The Netherlands
Osman Semih Kayhan	Bosch Security Systems, The Netherlands
Marcos Baptista Ríos	Alice Biometrics, Spain
Lorenzo Brigato	Sapienza University of Rome, Italy

W18 - Mobile Intelligent Photography and Imaging

| Chongyi Li | Nanyang Technological University, Singapore |
| Shangchen Zhou | Nanyang Technological University, Singapore |

Ruicheng Feng	Nanyang Technological University, Singapore
Jun Jiang	SenseBrain Research, USA
Wenxiu Sun	SenseTime Group Limited, China
Chen Change Loy	Nanyang Technological University, Singapore
Jinwei Gu	SenseBrain Research, USA

W19 - People Analysis: From Face, Body and Fashion to 3D Virtual Avatars

Alberto Del Bimbo	University of Florence, Italy
Mohamed Daoudi	IMT Nord Europe, France
Roberto Vezzani	University of Modena and Reggio Emilia, Italy
Xavier Alameda-Pineda	Inria Grenoble, France
Marcella Cornia	University of Modena and Reggio Emilia, Italy
Guido Borghi	University of Bologna, Italy
Claudio Ferrari	University of Parma, Italy
Federico Becattini	University of Florence, Italy
Andrea Pilzer	NVIDIA AI Technology Center, Italy
Zhiwen Chen	Alibaba Group, China
Xiangyu Zhu	Chinese Academy of Sciences, China
Ye Pan	Shanghai Jiao Tong University, China
Xiaoming Liu	Michigan State University, USA

W20 - Safe Artificial Intelligence for Automated Driving

Timo Saemann	Valeo, Germany
Oliver Wasenmüller	Hochschule Mannheim, Germany
Markus Enzweiler	Esslingen University of Applied Sciences, Germany
Peter Schlicht	CARIAD, Germany
Joachim Sicking	Fraunhofer IAIS, Germany
Stefan Milz	Spleenlab.ai and Technische Universität Ilmenau, Germany
Fabian Hüger	Volkswagen Group Research, Germany
Seyed Ghobadi	University of Applied Sciences Mittelhessen, Germany
Ruby Moritz	Volkswagen Group Research, Germany
Oliver Grau	Intel Labs, Germany
Frédérik Blank	Bosch, Germany
Thomas Stauner	BMW Group, Germany

W21 - Real-World Surveillance: Applications and Challenges

| Kamal Nasrollahi | Aalborg University, Denmark |
| Sergio Escalera | Universitat Autònoma de Barcelona, Spain |

Radu Tudor Ionescu	University of Bucharest, Romania
Fahad Shahbaz Khan	Mohamed bin Zayed University of Artificial Intelligence, United Arab Emirates
Thomas B. Moeslund	Aalborg University, Denmark
Anthony Hoogs	Kitware, USA
Shmuel Peleg	The Hebrew University, Israel
Mubarak Shah	University of Central Florida, USA

W22 - Affective Behavior Analysis In-the-Wild

Dimitrios Kollias	Queen Mary University of London, UK
Stefanos Zafeiriou	Imperial College London, UK
Elnar Hajiyev	Realeyes, UK
Viktoriia Sharmanska	University of Sussex, UK

W23 - Visual Perception for Navigation in Human Environments: The JackRabbot Human Body Pose Dataset and Benchmark

Hamid Rezatofighi	Monash University, Australia
Edward Vendrow	Stanford University, USA
Ian Reid	University of Adelaide, Australia
Silvio Savarese	Stanford University, USA

W24 - Distributed Smart Cameras

Niki Martinel	University of Udine, Italy
Ehsan Adeli	Stanford University, USA
Rita Pucci	University of Udine, Italy
Animashree Anandkumar	Caltech and NVIDIA, USA
Caifeng Shan	Shandong University of Science and Technology, China
Yue Gao	Tsinghua University, China
Christian Micheloni	University of Udine, Italy
Hamid Aghajan	Ghent University, Belgium
Li Fei-Fei	Stanford University, USA

W25 - Causality in Vision

Yulei Niu	Columbia University, USA
Hanwang Zhang	Nanyang Technological University, Singapore
Peng Cui	Tsinghua University, China
Song-Chun Zhu	University of California, Los Angeles, USA
Qianru Sun	Singapore Management University, Singapore
Mike Zheng Shou	National University of Singapore, Singapore
Kaihua Tang	Nanyang Technological University, Singapore

W26 - In-Vehicle Sensing and Monitorization

Jaime S. Cardoso	INESC TEC and Universidade do Porto, Portugal
Pedro M. Carvalho	INESC TEC and Polytechnic of Porto, Portugal
João Ribeiro Pinto	Bosch Car Multimedia and Universidade do Porto, Portugal
Paula Viana	INESC TEC and Polytechnic of Porto, Portugal
Christer Ahlström	Swedish National Road and Transport Research Institute, Sweden
Carolina Pinto	Bosch Car Multimedia, Portugal

W27 - Assistive Computer Vision and Robotics

Marco Leo	National Research Council of Italy, Italy
Giovanni Maria Farinella	University of Catania, Italy
Antonino Furnari	University of Catania, Italy
Mohan Trivedi	University of California, San Diego, USA
Gérard Medioni	Amazon, USA

W28 - Computational Aspects of Deep Learning

Iuri Frosio	NVIDIA, Italy
Sophia Shao	University of California, Berkeley, USA
Lorenzo Baraldi	University of Modena and Reggio Emilia, Italy
Claudio Baecchi	University of Florence, Italy
Frederic Pariente	NVIDIA, France
Giuseppe Fiameni	NVIDIA, Italy

W29 - Computer Vision for Civil and Infrastructure Engineering

Joakim Bruslund Haurum	Aalborg University, Denmark
Mingzhu Wang	Loughborough University, UK
Ajmal Mian	University of Western Australia, Australia
Thomas B. Moeslund	Aalborg University, Denmark

W30 - AI-Enabled Medical Image Analysis: Digital Pathology and Radiology/COVID-19

Jaime S. Cardoso	INESC TEC and Universidade do Porto, Portugal
Stefanos Kollias	National Technical University of Athens, Greece
Sara P. Oliveira	INESC TEC, Portugal
Mattias Rantalainen	Karolinska Institutet, Sweden
Jeroen van der Laak	Radboud University Medical Center, The Netherlands
Cameron Po-Hsuan Chen	Google Health, USA

Diana Felizardo	IMP Diagnostics, Portugal
Ana Monteiro	IMP Diagnostics, Portugal
Isabel M. Pinto	IMP Diagnostics, Portugal
Pedro C. Neto	INESC TEC, Portugal
Xujiong Ye	University of Lincoln, UK
Luc Bidaut	University of Lincoln, UK
Francesco Rundo	STMicroelectronics, Italy
Dimitrios Kollias	Queen Mary University of London, UK
Giuseppe Banna	Portsmouth Hospitals University, UK

W31 - Compositional and Multimodal Perception

Kazuki Kozuka	Panasonic Corporation, Japan
Zelun Luo	Stanford University, USA
Ehsan Adeli	Stanford University, USA
Ranjay Krishna	University of Washington, USA
Juan Carlos Niebles	Salesforce and Stanford University, USA
Li Fei-Fei	Stanford University, USA

W32 - Uncertainty Quantification for Computer Vision

Andrea Pilzer	NVIDIA, Italy
Martin Trapp	Aalto University, Finland
Arno Solin	Aalto University, Finland
Yingzhen Li	Imperial College London, UK
Neill D. F. Campbell	University of Bath, UK

W33 - Recovering 6D Object Pose

Martin Sundermeyer	DLR German Aerospace Center, Germany
Tomáš Hodaň	Reality Labs at Meta, USA
Yann Labbé	Inria Paris, France
Gu Wang	Tsinghua University, China
Lingni Ma	Reality Labs at Meta, USA
Eric Brachmann	Niantic, Germany
Bertram Drost	MVTec, Germany
Sindi Shkodrani	Reality Labs at Meta, USA
Rigas Kouskouridas	Scape Technologies, UK
Ales Leonardis	University of Birmingham, UK
Carsten Steger	Technical University of Munich and MVTec, Germany
Vincent Lepetit	École des Ponts ParisTech, France, and TU Graz, Austria
Jiří Matas	Czech Technical University in Prague, Czech Republic

W34 - Drawings and Abstract Imagery: Representation and Analysis

Diane Oyen	Los Alamos National Laboratory, USA
Kushal Kafle	Adobe Research, USA
Michal Kucer	Los Alamos National Laboratory, USA
Pradyumna Reddy	University College London, UK
Cory Scott	University of California, Irvine, USA

W35 - Sign Language Understanding

Liliane Momeni	University of Oxford, UK
Gül Varol	École des Ponts ParisTech, France
Hannah Bull	University of Paris-Saclay, France
Prajwal K. R.	University of Oxford, UK
Neil Fox	University College London, UK
Ben Saunders	University of Surrey, UK
Necati Cihan Camgöz	Meta Reality Labs, Switzerland
Richard Bowden	University of Surrey, UK
Andrew Zisserman	University of Oxford, UK
Bencie Woll	University College London, UK
Sergio Escalera	Universitat Autònoma de Barcelona, Spain
Jose L. Alba-Castro	Universidade de Vigo, Spain
Thomas B. Moeslund	Aalborg University, Denmark
Julio C. S. Jacques Junior	Universitat Autònoma de Barcelona, Spain
Manuel Vázquez Enríquez	Universidade de Vigo, Spain

W36 - A Challenge for Out-of-Distribution Generalization in Computer Vision

Adam Kortylewski	Max Planck Institute for Informatics, Germany
Bingchen Zhao	University of Edinburgh, UK
Jiahao Wang	Max Planck Institute for Informatics, Germany
Shaozuo Yu	The Chinese University of Hong Kong, China
Siwei Yang	Hong Kong University of Science and Technology, China
Dan Hendrycks	University of California, Berkeley, USA
Oliver Zendel	Austrian Institute of Technology, Austria
Dawn Song	University of California, Berkeley, USA
Alan Yuille	Johns Hopkins University, USA

W37 - Vision With Biased or Scarce Data

Kuan-Chuan Peng	Mitsubishi Electric Research Labs, USA
Ziyan Wu	United Imaging Intelligence, USA

W38 - Visual Object Tracking Challenge

Matej Kristan	University of Ljubljana, Slovenia
Aleš Leonardis	University of Birmingham, UK
Jiří Matas	Czech Technical University in Prague, Czech Republic
Hyung Jin Chang	University of Birmingham, UK
Joni-Kristian Kämäräinen	Tampere University, Finland
Roman Pflugfelder	Technical University of Munich, Germany, Technion, Israel, and Austrian Institute of Technology, Austria
Luka Čehovin Zajc	University of Ljubljana, Slovenia
Alan Lukežič	University of Ljubljana, Slovenia
Gustavo Fernández	Austrian Institute of Technology, Austria
Michael Felsberg	Linköping University, Sweden
Martin Danelljan	ETH Zurich, Switzerland

Contents – Part IV

W11 - Skin Image Analysis

W12 - Cross-Modal Human-Robot Interaction

W13 - Text in Everything

W14 - BioImage Computing

**W15 - Visual Object-Oriented Learning Meets Interaction: Discovery,
Representations, and Applications**

W16 - AI for Creative Video Editing and Understanding

W09 - Self-supervised Learning: What Is Next?

W09 - Self-supervised Learning: What Is Next?

The past two years have seen major advances in self-supervised learning, with many new methods reaching astounding performances on standard benchmarks. Moreover, many recent works have shown the large potential of coupled data sources such as image-text in producing even stronger models capable of zero-shot tasks, and often inspired by NLP. We have just witnessed a jump from the "default" single-modal pretraining with CNNs to transformer-based multi-modal training, and these early developments will surely mature in the coming months. However, despite this it is also apparent that there are still major unresolved challenges and it is not clear what the next step-change is going to be. In this workshop we aim to highlight and provide a forum to discuss potential research direction seeds, from radically new self-supervision tasks, data sources, and paradigms to surprising counter-intuitive results. Through invited speakers and paper oral talks, our goal is to provide a forum to discuss and exchange ideas where both the leaders in this field, as well as the new, younger generation can equally contribute to discussing the future of this field.

October 2022

Yuki M. Asano
Christian Rupprecht
Diane Larlus
Andrew Zisserman

Towards Self-Supervised and Weight-preserving Neural Architecture Search

Zhuowei Li[1]($\boxtimes$), Yibo Gao[2], Zhenzhou Zha[3], Zhiqiang Hu[4], Qing Xia[4], Shaoting Zhang[4], and Dimitris N. Metaxas[1]

[1] Rutgers Univeristy, New Jersey, USA
zl502@cs.rutgers.edu
[2] University of Electronic Science and Technology of China, Chengdu, China
[3] Zhejiang University, Hangzhou, China
[4] SenseTime Research, Shanghai, China

Abstract. Neural architecture search (NAS) techniques can discover outstanding network architecture while saving tremendous labor from human experts. Recent advancements further reduce the computational overhead to an affordable level. However, it is still cumbersome to deploy NAS in real-world applications due to the fussy procedures and the supervised learning paradigm. In this work, we propose the self-supervised and weight-preserving neural architecture search (SSWP-NAS) as an extension of the current NAS framework to allow the self-supervision and retain the concomitant weights discovered during the search stage. As such, we merge the process of architecture search and weight pre-training, and simplify the workflow of NAS to a one-stage and proxy-free procedure. The searched architectures can achieve state-of-the-art accuracy on CIFAR-10, CIFAR-100, and ImageNet datasets without using manual labels. Moreover, experiments demonstrate that using the concomitant weights as initialization consistently outperforms the random initialization and a separate weight pre-training process by a clear margin under semi-supervised learning scenarios. Codes are available at https://github.com/LzVv123456/SSWP-NAS.

Keywords: Self-Supervised learning · Neural architecture search

1 Introduction

The development of NAS algorithms save considerable time and efforts of experts through automating the neural architecture design process. It has achieved

---

Z. Li and Y. Gao—Equal contributions.
Y. Gao and Z. Zha—This work was done during the internship at SenseTime.

Supplementary Information The online version contains supplementary material available at https://doi.org/10.1007/978-3-031-25069-9_1.

© The Author(s), under exclusive license to Springer Nature Switzerland AG 2023
L. Karlinsky et al. (Eds.): ECCV 2022 Workshops, LNCS 13804, pp. 3–19, 2023.
https://doi.org/10.1007/978-3-031-25069-9_1

state-of-the-art performances in a series of vision tasks including image recognition [30,47,48], semantic segmentation [18,44] and object detection [8,37]. Recent advances on weight-sharing NAS [28] and differentiable NAS [22,39] further reduce the searching cost from thousands of GPU-days to a couple.

Despite the significant computational reduction made by current NAS methods, it is still cumbersome to deploy the NAS techniques in real-world applications due to the fussy procedures. As shown in Fig. 1, a typical workflow of NAS consists of a surrogate-structure search phase following by a final architecture selection process. Then a standalone weight pre-training procedure needs to be taken before transferring the architecture to downstream tasks. It is nontrivial to pre-train a network, taking even more time than the searching process. Besides, existing NAS workflows largely rely on manual annotations, making the domain-specific NAS even more unwieldy.

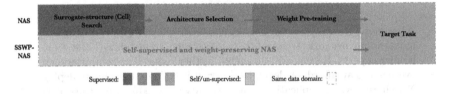

Fig. 1. Overview for the regime of general NAS and the proposed SSWP-NAS.

Driven by the inconvenience of the current NAS paradigm, we propose a new framework, namely self-supervised and weight-preserving neural architecture search (SSWP-NAS), as an extension of the current NAS methodology with the following two prominent properties: (1) SSWP-NAS is self-supervised so that it does not rely on manual signals to perform optimization. This property also removes the dependency on proxy-datasets (e.g. ImageNet [32]). (2) SSWP-NAS has the weight-preserving property, which means the concomitant weights generated during the search process can be retained and serve as initialization to benefit transfer learning. This property simplifies the current NAS workflow from a two-stage (i.e. search and pre-train) fashion to one-stage. To achieve weight-preserving, we align the dimensionality of the architecture used during the search and train stage and leverage stochastic operation sampling strategy to reduce the memory footprint. To remove the dependency on manual labels, we probe how the designed searching process copes with self-supervised learning objectives. We further observed a persistent optimization challenge dubbed *network inflation issue* due to the inconsistent optimization targets and the stochastic strategy. To overcome this challenge, we propose the forward progressive prune (FPP) operation that gradually bridges the gap between the optimization targets and reduces the extend of stochastic operations.

Empirically, SSWP-NAS achieves state-of-the-art accuracy on CIFAR-10 (2.41% error rate), CIFAR-100 (16.47% error rate), and ImageNet (24.3% top-1 error rate under restricted resources) when training on the searched architecture from scratch. When using concomitant weights as initialization, our method consistently outperforms two counterparts i.e. random initialization and two-stage weight pre-training method, by a clear margin under semi-supervised learning scenarios. Besides, we show that self-supervised learning objective surpasses the supervised counterpart in our framework and FPP is beneficial under both supervised and self/un-supervised learning objectives. Comprehensive ablation studies have also been conducted towards the proposed designs. Our main contributions can be summarized in three-fold:

- We propose SSWP-NAS that enjoys the self-supervised learning and weight-preserving property. It simplifies the current usage of NAS technique from a two-stage manner to one stage. As a by-product, SSWP-NAS is also proxy-free, which means it relies on neither the surrogate structure nor the proxy-dataset.
- We propose the FPP to address the network inflation issue that occurs during the designed weight-preserving search process. We empirically show that FPP is beneficial under both supervised and self/un-supervised signals.
- SSWP-NAS searches the architecture and generates the pre-train weights concurrently while achieving state-of-the-art performance regarding the quality of both the architecture and the pre-train weights.

2 Related Work

Neural Architecture Search. Early arts for NAS typically leverage on reinforcement learning (RL) [47,48] or evolutionary algorithms (EA) [21,30] to optimize a controller that samples a sequence of discrete operations to form the architecture. This straightforward implementation consumes tremendous computational resources. As a remedy, the following works propose the weight-sharing [22,28] strategy and surrogate structures [22,48] to reduce the computational overhead. DARTS [22] further simplify the NAS framework by relaxing the search space from discrete to continuous. This *One-shot* searching framework then becomes popular in the NAS domain due to its simplicity and efficiency. The proposed SSWP-NAS also inheres to this line of research. While the original DARTS design is observed to suffer from performance degeneration [4] and mode collapse [17] issue. To this end, DARTS+ [17] introduces the early stop to suppress the over-characterized non-parameterized operations. P-DARTS [4] tries to alleviate the performance drop by gradually increasing the depth of the surrogate structure. ProxylessNAS [2] first achieves the differentiable architecture search without a surrogate structure. Despite the advances up to date, existing methods still rely on manual supervision to perform optimization, and none of them are able to preserve the concomitant weights.

Self-supervised Learning. Not until recently, self-supervised learning mainly relies on heuristic pretext tasks [7,26,42,46] to form the supervision signal, and their performance lags behind a lot comparing with the supervised counterpart. Emerging of the contrastive-based self-supervised learning largely close the gap between the self-supervised and supervised weight pre-training [3,10,16,27,36].

It encourages a pulling force within the positive pair while pushing away negative pairs through minimizing a discriminative loss. BYOL [9] and SimSiam [5] also demonstrate that the negative pairs are not necessary. In this work, we investigate how state-of-the-art self-supervised learning methods cope with the weight-preserving network search.

Self/un-supervised NAS. Most recently, some other works also explore the self/un-supervised learning objective under the NAS framework. UnNAS [19] explores how different pretexts tasks can replace the supervised discrimination task. It shows that labels are not necessary for NAS, and metrics used in pretext tasks can be a good proxy for the structure selection. RLNAS [43] shifts from the performance-based evaluation metric to the convergence-based metric, and it further uses the random labels to generate supervision signals. Among existing self/un-supervised NAS works, SSNAS [14] and CSNAS [25] are most comparable to our method as they also explore the contrastive learning in NAS domain. Yet, they only consider from an architecture optimization perspective and their frameworks still fall into proxy-based searching. We here instead search for both the architecture and concomitant weights as an integrity.

3   Methodology

In this section, we first introduce the prior knowledge about differentiable NAS (DARTS) which serves as the foundation of our framework. Then we detail how to extend the DARTS towards the weight-preserving search and self-supervised optimization. Afterwards, we investigate the network inflation issue and propose a simple solution. Finally, we demonstrate how to search a network (architecture plus weights) using SSWP-NAS.

3.1   Preliminary: Differentiable NAS

Neural architecture search (NAS) task is generally formulated as a bi-level optimization task [1,6] where the upper-level variable α refers to the architecture parameters and lower-lever variable w represents the operation parameters:

$$\min_{\alpha} \mathcal{L}_{val}(w^*(\alpha), \alpha) \tag{1}$$

$$s.t. \quad w^*(\alpha) = argmin_w \, \mathcal{L}_{train}(w, \alpha) \tag{2}$$

In practice, two sets of parameters are optimized in an alternative manner. Differentiable NAS (DARTS) [22] further includes the architecture parameters to the computational graph via relaxing the search space from discrete to continuous. As such, it can effectively evolve both the architecture structure and operation weights leveraging the stochastic gradient descent [31] techniques.

Inspired by the success of manually-designed structural motifs, most NAS methods further transfer the searching target from the whole architecture to a cell structure [20,30,48] which is then stacked repetitively to build the final architecture. DARTS [22] also adopts this strategy to search a cell structure using a light-weight proxy architecture. Specifically, DARTS search space [22] represented by a cell can be interpreted as a directed acyclic graph (DAG) which consists of 7 nodes and 14 edges. Each edge between node m and n is associated with a collection of candidate operations $\mathcal{O}$ (e.g. 5×5 conv, 7×7 conv, average pooling) weighted by a real-valued vector $\boldsymbol{v}^{(m,n)} \in \mathbb{R}^{1 \times M}$ where $M = |\mathcal{O}|$. And the information flow $f_{m,n}$ between node $\boldsymbol{x}_m$ and $\boldsymbol{x}_n$ is defined as

$$f_{m,n}(\boldsymbol{x}_m) = \sum_{o \in \mathcal{O}} \frac{\exp(\boldsymbol{v}_o^{(m,n)})}{\sum_{o' \in \mathcal{O}} \exp(\boldsymbol{v}_{o'}^{(m,n)})} o(\boldsymbol{x}_m) \tag{3}$$

where $m < n$ and an intermediate node is summation of all its predecessors: $\boldsymbol{x}_n = \sum_{m<n} f_{m,n}(\boldsymbol{x}_m)$. The final output of a cell is the concatenation of 4 intermediate nodes (except for 2 input nodes and 1 output node) over the channel dimension. We here refer to DARTS [22] paper for more details.

3.2   Towards Weight-preserving

One of the most critical challenges in NAS is the surge of the memory footprint and the computational overhead. As a remedy, most existing methods search a light-weight surrogate architecture to find the cell structure. While reducing the computational overhead, the prevalence of proxy strategy excludes the possibility of the weight-preserving property from the very beginning. Due to the non-identical structures used during the search and the train stage, concomitant weights discovered in the search phase are abandoned, and only the searched cell structure is reserved for the downstream application.

To retain the concomitant weights, we first build our target architecture using the DARTS [22] search space, which is well-established and contains abundant sub-graphs [41]. Then we dispose the surrogate architecture and unify the architecture used during the search and train stage. We further allow different cell structures at each level of the architecture, i.e., we search a set of cells to directly build the final architecture instead of searching a single cell structure as the building unit. As clued in Sect. 3.1, this over-parameterized formation will consume $M = |\mathcal{O}|$ times memory comparing with a compact architecture. To this end, we leverage stochastic algorithms [2,39,40] to cut down the memory usage. Here, we adopt the path-binarization strategy, which is first proposed in ProxylessNAS [2] to overcome the accuracy degeneration issue caused by the depth-gap between the surrogate and the final architecture. Concretely, only a single operation on an edge is sampled and activated according to a learned distribution at each iteration. As such, the memory footprint during the search is reduced to the same magnitude as a compact architecture.

Through renouncing the proxy strategy and adopting the stochastic technique, we meet the indispensable conditions towards the weight-preserving. While in practice, we observe an undesirable edge gained by the skip-connection operation. Similar phenomenons have also been observed in previous supervised and proxy-based search [4,17,33]. Such over-ratings for non-parameterized operations not only hinder the general quality of the structure, but also result in insufficient updates of parameterized operations during the lower-level optimization. To this end, we introduce a non-parameterized operation, dropout ($p = 0.2$ across our experiments), during the lower-level optimization. By doing so, we implicitly regularize the importance of non-parameterized operation and increase the sampling odds for parameterized operation without direct interference with the learned upper-level parameter distribution.

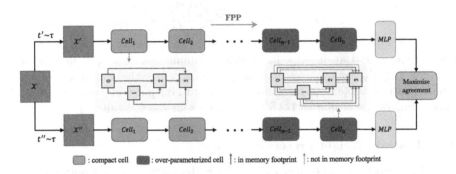

Fig. 2. Overview of the proposed SSWP-NAS framework. τ is a collection of data augmentations, t' and t'' are two transformations sampled from τ that transfer image x to two different views x' and x'', respectively. Detailed cell structure is omitted for the interpretation purpose.

3.3   Towards Self-supervised Learning

It is conceptually straightforward to remove the dependency of the manual annotation by replacing the supervised signal with a self/un-supervised counterpart. Following the common practice of NAS, we are seeking a favorable architecture through modeling the conditional probability (discriminative model) rather than learning the joint distribution (generative model). As such, we subscribe our exploration for the learning objective to the discriminative sub-field. And among the bag of self/un-supervised discriminative pretext tasks, contrastive learning based methodologies [3,9,10,27,36] demonstrate an outstanding representation learning ability over the peers. So we further focus our attention on probing how the contrastive learning copes with the designed weight-preserving search process.

Since the concept of positive and negative pair lies at the core of the contrastive learning, we study two representative methods, SimCLR [3] and BYOL [9] which are formulated with and without negative pairs. According to our pilot trials, BYOL only achieves 3.05% top-1 error rate on cifar-10 while SimCLR achieving 2.56%. We observe that the failure of the BYOL framework originates from the incompatible patterns between the stochastic operation sampling and exponential moving average (EMA). BYOL essentially forms a teacher-student pair using the current snapshot of the operation weights θ_t and its corresponding EMA counterpart $\theta'_t = \tau\theta'_{t-1} + (1 - \tau)\theta_t$ where τ is a smoothing factor. Then BYOL employees a symmetric consistency loss between the pair as the supervision signal. However, stochastic operation sampling turns the parameter distribution θ_t into ξ_t where ξ and θ are different operation distributions. And this inconsistency causes the mode collapse of the EMA, and thus fails the BYOL framework. So we take the strategy from SimCLR [3] where positive pairs are formed by two transformations of the same image and negative pairs are constructed using different images within a mini-batch. The whole architecture is then optimized in an end-to-end manner through minimizing the infoNCE loss [34,38]. With mini-batch size N, we have:

$$\mathcal{L}_{batch} = \sum_{i=1}^{2N} \mathcal{L}_i \tag{4}$$

$$\mathcal{L}_i = -\log \frac{\exp\left(sim(z_i, z_{j(i)})/\tau\right)}{\sum_{k=1}^{2N} \mathbb{1}_{[k \neq i]} \exp(sim(z_i, z_k))/\tau)} \tag{5}$$

where $z_i = P(E(\tilde{x}_i))$. $E(\cdot)$ denotes the searching encoder architecture and $P(\cdot)$ is a projection neck (multi-layer perception) added at the tail of searching structure. The Subscript i and $j(i)$ refers to two different views of the same image. And $sim(\cdot)$ measures the cosine similarity between two given vectors. Fig. 2 exhibits the overview of our framework. It is worth noting that $E_t(\cdot) \neq E_{t-1}(\cdot)$ at different time steps and $P(\cdot)$ is only employed during the search stage and being replaced by a linear classification layer at train stage. By leveraging this self-guided label generation method, SSWP-NAS gets rid of the dependency on manual labels.

3.4 Network Inflation Challenge

Despite a neat couple made by the weight-preserving search and the self-supervised optimization, it is challenging to optimize the proposed framework in practice. We observe that the difficulty originates from the combination of the over-parameterized structure and the stochastic operation sampling strategy, which in together we referred to as *network inflation* issue.

From a macro perspective, we are optimizing two sets of over-parameterized variable distributions α and w alternatively during the search phase. And at the end of search process, differentiable NAS [22] relies on the non-linear prune operation to approximate the compact variable $\tilde{\alpha}^*$ and $\tilde{w}^*$ using the over-parameterized variable α^* and w^*. This leaping relaxation essentially relies on the good generalization ability of the hierarchical convolutional structures. From a micro view, a hierarchical convolutional structure can be viewed as a sequential model. Given a random tie-breaking index i in a N-layer structures at time step t, we are maximizing posterior probability $\arg\max_{w_{i \sim N}} p(w_{i \sim N} | w_{1 \sim i-1}^t, D)$ where D is the data distribution and subscript $a \sim b$ denotes layer index from a to b. Due to the stochastic sampling strategy, conditions $w_{1 \sim i-1}^{t-1} \neq w_{1 \sim i-1}^t$ while $w_{i \sim N}$ can be deemed as unchanged. Even though stochastic algorithm theoretically guarantees the same global convergence, if it exists, this non-stationary condition increases the difficulty of optimization at each time step.

To handle the network inflation challenge, we propose a simple and effective solution, namely *forward progressive prune (FPP)*. Different from the general differentiable NAS pipeline where an architecture prune process is conducted for the target cell structure at the end of the search stage. We impose a cell-level progressive prune in the forward propagation direction during the search stage. At each prune step, we prune all edges contained in a cell, and for each edge, we only keep the operation with highest learned credit. By doing so, we reformulate the optimization target as $\arg\max_{w_{i\sim N}} p(w_{i\sim N}|\tilde{w}^*_{1\sim i-1}, D)$ where the condition $\tilde{w}^*_{1\sim i-1}$ is fixed and transfer the optimization target from $p(w_{1\sim N})$ to $p(w_{i\sim N}|\tilde{w}^*_{1\sim i-1})$ which is closer to the final goal $p(\tilde{w}^*_{1\sim N})$. As such, FPP gradually aligns the searching target with the final objective. And it fixes the non-stationary conditions at some point during the search and allows the following layers to adjust according to the fixed preceding layers. The forward propagation direction design also follows the common acknowledgment that shallow layers of a CNN architecture capture easier low-level features while deeper layers grab more complicated semantics.

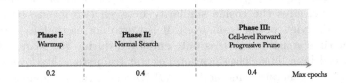

Fig. 3. Different phases during the search stage.

3.5   Searching with SSWP-NAS

SSWP-NAS is a proxy-free search, so one can search the target architecture directly on target data domain without using labels. As depicted in Fig. 3, the search stage of SSWP-NAS is divided into three phases. At the warm-up phase, we only update operation parameters to allow a better initialization of parameterized operations. Then we update both operation and architecture parameters alternatively as a standard differentiable NAS. Finally, we start the FPP phase, in which we progressively perform cell-level prune. The only extra hyperparameter introduced is the time span of different phases. We empirically suggest that the split $[0.2, 0.4, 0.4]$ works well in general. Given the proportion for phase III, the time interval T for pruning of two adjacent cells is calculated as $T = \frac{\text{max epochs} \times \text{FPP ratio}}{\text{cell num}}$.

4   Experiments

Experiments are organized into four sub-sections. We first provide the general settings used across our experiments. Then, we isolate the concomitant weights and benchmark the architecture quality searched using SSWP-NAS. Afterwards, we conduct comprehensive ablation studies towards our designs. Finally, we thoroughly study the effectiveness of concomitant weights. Detailed searched architecture are attached in *Appendix C*.

4.1 Experimental Settings

Following the well-established benchmark, we conduct our experiments on CIFAR-10/100 [15] and ImageNet [32] three datasets. We search an architecture consisting 20 cells with 36 initial channels for 300 epochs using mini-batch size 96 on CIFAR-10/100. For ImageNet, we search a structure that contains 14 cells with 48 initial channels for 100 epochs using mini-batch size 256. The magnitudes of searched architectures are kept the same as the final architecture trained in DARTS [22]. All hyper-parameters used during the search and train stage, except for our proposed designs, are inherited from DARTS [22]. See *Appendix A* for detailed specifications. And 1_{st} order optimization is employed across all settings. Experiments related to CIFAR-10/100 [15] are conducted on a single Nvidia A100 40GB and it scales up to 4 for ImageNet [32].

Table 1. Comparisons between SSWP-NAS and state-of-the-art methods on CIFAR-10/100. Here the search cost only counts the time used for the search. proxy-based methods commonly need another architecture selection procedure which generally costs another 1 GPU day [22]. While this procedure is not needed in our framework.

Architecture	Test Error (%)		Params	Search Cost	Search	Search
	CIFAR-10	CIFAR-100	(M)	(GPU days)	Type	Method
DenseNet-BC [13]	3.46	17.18	25.6	-	Supervised	Manual
NASNet-A [48]	2.65	-	3.3	1800	Supervised	RL
AmoebaNet-A [30]	3.34 ± 0.06	-	3.2	3150	Supervised	Evolution
AmoebaNet-B [30]	2.55 ± 0.05	-	2.8	3150	Supervised	Evolution
PNAS [20]	3.41 ± 0.09	-	3.2	225	Supervised	SMBO
Hireachical Evolution [21]	3.75 ± 0.12	-	15.7	300	Supervised	Evolution
ENAS [28]	2.89	-	4.6	0.5	Supervised	RL
NAONet [23]	3.18	15.67	10.6	200	Supervised	NAO
DARTS (1st order) [22]	3.0 ± 0.14	17.76	3.3	1.5	Supervised	Gradient
DARTS (2nd order) [22]	2.76 ± 0.09	17.54	3.3	4.0	Supervised	Gradient
SNAS (moderate) [39]	2.85 ± 0.02	-	2.8	1.5	Supervised	Gradient
ProxylessNAS-G [2]	2.08	-	5.7	4.0	Supervised	Gradient
P-DARTS [4]	2.50	16.55	3.4	0.3	Supervised	Gradient
PC-DARTS [40]	2.57 ± 0.07	-	3.6	0.1	Supervised	Gradient
BayesNAS [45]	2.81 ± 0.04	-	3.4	0.2	Supervised	Gradient
CSNAS$_{N=5}$ [25]	2.66 ± 0.07	-	3.4	1.0	Self-supervised	SMBO-TPE
SSNAS [14]	2.61	16.64	-	-	Self-supervised	Gradient
SSWP-NAS$_{e=300}$[†]	2.56 ± 0.07	17.27	4.0	1.0	Self-supervised	Gradient
SSWP-NAS$_{e=500}$[†]	**2.41 ± 0.07**	**16.47**	3.8	1.8	Self-supervised	Gradient

[†]: run 5 times with different seeds.

4.2 Benchmarking SSWP-NAS

CIFAR-10/100. We first search SSWP-NAS for 300 ($e = 300$) and 500 ($e = 500$) epochs on CIFAR dataset. Then we train the searched architecture from scratch with typical supervised learning. As shown in Table 1, SSWP-NAS$_{e=300}$ achieves state-of-the-art performance and outperforms DARTS (1_{st} order) significantly by searching the same epochs. When we extend the searching duration from 300 epochs to 500 epochs, SSWP-NAS surpasses existing methods by a clear margin.

Table 2. Comparison with state-of-the-art architectures on ImageNet (restricted resources)

Architecture	Test Error (%)		Params	×+	Search Cost	Search	Search
	Top-1	Top-5	(M)	(M)	(GPU days)	Type	Method
Inception-v2 [35]	25.2	7.8	11.2	-	-	-	Manual
MobileNet-v3 (Large 1.0) [12]	24.8	-	5.4	219	-	-	Manual
ShuffleNet(2×)-v2 [24]	25.1	-	≈ 5	591	-	-	Manual
NASNet-A [48]	26.0	8.4	5.3	564	2000	Supervised	RL
AmoebaNet-A [30]	25.5	8.0	5.1	555	3150	Supervised	Evolution
AmoebaNet-B [30]	26.0	8.5	5.3	555	3150	Supervised	Evolution
PNAS [20]	25.8	8.1	5.1	588	255	Supervised	SMBO
DARTS (2nd order) [22]	26.7	8.7	4.7	574	4	Supervised	Gradient
P-DARTS [4]	24.1	7.3	5.4	597	2.0	Supervised	Gradient
PC-DARTS [40]	24.2	7.3	5.3	597	3.8	Supervised	Gradient
ProxylessNAS [2]	24.9	7.5	7.1	465	8.3	Supervised	Gradient
SSNAS [14]	27.8	9.6	5.2	-	-	Self-supervised	Gradient
CSNAS$_{N=5}$ [25]	25.8	8.3	5.1	590	2.5	Self-supervised	SMBO-TPE
SSWP-NAS$_{e=50}$	24.8	7.8	5.0	597	3.5	Self-supervised	Gradient
SSWP-NAS$_{e=100}$	24.3	7.5	4.9	595	7	Self-supervised	Gradient

ImageNet. We search 50 and 100 epochs on ImageNet [32] and then train the searched structure for 250 epochs. As displayed in Table 2, SSWP-NAS also achieves state-of-the-art accuracy on ImageNet under limited budgets. One potential drawback is that SSWP-NAS takes a relative longer time to search when considering the architecture solely as we are directly searching for the final structure instead of a surrogate structure. However, the weight-preserving property offsets this computational overhead as other existing methods need an extra non-trivial pre-train step if they intend to initialize with pre-trained weights. By achieving state-of-the-art performance on both CIFAR and ImageNet datasets, it also manifests the generality of the proposed SSWP-NAS framework.

4.3 Ablation Study

For simplicity, we abbreviate self-supervised learning and supervised learning as SSL and SL, respectively, in the following sections.

SSL vs. SL. In order to substantiate the effectiveness of SSL, we compare it with the SL (cross-entropy loss function). By keeping all other settings the same, we switch between SL and SSL objectives. As shown in Table 3a, SSL outperforms SL significantly in our framework. This result not only exhibits the inessentiality of human annotations for architecture search, but also suggests that the manual interference may function adversely by limiting the optimization manifold. SSL objective may support learning a more generic structure for feature extraction. Beyond this frank improvement, we further show that the SSL searched architecture boosts the self-supervised weight pre-training in *Appendix B*.

Effectiveness of FPP. To demonstrate the effectiveness of the FPP module, we implement SSWP-NAS with and without FPP. When disregarding FPP, we remove phase III during the search and adjust the ratio of phases I and II to 0.2 and 0.8 accordingly. Then the architecture prune process is conducted once for all cells after the search stage. Besides the original FPP, we also implement a reversed version, namely backward progressive pruning (BPP), that starts pruning from the last cell and propagates backwardly. According to Table 3a, FPP consistently improves the qualities of the searched architectures under both SSL and SL scenarios. BPP, on the contrary, even degenerates the performances. This result coincides with our knowledge that shallow layers extract low-level features, which are easy to learn. In contrast, deeper layers capture higher-level semantic features and leverage the low-level features to compose.

Table 3. Ablation studies on learning objective, FPP, skip-connection dropout and proxy-free search.

(a) **SSL vs. SL and FPP.** Ablations for learning objectives and FPP.

Name	Test Error (%)	Dataset	Search Epoch
SL	2.92	CIFAR-10	300
SL+BPP	2.93	CIFAR-10	300
SL+FPP	2.77	CIFAR-10	300
SSL	2.65	CIFAR-10	300
SSL+BPP	2.73	CIFAR-10	300
SSL+FPP	**2.56**	CIFAR-10	300

(c) **Dropout for skip-connection.** Abations on different dropout rate for skip connections.

Name	Test Error (%)	Dataset	Seach Epoch
no drop	2.63	CIFAR-10	300
0.5 drop	2.68	CIFAR-10	300
0.2 drop	**2.56**	CIFAR-10	300

(b) **FPP time span.** Abations for different ratios of searching phases.

Name	Test Error (%)	Ratio	Dataset	Seach Epoch
shorter	2.65	$[0.2, 0.6, 0.2]$	CIFAR-10	300
longer	2.68	$[0.2, 0.2, 0.6]$	CIFAR-10	300
default	**2.56**	$[0.2, 0.4, 0.4]$	CIFAR-10	300

(d) **Proxy-free search.** Ablations for searching w/o proxy dataset.

Name	Accuracy (%) Top-1	Top-5	Search Epoch	Search Dataset	Train Dataset
transfer	69.6	88.45	600	CIFAR-10	ImageNet-tiny
proxy-free	**70.53**	88.54	300	ImageNet-tiny	ImageNet-tiny

Time span of FPP. In this subsection, we study the role of time span in the FPP process and recommend the default setting. We allow a longer and a shorter FPP time span by adjusting the ratio to $[0.2, 0.2, 0.6]$ and $[0.2, 0.6, 0.2]$, respectively. Table 3b compares the result of different time spans. It is shown that a balanced split between phase II and phase III strikes for a better result. Shorter FPP may result in the under-optimization of a single cell during the pruning phase. Longer FPP may squeeze the space of the normal bi-level optimization process, leading to potentially sub-optimal results. And a longer searching duration will relieve pressures of both phases by allowing more epochs at each phase and thus leads to a better result, as shown in Fig. 4.

Dropout for skip-connection. Here we add a dropout rate for skip-connection during the lower-level optimization process. As shown in Table 3c, using $p = 0.2$ (result from a coarse grid search) improves the overall quality of the architecture. However, the architecture quality is relatively sensitive to the dropout rate as the skip-connection is a crucial component to prevent gradient vanishing issue. Over-suppressing of the skip-connection can hamper the performance.

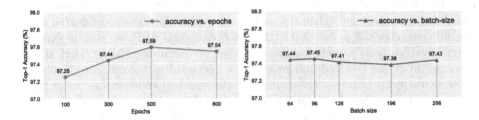

Fig. 4. For ablations of searching epochs, we fix the batch-size to 96 and increasing the searching epochs from 100 to 800. As for ablations of the batch-size, we fix the searching epoch as 300 and scale batch-size from 64 and 256.

Proxy-free search. It is well-established [3,11,29] that the performance of transfer learning drops when the gap between the target domain and the pre-trained domain is large. Given these priors, we focus on how proxy-free search can benefit the architecture quality. We use ImageNet-tiny [32] as the target domain and treat CIFAR-10 [15] as the proxy domain. Then we carry out transfer learning and proxy-free search. We doubled the search epochs on CIFAR-10 to keep the same iterations used during the search (ImageNet-tiny contains 10^5 images with size 64×64, and CIFAR-10 contains 5×10^4 images with size 32×32 in the training dataset). So the only difference between the two settings is the domain gap (image dimension, context, etc.) itself. As shown in Table 3d, proxy-free search results in a better architecture quality. This result suggests that given the existence of labeled proxy datasets, it is still favorable to search on the target domain directly without the label.

Impact of epochs and batch-size. As verified in self-supervised weight pre-training [3,9,10], network weights benefit from longer training and larger mini-batch size. Therefore, we examine how these two hyper-parameters affect the architecture searching process. As displayed in Fig. 4, in consent with SSL weight pre-training, the quality of searched architecture also improves as the searching duration increases. It saturates around 500 epochs on CIFAR-10/100. This result encourages a longer searching epoch and manifests that our design does not suffer from the mode collapse issue [2,17]. And even searching for only 100 epochs (cost 0.4 GPU-day) on CIFAR-10, SSWP-NAS still surpasses DARTS (1_{st} order) [22].

Yet, the searched architecture does not take advantage of a larger mini-batch size in our approach (see Fig. 4). As a result, one can use SSWP-NAS reliably under the limited GPU memory without worrying about the degeneration of architecture quality. However, this conclusion is only effective under the scenario where the magnitude of mini-batch size is a few hundred. We do not verify the impact of mini-batch size in thousands as used in several self-supervised weight-pretraining works [3,5,9,10].

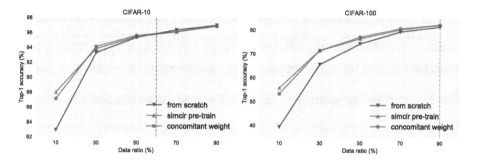

Fig. 5. Performances of the random initialization, simclr weight pre-train and concomitant weights on CIFAR-10/100 under different training data proportions.

4.4   Weight-preserving Benefits Semi-supervised Learning

In this section, we use both the searched architecture and concomitant weights from SSWP-NAS and probe their performances under semi-supervised scenarios. In particular, we first search using SSWP-NAS for 500 epochs with batch-size 96 on CIFAR dataset (train data). Then we gradually reduce the labeled training data from 90% to 10% with step-size 20% to mimic semi-supervised learning scenarios. Two baselines are included to demonstrate the effectiveness of concomitant weights. The first baseline noted as *random initialization* only uses the architecture. It is then trained from scratch using the given data ratio with a random initialization. The second baseline,*simclr pre-train*, in which we take the architecture searched by SSWP-NAS and pre-train it using SimCLR [3] for another 500 epochs using the same batch-size. The second baseline corresponds to a common two-stage framework without weight-preserving property. All three methods share the same architecture and the only difference is how they are initialized. We run the above settings on both CIFAR-10 and CIFAR-100.

As shown in Fig. 5, concomitant weights transfer clear positive information by outperforming the random initialization significantly. The gap between the concomitant weights and random initialization is bridged when using around 60% train data on CIFAR-10. However, this benefit diminishes not until around 90% of the train data on CIFAR-100. This result agrees with our intuition that self-supervised weight pre-training contributes more when labels are relatively scarce for each category, and the task is more challenging. More importantly, concomitant weights also surpass the two-stage framework pre-trained using Sim-CLR [3] except for the extremely scare data setting (with only 10% data). On the one hand, this result substantiates our contribution by merging the common two-stage pipeline into one-stage; on the other hand, it further suggests an interesting potential that evolving both architecture and weights simultaneously may serve as a better paradigm than the current isolated manner.

To better understand the differences between two-stage pre-trained weights and our concomitant weights, we further verify their reactions to different learning rates. Here we use 50% data from CIFAR-100 as a proxy-task. Table 4a shows that, unlike the typical two-stage pre-trained weights, concomitant weights consistently enjoy a larger learning rate. This result implies different statistical distributions between the concomitant weights and two-stage pre-trained weights.

Table 4. Ablation studies regarding concomitant weights.

(a) Comparisons of the two-stage pre-trained weights with the concomitant weights under different learning rates.

| Downstream task learning rate | Accuracy (%) | | Train data ratio (%) | Dataset |
	SimCLR Pre-train	Concomitant Weight		
0.01	75.82	73.18	50%	CIFAR-100
0.025	**76.1**	74.96	50%	CIFAR-100
0.1	76.0	77.2	50%	CIFAR-100
0.2	75.99	**77.54**	50%	CIFAR-100
0.5	74.18	75.7	50%	CIFAR-100

(b) Impact of dropout and FPP on concomitant weights.

FPP	Dropout	Relative gain	Train data ratio	Dataset
✗	✗	-	50%	CIFAR-100
✗	✓	0.23	50%	CIFAR-100
✓	✗	0.01	50%	CIFAR-100
✓	✓	0.19	50%	CIFAR-100

Finally, we also investigate the impact of proposed modules on the quality of concomitant weights. Since architecture and concomitant weights are evolved simultaneously, we can not drive to a conclusion by directly comparing the accuracy. To this end, we use *relative gain* to isolate the impact of the target operation towards the concomitant weights. $relative\ gain = (acc_f^{w/} - acc_s^{w/}) - (acc_f^{w/o} - acc_s^{w/o})$ where $w/$ and w/o denotes with and without concomitant weights. Subscript f and s represents fine-tuning and train from scratch, respectively. By doing so, we offset the influence of the architecture structure and focus on concomitant weights. As exhibited in Table 4b, dropout improves the quality of concomitant weights, this result coincides with our assumption that insufficient update of parameterized operation may hinder the quality of the concomitant weights. And according to experiments, FPP does not have a clear impact on overall quality of concomitant weights.

5 Conclusion

In this work, instead of trying to further reduce the computational overhead of the search process, the proposed SSWP-NAS strikes for a simplified workflow of NAS. It enjoys both self-supervising and weight-preserving two properties. Experiments show that self-supervised learning consistently benefits SSWP-NAS, and the concomitant weights successfully merge the two-stage framework into the one stage. Comprehensive ablation studies substantiate the effectiveness of our proposed designs. For the future work, it is important to probe why and how architecture and concomitant weights can boost each other. And it is also compelling to design the new self-supervised paradigm specifically for joint-optimization of architecture and concomitant weights. From the practical consideration, enabling multi-objective and hardware-aware learning will be useful.

References

1. Anandalingam, G., Friesz, T.L.: Hierarchical optimization: an introduction. Ann. Oper. Res. **34**(1), 1–11 (1992)
2. Cai, H., Zhu, L., Han, S.: ProxylessNAS: direct neural architecture search on target task and hardware. In: International Conference on Learning Representations (2019)
3. Chen, T., Kornblith, S., Norouzi, M., Hinton, G.: A simple framework for contrastive learning of visual representations. (2020) arXiv preprint arXiv:2002.05709
4. Chen, X., Xie, L., Wu, J., Tian, Q.: Progressive differentiable architecture search: bridging the depth gap between search and evaluation. In: Proceedings of the IEEE International Conference on Computer Vision, pp. 1294–1303 (2019)
5. Chen, X., He, K.: Exploring simple siamese representation learning. In: 2021 IEEE/CVF Conference on Computer Vision and Pattern Recognition (CVPR), pp. 15745–15753 (2021)
6. Colson, B., Marcotte, P., Savard, G.: An overview of bilevel optimization (2007).https://doi.org/10.1007/s10479-007-0176-2
7. Doersch, C., Gupta, A.K., Efros, A.A.: Unsupervised visual representation learning by context prediction. In: 2015 IEEE International Conference on Computer Vision (ICCV), pp. 1422–1430 (2015)
8. Ghiasi, G., Lin, T.Y., Pang, R., Le, Q.V.: NAS-FPN: Learning scalable feature pyramid architecture for object detection. In: 2019 IEEE/CVF Conference on Computer Vision and Pattern Recognition (CVPR), pp. 7029–7038 (2019)
9. Grill, J.B., et al.: Bootstrap your own latent - a new approach to self-supervised learning. In: Larochelle, H., Ranzato, M., Hadsell, R., Balcan, M.F., Lin, H. (eds.) Advances in Neural Information Processing Systems. Curran Associates, Inc. 33, pp. 21271–21284 (2020)
10. He, K., Fan, H., Wu, Y., Xie, S., Girshick, R.B.: Momentum contrast for unsupervised visual representation learning. In: 2020 IEEE/CVF Conference on Computer Vision and Pattern Recognition (CVPR), pp. 9726–9735 (2020)
11. He, K., Girshick, R.B., Dollár, P.: Rethinking imagenet pre-training. In: 2019 IEEE/CVF International Conference on Computer Vision (ICCV), pp. 4917–4926 (2019)
12. Howard, A.G., et al.: Searching for mobilenetv3. In: 2019 IEEE/CVF International Conference on Computer Vision (ICCV), pp. 1314–1324 (2019)
13. Huang, G., Liu, Z., Van Der Maaten, L., Weinberger, K.Q.: Densely connected convolutional networks. In: Proceedings of the IEEE conference on computer vision and pattern recognition, pp. 4700–4708 (2017)
14. Kaplan, S., Giryes, R.: Self-Supervised neural architecture search. CoRR abs/2007.01500 (2020)

15. Krizhevsky, A.: Learning multiple layers of features from tiny images. Tech. rep. (2009)
16. Li, J., Zhou, P., Xiong, C., Hoi, S.: Prototypical contrastive learning of unsupervised representations. In: International Conference on Learning Representations (2021)
17. Liang, H., et al.: DARTS+: improved differentiable architecture search with early stopping. CoRR abs/1909.06035 (2019)
18. Liu, C., et al.: Auto-DeepLab: Hierarchical neural architecture search for semantic image segmentation. In: 2019 IEEE/CVF Conference on Computer Vision and Pattern Recognition (CVPR), pp. 82–92 (2019)
19. Liu, C., Doll'ar, P., He, K., Girshick, R.B., Yuille, A.L., Xie, S.: Are labels necessary for neural architecture search? In: ECCV (2020)
20. Liu, C., et al.: Progressive neural architecture search. In: Proceedings of the European Conference on Computer Vision (ECCV), pp. 19–34 (2018)
21. Liu, H., Simonyan, K., Vinyals, O., Fernando, C., Kavukcuoglu, K.: Hierarchical representations for efficient architecture search. In: International Conference on Learning Representations (2018)
22. Liu, H., Simonyan, K., Yang, Y.: DARTS: differentiable architecture search. In: International Conference on Learning Representations (2019)
23. Luo, R., Tian, F., Qin, T., Chen, E., Liu, T.Y.: Neural architecture optimization. In: Proceedings of the 32nd International Conference on Neural Information Processing Systems, pp. 7827–7838 (2018)
24. Ma, N., Zhang, X., Zheng, H.T., Sun, J.: ShuffleNet v2: practical guidelines for efficient CNN architecture design. In: Proceedings of the European Conference on Computer Vision (ECCV) September 2018
25. Nguyen, N., Chang, J.M.: Contrastive self-supervised neural architecture search. CoRR abs/2102.10557 (2021)
26. Noroozi, M., Favaro, P.: Unsupervised learning of visual representations by solving jigsaw puzzles. In: ECCV (2016)
27. van den Oord, A., Li, Y., Vinyals, O.: Representation learning with contrastive predictive coding. ArXiv abs/1807.03748 (2018)
28. Pham, H., Guan, M., Zoph, B., Le, Q., Dean, J.: Efficient neural architecture search via parameters sharing. In: Dy, J., Krause, A. (eds.) Proceedings of the 35th International Conference on Machine Learning. Proceedings of Machine Learning Research, vol. 80, pp. 4095–4104. PMLR 10–15 Jul 2018
29. Raghu, M., Zhang, C., Kleinberg, J., Bengio, S.: Transfusion: understanding transfer learning for medical imaging. In: Advances in Neural Information Processing Systems. vol. 32. Curran Associates, Inc. (2019)
30. Real, E., Aggarwal, A., Huang, Y., Le, Q.V.: Regularized evolution for image classifier architecture search. Proceedings of the AAAI Conference on Artificial Intelligence 33(01), 4780–4789 (2019)
31. Ruder, S.: An overview of gradient descent optimization algorithms. CoRR abs/1609.04747 (2016)
32. Russakovsky, O., Deng, J., Su, H., Krause, J., Satheesh, S., Ma, S., Huang, Z., Karpathy, A., Khosla, A., Bernstein, M., Berg, A.C., Fei-Fei, L.: ImageNet Large Scale Visual Recognition Challenge. Int. J. Comput. Vision 115(3), 211–252 (2015). https://doi.org/10.1007/s11263-015-0816-y
33. Shu, Y., Wang, W., Cai, S.: Understanding architectures learnt by cell-based neural architecture search. In: International Conference on Learning Representations (2020)

34. Sohn, K.: Improved deep metric learning with multi-class n-pair loss objective. In: Lee, D., Sugiyama, M., Luxburg, U., Guyon, I., Garnett, R. (eds.) Advances in Neural Information Processing Systems. vol. 29. Curran Associates, Inc. (2016)
35. Szegedy, C., Vanhoucke, V., Ioffe, S., Shlens, J., Wojna, Z.: Rethinking the inception architecture for computer vision. In: Proceedings of the IEEE Conference on Computer Vision and Pattern Recognition (CVPR) June 2016
36. Tian, Y., Krishnan, D., Isola, P.: Contrastive multiview coding (2020)
37. Wang, N., et al.: NAS-FCOS: Fast neural architecture search for object detection. In: IEEE/CVF Conference on Computer Vision and Pattern Recognition (CVPR) June 2020
38. Wu, Z., Xiong, Y., Yu, S.X., Lin, D.: Unsupervised feature learning via non-parametric instance discrimination. In: 2018 IEEE/CVF Conference on Computer Vision and Pattern Recognition, pp. 3733–3742 (2018)
39. Xie, S., Zheng, H., Liu, C., Lin, L.: SNAS: stochastic neural architecture search. In: International Conference on Learning Representations (2019)
40. Xu, Y., et al.: Pc-darts: Partial channel connections for memory-efficient architecture search. In: International Conference on Learning Representations (2020)
41. Ying, C., Klein, A., Christiansen, E., Real, E., 0002, K.M., Hutter, F.: NAS-Bench-101: Towards Reproducible Neural Architecture Search. In: Proceedings of the 36th International Conference on Machine Learning, pp. 7105–7114. PMLR (2019)
42. Zhang, R., Isola, P., Efros, A.A.: Split-brain autoencoders: Unsupervised learning by cross-channel prediction. In: 2017 IEEE Conference on Computer Vision and Pattern Recognition (CVPR), pp. 645–654 (2017)
43. Zhang, X., Hou, P., Zhang, X., Sun, J.: Neural architecture search with random labels. In: 2021 IEEE/CVF Conference on Computer Vision and Pattern Recognition (CVPR), pp. 10902–10911 (2021)
44. Zhang, Y., Qiu, Z., Liu, J., Yao, T., Liu, D., Mei, T.: Customizable architecture search for semantic segmentation. In: 2019 IEEE/CVF Conference on Computer Vision and Pattern Recognition (CVPR), pp. 11633–11642 (2019)
45. Zhou, H., Yang, M., Wang, J., Pan, W.: BayesNAS: A Bayesian Approach for Neural Architecture Search. In: Proceedings of the 36th International Conference on Machine Learning, pp. 7603–7613. PMLR (2019)
46. Zhuang, C., Zhai, A., Yamins, D.: Local aggregation for unsupervised learning of visual embeddings. In: 2019 IEEE/CVF International Conference on Computer Vision (ICCV), pp. 6001–6011 (2019)
47. Zoph, B., Le, Q.V.: Neural architecture search with reinforcement learning. (2017) ArXiv abs/1611.01578
48. Zoph, B., Vasudevan, V., Shlens, J., Le, Q.V.: Learning transferable architectures for scalable image recognition. In: Proceedings of the IEEE conference on computer vision and pattern recognition, pp. 8697–8710 (2018)

MoQuad: Motion-focused Quadruple Construction for Video Contrastive Learning

Yuan Liu[1(✉)], Jiacheng Chen[2], and Hao Wu[3]

[1] The University of Hong Kong, Hong Kong, China
`3463423099@qq.com`
[2] Simon Fraser University, Burnaby, Canada
[3] Bytedance Inc, Beijing, China

Abstract. Learning effective motion features is an essential pursuit of video representation learning. This paper presents a simple yet effective sample construction strategy to boost the learning of motion features in video contrastive learning. The proposed method, dubbed **Mo**tion-focused **Quad**ruple Construction (MoQuad), augments the instance discrimination by meticulously disturbing the appearance and motion of both the positive and negative samples to create a quadruple for each video instance, such that the model is encouraged to exploit motion information. Unlike recent approaches that create extra auxiliary tasks for learning motion features or apply explicit temporal modelling, our method keeps the simple and clean contrastive learning paradigm (*i.e.*, SimCLR) without multi-task learning or extra modelling. In addition, we design two extra training strategies by analyzing initial MoQuad experiments. By simply applying MoQuad to SimCLR, extensive experiments show that we achieve superior performance on downstream tasks compared to the state of the arts. Notably, on the UCF-101 action recognition task, we achieve 93.7% accuracy after pre-training the model on Kinetics-400 for only 200 epochs, surpassing various previous methods.

Keywords: Video representation learning · Contrastive learning

1 Introduction

The recent progress of self-supervised visual representation learning has provided exciting opportunities for exploiting the huge amount of unlabelled web images and videos to train extremely powerful neural networks. In the image domain, seminal works based on contrastive learning (*e.g.*, SimCLR [7], MoCo [20]) have largely closed the performance gap between self-supervised learning methods and the supervised learning counterparts across various downstream tasks. However,

---

Y. Liu and J. Chen–Work done when interned at Bytedance Inc.

Supplementary Information The online version contains supplementary material available at https://doi.org/10.1007/978-3-031-25069-9_2.

ⓒ The Author(s), under exclusive license to Springer Nature Switzerland AG 2023
L. Karlinsky et al. (Eds.): ECCV 2022 Workshops, LNCS 13804, pp. 20–38, 2023.
https://doi.org/10.1007/978-3-031-25069-9_2

the progress in the video domain lags behind as videos contain rich motion information, making the learning of video representations more challenging.

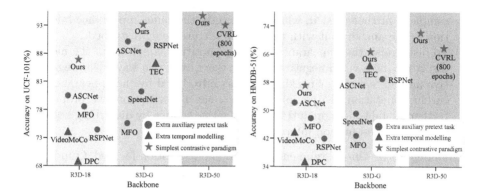

Fig. 1. Performance of MoQuad on UCF-101 and HMDB-51 action recognition (transfer learning pre-trained on Kinetics-400 [5] for only 200 epochs) with different video backbones, compared to other video self-supervised learning methods. The full results corresponding to this figure are available in Table 1.

Contrastive learning is still an effective framework for video self-supervised learning [12,38], but to further encourage the learning of motion-oriented features, extra designs are usually needed. For example, RspNet [6], ASCNet [23] and Pace [40] propose an extra motion branch and conduct multi-task learning, while DPC [16], MemDPC [16] and VideoMoCo [36] apply explicit temporal modeling for better modeling motion information.

This paper aims to keep the simplest video contrastive learning paradigm (*i.e.*, single-task, no extra modelling), but significantly improve the learning of motion features. Our main contribution is a carefully designed sample construction strategy that enriches the vanilla instance discrimination and enforces the model to capture effective motion information to finish the discrimination task.

Concretely, for each anchor clip, the instance discrimination [12] task requires the model to pull clips from the same video instance (positive sample) closer while pushing clips from different instances (negative sample) apart. Our method then builds upon three progressive analyses:

(1) Positive clip pairs from the same video instance share similar appearance and motion features, but ConvNets tend to excessively exploit the appearance clues (*e.g.*, background) to hack the instance discrimination task, thus neglecting the motion features [41]. To prevent the model from overly relying on the appearance clues, it is necessary to disturb the appearance of the positive sample.

(2) However, compared to the negative samples from other videos, the appearance disturbed positive sample still shares higher similarity with the anchor in terms of the appearance, thus still leaving space for the model to bypass the motion feature. To alleviate this, we argue that an intra-video negative sample is further needed, which has exactly the same appearance information as the anchor, but with the motion feature being disturbed.

(3) To distinguish the appearance-disturbed positive sample from the motion-disturbed intra-video negative sample, there exist two ways: learning effective motion features or hacking the bias introduced by the appearance disturbing operation. To completely block the latter option, we propose to further include one more intra-video negative sample by disturbing both the appearance and motion information.

Combining these three arguments, our method, **Mot**ion-focused **Quad**ruple Construction(MoQuad), creates a quadruple for each video instance consisting of 1) an anchor, 2) an appearance-disturbed positive sample, 3) a motion-disturbed intra-video negative sample, and 4) a motion-and-appearance-disturbed intra-video negative sample, and follows the standard contrastive learning paradigm to train the model. We also conduct detailed analyses to determine the appropriate disturbing operations for motion and appearance.

By simply applying MoQuad to SimCLR, experiments show that we considerably improve the transfer learning performance. In addition, by analyzing the initial MoQuad experiments, we further derive two training strategies that bring consistent improvements: 1) a warm-up strategy that warms up the model with an appearance-focused instance discrimination task, such that the model can concentrate more on learning motion features when training with MoQuad and being less distracted by appearance information; and 2) a hard negative sample mining strategy that pushes the model to better discriminate videos with potentially similar motion types.

Without any multi-tasking learning or extra temporal modelling, MoQuad surpasses state-of-the-art video self-supervised learning methods on various downstream tasks, including action recognition and video retrieval, with shorter pre-training schedules (See Fig. 1). Extensive ablation studies and analyses further justify the effectiveness of different components of our method.

2   Related Work

Self-supervised image representation learning Self-supervised image representation learning has become a hot topic in recent years. Previous studies focus on creating pretext tasks explicitly, such as predicting the rotation angle of an image [13], solving a jigsaw puzzle task [34] and solving a relative patch predicting task [47]. Recently, some works optimize clustering and representation learning jointly [4,29,45], or learning image visual representation by discriminating instances from each other through contrastive learning. [7] pulls two augmented crops from the same images together while pushing crops from different images apart, using InfoNCE [35]. [42] proposes to use a memory bank to store all image representations. To keep the representations consistent, [20] proposes MoCo, which stores image embeddings from a momentum update encoder in a queue and achieves superior performance. The above-mentioned methods usually rely on a large number of negative samples. Meanwhile, BYOL [15] and Siamese [8] learn meaningful representations by only maximizing the similarity of two augmented positive samples without using any negative samples.

Self-supervised video representation learning Similar to image- representation learning approaches, some video representation learning approaches create pretext tasks [1,9,10,22,25,26,31]. Recently, contrastive learning becomes the mainstream in self-supervised video representation learning [16–18,38]. For example, CVRL [38] pulls two augmented clips from the same video together while pushing clips from different videos apart. Since the motion information is very important in videos, some approaches, *e.g.*[41], focus on extracting the motion of videos by preventing the model from focusing on the appearance. To extract both the motion and appearance information of videos, [6,23,40] propose to create two branches and achieve satisfactory results.

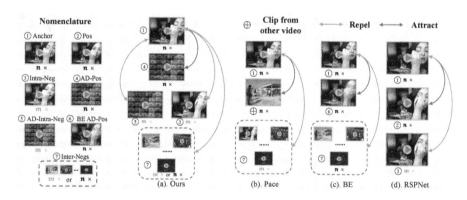

Fig. 2. Comparison between MoQuad with other motion-focused tasks from previous methods. (a): MoQuad (b): motion task of Pace [40] (c): motion task of BE [41] (d): motion task of RSPNet [6]. $n\times$ and $m\times$ denote different video playback speeds. We disturb the appearance of the positive sample to create the appearance-disturbed positive sample and introduce two intra-video negative samples, one with motion disturbed and the other with both motion and appearance disturbed. Inter-Negs are clips from other videos. Please see Sec. 3.1 for full details.

3   Method

Our method consists of 1) the MoQuad sample construction strategy, which is the key innovation of the paper (Sec 3.1), and 2) two extra training strategies derived by analyzing our initial experiments with MoQuad (Sec 3.2).

3.1   MoQuad Sample Construction

Definition of the motion-focused quadruple. Following the three-step analyses in Sec.1, MoQuad creates a quadruple from each video instance by carefully disturbing the motion and appearance information, aiming to enforce the learning of effective motion features. Given a video V_i, the quadruple consists of:

- An anchor (Anchor): a clip randomly sampled from V_i, denoted as A_i.
- An appearance-disturbed positive sample (AD-Pos): a clip randomly sampled from V_i, but with its appearance disturbed, denoted as $\bar{P}_i$.
- An intra-video negative sample (Intra-Neg): a clip randomly sampled from V_i, but with its motion information disturbed, denoted as N_i.
- An appearance-disturbed Intra-Neg (AD-Intra-Neg): a clip randomly sampled from V_i, but with both motion and appearance disturbed, denoted as $\bar{N}_i$.

We create the quadruple for each video in a batch, and clips from other videos serve as the inter-video negative samples (Inter-Negs) for the Anchor.

Instance Discrimination With MoQuad. We pass A_i, $\bar{P}_i$, N_i, and $\bar{N}_i$ to the feature extractor F and projection head H to get the clip features, denoted as z_{iA}, $\bar{z}_{iP}$, z_{iN}, and $\bar{z}_{iN}$, respectively. All the clip features are normalized. The vanilla instance discrimination task is augmented with per-instance quadruple:

$$L_m = -\sum_{i=0}^{B-1} \log \frac{\exp\left(z_{iA}\bar{z}_{iP}/\tau\right)}{\exp\left(z_{iA}\bar{z}_{iP}/\tau\right) + \sum S_{\text{Intra}} + \sum S_{\text{Inter}}} \tag{1}$$

τ is the temperature and B is the batch size. We use S_{Intra} and S_{Inter} to denote $\{\exp(z_{iA}z_{iN}/\tau), \exp(z_{iA}\bar{z}_{iN}/\tau)\}$ and $\{\exp\left(z_{iA}z_j/\tau\right)\}_{j\neq i}$, respectively. Note that z_j includes z_{jA}, $\bar{z}_{jP}$, z_{jN} and $\bar{z}_{jN}$, where i and j are the indices of two different videos.

Criteria for Choosing Disturbing Operations. Determining the concrete disturbing operations is a key step to complete the design of MoQuad, and we first set up the desiderata for them:

- Appearance: The appearance disturbing operation is to augment the positive samples (AD-Pos). It should inject sufficiently strong noise into the video clip on the premise of preserving the original motion information, such that the model can be pushed hard to learn motion-focused features. Note that this operation is also applied to AD-Intra-Neg to prevent the model from hacking the bias introduced by the operation (See the analyses in Sec. 1).
- Motion: The motion disturbing operation is to augment the negative samples (*i.e.*, Intra-Neg and AD-Intra-Neg). In contrast to the appearance disturbing operation, it should only introduce subtle change to the motion information of a clip to make sure that the discrimination task cannot be trivially solved by the model. Similar arguments are also discussed in [3].

With these criteria, we are now ready to choose the disturbing operations.

Appearance Disturbing Operation. Appearance information contains both low and high-frequency information. BE [41] (Fig. 3(a)) provides an off-the-shelf operation by adding to each video frame a noise image randomly picked from the current video. However, this operation still keeps most of the high-frequency information of the original video and is not aggressive enough according to our criteria. To break both the low and high-frequency signals and make the operation stronger, we propose a simple improvement for constructing the noise image. Given a video V_i, we create an empty noise image (with the same shape as V_i), denoted as D. We then slice D into $k \times k$ windows (we use $k = 5$), and randomly sample a frame from another video to replace each empty window in D. We insert this noise image D to each frame of V_i by a weighted average of $\bar{V}_i = (1 - \lambda)V_i + \lambda D$. Figure 3(c) illustrates this operation, and we name it as Repeated Appearance Disturbance (RAD). Note that the repeated frame can be sampled from the current video(Fig. 3(b)), but then it only introduces redundant signals to V_i and the RAD (intra) is not as strong as RAD (inter).

(a) BE   (b) RAD (Intra)   (c) RAD (Inter)

Fig. 3. Appearance disturbing operations. (a) the operation from BE [41] (b) RAD with the noise image from the current video (intra), and (c) RAD with noise image extracted from another video (inter).

Motion Disturbing Operation. The motion information in videos can be generally represented as the temporal gradients [41]. There are three common approaches to disturb the temporal information of a video in the literature [6, 27, 33]: 1) change the playback speed (Speed), 2) reverse the frame order (Reverse), or 3) shuffle the frame order (Shuffle). Based on our criteria above, Reverse and Shuffle could change the motion information too much, thus making the negative samples trivial to be discriminated by the model. On the contrary, Speed only modifies the motion information subtly and is a more appropriate choice for MoQuad. To be more concrete, given a video V_i, we extract two clips C_i^n and C_i^m with dilation rate n and m, respectively. C_i^m and C_i^n are considered as an Intra-Neg for each other. With all operations determined, Fig. 2(a) is an illustration for the instance discrimination task with MoQuad quadruples.

Relation to Previous Methods. Pace [40], RspNet [6] and BE [41] are related previous methods that construct extra samples to improve the learning of motion feature for instance discrimination. All of them demonstrate different properties with MoQuad, and we make an illustrative comparison in Fig. 2.

Pace (Fig. 2(b)) enforces the model to pull clips of the same playback speed closer, no matter if these clips are from the same video or not. RspNet (Fig. 2(d)) argues that it is inappropriate to compare the speed of clips of different videos, thus only requiring the model to distinguish the playback speed of clips from the same video. Both Pace and RspNet add an extra speed-discrimination task to the contrastive learning framework. However, distinguishing the speed requires the model to focus more on low-level motion features (*e.g.*, the scale of the temporal gradients), and the model could fail to capture high-level motion semantics (*e.g.* motion types) without other negative samples. BE (Fig. 2(c)) does not use any extra auxiliary task. Instead, it breaks the appearance information of the positive sample to alleviate the negative impact of background bias [41]. However, as we analyzed and visualized above, its disturbing operation keeps most of the original appearance clues in the positive sample, and the model might still bypass the learning of effective motion features.

MoQuad (Fig. 2(a)) does not suffer from the potential issues of these methods. Compared to Pace and RspNet, we consider playback speed as a way to disturb the motion feature of Intra-Neg, instead of making it the only discrimination target (*i.e.*, our Inter-Negs can have the same speed but different motion type as the Anchor). Compared to BE, we employ stronger appearance disturbance by RAD and create extra motion-disturbed negative samples to further prevent the model from bypassing the learning of motion features.

3.2   Extra Training Strategies for MoQuad

In this section, we will introduce the two extra training strategies.

Appearance Task Warm-Up. When training with MoQuad, the model exhibits two-stage learning progress: it starts with learning appearance-oriented features and gradually shifts to capture motion-oriented features (visualization in Fig. 4(left)), suggesting that the learning of appearance and motion can be better disentangled. Inspired by this discovery, we propose to use an appearance-focused task to warm up the model before training with MoQuad. We thus borrow the appearance task from RspNet [6] directly use it as the warm-up task. With the warm-up, MoQuad focuses on learning motion features from the very beginning of the training and learns better motion features (Fig. 4(right)). Algorithm 1 details the training schedule with the appearance task warm-up. The appearance task is simply an adapted instance discrimination task:

Given a batch of videos $V_B = \{V_i\}_{i=0}^{B-1}$, for each video V_i in V_B, we extract two clips with dilation n and m respectively, denoted as C_i^n and C_i^m. We pass C_i^n and C_i^m to feature extractor F and projection head H to get the features, and normalize these features to get the final feature vectors: z_i^n and z_i^m. The InfoNCE loss [35] again pulls clips from the same video together while pushing clips from different videos apart:

$$L_a = -\sum_{i=0}^{B-1} \log \frac{\exp\left(z_i^n z_i^m / \tau\right)}{\exp\left(z_i^n z_i^m / \tau\right) + \sum_{i \neq j, s \in \{n,m\}} \exp\left(z_i^n z_j^s / \tau\right)} \qquad (2)$$

Hard Negative Sample Mining. By visualizing the instance discrimination results of MoQuad, we find that the model sometimes has difficulty in distinguishing the positive samples from hard Inter-Negs. These hard inter-video negative samples turn out to share similar motion semantics as the Anchor (please check the Supplementary for the visualization and more analyses). Based on this observation, we propose a hard negative sample mining strategy to enforce the model to push these hard Inter-Negs farther away from the Anchor. For all the $\exp(z_{iA} z_j / \tau)$ in Eq. 1, we first get the top-K elements by:

$$S_{\text{Topks}} = TopK(\{\exp(z_{iA} z_j / \tau)\}_{i \neq j}) \tag{3}$$

where $K = \beta \times len(\{\exp(z_{iA} z_j / \tau)\}_{i \neq j})$ and β is the percentage of the total negative samples that are treated as hard negatives. We then assign a large weight, $\alpha(\alpha > 1)$, to these hard negative samples, and augment Eq. 1 by:

$$L_{\text{m}} = -\sum_{i=0}^{B-1} \log \frac{\exp(z_{iA} \bar{z}_{iP} / \tau)}{\exp(z_{iA} \bar{z}_{iP} / \tau) + \alpha \sum S_{\text{Intra}} + \alpha \sum S_{\text{Topks}} + \sum S_{\text{Easy}}} \tag{4}$$

For notation simplicity, S_{Easy} is defined as $S_{\text{Easy}} = S_{\text{Inter}} - S_{\text{Topks}}$ where $-$ is the difference between two sets.

Algorithm 1. Two-Stage Training Mechanism

Require: video set V, feature extractor F, projection head H, total epochs E, current epoch e and the percentage of total epochs p, used for the appearance task.
1: Initialize $e \leftarrow 0$
2: // *Appearance Task Warmup*
3: **while** $e < p \times E$ **do**
4:     Sample a batch of videos $V_B = \{V_i\}_{i=0}^{B-1}$ from video sets V.
5:     For each video in V_B, we extract two clips, C_i^n and C_i^m.
6:     Pass these clips to F and H to get z_i^n and z_i^m.
7:     Use Eq.2 to finish the instance discrimination task.
8:     $e \leftarrow e + 1$
9: **end while**
10: // *Instance Discrimination with MoQuad quadruple*
11: **while** $p \times E \leqslant e < E$ **do**
12:     Sample a batch of videos $V_B = \{V_i\}_{i=0}^{B-1}$ from video sets V.
13:     For each video in V_B, we create the quadruple: A_i, $\bar{P}_i$, N_i and $\bar{N}_i$
14:     Pass A_i, $\bar{P}_i$, N_i, $\bar{N}_i$ to F and H to get z_{iA}, $\bar{z}_{iP}$, z_{iN} and $\bar{z}_{iN}$.
15:     Use Eq.1 to finish the instance discrimination task.
16:     $e \leftarrow e + 1$
17: **end while**

---

4   Experiments

We first describe the experimental setups in Sec. 4.1, including the datasets, pre-training details, evaluation settings, *etc.*. In Sec. 4.2, we compare MoQuad with the state of the arts on two downstream tasks: action recognition and video retrieval. We then provide ablation studies in Sec. 4.3 to validate different design choices. In Sec. 4.4, we further present additional analyses to help understand the effectiveness of our method.

4.1   Experimental Settings

Datasets. Four video action recognition datasets are covered in our experiments: Kinetics-400 (K400) [5], UCF101 [39], HMDB51 [30] and Something-Something-V2 (SSv2) [14]. K400 consists of 240K training videos from 400 human action classes, and each video lasts about 10 s. UCF101 contains 13,320 YouTube videos from 101 realistic action categories. HMDB51 has 6,849 clips from 51 action classes. SSv2 provides 220,847 videos from 174 classes and the videos contain more complex action information compared to UCF101 and HMDB51.

Self-supervised Pre-training. There are various pre-training settings in the literature, we follow the same hyper-parameter choices as CVRL [38], including the image resolution (224×224) and the number of frames per clip (16). We evaluate our method with three different backbones: R3D-18 [19], S3D-G [43], and R3D-50 [38], so that we can compare with various previous works that use backbones with different model size. When pre-trained on K400, SSv2, and UCF101, the model is trained for 200, 200, and 800 epochs, respectively, which follows the most common pre-training schedule in the literature [6,22,38,40]. Due to limited computational resources, we conduct ablation studies by pre-training the small R3D-18 backbone on UCF101 by default, unless otherwise specified. We use LARS [46] as our optimizer with a mini-batch of 256, which is consistent with SimCLR [7]. The learning rate is initialized to 3.2, and we use a half-period cosine learning rate decay scheduling strategy [21]. For appearance disturbing, we set $k = 5$ and uniformly sample λ from $[0.1, 0.5]$. For the warmup with appearance task, the appearance task takes up the first 20% of the training epochs. β and α are set to be 0.01 and 1.5 respectively for the hard negative sample mining.

Supervised Fine-tuning for Action Recognition. Consistent with previous works [16,23,38], we fine-tune the pre-trained models on UCF101 or HMDB51 for 100 epochs and report the accuracy of action recognition. We sample 16 frames with dilation of 2 for each clip. The batch size is 32, and we use LARS [46] as the optimizer. The learning rate is 0.08, and we employ a half-period cosine learning rate decay strategy.

Linear Evaluation Protocol for Action Recognition. We also evaluate the video representations with the standard linear evaluation protocol [38]. We sample 32 frames for each clip with a temporal stride of 2 from each video. The linear classifier is trained for 100 epochs.

Evaluation Details for Action Recognition. For computing the accuracy of action recognition, we follow the common evaluation protocol [11], which densely samples 10 clips from each video and employs a 3-crop evaluation. The softmax probabilities of all the 10 clips are averaged for each video to get the final classification results.

4.2   Comparison with State of the Arts

We compare MoQuad with state-of-the-art self-supervised learning approaches on action recognition and video retrieval.

Evaluation on action recognition. As presented in Table 1, we compare our method with state of the arts under various experimental setups, with UCF101 and HMDB51 as the downstream datasets. Our method always outperforms the previous works with a clear margin when given the same pre-training settings. We note that even without the two extra training strategies, the vanilla MoQuad is still superior to previous methods.

As mentioned in Sec 4.1, UCF101 and HMDB51 are relatively small-scale datasets. To strengthen the comparison results, we further evaluate our model with the linear evaluation protocol on K400 and SSv2 datasets (Table 2). Our method can still consistently outperform previous works even with smaller effective batch size (*i.e.*, the number of different video instances in a batch).

Evaluation on video retrieval. We evaluate the learned video representation with nearest neighbour video retrieval. Following previous works [6,40], we sample 10 clips for each video uniformly and pass them to the pre-trained model to get the clip features. Then, we apply average-pooling over the 10 clips to get the video-level feature vector. We use each video in the test split as the query and look for the k nearest videos in the training split of UCF101. The feature backbone is R3D-18, and the model is pre-trained on UCF101. The top-k retrieval accuracy ($k = 1, 5, 10$) is the evaluation metric. As in Table 3, we outperform previous methods consistently on all three metrics. Note that the gap between our method and recent work, ASCNet [23], is small, especially considering ASCNet has a smaller resolution. However, we do outperform ASCNet in Table 1 under the same experimental settings (*i.e.*, resolution 224, pre-trained on K400, and S3D-G backbone). A potential explanation is that fine-tuning could maximize the benefits of MoQuad.

Table 1. Action recognition results on UCF101 and HMDB51. We report the fine-tuning results of MoQuad and various baselines under different setups. MoQuad † is our method with the two extra training strategies. ^{800}: pre-trained for 800 epochs, which is longer than the standard training schedule (*i.e.*, 200 epochs).

Method	Dataset	Backbone	Frame	Resolution	UCF101	HMDB51
CoCLR [18]$_{2020}$	UCF101	S3D	32	128	81.4	52.1
BE [41]$_{2021}$	UCF101	R3D-34	16	224	83.4	53.7
MoQuad (Ours)	UCF101	R3D-18	16	224	80.9	51.0
MoQuad † (Ours)	UCF101	R3D-18	16	224	82.7	53.0
MoQuad † (Ours)	UCF101	S3D-G	16	224	**87.4**	**57.3**
Pace [40]$_{2020}$	K400	R(2+1)D	16	112	77.1	36.6
MemoryDPC [17]$_{2020}$	K400	R3D-34	40	224	86.1	54.5
BE [41]$_{2021}$	K400	R3D-34	16	224	87.1	56.2
CMD [24]$_{2021}$	K400	R3D-26	16	112	83.7	55.2
DPC [16]$_{2019}$	K400	R3D-18	40	224	68.2	34.5
VideoMoCo [36]$_{2021}$	K400	R3D-18	32	112	74.1	43.1
RSPNet [6]$_{2021}$	K400	R3D-18	16	112	74.3	41.8
MFO [37]$_{2021}$	K400	R3D-18	16	112	79.1	47.6
ASCNet [23]$_{2021}$	K400	R3D-18	16	112	80.5	52.3
MoQuad (Ours)	K400	R3D-18	16	224	85.6	56.2
MoQuad † (Ours)	K400	R3D-18	16	224	**87.3**	**57.7**
MFO [37]$_{2021}$	K400	S3D	16	112	76.5	42.3
CoCLR [18]$_{2020}$	K400	S3D	32	128	87.9	54.6
SpeedNet [2]$_{2020}$	K400	S3D-G	64	224	81.1	48.8
RSPNet [6]$_{2021}$	K400	S3D-G	64	224	89.9	59.6
TEC [28]$_{2021}$	K400	S3D-G	32	128	86.9	63.5
ASCNet [23]$_{2021}$	K400	S3D-G	64	224	90.8	60.5
MoQuad (Ours)	K400	S3D-G	16	224	91.9	64.7
MoQuad † (Ours)	K400	S3D-G	16	224	**93.0**	**65.9**
CVRL [38]$_{2020}^{800}$	K400	R3D-50	16	224	92.9	67.9
MoQuad (Ours)	K400	R3D-50	16	224	93.0	66.9
MoQuad † (Ours)	K400	R3D-50	16	224	93.7	68.0
MoQuad $^{\dagger\ 800}$ (Ours)	K400	R3D-50	16	224	**94.7**	**71.5**

Table 2. Action recognition results on two large-scale datasets. We further provide linear evaluation results on K400 and SSv2 to demonstrate the effectiveness of our method. Following the two recent works: CVRL [38] and COP$_f$ [22], a R3D-50 backbone is pre-trained on the training split of the corresponding dataset for 200 epochs before conducting the standard linear evaluation. *: our re-implemented version.

Method	Backbone	Frame	Batch Size	Resolution	Linear eval
Pre-trained and evaluated on K400 dataset					
CVRL [38]$_{2021}$	R3D-50	16	512	224	62.9
COP$_f$ [22]$_{2021}$	R3D-50	16	512	224	63.4
MoQuad (Ours)	R3D-50	16	256	224	64.4
Pre-trained and evaluated on SSv2 dataset					
CVRL [38]$_{2021*}$	R3D-50	16	256	224	31.5
COP$_f$ [22]$_{2021}$	R3D-50	16	512	224	41.1
MoQuad (Ours)	R3D-50	16	256	224	44.0

Table 3. Evaluation on the video retrieval task.

Method	Backbone	Frame	Resolution	Top-k		
				$k = 1$	$k = 5$	$k = 10$
ClipOrder [44]$_{2019}$	R3D-18	16	112	14.1	30.3	40.0
SpeedNet [2]$_{2020}$	S3D-G	64	224	13.0	28.1	37.5
MemDPC [17]$_{2020}$	R(2+1)D	40	224	20.2	40.4	52.4
VCP [32]$_{2019}$	R3D-18	16	112	18.6	33.6	42.5
Pace [40]$_{2020}$	R(2+1)D	16	224	25.6	42.7	51.3
CoCLR-RGB [18]$_{2020}$	S3D-G	32	128	53.3	69.4	76.6
RSPNet [6]$_{2021}$	R3D-18	16	112	41.1	59.4	68.4
ASCNet [23]$_{2021}$	R3D-18	16	112	58.9	76.3	82.2
MoQuad (Ours)	R3D-18	16	224	**60.8**	**77.4**	**83.5**

4.3   Ablation Studies

This subsection provides ablation studies for different design choices of the paper. As described in Sec. 4.1, all ablation experiments use an R3D-18 backbone due to the limited computational resources and a large number of trials.

The design of quadruple. We validate the effectiveness of each component of our quadruple in Table 4. We use SimCLR [7] as the base method and progressively add the elements of the quadruple. Both the fine-tuning and linear evaluation results are reported for UCF101 and HMDB51. Regardless of the pre-training dataset, each element of MoQuad quadruple improves the performance consistently. We would like to point out that K400 is much larger than UCF101 (see Sec. 4.1 for dataset stats), indicating that the benefits of our method do not degenerate as the size of the pre-training dataset grows.

The choice of disturbing operations. We present the results of different motion and appearance disturbing operations in Table 5. The left sub-table shows that RAD (inter) is better than the simple trick proposed by [41] and RAD (intra), justifying our analysis in Sec. 3.1.

Table 4. Ablation studies for components of MoQuad quadruple. We pre-train the R3D-18 backbone on UCF101 and K400, respectively. The fine-tuning and linear evaluation results on UCF101 and HMDB51 are reported.

AD-Pos	Intra-Neg	AD-Intra-Neg	Fine-tuning		Linear eval.	
			UCF	HMDB	UCF	HMDB
Pre-trained on UCF101 dataset						
-	-	-	76.0	44.3	68.8	33.6
✓	-	-	78.2	47.2	70.0	37.0
✓	✓	-	79.6	49.3	71.5	38.8
✓	✓	✓	**80.9**	**51.0**	**73.9**	**40.3**
Pre-trained on K400 dataset						
-	-	-	80.9	50.2	75.0	44.3
✓	-	-	82.1	53.6	76.2	47.4
✓	✓	-	84.5	55.0	77.3	48.4
✓	✓	✓	**85.6**	**56.2**	**79.2**	**50.1**

Table 5. Ablation studies for the disturbing operations. **(Left)** Different appearance disturbing operations. **(Right)** Different motion disturbing operations. The models are pre-trained on UCF101 for 800 epochs.

Operations	Fine-tuning		Linear eval.		Operations	Fine-tuning		Linear eval.	
	UCF	HMDB	UCF	HMDB		UCF	HMDB	UCF	HMDB
BE [41]	79.7	48.5	71.8	39.4	Reverse	76.3	45.4	69.8	34.7
RAD (Intra)	80.1	49.0	72.0	39.4	Shuffle	77.4	46.2	70.3	36.4
RAD (Inter)	80.9	51.0	73.9	40.3	Speed	80.9	51.0	73.9	40.3

As also discussed in Sec. 3.1, reversing the frame order (Reverse), shuffling the frames (Shuffle), and changing video playback speed (Speed) are three commonly used tricks to break the motion information of a video. As shown by the comparison in Table 5 (right), Speed produces the best results, which is consistent with our analyses in Sec. 3.1. Reverse and Shuffle could change the motion feature of a video too aggressively, thus making the negative sample easy to be discriminated by the model.

Extra Training Strategies. Table 1 has shown that the two training strategies provide consistent improvements to MoQuad. We further validate the effectiveness of each strategy and the choice of the hyper-parameters in Table 6. The hard negative sampling mining and the warm-up with appearance task increase accuracy with proper hyper-parameters. To help better understand the warm-up strategy, we plot the ranking of different training samples from our quadruple in the pre-training process in Fig. 4. As shown in the left figure, the ranking of the Intra-Negs decreases with that of AD-Pos, implying the model first focuses on the appearance of the video. While the right figure shows the warm-up strategy could increase the average ranking of the Intra-Negs during the training process. Note that a higher average ranking of Intra-Negs suggests that the model can better distinguish the subtle differences in motion information injected by the motion disturbing operation.

Table 6. Ablation studies for the two extra training strategies. (Left) The hard negative sample mining strategy. $\beta = 0$ and $\alpha = 0$ is the best entry in Table 4(top). **(Right)** Appearance task warm-up. The entry of 0% is the best entry in the left subtable. All models are pre-trained on UCF101 for 800 epochs.

β	α	Fine-tuning		Linear eval.		Warmup ratio	Fine-tuning		Linear eval.	
		UCF	HMDB	UCF	HMDB		UCF	HMDB	UCF	HMDB
0	0	80.9	51.0	73.9	40.3	0%	81.7	52.3	74.9	41.4
0.01	1.5	**81.7**	**52.3**	**74.9**	41.4	10%	**82.8**	53.0	75.6	42.0
0.01	2.0	81.5	51.6	74.8	**41.6**	20%	82.7	**53.0**	**76.6**	**42.9**
0.01	3.0	80.9	50.8	74.2	41.0	40%	81.0	50.8	74.2	40.8
0.05	1.5	79.3	49.7	72.7	40.3	60%	80.0	49.7	72.1	39.9

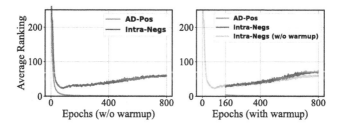

Fig. 4. Understanding the improvement brought by the warmup of appearance task. We plot the average rankings of the AD-Pos and the Intra-Negs. Intra-Negs includes Intra-Neg and the AD-Intra-Neg. Note that there are no Intra-Negs in the appearance task, thus we start plotting the right figure after the warm-up stage. The warm-up with the appearance task increases the average ranking of Intra-Negs.

4.4   More Analyses

Fine-grained action recognition results on SSv2. The paper's motivation is to improve the motion learning for contrastive learning. To further verify if MoQuad indeed learns better motion feature, we extend the linear evaluation on the SSv2 dataset to a fine-grained split of video categories. As presented in Table 7, we compare MoQuad with our base method, SimCLR [7], on different categories, under the linear evaluation protocol. All the models are pre-trained on the training split of SSv2. MoQuad and SimCLR get similar performances for the top categories, but MoQuad significantly outperforms SimCLR on the bottom ones. Specifically, our method improves SimCLR by more than 40% on "Move something down". Note that the bottom categories require accurate temporal features to be correctly recognized, indicating that MoQuad does teach the model to learn motion features better.

Table 7. Per-category accuracy on SSv2. We compare MoQuad with our base method, SimCLR [7], on fine-grained splits of SSv2 dataset to further demonstrate the capacity of our method to learn better motion features. The categories on the top can mostly be recognized using only the appearance information, while the bottom categories require deeper understanding of the motion in the video.

Video category in SSv2	SimCLR	MoQuad
Categories that do not need strong temporal information to classify		
Tear something into two pieces	75.2	83.9
Approach something with camera	60.1	88.9
Show something behind something	56.3	57.4
Plug something into two pieces	57.2	56.3
Hold something	27.4	26.3
Categories that need heavy temporal information to classify		
Move something and something closer	41.4	75.2
Move something and something away	29.2	74.5
Move something up	34.3	49.5
Move something down	24.3	68.6

Fig. 5. Visualizing the region of interest (RoI). We use CAM [48] to plot the RoI of the model. MoQuad focuses more on the moving regions than SimCLR.

RoI Visualization. We visualize the Region of Interest (RoI) of MoQuad and SimCLR in Fig. 5 using the tool provided by CAM [48]. Compared to SimCLR, MoQuad better constrains the attention to the moving regions, providing a clue that our method does improve the learning of motion-oriented features.

5 Conclusion

This paper proposes a simple yet effective method (MoQuad) to improve the learning of motion features in contrastive learning framework. A carefully designed sample construction strategy is the core of MoQuad, while two additional training strategies further improve the performance. By simply applying MoQuad to SimCLR, extensive experiments show that we outperform state-of-the-art self-supervised video representation learning approaches on both action recognition and video retrieval.

References

1. Agrawal, P., Carreira, J., Malik, J.: Learning to see by moving. In: 2015 IEEE International Conference on Computer Vision (ICCV), pp. 37–45 (2015)
2. Benaim, S., et al.: SpeedNet: Learning the speediness in videos. In: 2020 IEEE/CVF Conference on Computer Vision and Pattern Recognition (CVPR), pp. 9919–9928 (2020)
3. Cai, T.T., Frankle, J., Schwab, D.J., Morcos, A.S.: Are all negatives created equal in contrastive instance discrimination? (2020) arXiv preprint arXiv:2010.06682
4. Caron, M., Bojanowski, P., Mairal, J., Joulin, A.: Unsupervised pre-training of image features on non-curated data. In: 2019 IEEE/CVF International Conference on Computer Vision (ICCV), pp. 2959–2968 (2019)
5. Carreira, J., Zisserman, A.: Quo vadis, action recognition? a new model and the kinetics dataset. In: 2017 IEEE Conference on Computer Vision and Pattern Recognition (CVPR), pp. 4724–4733 (2017)
6. Chen, P., Huang, D., He, D., Long, X., Zeng, R., Wen, S., Tan, M., Gan, C.: RSPNet: Relative speed perception for unsupervised video representation learning. In: AAAI (2021)
7. Chen, T., Kornblith, S., Norouzi, M., Hinton, G.: A simple framework for contrastive learning of visual representations. In: III, H.D., Singh, A. (eds.) In: Proceedings of the 37th International Conference on Machine Learning. Proceedings of Machine Learning Research, vol. 119, pp. 1597–1607. PMLR (13–18 Jul 2020), https://proceedings.mlr.press/v119/chen20j.html
8. Chen, X., He, K.: Exploring simple siamese representation learning. (2020) ArXiv abs/2011.10566
9. Diba, A., Sharma, V., Gool, L., Stiefelhagen, R.: Dynamonet: dynamic action and motion network. In: 2019 IEEE/CVF International Conference on Computer Vision (ICCV), pp. 6191–6200 (2019)
10. Epstein, D., Chen, B., Vondrick, C.: Oops! predicting unintentional action in video. In: 2020 IEEE/CVF Conference on Computer Vision and Pattern Recognition (CVPR), pp. 916–926 (2020)

11. Feichtenhofer, C., Fan, H., Malik, J., He, K.: SlowFast networks for video recognition. In: 2019 IEEE/CVF International Conference on Computer Vision (ICCV), pp. 6201–6210 (2019)
12. Feichtenhofer, C., Fan, H., Xiong, B., Girshick, R.B., He, K.: A large-scale study on unsupervised spatiotemporal representation learning. In: 2021 IEEE/CVF Conference on Computer Vision and Pattern Recognition (CVPR), pp. 3298–3308 (2021)
13. Gidaris, S., Singh, P., Komodakis, N.: Unsupervised representation learning by predicting image rotations. (2018) arXiv preprint arXiv:1803.07728
14. Goyal, R., Kahou, S.E., Michalski, V., Materzynska, J., Westphal, S., Kim, H., Haenel, V., Fründ, I., Yianilos, P.N., Mueller-Freitag, M., Hoppe, F., Thurau, C., Bax, I., Memisevic, R.: The "something something" video database for learning and evaluating visual common sense. In: 2017 IEEE International Conference on Computer Vision (ICCV), pp. 5843–5851 (2017)
15. Grill, J.B., et al.: Bootstrap your own latent: a new approach to self-supervised learning. (2020) ArXiv abs/2006.07733
16. Han, T., Xie, W., Zisserman, A.: Video representation learning by dense predictive coding. In: 2019 IEEE/CVF International Conference on Computer Vision Workshop (ICCVW), pp. 1483–1492 (2019)
17. Han, T., Xie, W., Zisserman, A.: Memory-augmented dense predictive coding for video representation learning. In: ECCV (2020)
18. Han, T., Xie, W., Zisserman, A.: Self-supervised co-training for video representation learning. (2020) ArXiv abs/2010.09709
19. Hara, K., Kataoka, H., Satoh, Y.: Can spatiotemporal 3d CNNs retrace the history of 2d CNNs and imagenet? In: 2018 IEEE/CVF Conference on Computer Vision and Pattern Recognition, pp. 6546–6555 (2018)
20. He, K., Fan, H., Wu, Y., Xie, S., Girshick, R.B.: Momentum contrast for unsupervised visual representation learning. In: 2020 IEEE/CVF Conference on Computer Vision and Pattern Recognition (CVPR), pp. 9726–9735 (2020)
21. He, T., Zhang, Z., Zhang, H., Zhang, Z., Xie, J., Li, M.: Bag of tricks for image classification with convolutional neural networks. In: Proceedings of the IEEE/CVF Conference on Computer Vision and Pattern Recognition, pp. 558–567 (2019)
22. Hu, K., Shao, J., Liu, Y., Raj, B., Savvides, M., Shen, Z.: Contrast and order representations for video self-supervised learning. In: Proceedings of the IEEE/CVF International Conference on Computer Vision (ICCV), pp. 7939–7949 October 2021
23. Huang, D., et al.: ASCNet: Self-supervised video representation learning with appearance-speed consistency. In: Proceedings of the IEEE/CVF International Conference on Computer Vision (ICCV), pp. 8096–8105 October 2021
24. Huang, L., Liu, Y., Wang, B., Pan, P., Xu, Y., Jin, R.: Self-Supervised video representation learning by context and motion decoupling. In: CVPR (2021)
25. Isola, P., Zoran, D., Krishnan, D., Adelson, E.: Learning visual groups from co-occurrences in space and time. (2015) ArXiv abs/1511.06811
26. Jayaraman, D., Grauman, K.: Learning image representations tied to ego-motion. In: 2015 IEEE International Conference on Computer Vision (ICCV), pp. 1413–1421 (2015)
27. Jenni, S., Meishvili, G., Favaro, P.: Video representation learning by recognizing temporal transformations. (2020) ArXiv abs/2007.10730
28. Jenni, S., Jin, H.: Time-equivariant contrastive video representation learning. In: Proceedings of the IEEE/CVF International Conference on Computer Vision (ICCV), pp. 9970–9980 October 2021

29. Kolouri, S., Martin, C.E., Hoffmann, H.: Explaining distributed neural activations via unsupervised learning. In: 2017 IEEE Conference on Computer Vision and Pattern Recognition Workshops (CVPRW), pp. 1670–1678 (2017)
30. Kuehne, H., Jhuang, H., Garrote, E., Poggio, T., Serre, T.: HMDB: a large video database for human motion recognition. In: 2011 International conference on computer vision, pp. 2556–2563. IEEE (2011)
31. Lai, Z., Lu, E., Xie, W.: Mast: A memory-augmented self-supervised tracker. In: 2020 IEEE/CVF Conference on Computer Vision and Pattern Recognition (CVPR), pp. 6478–6487 (2020)
32. Luo, D., et al.: Video cloze procedure for self-supervised spatio-temporal learning. (2020) ArXiv abs/2001.00294
33. Misra, I., Zitnick, C.L., Hebert, M.: Shuffle and Learn: Unsupervised Learning Using Temporal Order Verification. In: Leibe, B., Matas, J., Sebe, N., Welling, M. (eds.) ECCV 2016. LNCS, vol. 9905, pp. 527–544. Springer, Cham (2016). https://doi.org/10.1007/978-3-319-46448-0_32
34. Noroozi, M., Favaro, P.: Unsupervised Learning of Visual Representations by Solving Jigsaw Puzzles. In: Leibe, B., Matas, J., Sebe, N., Welling, M. (eds.) ECCV 2016. LNCS, vol. 9910, pp. 69–84. Springer, Cham (2016). https://doi.org/10.1007/978-3-319-46466-4_5
35. van den Oord, A., Li, Y., Vinyals, O.: Representation learning with contrastive predictive coding. (2018) ArXiv abs/1807.03748
36. Pan, T., Song, Y., Yang, T., Jiang, W., Liu, W.: VideoMoCo: Contrastive video representation learning with temporally adversarial examples. (2021) ArXiv abs/2103.05905
37. Qian, R., et al.: Enhancing self-supervised video representation learning via multi-level feature optimization. In: Proceedings of the IEEE/CVF International Conference on Computer Vision (ICCV), pp. 7990–8001 October 2021
38. Qian, R., et al.: Spatiotemporal contrastive video representation learning. In: CVPR (2021)
39. Soomro, K., Zamir, A.R., Shah, M.: A dataset of 101 human action classes from videos in the wild. Center for Research in Computer Vision **2** (11) (2012)
40. Wang, J., Jiao, J., hui Liu, Y.: Self-supervised video representation learning by pace prediction. In: ECCV (2020)
41. Wang, J., Gao, Y., Li, K., Lin, Y., Ma, A.J., Sun, X.: Removing the background by adding the background: towards background robust self-supervised video representation learning. In: CVPR (2021)
42. Wu, Z., Xiong, Y., Yu, S.X., Lin, D.: Unsupervised feature learning via non-parametric instance discrimination. In: Proceedings of the IEEE Conference on Computer Vision and Pattern Recognition, pp. 3733–3742 (2018)
43. Xie, S., Sun, C., Huang, J., Tu, Z., Murphy, K.P.: Rethinking spatiotemporal feature learning: speed-accuracy trade-offs in video classification. In: ECCV (2018)
44. Xu, D., Xiao, J., Zhao, Z., Shao, J., Xie, D., Zhuang, Y.: Self-supervised spatiotemporal learning via video clip order prediction. In: 2019 IEEE/CVF Conference on Computer Vision and Pattern Recognition (CVPR), pp. 10326–10335 (2019)
45. YM., A., C., R., A., V.: Self-labelling via simultaneous clustering and representation learning. In: International Conference on Learning Representations (2020). https://openreview.net/forum?id=Hyx-jyBFPr
46. You, Y., Gitman, I., Ginsburg, B.: Large batch training of convolutional networks. (2017) arXiv preprint arXiv:1708.03888

47. Zhang, R., Isola, P., Efros, A.: Colorful Image Colorization. In: Leibe, B., Matas, J., Sebe, N., Welling, M. (eds.) ECCV 2016. LNCS, vol. 9907, pp. 649–666. Springer, Cham (2016). https://doi.org/10.1007/978-3-319-46487-9_40
48. Zhou, B., Khosla, A., Lapedriza, A., Oliva, A., Torralba, A.: Learning deep features for discriminative localization. In: Computer Vision and Pattern Recognition (2016)

On the Effectiveness of ViT Features
as Local Semantic Descriptors

Shir Amir[1($\boxtimes$)], Yossi Gandelsman[2], Shai Bagon[1], and Tali Dekel[1]

[1] Department of Computer Science and Applied Mathematics, The Weizmann
Institute of Science, Rehovot, Israel
shiramiremail@gmail.com
[2] Berkeley Artificial Intelligence Research (BAIR), Berkeley, USA

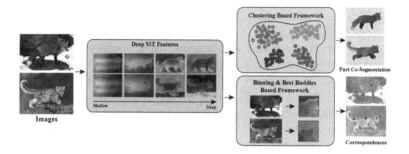

Fig. 1. Based on our new observations on deep ViT features, we devise *lightweight zero-shot* methods to solve fundamental vision tasks (e.g. part co-segmentation and semantic correspondences). Our methods are applicable even in challenging settings where the images belong to different classes (e.g. fox and leopard).

Abstract. We study the use of deep features extracted from a pretrained Vision Transformer (ViT) as dense visual descriptors. We observe and empirically demonstrate that such features, when extracted from a self-supervised ViT model (DINO-ViT), exhibit several striking properties, including: (i) the features encode powerful, well-localized semantic information, at high spatial granularity, such as object *parts*; (ii) the encoded semantic information is *shared across related, yet different object categories*, and (iii) positional bias changes gradually *throughout the layers*. These properties allow us to design simple methods for a variety of applications, including co-segmentation, part co-segmentation and semantic correspondences. To distill the power of ViT features from convoluted design choices, we restrict ourselves to *lightweight zero-shot* methodologies (e.g., binning and clustering) applied directly to the features. Since our methods require no additional training nor data, they are readily applicable across a variety of domains. We show by extensive

Supplementary Information The online version contains supplementary material available at https://doi.org/10.1007/978-3-031-25069-9_3.

© The Author(s), under exclusive license to Springer Nature Switzerland AG 2023
L. Karlinsky et al. (Eds.): ECCV 2022 Workshops, LNCS 13804, pp. 39–55, 2023.
https://doi.org/10.1007/978-3-031-25069-9_3

qualitative and quantitative evaluation that our simple methodologies achieve competitive results with recent state-of-the-art *supervised* methods, and outperform previous unsupervised methods by a large margin. Code is available in https://dino-vit-features.github.io/.

Keywords: ViT · Deep features · Zero-shot methods

1 Introduction

"Deep Features" – features extracted from the activations of layers in a pre-trained neural network – have been extensively used as visual descriptors in a variety of visual tasks, yet have been mostly explored for CNN-based models. For example, deep features extracted from CNN models that were pre-trained for visual classification (e.g., VGG [49]) have been utilized in numerous visual tasks including image generation and manipulation, correspondences, tracking and as a general perceptual quality measurement.

Recently, Vision Transformers (ViT) [13] have emerged as a powerful alternative architecture to CNNs. ViT-based models achieve impressive results in numerous visual tasks, while demonstrating better robustness to occlusions, adversarial attacks and texture bias compared to CNN-based models [37]. This raises the following questions: Do these properties reflect on the internal representations learned by ViTs? Should we consider deep ViT features as an alternative to deep CNN features? Aiming to answer these questions, we explore the use of deep ViT features as general dense visual descriptors: we empirically study their unique properties, and demonstrate their power through a number of real-world visual tasks.

In particular, we focus on two pre-trained ViT models: a supervised ViT, trained for image classification [13], and a self-supervised ViT (DINO-ViT), trained using a self-distillation approach [3]. In contrast to existing methods, which mostly focus on the features from the deepest layer [3,48,54], we dive into the self-attention modules, and consider the various facets (tokens, queries, keys, values) *across different layers*. We observe and empirically demonstrate that DINO-ViT features: (i) encode powerful high-level information at high spatial resolution, i.e., capture semantic object *parts*, (ii) this encoded semantic information is *shared across related, yet different object classes*, and (iii) positional information gradually decreases *throughout* layers, thus the intermediate layers encode position information as well as semantics. We demonstrate that these properties are not only due to the ViT architecture but also significantly influenced by the training supervision.

Relying on these observations, we unlock the effectiveness of DINO-ViT features by considering their use in a number of fundamental vision tasks: co-segmentation, part co-segmentation, and semantic point correspondences. Moreover, equipped with our new observations, we tackle the task of part co-segmentation in a *challenging unconstrained setting* where neither the number of input images, nor their domains are restricted. We further present how our part co-segmentation can be applied to videos. To the best of our knowledge, we are the first to show results of part co-segmentation in such challenging cases (Fig. 6). We apply *simple, zero-shot* methodologies to deep ViT features for all these tasks, which do not require further training. Deliberately avoiding large-scale learning-based models showcases the effectiveness of the learned DINO-ViT representations. We demonstrate that without bells and whistles, DINO-ViT features are already powerful enough to achieve competitive results compared to state-of-the-art models specifically designed and trained for each individual task. We thoroughly evaluate our performance qualitatively and quantitatively.

To conclude, our key contributions are: (i) We uncover surprising *localized semantic information*, far beyond saliency, readily available in ViT features. (ii) Our new observations give rise to *lightweighted zero-shot* methodologies for tackling co- and part co-segmentation as well as semantic correspondences. (iii) We are the first to show part co-segmentation in *extreme settings*, showing how objects can be consistently segmented into parts across different categories, and across a variety of image domains, for some of which training data is scarce.

2   Related Work

CNN-based Deep Features. Features of pre-trained CNNs are a cornerstone for various vision tasks from object detection and segmentation [5,19], to image generation [17,46]. These representations were shown to align well with human perception [17,26,35,58] and to encode a wide range of visual information - from low level features (e.g. edges and color) to high level semantic features (e.g. object parts) [4,40]. Nevertheless, they exhibit a strong bias towards texture [18], and lack positional information due to their shift equivariance [57]. Moreover, their restricted receptive field [34] makes them capture mostly local information and ignore long-range dependencies [53]. Here, we study the deep features of a less restrictive architecture - the Vision Transformer, as an alternative.

Vision Transformer (ViT). Vision Transformers [13] have recently been used as powerful CNN alternatives. ViT-based models achieve impressive results in a variety of visual tasks [2,7,13], while demonstrating better robustness to occlusions, adversarial attacks, and texture bias compared to CNN-based models [37].

In particular, Caron et al. [3] presented DINO-ViT – a ViT model trained without labels, using a self-distillation approach. They observed that the attention heads of this model attend to salient foreground regions in an image. They further showed the effectiveness of DINO-ViT features for several tasks that benefit from this property, including image retrieval and object segmentation.

Recent works follow this observation and utilize these features for object discovery [48,54], semantic segmentation [20] and category discovery [52]. All these works treat pre-trained DINO-ViT as a black-box, only considering features extracted from it's last layer, and their use as global or figure/ground-aware representations. In contrast, we examine the continuum of Deep ViT features *across layers*, and dive into the different representations inside each layer (e.g. the keys, values, queries of the attention layers). We observe *new* properties of these features besides being aware to foreground objects, and put these observations to use by solving fundamental vision tasks.

Concurrently, [11,37,42] study theoretical aspects of the underlying machinery, aiming to analyze how ViTs process visual data compared to CNN models. Our work aims to bridge the gap between better understanding Deep ViT representations and their use in real-world vision tasks in a zero-shot manner.

Co-segmentation. Co-segmentation aims to jointly segment objects common to all images in a given set. Several unsupervised methods used hand-crafted descriptors [15,44,45] for this task. Later, CNN-based methods applied supervised training [31] or fine-tuning [29,30,56] on *intra-class* co-segmentation datasets. The supervised methods obtain superior performance, yet their notion of "commonality" is restricted by their training data. Thus, they struggle generalizing to new *inter-class* scenarios. We, however, show a *lightweight unsupervised* approach that is competitive to *supervised* methods for intra-class co-segmentation *and* outperforms them in the inter-class setting.

Part Co-segmentation. Given a set of images with similar objects, the task is to discover common object *parts* among the images. Recent methods [9,24,32] train a CNN encoder-decoder in a self-supervised manner to solve this task, while [10] applies matrix factorization on pre-trained deep CNN features. In contrast, we utilize a pre-trained self-supervised ViT to solve this task, and achieve competitive performance to the methods above. Due to the zero-shot nature of our approach, we are able to apply part co-segmentation *across classes*, and on domains that lack training supervision (see Fig. 6). To the best of our knowledge, we are the first to address such challenging scenarios.

Semantic Correspondences. Given a pair of images, the task is to find semantically corresponding points between them. Aberman et al. [1] propose a sparse correspondence method for inter-class scenarios leveraging pre-trained CNN features. Recent *supervised* methods employ transformers for dense correspondence in images from the same scene [25,50]. Cho et al. [7] use transformers for semantic point correspondences by training directly on annotated point correspondences. We show that utilizing ViT features in a *zero-shot* manner can be competitive to *supervised* methods while being more robust to different pose and scale than previous unsupervised methods.

3   ViT Features as Local Patch Descriptors

We explore ViT features as *local patch descriptors*. In a ViT architecture, an image is split into n non-overlapping patches $\{p_i\}_{i \in 1..n}$ which are processed into *spatial tokens* by linearly projecting each patch to a d-dimensional space, and adding learned positional embeddings. An additional [CLS] token is inserted to capture global image properties. The set of tokens are then passed through L transformer encoder layers, each consists of normalization layers (LN), Multihead Self-Attention (MSA) modules, and MLP blocks (with skip connections):

$$\hat{T}^l = \mathsf{MSA}(\mathsf{LN}(T^{l-1})) + T^{l-1}, \quad T^l = \mathsf{MLP}(\mathsf{LN}(\hat{T}^l)) + \hat{T}^l \tag{1}$$

where $T^l = [t_0^l, \ldots, t_n^l]$ are the output tokens for layer l.

In each MSA block, tokens are linearly projected into queries, keys and values:

$$q_i^l = W_q^l \cdot t_i^{l-1}, \quad k_i^l = W_k^l \cdot t_i^{l-1}, \quad v_i^l = W_v^l \cdot t_i^{l-1} \tag{2}$$

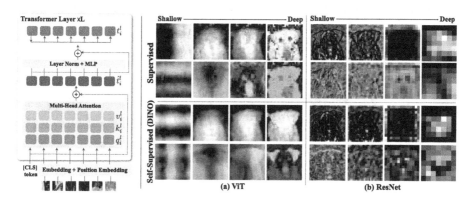

Fig. 2. *ViT Architecture (Left).* An image is split into n non-overlapping patches and gets a [CLS] token. These patches are embedded, added positional embeddings and passed through transformer layers. Each patch is directly associated with a set of features in each layer: a key, query, value and token; each can be used as patch descriptors. *Deep features visualization via PCA (Right):* Applied on supervised and self-supervised (a) ViTs and (b) CNN-ResNet models. We fed 18 images from AFHQ [8] to each model, extract features from a given layer, and perform PCA on them. For each model, we visualize PCA components at each layer, for an example image (Dalmatian dog in Fig. 7 left): the first component is shown on the top, while second-to-fourth components are shown as RGB images below. ResNet PCA is upsampled for visualization purposes. (Color figure online)

which are then fused using multihead self-attention. Figure 2 Left illustrates this process, for full details see [13]. Besides the initial image patches sampling, ViTs have no additional spatial sampling; hence, each image patch p_i is *directly* associated with a set of features: $\{q_i^l, k_i^l, v_i^l, t_i^l\}$, including its query, key, value and token, at each layer l, respectively. We next focus our analysis on using the *keys* as 'ViT features'. We justify this choice via ablation in Sects. 5.2 and 5.3.

3.1 Properties of ViT's Features

We focus on two pre-trained ViT models, both have the same architecture and training data, but differ in their training supervision: a *supervised ViT*, trained for image classification using ImageNet labels [13], and a *self-supervised ViT* (DINO-ViT), trained using a self-distillation approach [3]. We next provide qualitative analysis of the internal representations learned by both models, and empirically originate their properties to the *combination* of architecture and training supervision. In Sect. 5, we show these properties enable several applications, through which we quantitatively validate our observations.

Figure 2 Right (a) shows a simple visualization of the learned representation by supervised ViT and DINO-ViT: for each model, we extract deep features (keys) from a set of layers, perform PCA, and visualize the resulting leading components. Figure 2 Right (b) shows the same visualization for two respective CNN-ResNet [21] models trained using the same two supervisions as the ViT models: image classification, and DINO [3]. This simple visualization shows fundamental differences between the internal representations of each model.

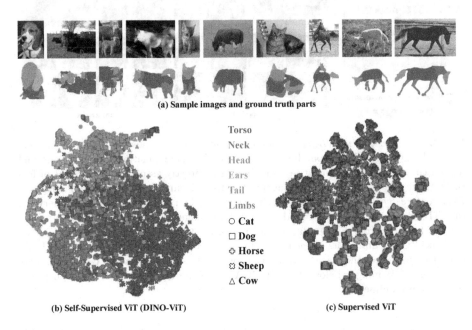

Fig. 3. *t-SNE visualization.* We take 10 images from 5 animal categories from PASCAL-Parts [6]. (a) shows representative images and ground-truth part segments. For each image we extract ViT features from DINO-ViT and a supervised ViT. For each model, all features are jointly projected to 2D using t-SNE [41]. Each 2D point is colored according to its ground-truth *part*, while its shape represents the class. In (b) DINO-ViT features are organized mainly by parts, across different object categories, while in (c) supervised ViT features are grouped mostly by class, regardless of object parts.

Semantics vs. Spatial Granularity. One noticeable difference between CNN-ResNet and ViT is that CNNs trade spatial resolution with semantic information in the deeper layers, as shown in Fig. 2 Right (b): the feature maps in the deepest layer have very low resolution ($\times 32$ smaller than the input image), and thus provide poorly localized semantic information. In contrast, ViT maintains the same spatial resolution through all layers. Also, the receptive field of ViT is the entire image in all layers – each token t_i^l attends to all other tokens t_j^l. Thus, ViT features provide fine-grained semantic information *and* higher spatial resolution.

Representations Across Layers. It is well known that the space of deep CNN-based features has a hierarchy of representation: early layers capture low-level elements such as edges or local textures (shallow layers in Fig. 2 Right (b)), while deeper layers gradually capture more high level concepts [4,40,46]. In contrast, we notice a different type of representation hierarchy in ViTs: *Shallow features mostly contain positional information*, while in deeper layers, this is reduced in favor of more semantic features. For example, in Fig. 2 Right (a) the deep features distinguish dog features from background features, while the shallow features are gathered mostly based on their spatial location. Interestingly, intermediate ViT features contain both positional and semantic information.

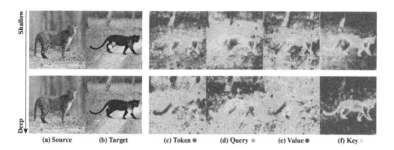

Fig. 4. *Facets of ViT:* We compute the similarity between a feature associated with the magenta point in the source image (a) to all features in the target image (b). We do this for intermediate features (top row) and features from the last layer (bottom row). (c-f) are the resulting similarity maps, when using different facets of DINO-ViT as features: tokens, queries, values and keys. Red indicates higher similarity. For each facet, the closest point in the target image is marked with a unique color, specified near the facet name. The keys (f) have cleaner similarity map compared to other facets. (Color figure online)

Semantic Information Across Super-classes. Figure 2 Right (b) exhibits the supervised ViT model (top) produces "noisier" features compared to DINO-ViT (bottom). To further contrast the two ViTs, we employ t-SNE [41] to the keys of the last layer $[k_i^{11}]$, extracted from 50 animal images from PASCAL-Parts [6]. Figure 3 presents the 2D-projected keys. Intriguingly, the keys from a DINO-ViT show semantic similarity of body parts across different classes (grouped by *color*), while the keys from a supervised ViT display similarity within each class regardless of body part (grouped by *shape*). This demonstrates that while supervised ViT spatial features emphasize *global* class information, DINO-ViT features have *local* semantic information resembling semantic object *parts*.

Different Facets of ViT Representation. So far, we focused on using keys as 'ViT features'. However, ViT provides different facets that are also directly associated with each image patch (Fig. 2 Left). We empirically observe slight differences in the representations of ViT facets, as shown in Fig. 4. In particular, we found the keys to provide a slightly better representation, e.g., they depict less sensitivity to background clutter than the other facets. In addition, both keys and queries posses more positional bias in intermediate layers than values and tokens.

4   Deep ViT Features Applied to Vision Tasks

We demonstrate the effectiveness of deep DINO-ViT features as local patch descriptors several visual tasks. We *deliberately* apply only simple, lightweight methodologies on the extracted features, without any additional training nor fine-tuning, to showcase the effectiveness of DINO-ViT representations. For full implementation details, see supplementary material (SM).

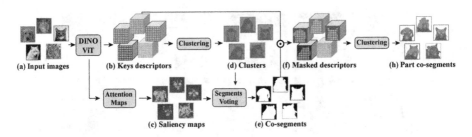

Fig. 5. *Co-segmentation and part co-segmentation pipeline.* Input images (a) are fed separately to DINO-ViT to obtain (b) spatial dense descriptors and (c) saliency maps (from the ViT's self-attention maps). All the extracted descriptors are clustered together (d). Each cluster is assigned as foreground or background via a saliency maps based voting process. Foreground segments form the co-segmentation results (e). The process is repeated on foreground features (f) alone to yield the common parts (h).

Co-segmentation. Our co-segmentation approach, applied to a set of N input images, comprises of two steps, followed by GrabCut [43] to refine the binary co-segmentation masks, as illustrated in Fig. 5(a-e):

1. *Clustering:* We treat the set of extracted descriptors across all images and all spatial locations as a bag-of-descriptors, and cluster them using k-means. At this stage, the descriptors are clustered into semantic common segments. As illustrated in Fig. 2 Right, the most prominent features' component distinguishes foreground and background, which ensures their separation. The result of this stage is K clusters that induce segments in all images.
2. *Voting:* We use a simple voting procedure to select clusters that are salient and common to most of the images. Let $\mathsf{Attn}_i^{\mathcal{I}}$ be the mean [CLS] attention of selected heads in the last layer in image $\mathcal{I}$ of patch i. Let $S_k^{\mathcal{I}}$ be the set of all patches in image $\mathcal{I}$ belonging to cluster k. The saliency of segment $S_k^{\mathcal{I}}$ is:

$$\mathsf{Sal}\left(S_k^{\mathcal{I}}\right) = \frac{1}{|S_k^{\mathcal{I}}|} \sum_{i \in S_k^{\mathcal{I}}} \mathsf{Attn}_i^{\mathcal{I}} \tag{3}$$

Each segment votes for the saliency of the cluster k:

$$\mathsf{Votes}\left(k\right) = \mathbb{1}_{\left[\sum_{\mathcal{I}} \mathsf{Sal}\left(S_k^{\mathcal{I}}\right) \geq \tau\right]} \tag{4}$$

For some threshold τ. A cluster k is considered "foreground" iff its Votes (k) is above percentage p of all the images.

Part Co-segmentation. To further co-segment the foreground objects into common *parts*, we repeat the clustering step only on foreground descriptors, see Fig. 5(f-h). By doing so, descriptors of common semantic parts across images are grouped together. We further refine the part masks using multi-label CRF [27]. In practice, we found k-means to perform well, but other clustering methods (e.g. [16,39]) can be easily plugged in. For co-segmentation, the number of clusters is automatically set using the elbow method [38], whereas for part co-segmentation, it is set to the desired number of object parts. Our method can be applied to a variety of object categories, and to arbitrary number of input images N, ranging from two to thousands of images. On small sets we apply random crop and flip augmentations for improved clustering stability (see SM for more details).

Point Correspondences. Semantic information is necessary yet insufficient for this task. For example, matching points on an animal's tail in Fig. 1, relying only on semantic information is ambiguous: all points on the tail are equally similar. We reduce this ambiguity in two manners:

1. *Positional Bias:* We want the descriptors to be position-aware. Features from earlier layers are more sensitive to their position in the image (see Sect. 3.1); hence we use mid-layer features which provide a good trade-off between position and semantic information.
2. *Binning:* We incorporate context into each descriptor by integrating information from adjacent spatial features. This is done by applying log-binning to each spatial feature, as illustrated in Fig. 1.

To automatically detect reliable matches between images, we adopt the notion of "Best Buddies Pairs"(BBPs) [12], i.e., we only keep descriptor pairs which are mutual nearest neighbors. Formally, let $M = \{m_i\}$ and $Q = \{q_i\}$ be sets of binned descriptors from images I_M and I_Q respectively. The set of BBPs is thus:

$$\mathsf{BB}(M,Q) = \{(m,q) \mid m \in M, \ q \in Q, \ \mathsf{NN}(m,Q) = q \wedge \mathsf{NN}(q,M) = m\} \quad (5)$$

Where $\mathsf{NN}(m,Q)$ is the nearest neighbor of m in Q under cosine similarity.

Resolution Increase. The spatial resolution of ViT features is inversely proportional to size of the *non-overlapping* patches, p_i. Our applications benefit from higher spatial feature resolution. We thus modify ViT to extract, at test time, *overlapping* patches, interpolating their positional encoding accordingly. Consequently, we get, without any additional training, ViT features at finer spatial resolution. Empirically, we found this method to work well in all our experiments.

5   Results

5.1   Part Co-segmentation

Challenging Small Sets. In Fig. 6, we present several image pairs collected from the web. These examples pose challenge due to different appearance (e.g.

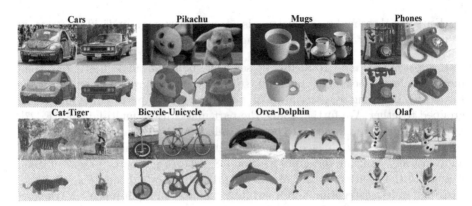

Fig. 6. *Part Co-segmentation of Image Pairs:* Our method semantically co-segments common object parts given as little as two input images. See the SM for more examples.

Fig. 7. *Part Co-segmentation on AFHQ:* We apply our method on the test set of AFHQ [8] containing 1.5 K images of different animal faces. More results are in SM.

cars, phones), different classes (e.g. bicycle-unicycle, cat-tiger) and belonging to domains that are difficult to accommodate training sets for (e.g. pikachu, olaf). Our zero-shot method manages to provide semantically consistent part segments for each image pair. For example, in the bicycle-unicycle example the tires, spokes, chassis and saddle parts are consistently found. To the best of our knowledge, we are the first to handle such challenging cases.

Video Part Co-segmentation. We extend our framework to work on videos by applying it to frames of a single video. Since DINO-ViT features are consistent across video frames, applying our observations to video co-segmentation yields temporally consistent parts. To the best of our knowledge, we are the first to apply part co-segmentation on videos. We include multiple examples in the SM.

Inter-class Results. In Fig. 7 we apply our part co-segmentation with $k = 10$ parts on AFHQ [8] test set, containing 1.5K images of different animal faces. Our method provides consistent parts across *different* animal classes, e.g. ears marked in orange, forehead marked in blue, whiskers marked in purple, etc.

CUB [55] evaluation. Following [9,24], we evaluate performance on CUB [55] test set, which contains 5K images of different bird species. Following [24], we

Fig. 8. *Part co-segmentation comparison on CUB:* We show results on randomly chosen images from CUB [55]. Our results are more semantically consistent across parts than the *supervised* SCOPS [24] and are competitive to the *supervised* Choudhury et al. [9].

Table 1. *Part Co-segmentation results:* We report mean error of landmark regression on three CUB [55] test sets, and NMI and ARI [9] measures on the entire CUB test set. All methods predict $k = 4$ parts. † method uses image-level supervision, ‡ methods use ground truth foreground masks as supervision.

Method	key-point regression ↓			FG-NMI ↑	FG-ARI ↑	NMI ↑	ARI ↑
	CUB-01	CUB-02	CUB-03				
supervised							
SCOPS [24]‡ (model)	18.3	17.7	17.0	39.1	17.9	24.4	7.1
Huang and Li [23]†	<u>15.1</u>	17.1	<u>15.7</u>	–	–	26.1	13.2
Choudhury et al.[9]‡	**11.3**	<u>15.0</u>	**10.6**	**46.0**	**21.0**	**43.5**	**19.6**
unsupervised							
ULD [51,59]	30.1	29.4	28.2	–	–	–	–
DFF [10]	22.4	21.6	22.0	32.4	14.3	25.9	12.4
SCOPS [24] (paper)	18.5	18.8	21.1	–	–	–	–
Ours	17.1	**14.7**	19.6	<u>39.4</u>	<u>19.2</u>	<u>38.9</u>	<u>16.1</u>

measure the key-point regression error between the predicted and ground truth landmarks in Table 1 on three test sets from CUB [55]. In addition, we follow [9] treating the part segments as clusters, and report NMI and ARI. FG-NMI and FG-ARI disregard the background part as a cluster. Our method surpasses unsupervised methods *by a large margin*, and is competitive to [9] which is *supervised* by foreground masks. Figure 8 shows our method produces more semantically coherent parts, with similar quality to [9]. Further evaluation on the CelebA [33] dataset is available in the SM.

5.2   Co-segmentation

We evaluate our performance on several *intra-class* co-segmentation datasets of varying sizes - MSRC7 [47], Internet300 [44] and PASCAL-VOC [14]. Further-

Table 2. *Co-segmentation evaluation:* We report mean Jaccard index $\mathcal{J}_m$ and precision $\mathcal{P}_m$ over all sets in each dataset. We compare to unsupervised methods [15,44] and methods supervised with ground truth segmentation masks [29–31,56].

Method	Training Set	MSRC [47]		Internet300 [44]		PASCAL -VOC [14]		PASCAL -CO	
		$\mathcal{J}_m$	$\mathcal{P}_m$	$\mathcal{J}_m$	$\mathcal{P}_m$	$\mathcal{J}_m$	$\mathcal{P}_m$	$\mathcal{J}_m$	$\mathcal{P}_m$
supervised									
SSNM [56]	COCO-SEG	81.9	95.2	74.1	93.6	71.0	<u>94.9</u>	<u>74.2</u>	<u>94.5</u>
DOCS [31]	VOC2012	82.9	95.4	72.5	93.5	<u>65.0</u>	94.2	34.9	53.7
CycleSegNet [30]	VOC2012	**87.2**	**97.9**	<u>80.4</u>	–	**75.4**	**95.8**	–	–
Li et al. [29]	COCO	–	–	**84.0**	**97.1**	63.0	94.1	–	–
unsupervised									
Hsu et al.[22]	–	–	–	69.8	92.3	60.0	91.0	–	–
DeepCO3 [28]	–	54.7	87.2	53.4	88.0	46.3	88.5	37.3	74.1
TokenCut [54]	–	81.2	94.9	65.2	91.3	57.8	90.6	75.8	93.0
Faktor et al.[15]	–	77.0	92.0	–	–	46.0	84.0	41.4	79.9
Rubinstein et al.[44]	–	74.0	92.2	57.3	85.4	–	–	–	–
Ours	–	<u>86.7</u>	<u>96.5</u>	79.5	<u>94.6</u>	60.7	88.2	**79.5**	**94.7**

Table 3. *Co-segmentation ablation:* on PASCAL-Co for saliency baselines and our method using different ViT facets. Our method surpasses all baselines, and our choice of keys yields better performance than default chosen DINO-ViT tokens.

	DINO Saliency Baselines		Sup. Saliency Baselines		Ours			
	<u>ViT</u>	ResNet	<u>ViT</u>	ResNet	Keys	Tokens	Queries	Values
$\mathcal{J}_m$	75.0	37.7	39.9	40.0	**79.5**	69.2	72.7	49.2
$\mathcal{P}_m$	93.1	78.1	69.7	78.9	**94.7**	90.68	91.7	83.3

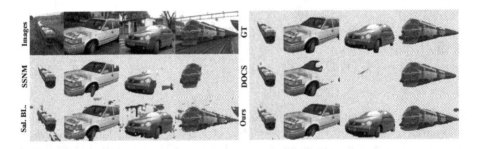

Fig. 9. *PASCAL-CO for inter-class co-segmentation:* Each set contains images from related classes. Our method captures regions of all common objects from different classes, contrary to supervised methods [31,56]. Saliency Baseline [3] results are noisy.

more, to evaluate *inter-class* co-segmentation, we compose a new dataset from PASCAL [14] images, named "PASCAL Co-segmentation" (PASCAL-CO). Our dataset has forty sets of six images, each from semantically related classes (e.g., car-bus-train, bird-plane). Figure 9 shows a sample set, the rest is in the SM.

Quantitative Evaluation. We compare our *unsupervised* approach to state-of-the-art *supervised* methods, trained on large datasets with ground truth segmentation masks, [29–31,56]; and *unsupervised* methods, [15,22,28,44,54]. We report Jaccard Index ($\mathcal{J}_m$), which reflects both precision (covering the foreground) and accuracy (no foreground "leakage"), and mean precision ($\mathcal{P}_m$). The results appear in Table 2. Our method surpasses the unsupervised methods *by a large margin*, and is competitive to the *supervised* methods. In the *inter-class* scenario (PASCAL-CO), our method surpasses all other methods.

Ablation. We conduct an ablation study to validate our observations in 3.1. As mentioned in Sect. 2, Caron et al. [3] observed DINO-ViT attention heads attend to salient regions in the image, and threshold them to perform object segmentation. We name their method "DINO-ViT Saliency Baseline" as mentioned in Table 3. We apply the same baseline with attention heads from a supervised ViT (Sup. ViT Saliency Baseline). To compare ViT with CNN representations, we also apply a similar method thresholding ResNet features (DINO / Sup. ResNet Saliency Baseline), implementation details are in the SM. We also ablate our method with different facets. The supervised ViT performs poorest, while both ResNet baselines perform similarly. The DINO-ViT baseline exceeds them and is closer to our performance. The remaining performance gap between our method and the DINO-ViT baseline can be attributed to one bias in the DINO-ViT baseline - it captures foreground salient objects regardless of their commonality to the other objects in the images. For example, the house behind the blue car in Fig. 9 is captured by DINO-ViT Saliency Baseline but is not captured by our method. This corroborates our observation that the properties of DINO-ViT stem from both architecture and training method. The facet ablation demonstrates our observation that keys are superior than other facets.

5.3   Point Correspondences

Qualitative Results. We test our method on numerous pairs, compared with the VGG-based method, NBB [1]. Figure 10 shows our results are more robust to changes of appearance, pose and scale on both intra- and inter-class pairs.

Quantitative Evaluation. We evaluate on 360 random Spair71k [36] pairs, and measure performance by Percentage of Correct Keypoint (PCK) - a predicted keypoint is considered correct if it lies within a $\alpha \cdot \max(h, w)$ radius from the annotated keypoint, where (h, w) is the image size. We modify our method to match this evaluation protocol to compute the binned descriptors for the given keypoints in the source image, and find their nearest-neighbors in the target image. We compare to NBB [1] (VGG19-based) and CATs [7] (ResNet101-based). Table 4 shows that our method outperforms NBB *by a large margin*, and closes the gap towards the *supervised* CATs [7].

Ablations. We ablate our method using on different facets and layers, with and without binning. Table 4 empirically corroborates our observation of keys being a better than other facets, and that features from earlier layers are more sensitive to their position in the image. The correspondence task benefits from these intermediate features more than plainly using the deepest features (Sect. 3.1).

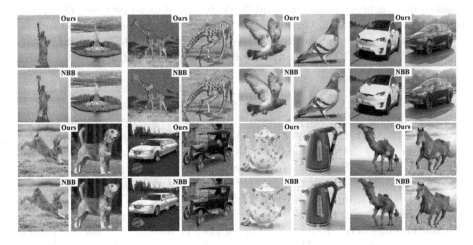

Fig. 10. *Correspondences Comparison to NBB* [1]: On intra-class (top-row) and inter-class (bottom-row) scenarios. Our method is more robust to appearance, pose and scale variations. Full size results are available in the SM.

Table 4. *Correspondence Evaluation on Spair71k:* We randomly sample 20 image pairs per category, and report the mean PCK across all categories ($\alpha = 0.1$); higher is better. We include a recent supervised method [7] for reference.

Method	Layer 9				Layer 11				NBB [1]	Supervised [7]
	Key	Query	Value	Token	Key	Query	Value	Token		
With bins	<u>56.48</u>	54.96	52.33	56.03	53.45	52.35	49.37	50.34	26.98	**61.43**
Without bins	52.27	49.35	43.97	50.14	47.08	42.64	41.56	46.09		

6   Conclusion

We provided new empirical observations on the internal features learned by ViTs under different supervisions, and harnessed them for several real-world vision tasks. We demonstrated the power of these observations by applying only lightweight zero-shot methodologies to these features, and still achieving competitive results to state-of-the-art supervised methods. We also presented new capabilities of part co-segmentation across classes, and on domains that lack available training sets. We believe that our results hold great promise for considering deep ViT features as an alternative to deep CNN features.

Acknowledgments. We thank Miki Rubinstein, Meirav Galun, Kfir Aberman and Niv Haim for their insightful comments and discussion. This project received funding from the Israeli Science Foundation (grant 2303/20), and the Carolito Stiftung. Dr Bagon is a Robin Chemers Neustein Artificial Intelligence Fellow.

References

1. Aberman, K., Liao, J., Shi, M., Lischinski, D., Chen, B., Cohen-Or, D.: Neural best-buddies: sparse cross-domain correspondence. In: TOG (2018)
2. Carion, N., Massa, F., Synnaeve, G., Usunier, N., Kirillov, A., Zagoruyko, S.: End-to-end object detection with transformers. In: Vedaldi, A., Bischof, H., Brox, T., Frahm, J.-M. (eds.) ECCV 2020. LNCS, vol. 12346, pp. 213–229. Springer, Cham (2020). https://doi.org/10.1007/978-3-030-58452-8_13
3. Caron, M., et al.: Emerging properties in self-supervised vision transformers. In: ICCV (2021)
4. Carter, S., Armstrong, Z., Schubert, L., Johnson, I., Olah, C.: Activation atlas. Distill (2019)
5. Chen, L.C., Papandreou, G., Kokkinos, I., Murphy, K., Yuille, A.L.: DeepLab: semantic image segmentation with deep convolutional nets, atrous convolution, and fully connected CRFs. In: IEEE Transactions on Pattern Analysis and Machine Intelligence (2017)
6. Chen, X., Mottaghi, R., Liu, X., Fidler, S., Urtasun, R., Yuille, A.: Detect what you can: detecting and representing objects using holistic models and body parts. In: CVPR (2014)
7. Cho, S., Hong, S., Jeon, S., Lee, Y., Sohn, K., Kim, S.: Semantic correspondence with transformers. In: NeurIPS (2021)
8. Choi, Y., Uh, Y., Yoo, J., Ha, J.W.: Stargan v2: diverse image synthesis for multiple domains. In: CVPR (2020)
9. Choudhury, S., Laina, I., Rupprecht, C., Vedaldi, A.: Unsupervised part discovery from contrastive reconstruction. In: NeurIPS (2021)
10. Collins, E., Achanta, R., Süsstrunk, S.: Deep feature factorization for concept discovery. In: Ferrari, V., Hebert, M., Sminchisescu, C., Weiss, Y. (eds.) Computer Vision – ECCV 2018. LNCS, vol. 11218, pp. 352–368. Springer, Cham (2018). https://doi.org/10.1007/978-3-030-01264-9_21
11. Cordonnier, J.B., Loukas, A., Jaggi, M.: On the relationship between self-attention and convolutional layers. In: ICLR (2019)
12. Dekel, T., Oron, S., Rubinstein, M., Avidan, S., Freeman, W.T.: Best-buddies similarity for robust template matching. In: CVPR (2015)
13. Dosovitskiy, A., et al.: An image is worth 16x16 words: transformers for image recognition at scale. In: ICLR (2021)
14. Everingham, M., Eslami, S.M.A., Van Gool, L., Williams, C.K.I., Winn, J., Zisserman, A.: The pascal visual object classes challenge: a retrospective. In: IJCV (2015)
15. Faktor, A., Irani, M.: Co-segmentation by composition. In: ICCV (2013)
16. Fowlkes, C., Belongie, S., Chung, F., Malik, J.: Spectral grouping using the nyström method. In: TPAMI (2004)
17. Gatys, L.A., Ecker, A.S., Bethge, M.: Image style transfer using convolutional neural networks. In: CVPR (2016)

18. Geirhos, R., Rubisch, P., Michaelis, C., Bethge, M., Wichmann, F.A., Brendel, W.: Imagenet-trained CNNs are biased towards texture; increasing shape bias improves accuracy and robustness. In: ICLR (2019)

19. Girshick, R.B., Donahue, J., Darrell, T., Malik, J.: Rich feature hierarchies for accurate object detection and semantic segmentation. In: CVPR (2014)

20. Hamilton, M., Zhang, Z., Hariharan, B., Snavely, N., Freeman, W.T.: Unsupervised semantic segmentation by distilling feature correspondences. In: ICLR (2022)

21. He, K., Zhang, X., Ren, S., Sun, J.: Deep residual learning for image recognition. In: CVPR (2016)

22. Hsu, K.J., Lin, Y.Y., Chuang, Y.Y.: Co-attention CNNs for unsupervised object co-segmentation. In: IJCAI (2018)

23. Huang, Z., Li, Y.: Interpretable and accurate fine-grained recognition via region grouping. In: CVPR (2020)

24. Hung, W.C., Jampani, V., Liu, S., Molchanov, P., Yang, M.H., Kautz, J.: SCOPS: self-supervised co-part segmentation. In: CVPR (2019)

25. Jiang, W., Trulls, E., Hosang, J., Tagliasacchi, A., Yi, K.M.: COTR: correspondence transformer for matching across images. In: ICCV (2021)

26. Johnson, J., Alahi, A., Fei-Fei, L.: Perceptual losses for real-time style transfer and super-resolution. In: Leibe, B., Matas, J., Sebe, N., Welling, M. (eds.) ECCV 2016. LNCS, vol. 9906, pp. 694–711. Springer, Cham (2016). https://doi.org/10.1007/978-3-319-46475-6_43

27. Krähenbühl, P., Koltun, V.: Efficient inference in fully connected CRFs with gaussian edge potentials. In: NeurIPS (2011)

28. Hsu, K.-J., Lin, Y.-Y., Chuang, Y.-Y.: DeepCO3: deep instance co-segmentation by co-peak search and co-saliency detection. In: CVPR (2019)

29. Li, B., Sun, Z., Li, Q., Wu, Y., Hu, A.: Group-wise deep object co-segmentation with co-attention recurrent neural network. In: ICCV (2019)

30. Li, G., Zhang, C., Lin, G.: CycleSegNet: object co-segmentation with cycle refinement and region correspondence. In: TIP (2021)

31. Li, W., Hosseini Jafari, O., Rother, C.: Deep object co-segmentation. In: Jawahar, C.V., Li, H., Mori, G., Schindler, K. (eds.) ACCV 2018. LNCS, vol. 11363, pp. 638–653. Springer, Cham (2019). https://doi.org/10.1007/978-3-030-20893-6_40

32. Liu, S., Zhang, L., Yang, X., Su, H., Zhu, J.: Unsupervised part segmentation through disentangling appearance and shape. In: CVPR (2021)

33. Liu, Z., Luo, P., Wang, X., Tang, X.: Deep learning face attributes in the wild. In: ICCV (2015)

34. Luo, W., Li, Y., Urtasun, R., Zemel, R.: Understanding the effective receptive field in deep convolutional neural networks. In: NeurIPS (2016)

35. Mechrez, R., Talmi, I., Zelnik-Manor, L.: The contextual loss for image transformation with non-aligned data. In: Ferrari, V., Hebert, M., Sminchisescu, C., Weiss, Y. (eds.) Computer Vision – ECCV 2018. LNCS, vol. 11218, pp. 800–815. Springer, Cham (2018). https://doi.org/10.1007/978-3-030-01264-9_47

36. Min, J., Lee, J., Ponce, J., Cho, M.: Spair-71k: a large-scale benchmark for semantic correspondence. CoRR (2019)

37. Naseer, M., Ranasinghe, K., Khan, S., Hayat, M., Khan, F.S., Yang, M.H.: Intriguing properties of vision transformers. In: NeurIPS (2021)

38. Ng, A.: Clustering with the k-means algorithm. In: Machine Learning (2012)

39. Ng, A., Jordan, M., Weiss, Y.: On spectral clustering: analysis and an algorithm. In: NeurIPS (2001)

40. Olah, C., Mordvintsev, A., Schubert, L.: Feature visualization. In: Distill (2017)

41. Poličar, P.G., Stražar, M., Zupan, B.: openTSNE: a modular python library for t-SNE dimensionality reduction and embedding. bioRxiv (2019)
42. Raghu, M., Unterthiner, T., Kornblith, S., Zhang, C., Dosovitskiy, A.: Do vision transformers see like convolutional neural networks? In: NeurIPS (2021)
43. Rother, C., Kolmogorov, V., Blake, A.: "GrabCut": interactive foreground extraction using iterated graph cuts. In: TOG (2004)
44. Rubinstein, M., Joulin, A., Kopf, J., Liu, C.: Unsupervised joint object discovery and segmentation in internet images. In: CVPR (2013)
45. Rubio, J.C., Serrat, J., López, A., Paragios, N.: Unsupervised co-segmentation through region matching. In: CVPR (2012)
46. Shocher, A., et al.: Semantic pyramid for image generation. In: CVPR (2020)
47. Shotton, J., Winn, J., Rother, C., Criminisi, A.: *TextonBoost*: joint appearance, shape and context modeling for multi-class object recognition and segmentation. In: Leonardis, A., Bischof, H., Pinz, A. (eds.) ECCV 2006. LNCS, vol. 3951, pp. 1–15. Springer, Heidelberg (2006). https://doi.org/10.1007/11744023_1
48. Siméoni, O., et al.: Localizing objects with self-supervised transformers and no labels. In: BMVC (2021)
49. Simonyan, K., Zisserman, A.: Very deep convolutional networks for large-scale image recognition. In: ICLR (2015)
50. Sun, J., Shen, Z., Wang, Y., Bao, H., Zhou, X.: LoFTR: detector-free local feature matching with transformers. In: CVPR (2021)
51. Thewlis, J., Bilen, H., Vedaldi, A.: Unsupervised learning of object landmarks by factorized spatial embeddings. In: ICCV (2017)
52. Vaze, S., Han, K., Vedaldi, A., Zisserman, A.: Generalized category discovery. In: ICLR (2022)
53. Wang, X., Girshick, R., Gupta, A., He, K.: Non-local neural networks. In: CVPR (2018)
54. Wang, Y., Shen, X., Hu, S.X., Yuan, Y., Crowley, J., Vaufreydaz, D.: Self-supervised transformers for unsupervised object discovery using normalized cut. In: CVPR (2022)
55. Welinder, P., et al.: Caltech-UCSD Birds 200. Tech. Rep. CNS-TR-2010-001, California Institute of Technology (2010)
56. Zhang, K., Chen, J., Liu, B., Liu, Q.: Deep object co-segmentation via spatial-semantic network modulation. In: AAAI (2020)
57. Zhang, R.: Making convolutional networks shift-invariant again. In: ICML (2019)
58. Zhang, R., Isola, P., Efros, A.A., Shechtman, E., Wang, O.: The unreasonable effectiveness of deep features as a perceptual metric. In: CVPR (2018)
59. Zhang, Y., Guo, Y., Jin, Y., Luo, Y., He, Z., Lee, H.: Unsupervised discovery of object landmarks as structural representations. In: CVPR (2018)

Anomaly Detection Requires Better Representations

Tal Reiss$^{(\boxtimes)}$, Niv Cohen, Eliahu Horwitz, Ron Abutbul, and Yedid Hoshen

School of Computer Science and Engineering, The Hebrew University of Jerusalem,
Jerusalem, Israel
tal.reiss@mail.huji.ac.il
http://www.vision.huji.ac.il/ssrl_ad/

Abstract. Anomaly detection seeks to identify unusual phenomena, a central task in science and industry. The task is inherently unsupervised as anomalies are unexpected and unknown during training. Recent advances in self-supervised representation learning have directly driven improvements in anomaly detection. In this position paper, we first explain how self-supervised representations can be easily used to achieve state-of-the-art performance in commonly reported anomaly detection benchmarks. We then argue that tackling the next generation of anomaly detection tasks requires new technical and conceptual improvements in representation learning.

Keywords: Anomaly detection · Self-Supervised learning ·
Representation learning

1 Introduction

Discovery commences with the awareness of anomaly, i.e., with the recognition that nature has somehow violated the paradigm-induced expectations that govern normal science.

———Kuhn, The Structure of Scientific Revolutions (1970)

I do not know what I may appear to the world, but to myself I seem to have been only like a boy playing on the seashore, and diverting myself in now and then finding a smoother pebble or a prettier shell than ordinary, whilst the great ocean of truth lay all undiscovered before me.

———Isaac Newton

Anomaly detection, discovering unusual patterns in data, is a core task for human and machine intelligence. The importance of the task stems from the centrality of discovering unique or unusual phenomena in science and industry. For example, the fields of particle physics and cosmology have, to large extent, been driven by the discovery of new fundamental particles and stellar objects.

© The Author(s), under exclusive license to Springer Nature Switzerland AG 2023
L. Karlinsky et al. (Eds.): ECCV 2022 Workshops, LNCS 13804, pp. 56–68, 2023.
https://doi.org/10.1007/978-3-031-25069-9_4

Similarly, the discovery of new, unknown, biological organisms and systems is a driving force behind biology. The task is also of significant economic potential. Anomaly detection methods are used to detect credit card fraud, faults on production lines, and unusual patterns in network communications.

Detecting anomalies is essentially unsupervised as only "normal" data, but no anomalies, are seen during training. While the field has been intensely researched for decades, the most successful recent approaches use a very simple two-stage paradigm: (i) each data point is transformed to a representation, often learned in a self-supervised manner. (ii) a density estimation model, often as simple as a k nearest neighbor estimator, is fitted to the normal data provided in a training set. To classify a new sample as normal or anomalous, its estimated probability density is computed - low likelihood samples are denoted as anomalies.

In this position paper, we first explain that advances in representation learning are the main explanatory factor for the performance of recent anomaly detection (AD) algorithms. We show that this paradigm essentially "solves" the most commonly reported image anomaly detection benchmark (Sect. 4). While this is encouraging, we argue that existing self-supervised representations are unable to solve the next generation of AD tasks (Sect. 5). In particular, we highlight the following issues: (i) masked-autoencoders are much worse for AD than earlier self-supervised representation learning (SSRL) methods (ii) current approaches perform poorly in for datasets with multiple objects per-image, complex background, fine-grained anomalies. (iii) in some cases SSRL performs worse than handcrafted representations (iv) for "tabular" datasets, no representation performed better than the original representation of the data (i.e. that data itself) (v) in the presence of nuisance factors of variation, it is unclear whether SSRL can *in-principle* identify the optimal representation for effective AD.

Anomaly detection presents both rich rewards as well as significant challenges for representation learning. Overcoming these issues will require significant progress, both technical and conceptual. We expect that increasing the involvement of the self-supervised representation learning community in anomaly detection will mutually benefit both fields.

2   Related Work

Classical AD approaches were typically based on either density estimation [9,20] or reconstruction [15]. With the advent of deep learning, classical methods were augmented by deep representations [19,23,24,38]. A prevalent way to learn these representations was to use self-supervised methods, e.g. autoencoder [30], rotation classification [10,13], and contrastive methods [35,36]. An alternative approach is to combine pretrained representations with anomaly scoring functions [25,27,28,32]. The best performing methods [27,28] combine pretraining on auxiliary datasets and a second finetuning stage on the provided normal samples in the training set. It was recently established [27] that given sufficiently powerful representations (e.g. ImageNet classification), a simple criterion based on the kNN distance to the normal training data achieves strong performance. We therefore limit the discussion of AD in this paper to this simple technique.

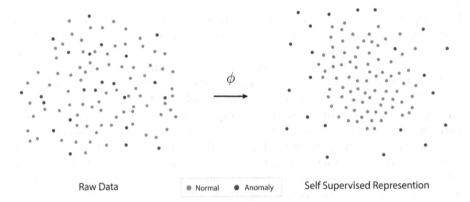

Raw Data   ● Normal   ● Anomaly   Self Supervised Representation

Fig. 1. *Normal and Anomalous Representations:* The self-supervised representations transform the raw data into a space in which normal and anomalous data can be easily separated using density estimation methods

3 Anomaly Detection as a Downstream Task for Representation Learning

In this section we describe the computational task, method, and evaluation setting for anomaly detection.

Task definition. We assume access to N random samples, denoted by $\mathcal{X}_{train} = \{x_1, x_2...x_N\}$, from the distribution of the normal data $p_{norm}(x)$. At test time, the algorithm observes a sample $\tilde{x}$ from the real-world distribution $p_{real}(x)$, which consists of a combination of the normal and anomalous data distributions: $p_{norm}(x)$ and $p_{anom}(x)$. The task is to classify the sample $\tilde{x}$ as normal or anomalous.

Representations for anomaly detection. In AD, it is typically assumed that anomalies $a \sim p_{anom}$ have a low likelihood under the normal data distribution, i.e. that $p_{norm}(a)$ is small. Under this assumption, the PDF of normal data p_{norm} acts as an effective anomaly classifier. In practice, however, training an estimator q for scoring anomalies using p_{norm} is a challenging statistical task. The challenge is greater when: (i) the data are high-dimensional (e.g. images) (ii) p_{norm} is sparse or irregular (iii) normal and anomalous data are not separable using simple functions. Representation learning may overcome these issues by transforming the sample x into a representation $\phi(x)$, which is of lower dimension, where p_{norm} is relatively smooth and where normal and anomalous data are more separable. As no anomaly labels are provided, self-supervised representation learning is needed.

A two-stage anomaly detection paradigm. Given a self-supervised representation ϕ, we follow a simple two stage anomaly detection paradigm: (i) *Representation encoder*: each sample during training or test is mapped to a feature descriptor using the mapping function ϕ. (ii) *Density estimation*: a probability estimator $q_{norm}(x)$ is fitted to the distribution of the normal sample features $\mathcal{X}_{train} = \{\phi(x_1), \phi(x_2)...\phi(x_N)\}$. A sample is scored at test time by first mapping it to the representation space $\phi(\tilde{x})$ and scoring it according to the density estimator. Given an estimator q_{norm} of the normal probability density p_{norm}, the anomaly score s is given by $s(\tilde{x}) = -q_{norm}(\phi(\tilde{x}))$. Normal data will typically obtain lower scores than anomalous samples. A user can then set a threshold for the prediction of anomalies based on an appropriate false positive rate.

4   Successful Representation Learning Enables Anomaly Detection

Detecting anomalies in images is probably the most researched task by the deep learning anomaly detection community. In this section, we show that the simple paradigm presented in Sect. 3 achieves state-of-the-art results. As most density estimators achieve very similar results, the anomaly detection performance is mostly determined by the quality of learned representation. This makes anomaly detection an excellent testing ground for representations. Furthermore, we discuss different approaches to finetune a representation on the normal train data and show significant gains.

Learning Representations From the Normal Data. Perhaps the most common approach taken by recent AD methods is to learn the representations in a self-supervised manner using solely the normal samples (i.e. the training dataset). Examples of such methods are RotNet [13], CSI [36] and others. The main disadvantage of such methods is that most of the datasets are of small size and hence do not suffice for learning powerful representations.

Extracting Representations From a Pretrained Model. A very simple alternative is to use an off-the-shelf pretrained model and extract features for the normal (i.e. training) data from it. The pretraining may be either supervised (e.g. using ImageNet labels [8,12]) or self-supervised (e.g. DINO), in both cases pretraining may be performed on ImageNet. These representations tend to perform much better than those extracted from models trained only on the normal data.

A hybrid approach. A natural extension to the above approaches is to combine the two. This is done by using the pretrained model as an initialization for a self-supervised finetuning phase (on the normal data). In this way, the powerful representation of the pretrained model can be used and refined within the context of the anomaly detection dataset and task. Multiple approaches [27,28] have been used for the self-supervised finetuning stage. However, in this paper we present what is possibly the simplest approach, using DINO's objective for the finetuning stage. In this approach, a pretrained DINO model is used as an initialization. During the finetuning phase, the model is trained on the target anomaly detection training dataset (i.e. only normal data) in a self-supervised manner by simply using the original DINO objective.

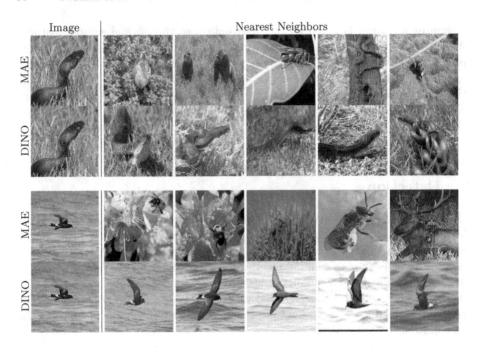

Fig. 2. *MAE vs. DINO nearest neighbors:* For each image, the top 5 nearest neighbors are shown according to their order. Note how MAE neighbors are chosen mostly based on the colors and not their semantic contents, in contrast, DINO neighbors are semantically accurate.

In Fig. 1 the above process is demonstrated with a toy example. Table 1 presents anomaly detection results on the CIFAR-10 [18] dataset, which is the most commonly used dataset for evaluation. As can be seen, using representations extracted from a recent self-supervised method (i.e. DINO) following the hybrid approach and coupled with a trivial kNN estimator for the density estimation phase nearly solves this dataset. Although a possible conclusion could have been that the anomaly detection task has been solved, in the next section we show this is not the case.

5   Gaps in Anomaly Detection Point to Bottlenecks in Representations Learning

While Sect. 4 presented a very optimistic view of the ability of representation learning to solve anomaly detection, in this section we paint a more complex picture. We use this to highlight several limitations of current self-Supervised representations.

Table 1. *Image anomaly detection results:* Mean ROC-AUC %. Bold denotes the best results, FT stands for finetuned

Approach	Self-Supervised		Pretrained		Hybrid		
Method	RotNet [13]	CSI [36]	ResNet	DINO	PANDA [27]	MSAD [28]	DINO-FT
CIFAR-10	90.1	94.3	92.5	97.1	96.2	97.2	**98.4**

5.1 Masked-Autoencoder: Advances in Self-Supervised Learning Do Not Always Imply Better Anomaly Detection

Recently, masked-autoencoder (MAE) based methods achieved significant improvements on several self-supervised representation learning benchmarks [11]. Yet, the representations learnt by MAE underperform contrastive self-supervised methods on unsupervised tasks such as anomaly detection. A comparison between MAE to contrastive self-Supervised method (DINO) is presented in Table 2 demonstrating the much better performance of DINO for AD. Fine-tuning on the normal training data improves both methods, however a large gap still remains. Implementation details for the experiments can be found in the App. A. In many papers, self-supervised methods are evaluated using supervised benchmarks, such as classification accuracy with finetuning. The key difference between anomaly detection and ordinary benchmarks where MAE excel is that anomaly detection is an unsupervised task. This is also suggested by MAE's worse performance with linear probing (as reported by the original paper), where the supervised labels cannot be used to improve the backbone representations.

MAE's optimization objective may explain why its strong representation does not translate into better anomaly detection capabilities. As MAE's objective is to reconstruct patches, it may learn a representation that encodes local information needed for reconstructing the image, overlooking semantic object properties. Consequently, the nearest neighbors may pay more attention to local similarity than to global semantic properties (See Fig. 2). In contrast, the goal of contrastive-based objectives is to map semantically similar images to nearby representations, disregarding some of the local properties.

Conclusion. Better performance on supervised downstream tasks does not necessarily imply better representations across the board. In some cases, while the representation may excel in a supervised downstream task, it may underperform in an unsupervised counterpart. Looking forward, we suggest that new self-supervised representation learning methods present evaluations on unsupervised anomaly detection tasks alongside the common supervised benchmarks.

5.2 Complex Datasets: Current Representations Struggle on Scenes, Finegrained Classes, Multiple Objects

Current representations are very effective for anomaly detection on datasets with a single object occupying a large portion of the image. Furthermore, these

Table 2. *Anomaly detection comparison of MAE and DINO:* Mean ROC-AUC %. Bold denotes the best results

Method	CIFAR-10	CUB-200	INet-S
MAE	78.1	73.1	83.2
DINO	**97.1**	**93.9**	**99.3**

methods typically perform well when the number of object categories in the normal train set is small and have coarse differences (e.g. "cat" and "ship"). A prime example is CIFAR-10, which is virtually solved. On the other hand, anomaly detection accuracy is much lower on more complex datasets containing multiple small objects, complex backgrounds; and when anomalies consist of related object categories (e.g. "sofa" and "armchair"). We modified the MS-COCO [21] dataset by using all images from a single super-category ('vehicles') as normal data, apart from a single category ('bicycle') which are used as anomalies. We experiment both with cropping just the object bounding boxes or using the entire image (including the background and other objects). Similarly, we report results for a multi-modal CUB-200 [37] anomaly detection benchmark. The results are presented in Table 3 (implementation details can be found in the Appendix). It is clear that these datasets are far from solved and that better representations are needed to achieve acceptable performance.

Conclusion. While current representations are effective for relatively easy datasets, more realistic cases with small objects, backgrounds and many object categories call for the development of new SSRL methods.

Table 3. *Multi-modal datasets:* Mean ROC-AUC %. "MS-COCO-I" / "MS-COCO-O" indicates MS-COCO image / object level benchmarks (respectively).

Method	MS-COCO-I	MS-COCO-O	CUB-200
PANDA [27]	61.5	77.0	78.4
MSAD [28]	61.7	76.9	80.1
DINO	61.5	73.4	74.5

5.3   Unidentifiability: Representations for Anomaly Detection May Be Ambiguous Without Further Guidance

In some settings, we would like our representation to focus on specific attributes (which we denote as *relevant*) while ignoring nuisance attributes that might bias the model. Consider two different companies interested in anomaly detection in cars. The first company may be interested in detecting novel car models, while the second is interested in unusual driving behaviors. Although both may wish to apply density estimation using a state-of-the-art self-supervised representation,

Table 4. *Summary of the findings of from Horwitz and Hoshen* [14]: Average metrics across all MVTec3D-AD classes, "INet" indicates ImageNet [8] pretrained features, PC indicates point cloud. I-ROC indicates image level ROC-AUC % [4], P-ROC indicates pixel level ROC-AUC %. Higher score indicates better the results

Modality	RGB	Depth	Depth	Depth	Depth	Depth	PC	RGB+PC
Method	INet	INet	NSA [33]	Raw	HoG [7]	SIFT [22]	FPFH [31]	RGB+FPFH
PRO [2]	87.6	58.6	57.2	19.1	61.4	86.6	92.4	**96.4**
I-ROC	78.5	63.7	69.6	52.8	56.0	71.4	75.3	**86.5**
P-ROC	96.6	82.1	81.7	54.8	84.5	95.4	98.0	**99.3**

each will view the ground truth anomalies of the other company as a false-positive case. As each company is interested in different anomalies, they may require different representations. One company would require the representation to contain only the driving patterns and be agnostic to the car model, at the same time, the other company would strive for the opposite. As these preferences are not present at the time of pretraining the self-supervised backbone, the correct solution is often unidentifiable.

One initial effort is RedPANDA [6] that proposed providing labels for nuisance attributes. We note that only the attributes to be ignored are labeled, while the other attributes (the ones in which characterize anomalies) are not provided. Representation learning is then performed using domain-supervised disentanglement [16], resulting in a representation only describing the unlabelled attributes. Yet, the field of domain-supervised disentanglement is still in its infancy, and the assumption of nuisance attribute labels is often not applicable.

Conclusion. Self-supervised representation learning methods are designed to focus on semantic attributes of images, but choosing the most relevant ones is unidentifiable without further guidance. Incorporating guidance may be achieved by a careful choice of inductive bias [16] (e.g. augmentations) or using concept-based representation techniques [17].

5.4   3D Point Clouds: Self-supervised Representations Do Not Always Improve over Handcrafted Ones

In an empirical investigation [14], we evaluated representative methods designed for different modalities on the MVTec3D-AD dataset [3]. The paper showed that currently, handcrafted features for 3D surface matching outperform learning-based methods designed either for images or for 3D point clouds. A key insight was that rotation invariance is very beneficial in this modality, and is often overlooked. A summary of the findings, taken from the original paper, is found in Table 4.

Conclusion. When dealing with 3D point-cloud, self-supervised representations are yet to outperform handcrafted features for anomaly detection. For modalities less mature than images, domain specific priors may still need to be integrated into the architecture or objective. This stresses the need for better 3D point-cloud representations.

5.5 Tabular Data: When Representations Do Not Improve over the Original Data

The tabular setting is probably the most general anomaly detection setting, where each sample in the dataset consists of a set of numerical and categorical variables. This is strictly harder than any other setting as no regularity in the data can be assumed. Such data are frequently encountered, as unstructured databases are very common. In recent years, self-supervised methods have been proposed for tabular anomaly detection [1,26,34,39]. These methods differ by the auxiliary task that they use for representation learning (and potentially also for anomaly scoring). Two representative deep learning approaches are GOAD [1] which predicts geometric transformations, and use the prediction errors to detect anomalies, and ICL [34] which adopts the contrastive learning task for training and for anomaly scoring by differentiating between in-window and out-window features. As part of our evaluation, we used both their standard pipeline (i.e. their auxiliary tasks for anomaly scoring) and our AD density estimation paradigm (see Appendix). These results were then compared with kNN on the original raw features without any modifications. The results are presented in Table 5. Self-supervised representation learning did not improve performance in comparison with the original raw features.

Conclusion. Representation learning for general datasets is an open research question. Some prior knowledge of the dataset must be used in order to learn non-trivial data representations, at least in the context of anomaly detection.

Table 5. *Tabular results:* Mean F1 & ROC-AUC % from the ODDS benchmarks results. Bold denotes the best results

Method	GOAD [1]		ICL [34]		Raw
Scoring	Auxiliary	kNN	Auxiliary	kNN	kNN
F1	54.4	63.2	68.1	69.8	**69.9**
ROC-AUC	78.2	87.6	88.9	89.4	**90.2**

6 Final Remarks

In this position paper, we advocated the study of self-supervised representations for the task of anomaly detection. We explained that advances in representation learning have been the main driving force behind progress in anomaly detection. On the other hand, we demonstrated that current self-supervised representation learning methods often fall short in challenging anomaly detection settings. Our hope is that interplay between the self-supervised representation learning and anomaly detection fields will result in mutual benefits for both communities.

Acknowledgements. This work was partially supported by the Malvina and Solomon Pollack Scholarship, a Facebook award, the Israeli Cyber Directorate, the Israeli Higher Council and the Israeli Science Foundation. We also acknowledge support of Oracle Cloud credits and related resources provided by the Oracle for Research program.

A   Appendix

In this paper we report anomaly detection results using the standard uni-modal protocol, which is widely used in the anomaly detection community. In the uni-modal protocol, multi-class datasets are converted to anomaly detection by setting a class as normal and all other classes as anomalies. The process is repeated for all classes, converting a dataset with C classes into C datasets. Finally, we report the mean ROC-AUC % over all C datasets as the anomaly detection results.

A.1   Anomaly detection comparison of MAE and DINO

We compare between DINO [5] and MAE [11] as a representation for a kNN based anomaly detection algorithm. For MAE, we experimented both with kNN and reconstruction error for anomaly scoring and found that the latter works badly, therefore we report just the kNN results. We evaluate using a variety of datasets, in the uni-modal setting described above. We used the following datasets:

INet-S [29]: The dataset is subset of 10 animal classes taken from ImageNet21k (e.g. "petrel", "tyrannosaur", "rat snake", "duck", "bee fly", "sheep", "beer cub", "red deer", "silverback", "opossum rat") that do not appear in ImageNet1K dataset. The dataset is coarse-grained and contains images relatively close to ImageNet1K dataset. It intended to convey that even for easy tasks the MAE doesn't achieve as good results as DINO.

CIFAR-10 [18]: Consists of low-resolution 32×32 images from 10 different classes.

CUB-200 [37]: Bird species image dataset which contains 11,788 images of 200 subcategories. In the experiment we calculated mean ROC-AUC% over the 20 first categories.

A.2   Multi-modal datasets

In these experiment we specify a single class as anomalous, and treat all images which does not contain it as normal.

MS-COCO-I [21]: We build a multi-modal anomaly detection dataset comprised of scenes benchmarks, where each image is evaluated against other images featuring similar scenes. We choose 10 object categories ("bicycle", "traffic light", "bird" , "backpack", "frisbee", "bottle", "banana", "chair", "tv", "microwave", "book") from different MS-COCO super-categories. To construct a multi-modal anomaly detection benchmark, we designate an object category from the list as the anomalous class, and training images of a similar super-category that do not contain it as our normal train set. Our test set contains all the test images from that super-category, where images containing the anomalous object are labelled as anomalies. This process is repeated for the 10 object categories resulting in 10 different evaluations. We report their average ROC-AUC %.

MS-COCO-O: We introduce a similar benchmark to MS-COCO-I, focusing on single objects rather than scenes. We crop all objects from our 10 super-categories (described above) according to the MS-COCO supplied bounding boxes. We repeat a similar process, using a similar object category as normal and the rest as anomalies.

CUB-200 [37]: We create a multi-modal anomaly detection benchmark based on the CUB-200 dataset. We focus on the 20 first categories, designating only one as an anomaly each time.

A.3   Tabular domain

Various datasets used for tabular data anomaly detection were used for the experiments. A total of 31 datasets from Outlier Detection DataSets (ODDS)[1] are employed. For the evaluation of GOAD and ICL we used the official repositories and made an effort to select the best configuration available. For all density estimation evaluations we used kNN with $k = 5$ nearest neighbors. To convert GOAD and ICL into the standard paradigm of representation learning followed by density estimation: i) we use the original approaches to train a feature encoder (followed by a classifier which we discard) ii) we use the feature encoder to represent each sample iii) density estimation is performed on the representations using kNN exactly as in Sect. 3.

References

1. Bergman, L., Hoshen, Y.: Classification-based anomaly detection for general data. In: ICLR (2020)
2. Bergmann, P., Fauser, M., Sattlegger, D., Steger, C.: MVTec AD-A comprehensive real-world dataset for unsupervised anomaly detection. In: Proceedings of the IEEE/CVF Conference on Computer Vision and Pattern Recognition, pp. 9592–9600 (2019)
3. Bergmann, P., Jin, X., Sattlegger, D., Steger, C.: The MVTec 3D-AD dataset for unsupervised 3D anomaly detection and localization (2021). arXiv preprint arXiv:2112.09045
4. Bradley, A.P.: The use of the area under the ROC curve in the evaluation of machine learning algorithms. Pattern Recogn. **30**(7), 1145–1159 (1997)
5. Caron, M., et al.: Emerging properties in self-supervised vision transformers. In: Proceedings of the IEEE/CVF International Conference on Computer Vision, pp. 9650–9660 (2021)
6. Cohen, N., Kahana, J., Hoshen, Y.: Red PANDA: disambiguating anomaly detection by removing nuisance factors (2022) arXiv preprint arXiv
7. Dalal, N., Triggs, B.: Histograms of oriented gradients for human detection. In: 2005 IEEE Computer Society Conference on Computer Vision and Pattern Recognition (CVPR 2005), vol. 1, pp. 886–893. IEEE (2005)

---

[1] http://odds.cs.stonybrook.edu/.

8. Deng, J., Dong, W., Socher, R., Li, L.J., Li, K., Fei-Fei, L.: ImageNet: a large-scale hierarchical image database. In: 2009 IEEE Conference on Computer Vision and Pattern Recognition, pp. 248–255. IEEE (2009)

9. Eskin, E., Arnold, A., Prerau, M., Portnoy, L., Stolfo, S.: A geometric framework for unsupervised anomaly detection. In: Barbará, D., Jajodia, S. (eds.) Applications of Data Mining in Computer Security. Advances in Information Security, vol. 6, pp. 77–101. Springer, Boston (2002). https://doi.org/10.1007/978-1-4615-0953-0_4

10. Golan, I., El-Yaniv, R.: Deep anomaly detection using geometric transformations. In: NeurIPS (2018)

11. He, K., Chen, X., Xie, S., Li, Y., Dollár, P., Girshick, R.: Masked autoencoders are scalable vision learners. In: Proceedings of the IEEE/CVF Conference on Computer Vision and Pattern Recognition, pp. 16000–16009 (2022)

12. He, K., Zhang, X., Ren, S., Sun, J.: Deep residual learning for image recognition. In: Proceedings of the IEEE Conference on Computer Vision and Pattern Recognition, pp. 770–778 (2016)

13. Hendrycks, D., Mazeika, M., Kadavath, S., Song, D.: Using self-supervised learning can improve model robustness and uncertainty. In: NeurIPS (2019)

14. Horwitz, E., Hoshen, Y.: An empirical investigation of 3D anomaly detection and segmentation (2022)

15. Jolliffe, I.: Principal component analysis. Springer (2011) https://doi.org/10.1007/978-1-4757-1904-8

16. Kahana, J., Hoshen, Y.: A contrastive objective for learning disentangled representations (2022) arXiv preprint arXiv:2203.11284

17. Koh, P.W., et al.: Concept bottleneck models. In: International Conference on Machine Learning, pp. 5338–5348. PMLR (2020)

18. Krizhevsky, A., Hinton, G., et al.: Learning multiple layers of features from tiny images (2009)

19. Larsson, G., Maire, M., Shakhnarovich, G.: Learning representations for automatic colorization. In: Leibe, B., Matas, J., Sebe, N., Welling, M. (eds.) ECCV 2016. LNCS, vol. 9908, pp. 577–593. Springer, Cham (2016). https://doi.org/10.1007/978-3-319-46493-0_35

20. Latecki, L.J., Lazarevic, A., Pokrajac, D.: Outlier detection with kernel density functions. In: Perner, P. (ed.) MLDM 2007. LNCS (LNAI), vol. 4571, pp. 61–75. Springer, Heidelberg (2007). https://doi.org/10.1007/978-3-540-73499-4_6

21. Lin, T.-Y., et al.: Microsoft COCO: common objects in context. In: Fleet, D., Pajdla, T., Schiele, B., Tuytelaars, T. (eds.) ECCV 2014. LNCS, vol. 8693, pp. 740–755. Springer, Cham (2014). https://doi.org/10.1007/978-3-319-10602-1_48

22. Lowe, D.G.: Distinctive image features from scale-invariant keypoints. Int. J. Comput. vis. **60**(2), 91–110 (2004)

23. Mathieu, M., Couprie, C., LeCun, Y.: Deep multi-scale video prediction beyond mean square error. In: ICLR (2016)

24. Noroozi, M., Favaro, P.: Unsupervised learning of visual representations by solving Jigsaw Puzzles. In: Leibe, B., Matas, J., Sebe, N., Welling, M. (eds.) ECCV 2016. LNCS, vol. 9910, pp. 69–84. Springer, Cham (2016). https://doi.org/10.1007/978-3-319-46466-4_5

25. Perera, P., Patel, V.M.: Learning deep features for one-class classification. IEEE Trans. Image Process. **28**(11), 5450–5463 (2019)

26. Qiu, C., Pfrommer, T., Kloft, M., Mandt, S., Rudolph, M.: Neural transformation learning for deep anomaly detection beyond images. In: International Conference on Machine Learning, pp. 8703–8714. PMLR (2021)

27. Reiss, T., Cohen, N., Bergman, L., Hoshen, Y.: Panda: adapting pretrained features for anomaly detection and segmentation. In: Proceedings of the IEEE/CVF Conference on Computer Vision and Pattern Recognition, pp. 2806–2814 (2021)
28. Reiss, T., Hoshen, Y.: Mean-shifted contrastive loss for anomaly detection (2021). arXiv preprint arXiv:2106.03844
29. Ridnik, T., Ben-Baruch, E., Noy, A., Zelnik-Manor, L.: ImageNet-21k pretraining for the masses (2021)
30. Ruff, L., et al.: Deep one-class classification. In: ICML (2018)
31. Rusu, R.B., Blodow, N., Beetz, M.: Fast point feature histograms (FPFH) for 3D registration. In: 2009 IEEE International Conference on Robotics and Automation, pp. 3212–3217 (2009)
32. Salehi, M., Sadjadi, N., Baselizadeh, S., Rohban, M.H., Rabiee, H.R.: Multiresolution knowledge distillation for anomaly detection. In: Proceedings of the IEEE/CVF Conference on Computer Vision and Pattern Recognition, pp. 14902–14912 (2021)
33. Schlüter, H.M., Tan, J., Hou, B., Kainz, B.: Self-Supervised out-of-distribution detection and localization with natural synthetic anomalies (NSA) (2021). arXiv preprint arXiv:2109.15222
34. Shenkar, T., Wolf, L.: Anomaly detection for tabular data with internal contrastive learning. In: International Conference on Learning Representations (2021)
35. Sohn, K., Li, C.L., Yoon, J., Jin, M., Pfister, T.: Learning and evaluating representations for deep one-class classification (2020). arXiv preprint arXiv:2011.02578
36. Tack, J., Mo, S., Jeong, J., Shin, J.: CSI: Novelty detection via contrastive learning on distributionally shifted instances. In: NeurIPS (2020)
37. Welinder, P., et al.: Caltech-UCSD birds 200. **2**(5), 11 (2010)
38. Zhang, R., Isola, P., Efros, A.A.: Colorful image colorization. In: Leibe, B., Matas, J., Sebe, N., Welling, M. (eds.) ECCV 2016. LNCS, vol. 9907, pp. 649–666. Springer, Cham (2016). https://doi.org/10.1007/978-3-319-46487-9_40
39. Zong, B., et al.: Deep autoencoding gaussian mixture model for unsupervised anomaly detection. In: International Conference on Learning Representations (2018)

Leveraging Self-Supervised Training for Unintentional Action Recognition

Enea Duka$^{(\boxtimes)}$, Anna Kukleva, and Bernt Schiele

MPI for Informatics, Saarbrücken, Germany
{enea.duka,akukleva,schiele}@mpi-inf.mpg.de
https://www.mpi-inf.mpg.de

Abstract. Unintentional actions are rare occurrences that are difficult to define precisely and that are highly dependent on the temporal context of the action. In this work, we explore such actions and seek to identify the points in videos where the actions transition from intentional to unintentional. We propose a multi-stage framework that exploits inherent biases such as motion speed, motion direction, and order to recognize unintentional actions. To enhance representations via self-supervised training for the task of unintentional action recognition we propose temporal transformations, called **T**emporal **T**ransformations of **I**nherent **B**iases of **U**nintentional **A**ctions (T^{2}IBUA). The multi-stage approach models the temporal information on both the level of individual frames and full clips. These enhanced representations show strong performance for unintentional action recognition tasks. We provide an extensive ablation study of our framework and report results that significantly improve over the state-of-the-art.

Keywords: Unintentional action recognition · Multi-stage temporal modeling · Self-supervised representation enhancement

1 Introduction

Video action understanding has witnessed great progress over the past several years in the fields of action detection, recognition, segmentation, caption generation, tracking and many others [5,13,22,37,42]. However, these methods implicitly rely on the continuity of the underlying intentional action, such as one starts and then continues the same or related activity for some time. In this paper, we study unintentional actions, for example, when one falls down by accident during jogging. This is a challenging task due to the difficulty in defining precisely the intentionality of an action while largely depending on the temporal context of the action. Moreover,

---

E. Duka and A. Kukleva—Equal contribution.

Supplementary Information The online version contains supplementary material available at https://doi.org/10.1007/978-3-031-25069-9_5.

ⓒ The Author(s), under exclusive license to Springer Nature Switzerland AG 2023
L. Karlinsky et al. (Eds.): ECCV 2022 Workshops, LNCS 13804, pp. 69–85, 2023.
https://doi.org/10.1007/978-3-031-25069-9_5

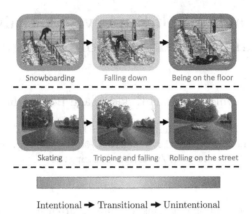

Intentional → Transitional → Unintentional

Fig. 1. Transition from intentional to unintentional action. We can notice an abrupt change of motion in the first example, when the snowboarder starts to fall down the stairs, and a sudden change in speed in the second example when the skater starts rolling on the ground. In both cases, there is a strict order from intentional to unintentional action.

annotating such videos is both costly and difficult with human accuracy being 88% in localising the transition between intentional and unintentional actions [8]. The potential of this research covers assistance robotics, health care devices, or public video surveillance, where detecting such actions can be critical. In this work, we propose to look at unintentional actions from the perspective of the discontinuity of the intentionality and exploit inherent biases in this type of videos.

We design our framework to explore videos where a transition from intentional to unintentional action occurs. We are the first to explicitly leverage the inherent biases of unintentional videos in the form of motion cues and high-level action ordering within the same video, specifically the transition from unintentional to intentional and vice-versa.

To this end, we propose a three-stage framework where, starting from a pretrained representation, the first two stages leverage self-supervised training to address unintentional biases on frame and video clip levels and further enhance these representations. We regularly observe abrupt motion changes in videos of unintentional actions when an unwitting event happens. Therefore, we formulate **T**emporal **T**ransformations to exploit **I**nherent **B**iases in videos that resemble **U**nintentional **A**ction motion changes (T^2IBUA). Using T^2IBUA we learn intra and inter-clip information in a multi-stage manner. In the first stage, we capture local information from the neighbouring frames by predicting the discrete label of the applied T^2IBUA on the frame level. Then, in the second stage, we train the model to integrate information from the entire video by predicting T^2IBUA on the clip level. See Fig. 4 for different T^2IBUA levels. Inspired by the growing popularity of transformer models in vision [10,14,28,32], we utilize a multi-level transformer architecture that serves as a temporal encoder in our framework to model long-term relations. In the last stage of supervised downstream learning, we particularly benefit from the high-level ordering by deploying conditional random fields (CRF) to explicitly enforce smooth transitions within the same video from intentional to unintentional actions. By modelling these explicit global temporal relations, we enhance our results on various downstream tasks such as unintentional action classification, localisation and anticipation.

Our work makes the following contributions. Firstly, we propose a framework that includes three learning stages for unintentional action recognition: by leveraging inherent biases in the first and in the second stages with self-supervised feature enhancement, the third stage employs supervised learning for unintentional action recognition on the top of these amplified representations. In addition, we present a multi-stage temporal encoder as a transformer-based architecture in combination with conditional random fields to capture local and global temporal relations between intentional and unintentional actions. Finally, we show state-of-the-art results for classification, detection, and anticipation tasks and perform various ablation studies that evaluate the effectiveness of each of the components of the proposed framework.

2 Related Work

Unintentional Action Recognition. The topic of unintentional action recognition is recently introduced by Epstein et al. [8]. The authors collect the dataset Oops! and propose a framework to study unintentional action recognition based on three directions such as classification, localization, and anticipation. The framework consists of two consecutive parts to learn the representations in a self-supervised way and then adapt the model to unintentional actions. In contrast to [8] we leverage self-supervised training to tailor pre-trained representations to unintentional actions and model temporal information in a multi-stage fashion through the multiple learning stages. Han et al. [16] predict the future steps of the video based on the feature space on the given history frames. Epstein et al. [9] utilize a 3D CNN in combination with Transformer blocks to enhance representation learning and recognize discontinuities in the videos. As opposed to [9] we follow a two-step approach with a self-supervised enhancement of the pre-trained features followed by a fully-supervised fine-tuning on unintentional action recognition tasks as in [8,15].

Video Representation Learning. Nowadays, video representation learning is an active research topic. Various types of architectures are developed, such as two-stream networks [3,4,11], 3D convolution networks [3,33,34], Transformer based architectures [10,25,28] that rely on supervised training. Many other works explore self-supervised learning to utilize more widely available resources [1,20,21,35]. The first methods [12,24,27] focus on frame-level video augmentations such as frame shuffling and odd frame insertion in an existing frame sequence. Then they train their model to distinguish these augmentations [27]. Xu et al. [41] modifies this paradigm by augmenting the videos on the clip level instead of frame level. The authors shuffle the video clips and predict the correct order of the clips, thus they employ spatial-temporal information from videos. Han et al. [15] formulate the problem as future clip prediction. The recent work [6,29] apply contrastive learning between different augmentation of videos. Some other directions for the self-supervised learning include colourization of greyscale images as a proxy task [23,36], or leveraging temporal co-occurrence as a learning signal [18,38]. Wei et al. [40] utilize the arrow of time by detecting the direction of video playback. Another popular direction for the video representation learning is to learn temporal or spatial-temporal cycle-consistency as a proxy task [19,39]. While in this work, we introduce a self-supervised feature enhancement that exploits the inherent biases of unintentional actions. We show that this amplification improves the unintentional action recognition by large margin with respect to previous work and the baselines.

3 Approach to Exploit the UA Inherent Biases

In this and the next sections, we present our framework to study unintentional actions (UA) in videos. First, we provide an overview of our approach in Sect. 3.1. In Sect. 3.2 we detail T^{2}IBUA for self-supervised training, and then in Sect. 4 we describe the learning stages for our framework.

Notation: Let $X \in \mathcal{R}^{T \times W \times H \times 3}$ be an RGB video, where T, W and H are the number of frames, width and height respectively. We denote a clip sampled from this video as $x \in \mathcal{R}^{t \times W \times H \times 3}$ where t is the number of frames in the clip and $t \leq T$. As f we further denote individual frames from the video or clip. We denote T^{2}IBUA as a function $\mathcal{T}(\cdot)$, the spatial encoder as $\mathfrak{S}(\cdot)$ and the temporal encoder as $\mathfrak{T}(\cdot)$, and a linear classification layer as $MLP_i(\cdot)$, where i is a learning stage indicator.

3.1 Framework Overview

Unintentional actions (UAs) are highly diverse and additionally happen rarely in daily life, making it difficult to collect representative and large-scale datasets. To overcome this issue, we propose a three-stage learning framework for unintentional action recognition. Starting from the observation that unintentional actions are often related to changes in motion such as speed or direction, we aim to leverage such inherent biases of unintentional actions in our framework. In particular, simulating these inherent biases with a set of temporal video transformations, the first two stages of our framework greatly enhance pre-trained representation in a self-supervised fashion. More specifically, the first stage uses these transformations to learn intra-clip information (frame2clip) whereas the second stage uses the same transformations to address inter-clip information (frame2clip2video). The third stage then refines this representation via finetuning on the downstream unintentional action tasks using labelled data. We additionally enforce smooth transitions from intentional to unintentional action during the third stage by employing a conditional random field. While simulating the above-mentioned inherent motion biases cannot possibly cover the entire

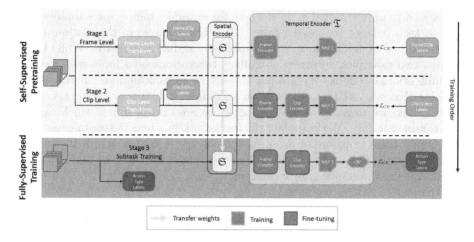

Fig. 2. Framework overview. In the first and the second stages, we use self-supervised feature enhancement by predicting T^2IBUA . In the **first stage**, we enhance the representations so that they encode short-term dependencies based on neighbouring frames by training the frame encoder $\mathfrak{T}_{frame}$ and MLP_1. During the **second stage**, we further fine-tune representations so that they encode long-term dependencies using inter clip information by fine-tuning the frame encoder $\mathfrak{T}_{frame}$ and training the clip encoder $\mathfrak{T}_{clip}$ together with MLP_2. During the **stage three**, we train in a fully-supervised way for downstream tasks by fine-tuning the frame $\mathfrak{T}_{frame}$ and clip $\mathfrak{T}_{clip}$ encoders, while we train MLP_3 and the CRF parameters.

diversity of unintentional actions, it results in a powerful representation leveraged by the third stage of our approach and achieves a new state of the art for unintentional action recognition tasks.

For the first and the second stages, we generate labels based on the selected transformations. We extract frame features with a fixed pretrained spatial encoder, in particular using a pretrained ViT [7] model. Furthermore, for fair comparison to previous work we also explore random initialization without any additional pre-training employing a Resnet3D architecture. Then we pass the frame features to the temporal encoder. The architecture of the temporal encoder changes from stage to stage that we discuss in detail in Sect. 4. For the first and the second stages, we use cross entropy loss with self-generated labels. The overview of the framework is shown in Fig. 2.

3.2 Temporal Transformations of Inherent Biases of Unintentional Actions (T^2IBUA)

The potential diversity of unintentional actions encompasses all possible types of activities, since it is human nature to have failures during executing intentional actions. We aim to grasp the motion inherent biases of unintentional actions and propose several temporal transformations grouped in two categories: motion speed and motion direction transformations. Changes in the motion speed can correspond to unintentional actions when, for example, the person stumbles and tries to keep balance by moving faster. Whereas, changes in the motion direction can occur by falling down or unexpectedly sliding backwards. While these are just two examples, in practice similar observations hold for a wide variety of unintentional actions. We ground the set of transformations on the above intuition that connect temporal motion and inherent biases of unintentionality. In this work, we consider each video as an ordered set of units, which can be either frames or clips. We formulate the framework in a multi-stage manner and, thus, apply these transformations in the first stage to frames and in the second stage to clips. In Fig. 4 we show the difference between frame-level and clip-level T^2IBUA using the shuffle transformation.

We define T^2IBUA as follows:

- *Speed-up*: We synthetically vary the number of units of the video by uniformly subsampling them with the ratios $\{1/2, 1/4, 1/8\}$
- *Random point speed-up*: We sample a random index $ri \in \{1 \cdots t\}$ to indicate the unit from which we are synthetically speeding up the video. Specifically, we start subsampling of the video units after ri unit with ratio $1/\rho$:

$$[1, 2, \cdots, t-1, t] \rightarrow [1, 2, \cdots, ri, ri + \rho, \cdots, t - \rho, t].$$

  where t is the length of the video.
- *Double flip*: We mirror the sequence of the video units and concatenate them to the original counterpart:

$$[1, 2, \cdots, t-1, t] \rightarrow [1, 2, \cdots, t-1, t, t-1, \cdots, 2, 1].$$

- *Shuffle*: The video units are sampled in a random, non-repeating and non-sorted manner.
- *Warp*: The video units are sampled randomly and sorted increasingly by their original index.

All transformations are depicted in Fig. 3. We group T^2IBUA in the motion speed group where we include the *Speed-up* variations and the motion direction group where we include the rest of the proposed T^2IBUA.

T^2IBUA **prediction:** For the feature enhancement, we associate a discrete label to each T^2IBUA and train our framework in a self-supervised way to predict the label of the T^2IBUA that we apply either on frame or clip level. Each input sequence is transformed into a set of 6 sequences. The correspondences between the self-generated labels and the 6 transformed sequences are as follows: 1 - Initial sequence without T^2IBUA; {2, 3, 4} - *Speed-up* with sampled ratio $\in$ $\{1/2, 1/4, 1/8\}$; 5 - *Random point speed-up*; 6 - *Double flip*; 7 - *Shuffle*; 8 - *Warp*. Specifically, we predict the transformation applied to each of the 6 resulting sequences correspondingly. Therefore, the model learns to distinguish various simulated inherent biases of unintentional actions on the same sequence.

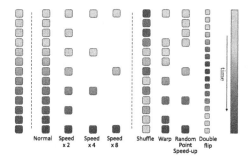

Fig. 3. We group T^{2}IBUA in the **speed** group, where the speed of the video changes uniformly, and the **direction** group, where the speed changes non-uniformly, or we permute the units of the video.

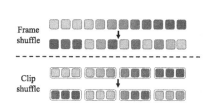

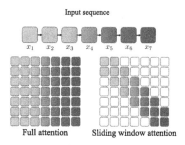

Fig. 4. Shuffle transformation. **Top**: Frame level. Each frame moves randomly to any position. **Bottom**: Clip level. We group frames into clips. Frames cannot change the position within a clip, while clips are shuffled randomly.

Fig. 5. Attention computation. **Left**: Full attention. Required memory and processing time scale quadratically. **Right**: Sliding window attention mechanism. Required memory and processing time scale linearly with the input sequence length.

4   Multi-Stage Learning for Unintentional Action Recognition

In this section, we first introduce the Transformer block, a building element for the different modules of our temporal encoder in Sect. 4.1. Then we discuss in details each learning stage of our framework in Sect. 4.2, 4.3 and 4.4.

4.1   Transformer Block

In this work, to process long sequences, we utilize a transformer architecture. We integrate intra and inter clip information and capture information for unintentional action recognition from different granularity levels, frame and clip levels. To aggregate information over the whole video, we need to process a long sequence of consecutive frames or clips from the video. Further, we refer to the elements of frame or clip sequences as units. In the datasets, all videos are usually of different lengths, and the vanilla transformer model [7] scales quadratically with the sequence length in the required memory and processing time. The Longformer [2] addresses this problem by attending only to a fixed-sized window for each unit. The difference in the attention computation is shown in Fig. 5. We apply a window of size w so that each unit attends to $\frac{1}{2}w$ units from the past and $\frac{1}{2}w$ from the future. All building blocks for the temporal encoder in our work follow the structure of Longformer attention blocks. The transformer blocks can update the input sequence based on the temporal context or aggregate the temporal information from the entire sequence into one representation vector. For the former, for a given input sequence of units $U = \{u_1, \cdots, u_K\}$ we update the sequence and output $U' = \{u'_1, \cdots, u'_K\}$. For the latter, we expand the given input sequence with an additional classification unit cls that serves as aggregation unit across the whole sequence [7,28]. Therefore, we map the expanded input sequence $U \cup \{cls\}$ to cls' that aggregates the sequence into one representation. For more details we refer to the original paper [2].

4.2   [Stage 1] Frame2Clip (F2C) Learning

In the first stage of our framework, we enhance the pre-trained features such they encode intra clip information based on neighboring frames. We operate within short clips and, thus, capture local motion information. First, we sample a short clip x of the length n from a long input video X, where $n < |X|$ and $|X|$ is the length of the input video. The clip x consists of sequential frames $x = \{f_1, \cdots, f_n\}$ from the same video X. We apply T^2IBUA T to this sequence of frames of the clip x as we show schematically in the shuffle transformation on the top of Fig. 4. As a label for the transformed clip $T(x)$ we use the index that corresponds to the respective T^2IBUA, e.g. the shuffle transformation has index 7. Then, from the spatial encoder $\mathfrak{S}$ we obtain the spatial frame features for the sequence. Thereafter, to impose the intra connections between the frames of the same sequence, we use the temporal frame encoder $\mathfrak{T}_{frame}$ that is composed of the transformer blocks. The output of this encoder is one representation vector for the clip, hence we expand the sequence of frame features with an additional unit cls_1 to aggregate the temporal information from the clip x into one representation $\tilde{x}$. Finally, we predict the T^2IBUA index with a classification layer MLP_1. We compute cross-entropy loss and optimize parameters of the temporal frame encoder $\mathfrak{T}_{frame}$ and a classification layer MLP_1. Formally, we apply the following pipeline in the first stage:

$$\hat{y}_{clip} = MLP_1(\mathfrak{T}_{frame}(\mathfrak{S}(T(f_1, \cdots, f_n)), cls_1)), \tag{1}$$

where $\hat{y}_{clip}$ is the predicted label of the transformation applied to the clip x. Note that at this stage of the training, we process all sampled clips independently of each other. Our F2C learning implies intra-clip encoding, therefore, it can be also substituted with the more common spatio-temporal networks such as ResNet3D that learn representations of short clips.

4.3   [Stage 2] Frame2Clip2Video (F2C2V) Learning

In the second stage of our framework, we further enhance the representations based on the clip level transformations. At this stage, we integrate inter clip information into the video representation to model long-term dependencies. The input video X of length $|X|$ we split into overlapping clips $z = \{x_1, x_2, \cdots, x_N\}$ that we sample with stride k from the video, where $N = \frac{|X|-n}{k} + 1$. Each clip consists of n frames $x_i = \{f_1^i, \cdots, f_n^i\}$. During F2C2V learning stage we apply T^{2}IBUA $\mathcal{T}$ to the sequence of clips of the whole video X as we show schematically on the bottom of Fig. 4. Specifically, the order of consecutive frames within each clip remains fixed, whereas the sequence of the clips is transformed $\mathcal{T}(z)$. As in the first stage, we first pass frames of the clips through the spatial encoder $\mathfrak{S}$ to obtain frame-level features, then the sequence of frames for each clip $x_i = \{f_1^i, \cdots, f_n^i\}$ is aggregated into one clip representation $\tilde{x}_i$ by the frame encoder $\mathfrak{T}_{frame}$. Note that the frame sequence during this stage follows the original order. Then we aggregate the transformed sequence of clip representations $\mathcal{T}(\tilde{x}_1, \tilde{x}_2, \cdots, \tilde{x}_N)$ into video representation vector $\tilde{X}$ with the temporal clip encoder $\mathfrak{T}_{clip}$. We predict T^{2}IBUA that was applied to the sequence of clips with the classification layer MLP_2. Formally, the second stage is as follows:

$$\tilde{x}_i = \mathfrak{T}_{frame}(\mathfrak{S}(f_1^i, \cdots, f_n^i), cls_1), \ \forall i \in \{1, \ldots, N\}; \tag{2}$$

$$\hat{y}_{video} = MLP_2(\mathfrak{T}_{clip}(\mathcal{T}(\tilde{x}_1, \cdots, \tilde{x}_N), cls_2)), \tag{3}$$

where $\hat{y}_{video}$ is the predicted T^{2}IBUA label applied to a sequence of clips from video X. To aggregate all the clips into one video vector we similarly use an additional classification token cls_2 as in the previous stage. At this stage, we optimize the parameters of the temporal clip encoder $\mathfrak{T}_{clip}$ and the classification layer MLP_2, the parameters of the temporal frame encoder $\mathfrak{T}_{frame}$ we transfer from the previous stage and fine-tune. Note that at each stage we utilize a new classification layer and discard the one from the previous stage.

4.4   [Stage 3] Downstream Transfer to Unintentional Action Tasks

In the last stage, we extend our framework to supervised unintentional action recognition tasks. For these tasks, we predict the unintentionality for each short clip rather than for the entire video. This stage is completely supervised with clip level labels, therefore we do not apply T^{2}IBUA. The input video X is divided into temporally ordered overlapping clips $z = \{x_1, x_2, \cdots, x_N\}$. We follow the pipeline of the second stage and extract clip level representation vectors $\{\tilde{x}_1, \tilde{x}_2, \cdots, \tilde{x}_N\}$ with the temporal frame encoder $\mathfrak{T}_{frame}$. Then we use the temporal clip encoder $\mathfrak{T}_{clip}$ without the classification aggregation unit cls_2 given that for the supervised tasks we use clip-level classification labels. The output of $\mathfrak{T}_{clip}$ then is the sequence of clip representations instead of one video representation vector. By using the temporal clip encoder $\mathfrak{T}_{clip}$ each clip representation vector is updated with the inter clip information from the whole video X. Finally, we pass the clip vectors through the classification layer MLP_3 to obtain the emission scores that we use in the CRF layer $\mathfrak{T}_{CRF}$. The overall downstream transfer stage is as follows:

$$\tilde{x}_i = \mathfrak{T}_{frame}(\mathfrak{S}(f_{i,1}, \cdots, f_{i,n}), cls_1), \ \forall i \in \{1, \ldots, N\}; \tag{4}$$

$$\{\hat{y}_1, \cdots, \hat{y}_N\} = \mathfrak{T}_{CRF}(MLP_3(\mathfrak{T}_{clip}(\tilde{x}_1, \cdots, \tilde{x}_N)), T), \tag{5}$$

where $\{\hat{y}_1, \cdots, \hat{y}_N\}$ predicted clip level labels and T is the transition matrix of the CRF layer.

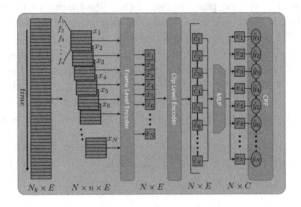

Fig. 6. Multi-stage temporal encoder. The $\mathfrak{T}_{frame}$ encoder aggregates the frames of each input clip x_1 into single-vector clip representation $\tilde{x}_i$. $\mathfrak{T}_{clip}$ updates $\tilde{x}_i$ with inter-clip information resulting into $\tilde{c}_i$. MLP turns $\tilde{c}_1$ into emission scores e_i for the CRF layer, which predicts clip-level labels.

The CRF layer aims to explicitly model the high-level temporal structure of the unintentional videos between the clips as shown in Fig. 1. For videos containing unintentional actions the high-level order imposes a smooth transition from normal to transition to unintentional clips, therefore we can incorporate this prior information on the order of the clip labels. The CRF layer facilitates dependencies between the current clip and the previously predicted labels and enforces neighbouring clips to have the same label. This layer is parametrized by the transition matrix $T \in \mathcal{R}^{C \times C}$, where $T_{i,j}$ is the score of transitioning from label i to label j and C stands for the number of classes. We denote by θ the parameters of the temporal encoder $\mathfrak{T}$ and MLP_3 together that we optimize during training, by $L \in \mathcal{R}^N$ the vector of clip labels for the input sequence of clips z and by $E \in \mathcal{R}^{N \times C}$ an emission matrix that consists of logits for each unit of the input sequence generated by MLP_3 layer. Then, we can calculate the scores for a label sequence $\hat{L}$ given the input sequence $z = \{x_1, \cdots, x_N\}$, trainable weights θ and the transition matrix T as

$$s(\hat{L}, z, \theta, T) = \sum_{t=1}^{n} \left(T_{L_{t-1}, Lt} + E_{t, L_t} \right).$$ (6)

Let L_{all} be the set of all possible vectors of labels for the input clip sequence. Then the loss function is the negative log-likelihood:

$$\mathcal{L}(\theta, T) = -log \frac{exp(s(L_{gt}, z, \theta, T))}{\sum_{L \in L_{all}} exp(s(L, z, \theta, T))},$$ (7)

where L_{gt} the ground truth labels for each clip of the input sequence z. The total loss is the average over all training sequences.

During the inference, we compute the optimal labels L_{opt} for the input sequence of clips z_{test} as

$$L_{opt} = \underset{L \in L_{all}}{argmax} \, s(L, z_{test}, \theta^{\star}, T^{\star}),$$ (8)

where $\theta^{\star}$ and $T^{\star}$ are the optimized model parameters. Note that Eq. 8 can be solved efficiently with the Viterbi algorithm.

5   Experimental Results

In this section, we present experimental findings that validate our framework. First, we introduce the Oops! dataset, present different backbone models used in the framework and discuss the implementation details of our method. We further compare our framework to the state-of-the-art on the three subtasks in Sect. 5.1 and then we present extensive ablation experiments to validate the impact of each of the components in Sect. 5.2.

Dataset: Oops! [8] is a collection of 20,338 amateur fail videos from YouTube. The dataset is split into three sets: 7,368 labelled videos for supervised training, 6,739 labelled videos for validation, and 6,231 unlabelled videos for pretraining. Each label in the first two sets consists of the timestamp where the action transitions from intentional to unintentional. Following the clip sampling procedure described in Sect. 4 we get $18,069$ intentional clips, $4,137$ transitional clips and $19,679$ unintentional clips.

Spatial Encoder: To disentangle the influence of different components of the framework, to provide a fair comparison to previous methods and to follow the trend of performant architectures, we use two backbones for the main experiments. Similarly to prior work [8,16], we employ ResNet3D (R3D) [17]. This backbone is spatio-temporal and, therefore, substitutes both spatial $\mathfrak{S}$ and frame level $\mathfrak{T}_{frame}$ encoder as we discuss in Sect. 4.2, we refer to it as F2C level for R3D backbone. In this setup, we learn the representations from scratch instead of enhancing pre-trained representations to fairly compare to previous works. To decouple spatial and temporal dimensions we use ViT model [7] pretrained on ImageNet-21K [30] (IN-21K). In this setup, we leverage the pre-trained image representations and further enhance them for unintentional actions. We note that we freeze ViT model, while R3D model we train from scratch with a random initialization. Each encoder consists of three stacked transformer blocks, each block constitutes of 16 parallel heads. Additional backbones are in the supplement.

Implementation details: We sample clips of 16 frames from the input video with stride $k = 4$. We train for 100 epochs for the first and second stage and for 50 epochs for the third stage with the AdamW [26] optimizer with weight decay of $1e - 4$. The starting learning rate for all stages is $1e - 4$ and is decreased to $1e - 6$ using a cosine decay policy. During the third stage, the loss function is weighted as $\omega_i = {}^{max(\eta_1, \eta_2, \cdots, \eta_c)}/\eta_i$, where η_1 is the number of labels in class i and c is the number of classes. We recalculate the weights during the anticipation task as class sample distribution changes over time. More details can be found in the supplement.

Table 1. Comparison of our approach to state-of-the-art for UA classification, localisation and anticipation. F2C denotes the first (frame2clip) learning stage, F2C2V denotes the second (frame2clip2video) learning stage. τ_L indicates a time window in seconds for ground truth assignment for the localization task. τ_A indicates the time step in the future that we predict. Init. indicates fully supervised initialization of the backbone if applies. * indicates that the backbone is frozen during all stages.

Method	Backbone	Init.	Pretrain Dataset	Cls Acc	Loc. $\tau_L : 0.25$	$\tau_L : 1$	Ant $\tau_A : 1$
K700 Supervision [8]	R(18)3D	K700	–	64.0	46.7	75.9	59.7
Epstein et al. [8]	R(18)3D	–	Oops!	61.6	36.6	65.3	56.7
Han et al. [16]	(2+3D)R18	–	(K400+Oops!)	64.4	–	–	–
Ours (F2C)	R(18)3D	–	Oops!	65.3	37.7	67.8	66.7
Ours (F2C2V)	R(18)3D	–	Oops!	**74.0**	39.4	**69.5**	**76.1**
Ours (F2C)	ViT*	IN-21K	Oops!	65.5	41.4	72.2	69.2
Ours (F2C2V)	ViT*	IN-21K	Oops!	**76.9**	**42.8**	**72.8**	**78.1**

5.1   Comparison to State-of-the-art

In this section we compare the performance of our framework on the three downstream tasks as classification, localization and anticipation. In Table 1 we provide a comparison to the previous state-of-the-art method across all three tasks. The first row in the table corresponds to supervised pretraining on the Kinetics 700 [31]. The benchmark that we follow for the fair comparison across all the tasks is defined in Oops! dataset [8].

Unintentional Action Classification. We first compare our framework to the recent state-of-the-art work on unintentional action classification. We divide Table 1 into blocks with comparable backbones and pre-training methods. We can observe a significant improvement over the previous state-of-the-art method by approximately 10 points with R3D backbone and training from scratch. In contrast to Han et al. [16] we pretrain our model only on the Oops! dataset without additional external data. With our temporal encoder on the frame level, we gain minor improvement by 0.9 points indicating that T^2IBUA helps to improve even on the F2C level, while our two stage temporal encoder allows us to achieve substantial increase in the performance by 9.6 points. We notice further increase in performance when we use pre-trained frozen ViT as our backbone. For this setup, we enhance the pre-trained features and gain an improvement of 1.1 points with our frame level encoder and 12.5 points with our two stage temporal encoder. These results indicate the importance of connections between intra and inter clip information in the video

Unintentional Action Localisation. For this downstream task, we localize the transition point from intentional to unintentional action in the video. We directly validate the network trained on unintentional action classification task and detect the transition point from intentional action to unintentional. The transitional clip for this task we define as the clip with the maximum output score of being transitional. In Table 1 the localisation column shows the performance of our framework for two temporal localisation thresholds. By using R3D as our backbone, we improve by 2.8 and 4.2 points for the $\tau_L = 0.25$ and $\tau_L = 1$ respectively compared to Epstein et al. [8].We imporove by 6.2 and 7.5 points for each threshold respectively when using ViT as our backbone. Additionally, we can observe the same trend as for the classification task, specifically, the clip level model outperforms the frame level model. Note that for this task we reuse the model that is trained for the classification task where the number of normal and unintentional clips is notably higher than the number of transitional clips. The localisation task requires the model to be very specific on the boundaries when the unintentionality starts. We suppose that this influences the smaller improvement for the localization task than for the other tasks.

Unintentional Action Anticipation. Further, we validate our framework on anticipation of unintentional actions. During the supervised training, we train our model to predict the label of the future clip. To directly compare to the previous work, we anticipate an action 1.5 seconds into the future as in previous work. Our model achieves new state-of-the-art results with a considerable improvement by 19.4 and 21.4 points for the R3D and ViT backbones respectively. Note that the performance of our framework is better for the anticipation task than for classification. As mentioned, the dataset includes more unintentional clips than normal clips, and in combination that we are able to predict unintentional clips more accurately, it leads to the difference in the performance between the downstream tasks.

5.2 Ablation Study

We perform ablation studies on different components and backbones of our framework to assess their influence on the overall performance. We use the unintentional action classification task for all the evaluations in this section. Extended ablations and qualitative results can be found in the supplement.

Feature Enhancement Stages. In this section we analyse the influence of the self-supervised training to enhance the representations with T^2IBUA on the UA classification task. In Table 2 the first and the second rows show the performance with and without T^2IBUA respectively. Specifically, for the first row, we skip the self-supervised procedure for the pretraining of the parameters and directly optimize from scratch the respective temporal encoders with the supervised task. For the second row, we include the self-supervised feature enhancement corresponding to the respective learning stages. We can observe that for each temporal encoder (F2C and F2C2V) we have a significant increase with T^2IBUA representation enhancing (learning). Considering only intra clip information (F2C) we improve by 4.6 and 4.5 points for the ViT and R3D backbones respectively while using additionally inter clip information we improve by 4.7 and 4.8 points. These results confirm the importance of the self-supervised feature enhancement for UA recognition performance.

Table 2. Ablation results: Influence of T^2IBUA and CRF layer on classification performance for two representations, pretrained ViT representation and R3D.

T^2IBUA	CRF	ViT		R3D	
		F2C	F2C2V	F2C	F2C2V
–	–	60.9	69.6	59.3	65.6
✔	–	65.5	74.3	63.8	70.4
✔	✔	65.0	**76.9**	65.3	**74.0**

Table 3. Influence of T^2IBUA groups on UA classification task for different stages of learning of the temporal encoder.

T^2IBUA	F2C	F2C2V
–	60.9	73.2
Speed	62.5	73.8
Direction	63.3	75.1
All	**65.5**	**76.9**

Learning Stages for Temporal Encoder. In this section, we evaluate the impact of multi-stage learning of our temporal encoder. In Table 2 we show in the F2C column the performance across different settings for the corresponding encoder, see the detailed structure in Fig. 6 for ViT backbone, whereas R3D backbone comprises this stage. In the second column with the F2C2V encoder, we observe a significant improvement by about $5 - 10$ points consistently across all the settings. Furthermore, we evaluate the performance of the spatial features from the fixed pretrained ViT model [7] that we use as input features to our temporal encoder. We obtain 58.4 points without T^2IBUA and without CRF. It supports the importance of the temporal encoder that is able to capture successfully inter and intra clip information.

T^2IBUA **Groups.** We assess the influence of the T^2IBUA groups on the overall performance. Table 3 shows that both groups improve the performance for both encoder levels. We notice that the direction T^2IBUA have a more significant impact than the speed T^{2}IBUA. Whereas, in contrast to the separate groups, the combination of the speed and the direction transformations leads to greatly enhanced representations that effectively capture the inherent biases of UA. We additionally study an influence of each transformation separately that shows random point speed-up to be the most important, while the combination of all T^{2}IBUA is still predominant. More detailed study is in the supplement.

6   Conclusion

In this paper, we propose a multi-stage framework to exploit inherent biases that exist in videos of unintentional actions. First, we simulate these inherent biases with the temporal transformations that we employ for self-supervised training to enhance pre-trained representations. We formulate the representation enhancement task in a multi-stage way to integrate inter and intra clip video information. This leads to powerful representations that significantly enhance the performance on the downstream UA tasks. Finally, we employ CRF to explicitly model global dependencies. We evaluate our model on the three unintentional action recognition tasks such as classification, localisation, and anticipation, and achieve state-of-the-art performance across all of them.

References

1. Agrawal, P., Carreira, J., Malik, J.: Learning to see by moving. In: 2015 IEEE International Conference on Computer Vision (ICCV) (2015)
2. Beltagy, I., Peters, M.E., Cohan, A.: Longformer: the long-document transformer. arXiv preprint arXiv:2004.05150 (2020)
3. Carreira, J., Zisserman, A.: Quo vadis, action recognition? A new model and the kinetics dataset. In: 2017 IEEE Conference on Computer Vision and Pattern Recognition (CVPR) (2017)
4. Chen, J., Xu, Y., Zhang, C., Xu, Z., Meng, X., Wang, J.: An improved two-stream 3D convolutional neural network for human action recognition. In: International Conference on Automation and Computing (ICAC) (2019)
5. Dai, R., Das, S., Minciullo, L., Garattoni, L., Francesca, G., Bremond, F.: PDAN: pyramid dilated attention network for action detection. In: Proceedings of the IEEE/CVF Winter Conference on Applications of Computer Vision (2021)
6. Dave, I., Gupta, R., Rizve, M.N., Shah, M.: TCLR: temporal contrastive learning for video representation. arXiv preprint arXiv:2101.07974 (2021)
7. Dosovitskiy, A., et al.: An image is worth 16 × 16 words: Transformers for image recognition at scale. In: International Conference on Learning Representations (ICLR) (2020)
8. Epstein, D., Chen, B., Vondrick, C.: Oops! Predicting unintentional action in video. In: 2020 IEEE/CVF Conference on Computer Vision and Pattern Recognition (CVPR) (2020)
9. Epstein, D., Vondrick, C.: Video representations of goals emerge from watching failure. arXiv preprint arXiv:2006.15657 (2020)
10. Fan, H., et al.: Multiscale vision transformers. arXiv preprint arXiv:2104.11227 (2021)

11. Feichtenhofer, C., Pinz, A., Zisserman, A.: Convolutional two-stream network fusion for video action recognition. In: Conference on Computer Vision and Pattern Recognition (CVPR) (2016)
12. Fernando, B., Bilen, H., Gavves, E., Gould, S.: Self-supervised video representation learning with odd-one-out networks. In: IEEE International Conference on Computer Vision and Pattern Recognition (CVPR) (2017)
13. Ging, S., Zolfaghari, M., Pirsiavash, H., Brox, T.: COOT: cooperative hierarchical transformer for video-text representation learning. arXiv preprint arXiv:2011.00597 (2020)
14. Girdhar, R., Carreira, J., Doersch, C., Zisserman, A.: Video action transformer network. In: 2019 IEEE/CVF Conference on Computer Vision and Pattern Recognition (CVPR) (2019)
15. Han, T., Xie, W., Zisserman, A.: Video representation learning by dense predictive coding. In: 2019 IEEE/CVF Conference on Computer Vision and Pattern Recognition Workshops (CVPRW) (2019)
16. Han, T., Xie, W., Zisserman, A.: Memory-augmented dense predictive coding for video representation learning. In: European Conference on Computer Vision (ECCV) (2020)
17. Hara, K., Kataoka, H., Satoh, Y.: Can spatiotemporal 3D CNNs retrace the history of 2D CNNs and imagenet? In: 2018 IEEE/CVF Conference on Computer Vision and Pattern Recognition (CVPR) (2018)
18. Isola, P., Zoran, D., Krishnan, D., Adelson, E.H.: Learning visual groups from co-occurrences in space and time. arXiv preprint arXiv:1511.06811 (2015)
19. Jabri, A., Owens, A., Efros, A.A.: Space-time correspondence as a contrastive random walk. arXiv preprint arXiv:2006.14613 (2020)
20. Jayaraman, D., Grauman, K.: Learning image representations equivariant to ego-motion. In: International Conference on Computer Vision (ICCV) (2015)
21. Kim, D., Cho, D., Kweon, I.S.: Self-supervised video representation learning with space-time cubic puzzles. In: Proceedings of the AAAI Conference on Artificial Intelligence (AAAI) (2019)
22. Kukleva, A., Kuehne, H., Sener, F., Gall, J.: Unsupervised learning of action classes with continuous temporal embedding. In: 2019 IEEE/CVF Conference on Computer Vision and Pattern Recognition (CVPR) (2019)
23. Larsson, G., Maire, M., Shakhnarovich, G.: Colorization as a proxy task for visual understanding. In: 2017 IEEE Conference on Computer Vision and Pattern Recognition (CVPR) (2017)
24. Lee, H.Y., Huang, J.B., Singh, M., Yang, M.H.: Unsupervised representation learning by sorting sequences. In: 2017 IEEE International Conference on Computer Vision (ICCV) (2017)
25. Liu, Z., et al.: Video swin transformer. arXiv preprint arXiv:2106.13230 (2021)
26. Loshchilov, I., Hutter, F.: Fixing weight decay regularization in adam. arXiv preprint arXiv:1711.05101 (2017)
27. Misra, I., Zitnick, C.L., Hebert, M.: Unsupervised learning using sequential verification for action recognition. arXiv preprint arXiv:1603.08561 **2**, 8 (2016)
28. Neimark, D., Bar, O., Zohar, M., Asselmann, D.: Video transformer network. arXiv preprint arXiv:2102.00719 (2021)
29. Qian, R., et al.: Spatiotemporal contrastive video representation learning. In: 2021 IEEE/CVF Conference on Computer Vision and Pattern Recognition (CVPR) (2021)
30. Ridnik, T., Ben-Baruch, E., Noy, A., Zelnik-Manor, L.: ImageNet-21k pretraining for the masses. arXiv preprint arXiv:2104.10972 (2021)

31. Smaira, L., Carreira, J., Noland, E., Clancy, E., Wu, A., Zisserman, A.: A short note on the kinetics-700-2020 human action dataset. arXiv preprint arXiv:2010.10864 (2020)
32. Sun, C., Myers, A., Vondrick, C., Murphy, K., Schmid, C.: VideoBERT: a joint model for video and language representation learning. In: 2019 IEEE/CVF International Conference on Computer Vision (ICCV) (2019)
33. Tran, D., Bourdev, L.D., Fergus, R., Torresani, L., Paluri, M.: C3D: generic features for video analysis. arXiv preprint arXiv:1412.0767 (2014)
34. Tran, D., Wang, H., Torresani, L., Ray, J., LeCun, Y., Paluri, M.: A closer look at spatiotemporal convolutions for action recognition. In: 2018 IEEE/CVF Conference on Computer Vision and Pattern Recognition (CVPR) (2018)
35. Vondrick, C., Pirsiavash, H., Torralba, A.: Generating videos with scene dynamics. In: Advances in Neural Information Processing Systems (NIPS) (2016)
36. Vondrick, C., Shrivastava, A., Fathi, A., Guadarrama, S., Murphy, K.: Tracking emerges by colorizing videos. In: European Conference on Computer Vision (ECCV) (2018)
37. Wang, N., Zhou, W., Wang, J., Li, H.: Transformer meets tracker: exploiting temporal context for robust visual tracking. In: 2021 IEEE/CVF Conference on Computer Vision and Pattern Recognition (CVPR) (2021)
38. Wang, X., Gupta, A.: Unsupervised learning of visual representations using videos. In: 2015 IEEE International Conference on Computer Vision (ICCV) (2015)
39. Wang, X., Jabri, A., Efros, A.A.: Learning correspondence from the cycle-consistency of time. In: 2019 IEEE/CVF Conference on Computer Vision and Pattern Recognition (CVPR) (2019)
40. Wei, D., Lim, J., Zisserman, A., Freeman, W.T.: Learning and using the arrow of time. In: 2018 IEEE/CVF Conference on Computer Vision and Pattern Recognition (CVPR) (2018)
41. Xu, D., Xiao, J., Zhao, Z., Shao, J., Xie, D., Zhuang, Y.: Self-supervised spatiotemporal learning via video clip order prediction. In: 2019 IEEE/CVF Conference on Computer Vision and Pattern Recognition (CVPR) (2019)
42. Yang, C., Xu, Y., Shi, J., Dai, B., Zhou, B.: Temporal pyramid network for action recognition. In: 2020 IEEE/CVF Conference on Computer Vision and Pattern Recognition (CVPR) (2020)

A Study on Self-Supervised Object Detection Pretraining

Trung Dang[1]([✉]), Simon Kornblith[2], Huy Thong Nguyen[2], Peter Chin[3], and Maryam Khademi[2]

[1] Boston University, Boston, US
trungvd@bu.edu
[2] Google LLC, Mountain View, US
[3] Dartmouth College, Hanover, US

Abstract. In this work, we study different approaches to self-supervised pretraining of object detection models. We first design a general framework to learn a spatially consistent dense representation from an image, by randomly sampling and projecting boxes to each augmented view and maximizing the similarity between corresponding box features. We study existing design choices in the literature, such as box generation, feature extraction strategies, and using multiple views inspired by its success on instance-level image representation learning techniques [6,7]. Our results suggest that the method is robust to different choices of hyperparameters, and using multiple views is not as effective as shown for instance-level image representation learning. We also design two auxiliary tasks to predict boxes in one view from their features in the other view, by (1) predicting boxes from the sampled set by using a contrastive loss, and (2) predicting box coordinates using a transformer, which potentially benefits downstream object detection tasks. We found that these tasks do not lead to better object detection performance when finetuning the pretrained model on labeled data.

Keywords: Self-supervised · Object detection

1 Introduction

Pretraining a model on a large amount of labeled images and finetuning on a downstream task, such as object detection or instance segmentation, has long been known to improve both performance and convergence speed on the downstream task. Recently, self-supervised pretraining has gained popularity since it significantly reduces the cost of annotating large-scale datasets, while providing superior performance compared to supervised pretraining. Although a number of prior works obtain semantic representations from unlabeled images via the use of proxy tasks [14,23,25], recent works focus on instance-level contrastive learning [8–10,17] or self-distillation [7,15] methods. These methods learn an instance-level representation for each image that performs competitively with using only a linear [8] or K-nearest neighbor [7] classifier, closing the gap to the performance of supervised baselines.

© The Author(s), under exclusive license to Springer Nature Switzerland AG 2023
L. Karlinsky et al. (Eds.): ECCV 2022 Workshops, LNCS 13804, pp. 86–99, 2023.
https://doi.org/10.1007/978-3-031-25069-9_6

However, most of these contrastive and self-distillation methods focus on learning an instance-level representation. They are likely to entangle information about different image pixels, and are thus sub-optimal for transfer learning to dense prediction tasks. A recent line of work aims to improve self-supervised pretraining for object detection by incorporating modules in the detection pipeline [31], and taking into account spatial consistency [26,28,30,31,34,35]. Rather than comparing representation at instance-level, these methods propose to leverage view correspondence, by comparing feature vectors [26,30] or RoIAlign features of sampled boxes [28,31] at the same location from two augmented view of the same image. These pretraining strategies have been shown to benefit downstream dense prediction tasks. However, there may still be a discrepancy between the pretraining and the downstream task, since they are optimized towards different objectives.

Following prior works that pretrain SSL models for object detection models by sampling object bounding boxes and leveraging view correspondence [28,31], we study how different design choices affect performance. Specifically, we investigate strategies for box sampling, extracting box features, the effect of multiple views (inspired by `multi-crop` [6]), and the effect of box localization auxiliary tasks. We evaluate these proposals by pretraining on ImageNet dataset and finetuning on COCO dataset. Our results suggest that (1) the approach is robust to different hyperparameters and design choices, (2) the application of `multi-crop` and box localization pretext tasks in our framework, as inspired by their success in the literature, does not lead to better object detection performance when finetuning the pretrained model on labeled data.

2 Related Work

2.1 Self-Supervised Learning from Images

A large number of recent work on self-supervised learning focuses on constrastive learning, which learns the general feature of an image by using data augmentations and train the model to discriminate views coming from the same image (positive pairs) and other images (negative pairs), usually by using the InfoNCE loss [16, 29,32]. In practice, contrastive learning requires simultaneous comparison among a large number of sampled images, and benefits from large batches [8,9], memory banks [10,17], false negative cancellation [22], or clustering [1,5,21]. Recently, noncontrastive methods such as BYOL [15] and its variants [7,11] obtain competitive results without using negative pairs, by training a student network to predict the representations obtained from a teacher network.

While instance-level representation learning methods show promising results on transfer learning for image classification benchmarks, they are sub-optimal for dense prediction tasks. A recent line of works focus on pre-training a backbone for object detection. Similar to instance-level representation learning, these pre-training methods can be based on contrastive learning (e.g. VADeR [26], DenseCL [30], and PixPro [34]), or self-distillation, (e.g. SoCo [31] and SCRL [28]). These methods share a general idea of leveraging view correspondence, which is available from the spatial relation between augmented views and the original image. Beyond pre-training the backbone, UP-DETR [12] and DETReg [3] propose a way to pre-train a whole object detection pipeline, using a frozen backbone trained with SSL on object-centric images to extract object features. Xie et al. [33] leverage image-level self-supervised pre-training to discover object correspondence and perform object-level representation learning from scene images.

Many of these object detection pre-training techniques rely on heuristic box proposals [3,31] or frozen backbone trained on ImageNet [3,12]. Our work, however, aims to study the potential of end-to-end object detection pre-training without them. The framework we study is closest to SCRL [28], which adopts the approach from BYOL [15].

2.2   Object Detection

Faster R-CNN [27] is a popular object detector, which operates in two stages. In the first stage, a single or multi-scale features extracted by the backbone are fed into the Region Proposal Network to get object proposals. In the second stage, the pooled feature maps inside each bounding box are used to predict objects. Low-quality object proposals and predictions are filtered out with Non-Maximum Suppression (NMS), which is heuristic and non-differentiable.

Detection Transformer (DETR) [4] is a simpler architecture than Faster R-CNN and operates in a single stage. The features retrieved by the backbone are encoded by a transformer encoder. The transformer decoder attends to the encoded features and uses a fixed number of query embeddings to output a set of box locations and object categories. DETR can learn to remove redundant detections without relying on NMS; however, the set-based loss and the transformer architecture are notoriously difficult to train. Deformable DETR [36] with multi-scale deformable attention modules has been shown to improve over DETR in both performance and training time.

3   Approach

In this section, we describe the general framework and the notations we use in our study. We first generate multiple views of an image via a sequence of image augmentations. Our framework aims to learn a spatially consistent representation by matching features of boxes covering the same region across views. To avoid mode collapse, we train a student network to predict the output of a teacher network, following BYOL [15]. At the end of this section, we compare the proposed framework with a number of existing pretraining techniques for object detection.

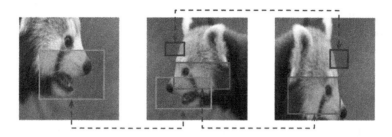

Fig. 1. Example of view and random box generation with number of views $V = 3$ and number of boxes $K = 3$.

3.1 View Construction and Box Sampling

We first randomly crop the original image to obtain a base view, with minimum scale of s_{base}. Next, V augmented views $v_1, \cdots v_V$ are constructed from the base view using V different image augmentations $t_1, \cdots, t_V \sim \mathcal{T}$, each of which is a sequence of random cropping with minimum scale of s_{view}, color jittering, and Gaussian blurring. Here, $\mathcal{T}$ is the distribution of image augmentations. The minimum scale of these V views with regards to the original image is $s_{\text{base}} \times s_{\text{view}}$. We separate s_{base} and s_{view} and choose $s_{\text{view}} > 0.5$ to make sure that views are pairwise overlapped. The views are also resized to a fixed size to be processed in batches.

Next, we sample K boxes $b^1, \cdots, b^K \sim \mathcal{B}$ relative to the base view, where $b^k \in \mathbb{R}^4$ is the box coordinate of the top left and the bottom right corner, $\mathcal{B}$ is the box distribution. We transform these boxes from the base view to each augmented view v_i, based on the transformation t_i used to obtain v_i. We keep only valid boxes that are completely fitted inside the view. Let τ_{t_i} be the box transformation, B_i be the set of valid box indices by this transformation, the set of boxes sampled for view i is denoted as $\{b_i^k = \tau_{t_i}(b^k) | k \in B_i\}$. During this sampling process, we make sure that each box in the base view is completely inside at least two augmented views by over-sampling and removing boxes that do not satisfy this requirement.

The result of this view construction and box sampling process is V augmented views of the original image with a set of at most K boxes for each view. Figure 1 describes the view and random box generation with $V = 3$ and $K = 3$.

3.2 SSL Backbone

Our strategy for training the backbone follows BYOL [15]. We train an online network f_θ, whose output can be used to predict the output of a target network f_ξ, where θ and ξ are weights of the online and target network, respectively. The target network is built by taking an exponential moving average of the weights of the online network, which is analogous to model ensembling and has been previously shown to improve performance [7,15].

Specifically, let $y_i = f_\theta(v_i), y'_i = f_\xi(v_i)$ be the output of the backbone for each view i. Different from instance-level representation learning [7,15] where global representations are compared, we compare local regions of interest refined by sampled boxes. Let ϕ be a function that takes a feature map y, and box coordinates b to output a box representation $\phi(y, b)$ (e.g., RoIAlign). Following [15], we add projection layers g_θ, g_ξ for both networks and a prediction layer q_θ for the online network. We obtain box representations $u_i^k = g_\theta(\phi(y_i, b_i^k))$ from the online network, and $u'^k_i = g_\xi(\phi(y'_i, b_i^k))$ from the target network. The SSL loss is computed as:

$$\mathcal{L}_{\text{BYOL}}(y_i, y'_i, b_i; \theta) = \sum_{\substack{i=1,j=1 \\ j \neq i}}^{V} \frac{1}{|B_i \cap B_j|} \sum_{k \in B_i \cap B_j} \|q_\theta(u_i^k) - u'^k_j\|^2 \tag{1}$$

3.3 Comparison with Prior Work

Our framework reduces to BYOL when $V = 2$, $K = 1$, and τ_{t_i} returns the whole view boundary regardless of the transformation t_i. In this case, only global representation of two augmented views are compared.

The framework is similar to SCRL when $V = 2$, $\mathcal{B}$ is a uniformly random box distribution, and $\phi(y, b)$ outputs the 1×1 RoIAlign of the feature map y with regards to box b. We do not remove overlapping boxes as in SCRL, since object bounding boxes are not necessarily separated.

In SoCo [31], boxes are sampled from box proposals via the selective search algorithm. A third view (and fourth view), which is a resizing of one of the two views is also included to encourage learning object scale variance. We generalize it to V views in our framework. SoCo offers pretraining FPN layers in Faster R-CNN for better transfer learning efficiency. We however only focus on pretraining the ResNet backbone, which is included in both Faster R-CNN and DETR (Fig. 2).

4 Experiments

4.1 Experimental Setup

Dataset. We pretrain the ResNet50 backbone on ImageNet [13] (~ 1.28m images), and finetune the object detection model on MS-COCO [24] (~ 118k images).

Image Augmentation. After we crop a view from the image as described in Sect. 3, we resize it to 256×256, and follow previous work [8,15,28] in applying random horizontal flipping, color jittering, and Gaussian blurring.

Network Architecture. We use ResNet-50 [19] as the backbone, which outputs a feature map of shape $(7, 7, 2048)$.

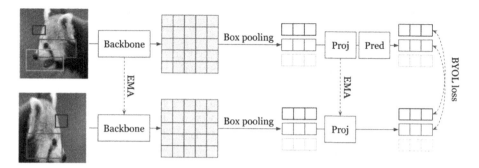

Fig. 2. General framework: we train an online network to predict the output of a target network under a different augmented view and update the target network with a moving average of the online network. The representations are compared at regional level via random box sampling.

Pretraining Setup. For the baseline, we follow [15] to train a BYOL model for 300 epochs (top-1: 73.0) and 1,000 epochs (top-1: 74.3). For our framework, if not explicitly stated otherwise, we use $V = 2, K = 8, s_{base} = 0.9, s_{view} = 0.6$. We use LARS optimizer with a base initial learning rate $0.3 \times$ batch size/256 for 300 epochs and $0.2 \times$ batch size/256 for 1,000 epochs, with a cosine learning rate decay, and a warm up period of 10 epochs.

Evaluation. We evaluate pretrained models on the COCO object detection task. We fine-tune a Mask R-CNN detector with an FPN backbone on the COCO train2017 split with the standard 1× schedule, following [18]. For DETR fine-tuning, we use the initial learning rate 1×10^{-4} for transformers and 5×10^{-5} for the CNN backbone, following UP-DETR [12]. The model is trained with 150 and 300 epoch schedules, with the learning rate multiplied by 0.1 at 100 and 200 epochs, respectively.

Table 1 compares our framework with both instance-level and dense pretraining methods. Our proposed framework shows a clear performance improvement over methods that only consider instance-level contrasting. Among methods that leverage view correspondence to learn dense representation, our results are comparable with SCRL. Note that some methods, for example DetCon and SoCo, use unsupervised heuristic in obtaining object bounding boxes or segmentation masks, thus are not directly comparable.

In the following sections, we explore different settings and techniques built on top of this framework to study if they improve the performance.

4.2 Effect of Box Sampling Strategies

We focus on a general random box sampling strategies, as in [12,28]. While some box proposal algorithms (e.g., selective search) have been shown to produce sufficiently good object boundaries to improve SSL performance [12,31], we want

Table 1. Results of object detection fine-tuned on COCO with Faster R-CNN + FPN.

Method	Epoch	AP^b	AP^b_{50}	AP^b_{75}
Random init	–	32.8	50.9	35.3
Supervised	–	39.7	59.5	43.3
MoCo	200	38.5	58.3	41.6
SimCLR [8]	–	38.5	58.0	42.0
BYOL* [15]	300	39.1	59.9	42.4
BYOL* [15]	1k	40.1	61.3	43.9
MoCo-v2	–	39.8	59.8	43.6
$DetCon_S$ [20]	300	41.8	–	–
$DetCon_B$ [20]	300	42.0	–	–
SCRL [28]	1k	40.9	62.5	44.5
SoCo [31]	100	42.3	62.5	46.5
SoCo [31]	400	43.0	63.3	47.4
DenseCL [30]	200	40.3	59.9	44.3
PLRC [2]	200	40.7	60.4	44.7
Ours	300	39.9	60.7	43.8
Ours	1k	40.8	62.1	44.7

to avoid incorporating additional inductive bias in the form of rules to generate boxes, since the efficacy of such rules could depend on the dataset.

We study the effect of four hyperparameters of the random box sampling strategy: (1) number of boxes per image (K), (2) box coordinate jittering rate (relative to each coordinate value) ($\%n$), (3) minimum box size (S_{min}), and (4) minimum scale for each view (s_{view}). For each attribute, we report results on several chosen values as in Table 2.

The results are shown in Table 2. For the number of boxes per image K, it can be seen that increasing the number of boxes does not have a large effect on the finetuning performance. The results slightly drop when introducing box jittering, which was proposed in [31]. The approach is pretty robust against changing the minimum box size S_{min}. For the minimum scale for each view s_{view}, it can be observed that having a larger scale (i.e. larger overlapping area between views, which boxes are sampled from) does not help to increase the pretraining efficacy.

4.3   Effect of Methods to Extract Box Features

We explore three different ways to extract features for each box (choices of $\phi(y, b)$): (1) RoIAlign 1×1 (denoted as ra1), (2) RoIAlign $c \times c$ with crop size $c > 1$ (denoted as ra3, ra7, etc.), and (3) averaging cells in the feature map that overlap with the box, similar to 1×1 RoIPooling (denoted as avg). While SCRL and SOCO use ra1 [28,31], ra7 offers more precise features and is used in Faster R-CNN [27] to extract object features. avg shifts box coordinates slightly, introducing variance in the scale and location.

Table 2. Effect of Box Sampling by number of boxes K, box jittering rate %n, minimum box size S_{min}, and minimum scale for each view s_{view}. Underlined numbers are results of the default setting (Sect. 4.1).

K	AP^b	AP^b_{50}	AP^b_{75}	%n	AP^b	AP^b_{50}	AP^b_{75}	S_{min}	AP^b	AP^b_{50}	AP^b_{75}	s_{view}	AP^b	AP^b_{50}	AP^b_{75}
4	40.0	60.8	43.7	0	**39.9**	60.7	43.8	0	39.9	60.7	43.8	0.5	40.1	60.9	43.5
8	39.9	60.7	43.8	0.05	39.7	60.5	43.5	0.05	40.0	60.7	43.8	0.6	39.9	60.7	43.8
16	**40.2**	61.0	44.1	0.10	39.7	60.6	43.1	0.10	39.9	60.6	43.3	0.7	39.6	60.1	43.2
32	39.9	60.7	43.5	0.20	39.8	60.7	43.4	0.20	39.8	60.4	43.5	0.8	39.7	60.2	43.6

Additionally, with the use of RoIAlign $c \times c$, we want to examine the necessity of random box sampling. Specifically, we compare the dense features of the shared area of two views, which is similar to comparing $c \times c$ identical boxes forming a grid in the shared area.

The results are shown in Table 3. We observe that ra1 achieves the best performance, although the differences are marginal. Moreover, when not using random box sampling, AP scores drop significantly, hinting that comparing random boxes with diversified locations and scales is necessary for a good pretrained model.

Table 3. Effect of extracting box features.

	box sampling			shared area		
Feature extraction	AP^b	AP^b_{50}	AP^b_{75}	AP^b	AP^b_{50}	AP^b_{75}
ra1	**39.9**	60.7	43.8	–	–	–
ra3	39.7	60.4	43.4	39.3	59.9	42.8
ra7	39.6	60.4	43.4	39.4	60.0	42.8
avg	39.8	60.7	43.3	–	–	–

4.4   Effect of Multiple Views

multi-crop has been shown to be an effective strategy in instance-level representation learning [6,7,22], with both contrastive and non-contrastive methods. In the context of dense representation learning, the only similar adoption of multi-crop we found is in SoCo [31]; however, their third view is only a resize of one of two main views. We are interested in examining if an adaptation of multi-crop that more closely resembles the original proposal of [6] provides meaningful improvements for our task. We consider two settings for our experiments:

Using Multiple Views. In the 2-view ($V = 2$) setting, since only features corresponding to the shared area of two views are considered for training, the computation related to the non-overlapping area may be useless. For better efficiency, we consider using more than two views $(V > 2)$. The similarity between each pair of views is included in the loss as in Eq. 1. We expect increasing the number of views improves the pretrained model since the the model is trained using more augmented views within the same number of epochs.

Using Local-to-Global Correspondences. Instead of obtaining RoIAlign from the view's dense representation for each box, we crop the image specified by each box and obtain the features with a forward pass through the backbone. For example, if the number of boxes $K = 8$, we will perform 8 forward passes through the network on the 8 crops to obtain box features. These features will be compared against box features obtained with RoIAlign from the *global view*. This is similar to the adoption of local views in DINO [7], except the local views are compared only against the corresponding regions in the feature map obtained from the global view, rather than a representation of the entire image. Figure 3 shows how local views can be leveraged in SSL pretraining.

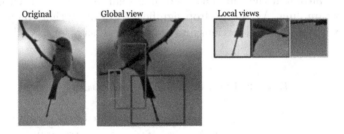

Fig. 3. Leveraging local views in SSL pretraining. Each sampled box in a global view is compared with the representation of the image cropped from the box.

Table 4. Results with multiple views.

#views	AP^b	AP^b_{50}	AP^b_{75}
2	**39.9**	60.5	43.6
3	39.8	60.8	43.4
4	39.8	60.7	43.5

Table 4 shows the results of using multiple views for $V = 3$ or $V = 4$, where we do not observe a significant performance gain despite more computation at each step. This suggests that constructing more augmented views at the same scale does not necessarily lead to an increase in the performance. Note that in this first design, we did not downscale views or use local views as in [6,7]. For the second design, table 5 shows the results of using global and local views, with each local view covering an smaller area inside the global view and resized to 96×96. We observe that the performance drops significantly. These results suggest that

Table 5. Results with global and local views.

#views	ra1			avg		
	AP^b	AP^b_{50}	AP^b_{75}	AP^b	AP^b_{50}	AP^b_{75}
2 global	**39.9**	60.7	43.8	**39.8**	60.7	43.3
1 global + 8 local	38.1	58.7	41.5	38.3	58.8	41.9
2 global + 8 local	38.3	58.8	41.7	39.0	59.7	42.7

although the `multi-crop` strategy, either with or without considering local-to-global correspondences, is effective for learning global image features [6,7], it is not effective for learning dense features.

4.5   Effect of Box Localization Auxiliary Task

In addition to the SSL loss, we consider a box localization loss to match the objective of SSL pretraining with that of an object detection model. Existing methods usually improve pretraining for dense prediction tasks by leveraging spatial consistency; however, a self-supervised pretext task designed specifically for the object detection tasked has been less studied. UP-DETR [12] has demonstrated that using box features extracted by a well-trained vision backbone to predict box location helps DETR pretraining. In this section, we present our effort to incorporate such object detection pretext tasks into our pretraining framework. Two types of box localization loss $\mathcal{L}_{box}$ are considered. Given a box feature from a view, we can compute either (1) a box prediction loss, i.e. a contrastive loss that helps predict the corresponding box among up to K boxes from another view; or (2) a box regression loss, an L1 distance and general IoU loss of box coordinates from another view predicted by using a transformer. The final loss is defined as $\mathcal{L} = \mathcal{L}_{BYOL} + \lambda\mathcal{L}_{box}$, where λ is the weight of the box localization term.

Box Prediction Loss. Given a box feature u_i^k from a view i, we want to predict which box from $\{u_j^1, \cdots u_j^K\}$ in view j corresponds to u_i^k. This can be done by comparing the similarity between feature from each of these K boxes with u_i^k. We use a contrastive loss to minimize the distance between positive box pairs, and maximize the distance between negative box pairs.

$$\mathcal{L}_{i,j}^{box\_pred} = -\sum_{k=1}^{K} \log \frac{\exp(\text{sim}(u_i^k, u_j^k)/\tau)}{\sum_{k'=1}^{K} \exp(\text{sim}(u_i^k, u_j^{k'})/\tau)}$$

Box Regression Loss. Inspired by the DETR architecture [4], we employ a transformer, which takes u_j^k as the query and looks over the representation of the i-th view to predict the box location u_i^k. The output of the transformer is a

Table 6. Results of object detection fine-tuned on COCO with Faster R-CNN + FPN.

Method	λ	AP^b	AP^b_{50}	AP^b_{75}
Random init	–	32.8	50.9	35.3
BYOL	–	39.1	59.9	42.4
No box loss	0.00	**39.9**	60.5	43.6
Box prediction	0.01	39.8	60.3	43.6
	0.05	39.5	59.9	43.1
	0.10	39.5	60.0	43.4
Box regression	0.01	39.6	60.4	43.3
	0.05	39.1	59.4	42.9

Table 7. Results of object detection fine-tuned on COCO with DETR.

Method	AP^b	AP^b_{50}	AP^b_{75}
supervised	37.7^5	59.0	39.2
BYOL [15]	36.0^6	56.9	37.2
Ours	35.6	56.0	37.0
Ours + regression	35.7	55.8	37.5

vector of size 4 for each box, representing the coordinate of the box center, its height and width. In addition to L1 loss, we also use the generalized IoU loss, which is invariant to box scales. The bounding box loss is defined as

$$\mathcal{L}_{\text{box}}(\tilde{b}_i^k, b_i^k) = \lambda_{\text{giou}} \mathcal{L}_{\text{giou}}(\tilde{b}_i^k, b_i^k) + \lambda_{\text{box}} \|\tilde{b}_i^k - b_i^k\|_1$$

where λ_{giou} and λ_{box} are weights in the loss, $\tilde{b}_i^k = \text{Decoder}(y_i, u_j^k)$ is a predicted box. The box loss is defined as

$$\mathcal{L}_{\text{box}} = \sum_{\substack{i=1 \\ j=1, j \neq i}}^{V} \frac{1}{|B_i \cap B_j|} \sum_{k \in B_i \cap B_j} \left(\mathcal{L}_{\text{box}}(\tilde{b}_i^k, b_i^k) + \mathcal{L}_{\text{box}}(\tilde{b}_j^k, b_j^k) \right)$$

Table 6 shows the results with two proposed losses when fine-tuning the Faster R-CNN model on the COCO dataset. It can be seen that these auxiliary losses, despite our expectation, have an adverse effect on the finetuning performance. We suggest that although these tasks encourage learning a representation that is useful for box prediction, the gap between these and a supervised task on labeled data is still significant that finetuning is not very effective.

Table 7 shows the results when fine-tuning DETR, which shares the decoder architecture with the decoder used to obtain box regression loss. While our framework improves the fine-tuning performance ($+0.8$ AP^b) as shown in Table 1, it does not improve the results in the case of fine-tuning DETR (-0.4 AP^b).

5   Conclusion

We studied a self-supervised pretraining approach for object detection based on sampling random boxes and maximizing spatial consistency. We investigated the effect of different box generation and feature extraction strategies. Moreover, we tried incorporating multi-crop and additional self-supervised object detection pretext tasks to the proposed framework. We found that the method is robust against different design choices.

Acknowledgement. We thank our colleagues from Google Brain Toronto and Brain AutoML, Ting Chen and Golnaz Ghiasi who provided insight and expertise that greatly assisted this research.

References

1. Asano, Y.M., Rupprecht, C., Vedaldi, A.: Self-labelling via simultaneous clustering and representation learning. arXiv preprint arXiv:1911.05371 (2019)
2. Bai, Y., Chen, X., Kirillov, A., Yuille, A., Berg, A.C.: Point-level region contrast for object detection pre-training. arXiv preprint arXiv:2202.04639 (2022)
3. Bar, A.,et al.: DETReg: unsupervised pretraining with region priors for object detection. arXiv preprint arXiv:2106.04550 (2021)
4. Carion, N., Massa, F., Synnaeve, G., Usunier, N., Kirillov, A., Zagoruyko, S.: End-to-end object detection with transformers. In: Vedaldi, A., Bischof, H., Brox, T., Frahm, J.-M. (eds.) ECCV 2020. LNCS, vol. 12346, pp. 213–229. Springer, Cham (2020). https://doi.org/10.1007/978-3-030-58452-8_13
5. Caron, M., Bojanowski, P., Joulin, A., Douze, M.: Deep clustering for unsupervised learning of visual features. In: Proceedings of the European Conference on Computer Vision (ECCV), pp. 132–149 (2018)
6. Caron, M., Misra, I., Mairal, J., Goyal, P., Bojanowski, P., Joulin, A.: Unsupervised learning of visual features by contrasting cluster assignments. Advances in Neural Information Processing Systems 33, 9912–9924 (2020)
7. Caron, M., et al.: Emerging properties in self-supervised vision transformers. In: Proceedings of the IEEE/CVF International Conference on Computer Vision, pp. 9650–9660 (2021)
8. Chen, T., Kornblith, S., Norouzi, M., Hinton, G.: A simple framework for contrastive learning of visual representations. In: International Conference on Machine Learning, pp. 1597–1607. PMLR (2020)
9. Chen, T., Kornblith, S., Swersky, K., Norouzi, M., Hinton, G.E.: Big self-supervised models are strong semi-supervised learners. Advances in neural information processing systems 33, 22243–22255 (2020)
10. Chen, X., Fan, H., Girshick, R., He, K.: Improved baselines with momentum contrastive learning. arXiv preprint arXiv:2003.04297 (2020)
11. Chen, X., He, K.: Exploring simple siamese representation learning. In: Proceedings of the IEEE/CVF Conference on Computer Vision and Pattern Recognition, pp. 15750–15758 (2021)
12. Dai, Z., Cai, B., Lin, Y., Chen, J.: UP-DETR: unsupervised pre-training for object detection with transformers. In: Proceedings of the IEEE/CVF Conference on Computer Vision and Pattern Recognition, pp. 1601–1610 (2021)

13. Deng, J., Dong, W., Socher, R., Li, L.J., Li, K., Fei-Fei, L.: ImageNet: a large-scale hierarchical image database. In: 2009 IEEE Conference on Computer Vision and Pattern Recognition, pp. 248–255. IEEE (2009)
14. Doersch, C., Zisserman, A.: Multi-task self-supervised visual learning. In: Proceedings of the IEEE International Conference on Computer Vision, pp. 2051–2060 (2017)
15. Grill, J.B., Strub, F., Altché, F., Tallec, C., Richemond, P., Buchatskaya, E., Doersch, C., Avila Pires, B., Guo, Z., Gheshlaghi Azar, M., et al.: Bootstrap your own latent-a new approach to self-supervised learning. Adv. Neural Inf. Process. Syst. **33**, 21271–21284 (2020)
16. Gutmann, M., Hyvärinen, A.: Noise-contrastive estimation: a new estimation principle for unnormalized statistical models. In: Proceedings of the Thirteenth International Conference on Artificial Intelligence and Statistics, pp. 297–304. JMLR Workshop and Conference Proceedings (2010)
17. He, K., Fan, H., Wu, Y., Xie, S., Girshick, R.: Momentum contrast for unsupervised visual representation learning. In: Proceedings of the IEEE/CVF Conference on Computer Vision and Pattern Recognition, pp. 9729–9738 (2020)
18. He, K., Girshick, R., Dollár, P.: Rethinking imageNet pre-training. In: Proceedings of the IEEE/CVF International Conference on Computer Vision, pp. 4918–4927 (2019)
19. He, K., Zhang, X., Ren, S., Sun, J.: Deep residual learning for image recognition. In: Proceedings of the IEEE Conference on Computer Vision and Pattern Recognition, pp. 770–778 (2016)
20. Hénaff, O.J., Koppula, S., Alayrac, J.B., van den Oord, A., Vinyals, O., Carreira, J.: Efficient visual pretraining with contrastive detection. In: Proceedings of the IEEE/CVF International Conference on Computer Vision, pp. 10086–10096 (2021)
21. Huang, J., Dong, Q., Gong, S., Zhu, X.: Unsupervised deep learning by neighbourhood discovery. In: International Conference on Machine Learning, pp. 2849–2858. PMLR (2019)
22. Huynh, T., Kornblith, S., Walter, M.R., Maire, M., Khademi, M.: Boosting contrastive self-supervised learning with false negative cancellation. In: Proceedings of the IEEE/CVF Winter Conference on Applications of Computer Vision (WACV), pp. 2785–2795 (January 2022)
23. Larsson, G., Maire, M., Shakhnarovich, G.: Learning representations for automatic colorization. In: Leibe, B., Matas, J., Sebe, N., Welling, M. (eds.) ECCV 2016. LNCS, vol. 9908, pp. 577–593. Springer, Cham (2016). https://doi.org/10.1007/978-3-319-46493-0_35
24. Lin, T.-Y., Maire, M., Belongie, S., Hays, J., Perona, P., Ramanan, D., Dollár, P., Zitnick, C.L.: Microsoft COCO: common objects in context. In: Fleet, D., Pajdla, T., Schiele, B., Tuytelaars, T. (eds.) ECCV 2014. LNCS, vol. 8693, pp. 740–755. Springer, Cham (2014). https://doi.org/10.1007/978-3-319-10602-1_48
25. Noroozi, M., Favaro, P.: Unsupervised learning of visual representations by solving jigsaw puzzles. In: Leibe, B., Matas, J., Sebe, N., Welling, M. (eds.) ECCV 2016. LNCS, vol. 9910, pp. 69–84. Springer, Cham (2016). https://doi.org/10.1007/978-3-319-46466-4_5
26. O Pinheiro, P.O., Almahairi, A., Benmalek, R., Golemo, F., Courville, A.C.: Unsupervised learning of dense visual representations. Adv. Neural Inf. Process. Syst. **33**, 4489–4500 (2020)
27. Ren, S., He, K., Girshick, R., Sun, J.: Faster R-CNN: towards real-time object detection with region proposal networks. In: Advances in neural information processing systems **28** (2015)

28. Roh, B., Shin, W., Kim, I., Kim, S.: Spatially consistent representation learning. In: Proceedings of the IEEE/CVF Conference on Computer Vision and Pattern Recognition, pp. 1144–1153 (2021)
29. Sohn, K.: Improved deep metric learning with multi-class N-pair loss objective. In: Lee, D., Sugiyama, M., Luxburg, U., Guyon, I., Garnett, R. (eds.) Advances in Neural Information Processing Systems. vol. 29. Curran Associates, Inc. (2016). https://proceedings.neurips.cc/paper/2016/file/6b180037abbebea991d8b1232f8a8ca9-Paper.pdf
30. Wang, X., Zhang, R., Shen, C., Kong, T., Li, L.: Dense contrastive learning for self-supervised visual pre-training. In: Proceedings of the IEEE/CVF Conference on Computer Vision and Pattern Recognition, pp. 3024–3033 (2021)
31. Wei, F., Gao, Y., Wu, Z., Hu, H., Lin, S.: Aligning pretraining for detection via object-level contrastive learning. In: Advances in Neural Information Processing Systems **34** (2021)
32. Wu, Z., Xiong, Y., Yu, S.X., Lin, D.: Unsupervised feature learning via non-parametric instance discrimination. In: Proceedings of the IEEE Conference on Computer Vision and Pattern Recognition, pp. 3733–3742 (2018)
33. Xie, J., Zhan, X., Liu, Z., Ong, Y., Loy, C.C.: Unsupervised object-level representation learning from scene images. In: Advances in Neural Information Processing Systems **34** (2021)
34. Xie, Z., Lin, Y., Zhang, Z., Cao, Y., Lin, S., Hu, H.: Propagate yourself: exploring pixel-level consistency for unsupervised visual representation learning. In: Proceedings of the IEEE/CVF Conference on Computer Vision and Pattern Recognition, pp. 16684–16693 (2021)
35. Yang, C., Wu, Z., Zhou, B., Lin, S.: Instance localization for self-supervised detection pretraining. In: Proceedings of the IEEE/CVF Conference on Computer Vision and Pattern Recognition, pp. 3987–3996 (2021)
36. Zhu, X., Su, W., Lu, L., Li, B., Wang, X., Dai, J.: Deformable DETR: deformable transformers for end-to-end object detection. arXiv preprint arXiv:2010.04159 (2020)

Internet Curiosity: Directed Unsupervised Learning on Uncurated Internet Data

Alexander C. Li[1]([✉])[ID], Ellis Brown[1][ID], Alexei A. Efros[2][ID], and Deepak Pathak[1][ID]

[1] Carnegie Mellon University, Pittsburgh, US
alexanderli@cmu.edu
[2] University of California, Berkeley, US

Abstract. We show that a curiosity-driven computer vision algorithm can learn to efficiently query Internet text-to-image search engines for images that improve the model's performance on a specified dataset. In contrast to typical self-supervised computer vision algorithms, which learn from static datasets, our model actively expands its training set with the most relevant images. First, we calculate an image-level curiosity reward that encourages our model to find the most useful images for pre-training. Second, we use text similarity scores to propagate observed curiosity rewards to untried text queries. This efficiently identifies relevant semantic clusters without any need for class labels or label names from the targeted dataset. Our method significantly outperforms models that require 1–2 orders of magnitude more compute and data.

Keywords: Self-supervised learning · Active learning · Curiosity

1 Introduction

Billions of photos are uploaded to the Internet every day, capturing an incredible diversity of real-world objects and scenes. These images offer self-supervised models, which do not require label information, a valuable opportunity to learn useful representations that can improve performance on downstream tasks [3,7,8]. However, today's computer vision models do not fully take advantage of the breadth of visual knowledge on the Internet. These models are typically trained on hand-curated datasets (e.g., ImageNet [2]), which typically yield suboptimal transfer performance when they do not contain much data relevant to the downstream task. Training on a broader slice of Internet images, such as Instagram-1B [6], can potentially alleviate this issue by sampling a wider variety of images. However, this undirected approach still often fails to find relevant data and wastes significant training time on irrelevant data that do not contribute to improved performance on the downstream dataset.

We argue that the best approach to handle the scale of Internet data is *directed curiosity*: our models should seek out data that improve their performance on a particular target task. This is different from the (typical) undirected

---

Alexander C. Li and E. Brown—Authors contributed equally to this work.

© The Author(s), under exclusive license to Springer Nature Switzerland AG 2023
L. Karlinsky et al. (Eds.): ECCV 2022 Workshops, LNCS 13804, pp. 100–104, 2023.
https://doi.org/10.1007/978-3-031-25069-9_7

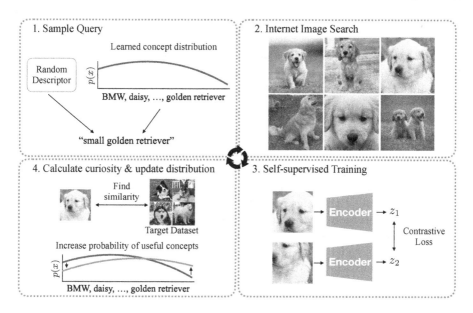

Fig. 1. Overview of Internet Curiosity. Our goal is to efficiently search the Internet for images that improve our performance on a target dataset. In each iteration, we generate a text query by combining a random descriptor with a concept sampled from a learned distribution. We query Google Images with the resulting phrase and download the top 100 image results. We add these images to the set of previously downloaded images and perform self-supervised learning on the combined dataset. Finally, we evaluate the relevance of the new images and increase the likelihood of the query and other related queries if the new images were similar to the target dataset.

self-supervised learning setting, where the goal is to learn general representations that broadly transfer to other computer vision tasks. Here, we only care about performance on a *single* dataset that is known beforehand.

We propose "Internet Curiosity", which continually improves a model with relevant images found on the Internet. This method alternates between searching for images on the Internet with text queries, training on acquired images, determining which images were useful, and prioritizing what to search for next (see Fig. 1). Our setting is different from active learning [11], where the goal is to selectively obtain labels for data points from a fixed dataset. In contrast, Internet Curiosity continually expands the size of its dataset and requires no labels for training, even from the target dataset. To summarize our contributions:

1. We introduce the paradigm of "directed curiosity," where a self-supervised algorithm actively seeks out data that is relevant to its target task.
2. We show that weak supervision in the form of text can be used to efficiently sift through the vast ocean of Internet image data.
3. We use nearest-neighbor representation distance to identify relevant images.
4. We use text similarity to estimate the quality of untried, similar queries.
5. Compared to pre-trained models, Internet Curiosity is far more compute- and data-efficient, while matching or outperforming them on accuracy.

2   Internet Curiosity

Since we have a known target dataset, we can prioritize collecting data that is expected to be helpful for this task. We make as few assumptions as possible and assume that we have only unlabeled training data from the target domain, without any labels or information about what the dataset is about.

2.1   Image Search

We use text-based image search to find images online. Image search is fast, returns diverse images from across the internet, and enables searches for vastly different queries all at the same time. Note that image search is noisy and makes use of weak supervision (the image-text pairing on webpages). For this reason, we only perform self-supervised training on the downloaded images. We use a publicly available codebase to query Google Images, which can download the top 100 images for each query. We turn on SafeSearch, mark preference for photographs, and set the minimum image size to 350, the smallest setting available.

2.2   Self-Supervised Training

Self-supervised learning allows us to learn from the unlabeled images that we download from Google Image Search. In each iteration, we train for 10 epochs using MoCo-v3 [1] to fine-tune a ResNet-50 model [4] on a mixture of newly downloaded images, previous images, and target dataset images. We use a batch size of 200 and a base learning rate of 0.15. At iteration 0, we initialize our model using a MoCo-v3 checkpoint trained for 100 epochs on ImageNet.

2.3   "Densifying" the Target Dataset via Curiosity

Our curiosity reward is formulated to find images similar in representation space to the target dataset images. These images act as "hard negatives" for the target dataset and should result in the biggest improvement in representation quality. Thus, we compute the curiosity reward for a particular image as its representation's negative distance to its closest neighbor in the target dataset.

2.4   Generating Queries

We construct queries by combining two components:

1. *Concepts* specify semantic categories such as people, places, or objects.
2. *Descriptors* are modifiers that are generally applicable for any concept. They provide a simple way to generate semantic variations in appearance.

We use a large vocabulary of concepts that is explicitly constructed to contain a variety of visual concepts. We draw our concepts from the Britannica Concise Encyclopedia and use 41 descriptors, encompassing size, shape, and color.

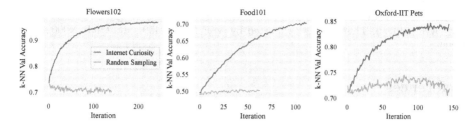

Fig. 2. Learning curves for 3 targeted datasets. Internet Curiosity outperforms a baseline that randomly samples queries uniformly from the concept vocabulary.

Table 1. Our method significantly improves k-NN accuracy for each dataset while using only 10% the training time of algorithms with static, curated datasets.

Model	Flowers	Food	Pets	Images	GPU-hours
CLIP ResNet-50 [9]	88.3	**79.4**	70.8	400×10^6	10^3
MoCo-v3 (ImageNet pre-train)	74.6	49.9	71.5	1.2×10^6	72
MoCo-v3 (ImageNet + target)	87.0	51.0	73.0	1.2×10^6	74
Random exploration	70.6	50.3	71.7	2×10^6	72 + 72
Ours	**97.1**	70.4	**84.2**	2×10^6	72 + 72

It is inefficient to try every possible query given our large vocabulary; instead, we use a pre-trained sentence similarity model [10] to build a similarity graph for our concepts. Each pair of concepts has edge weight equal to their predicted text similarity. To ensure large cliques of concepts do not dominate, we only keep the largest 50 connections per concept. After searching for a query and evaluating curiosity rewards for each resulting image, we estimate rewards for unseen concepts via the similarity graph using a weighted sum. Our concept distribution for the next iteration is then the softmax of each reward estimate.

3 Results

We evaluate Internet Curiosity on Flowers-102, Food101, and Oxford-IIT Pets, which are 3 small-scale fine-grained classification datasets commonly used to evaluate transfer learning performance for large pre-trained models [5,9]. Figure 2 shows that directed exploration with Internet Curiosity is far better than a baseline that samples queries uniformly at random from the concept vocabulary. In fact, random sampling typically decreases accuracy, likely due to the fact that Internet images can be unsuited for general pre-training (e.g. watermarks, images of text, large distribution shift). Table 1 shows that our method universally improves on the starting MoCo model and can outperform a CLIP [9] model of the same size while using much less compute and data.

Acknowledgements. AL is supported by the NSF GRFP under grants DGE1745016 and DGE2140739. This work is supported by NSF IIS-2024594 and ONR N00014-22-1-2096.

References

1. Chen, X., Xie, S., He, K.: An empirical study of training self-supervised vision transformers. In: Proceedings of the IEEE/CVF International Conference on Computer Vision, pp. 9640–9649 (2021)
2. Deng, J., Dong, W., Socher, R., Li, L.J., Li, K., Fei-Fei, L.: ImageNet: a large-scale hierarchical image database. In: 2009 IEEE Conference on Computer Vision and Pattern Recognition, pp. 248–255. IEEE (2009)
3. He, K., Fan, H., Wu, Y., Xie, S., Girshick, R.: Momentum contrast for unsupervised visual representation learning. In: 2020 IEEE/CVF Conference on Computer Vision and Pattern Recognition (CVPR) (2020)
4. He, K., Zhang, X., Ren, S., Sun, J.: Deep residual learning for image recognition. In: 2016 IEEE Conference on Computer Vision and Pattern Recognition (CVPR) (2016)
5. Kornblith, S., Shlens, J., Le, Q.V.: Do better imageNet models transfer better? In: Proceedings of the IEEE/CVF Conference on Computer Vision and Pattern Recognition, pp. 2661–2671 (2019)
6. Mahajan, D., et al.: Exploring the limits of weakly supervised pretraining. In: Proceedings of the European conference on computer vision (ECCV) (2018)
7. Oord, A.v.d., Li, Y., Vinyals, O.: Representation learning with contrastive predictive coding. preprint arXiv:1807.03748 (2018)
8. Pathak, D., Krahenbuhl, P., Donahue, J., Darrell, T., Efros, A.A.: Context encoders: Feature learning by inpainting. In: Proceedings of the IEEE Conference on Computer Vision and Pattern Recognition, pp. 2536–2544 (2016)
9. Radford, A., et al.: Learning transferable visual models from natural language supervision. In: International Conference on Machine Learning, pp. 8748–8763. PMLR (2021)
10. Reimers, N., Gurevych, I.: Sentence-BERT: sentence embeddings using siamese BERT-networks. arXiv preprint arXiv:1908.10084 (2019)
11. Settles, B.: Active learning literature survey (2009)

W10 - Self-supervised Learning for Next-Generation Industry-Level Autonomous Driving

W10 - Self-supervised Learning for Next-Generation Industry-Level Autonomous Driving

Self-supervised learning for next-generation industry-level autonomous driving refers to a variety of studies that attempt to refresh the solutions for challenging real-world perception tasks by learning from unlabeled or semi-supervised large-scale collected data to incrementally self-train powerful recognition models. Thanks to the rise of large-scale annotated data sets and advances in computing hardware, various supervised learning methods have significantly improved the performance in many problems (e.g. 2D detection, instance segmentation, and 3D Lidar Detection) in the field of self-driving. However, these supervised learning approaches are notoriously "data hungry", especially in the current autonomous driving fields. To facilitate an industry-level autonomous driving system in the future, the desired visual recognition model should be equipped with the ability of self-exploring, self-training, and self-adapting across diverse new-appearing geographies, streets, cities, weather conditions, object labels, viewpoints, or abnormal scenarios. To address this problem, many recent efforts in self-supervised learning, large-scale pretraining, weakly supervised learning, and incremental/continual learning have been made to improve the perception systems to deviate from traditional paths of supervised learning for self-driving solutions. This workshop aims to investigate advanced ways of building next-generation industry level autonomous driving systems by resorting to self-supervised/semi-supervised learning.

October 2022

Xiaodan Liang
Hang Xu
Fisher Yu
Wei Zhang
Michael C. Kampffmeyer
Ping Luo

Towards Autonomous Grading in the Real World

Yakov Miron[1,2], Yuval Goldfracht[1(✉)], and Dotan Di Castro[1]

[1] Bosch Center for Artificial Intelligence (BCAI), Haifa, Israel
yuval.goldfracht@il.bosch.com
[2] The Autonomous Navigation and Sensor Fusion Lab, The Hatter Department of Marine Technologies, University of Haifa, Haifa, Israel

Abstract. Surface grading is an integral part of the construction pipeline. Here, a dozer, which is a key machinery tool in any construction site, is required to level an uneven area containing pre-dumped sand piles. In this work, we aim to tackle the problem of autonomous surface grading on real-world scenarios. We design both a realistic physical simulation and a scaled real-world prototype environment mimicking the real dozer dynamics and sensory information. We establish heuristics and learning strategies in order to solve the problem. Through extensive experimentation, we show that although heuristics are capable of tackling the

Fig. 1. A comparison of grading policy on a **simulated dozer** and a **real-world scaled prototype**. Here, the agent is provided with an initial graded area and is required to extend it. **Top Row:** Our experimental setup shown in Sect. 3.5 showing the scaled dozer prototype facing the sand piles. **Middle & Bottom Rows:** Heightmaps extracted from our simulation and scaled experimental setup respectively. Sand piles appear as dark blobs that indicate their height. Columns compare the grading policy on a similar scenario in simulation and experimental setup.

Supplementary Information The online version contains supplementary material available at https://doi.org/10.1007/978-3-031-25069-9_8.

ⓒ The Author(s), under exclusive license to Springer Nature Switzerland AG 2023
L. Karlinsky et al. (Eds.): ECCV 2022 Workshops, LNCS 13804, pp. 107–117, 2023.
https://doi.org/10.1007/978-3-031-25069-9_8

problem in a clean and noise-free simulated environment, they fail catastrophically when facing real-world scenarios. However, we show that the simulation can be leveraged to guide a learning agent, which can generalize and solve the task both in simulation and in a scaled prototype environment.

1   Introduction

Recent years have seen a dramatic rising demand for automation in the construction industry, which suffer from labor shortage due to various reasons. Unfortunately, construction sites are unpredictable and unstructured environments by nature, where multiple machines work simultaneously, on a variety of challenging tasks. For these reasons, automation in the field is considered to be an extremely difficult task, which has not yet been solved.

Autonomous grading, similarly to many other robotics tasks, suffers from several challenges. First, partial observability of the environment, which hinders the decision making process, is caused by limited field-of-view and missing or noisy sensor information. Second, data collection is extremely challenging as it requires significant amount of time and labor, required to physically design the grading scenario. In addition, it can also be dangerous as it involves operating heavy machinery in a complex environment.

In this work, we aim to learn a robust behavior policy capable of autonomous grading in the real-world (see Fig. 1). Specifically, we make use of *privileged learning* [20], where the agent is trained on noisy data to imitate an expert, which has direct access to perfect and noise-free measurements. In addition, we design both simulator and scaled real-world prototype in order to rapidly explore and deploy the learned policies. This mode of action aims to overcome the challenges of learning a behaviour policy for robotics applications. Our main contributions are as follows:
(1) We create a physically realistic simulation environment for training and evaluation of grading policies.
(2) We train an agent to learn a robust policy capable of operating in real-world scenarios, using *privileged learning*.
Finally, we validate our methods and assumptions on a scaled prototype environment, which mimics real-world vehicle dynamics, sensors and soil interaction.

2   Related Work

2.1   Bulldozer Automation

The field of autonomous construction vehicles, has attracted a growing interest over the past few years, Here, the majority of work focuses around bulldozers, excavators and wheel-loaders on different aspects of the autonomous construction task. [4] implemented a heuristic approach to grade sand piles. [16], developed a pioneering software package for automation of a construction vehicle, where their excavator was one of the first to operate autonomously. [6] trained an agent using a model-free reinforcement learning approach and showed impressive results in simulation.

2.2   Sand Simulation

Precise particle simulation, e.g sand, is an active research field for both classic and modern tools. Classic methods describe soil using solid mechanical equations [12], discrete element methods [2], or fluid mechanics [17]; whereas newer modern methods utilize deep neural networks [15] to simulate the reaction of particles to forces. While these methods achieve outstanding results and have impacted both the gaming and cinematic industries, they require high computational cost and long run times. As opposed to such methods, [5] suggested a simple and efficient heightmap approach aimed at robotic applications. [14] created a simulator, aimed towards policy evaluation for bulldozers, that is fast, fairly accurate and captures the main aspects of the interaction between the sand and the vehicle.

2.3   Sim-to-Real

The *sim-to-real* gap is an active research field, which can be divided into two main categories *dynamics* and *perception*. The *dynamics* gap stems from the inability to precisely model the environment and is tackled by solving a robust objective using dynamic randomization [9] or adversarial training [10, 19]. Overcoming the *perception* gap is done by learning a mapping between the simulation and the real-world [11], [8] or learning robust feature extractors [7].

3   Method

In this work, we focus our efforts on several fronts: (*i*) creating a realfistic simulation environment (*ii*) exploring behavioral cloning (**BC**) techniques that enable learning of a robust policy for real-world scenarios (*iii*) rapid policy evaluation on a real-world scaled prototype.

3.1   Problem Formulation

We formalize the problem as a POMDP/R [18] and define a 4-tuple. **State:** consists of all the information required to obtain the optimal policy and determine the outcome of each action (see Fig. 2a). **Observation:** contains an *ego-view* of the current state. Meaning, a bounding box view around the current location of the dozer (see Fig. 2a) **Actions:** We use the "Start point and Push point" action-set [4], shown in Fig. 3. At each state, the agent selects two coordinates *push point* and *start point* denoted by (P, S), respectively. The dozer drives towards P, then reverses slightly further than it's initial location, marked as B, and finally moves to the next starting position S. **Transitions:** Transitions are determined by a dozer's dynamics and physical properties of soil and the environment.

3.2 Dozer Simulation

The complexity of surface grading lies in the interaction between the vehicle and the soil. However, fast and efficient training is crucial for rapid evaluation and development of control algorithms. To mitigate these two aspects, we created a physically realistic simulated environment taking these considerations into account. In our simulation, each sand pile is initially modeled as a multivariate gaussian distribution as follows:

$$f(x,y) = \frac{V}{2\pi\sigma_x\sigma_y} \cdot \exp\left(-\frac{1}{2}\left[\left(\frac{x-\mu_x}{\sigma_x}\right)^2 + \left(\frac{y-\mu_y}{\sigma_y}\right)^2\right]\right) \quad (1)$$

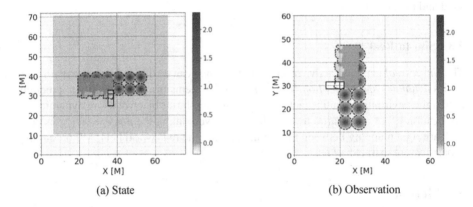

(a) State   (b) Observation

Fig. 2. An example of the state and observation (marked in gray) used in our problem. The state includes the entire information for the task (full heightmap of the area and dozer position). The observation is a subsection bounding box of the state around the dozer in dozer *ego-view* axis.

Where x, y are the Cartesian coordinates of the heightmap, $f(x,y)$ is the height of the soil at each point, $V[cm^3]$ is the volume, and σ_x, σ_y define the footprint of the sand pile. Similarly to [1], we assume linear mechanics of a dozer, i.e. a linear relationship between the dozer's velocity and the load on the blade. In our implementation, the simulator receives an action tuple (P, S), as defined in Sect. 3.1, and generates the low-level control and movement. Then, the simulator evaluates the time it took to perform these actions, taking into consideration the volume of sand moved and the distance traveled. After an action is performed. Finally, the simulator returns the updated observation.

3.3 Baseline Algorithm

The *baseline* algorithm, inspired by the behavior of a human driver, relies on edge detection techniques for sand-pile recognition and is presented in Fig. 3. After recognizing the sand piles it plans 2 way-points – P and S. P is selected based on the nearest sand pile while ensuring that sand fills approximately 50% of the blade. Thus, no side spillage or slippage of the tracks occurs. As the dozer is incapable of reversing on uneven ground, the B-point is reached by reversing in the same path it previously drove,

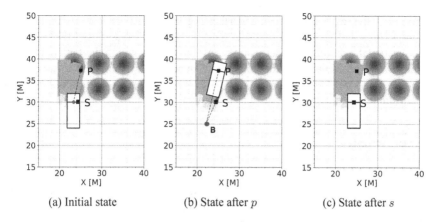

(a) Initial state   (b) State after p (c) State after s

Fig. 3. Example of actions (P, S) and the trajectory between these actions. In Fig. 3a, the red dot describes the initial position. The action's order is as follows: (i) From origin rotate to face P. (ii) Drive forward to P. This action is the only one where the dozer interacts with sand and grades it. (iii) Reverse back to B (blue dot in Fig. 3b). (iv) Rotate to face next S. (v) Drive forward to next S. (Color figure online)

yet slightly further than where it started. After reaching B, the dozer rotates and moves towards the the S-point. This point is selected at the frontier of the nearest sand pile. This process repeats until the entire area is graded.

3.4   Privileged Behavioral Cloning

The basis for utilizing privileged learning techniques [20], is the fact that the *baseline* performs very well in a clean environment (i.e., no noise or inaccuracies). As such, we provide the *baseline* with a clean observation o_t and collect two types of data tuples $\{o_t, a_t\}$ and $\{N(o_t), a_t\}$ for the **BC** and **privileged BC** agents, respectively, where a_t is the action performed by the *baseline* and $N(o_t)$ is the augmented observation. In our case, augmentations include low-volume piles that are spread randomly around the target area, depth measurement noise and occlusions. The **privileged BC** loss is as follows:

$$L(\theta; D) = \mathop{\mathbb{E}}_{(o,a)\sim D}\Big[l\big(\pi_\theta(N(o)), a\big)\Big],$$

Here, D is the training dataset generated by an expert demonstrator presented with clean observations. D contains corresponding actions a, and clean observations o. $N(o)$ is the augmented observations, π_θ is the agent's learned policy and $l(\cdot)$ is a cross entropy metric. Figure 4 illustrate this concept.

3.5   Scaled Prototype Environment

While the simulated environment enables rapid testing, it is not physically precise. As such, we built a $1 : 9$ scaled prototype environment that includes a RGBD camera, a scaled dozer prototype of size 60×40 cm, and a sandbox of size 250×250 cm in order to mimic the real-world inaccuracies (see Fig. 1). The RGBD camera captures the entire target area and provides a realistic dense heightmap. In addition, the camera is used to localize the prototype dozer using an ArUco marker within the target area [13]. The images and positions from the camera are then used by the agent to predict the next action. The dozer then implements a low-level controller, moving according to the chosen destinations in a closed control loop manner.

4   Experiments

We conduct rigorous experimentation in both simulated and scaled prototype environments in order to validate our assumptions. First, we show that although the *baseline* performs exceptionally well in a clean simulation, it fails catastrophically when presented with real-world observations. Second, we train two DNN based agents according to Sect. 3.4, denoted as **BC** and **privileged BC**, and show their ability to imitate the *baseline*'s decision making process. Finally, we show the advantage of the **privileged BC** agent and it's ability to generalize to real-world observations, despite training only in simulation

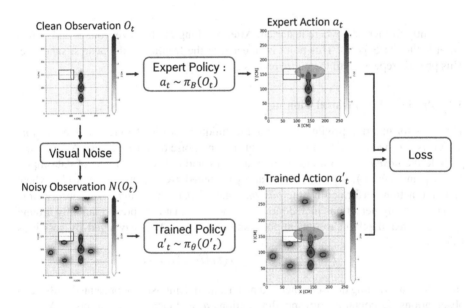

Fig. 4. Illustration of the suggested *privileged learning* algorithm. A clean observation o_t is augmented to create a noisy observation $N(o_t)$. The expert policy, i.e. *baseline*, uses o_t in order to produce an ideal action a_t - showm as green and red way-points. The **privileged BC** agent aims to learn a robust policy π_θ that takes the augmented observation $N(o_t)$ as input an outputs an action a'_t similar as possible to a_t. (Color figure online)

4.1 Simulation Results

As our policy model, we use a ResNet based [3] end-to-end fully convolutional neural network with dilated convolutions where the input size is $(H \times W)$, and the the output size is $(H \times W \times 2)$ - a channel for each of the actions (P, S) (See Fig. 3). we train both agents using 200 random initial states and solve those tasks using the *baseline* while recording the observation-action sets (o_t, a_t). In each episode, we randomize the number of sand piles, shapes, locations, volumes and initial location.

Table 1. Simulation results of our algorithms. Each result shows an average of **50** episodes. Volume cleared is measured in % ($\uparrow$ is better) and time in minutes ($\downarrow$ is better).

Algorithm	Volume cleared [%]	Time [minutes]
baseline	98.2	11.8
BC	88.6	16.4
Privileged BC	87.3	12.9

Table 2. Percentage of successful actions performed on the scaled prototype environment. We examined **35** states and the selected actions of all three agents, i.e., **BC**, **privileged BC** and the *baseline* on all states. An action is classified as successful if it fits the distribution of an expert on a real (noisy) environment. Here, **privileged BC** outperforms other agents thus supports our claim. Please refer to Fig. 7 for an illustration of the above.

	baseline	BC	Privileged BC
Success [%]	14.3	60	**91.4**

Observing the behavior of the *baseline* method, we conclude that it performs exceptionally well when given access to clean measurements (see Fig. 5a). However, it is highly sensitive when provided with noisy observations (see Fig. 5b) where it often fails. Regarding our agents, we can conclude that both **BC** and **privileged BC** agents can indeed imitate the *baseline* when presented with clean observations but do not outperform it (see Fig. 5a and Table 1). In addition, we found that the **BC** agent outperforms the **privileged BC** agent when presented with clean observations. However, as opposed to the privileged BC paradigm, it is unable to generalize when confronted with noisy observations. This sits in line with well know theory on robustness and generalization [21].

4.2 Scaled Prototype Environment Results

Motivated by the results of Sect. 4.1, we continue and compare all methods on a scaled prototype environment. This environment addresses two key aspects of the *sim-to-real* gap. First, the underlying dynamics of our simulation are put to the test, as we cannot fully model the intricate interactions between the soil and the dozer. Second, the usage of a RGBD camera introduces noise and other inaccuracies into the observation space. It is important to notice that **we do not train** our agents in this environment but merely deploy the agents trained in Sect. 4.1 and test whether they can generalize well to this environment.

We present examples of predictions in Fig. 7, and a full trajectory performed by the **privileged BC** agent in Fig. 1 and 6. In addition, we present a quantitative comparison of successful actions performed on the scaled prototype environment in Table 2. These results re-validate our conclusion from Sect. 4.1 that the *baseline* algorithm under-preforms when presented with real noisy observations.

In our quantitative experiments (see Table 2), both **BC** agents outperformed the *baseline*. However, only the **privileged BC** agent learned a robust feature extractor. This enabled it to solve the task with satisfying performance (over 90%). and empha-sizes the importance and benefit of leveraging *privileged learning* within our training procedure.

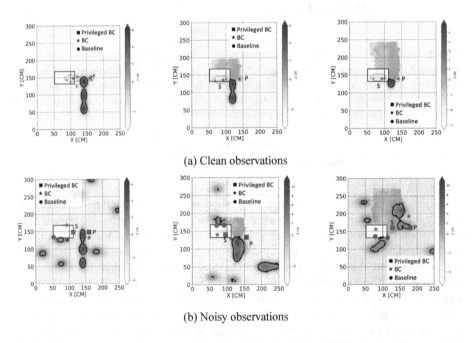

(a) Clean observations

(b) Noisy observations

Fig. 5. We compare the various agents, (*baseline*, **BC** and **privileged BC**), in simulation, on both clean and noisy observations. We observe that when evaluated on clean observations (Fig. 5a), all agents are capable of predicting actions with a similar intention to that of the *baseline*'s. On the other hand, when provided with noisy observations (Fig. 5b) both the *baseline* and **BC** agents fail to detect the sand piles as opposed to the **privileged BC** agent.

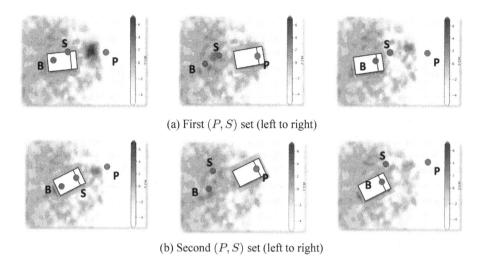

(a) First (P, S) set (left to right)

(b) Second (P, S) set (left to right)

Fig. 6. A trajectory collected from our scaled prototype environment including one sand pile using our **privileged BC** agent. Each row is a different sequence of (P, S) actions chosen by our agent. The sequence is shown from left to right. The P, S and B points shown in Fig. 3 are marked in red, green and blue respectively (Color figure online)

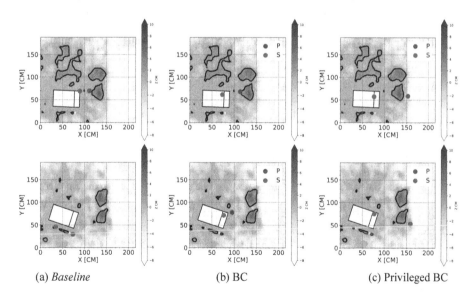

(a) *Baseline*   (b) BC   (c) Privileged BC

Fig. 7. Evaluation of various methods on real-world data. The dozer's location appears as a white element, the heightmap represents the height of each coordinate, sand pile are inscribed by black contours and the green and red dots are the actions predicted by the agent. **(a)** the *baseline* fails to generalize and differentiate between sand piles. **(b)** The **BC** agent learns to imitate the *baseline* on the clean data, and is thus prone to the same mistakes. **(c)** the **privileged BC** agent learns a robust policy that is capable of operating on the real-world data. (Color figure online)

5   Conclusions

In this work, we showed the importance of automating tasks in the field of construction, specifically the grading task. We argued that data collection in the real-world is not only expensive but often infeasible. Our proposed simulation is beneficial for two reasons. First, it enables generation of diverse scenarios. Second, it allows evaluation of planning policies prior to deployment in the real -world. Our suggested *baseline* approach, performed well in an ideal environment. However, it fails when presented with real observations that include inaccuracies. A similar behaviour was observed by the **BC** agent trained with clean observations. By combining the *baseline* and our simulation environment, using privileged learning, we were able to learn a robust behavior policy capable of solving the task. The **privileged BC** agent was the only one able to solve a complete grading task in our scaled prototype environment.

References

1. Caterpillar: D6/D6 XE track-type tractors specifications. https://s7d2.scene7.com/is/content/Caterpillar/CM20181217-51568-10948
2. Cundall, P.A., Strack, O.D.L.: A discrete numerical model for granular assemblies. Géotechnique **29**(1), 47–65 (1979)
3. He, K., Zhang, X., Ren, S., Sun, J.: Deep residual learning for image recognition. In: Proceedings of the IEEE Conference on Computer Vision and Pattern Recognition, pp. 770–778 (2016)
4. Hirayama, M., Guivant, J., Katupitiya, J., Whitty, M.: Path planning for autonomous bulldozers. Mechatronics **58**, 20–38 (2019)
5. Kim, W., Pavlov, C., Johnson, A.M.: Developing a simple model for sand-tool interaction and autonomously shaping sand. arXiv:1908.02745 (2019)
6. Kurinov, I., Orzechowski, G., Hämäläinen, P., Mikkola, A.: Automated excavator based on reinforcement learning and multibody system dynamics. IEEE Access **8**, 213998–214006 (2020)
7. Loquercio, A., Kaufmann, E., Ranftl, R., Müller, M., Koltun, V., Scaramuzza, D.: Learning high-speed flight in the wild. Sci. Robot. **6**(59), eabg5810 (2021)
8. Miron, Y., Coscas, Y.: S-flow GAN (2019)
9. Peng, X.B., Andrychowicz, M., Zaremba, W., Abbeel, P.: Sim-to-real transfer of robotic control with dynamics randomization. In: 2018 IEEE International Conference on Robotics and Automation (ICRA), pp. 3803–3810. IEEE (2018)
10. Pinto, L., Davidson, J., Sukthankar, R., Gupta, A.: Robust adversarial reinforcement learning. In: International Conference on Machine Learning, pp. 2817–2826. PMLR (2017)
11. Rao, K., Harris, C., Irpan, A., Levine, S., Ibarz, J., Khansari, M.: RL-CycleGAN: reinforcement learning aware simulation-to-real. In: Proceedings of the IEEE/CVF CVPR, pp. 11157–11166 (2020)
12. Reece, A.R.: Introduction to terrain vehicle systems: M. G. Bekker, Univ. of Michigan Press, Ann Arbor (1969). $27.50. J. Terramech. **7**, 75–77 (1970)
13. Romero-Ramirez, F.J., Muñoz-Salinas, R., Medina-Carnicer, R.: Speeded up detection of squared fiducial markers. Image Vis. Comput. **76**, 38–47 (2018)
14. Ross, C., Miron, Y., Goldfracht, Y., Di Castro, D.: AGPNet-autonomous grading policy network. arXiv preprint arXiv:2112.10877 (2021)

15. Sanchez-Gonzalez, A., Godwin, J., Pfaff, T., Ying, R., Leskovec, J., Battaglia, P.: Learning to simulate complex physics with graph networks. In: International Conference on Machine Learning, pp. 8459–8468. PMLR (2020)
16. Stentz, A., Bares, J., Singh, S., Rowe, P.: A robotic excavator for autonomous truck loading. Auton. Robot. **7**(2), 175–186 (1999)
17. Sulsky, D., Chen, Z., Schreyer, H.: A particle method for history-dependent materials. Comput. Methods Appl. Mech. Eng. **118**(1), 179–196 (1994)
18. Sutton, R.S., Barto, A.G.: Reinforcement learning: an introduction. MIT press (2018)
19. Tessler, C., Efroni, Y., Mannor, S.: Action robust reinforcement learning and applications in continuous control. In: ICML, pp. 6215–6224. PMLR (2019)
20. Vapnik, V., Vashist, A.: A new learning paradigm: learning using privileged information. Neural Netw. **22**(5–6), 544–557 (2009)
21. Wiesemann, W., Kuhn, D., Rustem, B.: Robust Markov decision processes. Math. Oper. Res. **38**(1), 1–208 (2013)

Bootstrapping Autonomous Lane Changes with Self-supervised Augmented Runs

Xiang Xiang[(✉)] [ID]

Key Lab of Image Processing and Intelligent Control, Ministry of Education
School of Artificial Intelligence and Automation, Huazhong University of Science
and Technology, Wuhan, China
xex@hust.edu.cn

Abstract. In this paper, we want to strengthen an autonomous vehicle's lane-change ability with limited lane changes performed by the autonomous system. In other words, our task is bootstrapping the predictability of lane-change feasibility for the autonomous vehicle. Unfortunately, autonomous lane changes happen much less frequently in autonomous runs than in manual-driving runs. Augmented runs serve well in terms of data augmentation: the number of samples generated from augmented runs in a single one is comparable with that of samples retrieved from real runs in a month. In this paper, we formulate the Lane-Change Feasibility Prediction problem and also propose a data-driven learning approach to solve it. Experimental results are also presented to show the effectiveness of learned lane-change patterns for the decision making.

Keywords: Perceptual signal · Signal fusion · Data augmentation · Auto labeling · Planning

1 Introduction

Motion planning is the brain of an autonomous vehicle. It handles the reasoning for driving behaviors and is one of the core challenges for making a complete self-driving a reality: to make safe, reliable, smooth, comfortable and efficient plans. It is closely related to decision making: taking inputs from a variety of perception and localization modules, the planner's job is to figure out where the autonomous vehicle should be going and how to get there safely and efficiently. For example, navigation, obstacle avoidance, being predictable to surrounding vehicles' drivers, and so on [9]. However, it is complex to act in the environment of road, vehicles, among other objects. Thus, it may need a data-driven model to plan such action actively: an autonomous vehicle's decision making should be better and better, along with increasing run miles and thus seeing more and more cases with more variety. In this paper, we are interested in a learning-based approach to autonomous vehicle's lane-change planning in particular, with the expectation to supplement the rule-based approach.

© The Author(s), under exclusive license to Springer Nature Switzerland AG 2023
L. Karlinsky et al. (Eds.): ECCV 2022 Workshops, LNCS 13804, pp. 118–130, 2023.
https://doi.org/10.1007/978-3-031-25069-9_9

Specifically speaking, we are interested in answering the question 'Can ego make a lane change (LC)?' at any given time during an autonomous run. Notably, our task is about prediction instead of recognition - simply telling if a happened LC action is safe or not, namely to answer 'Was it a safe LC?' Moreover, this task is not about the ego vehicle's intention to make LC, namely to answer 'Should ego make LC?' at any given time during an autonomous run. Namely, we take the LC intention as granted: 'given a LC intention, can ego make LC?' As a prediction problem, it can be naturally abstracted as a learning problem.

We want an autonomous vehicle to bootstrap itself from its own LC attempts, which are realistically operable to itself. While LC is a commonly-seen maneuver in manual driving, a trivial successful manual LC can be challenging to an autonomous vehicle: although both a driver and a vehicle observe the external road, their internal decision-making mechanisms are quite different. For the same reason, simulated run is also opted out. Our preference is to take advantage of the autonomous lane changes.

Contributions of this paper are summarized as follows:

- We formulate the Lane- Change Feasibility Prediction problem.
- We propose a data-driven learning approach to solve the proposed problem.
- We present experimental results that show the effectiveness of learned lane-change patterns for decision-making.

1.1 Challenge

Unfortunately, autonomous lane changes happen much less frequently in autonomous runs than in manual-driving runs. Normally, an autonomous system's strategy is conservative and the rules are strict, because the action of steering is generally more risky to take and more challenging to succeed than cruising or adjusting the speed in the same lane. In an hour-long run, there are usually only a very few autonomous LC attempts, some of which are even disengaged by the driver. Actually, it is always legitimate to pose the question 'Can ego make LC?'. If we keep asking that question every half a minute, our preference is to have over a hundred of potential autonomous LC attempts in an hour-long run.

Obviously, that is not practical in real runs. As we have excluded virtual runs or namely simulations in analogy to virtual reality, we prefer following the idea of augmented reality to replay the recorded real runs and yet alter the behavior of the ego vehicle - command it to make LC at arbitrary time to see if it will collide with approximately real surrounding vehicles in the records. If it collides, then this autonomous LC attempt fails and vice versa. We call this type of augmented records of autonomous runs **Augmented Runs**. The collision test will become one of the metrics to assign labels for samples that is associated with a specific augmented LC.

1.2 Related Works

There are quite a few published works that share the same task with us, while many share a similar motivation of boostrapping the decision-making ability.

Among those, there exist a series of recent literature from the perspective of reinforcement learning, such as [2,4,5,10,11,14]. While being leaning-based approaches, some of them still make explicit use of rules such as [6,14]. Some works learn to imitate human drivers [8,15]. Bevly *et. al.* has summarized a survey dedicated to lane changing and merging [3] yet not in the sense of autonomous lane-changing. Paden *et. al.* has presented a comprehensive survey on the broader topics of autonomous motion planning [12,13], where the concept of lane graph is examined. Althoff *et. al.* use a simple method for the lane-change prediction in a provable framework of cruise control [1], while Khelfa *et. al.* performs it in German two-lane highways [7], a realistic environment as we have in this work. In addition, instead of evaluating whether it is safe to change lane, the focus in [16,17] is detecting the intention of lane changing.

2   Problem Formulation

In this section, we first get warmed up with some representations of the driving environment and then formulate our task as a learning problem. The representations will become a source of evidences or namely features for learning the LC feasibility.

2.1   States of Lanes and Surrounding Vehicles

In this paper, we use the word **Tunnel** to represent the ego vehicle's situation in the local lanes in the map. Cues related to the map structure are called **Tunnel Features**. Other than the situations on the road, the ego vehicle must also have access to certain information about the environment around it in order to be able to predict the motion of moving objects. The topology of the ego vehicle with its surrounding vehicles is called **Anchor** in this paper. For all dynamic objects, the ego vehicle must have access to the class of the object. This information is vitally important as most prediction models have different algorithmic approaches to vehicles as opposed to pedestrians. Next, the ego need to have information regarding the dynamic objects' current state as well as their position and velocity. We call all those cues **Anchor Features** $\mathcal{F}$ which also include the tunnel features as anchors are built upon tunnels. Notably, there usually are several candidate anchors at a specific time stamp.

2.2   Formulation as a Learning Problem

We define a time interval as a sample. For our task of predicting LC feasibility, we call it a LC **prior interval** I, which does not contain the action of LC. Instead, if its label is positive, there is a subsequent LC action right after the interval. In practice, I is a set of time stamps $I := [t_s, t_e]$ where t_s is the starting time and t_e is the ending time. Particularly, we call the time interval containing a LC action a LC **segment** S. Similarly, S is also a set of time stamps. The boundary time between a LC prior interval and its subsequent LC segment is the LC **action start time** t_{as}.

In an autonomous run, our task of predicting a future LC's feasibility from historic observations is formulated as

Predicting LC Feasibility (PLCF) Problem: at arbitrary time T with the observation I that ends at T and namely given a sample $\mathcal{X} := \mathcal{F}(I)$, suppose $t_{as} := T$, output a label $\mathcal{Y} \in \{True, False\}$ to tell if the ego vehicle is able to complete a safe lane change in S that starts at T.

In the following sections, we will propose our solution to the PLCF Problem using supervised learning as it is a typical prediction problem, while the blocker is assigning the labels. We will use $1.0, -1.0$ to represent $True, False$, respectively.

3 Sample Preparation by Augmented Run

In order to train a supervised model to solve the PLCF problem, first of all we need to build a dataset with samples $(\mathcal{X}, \mathcal{Y})$. In the following, we first explain how features $\mathcal{X}$ can be automatically extracted through a pipeline for a prior interval I, given a time stamp T, from the raw records of real autonomous runs. Then, we elaborate on how the label $\mathcal{Y}$ can also be automatically assigned for I and also $\mathcal{X}$.

3.1 Extracting Anchor Features from Real Runs

This section presents the realization of per-time-stamp part of the feature-extracting function $\mathcal{F}(\cdot)$. We build a raw data processing pipeline as shown in Fig. 1 to extract anchor features including the ego vehicle's cues, the surrounding vehicles' cues, and also certain road structure cues. The raw data are recorded from online motion planning at real autonomous runs. The typical surrounding vehicles include the current-lane or namely the source-lane's front cars (FC) as well as LC target-lane's front cars (TF) and back cars (TB). Note that there might be more than one FC. As we use the centroid of a vehicle to represent it, naturally there is a vehicle *front length* and *back length*. Table 1 summaries those vehicles' features.

Table 1. Vehicle-cues part of the anchor feature list.

Vehicle-related Feature	Ego	TB	TF	FC
Speed	✓	✓	✓	✓
Acceleration	✓	✓	✓	✓
Heading angle	✓	✓	✓	✓
Vehicle front length	✓	✓	✓	✓
Vehicle back length	✓	✓	✓	✓
Distance to ego	×	✓	✓	✓
Relative speed w.r.t. ego	×	✓	✓	✓

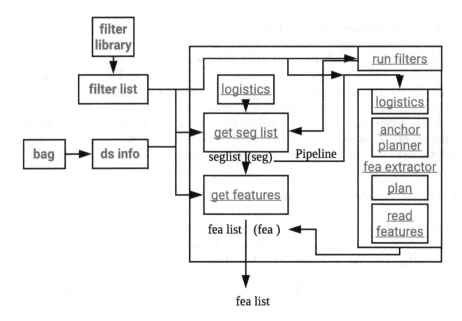

Fig. 1. Visualization of real-run raw data processing pipeline.

Also included as a feature is the time from the current time to *action start time*, which is given as a input. Table 2 summaries the map and lane related features. In correspondence to the *action start time*, we define the *action start distance* as the distance to the starting point after which a LC is allowed; similarly, the *action end distance* is defined as distance to the end point before which a LC is allowed.

Now, we take a closer look at Fig. 1, with Table 3 summarizing the important terms shown in it. A *filter library* is a generic base filter that specific filters can inherit from. Then, a *filter list* is an ordered list of filters aimed at getting a specific class of samples. Moreover, a *seg list* is a generic list of segments containing the core common task-dependent attributes and having list-like operators, say, merge, intersection, difference, union, etc. Lastly, an *anchor planner* is an offline planner that takes in the recorded ego motion at a single time stamp and re-runs offline in a single time stamp. Unlike the online planner running on the fly in real runs, the anchor planner's planned motion for the next time stamp will not affect the ego motion that already has been stored.

3.2   Auto-labeling from Augmented Runs

With $\mathcal{X}$ obtained for I, we propose an auto-labeling algorithm as illustrated in Algorithm 1, which implements the labeling function $\mathcal{L}(\cdot)$ by commanding an augmented LC run in the subsequent segment S and checking if the LC is safe in order to assign a label to I as well as $\mathcal{X}$.

Table 2. Tunnel-features part of the anchor feature list.

Map-related Feature
Current road's speed limit
Estimated source-lane average speed
Estimated target-lane average speed
Action start distance
Action end distance: float
Ahead distance until change if ego take action
Distance to the end of current lane
If there is crossing solid line?

Table 3. Non-rigorous glossary of the data pipeline. Presented are terms that are critical to the understanding the pipeline.

Term	Meaning
Bag	All raw records of a real run
Ds info	A package wraping up a bag's meta data
Filter list	An ordered list of filters
Filter library	A generic base filter
Seg list	A generic list of segments
Anchor planner	An offline planner
Feature list	A generic list of cues of an anchor

Algorithm 1: Auto-labeling $\mathcal{Y} = \mathcal{L}(\mathcal{X}, T)$ where $\mathcal{X} = \mathcal{F}(I)$.

Input : measurement matrix $\mathcal{X} \in \mathbb{R}^{d \times N}$,
time stamp $T \in \mathbb{R}$,
prior interval $I := \{t | t_s \leq t \leq t_e\}$,
anchor feature extractor $\mathcal{F} : I \to \mathcal{X}$

Output: label $\mathcal{Y} \in \{1.0, -1.0\}$

1 Construct a segment $S := \{t | T \leq t \leq T_e\}$ by setting a reasonable segment ending time T_e for S to be able to contain a LC that starts at T.

2 Augmented run by giving the ego vehicle a command to make LC at T and obtain action ending time t_{ae} when the LC completes.

3 Assign a label $\mathcal{Y}$ through checking safety metrics at sampled t_c where $t_{as} \leq t_c \leq t_{ae}$.

Augmented Run. Unlike the anchor planner, the *augmented run planner* is an online planner whose planned motion for the next time stamp will be indeed used as the ego motion at the next time stamp. However, it is still post-processing and thus also different from the online planner running on the fly at real runs. The augmented run planner iterates the following three steps:

1. In the first time stamp, the planner still takes in the recorded ego motion for this time stamp.
2. In the subsequent time stamp, it takes in the previously predicted next-time-stamp motion recorded in the trajectory planned in the previous time stamp.
3. In this online manner, the ego vehicle's motion will be updated incrementally from each time stamp's planned trajectory.

A visualization of an augmented of run of the lane change is shown in Fig. 2 where the yellow point moves across lanes, which does not happen in the recorded real run.

Safety Checking. The safety metrics are two-fold: one is the collision checking; the other is the braking difficulty check for the target-back vehicle. As a necessarily conservative convention, passing both safety checks for all sampled time stamps of LC in S induces $\mathcal{Y} = 1.0$ that denotes a positive sample and thus 'the ego can make LC'; in the other way around, failing any check in any time stamp induces $\mathcal{Y} = 1.0$ which denotes a negative sample and thus 'the ego cannot make LC'.

In summary, given a time stamp T, a sample $(\mathcal{X}, \mathcal{Y})$ can be obtained as $\left(\mathcal{F}(I), \mathcal{L}(\mathcal{F}(I), T)\right)$ through this two-stepped process $\mathcal{L}(\mathcal{F}(\cdot))$ consisting of the anchor feature extraction $\mathcal{X} = \mathcal{F}(I)$ and the auto-labeling $\mathcal{Y} = \mathcal{L}(\mathcal{X}, T)$. Notably, we do not need any anchor feature from the augmented run. Per time stamp T, we just need to prepare features for time stamps in the prior interval I instead of time stamps in the augmented-run segment S. In order to generate a dataset for

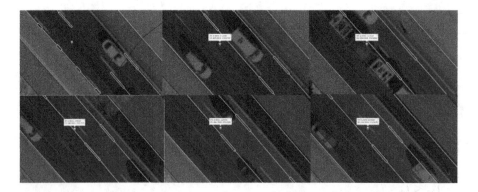

Fig. 2. Visualization of the ego vehicle's position in an augmented lane change. Yellow point denotes the centroid of our ego vehicle; the surrounding vehicles in the pictures are in the map record and do not reflect the real scenarios. The traffic goes from northwest to southeast in the pictures. Ordered from left to right, top to bottom. Observing from the traffic flow direction, the yellow point has moved from being close to the left of one lane center line (orange color) in the first presented frame to being close to the right of its left adjacent lane's center line (orange color as well). Thus, this ego vehicles completes a left LC from one lane center to its left lane center. (Color figure online)

training a model, we can iterate this two-stepped process by inputting a series of T per run and also do the same for a number of runs.

4 Supervised Learning

After the per-time-stamp anchor features are extracted, there still leaves one more step to fully realize the function $\mathcal{F}(\cdot)$ such that $\mathcal{X} = \mathcal{F}(I)$.

4.1 Interval-Level Feature Aggregation

After discretizing a prior interval $I := \{t | t_s \leq t \leq t_e\}$ through equally-spaced sampling time stamps, we will have $I := \{t_1, t_2, ..., t_i, ..., t_{N-1}, t_N\}$ that contains a number of equally-spaced discrete time stamps t_i where $i \in 1, 2, ..., N$, $t_1 := t_s$ and $t_N := t_e$. The exact number $N = \frac{t_e - t_s}{\Delta t} + 1$, depending on the interval duration $(t_e - t_s)$ and sampling rate Δt of time stamps.

Then, each time stamp gives a feature vector. Thus, we have a set of features $\mathcal{F} := \{\mathbf{f}_1, \mathbf{f}_2, ..., \mathbf{f}_i, ..., \mathbf{f}_{N-1}, \mathbf{f}_N\}$ where $\mathbf{f}_i \in \mathbb{R}^d$. In Algorithm 1, to be generic, we represent such a feature set as an column-wise arrangement of feature vectors so we form a measurement matrix $\mathcal{X} \in \mathbb{R}^{d \times N}$ and leave the specific implementation internally for $\mathcal{X}$ open. As the focus of this paper is not the time-series feature aggregation, in or experiments we choose the simple approach of feature averaging:

$$\bar{\mathbf{f}} = \frac{1}{N} \sum_{i=1}^{N} \mathbf{f}_i \tag{1}$$

no matter how many time stamps one interval has. In the literature, mean pooling has been widely used for temporal feature aggregation and turns out to effective. The underlying premise is that the features are homogeneous. Then, the mean vector of a set of feature vectors can be a reasonable summarization. The anchor features in a short interval are likely to be similar and then the feature vectors are expected to be homogenous. Till now, we have given a full realization of $\mathcal{F}(\cdot)$ and now we have $\mathcal{X} = \bar{\mathbf{f}}$.

4.2 Classification

We expect decision tree based classifier models to be effective to learn the discriminative information that separates positive samples from negative ones. The variations include Extra Tree, Emsemble Extra Tree, Random Forest, AdaBoost and Gradient Boost, among others. At training, we solve a set of equations

$$\left\{ \mathcal{Y}_k = \mathcal{C}(\bar{\mathbf{f}}_k) \right\}_{k=1}^{M} \tag{2}$$

for the parameters of a parameterized model $\mathcal{C}$ where $k \in \{1, 2, ..., M\}$ and M is the number of total training samples we have. Once $\mathcal{C}(\cdot)$ is realized, at testing, given a test sample $\bar{f}^{tst}$, our model's prediction is $\mathcal{Y}^{tst} := \mathcal{C}(\bar{f}^{tst})$.

Connecting to our PLCF Problem detailed in Sect. 2.2, given the k-th time stamp T_k^{trn} in the training set with the observation I_k^{trn}, we solve

$$\left\{ \mathcal{L}\left(\mathcal{F}(I_k^{trn}), T_k^{trn}\right) = \mathcal{C}\left(\mathcal{F}(I_k^{trn})\right) \right\}_{k=1}^{M} \tag{3}$$

for $\mathcal{C}$, the only unknown, with $\mathcal{F}$ and $\mathcal{L}$ already realized. Once $\mathcal{C}$ is realized, given a testing time stamp, our model only takes in its observation I_k^{tst} and predicts

$$\mathcal{Y}^{tst} := \mathcal{C}\left(\mathcal{F}(I^{tst})\right) \tag{4}$$

to tell if the ego vehicle is able to start a safe lane change at that testing time stamp. Till now, we have presented our approach to solve the PLCF Problem.

5   Experiments

In this section, we train various supervised models on a gently-sized example dataset, present the experimental results and compare them with training on manual LC runs. We build a dataset with 282 samples drawn from 6 augmented runs (left and right LC) on 3 bags (*i.e.*, recorded autonomous runs). Details of the dataset is summarized in Table 4.

5.1   Performance of Proposed Approach

Table 4. Summary of the example dataset of augmented LC cases. This dataset is built from recorded runs almost in one day.

Category	Full set size	Partial set size
Total	832	324
Positive	140	109
Negative	692	215
Left LC	596	223
Right LC	236	101

Table 5. Comparison of the performance of different classifier when **mixing** samples of left and right LC. The full dataset contains 832 samples in total with 140 positive ones and 692 negative ones. A partial dataset contains 324 samples in total with 109 positive ones and 215 negative ones. Results are averaged over stratified 5-fold cross validation.

Mixed dataset size	832	324
Classifier	Full-set Accuracy	Partial-set Accuracy
Decision Tree	$0.84(+/-0.10)$	$0.78(+/-0.10)$
Linear SVM	$0.71(+/-0.33)$	$0.61(+/-0.13)$
Extra Tree	$0.78(+/-0.19)$	$0.67(+/-0.08)$
Ensemble extra trees	$0.84(+/-0.14)$	$0.78(+/-0.15)$
Random forest	$\mathbf{0.88}(+/-0.07)$	$0.81(+/-0.11)$
AdaBoost	$0.86(+/-0.04)$	$0.81(+/-0.08)$
Gradient Boost	$\mathbf{0.88}(+/-0.05)$	$\mathbf{0.83}(+/-\mathbf{0.11})$

Table 5 presents the results when we mix the samples of left LC and right LC, while we think it makes more sense to separately treat the two LC tasks, given the normal traffic rule of slow vehicles on the right. Therefore, we also conduct the experiments on left-LC samples only and right-LC samples only, with results shown in Table 6 and Table 7, respectively.

5.2   Performance of Alternative Approach

As introduced in Sect. 1, we prefer learning on the augmented autonomous runs to learning on the manual-driving runs. However, regardless of our preference, we also collect a dataset from real manual-LC cases: we treat a manual LC case as

Table 6. Comparison of the performance of different classifier on only **left**-LC samples. The full left-LC dataset 596 samples in total with 87 positive ones and 509 negative ones. The partial left-LC dataset contains 223 samples in total with 69 positive ones and 154 negative ones. Results are averaged over stratified 5-fold cross validation.

Left-LC dataset size	596	324
Classifier	Full-set Acc	Partial-set Acc
Decision Tree	$0.88(+/-0.04)$	$0.78(+/-0.17)$
Linear SVM	$0.72(+/-0.58)$	$0.50(+/-0.32)$
Extra Tree	$0.82(+/-0.12)$	$0.65(+/-0.10)$
Ensemble extra trees	$0.85(+/-0.10)$	$0.74(+/-0.20)$
Random forest	$0.88(+/-0.05)$	$0.78(+/-0.19)$
AdaBoost	$\mathbf{0.90}(+/-0.02)$	$0.81(+/-0.05)$
Gradient Boost	$\mathbf{0.90}(+/-0.03)$	$\mathbf{0.83}(+/-0.12)$

Table 7. Comparison of the performance of different classifier on only **right**-LC samples. The full right-LC dataset contains 236 samples in total with 53 positive ones and 183 negative ones. The partial right-LC dataset contains 101 samples in total with 40 positive ones and 61 negative ones. Results are averaged over stratified 5-fold cross validation.

Right-LC dataset size	236	101
Classifier	Full-set Acc	Patial-set Acc
Decision Tree	$0.83(+/-0.13)$	$0.73(+/-0.16)$
Linear SVM	$0.69(+/-0.38)$	$0.59(+/-0.27)$
Extra Tree	$0.72(+/-0.27)$	$0.68(+/-0.17)$
Ensemble extra trees	$0.81(+/-0.18)$	$0.72(+/-0.18)$
Random forest	$\mathbf{0.87}(+/-0.08)$	$0.75(+/-0.20)$
AdaBoost	$0.86(+/-0.09)$	$\mathbf{0.75}(+/-0.11)$
Gradient Boost	$0.86(+/-0.11)$	$\mathbf{0.75}(+/-0.18)$

a positive sample and a disengaged LC case with intention from the autonomous system as a nagative sample, both based on the premise that the human driver sets a golden standard. Different from augmented runs, LC is occasional in real runs. For positive samples, there are averagely 1 manual LC case per run; for negative samples, there are averagely 2 disengaged autonomous LC cases per run. Table 8 the statistics of the full and monthly partial dataset.

We also follow the same protocol of extracting features from the intervals prior to the lane changes, which are either real LC or LC taken over by the driver in this approach. Table 9 summaries the performances on the full dataset and partial datasets as well. However, as the datasets are different, we cannot compare apple to apple.

Table 8. Summary of the dataset of the real LC-intended cases. The full dataset is built from recorded runs in almost three months: February, March and April of 2020. We do not distinguish left and right LC cases when collecting this dataset.

Category	Full-set size	March-set size	April-set size
Total	2116	864	924
Positive	749	299	328
Negative	1367	565	596

Table 9. Comparison of the performance of different classifier on the real LC-intended dataset. 2116 samples in total with 749 positive ones and 1367 negative ones. Results are averaged over stratified 5-fold cross validation.

Dataset	Full dataset	March set	April set
Dataset size	2116	864	924
Classifier	Accuracy	Accuracy	Accuracy
Decisi. Tree	0.83 (+/- 0.02)	0.81 (+/- 0.04)	0.81 (+/- 0.05)
Linear SVM	0.64 (+/- 0.16)	0.65 (+/- 0.12)	0.52 (+/- 0.23)
Extra Tree	0.65 (+/- 0.06)	0.62 (+/- 0.12)	0.66 (+/- 0.06)
En. extra trees	0.75 (+/- 0.02)	0.75 (+/- 0.01)	0.77 (+/- 0.06)
Rand. forest	0.87 (+/- 0.03)	0.84 (+/- 0.05)	0.84 (+/- 0.05)
AdaBoost	0.86 (+/- 0.05)	**0.86** (+/- 0.02)	0.85 (+/- 0.05)
Grad. Boost	**0.89** (+/- 0.04)	**0.86** (+/- 0.04)	**0.86** (+/- 0.03)

6 Conclusion

In this paper, we have formulated the Lane-Change Feasibility Prediction problem and also proposed a data-driven learning approach to solve it. Experimental results are also presented. As shown by the performances, we see the more samples, the higher accuracy. Augmented runs serves well in terms of data augmentation: the number of samples generated from augmented runs in a single one are comparable with that of samples retrieved from real runs in a month.

Notably, the problem formulation and proposed approach presented in this paper for lane change remain generic and can be adapted to other ego vehicle actions such as merging, accepting other vehicle to merge, and so on.

Acknowledgement. This research was supported by HUST Independent Innovation Research Fund (2021XXJS096), Sichuan University Interdisciplinary Innovation Research Fund (RD-03-202108), IPRAI Key Lab Fund (6142113220309), and the Key Lab of Image Processing and Intelligent Control, Ministry of Education, China.

References

1. Althoff, M., Maierhofer, S., Pek, C.: Provably-correct and comfortable adaptive cruise control. IEEE Trans. Intell. Veh. **6**(1), 159–174 (2020)
2. Bae, S., Saxena, D., Nakhaei, A., Choi, C., Fujimura, K., Moura, S.: Cooperation-aware lane change control in dense traffic. arXiv preprint arXiv:1909.05665 (2019)
3. Bevly, D., et al.: Lane change and merge maneuvers for connected and automated vehicles: a survey. IEEE Trans. Intell. Veh. **1**(1), 105–120 (2016)
4. Bouton, M., Nakhaei, A., Fujimura, K., Kochenderfer, M.J.: Cooperation-aware reinforcement learning for merging in dense traffic. In: 2019 IEEE Intelligent Transportation Systems Conference (ITSC), pp. 3441–3447. IEEE (2019)

5. Bouton, M., Nakhaei, A., Fujimura, K., Kochenderfer, M.J.: Safe reinforcement learning with scene decomposition for navigating complex urban environments. In: 2019 IEEE Intelligent Vehicles Symposium (IV), pp. 1469–1476. IEEE (2019)
6. Hong, J., Sapp, B., Philbin, J.: Rules of the road: predicting driving behavior with a convolutional model of semantic interactions. In: Proceedings of the IEEE Conference on Computer Vision and Pattern Recognition, pp. 8454–8462 (2019)
7. Khelfa, B., Tordeux, A.: Understanding and predicting overtaking and fold-down lane-changing maneuvers on european highways using naturalistic road user data. In: 2021 IEEE Intelligent Vehicles Symposium Workshops (IV Workshops), pp. 168–173. IEEE (2021)
8. Kuefler, A., Morton, J., Wheeler, T., Kochenderfer, M.: Imitating driver behavior with generative adversarial networks. In: 2017 IEEE Intelligent Vehicles Symposium (IV), pp. 204–211. IEEE (2017)
9. LaChapelle, D., Humphreys, T., Narula, L., Iannucci, P., Moradi-Pari, E.: Automotive collision risk estimation under cooperative sensing. In: IEEE International Conference on Acoustics, Speech and Signal Processing (ICASSP), pp. 9200–9204. IEEE (2020)
10. Mirchevska, B., Pek, C., Werling, M., Althoff, M., Boedecker, J.: High-level decision making for safe and reasonable autonomous lane changing using reinforcement learning. In: 21st International Conference on Intelligent Transportation Systems (ITSC), pp. 2156–2162. IEEE (2018)
11. Naumann, M., Königshof, H., Stiller, C.: Provably safe and smooth lane changes in mixed trafic. In: 2019 IEEE Intelligent Transportation Systems Conference (ITSC), pp. 1832–1837. IEEE (2019)
12. Paden, B., Čáp, M., Yong, S.Z., Yershov, D., Frazzoli, E.: A survey of motion planning and control techniques for self-driving urban vehicles. IEEE Trans. Intell. Veh. **1**(1), 33–55 (2016)
13. Sadat, A., Casas, S., Ren, M., Wu, X., Dhawan, P., Urtasun, R.: Perceive, Predict, and Plan: safe motion planning through interpretable semantic representations. In: Vedaldi, A., Bischof, H., Brox, T., Frahm, J.-M. (eds.) ECCV 2020. LNCS, vol. 12368, pp. 414–430. Springer, Cham (2020). https://doi.org/10.1007/978-3-030-58592-1_25
14. Wang, J., Zhang, Q., Zhao, D., Chen, Y.: Lane change decision-making through deep reinforcement learning with rule-based constraints. In: 2019 International Joint Conference on Neural Networks (IJCNN), pp. 1–6. IEEE (2019)
15. Xiang, X.: A brief review on visual tracking methods. In: 2011 Third Chinese Conference on Intelligent Visual Surveillance, pp. 41–44. IEEE (2011)
16. Yan, Z., Yang, K., Wang, Z., Yang, B., Kaizuka, T., Nakano, K.: Intention-based lane changing and lane keeping haptic guidance steering system. IEEE Trans. Intell. Veh. **6**(4), 622–633 (2020)
17. Zhang, Y., Lin, Q., Wang, J., Verwer, S., Dolan, J.M.: Lane-change intention estimation for car-following control in autonomous driving. IEEE Trans. Intell. Veh. **3**(3), 276–286 (2018)

W11 - Skin Image Analysis

W11 - ISIC Skin Image Analysis

Skin is the largest organ of the human body, and is the first area of a patient assessed by clinical staff. The skin delivers numerous insights into a patient's underlying health; for example, pale or blue skin suggests respiratory issues, unusually yellowish skin can signal hepatic issues, or certain rashes can be indicative of autoimmune issues. In addition, dermatological complaints are also among the most prevalent in primary care. Images of the skin are the most easily captured form of medical image in healthcare, and the domain shares qualities to standard computer vision datasets, serving as a natural bridge between standard computer vision tasks and medical applications. However, significant and unique challenges still exist in this domain. For example, there is remarkable visual similarity across disease conditions, and compared to other medical imaging domains, varying genetics, disease states, imaging equipment, and imaging conditions can significantly change the appearance of the skin, making localization and classification in this domain unsolved tasks. This workshop served as a venue to facilitate advancements and knowledge dissemination in the field of skin image analysis, raising awareness and interest for these socially valuable tasks. Invited speakers included major influencers in computer vision and skin imaging, along with authors of accepted papers.

October 2022

M. Emre Celebi
Catarina Barata
Allan Halpern
Philipp Tschandl
Marc Combalia
Yuan Liu

Artifact-Based Domain Generalization of Skin Lesion Models

Alceu Bissoto[1,4]([✉]) [iD], Catarina Barata[2] [iD], Eduardo Valle[3,4] [iD], and Sandra Avila[1,4] [iD]

[1] Institute of Computing, University of Campinas, Campinas, Brazil
{alceubissoto,sandra}@ic.unicamp.br
[2] Institute for Systems and Robotics, Instituto Superior Técnico, Lisbon, Portugal
ana.c.fidalgo.barata@tecnico.ulisboa.pt
[3] School of Electrical and Computing Engineering, University of Campinas, Campinas, Brazil
dovalle@dca.fee.unicamp.br
[4] Recod.ai Lab, University of Campinas, Campinas, Brazil

Abstract. Deep Learning failure cases are abundant, particularly in the medical area. Recent studies in out-of-distribution generalization have advanced considerably on well-controlled synthetic datasets, but they do not represent medical imaging contexts. We propose a pipeline that relies on artifacts annotation to enable generalization evaluation and debiasing for the challenging skin lesion analysis context. First, we partition the data into levels of increasingly higher biased training and test sets for better generalization assessment. Then, we create environments based on skin lesion artifacts to enable domain generalization methods. Finally, after robust training, we perform a test-time debiasing procedure, reducing spurious features in inference images. Our experiments show our pipeline improves performance metrics in biased cases, and avoids artifacts when using explanation methods. Still, when evaluating such models in out-of-distribution data, they did not prefer clinically-meaningful features. Instead, performance only improved in test sets that present similar artifacts from training, suggesting models learned to ignore the known set of artifacts. Our results raise a concern that debiasing models towards a single aspect may not be enough for fair skin lesion analysis.

Keywords: Skin lesions · Artifacts · Debiasing · Domain generalization

1 Introduction

Despite Deep Learning's superhuman performance on many tasks, models still struggle to generalize, stalling the adoption of AI for critical decisions such as medical diagnosis.

Skin lesion analysis is no exception. Recent works exposed concerning model behaviors, such as achieving high performances with the lesions fully occluded

© The Author(s), under exclusive license to Springer Nature Switzerland AG 2023
L. Karlinsky et al. (Eds.): ECCV 2022 Workshops, LNCS 13804, pp. 133–149, 2023.
https://doi.org/10.1007/978-3-031-25069-9_10

on the image [6], or exploiting the presence of artifacts (e.g., rulers positioned by dermatologists to measure lesions) to shortcut learning [7]. Moreover, current models fail to cope with underrepresented populations such as Black, Hispanic, and Asian people. Those shortcomings prevent automated skin analysis solutions from wider adoption and from realizing their potential public health benefits.

Domain Generalization (DG), which in computer vision studies how models fall prey to spurious correlations, is yet to be adequately adopted by the medical image analysis literature, partly because medical data often lack the *labeled environments* which are a critical input to most DG techniques. Within a corpus of data, Environments are groups or domains that share a common characteristic (e.g., predominant image color, image capturing device, demographic similarities). In DG research, datasets are often synthetic, creating environments on demand, or multi-sourced, with an environment for each source. Medical data, however, pose special challenges due to their complexity and multi-faceted nature, presenting multiple ways of grouping data, or latent environments whose full annotation is next to impossible. We are interested in adapting DG techniques to benefit those complex and rich tasks, considering those challenges.

We start by allowing the assessment of generalization performance, even when out-of-distribution data are unavailable, using a tunable version of "trap sets" [7]. Next, we infer existing, latent environments from available data, enabling the adoption of robust learning methods developed in the DG literature. Finally, after model training, we select robust features during test-time, censoring irrelevant information. Our extensive experiments show that it is possible to obtain models that are resilient to training with highly biased data. Code to reproduce our experiments is available at https://github.com/alceubissoto/artifact-generalization-skin.

Our main contributions are:

- We propose a method to adapt existing annotations into environments, successfully increasing the robustness of skin lesion analysis models;
- We propose a test-time procedure that consistently improves biased models' performance;
- We show that model debiasing is insufficient to increase out-of-distribution performance. Better characterization of out-of-distribution spurious sources is necessary to train more robust models.

2   Background

Domain Generalization (DG) and Domain Adaptation (DA) aim to study and mitigate distribution shifts between training and test data for a known (in the case of DA) or an unknown (in the case of DG) distribution test data distribution. Here we focus on DG techniques since the test distribution is almost always unknown for medical analysis.

A complete review of the extensive literature on DG is outside the scope of this work. We point the reader to two recent surveys of the area [33,43]. In this section, we will briefly review the two techniques directly used in this work.

DG techniques are contrasted with the classical **Empirical Risk Minimization** [37] **(ERM)** learning criterion, which assumes that the samples are independent and identically distributed (i.i.d.) and that train and test sets are sampled from the same distribution. For the sake of completeness, the ERM minimization goal is defined as $R_{\mathrm{ERM}}(\theta) = \frac{1}{n}\sum_{i=1}^{n}\ell(x_i, y_i; \theta)$, where ℓ is the classification loss, θ is the model's parameters, and n is the number of samples (x, y). DG techniques deal with train-test distribution shifts. We present two of them below.

Distributional Robust Optimization (DRO) [19,30] methods minimize the maximum risk for all groups (while ERM minimizes the global average risk). That way, the model focuses on high-risk groups, which usually comprise those with correlations underrepresented in the dataset. The risk is calculated as:

$$R_{\mathrm{DRO}}(\theta) := \max_{e \in \mathcal{E}_{\mathrm{tr}}} \hat{\mathbb{E}}_{P^e}[\ell(x, y; \theta)], \tag{1}$$

where we evaluate the expectation separately for each environment distribution P^e, and the data is separated into environments e, sampled from the set of all environments available for training $\mathcal{E}_{\mathrm{tr}}$.

DRO can prevent models from exploiting spurious correlations, for example, if the risk is low for a biased group and high for an unbiased group. In that case, success depends on groups being separated by bias. DRO can also raise the importance of small groups (e.g., rare animal subspecies, rare pathological conditions), which would be obliterated by averaging. DRO techniques require explicitly labeled environments, and one of our main contributions is evaluating one of them (GroupDRO) on inferred environments.

Representation Self-challenging (RSC) [20] is a three-step robust deep learning training method. At each training iteration, RSC sets to zero the most predictive part of the model representation, according to the gradients. More specifically, the model representations with the highest gradients will be set to zero before the model update. Such feature selection causes less dominant features in the training set to be learned by the model, potentially discarding easy-to-learn spurious correlations and thus preventing the so-called *shortcut learning* [32]. We use RSC as a strong baseline for comparing with our proposed pipeline, since this technique does not require environment labels, being adaptable to any classification problem. In a recent benchmark [40], RSC appears as one of the few effective methods, including for the PatchCamelyon histopathology dataset [4,22].

3 Methodology

The main objective is to learn more robust skin lesion representations using deep learning for skin lesion analysis, considering the binary problem of melanoma *vs.* benign. To achieve this, we present a pipeline (Fig. 1) that proposes 1) partitioning data into train/test trap sets that simulate a highly biased scenario; 2) crafting and exploiting training data partitions (environments) to learn robust representations through GroupDRO [30]; 3) selecting task-relevant features for inference, avoiding spurious ones.

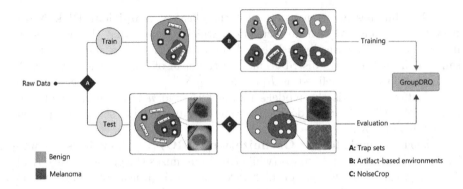

Fig. 1. Our proposed pipeline for debiasing. Images with artifacts are represented by the drawings of rulers, dark corners, and the combination of both. White circles represent samples where artifacts are absent. Our first step (A) partitions data into challenging train and test sets, called Trap sets. The training set is divided into environments (B). Each environment groups samples containing the same set of artifacts. These environments are used to train a robust learning algorithm, such as GroupDRO. In the last step (C), we select features of our trap-test set, censuring the background which may provide spurious correlations.

3.1 Trap Sets

Since spurious correlations inflate metrics, DG methods require carefully crafted protocols to measure generalization. Often, datasets introduce correlations with the class labels (using an extraneous feature, such as color in ColorMNIST [2]), which purposefully differ between the training and the test split.

Here, we follow the "trap set" procedure [7] to craft training and test with **amplified** correlations between artifacts and class labels (malignant *vs.* benign), which appear in **opposite** directions in the dataset splits. We adapt the trap set protocol, introducing a tunable level of bias, from 0 (randomly selected sets) to 1 (highly biased). This level controls, for each sample in the split, the probability of selecting it at random *versus* following the trap set procedure. Table 1 illustrates the correlations between artifact and class labels on the splits for the bias levels used in this work. We think our trap sets can expand generalization measurements to be used outside of specialized literature, reaching problems that urgently need out-of-distribution performance assessment.

3.2 Artifact-Based Environments

In DG, environments divide data according to spurious characteristics. For example, ColorMNIST [2] is divided into two environments: one that correlates colors with values (one color for each digit), and another with colors chosen randomly. Some environments correspond to data sources, as in PatchCamelyon [4,22], where each environment comprises data collected at the same hospital.

Table 1. Spearman correlations between diagnostic and each of the 7 considered artifacts to build the trap sets. As the factor increase, so does the correlations and differences between train and test.

factor	set	dark corner	hair	gel border	gel bubble	ruler	ink	patches
0	train	0.119	-0.104	0.003	0.055	0.142	0.023	-0.138
	test	0.135	-0.112	0.023	0.047	0.162	0.030	-0.149
0.5	train	0.233	-0.185	0.083	0.052	0.246	0.048	-0.110
	test	-0.129	0.083	-0.156	0.038	-0.074	-0.025	-0.217
1.0	train	0.36	-0.282	0.178	0.056	0.352	0.096	-0.062
	test	-0.438	0.296	-0.35	0.049	-0.335	-0.113	-0.319

Arjovsky et al. [2] mention that environments act to "reduce degrees of freedom in the space of invariant solutions". Thus, more environments help discard spurious features during training [29]. The plethora of concepts available enables multiple ways of dividing the dataset into environments, some of which will be more successful than others at achieving robust representations.

Many annotated concepts could be used for environment generation for skin lesion datasets. Recently, Daneshjou et al. [15] released a clinical skin lesion dataset presenting per-image specialist annotated information on Fitzpatrick skin types. Other metadata such as anatomical location, patient sex, and age are available for some datasets, such as the ISIC2019 [12]. In this work, we use the presence of artifacts in the image capture process to create the environment. The presence of those artifacts gives models the opportunity to exploit spurious correlations in order to shortcut learning [7,12,39]. We aim to prevent that, creating more robust models.

The 7 artifact types (see Table 1) may co-occur in lesions, with $2^7 = 128$ combinations. Adding the binary class label, that gives 256 potential environments (e.g., benign with no artifacts, benign with dark-corners, malignant with dark-corners and rulers, etc.), although some of those may contain very few (or zero) images. We use non-empty environments to train a robust learning algorithm.

Our risk minimization of choice is Group Distributionally Robust Optimization (GroupDRO) [30], a variation of DRO that includes more aggressive regularization, in the form of a hyperparameter to encourage fitting smaller groups, higher ℓ_2 regularization, and early stopping. It is a good fit due to our setting with many few-samples environments.

3.3   NoiseCrop: Test-Time Feature Selection

The last step in the pipeline is selecting robust features for inference. Recent work [8] shows that test-time feature selection yields considerable gains in performance, even when spurious correlations are learned.

In this step, we censor the input images' information to prevent models from using spurious features. We employ segmentation masks to separate foreground lesions, which host robust features, from background skin areas, which concentrate spurious information (e.g., skin tones, patches, and image artifacts).

We employ the ground-truth segmentation masks when available, and infer the segmentation (with a Deep Learning model [10]) when they are not. Since we post-process all masks through a convex hull operation, masks do not need to be pixel-perfect, instead they must roughly cover the whole lesion. To minimize the effect of the background pixels on the models, we replace them with a noisy background sampled uniformly from 0 to 255 in each RGB channel. We also eliminate *lesion size* information since the lack of scale guidelines for image capture makes size an unreliable feature subjected to spurious correlations. The convex hull of the segmentation mask is used to crop and re-scale the image such that lesion occupies the largest possible area while keeping the aspect ratio. We call those censoring procedures **NoiseCrop** (Fig. 2). Again, we stress, this censoring is applied only to test images.

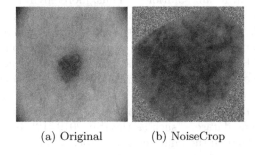

(a) Original   (b) NoiseCrop

Fig. 2. Comparison between Original and NoiseCrop images. In NoiseCrop, we remove the background information, replace it with a uniform noise, and resize the lesion to occupy the whole image.

4   Results

4.1   Data

We employ several high-quality datasets in this study (Table 2). The class labels are selected and grouped such that the task is always a binary classification of melanoma *vs.* benign (other, except for carcinomas). We removed from all analysis samples labeled basal cell carcinoma or squamous cell carcinoma. In the out-of-distribution test sets, we kept only samples labeled melanoma, nevus, and benign/seborrheic keratosis.

The artifact annotations [7] comprise 7 types: *dark corners (vignetting)*, *hair, gel borders, gel bubbles, rulers, ink markings/staining*, and *patches* applied to the patient skin. Ground-truth labels for those are available for the ISIC2018 [36] and

Table 2. Datasets used in our work

Dataset	# Samples	Classes	Set	Type
ISIC2019 (train) [11]	12,360	melanoma *vs.* nevus, actinic keratosis, benign keratosis, dermatofibroma, vascular lesion	training	dermoscopic
ISIC2019 (val) [11]	2,060	as above	validation	dermoscopic
ISIC2019 (test) [11]	6,182	as above	test	dermoscopic
PH2 [24]	200	melanoma *vs.* nevus, benign keratosis	test	dermoscopic
Derm7pt-Dermoscopic [21]	872	as above	test	dermoscopic
Derm7pt-Clinical [21]	839	as above	test	clinical
PAD-UFES-20 [27]	531	as above	test	clinical

Derm7pt [21]. For the larger ISIC2019, we infer those labels using independent binary per-artifact classifiers fine-tuned on the ISIC2018 annotations[1].

4.2 Model Selection and Implementation Details

Hyperparameter selection is crucial for DG. Following GroupDRO [30] protocol, we first performed a grid-search over learning rate (values 0.00001, 0.0001, 0.001), and weight-decay (0.001, 0.01, 0.1, 1.0), for 2 runs, on a validation set randomly split from the training set. Although GroupDRO suggests an unbiased (equal presence of all artifacts) validation set, we found such constraint unrealistic, since a perfectly unbiased data distribution is impossible to predict at training time. We follow the same hyperparameter search procedures for all techniques, including the baselines. Given the best combination on the validation set, we searched for GroupDRO's generalization adjustment argument among the values [0..5]. Sagawa et al. [30] added that hyperparameter to encourage fitting smaller groups. We provide, to illustrate an upper-bound of GroupDRO's performance, an **oracle** version whose hyperparameters were selected with privileged information from test time.

All models employ a ResNet-50 [17] backbone, fine-tuned for up to 100 epochs with SGD with momentum and patience of 22 epochs. Conventional data augmentation (shifts, rotations, color) is used on training and testing, with 50 replicas for the latter. On all plots, lines refer to the average of 10 runs, with shaded areas showing the standard error. Each run has a different training/validation partition and random seed.

---

[1] Each model is an ImageNet-pretrained Inceptionv4 [35] fine tuned with stochastic gradient descent, with momentum 0.9, weight decay 10^{-3}, and learning rate 10^{-3}, reduced to 10^{-4} after epoch 25. Batch size is 32, with reshuffling before each epoch. Data augmented with random crops, rotations, flips, and color transformations.

4.3   Debiasing of Skin Lesion Models

The trap set protocol partitions train and test in an intentional challenging way that is catastrophic for naive models. Models that exploit spurious correlations in the train "fall in the trap" resulting in very low performance (Table 3).

ERM achieves a ROC AUC of only 0.58, showing that trap sets successfully creates challenging biased train and test sets [7]. Our pipeline consider Group-DRO enabled by our artifact-based environments, followed by the application of NoiseCrop in test images. Debiased methods should produce solutions that are more invariant to the training bias, varying less from low to high bias scenarios.

Table 3. Results for different pipelines on a strong trap test (training bias = 1). Our results considerably surpass the state of the art in that scenario. †Reported from the original, using a ResNet-152 model on the ISIC2018 dataset.

Method	ROC AUC
ERM [37]	0.58
RSC [20]	0.59
Bissoto et al. [7]†	0.54
GroupDRO (Ours)	0.68
Full Pipeline (Ours)	**0.74**

Our solution reaches 0.74 AUC in the most biased scenario, while the ERM baseline performs not much better than chance—a difference of 16 percentage points. Other robust methods that do not make use of environments (RSC and Bissoto et al. [7]) failed to improve over ERM. To the best of our knowledge, this is the first time debiasing solutions succeed for skin lesion analysis.

Summary: Our pipeline is an effective strategy for debiasing, surpassing baselines and previous works by 16% points in high-bias scenarios.

4.4   Ablation Study

Next, we provide an ablation of our pipeline, individually evaluating the effects of the robust training enabled by our artifact-based environments and Noise-Crop. We consider increasingly high training biases to check the differences of performances in low and high biased scenarios. We show our results in Fig. 3: performances on the right inferred over NoiseCrop images, and without it (original images) on the left.

Artifact-Based GroupDRO. GroupDRO increases the robustness to artifacts, yielding an improvement of around 10 percentage points in the AUC metric for high-bias scenarios. In such biased contexts, trap sets punish the model for relying on the artifacts, causing both ERM and RSC to fall under 0.6 AUC. In low-bias scenarios, GroupDRO prevents models from relying on artifacts,

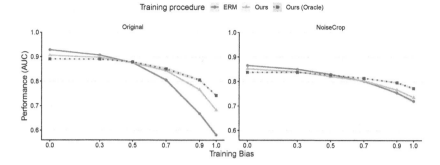

Fig. 3. Ablation study of our method. Each line represent a training method. On the left, we perform inference with original test samples, and in the right, we use NoiseCrop for inference. The Oracle serves as an upper-bound, where we ran our pipeline with access to the test distribution for hyperparameter decision. All methods are evaluated at our trap sets with increasing bias. Our artifact-based environments enable GroupDRO to improve robustness, and NoiseCrop improved robustness of all methods.

causing the performance to drop compared to the baseline. When using privileged information to select hyperparameters for GroupDRO, our oracle reached 0.77 AUC. In the DG literature, deciding hyperparameters is a crucial step, and it is not uncommon to see methods completely fail when hyperparameters are chosen without privileged information over unbiased sets [1,16]. We believe that our fine-grained environments considering each possible combination of artifacts allowed for more robustness to hyperparameter decision.

Test-Time Debiasing. To complete our proposed pipeline (as Fig. 1), we perform feature selection on inference-time. Unlike the direction usually pursued in the literature [2,30], our debiasing method does not require altering any procedure during training. The idea is to select the features present in the image *during test evaluation*, forcing the network to use the correct correlations learned to make the prediction. In Fig. 3 (right), the scenario drastically changes when the same networks from the left of the figure are tested with NoiseCrop images, especially for the most biased scenario. The ERM model, which was slightly better than chance when classifying skin lesions with unchanged images, surpassed 0.72 AUC when evaluated with NoiseCrop images. Composing this test procedure alongside robust training methods further improves performance, achieving our best result. The reported harm in the performance for less biased scenarios can be illusory since exploiting biases naturally translates to better in-distribution performance but less generalization power.

The steep increase in performance when using NoiseCrop test samples with the baseline model suggests that the network learns correct correlations even when training is heavily contaminated with spurious correlations, contrary to previous belief [28]. Still, we achieve our highest performance by using the debiasing procedure (through GroupDRO) and the NoiseCrop test. As in our pipeline, training and test-time debiasing are necessary to create more robust models.

Test-time debiasing appears as a quick effective method to increase robustness at the cost of using domain knowledge of the task. The main challenge is to make test-time debiasing more general, relying less on existing annotations, such as the segmentation masks we use for skin lesion images.

Summary: Artifact-based GroupDRO is an effective strategy for debiasing, and masking artifacts (spurious correlations) during test enable correct features to be used for inference. Our ablation suggests that models still learn robust predictive features even when trained on highly-biased data, but are ignored when known spurious correlations appear during test-time.

4.5   Out-of-Distribution Evaluation

We have previously shown the increased robustness of skin lesion analysis models when training with our artifact-based environments and NoiseCrop test samples. Now, we investigate the effect of the acquired robustness on out-of-distribution sets, which present different artifacts and attributes. Does robustness to the annotated artifacts cause models to rely more on robust features in general? We show our results in Fig. 4.

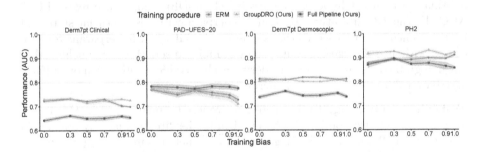

Fig. 4. The different lines compare the ERM baseline, our environment-enabled Group-DRO, and our full pipeline. We train the models with increasingly high biased sets (trap train). We evaluate the performance on 4 out-of-distribution test sets comprising clinical and dermoscopic samples. Unlike the plots using trap test for evaluation, trends here are subtler. The debiasing procedure improves performances on PH2, PAD-UFES-20, and for biased models on Derm7pt-Clinical. On Derm7pt-Dermoscopic, baselines still perform better, despite all the bias in train.

The performances on out-of-distribution test sets are more stable than on trap tests across training biases. This is because trap-test contains opposite correlations from training, punishing the model for learning the encouraged spurious correlations. Still, PH2 and PAD-UFES-20 lines show slight negative and positive trends, respectively, indicating the presence and exploitation of biases.

Our results show very noisy out-of-distribution performance according to the technique used. Our full pipeline present consistent advantage for PAD-UFES-20, while presenting lower performances at all other cases. Interestingly, when we skip NoiseCrop, using only GroupDRO for debiasing, we achieve positive results for all training biases in PH2, and for high training biases in Derm7pt Clinical. For Derm7pt-Dermato, the robust training procedure yielded no gains.

The differences between artifacts present in training (which are increasingly reinforced as training bias increases) and test may explain such irregular behavior. Analyzing the artifacts of each out-of-distribution test-set, we verified that the datasets most affected by the debiasing procedures reliably display a subset of the artifacts present on training. Specifically, PH2 presents dark corners, while PAD-UFES-20 display ink-markings. Derm7pt present rare cases of dark corners, and different style of rulers. Hair is the only artifact in all 4 test sets, while patches, and gel borders are absent in all sets. In Fig. 5, we show a selection of the artifacts from each considered out-of-distribution test-set. In such scenario, the models appear to learn to avoid known artifacts from training environments instead of learning to rely on clinically-relevant features.

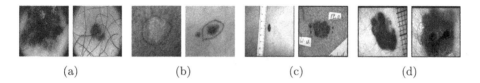

(a)   (b)   (c)   (d)

Fig. 5. Artifacts from the out-of-distribution test sets. While (a) PH2 and (b) PAD-UFES-20 present similar artifacts to ISIC2019 (our training set), Derm7pt ((c) Clinical and (d) Dermoscopic) present different ones. We hypothesize this caused debiasing solutions to be more effective in PH2 and PAD-UFES-20.

Another possible explanation for such variation is hinted by the overall low performance of NoiseCrop (except for PAD-UFES-20). There is a chance that the low performances are due to the domain shift introduced by the background noise, but this is unlikely since such shift did not affect our ISIC2019 experiments, where NoiseCrop reliably achieved our best performances. A more concerning and plausible explanation is that when censored of background information, models can not exploit other available sources of spurious correlations. Such spurious correlations are present in training and may even have very low correlations to the label. In addition, the natural distribution shift of correct features that happen in out-of-distribution sets, cause performances to drop. It is possible then, that the performance achieved by ERM and GroupDRO are overoptimistic. This shows the challenges of debiasing skin lesion models, agreeing with previous works [7] that suspected models combine weak correlations from several sources that may be hard to detect. For further advancing debiasing, future datasets must explicitly describe possible sources of spurious correlations [14].

Summary: When considering biased training scenarios, our proposed debiasing solutions surpassed baselines in 3 out of 4 test sets. Still, improvements depend on the similarity between the confounders used to partition environments and the ones present in test. Models fail when background is censored.

4.6 Qualitative Analysis

To inspect the effects from another angle, we used ScoreCAM [38] to create saliency maps[2]. We contrast our robust trained model with the ERM solution on the most biased scenario (training bias 1.0). In Fig. 6, we show cherry-picked malignant cases from the trap-test set that were misclassified by the ERM or GroupDRO, and that focused on an artifact. There are numerous samples in which the saliency maps indicate that ERM models focus on rulers. When trained with GroupDRO, models often correctly shift their attention to the lesion, causing the prediction to be correct. There are also cases where the baseline's attention correctly focuses on a lesion (even though the prediction is erroneous) and the robust model focuses on the artifact, but these are considerably less frequent.

Fig. 6. Qualitative analysis of malignant samples from the trap-test. We show three sets, each showing the original image followed by ScoreCam saliency maps of ERM and GroupDRO models (ours), in this order. Red (dashed) and blue (solid) borders mark wrong and correct predictions, respectively. In most scenarios, GroupDRO can shift the focus of the model from the artifact to the lesion (first two cases). However, there are still failure cases where the opposite happens (last case). (Color figure online)

5 Related Work

Artifacts on Skin Lesion Datasets. Artifacts affect skin-lesion-analysis models, which achieve a performance considerably higher than chance in images with the lesion fully occluded [6]. Generative models can amplify such biases [25]. Further investigation [7] analyzed the correlations between artifacts and labels, showing that, even with modest correlations, artifacts harmed performances. An analysis of the ISIC 2019 challenge [12] quantified the error rates of the top-ranked models when artifacts were present, finding that ink markings were particularly harmful for melanoma classification. Another work [14] recommended that future skin-lesion datasets describe artifacts and other potential confounders as metadata.

Evaluation of Generalization Performance. Out-of-distribution performance must be measured in challenging protocols, whose craft is laborious requiring attention to class proportions, correlations to other objects, or to background

---

[2] To minimize stochastic effects in the saliency maps, we compare models trained with the same random seed.

colors, textures, and scenes (e.g., ObjectNet [5], ImageNet-A [18]). Partially or fully synthetic datasets (e.g., natural images on artificial backgrounds) allow fine control of the spurious correlations and are often employed in more theoretical works, or as a first round of evaluations elsewhere [1,2]. An alternative to synthetic or handcrafted datasets is to employ naturally occurring environments (such as data source) and split the data holding out some environments exclusively for testing [16,34], e.g., in PatchCamelyon17-WILDS [4], one of the five source hospitals is used for testing, while the others are used for training.

Our assessment scheme forgoes either synthetic data or splitting the sets by hand. Our scheme requires (ground-truth or inferred) annotations for potential bias sources (such as the artifacts we use in this work), but once those are available, the trap sets automatically amplify their effect by creating train and test splits with inverse correlations. The tunable trap sets proposed in this work allow controlling a level of bias.

Debiasing Medical Imaging. Environment-dependent debiasing techniques seldom appear in the literature on medical image analysis. That is partly due to the lack of environment annotations, e.g., potential biasing attributes or artifacts. One way to see environments is through the lenses of causality, where they could be thought of as interventions in data [2]. Direct interventions in real-world data can be unfeasible or at least uncommon. For example, collecting the same image under different acquisition devices is uncommon if not for domain generalization purpose studies. Other types of shifts, such as the ones characterized by physical attributes, are impossible to intervene upon. It is impossible, for example, to see how a lesion on the face would be if it were in the palms and soles. Still, ideally, we would have enough environments to explain every source of noise in data, with slight differences between them. When environments are not annotated *a priori*, works develop mechanisms to create them. A common strategy is to assign whole data sources as environments [3,4,22]. However, when this strategy is successful, the different data sources (and environments) characterize only changes in a few aspects, such as the acquisition device. When differences across data sources are considerable, environments differ in many aspects simultaneously, harming debiasing performance. Other methods to generate environments rely on using differences in classes distributions [41], data augmentation procedures [9,42], or generative modeling [23]. After environments or domains are artificially generated through one of the techniques above, robust training use environments for feature alignment.

In our work, we use annotations of artifacts to create environments. Each environment presents a unique combination of artifact and label, yielding over 90 in training environments. Models trained with our environments successfully learned to avoid using artifacts for inference, improving performance in high-bias setups.

6   Conclusion

Debiasing skin lesion models is possible. In this paper, we introduced a pipeline that enables bias generalization assessment without access to out-of-distribution sets, followed by a strategy to create environments from available metadata, and finally, a test-set debiasing procedure. We evaluated our pipeline using a large challenging training dataset and noisy (inferred) artifact annotations.

Our findings suggest that domain generalization techniques, such as Group-DRO, can be employed for debiasing, as long as the environments represent spurious fine-grained differences, such as the presence of artifacts. Also, we showed that models learn a diverse set of features (spurious and robust), even in biased scenarios, and that removing spurious ones during test yields surprisingly good results without any training procedure changes. When we use training and test-time debiasing, we achieve our best result—GroupDRO enabled learning more robust features, while NoiseCrop allows using them during inference. For out-of-distribution sets, the debiasing success depends on the similarity between the artifacts they display and those in training, used to partition environments. Despite potentially learning more robust features with GroupDRO, the presence of different artifacts and spurious correlations in test-time can still bias predictions.

In future work, we envision methods that are less reliant on labels for both environment partition and test-time debiasing. The domain generalization literature is evolving, proposing methods that learn to separate environments solely from data [1,13], but are still to see the same success of supervised approaches. Alternatively to test-time debiasing, methods for model editing [26,31] could enable practitioners to guide models from a few annotated images by making explicit the presence of artifacts and other spurious features.

Acknowledgments. A. Bissoto is funded by FAPESP 2019/19619-7. C. Barata is funded by the FCT projects LARSyS (UID/50009/2020) and CEECIND/00326/2017. E. Valle is partially funded by CNPq 315168/2020-0. S. Avila is partially funded by CNPq 315231/2020-3, FAPESP 2013/08293-7, 2020/09838-0, and Google LARA 2021. The Recod.ai lab is supported by projects from FAPESP, CNPq, and CAPES.

References

1. Ahmed, F., Bengio, Y., van Seijen, H., Courville, A.: Systematic generalisation with group invariant predictions. In: International Conference on Learning Representations (ICLR) (2021)
2. Arjovsky, M., Bottou, L., Gulrajani, I., Lopez-Paz, D.: Invariant risk minimization. arXiv:1907.02893 (2019)
3. Aubreville, M., et al.: Mitosis domain generalization in histopathology images-the midog challenge. arXiv preprint arXiv:2204.03742 (2022)
4. Bandi, P., et al.: From detection of individual metastases to classification of lymph node status at the patient level: the camelyon17 challenge. IEEE Trans. Med. Imaging **38**(2), 550–560 (2018)

5. Barbu, A., et al.: Objectnet: a large-scale bias-controlled dataset for pushing the limits of object recognition models. In: Advances in Neural Information Processing Systems (NeurIPS) (2019)

6. Bissoto, A., Fornaciali, M., Valle, E., Avila, S.: (De)Constructing bias on skin lesion datasets. In: IEEE Conference on Computer Vision and Pattern Recognition Workshops (CVPRW) (2019)

7. Bissoto, A., Valle, E., Avila, S.: Debiasing skin lesion datasets and models? not so fast. In: IEEE Conference on Computer Vision and Pattern Recognition Workshops (CVPRW) (2020)

8. Borji, A.: Contemplating real-world object classification. In: International Conference on Learning Representations (ICLR) (2021)

9. Chang, J.-R., et al.: Stain Mix-Up: unsupervised domain generalization for histopathology images. In: de Bruijne, M., et al. (eds.) MICCAI 2021. LNCS, vol. 12903, pp. 117–126. Springer, Cham (2021). https://doi.org/10.1007/978-3-030-87199-4_11

10. Chen, L.-C., Zhu, Y., Papandreou, G., Schroff, F., Adam, H.: Encoder-decoder with atrous separable convolution for semantic image segmentation. In: Ferrari, V., Hebert, M., Sminchisescu, C., Weiss, Y. (eds.) ECCV 2018. LNCS, vol. 11211, pp. 833–851. Springer, Cham (2018). https://doi.org/10.1007/978-3-030-01234-2_49

11. Codella, N., et al.: Skin lesion analysis toward melanoma detection 2018: a challenge hosted by the international skin imaging collaboration (ISIC). arXiv preprint arXiv:1902.03368 (2019)

12. Combalia, M., et al.: Validation of artificial intelligence prediction models for skin cancer diagnosis using dermoscopy images: the 2019 international skin imaging collaboration grand challenge. Lancet Digital Health 4(5), e330–e339 (2022)

13. Creager, E., Jacobsen, J.H., Zemel, R.: Environment inference for invariant learning. In: International Conference on Machine Learning (ICML) (2021)

14. Daneshjou, R., et al.: Checklist for evaluation of image-based artificial intelligence reports in dermatology: clear derm consensus guidelines from the international skin imaging collaboration artificial intelligence working group. JAMA Dermatol. 158(1), 90–96 (2022)

15. Daneshjou, R., et al.: Disparities in dermatology AI: assessments using diverse clinical images. arXiv preprint arXiv:2111.08006 (2021)

16. Gulrajani, I., Lopez-Paz, D.: In search of lost domain generalization. In: International Conference on Learning Representations (ICLR) (2021)

17. He, K., Zhang, X., Ren, S., Sun, J.: Deep residual learning for image recognition. In: IEEE Conference on Computer Vision and Pattern Recognition (CVPR) (2016)

18. Hendrycks, D., Zhao, K., Basart, S., Steinhardt, J., Song, D.: Natural adversarial examples. In: IEEE Conference on Computer Vision and Pattern Recognition (CVPR) (2021)

19. Hu, W., Niu, G., Sato, I., Sugiyama, M.: Does distributionally robust supervised learning give robust classifiers? In: International Conference on Machine Learning (ICML) (2018)

20. Huang, Z., Wang, H., Xing, E.P., Huang, D.: Self-challenging improves cross-domain generalization. In: Vedaldi, A., Bischof, H., Brox, T., Frahm, J.-M. (eds.) ECCV 2020. LNCS, vol. 12347, pp. 124–140. Springer, Cham (2020). https://doi.org/10.1007/978-3-030-58536-5_8

21. Kawahara, J., Daneshvar, S., Argenziano, G., Hamarneh, G.: Seven-point checklist and skin lesion classification using multitask multimodal neural nets. IEEE J. Biomed. Health Inform. 23(2), 538–546 (2019)

22. Koh, P.W., et al.: Wilds: a benchmark of in-the-wild distribution shifts. arXiv:2012.07421 (2020)

23. Liu, H., et al.: Domain generalization in restoration of cataract fundus images via high-frequency components. In: International Symposium on Biomedical Imaging (ISBI) (2022)

24. Mendonça, T., Celebi, M., Mendonca, T., Marques, J.: Ph2: a public database for the analysis of dermoscopic images. In: Dermoscopy Image Analysis (2015)

25. Mikołajczyk, A., Majchrowska, S., Limeros, S.C.: The (de) biasing effect of gan-based augmentation methods on skin lesion images. In: Wang, L., Dou, Q., Fletcher, P.T., Speidel, S., Li, S. (eds.) MICCAI 2022. LNCS, vol. 13438, pp. 437–447. Springer, Cham (2022). https://doi.org/10.1007/978-3-031-16452-1_42

26. Mitchell, E., Lin, C., Bosselut, A., Finn, C., Manning, C.D.: Fast model editing at scale. In: International Conference on Learning Representations (ICLR) (2022)

27. Pacheco, A.G., et al.: Pad-ufes-20: a skin lesion dataset composed of patient data and clinical images collected from smartphones. Data Brief **32**, 106221 (2020)

28. Pezeshki, M., Kaba, O., Bengio, Y., Courville, A.C., Precup, D., Lajoie, G.: Gradient starvation: a learning proclivity in neural networks. In: Advances in Neural Information Processing Systems (NeurIPS) (2021)

29. Rosenfeld, E., Ravikumar, P., Risteski, A.: The risks of invariant risk minimization. In: International Conference on Learning Representations (ICLR) (2021)

30. Sagawa, S., Koh, P.W., Hashimoto, T.B., Liang, P.: Distributionally robust neural networks for group shifts: on the importance of regularization for worst-case generalization. In: International Conference on Learning Representations (ICLR) (2020)

31. Santurkar, S., Tsipras, D., Elango, M., Bau, D., Torralba, A., Madry, A.: Editing a classifier by rewriting its prediction rules. In: Advances in Neural Information Processing Systems (NeurIPS) (2021)

32. Shah, H., Tamuly, K., Raghunathan, A., Jain, P., Netrapalli, P.: The pitfalls of simplicity bias in neural networks. In: Advances in Neural Information Processing Systems (NeurIPS) (2020)

33. Shen, Z., et al.: Towards out-of-distribution generalization: a survey. arXiv:2108.13624 (2021)

34. Shrestha, R., Kafle, K., Kanan, C.: An investigation of critical issues in bias mitigation techniques. In: IEEE Winter Conference on Applications of Computer Vision (WACV) (2022)

35. Szegedy, C., Ioffe, S., Vanhoucke, V., Alemi, A.A.: Inception-v4, inception-resnet and the impact of residual connections on learning. In: AAAI Conference on Artificial Intelligence (AAAI) (2017)

36. Tschandl, P., Rosendahl, C., Kittler, H.: The ham10000 dataset, a large collection of multi-source dermatoscopic images of common pigmented skin lesions. Sci. Data **5**(1), 1–9 (2018)

37. Vapnik, V.: Principles of risk minimization for learning theory. In: Advances in Neural Information Processing Systems (NeurIPS) (1992)

38. Wang, H., et al.: Score-cam: Score-weighted visual explanations for convolutional neural networks. In: IEEE Conference on Computer Vision and Pattern Recognition Workshops (CVPRW) (2020)

39. Winkler, J., et al.: Association between surgical skin markings in dermoscopic images and diagnostic performance of a deep learning convolutional neural network for melanoma recognition. JAMA Dermatol. **155**(10), 1135–1141 (2019)

40. Ye, N., et al.: Ood-bench: quantifying and understanding two dimensions of out-of-distribution generalization. In: IEEE Conference on Computer Vision and Pattern Recognition (CVPR) (2022)
41. Yoon, C., Hamarneh, G., Garbi, R.: Generalizable feature learning in the presence of data bias and domain class imbalance with application to skin lesion classification. In: Shen, D., et al. (eds.) MICCAI 2019. LNCS, vol. 11767, pp. 365–373. Springer, Cham (2019). https://doi.org/10.1007/978-3-030-32251-9_40
42. Zhang, R., Xu, Q., Huang, C., Zhang, Y., Wang, Y.: Semi-supervised domain generalization for medical image analysis. In: International Symposium on Biomedical Imaging (ISBI) (2022)
43. Zhou, K., Liu, Z., Qiao, Y., Xiang, T., Loy, C.C.: Domain generalization: a survey. arXiv:2103.02503 (2021)

An Evaluation of Self-supervised Pre-training for Skin-Lesion Analysis

Levy Chaves[1,3]([✉]) [iD], Alceu Bissoto[1,3] [iD], Eduardo Valle[2,3] [iD],
and Sandra Avila[1,3] [iD]

[1] Institute of Computing (IC), Campinas, Brazil
{levy.chaves,alceubissoto,sandra}@ic.unicamp.br
[2] School of Electrical and Computing Engineering (FEEC), Campinas, Brazil
dovalle@dca.fee.unicamp.br
[3] Recod.ai Laboratory, University of Campinas, Campinas, Brazil

Abstract. Self-supervised pre-training appears as an advantageous alternative to supervised pre-trained for transfer learning. By synthesizing annotations on pretext tasks, self-supervision allows pre-training models on large amounts of pseudo-labels before fine-tuning them on the target task. In this work, we assess self-supervision for diagnosing skin lesions, comparing three self-supervised pipelines to a challenging supervised baseline, on five test datasets comprising in- and out-of-distribution samples. Our results show that self-supervision is competitive both in improving accuracies and in reducing the variability of outcomes. Self-supervision proves particularly useful for low training data scenarios (<1500 and <150 samples), where its ability to stabilize the outcomes is essential to provide sound results.

Keywords: Self-supervision · Out-of-distribution · Skin lesions · Melanoma · Classification · Small datasets

1 Introduction

Self-supervised learning bridges the gap between supervised learning, which leads to the most accurate models but requires human-annotated samples, and unsupervised learning, which can exploit non-annotated samples but often leads to disappointing accuracies. By using synthesized annotations on so-called *pretext tasks*, self-supervision is able to *pre-train* models on abundant pseudo-labels before tuning them for the downstream target task.

Applications for which annotated data is expensive or scarce—often the case for medical applications—especially benefit from self-supervision [2,32, 43]. Training state-of-the-art Deep Learning models require extensive training datasets, which are seldom available for medical applications. We can mitigate the issue by applying transfer learning, i.e., pre-training the models (with classical supervised learning) on a large, unrelated dataset, and fine-tuning them on the target dataset, but there is a risk that the representations learned during

© The Author(s), under exclusive license to Springer Nature Switzerland AG 2023
L. Karlinsky et al. (Eds.): ECCV 2022 Workshops, LNCS 13804, pp. 150–166, 2023.
https://doi.org/10.1007/978-3-031-25069-9_11

pre-training will not fully adapt to the downstream task [31]. Self-supervised pre-training has proved, thus, advantageous for transfer learning in many tasks, such as object localization [22], speech representation [26], and medical image classification [1,40,49].

In this work, we assess self-supervision pre-training for the automated diagnosis of skin lesions, an application for which traditionally transfer learning from models supervised on ImageNet is employed to mitigate the scarcity of data [31,41]. Our work improves on existing self-supervised applications to medical applications [2,21,40,42] by evaluating performances on out-of-distribution and low data (less than 150 samples in the stringiest case) regimens. We also evaluate adding an intermediate contrastive learning pre-training before performing the traditional fine-tuning protocol.

The main contributions of this work are:

- We assess five self-supervision learning candidates (BYOL, InfoMin, MoCo, SimCLR, and SwAV) against a competitive supervised baseline;
- We perform a systematic assessment of four transfer learning pipelines (the supervised baseline and three self-supervised contenders) in five publicly accessible test datasets, comprising in-distribution and out-distribution scenarios. Our results suggest that self-supervised models present superior performance in both in-distribution and out-of-distribution in almost all evaluated datasets;
- We assess the performance of our pipelines/datasets in a low-data training scenario (with as few as 148 samples). Again, we find performance improvement in favor of self-supervised pre-training.

We organized the remaining text as follows. We discuss the state-of-the-art on self-supervision in Sect. 2, comprising both general works and those dedicated to medical images, and skin lesions in particular. We detail our goals, datasets, pipelines, protocols, experimental design, and implementation details in Sect. 3. Experimental results and analyses appear in Sect. 4. Finally, we discuss our main findings, along with future research directions in Sect. 5.

2   Related Work

Self-supervised learning has attracted growing attention in the past decade, with hundreds of papers published in the past few years. For a comprehensive review, we refer the reader to the survey of Jing et al. [24]. We recommend the survey by Liu et al. [29] for a more fundamental/theoretical viewpoint on a broad scope of techniques. In this section, we will limit ourselves to the methods directly relevant to this work, and to a selection of methods used for medical images and, in particular, for skin-lesion analysis.

2.1   Self-supervised Learning for Visual Tasks

Self-supervised learning pre-trains models on auxiliary **pretext tasks** such as colorizing [48], predicting rotation angles [15], and in-painting [34], before fine-tuning them on the **downstream task**, i.e., the **target task**. This allows pre-learning representations on unlabeled data and then refining those representations on labeled data. The base model in self-supervised learning, called the **encoder**, transforms the input image into the **(latent) representations**. ResNet-50 is often employed as a backbone, due to its ability to conciliating simplicity and accuracy [2,7,9,16,18,47].

A critical breakthrough in self-supervised learning was the adoption of contrastive losses [24,38], which explicitly organize the feature space by bringing together the representations for related (positive) pairs of samples, while pushing apart the representations for unrelated (negative) pairs.

InstDisc [47] is the seminal work on contrastive self-supervised image classification. InstDisc reframes class-level classification as instance-level discrimination: each training sample becomes one label, whose data-augmented views must be recognized against data-augmented views from all other training samples. The challenge is extending the loss for so many labels (millions, in ImageNet), which is conquered by reformulating the softmax loss. An ℓ_2-normalization turns the dot products into cosine similarities. A temperature hyperparameter τ allows regulating the loss concentration. A memory bank caches the parameters for each label/instance. Finally, the softmax is approximated using the Noise-Contrastive Estimation [17], previously successful for training very large word embeddings. The technique creates very compact (128-d) representations, thus making storage and computation for the memory bank feasible, despite the large number of entries.

Instead of using a memory bank, SimCLR [9] employs end-to-end learning [18], adding an auxiliary dimension-reducing network (projection network) after the encoder and generating the representations on the fly for each batch. The pretext task and loss are very similar to InstDisc's, but only the samples present in the batch are considered in the computation of the loss, without resorting to the memory bank. Thus, SimCLR requires very large batch sizes (4096 samples *vs.* InstDisc' 256), and strong data-augmented views in order to be effective.

MoCo [18] proposes a dictionary of representations whose size is a hyperparameter that may be much larger than the batch size (which is limited to the GPU memory) while still being much smaller than the training set as in InstDisc. The entries in the dictionary are the past few batches, updated in a FIFO scheme. Pretext task and loss still work similar to InstDisc and SimCLR, but negative examples are now taken from the dictionary. The parameters for each label are updated using a "momentum" update, which prevents the representations from fluctuating too much. That is reminiscent of the proximal regularization of InstDisc, but the latter acts on the loss instead of the representations. Further, MoCo-V2 [11] added the projection network and strong data-augmented views as in SimCLR's into the original MoCo formulation.

In BYOL [16], one slow network creates targets for a fast network. The parameters of the fast network are learned by backpropagation, and the parameters of the slow network are the exponential moving average of the parameters of the fast network. In that manner, BYOL bootstraps its own target representations.

BYOL still matches data-augmented views between positive pairs as pretext, but without resorting to negative pairs. Instead, it feeds one view to the fast, and the other to the slow network, and uses the cosine distance between the two outputs as loss.

SwAV [7] is an interesting technique that, instead of using instance-based pairwise positive/negative examples, creates pseudo-labels by clustering the representations online, batch by batch. The pretext task is assigning data-augmented views of the same training sample to the same cluster, with an equipartitioning constraint preventing the trivial solution of a single cluster.

In contrast to the techniques above, which use standard data augmentation techniques to create the views of the samples for contrastive learning, InfoMin [39] *learns* how to create the views, using a criterion of minimizing the mutual information between views. The motivation is creating a challenging but feasible pretext task for the model.

2.2   Self-supervised Learning on Medical Tasks

Currently, there are two paths to follow when the matter is using self-supervised learning in medical applications. One is simply to use the same pretext task designed for general purpose computer vision or propose a slightly adapted version of such tasks that fit best into the current medical application. Early medical applications leveraged self-supervised pretext tasks of reconstructing distorted or damaged inputs [6,8,20,32], such as image reconstruction in retinal images [20], context restoration in fetal MRI [8], or depth estimation in monocular endoscopy [30]. Zhou et al. [49], working on X-ray images, employ a domain-general pretext task (the matching of data-augment views of instances of most methods of Sect. 2.1), and uses stronger baselines: both supervised pre-training on ImageNet, and self-supervised pre-training with MoCo [18], still showcasing improvements in downstream image classification. Their technique, *Comparing to Learn*, uses two networks, in knowledge-distillation teacher-student pair, for a momentum encoding scheme somewhat reminiscent of BYOL [16]. MoCo pre-training appears to be widely used in the medical field, bringing superior performance to other medical applications compared to their supervised counterpart for COVID diagnosis [37,43], and pleural effusion classification [10].

On the other hand, the second way to explore self-supervised is to leverage knowledge about the medical domain - by experience or any domain expert involved - and computer vision to design a custom-built pretext task for the target medical application. Suitable pretext tasks are crucial for learning predictive representations, motivating some works to evaluate whether domain-specific might improve self-supervised learning for medical images. For instance, Jamaludin et al. [23] pre-train a Siamese Net with a contrastive loss in which the positive pairs are patches of spinal magnetic resonance images depicting the same vertebrae of a patient across exams, and the negative pairs are corresponding vertebrae in different patients. They found that the scheme improves the prediction of intervertebral disc degeneration. Wenjia et al. [3] use a pretext in which the model has to predict the bounding boxes of anatomic features in heart magnetic resonance images, metadata ordinarily available in the DICOM files. They found improvements in the downstream task of segmenting the heart in the images. An issue with all those works is their choice of baseline, networks initialized with random weights, instead of stronger baselines such as models fine-tuned on ImageNet, or other schemes for self-supervision.

Most close to our work of performing a systematic evaluation, Truong et al. [40] assess four medical classifications and three distinct self-supervised pre-training in similar training regimens that ours but lacks evaluation regarding out-of-distribution performances. Hosseinzadeh et al. [21] evaluate the performance of fourteen self-supervised ImageNet models to a diverse set of tasks in medical image classification and segmentation. Again, out-of-distribution and low-data performance evaluations remained uncovered.

2.3 Self-supervised Learning on Skin Lesion Analysis

Wang et al. [44] employ a clustering pretext-task reminiscent of SwAV [7], but accumulating samples from several small batches, and employing different clustering and losses. Since the downstream task is *unsupervised* learning on the same clusters, although evaluated on the classes of ISIC 2018 Lesion Diagnostic Challenge, this work is in a gray zone between self-supervised and purely unsupervised learning. They found favorable results compared to other clustering techniques, but, not surprisingly, a large penalty compared to works that employ supervised fine-tuning. Segmentation tasks also benefits from self-supervision, such as in Li et al. [28], and Wang et al. [46], which both applied the self-supervised with color-based pretext tasks for segmenting skin lesions.

Most related to our work, Azizi et al. [2] performed a well-designed, systematic evaluation of SimCLR [9], for two medical tasks: skin-lesion analysis on a private dataset of $> 450,000$ teledermatology clinical images, and X-rays on the publicly available CheXpert dataset. Contrasting SimCLR pre-training to two strong supervised pre-training baselines, they find it advantageous for the skin-lesion task, and similar for the X-rays task. Their study is complementary to ours, with two medical tasks, three encoder architectures, three pre-training datasets, and the evaluation of a novel pretext technique for exploiting multiple images of the same clinical case they call Multi-Instance Contrastive Learning. Our study, whose focal point is skin-lesion diagnosis, evaluates five test datasets with in- and out-of-distribution images, three pipelines for self-supervision pre-training (in contrast to a challenging supervised pipeline), and five candidate self-supervision schemes. Verdelho et al. [42] compare only two self-supervised learning approaches both quantitatively and qualitatively, but lacks low-data and out-of-distribution evaluation.

We stand out our work from the ones available in the literature in Table 1 by highlighting the contributions of our experimental design.

Table 1. Overview of related works that evaluate self-supervised *vs.* supervised pre-training.

Work$_{year}$	#Evaluated Methods	Out-of-distribution Evaluation	Low-data Evaluation
Azizi et al. [2]$_{2021}$	2	No	Yes
Hosseinzadeh et al. [21]$_{2021}$	15	No	No
Truong et al. [40]$_{2021}$	5	No	Yes
Verdelho et al. [42]$_{2022}$	2	No	No
Ours$_{2022}$	6	Yes	Yes

3   Materials and Methods

This section details the methodology, comprising the datasets and factors in our experimental design. We also discuss how we conduct the experimental evaluation of all pipelines.

3.1   Datasets

Following ISIC 2020 Challenge [35], our task is melanoma *vs.* benign lesions classification. We evaluate our experiments in five, high-quality, publicly available datasets (Table 2).

Table 2. Description of the datasets used in this work. Mel.: number of melanomas. †Split used for test if omitted

Dataset (split†)	Size	Mel.	Lesion Diagnoses	Other information
isic19 [12] (train)	14 805	3121	Melanoma *vs.* actinic keratosis, benign keratosis, dermatofibroma, melanocytic nevus, vascular lesion	Dermoscopic images
isic19 (validation)	1 931	224	Idem	Dermoscopic images, in-distribution
isic19 (test)	3 863	396	Idem	Idem
isic20 [35]	1 743	581	Melanoma *vs.* actinic keratosis, benign keratosis, lentigo, melanocytic nevus, unknown (benign)	Dermoscopic images, out-of-distribution, additional unknown diagnosis
derm7pt-derm [25]	872	252	Melanoma *vs.* melanocytic nevus, seborrhoeic keratosis	Dermoscopic images, out-of-distribution
derm7pt-clinic [25]	839	248	Idem	Clinical images, out-of-distribution
pad-ufes-20 [33]	1 261	52	Melanoma *vs.* actinic keratosis, Bowen's disease, nevus, seborrheic keratosis	Clinical images, out-of-distribution, additional Bowen's disease diagnosis

We performed all training and validation in splits of the isic19 dataset. We removed samples from isic20 present in the isic19 train/validation splits to avoid contaminating the former. We removed basal cell carcinomas and squamous cell carcinomas from all datasets, leaving melanoma as the only malignant class.

The diversity of test datasets aimed at mitigating bias in evaluation [4,5,14], providing both in-distribution (same dataset, same type of image, same classes) and out-of-distribution (cross-dataset, different types of image, different classes) scenarios.

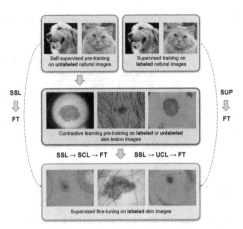

Fig. 1. An overview of our evaluated pipelines. In SSL → FT scheme we contrast the result of five fine-tuned SSL ImageNet pre-trained models on isic19 dataset (see Sect. 3.2) with the supervised counterpart. The SSL → SCL → FT pipeline differs from SSL → UCL → FT according to the employed contrastive loss. They both go through a pre-training stage (see Sect. 3.3)—which can be supervised (SCL) or unsupervised (UCL)—using the isic19 dataset and then performing a supervised fine-tuning. Figure inspired from Azizi et al. [2].

3.2   Experimental Design

We evaluate four alternative pipelines (Fig. 1), which vary in the pre-training and fine-tuning of the model. First, we wish to compare the baseline (supervised pre-training) pipeline (SUP → FT) with the basic self-supervision pipeline (SSL → FT) to establish whether self-supervision is advantageous. In addition, we wish to select a self-supervision scheme among five candidates (BYOL, InfoMin, MoCo-V2, SimCLR, and SwAV) to perform the remainder of the experiments. Selecting the most promising scheme at this stage is necessary for managing the number of experiments, as the next round of experiments will be extensive and, thus, expensive. The SSL → * → FT pipelines have an additional, intermediate pre-training step on the isic19 train split using supervised (SCL) or unsupervised (UCL) contrastive loss (Sect. 3.3).

In the first round of experiments, we attempt a few combinations of hyperparameters for each self-supervision scheme. We purposefully optimize the baseline pipeline more thoroughly to make it challenging. The exact search space appears in Sect. 3.4. We perform all searches on the isic19 validation split to avoid using privileged test information on this step [41]. To estimate the statistical variability of those experiments, we perform five replicates for every experiment, reflecting different random initializations for the training procedures (optimizer, scheduler, and augmentations).

The next round of experiments is a systematic evaluation of all pipelines (Fig. 1) under three data regimens: full training data with 100% of the samples, and low training data with 10, and 1% of the samples. The latter intends to simulate the frequent scenario on medical images of insufficient training data.

For each combination of pipeline and hyperparameter (see Subsect. 3.4), we measure their performance on the isic19 validation split five times, reflecting different random initializations for the training procedures, and, on the low-data experiments, also different random training subsets. We pick the five non-unique best combinations of hyperparameters for each pipeline. For each combination, we perform five replicates on the isic19 test split, resulting in 25 measurements for each pipeline.

3.3   SSL → UCL/SCL → FT Pipelines

These two pipelines investigate the benefit of introducing an additional contrastive learning pre-training step before the traditional fine-tuning. Even though it adds an additional computational cost, works in the literature report some advantages [1,2,13], but their evaluation only consider domains with abundant data availability (compared to ours). To this end, we investigate if the same observed improvements also translate to domains with only a few hundred data available.

We evaluated two contrastive losses:

Unsupervised Contrastive Loss (UCL): we performed the pre-training on the isic19 training set using the self-supervised NT-Xent contrastive loss [9]:

$$\mathcal{L}_{UCL} = \frac{-1}{2N} \sum_{i=1}^{2N} \log \frac{\exp(z_i \cdot z_i^+)/\tau}{\sum_{k \neq i}^{2N} \exp(z_i \cdot z_k)/\tau}, \tag{1}$$

where $z_* = f(x_*)$ is the representation for input x_i output by the encoder f, z_i^+ is the positive pair to z_i, and all vectors z_* are ℓ_2-normalized. The scalar temperature τ hyperparameter regulates the concentration/spreading of the loss. Only the input data is exploited, and class labels are ignored.

Supervised Contrastive Loss (SCL): we performed pre-training using a straightforward extension of the loss above [27], which incorporates class labels by grouping as positive all examples in the same class (instead of just the augmented pair coming from the same instance):

$$\mathcal{L}_{SCL} = \frac{-1}{2N} \sum_{i=1}^{2N} \frac{1}{|Z_i^+|} \sum_{z^+ \in Z_i^+} \log \frac{\exp(z_i \cdot z^+)/\tau}{\sum_{k \neq i}^{2N} \exp(z_i \cdot z_k)/\tau}, \tag{2}$$

where Z_i^+ is the set of all representations that are positive to z_i, and the other symbols are the same as in Eq. 1.

3.4   Implementation Details

We follow standard guidelines for self-supervised learning literature [36] and use ResNet-50 (1×) [19] as base encoder for all experiments. In SUP → FT scheme, we strive to make the baseline challenging, by performing, on the isic19 validation split, a thorough a grid search comprising batch size $(32, 128, 512)$, balanced batches (yes or no), starting learning rate $(0.1, 0.05, 0.005, 0.009, 0.0001)$, and learning rate scheduler (plateau, cosine). The optimizer is the SGD with a momentum of 0.9 and weight decay of 0.001. The plateau scheduler has patience of 10 epochs and a reduction factor of 10.

For the SSL $\rightarrow * \rightarrow$ FT pipelines, we employ two fully-connected layers to embed the ResNet-50 onto 128-dimensional representations, fed to the contrastive loss. We resized the input images to 224×224 and used SimCLR's recommended heavy image augmentation pretexts—color jitter, horizontal and vertical flips, random resized crop, and grayscale. We omitted the Gaussian blur because, for skin lesions, it leads many images to be very similar, harming the results. We used a learning rate of 0.001 with a cosine decay on an Adam optimizer. We perform a grid search through pre-training batch size (80 or 512), balanced batches (yes or no), temperature scale (0.1, 0.5, 1.0), and pre-training epochs (50 or 200).

For all pipelines, the fine-tuning lasts for 100 epochs with early stopping with patience of 22 epochs, monitored on the validation loss. Both schedulers have a minimum learning rate of 10^{-5}. All experiments ran in a single RTX 5000 GPU, except for the SSL $\rightarrow$ UCL/SCL $\rightarrow$ FT pipelines which required two Quadro RTX 8000 GPUs. The source code to reproduce our work in addition to detailed descriptions about the data is available on our public repository[1].

4   Results

As explained in Sect. 3.2, we organized our extensive experimental design in two rounds, corresponding to the next two subsections. In a third subsection, we analyze the second round of experiments in a low training data scenario.

4.1   Self-supervision Schemes *vs.* Baseline Comparison

In this first round of experiments, we compared the baseline pipeline (SUP $\rightarrow$ FT) to the basic self-supervision pipeline (SSL $\rightarrow$ FT) with five self-supervision schemes (BYOL, InfoMin, MoCo-V2, SimCLR, and SwAV). We optimized the baseline and self-supervised pipelines as explained in Sect. 3.4.

The results (Table 3) show that, despite having no access to the labels during the pre-training, and being less thoroughly optimized during the final fine-tuning, the models with self-supervised pre-training are very competitive. Indeed, two of the pipelines (SimCLR and SwAV) had averages above the ones in the baseline.

This first round of experiments intended to validate the applicability of self-supervised learning, and to select one self-supervised scheme for the expensive round of systematic evaluations in the next round. Thus, it comes with the important caveat that both optimization and evaluation were conducted in the isic19 validation set. The second round of experiments will evaluate the ability of the pipelines to generalize performance in the rigorous setting of a held-out test set.

---

[1] https://github.com/VirtualSpaceman/ssl-skin-lesions

Table 3. Results for the first round of experiments, comparing the supervised SUP → FT baseline to the basic SSL → FT pipeline with five SSL schemes. The metric is the AUC on the isic19 validation split. Despite the baseline using label information on pre-training, and being more thoroughly optimized, self-supervision pre-training is still very competitive with it.

Method	AUC (%)	Hyperparameters			
		learning rate	batch size	batches	scheduler
Sup. baseline	94.8 ± 0.6	0.009	128	balanced	plateau
SimCLR [9]	**95.6 ± 0.3**	0.01	32	unbalanced	plateau
SwAV [7]	95.3 ± 0.6	0.01	32	unbalanced	plateau
BYOL [16]	94.6 ± 0.5	0.01	32	unbalanced	plateau
InfoMin [39]	94.4 ± 0.5	0.001	32	unbalanced	plateau
MoCo-V2 [11]	93.9 ± 0.7	0.001	32	unbalanced	plateau

4.2 Systematic Evaluation of Pipelines

In the second round of experiments, we performed a systematic evaluation of the baseline pipeline, pre-trained with supervision (SUP → FT) against the three pipelines pre-trained with self-supervision (SSL → FT, SSL → UCL → FT, and SSL → SCL → FT). In this round, we only evaluated SimCLR as the self-supervision scheme for several reasons: it showed the best performance in the preliminary experiments (Sect. 4.1), it allows introducing annotation information easily with a supervised contrastive loss, it had one hyperparameter less than SwAV to optimize (number of clusters), and the ablation studies in the original papers helped to decide on a range of reasonable values for the temperature value.

As explained in Sect. 3.2, this round of experiments simulates a realistic machine-learning protocol, in which first we optimize the hyperparameters for each pipeline on the isic19 validation split, then evaluate the performance on a held-out test set. The test sets considered as in-distribution is isic19 test split; and isic20, derm7pt-derm, derm7pt-clinic, and pad-ufes20 as out-of-distribution. Those cross-dataset evaluations are critical to evaluate how well the pipelines generalize to different classes, image acquisition techniques, or even to subtle dataset variations across institutions.

The results appear in the topmost plot of Fig. 2, where each boxplot shows the distribution of 25 individual measurements (small black dots), corresponding to the best five non-unique hyperparameterizations, with five replicates for each of them. The boxplots show, as usual, the three quartiles (box), and the range of the data (whiskers) up to 1.5× the interquartile range (samples outside that range are plotted individually as "outliers"). The large red dots show the means for each experiment. The metric is the AUC on the test datasets labeled on the right vertical axis. To make the horizontal axis comparable across its domain, we linearize the AUC using the logit (i.e., the logarithm of the odds) in base 2, shown on the bottom axis. The original AUC values appear on the top axis.

The plots reveal two advantages for the self-supervised pipelines: first, performances (means and medians) tend to be higher; second, the variability (width of the boxes) tended to be smaller. That shows the ability of the self-supervised pre-training in improving the results and in making them more stable.

No consistent advantage in terms of trend improvement (mean, median) is evident among the different self-supervised pipelines, but in terms of variability reduction, the double-pre-trained pipelines (SSL → SCL/UCL → FT) appear to have a slight advantage.

4.3   Low Training Data Scenario

These results follow the same protocol as those in the previous section but with drastically reduced train datasets. The results appear in the middle and bottom-most plots in Fig. 2, for 10% (1480 samples) and 1% (148 samples), respectively, of the original train dataset. Other than for this restriction, the interpretation of the plots is the same as in the previous section.

The results are much noisier than the full-data experiments: in part, this is intrinsic to the smaller training sets, but the random choice of training subsets also contributes to increased variability.

Again, the self-supervised pipelines appear advantageous, both in terms of trend improvement (mean, median) and in terms of variability reduction, but here the advantage of the double-pre-trained pipelines (SSL → SCL/UCL → FT) seems more decisive, especially for the lowest data regimen, where it brings a clear improvement both in trend and variability. As we will discuss in the conclusions, such variability reduction is critical for the soundness of the deployment of low-data models.

4.4   Qualitative Analysis

We performed a qualitative analysis, since we would like to have clues of what different pre-training methods learned to make the decision. Therefore, to analyze the differences between the learned presentations for each pre-training, we performed the Score-CAM [45], method for visualizing the features learned by a neural network and the regions that activate a certain label. We preferred a gradient-free method which overcome both saturation, and false confidence issues [45] from gradient-based techniques.

Figure 3 shows the Score-CAM results for the top-3 self-supervised models (according to Table 3) *vs.* the supervised baseline for each training regime (100%, 10%, and 1% of training data). We randomly sampled three images from the isic19-test to show true positive, true negative, false positive, and false negatives cases. Apart from having different confidences about the prediction of each sample (bottom left rectangle) each model appears to highlight distinct regions. In general, both confidences and activation maps in true positive, and true negative seems to focus on similar regions in the 100% training regime, but the attention appears sparser in the 10 and 1% training data regime.

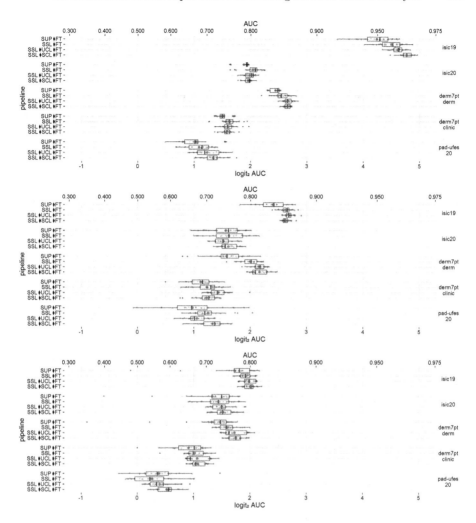

Fig. 2. Results for the second round of experiments, with a systematic comparison of the pipelines labeled on the left vertical axis at the datasets labeled on the right vertical axis. The top, middle, and bottom plots show results for 100, 10 and, 1% of the training data, respectively. Individual measurements represented by each boxplot appear as small black dots, whose means appear as larger red dots. In general, self-supervised pre-trained improves trends (medians, means) and reduces variability, both in the full-data and the low-data scenarios. (Color figure online)

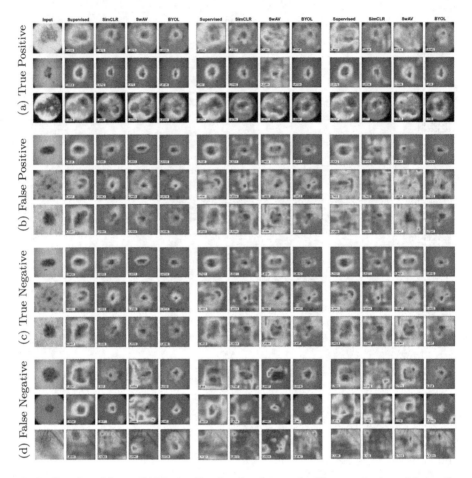

Fig. 3. Results of Score-CAM visualization for the top-3 self-supervised models *vs.* the supervised baseline for each training data regime. The brighter the more relevant the corresponding region in the image is for the model prediction. We randomly sampled three images from the isic19 test set to show true positive, false positive, true negative, and false negative scenarios; considering the full-data training case. For example, in a true positive scenario, we take three random images in which all images are classified as malignant correctly.

5   Conclusions

Our experiments show that self-supervised pre-training makes models easier to deploy than classical supervised pre-training. Even with a less thorough hyperoptimization, the former outperformed the latter in general trends and—especially—in variability.

It is hard to quantify this impression, but the models pre-trained with self-supervised also "felt" easier, faster, and more "ready-to-use" than the baseline models during training.

The advantage of the self-supervised pipelines was particularly prominent in the low-data scenarios, where their ability to stabilize the results, reducing variability, was even more noticeable. In those scenarios, especially the very-low data one, the double-pre-trained models appeared advantageous. We conjecture that self-supervised pre-training improved the mean performance due to the task granularity [1,13], in which self-supervised models showed to perform particularly better when the pre-training is empowered by contrastive learning. However, those findings are shown experimentally, without any theoretical formulation. Understanding what circumstances make self-supervised competitive (or even superior) from a theoretical perspective is a promising research area.

Very low-data scenarios are not, unfortunately, rare in medical applications. Models trained on such regimens will experience large variances in performance in contrast to models trained with adequate samples, but model designers will often be unaware of such variance (since they cannot run a simulation such as ours, comparing their model to others trained in different datasets of the same size). Our results suggest that self-supervised pre-training may reduce that variability, leading to saner models. Of course, models trained on small samples may suffer from severe biases, and extensive exploration is necessary to evaluate whether self-supervised might reinforce those biases [4]. In addition, we performed qualitative analysis using Score-CAM to visualize the network's attention.

Self-supervised learning is a thriving research area, and the possibility of creating domain-specific or at least domain-aware pretext tasks for skin-lesion analysis is an exciting avenue of continuation for this work. Domain-aware sampling methods for selecting the positive and negative pairs in contrastive learning—even while using commodity pretext tasks—also instigate the possibility of incorporating domain knowledge into self-supervision learning.

Acknowledgements. L. Chaves is partially funded by Santander and Google LARA 2021. A. Bissoto is funded by FAPESP 2019/19619-7. E. Valle is partially funded by CNPq 315168/2020-0. S. Avila is partially funded by CNPq 315231/2020-3, FAPESP 2013/08293-7, 2020/09838-0, and Google LARA 2021. This study was financed in part by the Coordenação de Aperfeiçoamento de Pessoal de Nível Superior – Brasil (CAPES) – Finance Code 001. The Recod.ai lab is supported by projects from FAPESP, CNPq, and CAPES.

References

1. Azizi, S., et al.: Robust and efficient medical imaging with self-supervision. arXiv preprint arXiv:2205.09723 (2022)
2. Azizi, S., et al.: Big self-supervised models advance medical image classification. In: International Conference on Computer Vision (ICCV) (2021)

3. Bai, W., et al.: Self-supervised learning for cardiac MR image segmentation by anatomical position prediction. In: Shen, D., et al. (eds.) MICCAI 2019. LNCS, vol. 11765, pp. 541–549. Springer, Cham (2019). https://doi.org/10.1007/978-3-030-32245-8_60

4. Bissoto, A., Fornaciali, M., Valle, E., Avila, S.: (De)Constructing bias on skin lesion datasets. In: Conference on Computer Vision and Pattern Recognition Workshops (CVPRW) (2019)

5. Bissoto, A., Valle, E., Avila, S.: Debiasing skin lesion datasets and models? Not so fast. In: Conference on Computer Vision and Pattern Recognition Workshops (CVPRW) (2020)

6. Boyd, J., Liashuha, M., Deutsch, E., Paragios, N., Christodoulidis, S., Vakalopoulou, M.: Self-supervised representation learning using visual field expansion on digital pathology. In: International Conference on Computer Vision (ICCV) (2021)

7. Caron, M., Misra, I., Mairal, J., Goyal, P., Bojanowski, P., Joulin, A.: Unsupervised learning of visual features by contrasting cluster assignments. In: Advances in Neural Information Processing Systems (NeurIPS) (2020)

8. Chen, L., Bentley, P., Mori, K., Misawa, K., Fujiwara, M., Rueckert, D.: Self-supervised learning for medical image analysis using image context restoration. Med. Image Anal. **58**, 101539 (2019)

9. Chen, T., Kornblith, S., Norouzi, M., Hinton, G.: A simple framework for contrastive learning of visual representations. In: International Conference on Machine Learning (ICML) (2020)

10. Chen, X., Yao, L., Zhou, T., Dong, J., Zhang, Y.: Momentum contrastive learning for few-shot covid-19 diagnosis from chest ct images. Pattern Recogn. **113**, 107826 (2021)

11. Chen, X., Fan, H., Girshick, R., He, K.: Improved baselines with momentum contrastive learning. arXiv preprint arXiv:2003.04297 (2020)

12. Codella, N., Gutman, D., Celebi, M.E., Helba, B., Marchetti, M.A., et al.: Skin lesion analysis toward melanoma detection: A challenge at the 2017 international symposium on biomedical imaging (ISBI), hosted by the international skin imaging collaboration (ISIC). In: International Symposium on Biomedical Imaging (ISBI) (2018)

13. Cole, E., Yang, X., Wilber, K., Mac Aodha, O., Belongie, S.: When does contrastive visual representation learning work? In: Conference on Computer Vision and Pattern Recognition (CVPR) (2022)

14. Geirhos, R., et al.: Shortcut learning in deep neural networks. Nat. Mach. Intelli. **2**(11), 665–673 (2020)

15. Gidaris, S., Singh, P., Komodakis, N.: Unsupervised representation learning by predicting image rotations. In: International Conference on Learning Representations (ICML) (2018)

16. Grill, J.B., et al.: Bootstrap your own latent - a new approach to self-supervised learning. In: Advances in Neural Information Processing Systems (NeurIPS) (2020)

17. Gutmann, M., Hyvärinen, A.: Noise-contrastive estimation: a new estimation principle for unnormalized statistical models. In: International Conference on Artificial Intelligence and Statistics (AISTATS) (2010)

18. He, K., Fan, H., Wu, Y., Xie, S., Girshick, R.: Momentum contrast for unsupervised visual representation learning. In: Conference on Computer Vision and Pattern Recognition (CVPR) (2020)

19. He, K., Zhang, X., Ren, S., Sun, J.: Deep residual learning for image recognition. In: Conference on Computer Vision and Pattern Recognition (CVPR) (2016)

20. Hervella, Á.S., Rouco, J., Novo, J., Ortega, M.: Retinal image understanding emerges from self-supervised multimodal reconstruction. In: Frangi, A.F., Schnabel, J.A., Davatzikos, C., Alberola-López, C., Fichtinger, G. (eds.) MICCAI 2018. LNCS, vol. 11070, pp. 321–328. Springer, Cham (2018). https://doi.org/10.1007/978-3-030-00928-1_37

21. Hosseinzadeh Taher, M.R., Haghighi, F., Feng, R., Gotway, M.B., Liang, J.: A systematic benchmarking analysis of transfer learning for medical image analysis. In: Albarqouni, S., et al. (eds.) DART/FAIR -2021. LNCS, vol. 12968, pp. 3–13. Springer, Cham (2021). https://doi.org/10.1007/978-3-030-87722-4_1

22. Hu, D., et al.: Discriminative sounding objects localization via self-supervised audiovisual matching. In: Advances in Neural Information Processing Systems (NeurIPS), vol. 33 (2020)

23. Jamaludin, A., Kadir, T., Zisserman, A.: Self-supervised learning for spinal MRIs. In: Cardoso, M.J., et al. (eds.) DLMIA/ML-CDS -2017. LNCS, vol. 10553, pp. 294–302. Springer, Cham (2017). https://doi.org/10.1007/978-3-319-67558-9_34

24. Jing, L., Tian, Y.: Self-supervised visual feature learning with deep neural networks: a survey. IEEE Trans. Pattern Anal. Mach. Intell. **43**, 4037–4058 (2020)

25. Kawahara, J., Daneshvar, S., Argenziano, G., Hamarneh, G.: Seven-point checklist and skin lesion classification using multitask multimodal neural nets. IEEE J. Biomed. Health Inform. **23**(2), 538–546 (2019)

26. Kawakami, K., Wang, L., Dyer, C., Blunsom, P., van den Oord, A.: Learning robust and multilingual speech representations. In: Conference on Empirical Methods in Natural Language Processing (EMNLP) (2020)

27. Khosla, P., et al.: Supervised contrastive learning. In: Advances in Neural Information Processing Systems (NeurIPS) (2020)

28. Li, Y., Chen, J., Zheng, Y.: A multi-task self-supervised learning framework for scopy images. In: International Symposium on Biomedical Imaging (ISBI) (2020)

29. Liu, X., et al.: Self-supervised learning: generative or contrastive. IEEE Trans. Knowl. Data Eng. **35**, 857–876 (2021)

30. Liu, X., et al.: Self-supervised learning for dense depth estimation in monocular endoscopy. In: Stoyanov, D., et al. (eds.) CARE/CLIP/OR 2.0/ISIC -2018. LNCS, vol. 11041, pp. 128–138. Springer, Cham (2018). https://doi.org/10.1007/978-3-030-01201-4_15

31. Menegola, A., Fornaciali, M., Pires, R., Bittencourt, F.V., Avila, S., Valle, E.: Knowledge transfer for melanoma screening with deep learning. In: International Symposium on Biomedical Imaging (ISBI) (2017)

32. Morís, D.I., Hervella, Á.S., Rouco, J., Novo, J., Ortega, M.: Context encoder self-supervised approaches for eye fundus analysis. In: International Joint Conference on Neural Networks (IJCNN) (2021)

33. Pacheco, A.G., et al.: Pad-ufes-20: a skin lesion dataset composed of patient data and clinical images collected from smartphones. Data Brief **32**, 106221 (2020)

34. Pathak, D., Krahenbuhl, P., Donahue, J., Darrell, T., Efros, A.A.: Context encoders: feature learning by inpainting. In: Conference on Computer Vision and Pattern Recognition (CVPR) (2016)

35. Rotemberg, V., et al.: A patient-centric dataset of images and metadata for identifying melanomas using clinical context. Scientific Data **8**(1), 1–8 (2021)

36. Srikar Appalaraju, Yi Zhu, Y.X., Fehervari, I.: Towards good practices in self-supervised representation learning. In: Advances in Neural Information Processing Systems Workshops (NeurIPSW) (2020)

37. Sriram, A., et al.: Covid-19 prognosis via self-supervised representation learning and multi-image prediction. arXiv preprint arXiv:2101.04909 (2021)

38. Tack, J., Mo, S., Jeong, J., Shin, J.: CSI: novelty detection via contrastive learning on distributionally shifted instances. In: Advances in Neural Information Processing Systems (NeurIPS) (2020)
39. Tian, Y., Sun, C., Poole, B., Krishnan, D., Schmid, C., Isola, P.: What makes for good views for contrastive learning? In: Advances in Neural Information Processing Systems (NeurIPS) (2020)
40. Truong, T., Mohammadi, S., Lenga, M.: How transferable are self-supervised features in medical image classification tasks? In: Machine Learning for Health, pp. 54–74. PMLR (2021)
41. Valle, E., et al.: Data, depth, and design: learning reliable models for skin lesion analysis. Neurocomputing **383**, 303–313 (2020)
42. Verdelho, M.R., Barata, C.: On the impact of self-supervised learning in skin cancer diagnosis. In: International Symposium on Biomedical Imaging (2022)
43. Vu, Y.N.T., Wang, R., Balachandar, N., Liu, C., Ng, A.Y., Rajpurkar, P.: Medaug: contrastive learning leveraging patient metadata improves representations for chest x-ray interpretation. In: Machine Learning for Healthcare Conference, pp. 755–769 (2021)
44. Wang, D., Pang, N., Wang, Y., Zhao, H.: Unlabeled skin lesion classification by self-supervised topology clustering network. Biomed. Signal Process. Control **66**, 102428 (2021)
45. Wang, H., et al.: Score-cam: score-weighted visual explanations for convolutional neural networks. In: Conference on Computer Vision and Pattern Recognition Workshops (CVPRW) (2020)
46. Wang, Z., Lyu, J., Luo, W., Tang, X.: Superpixel inpainting for self-supervised skin lesion segmentation from dermoscopic images. In: International Symposium on Biomedical Imaging (ISBI) (2022)
47. Wu, Z., Xiong, Y., Yu, S.X., Lin, D.: Unsupervised feature learning via non-parametric instance discrimination. In: Conference on Computer Vision and Pattern Recognition (CVPR) (2018)
48. Zhang, R., Isola, P., Efros, A.A.: Colorful image colorization. In: Leibe, B., Matas, J., Sebe, N., Welling, M. (eds.) ECCV 2016. LNCS, vol. 9907, pp. 649–666. Springer, Cham (2016). https://doi.org/10.1007/978-3-319-46487-9_40
49. Zhou, H.-Y., Yu, S., Bian, C., Hu, Y., Ma, K., Zheng, Y.: Comparing to learn: surpassing imagenet pretraining on radiographs by comparing image representations. In: Martel, A.L., et al. (eds.) MICCAI 2020. LNCS, vol. 12261, pp. 398–407. Springer, Cham (2020). https://doi.org/10.1007/978-3-030-59710-8_39

Skin_Hair Dataset: Setting the Benchmark for Effective Hair Inpainting Methods for Improving the Image Quality of Dermoscopic Images

Joanna Jaworek-Korjakowska[1,2]([⊠]) [iD], Anna Wojcicka[1] [iD],
Dariusz Kucharski[1] [iD], Andrzej Brodzicki[1] [iD], Connah Kendrick[3] [iD],
Bill Cassidy[3] [iD], and Moi Hoon Yap[3] [iD]

[1] AGH University of Science and Technology, Al Mickiewicza 30,
30-059 Krakow, Poland
{jaworek,wojcicka,darekk,brodzicki}@agh.edu.pl

[2] Department of Pathology, Stanford School of Medicine, 300 Pasteur Drive,
Stanford, CA 94305-5324, USA

[3] Manchester Metropolitan University, John Dalton Building, Chester Street,
Manchester M1 5GD, UK
{Connah.Kendrick,B.Cassidy,M.Yap}@mmu.ac.uk
http://www.springer.com/gp/computer-science/lncs

Abstract. Dermoscopic images are often contaminated by artifacts including clinical pen markings, immersion fluid air bubbles, dark corners, and most importantly hair, which makes interpreting them more challenging for clinicians and computer-aided diagnostic algorithms. Hence, automated artifact recognition and inpainting systems have the potential to aid the clinical workflow as well as serve as an preprocessing step in the automated classification of dermoscopic images. In this paper, we share the first release of a public dermoscopic image dataset with hair artifacts which can be accessed here https://skin-hairdataset. github.io/SHD/. The Skin_Hair dataset contains over 252 dermoscopic images including artificial hair and will be expanded over time. Furthermore, we present the primary results of applying machine learning algorithms and GAN based architectures to the hair inpainting problem in dermoscopic images. We envision that these results will serve as a benchmark for researchers who might work on the hair detection and reconstruction tasks with this dataset in the future. In this work, we present a skin lesion image dataset based on the ISIC dataset containing dermoscopic images, images containing artificial hairs and the corresponding ground-truth masks. Furthermore, we use four hair inpainting methods including Navier-Stokes, Telea, Hair_SinGAN and R-MNet architectures which we evaluate using image quality assessment metrics MSE, PSNR, UQI and SSIM. The R-MNet architecture achieved the highest SSIM score of 0.960.

Supplementary Information The online version contains supplementary material available at https://doi.org/10.1007/978-3-031-25069-9_12.

© The Author(s), under exclusive license to Springer Nature Switzerland AG 2023
L. Karlinsky et al. (Eds.): ECCV 2022 Workshops, LNCS 13804, pp. 167–184, 2023.
https://doi.org/10.1007/978-3-031-25069-9_12

Keywords: Melanoma · Dermoscopy · Hair inpainting · Artifacts · Hair removal · Image quality · GAN

1   Introduction

Removal of artifacts from dermoscopic images is a necessary step in classifying skin lesions since artifacts can lead to severe misinterpretation of the global and local structures both for clinical and computer-aided diagnosis. Automated analysis of dermoscopic images is a challenge task [12], with one of the main difficulties being the existence of a variety of artifacts including clinical pen markings, rulers, immersion fluid air bubbles, size-reference stickers, dark corners, and most commonly - hair (see Fig. 1). These artifacts are strikingly different when compared to the rest of the image in both color, shape, and features. As the unique patterns in human skin are often very subtle, those unwanted artifacts often draw the attention of the deep neural network, leading to a falsified diagnosis. Furthermore, the presence of hair may obscure and distort important areas that could determine the final classification.

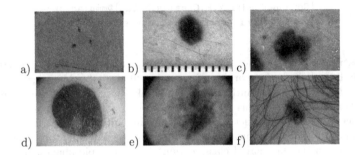

Fig. 1. Sample images from the ISIC database with the following artifacts: a) clinical pen markings, b) rulers, c) immersion fluid air bubbles, d) lens measurement reference, e) dark corners, e) hair [21]

One of the advantages of deep learning methods is the relative lack of pre-processing needed. In most computer vision tasks, including segmentation and classification which are mostly based on CNNs, datasets without any preparation or preprocessing are directly passed to the backbone of the CNN network in order to learn the features. However, prior research [28–30,38] indicates that, in the case of dermoscopic image analysis, most of the algorithms perform better when the artifacts are removed or inpainted.

The presence of hair in dermoscopic images poses a significant challenge as they may occlude some of the information of the lesion such as its boundary and texture. Hence, the removal of hair is an important preprocessing step which, due to its diverse appearances, causes significant problems. We propose a dermoscopic image dataset which gives the possibility to work in the area of removing artifacts and can be used as a benchmark for researchers working on hair detection and inpainting. The dataset uses images from the ISIC datasets [14–16,18,31,41] and consists of dermoscopic images with artificially added hairs as well as corresponding binary masks.

Based on the proposed dataset, which consist of 252 dermoscopic images, we have trained and evaluated two traditional inpainting methods, Telea [39] and Navier-Stokes [7]). In addition deep learning based methods Hair_SinGAN and R-MNet have been proposed. The hair inpainting algorithms have been evaluate using image quality assessment metrics MSE, PSNR, UQI and SSIM. The R-MNet architecture achieved the highest SSIM score 0.960.

The main novelty of this paper can be summarised as follows:

- We introduce a benchmark dataset with consists of 252 cases including: raw dermoscopic images (reference images), corresponding images with overlaying artificial hairs and binary masks which serve as ground-truth.
- We use state-of-the-art Reverse-Masking networks for the inpainting of hairs in dermoscopic images which applies changes only to the target region.
- We propose the Hair_SinGAN architecture, based on the work of [32] et al. which is trained on a single image.
- We statistically evaluate our hair inpainting methods using image quality assessment metrics MSE, PSNR, UQI and SSIM and suggest the R-MNet architecture to serve as the pre-processing method for dermoscopic images.

2   Related Work

Research focusing on automated skin lesion analysis often observe the occurrence of artifacts, but do not discuss how to circumvent the possible negative effects of their presence [23], or do not investigate the effects of their removal [45]. Early attempts to remove hair from skin lesion images were conducted by Lee et al. [25] who created the Dullrazor software. They used grayscale morphological closing to perform hair segmentation and bilinear interpolation to remove hairs from melanoma images. However, this approach is limited in that it is only effective in removing thick dark hairs from skin lesion images. This method would later be improved by [24] et al. who developed the E-shaver application which used an edge detector with color averaging making it more effective on different types of hairs. However, in their experiments, they tested on only 50 images. In the same year, Fiorese et al. [17] proposed the VirtualShave tool which used partial differential equation inpainting, and claimed performance comparable to human operators removing hair manually, with resulting images being almost indistinguishable from hair-free skin.

Later, Xie et al. [43] used a top-hat operator to segment and anisotropic diffusion to remove hair from skin lesion images. As per previous works, this study was not able to handle all types of hair. Additionally, this method was only tested on a very small dataset of 40 just images. Limited dataset testing is a common theme in many prior research projects in this domain [1,10,17,20,37,40].

Maglogiannis [27] et al. used combinations of Bottom-hat, Laplacian, and Sobel methods to identify and remove hair from dermoscopic images. They observed that the Laplacian of Gaussian and Sobel edge detection methods combined, together with a 3×3 wiener noise reduction filter, provided the best results.

Salido et al. [33] performed hair removal on dermoscopic images using morphological bottom-hat filtering with erosion and dilation by morphological opening. Inpainting was completed using a nonlinear model based on curvature-driven diffusions for nontexture images, originally proposed by [13].

Bardou et al. [6] used a variational autoencoder to remove hair from dermoscopic skin lesion images in the HAM10000 dataset without the need for paired samples. The encoder uses dermoscope images as input and builds a latent distribution which ignores hair as noise, while the decoder reconstructs a hair-free image. Their results show high quality inpainting, the reconstructed images are not identical to the input images as they look blurry and often distort the features of lesions.

In 2019, Talavera et al. [38] identified that there are currently no methods to benchmark the effectiveness of hair removal algorithms. They extracted 13 hairless images from the PH2 dataset and overlaid artificial hairs to test the effectiveness of 6 state-of-the-art algorithms and compared the results.

Li et al. [26] trained a U-Net with ISIC data to obtain hair masks, and propose an inpainting architecture comprising a gated convolution and SN-PatchGAN. They categorised hair in ISIC images as: thin; overlapping; faded; of similar contrast or colour to the underlying skin; and obscuring lesions. They observe that traditional hard-coded threshold-based hair removal methods are ineffective, and can result in over-removal which can cause loss of important lesion details, or under-removal where the hair cannot be removed effectively. They also propose an evaluation method (intra-structural similarity) to analyse the effect of hair removal based on a single dermoscopic image.

Song et al. [36] proposed a novel hair extraction method which utilised maximum variance fuzzy clustering, with a Criminisi algorithm used for repairing image regions where hair had been removed. This method is capable of fast hair extraction and segmentation with reduced computational complexity. The implementation does not require extensive learning based on a large number of parameters and training images, resulting in high execution efficiency.

More recently, Nauta et al. [29] found that CNN classifiers partly based predictions of benign images on the presence of colour calibration patches placed onto the skin during examinations. By artificially inserting colour calibration patches into malignant images, they showed that shortcut learning results in a significant increase in misdiagnoses. This work indicates that other artifact types may present similar issues.

We surveyed 38 state-of-the-art papers from the field of computer aided diagnosis in skin lesion tasks and checked if the authors mention any techniques for hair removal. Although most of the papers describe the difficulties of dealing with artifacts they often state that it is an issue for computer vision processing methods. The deep learning methodology hasn't been explored in this area, yet. Only in 7 papers researchers indicated the artifact removal or enhancement stage [2–5, 8, 9, 44].

3   Skin_Hair Dataset

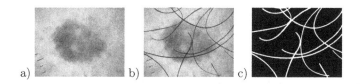

Fig. 2. Illustration of the dataset creation process: (a) clear image without hair, (b) hair extracted from different image placed over the clear image, and (c) ground truth reference mask.

The main issue in the process of detection, reconstruction and assessment of artifacts is mostly due to the lack of properly prepared datasets which do not include ground truth masks and reference images. Due to the artifact removal evaluation process we propose a novel Skin_Hair dataset that includes the raw dermoscopic images, images containing artificial hairs as well as ground-truth masks for evaluation purpose. This dataset is created by taking raw dermoscopic images without hair artifacts from the ISIC dataset that serve as a reference ground-truth image and for applying manually extracted hairs from other dermoscopic images from the ISIC dataset. The dataset can be obtained from the following repository: https://skin-hairdataset.github.io/SHD/.

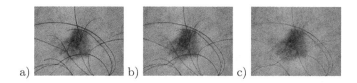

Fig. 3. Illustration of three types of hair colour: (a) dark, (b) brown, and (c) light. (Color figure online)

Raw dermoscopic images without hair, as well as hair patterns, were taken from the ISIC database [21], which is the largest publicly available dataset containing dermoscopic images with metadata. To successfully determine the effectiveness of the hair removal methods, we use the balanced ISIC dataset as presented in [11]. Pewton and Yap [30] annotated the dataset for numerous artifacts, including hair. Based on the provided information we divide the dataset into two separate parts containing hair and without hair, respectively. Due to the very large variety and complexity of the hair patterns, we decided to transfer the hair from other dermoscopic images, which allowed us to maintain their natural appearance (Fig. 2). The process of creating a dermoscopic image with hairs consists of the following steps: 1) Choosing a raw image without artifacts from the ISIC dataset, 2) Choosing an image including hairs from the ISIC dataset, 3) Manually marking the hair areas using Photoshop quick mask with alpha channel, soft, round brush with full opacity and size adapted to the size of the marked hair, 4) Cutting out the hair to a new transparent layer and clearing any additional areas of skin visible on this layer, 5) Applying the hair mask to the dermatoscopic image.

The extracted hair patterns have been augmented using the following methods: 1) randomly moving and rotating the mask, 2) modification of the selection with small, medium and large number of hairs; 3) changing the color of the hair into three main categories (light, brown, and dark - defined based on the analysis of the dataset) using brightness, contrast tool and color blending mode; 4) randomly applying different masks onto different clean images, without hair; and 5) for each modified pattern, a reference mask was created using a threshold tool (Fig. 4).

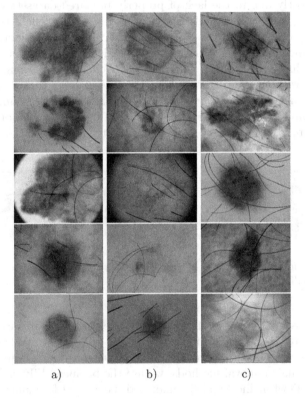

a)   b)   c)

Fig. 4. Illustrations of three types of hair size: (a) small, (b) medium, and (c) large.

The method is repeated for three different hair colours - dark, brown and light as presented in Fig. 3. In total, we used 77 non-hair images as the basis for applying different hair configurations. We augmented the extracted hair by changing the size, amount and colour. In total 252 images were generated with 84 unique masks to cover the different hair types. The Skin_Hair dataset contains: 35 images with small density (each in three colours - light, brown and dark), 27 images with medium density (each in three colours - light, brown and dark) and 22 images with high density (each in three colours - light, brown and dark). The process of mask extraction and hair addition was performed using Adobe Photoshop 23.4.1.

4 Effective Hair Inpainting Algorithms

As the artifact removal process is an obligatory step in image preprocessing we have considered 5 different inpainting techniques in order to compare the traditional computer vision inpainting methods including Navier-Stokes and Telea with two state-of-the art deep learning techniques - SinGAN and R-MNet.

4.1 Navier-Stokes

Bertalmio et al. found an analogy between the image inpainting problem and the stream function in a two-dimensional (2D) incompressible fluid. An approximate solution to the inpainting problem is obtained by numerically approximating the steady state solution of the 2D NSE (Navier-Stokes Equations) vorticity transport equation, and simultaneously solving the Poisson equation between the vorticity and stream function, in the region to be inpainted [7]. Image intensity is changed via a 'stream function'. Isophote lines (lines of equal brightness intensity) are propagated along the edges from the outside into the region that is being inpainted. Instead of using the vorticity of the fluid, the method uses the laplacian of the intensity. The direction of the flow is a vector field defined by the stream function. The algorithm continues the isophote lines and matches gradient vectors at the boundary of the inpainting region [7]. Results of the Navier-Stokes algorithm are presented in Fig. 5.

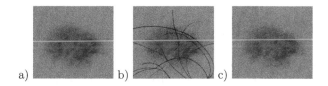

a)   b)   c)

Fig. 5. Illustration of the effects of the Navier-Stokes inpainting method: a) original dermoscopic image, b) dermoscopic image containing artificial hair, and c) Navier-Stokes inpainting outcomes.

4.2 Telea

Telea [39], proposed an inpainting algorithm using a Fast Marching Method (FMM). This method is considered faster and less complex to compute than other typical inpainting methods [39]. The algorithm uses known regions to grow inpainted regions into the target regions, and is enforced by the use of the FMM. The Fast Marching Method itself is a numerical technique used for solving a boundary value problem [34]. Here it is used to ensure that pixels closer to the known neighbours are inpainted before pixels with unknown neighbours. It performs a similar role to a distance transform but has an advantage of maintaining narrow bands - a boundary between known and unknown areas. The algorithm defines three types of pixels: $BAND$: the pixel belongs to the narrow band, $KNOWN$: the pixel is outside the inpainting boundary (known) and $INSIDE$: the pixel is inside the inpainting boundary (unknown). For each pixel, there are two values - T (distance to the edge) and I (grey-level intensity). The algorithm works in the following steps: 1) extract the $BAND$ point with the smallest T, 2) march the boundary inward by adding new points to it, 3) perform the inpainting: iterate over the $KNOWN$ points in the neighborhood of the current point (i, j) and compute $I(i, j)$ and the image gradient is estimated by central differences, 4) propagates the value T of point (i, j) to its neighbors (k, l) by solving the finite difference discretization problem, 5) inserts (k, l) with its new T in the heap. Results of the Telea algorithm are presented in Fig. 6.

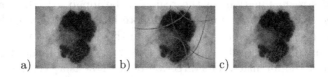

a)   b)   c)

Fig. 6. Illustration of the effects of the Telea inpainting algorithm: a) original dermoscopic image, b) dermoscopic image containing artificial hair, and c) Telea inpainting outcomes.

4.3 Hair_SinGAN Architecture

While most of deep learning models require large numbers of examples in order to be trained effectively, we tried to design an approach which works on as few examples as possible. In practice, this is technically difficult as the dataset needs to represent the underlying distribution, and the more examples the dataset consists of, the more accurate the representation is. However, considering a single image as a dataset itself, it can represent its own distribution. The general idea behind our approach is to analyze parts of an image which are not hidden behind artifacts that we want to remove, train the model on those parts, and then use the model in order to reconstruct areas hidden behind (Fig. 7). The algorithm starts with dividing the image into a set of smaller training rectangular regions. Those areas where the GT image mask shows no hair are training regions and a the rest of the image becomes a reconstruction region (see steps T1 and R1 on Fig. 7). Then, from the proposed image parts, multiple rectangles are extracted, which constitute inputs to a model based on Generative Adversarial Networks (T2 and R2). Additionally, in training phase (T4), those inputs are enriched with artificial artifacts (T3). On the other hand, original snippets serves as model's reconstruction goal (network output). After the training process is finished, snippets prepared from the reconstruction region (R2) are then fed into the network (R3). The outputs, with the hair removed, replaces fragments on the original image.

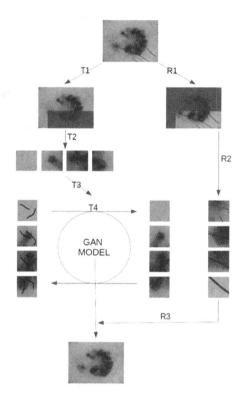

Fig. 7. SinGAN [35] algorithm pipeline, consisting of two independent branches. The T branch represents a model training process on artificially generated examples. Such a model is then used for inpainting on fragments covered with artifacts (showed on the branch R). In the final step, reconstructed fragments are replaced with those on the original image.

Given that this method requires only a single image for reconstruction, the results represent a valid alternative compared to prior traditional methods, especially in the case of poor quality datasets (Table 1). The main drawback of the algorithm is the ratio between the training area and the reconstruction area. When the ratio is below 1, reconstruction starts to disclose insufficient dataset problem.

4.4   R-MNet Method

For hair inpainting we employ the use of Reverse-Masking networks [22]. The advantage of this method is that it is similar to the traditional methods and the GAN focuses only on target regions, making no change to the surrounding regions. The network does this by importing the mask and then feeding into the network as in traditional structures, as illustrated in Fig. 8.

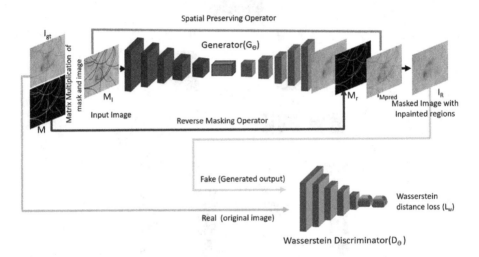

Fig. 8. Illustration of R-MNet [22] on skin lesions inpainting.

The mask is applied internally, using matrix operations which inverse the mask which is used to reapply the undamaged areas to the inpainted images. The network then uses surrounding regions to inpaint the selected areas, as with traditional prior techniques. The network uses an encoder-decoder structure, to focus on non-damaged regions, and reconstructs the inpainted image during the decode stage. However, owing to the network structure a custom loss function is used which focuses on damaged regions to assess reconstruction. The decoder uses a series of 5×5 convolutions of increasing filter depth. Each convolution is followed by a LeakyReLU activation, using an alpha of 0.2, a dropout of 0.5, followed by a max-pooling layer. The LeakyReLU is used to prevent the non activation of the neurons, as instead of the function being zero when $x < 0$, the leakyReLU will return some small negative number instead. The LeakyReLU is used during the encoding stage, allowing a diverse set of features to be captured, the dropout aids the network to deal with less features and the pooling focuses the network onto the core features. The decoder follows a series of up-sampling layers, with transposed convolutions, standard ReLU and batch normalisation, with a final tanh layer. The encoder mixes both 2D up-sampling and transpose convs to resoles the feature back to the original size while having some convolutional functions, the batch normalisation aids the network to generalise during training. To train our network we used the generated hair masks on the hairless images to ensure the network learned skin features only. Owing to this the network requires hairless examples as ground truth to ensure the system only learns to inpaint skin and the skin lesions. When training using R-MNet the network manages to inpaint the hair masks, as illustrated in Fig. 9 which shows that the network manages to successfully inpaint skin regions and parts of the lesions However, limitations, due to the limited umber of input images, are apparent.

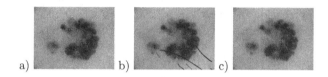

Fig. 9. Illustration of the visual appearances on skin lesions: a) original dermoscopic image, b) dermoscopic image containing artificial hair, and c) R-MNet inpainted image.

5 Result Analysis

Image Quality Assessment (IQA) is considered as a characteristic property of an image and describes the degradation of the perceived image. Quality of an image can be described technically with statistical metrics as well as objectively to indicate the deviation from the ideal or reference model. In our case the ideal is an image without the hair mask. There are several techniques and metrics available that can be used for objective image quality assessment. Here, we use the full-Reference (FR) approach, as we assess the quality of a test image in comparison with a reference image which is considered to be of perfect quality. We take advantage of image quality techniques to compare the outcomes of our proposed inpainting algorithms for the hair regions such as MSE (Mean Square Error), PSNR (Peak Signal to Noise Ratio), SSIM (Structured Similarity Index Measure), and UQI (Universal Quality Index).

MSE is the most common estimator of image quality measurement and refers to the second moment of error. The error is the difference between the estimator and the estimated outcome. It is a function of risk, considering the expected value of the squared error loss or quadratic loss, with a bias towards large deviation from the ground truth. MSE is a full reference metric with values closer to zero indicating higher similarity. MSE between two images such as $I(x, y)$ and $K(x, y)$ is defined as [42]:

$$MSE = \frac{1}{m\,n} \sum_{i=0}^{m-1} \sum_{j=0}^{n-1} [I(i,j) - K(i,j)]^2. \tag{1}$$

PSNR is used to calculate the ratio between the maximum possible signal power and the power of the distorting noise which affects the quality of its representation. This ratio between two images is computed in decibel form and is usually calculated as the logarithm term of decibel scale where the dynamic range varies between the largest and the smallest possible values which are changeable by their quality. The PSNR is defined as [19]:

$$PSNR = 10 \cdot \log_{10} \left(\frac{MAX_I^2}{MSE} \right) \tag{2}$$

SSIM is a perception based model where image degradation is considered as the change of perception in structural information [46]. It also takes into consideration other important perception based elements such as luminance masking and contrast masking. The difference of this method when compared to other techniques, such as MSE or PSNR, is that the other approaches estimate absolute errors. Structural information is the idea that the pixels have strong inter-dependencies especially when they are spatially close. The SSIM index is calculated on various windows of an image. The measure between two windows x and y of common size $N \times N$ is defined as [46]:

$$SSIM(x,y) = \frac{(2\mu_x\mu_y + c_1)(2\sigma_{xy} + c_2)}{(\mu_x^2 + \mu_y^2 + c_1)(\sigma_x^2 + \sigma_y^2 + c_2)} \tag{3}$$

where L is the dynamic range of the pixel-values (typically this is $2^{\#bits\ per\ pixel} - 1$, $k_1 = 0.01$ and $k_2 = 0.03$ by default).

UQI was the predecessor of SSIM and evaluates quality of an image using loss of correlation, luminance distortion, and contrast distortion. UQI is global rather than being local or specially intended to the images being tested or on the individual observers. The quality index is defined as:

$$Q = \frac{4\sigma_{xy}\overline{xy}}{(\sigma_x^2 + \sigma_y^2) + ((\overline{x})^2 + (\overline{y})^2)} \tag{4}$$

where, $\overline{x}$ and $\overline{y}$ are the mean values of the original and distorted images respectively. σ_x^2 and σ_y^2 are the variances. σ_{xy} is the covariance. The range of UQI is $[-1, 1]$ where 1 is achieved when the two images are identical.

Due to the R-MNet algorithm which requires the training set the proposed Skin_Hair dataset has been divided into training and testing sets including 170 and 82 images respectively. After inpainting (reconstructing) skin lesion images, we estimated the quality by using MSE, PSNR, UQI and SSIM metrics. The summary of quality matrices calculations is shown in Table 1. From this table, we observe that all metrics have given almost consistent results. From a representation perspective, SSIM and UQI is normalized, but MSE and PSNR are not. Therefore, SSIM and UQI can be treated as more understandable than MSE

Table 1. Summary of the IQA metrics including MSE, PSNR, SSIM and UQI for hair inpaining methods including Navier-Stokes, Telea, Hair_SinGAN and R-MNet. † = higher value is better; ⊎ = lower value is better.

Hair inpainting method	MSE ⊎	PSNR †	SSIM †	UQI †
Navier-Stokes	7.380	40.305	0.959	0.9984
Telea	**7.114**	**40.558**	0.959	0.9984
Hair_SinGAN	53.735	34.489	0.881	0.9976
R-MNet	23.743	**40.655**	**0.960**	**0.9985**

and PSNR. This is due to MSE and PSNR being absolute errors, however, SSIM provides perception and saliency-based errors. The highest SSIM value has been achieved by the R-MNet GAN based architecture (Fig. 10).

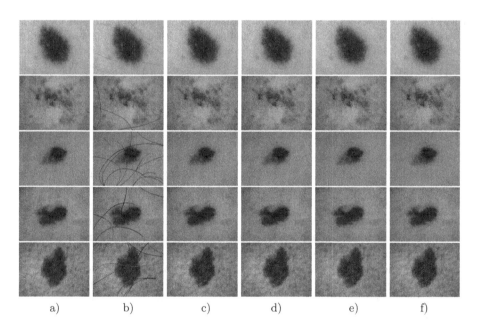

a)   b)   c)   d)   e)   f)

Fig. 10. Visual comparison of the results of different methods: a) original dermoscopic image, b) dermoscopic image containing artificial hair, c) Navier-Stokes inpainted image, d) Telea inpainted image, e) Hair_SinGAN inpainted image, and f) R-MNet inpainted image. The regions inpainted with Hair_SinGAN are the least distinguishable to the human eye. We can also observe, that the regions containing hair overlapping the lesions were the hardest to inpaint.

When artifact levels increase, the recovery quality of the output image is also shown to deteriorate, which we can be observed in Fig. 11.

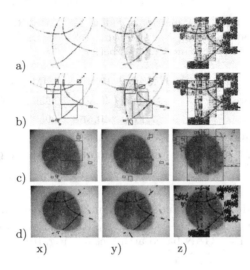

Fig. 11. Illustrations of the SSIM algorithm for three hair inpainting methods: x) Telea, y) R-MNet and z) Hair_SinGAN: a) image differences with darker regions show more disparity, b–c) filter using a minimum threshold area to remove the gray noise, and highlight the differences with a bounding box, d) visualisation of the exact differences. (Color figure online)

6   Conclusions

We demonstrate the application of image inpainting onto skin lesions for hair removal, and highlight key issues in this field. Namely the lack of ground truth data, and present a novel hair inpainting dataset for qualitative evaluation of inpainting techniques. The most important contribution is the release of a dataset of skin images with added hair. Although the dataset is limited in size it can provide valuable benchmarking on future hair removal techniques. However, we continue to work on extending the number of images and will release a larger second version at a later date. We hope that the existence of a large collection of corresponding images with their reference ground truths, will be a useful addition to the ISIC database, helpful for researchers wishing to work on skin lesions. Furthermore, we plan to add different artifacts such as measuring tools, such as air bubbles and dark corner. We have used this dataset to test four inpainting algorithms - two classical approaches (Navier-Stokes and Telea) and two of our own implementations based on GANs. Available ground truth images allowed us to statistically evaluate those methods. The lowest MSE was achieved by the Telea method. However, in terms of other metrics the best performing method proved to be R-MNet. We note however that while Hair_SinGAN achieved slightly inferior results, it was only trained on a single image for each example.

Acknowledgments. We gratefully acknowledge the funding support of the research project by the program "Excellence initiative—research university" for the AGH UST and the NAWA Bekker Scholarship for J. Jaworek-Korjakowska.

References

1. Abbas, Q., Celebi, M., García, I.F.: Hair removal methods: a comparative study for dermoscopy images. Biomed. Signal Process. Control **6**(4), 395–404 (2011). https://doi.org/10.1016/j.bspc.2011.01.003, https://www.sciencedirect.com/science/article/pii/S1746809411000048
2. Adegun, A.A., Viriri, S.: Deep learning-based system for automatic melanoma detection. IEEE Access **8**, 7160–7172 (2020)
3. Adegun, A.A., Viriri, S.: Deep learning techniques for skin lesion analysis and melanoma cancer detection: a survey of state-of-the-art. Artif. Intell. Rev. **54**, 811–841 (2020)
4. Almaraz-Damian, J.A., Ponomaryov, V., Sadovnychiy, S., Castillejos-Fernandez, H.: Melanoma and nevus skin lesion classification using handcraft and deep learning feature fusion via mutual information measures. Entropy **22**(4) (2020). https://doi.org/10.3390/e22040484, https://www.mdpi.com/1099-4300/22/4/484
5. Barbosa, J., Baleiras, M.: Melanoma detection using deep learning methods (2019)
6. Bardou, D., Bouaziz, H., Lv, L., Zhang, T.: Hair removal in dermoscopy images using variational autoencoders. Skin Res. Technol. **28**(3), 445–454 (2022). https://doi.org/10.1111/srt.13145, https://onlinelibrary.wiley.com/doi/abs/10.1111/srt.13145
7. Bertalmio, M., Bertozzi, A.L., Sapiro, G.: Navier-stokes, fluid dynamics, and image and video inpainting. In: Proceedings of the 2001 IEEE Computer Society Conference on Computer Vision and Pattern Recognition. CVPR 2001, vol. 1, pp. I–I. IEEE (2001)
8. Bisla, D., Choromanska, A., Berman, R., Stein, J., Polsky, D.: Towards automated melanoma detection with deep learning: data purification and augmentation. In: 2019 IEEE/CVF Conference on Computer Vision and Pattern Recognition Workshops (CVPRW), pp. 2720–2728 (2019)
9. Bisla, D., Choromanska, A., Stein, J., Polsky, D., Berman, R.: Skin lesion segmentation and classification with deep learning system. arXiv:abs/1902.06061 (2019)
10. Borys, D., Kowalska, P., Frackiewicz, M., Ostrowski, Z.: A simple hair removal algorithm from dermoscopic images. In: Ortuño, F., Rojas, I. (eds.) IWBBIO 2015. LNCS, vol. 9043, pp. 262–273. Springer, Cham (2015). https://doi.org/10.1007/978-3-319-16483-0_27
11. Cassidy, B., Kendrick, C., Brodzicki, A., Jaworek-Korjakowska, J., Yap, M.H.: Analysis of the ISIC image datasets: usage, benchmarks and recommendations. Med. Image Anal. **75**, 102305 (2022)
12. Celebi, M.E., Barata, C., Halpern, A., Tschandl, P., Combalia, M., Liu, Y.: Guest editorial: image analysis in dermatology. Med. Image Anal. **79**, 102468 (2022). https://doi.org/10.1016/j.media.2022.102468
13. Chan, T.F., Shen, J.: Nontexture inpainting by curvature-driven diffusions. J. Visual Commun. Image Represent. **12**(4), 436–449 (2001). https://doi.org/10.1006/jvci.2001.0487, https://www.sciencedirect.com/science/article/pii/S1047320301904870
14. Codella, N., et al.: Skin lesion analysis toward melanoma detection 2018: a challenge hosted by the international skin imaging collaboration (ISIC) (2018)

15. Codella, N.C., et al.: Skin lesion analysis toward melanoma detection: a challenge at the 2017 international symposium on biomedical imaging (ISBI), hosted by the international skin imaging collaboration (ISIC). In: 2018 IEEE 15th International Symposium on Biomedical Imaging (ISBI 2018), pp. 168–172. IEEE (2018)
16. Combalia, M., et al.: Bcn20000: Dermoscopic lesions in the wild (2019)
17. Fiorese, M., Peserico, E., Silletti, A.: Virtualshave: automated hair removal from digital dermatoscopic images. In: 2011 Annual International Conference of the IEEE Engineering in Medicine and Biology Society, pp. 5145–5148 (2011). https://doi.org/10.1109/IEMBS.2011.6091274
18. Gutman, D., et al.: Skin Lesion Analysis Toward Melanoma Detection: A Challenge at the International Symposium on Biomedical Imaging (ISBI) 2016, Hosted by the International Skin Imaging Collaboration (ISIC) (2016)
19. Horé, A., Ziou, D.: Image quality metrics: PSNR vs. SSIM. In: 2010 20th International Conference on Pattern Recognition, pp. 2366–2369 (2010). https://doi.org/10.1109/ICPR.2010.579
20. Huang, A., Kwan, S.Y., Chang, W.Y., Liu, M.Y., Chi, M.H., Chen, G.S.: A robust hair segmentation and removal approach for clinical images of skin lesions. In: 2013 35th Annual International Conference of the IEEE Engineering in Medicine and Biology Society (EMBC), pp. 3315–3318 (2013). https://doi.org/10.1109/EMBC.2013.6610250
21. ISIC: Isic archive gallery. Online, July 2020. https://www.isic-archive.com
22. Jam, J., Kendrick, C., Drouard, V., Walker, K., Hsu, G.S., Yap, M.H.: R-MNET: a perceptual adversarial network for image inpainting. In: Proceedings of the IEEE/CVF Winter Conference on Applications of Computer Vision, pp. 2714–2723 (2021)
23. Kassem, M.A., Hosny, K.M., Fouad, M.M.: Skin lesions classification into eight classes for ISIC 2019 using deep convolutional neural network and transfer learning. IEEE Access 8, 114822–114832 (2020). https://doi.org/10.1109/ACCESS.2020.3003890
24. Kiani, K., Sharafat, A.R.: E-shaver: an improved dullrazor® for digitally removing dark and light-colored hairs in dermoscopic images. Comput. Biol. Med. 41(3), 139–145 (2011). https://doi.org/10.1016/j.compbiomed.2011.01.003, https://www.sciencedirect.com/science/article/pii/S0010482511000047
25. Lee, T., Ng, V., Gallagher, R., Coldman, A., McLean, D.: Dullrazor®: A software approach to hair removal from images. Comput. Biol. Med. 27(6), 533–543 (1997). https://doi.org/10.1016/S0010-4825(97)00020-6, https://www.sciencedirect.com/science/article/pii/S0010482597000206
26. Li, W., Joseph Raj, A.N., Tjahjadi, T., Zhuang, Z.: Digital hair removal by deep learning for skin lesion segmentation. Pattern Recogn. 117, 107994 (2021). https://doi.org/10.1016/j.patcog.2021.107994, https://www.sciencedirect.com/science/article/pii/S0031320321001813
27. Maglogiannis, I., Delibasis, K.: Hair removal on dermoscopy images. In: 2015 37th Annual International Conference of the IEEE Engineering in Medicine and Biology Society (EMBC), pp. 2960–2963 (2015). https://doi.org/10.1109/EMBC.2015.7319013
28. Maron, R.C., et al.: Reducing the impact of confounding factors on skin cancer classification via image segmentation: technical model study. J. Med. Internet Res. 23 (2021)
29. Nauta, M., Walsh, R., Dubowski, A., Seifert, C.: Uncovering and correcting shortcut learning in machine learning models for skin cancer diagnosis. Diag-

nostics **12**(1) (2022). https://doi.org/10.3390/diagnostics12010040, https://www. mdpi.com/2075-4418/12/1/40

30. Pewton, S.W., Yap, M.H.: Dark corner on skin lesion image dataset: does it matter? In: Proceedings of the IEEE/CVF Conference on Computer Vision and Pattern Recognition (CVPR) Workshops, pp. 4831–4839, June 2022

31. Rotemberg, V., et al.: A patient-centric dataset of images and metadata for identifying melanomas using clinical context. Sci. Data 8, 34 (2021). https://doi.org/ 10.1038/s41597-021-00815-z

32. Rott Shaham, T., Dekel, T., Michaeli, T.: SinGAN: Learning a generative model from a single natural image. In: IEEE International Conference on Computer Vision (ICCV)(2019)

33. Salido, J.A.A., Ruiz, C.: Using morphological operators and inpainting for hair removal in dermoscopic images. In: Proceedings of the Computer Graphics International Conference. CGI 2017, Association for Computing Machinery, New York, NY, USA (2017). https://doi.org/10.1145/3095140.3095142, https://doi.org/10. 1145/3095140.3095142

34. Sethian, J.A.: A fast marching level set method for monotonically advancing fronts. Proc. Natl. Acad. Sci. **93**(4), 1591–1595 (1996)

35. Shaham, T.R., Dekel, T., Michaeli, T.: SinGAN: learning a generative model from a single natural image. In: Proceedings of the IEEE/CVF International Conference on Computer Vision, pp. 4570–4580 (2019)

36. Song, X., et al.: Research on hair removal algorithm of dermatoscopic images based on maximum variance fuzzy clustering and optimization criminisi algorithm. Biomed. Signal Process. Control **78**, 103967 (2022). https://doi. org/10.1016/j.bspc.2022.103967, https://www.sciencedirect.com/science/article/ pii/S1746809422004669

37. Sultana, A., Dumitrache, I., Vocurek, M., Ciuc, M.: Removal of artifacts from dermatoscopic images. In: 2014 10th International Conference on Communications (COMM), pp. 1–4 (2014). https://doi.org/10.1109/ICComm.2014.6866757

38. Talavera-Martínez, L., Bibiloni, P., González-Hidalgo, M.: Comparative study of dermoscopic hair removal methods. In: Tavares, J.M.R.S., Natal Jorge, R.M. (eds.) VipIMAGE 2019. LNCVB, vol. 34, pp. 12–21. Springer, Cham (2019). https://doi. org/10.1007/978-3-030-32040-9_2

39. Telea, A.: An image inpainting technique based on the fast marching method. J. Graphics Tools **9**(1), 23–34 (2004)

40. Toossi, M.T.B., Pourreza, H.R., Zare, H., Sigari, M.H., Layegh, P., Azimi, A.: An effective hair removal algorithm for dermoscopy images. Skin Res. Technol. **19** (2013)

41. Tschandl, P.: The HAM10000 dataset, a large collection of multi-source dermatoscopic images of common pigmented skin lesions (2018). https://doi.org/10.7910/ DVN/DBW86T, https://doi.org/10.7910/DVN/DBW86T

42. Wang, Z., Bovik, A.C.: Mean squared error: love it or leave it? A new look at signal fidelity measures. IEEE Signal Process. Mag. **26**(1), 98–117 (2009). https://doi. org/10.1109/MSP.2008.930649

43. Xie, F.Y., Qin, S.Y., Jiang, Z.G., Meng, R.S.: PDE-based unsupervised repair of hair-occluded information in dermoscopy images of melanoma. Comput. Med. Imaging Graph. **33**(4), 275–282 (2009). https://doi.org/10.1016/ j.compmedimag.2009.01.003, https://www.sciencedirect.com/science/article/pii/ S0895611109000056

44. Xie, Y., Zhang, J., Xia, Y., Shen, C.: A mutual bootstrapping model for automated skin lesion segmentation and classification. IEEE Trans. Med. Imaging **39**, 2482–2493 (2020)
45. Zanddizari, H., Nguyen, N., Zeinali, B., Chang, J.M.: A new preprocessing approach to improve the performance of CNN-based skin lesion classification. Med. Biol. Eng. Comput. **59**(5), 1123–1131 (2021). https://doi.org/10.1007/s11517-021-02355-5
46. Wang, Z., Bovik, A.C., Sheikh, H.R., Simoncelli, E.P.: Image quality assessment: from error visibility to structural similarity. IEEE Trans. Image Process. **13**(4), 600–612 (2004). https://doi.org/10.1109/TIP.2003.819861

FairDisCo: Fairer AI in Dermatology via Disentanglement Contrastive Learning

Siyi Du[1]([✉])[ID], Ben Hers[1], Nourhan Bayasi[1][ID], Ghassan Hamarneh[2][ID], and Rafeef Garbi[1][ID]

[1] University of British Columbia, Vancouver, BC, Canada
{siyi,bhers,nourhanb,rafeef}@ece.ubc.ca
[2] Simon Fraser University, Burnaby, BC, Canada
hamarneh@sfu.ca

Abstract. Deep learning models have achieved great success in automating skin lesion diagnosis. However, the ethnic disparity in these models' predictions, where lesions on darker skin types are usually underrepresented and have lower diagnosis accuracy, receives little attention. In this paper, we propose FairDisCo, a disentanglement deep learning framework with contrastive learning that utilizes an additional network branch to remove sensitive attributes, i.e. skin-type information from representations for fairness and another contrastive branch to enhance feature extraction. We compare FairDisCo to three fairness methods, namely, resampling, reweighting, and attribute-aware, on two newly released skin lesion datasets with different skin types: Fitzpatrick17k and Diverse Dermatology Images (DDI). We adapt two fairness-based metrics *DPM* and *EOM* for our multiple classes and sensitive attributes task, highlighting the skin-type bias in skin lesion classification. Extensive experimental evaluation demonstrates the effectiveness of FairDisCo, with fairer and superior performance on skin lesion classification tasks.

Keywords: Fairness · Skin lesion diagnosis · Medical imaging

1 Introduction

Cancer is the leading cause of death worldwide, accounting for nearly 10 million deaths in 2020, or about one in six deaths, and the skin is among the 6 most common organs invaded by cancer [40,47]. Nevertheless, the survival rate of patients can be considerably increased by early detection and treatment of skin lesions [4]. Traditional diagnosis and detection of skin cancer have been carried out by dermatologists via manual screening and visual inspection, which is time-consuming, complex and error-prone. With the advancement in computer vision, tedious examination procedures may be avoided by an automatic diagnosis of identifying possibly cancerous lesions. End-to-end deep neural networks (DNNs) were developed to capture more powerful features and output predictions based directly on the input images [14,19]. Despite their success, DNNs are vulnerable

© The Author(s), under exclusive license to Springer Nature Switzerland AG 2023
L. Karlinsky et al. (Eds.): ECCV 2022 Workshops, LNCS 13804, pp. 185–202, 2023.
https://doi.org/10.1007/978-3-031-25069-9_13

to biases that render their decisions "unfair". For example, a recent study has shown that patients with darker skin types (tones or colors) experience a significant drop-off in diagnosis accuracy compared to those of light types [12], which limits the successful clinical deployment of DNNs [21,33].

Fairness is challenging to address for skin lesion classification, given the lack of annotated data with diverse skin types. Kinyanjui et al. [32] first proposed a pipeline for automatic estimation of skin-type labels based on the individual typology angle (ITA) for skin disease datasets. Other works further adopted the ITA to generate skin-type labels and proposed skin-type bias mitigating strategies [9,11,33]. However, they validated their approaches mainly on datasets with a small fraction of dark skin-type samples, limiting their conclusions' statistical power and generalizability. Furthermore, Groh et al. [20] found that ITA might be unreliable for annotating large-scale image datasets. Recently, Groh et al. [21] and Daneshjou et al. [12] proposed two new datasets with varied skin types and manually annotated Fitzpatrick skin-type labels 1 through 6, named Fitzpatrick17k and Diverse Dermatology Images (DDI). Some researchers [9,48] used these datasets to test the efficacy of their proposed"fairness" approaches. However, they simplified the experimental settings by only testing generalizability on a subset of the data, i.e. training models on intermediate skin types (merging types 3 and 4) and testing on dark skins (merging types 5 and 6) without using data of fairer skin types 1 and 2 [9] or splitting multiple skin types into two sensitive groups; i.e. binary sensitive attribute classification task [48].

Fairness through blindness is a recent promising direction for mitigating unfairness in DNNs. Blindness to sensitive attributes is achieved by complementing the target DNN branch with a branch dedicated to classifying the sensitive attribute, e.g. gender. The framework captures semantic information to accurately perform the target task, e.g. classification, while simultaneously *minimizing* the accuracy of the sensitive-attribute classifier, learning sensitive-attribute agnostic representations. This can be achieved through adversarial learning [8,15,44] or disentangled representation learning [3,37,49]. Nevertheless, as noted in Elazar et al. [15] and Wang et al. [45], one of the main limitations of these methods is that they may hurt the representation learning of the target task. In particular, after sensitive attribute information has been removed from representations, the model might consider combinations of other attributes as a proxy and discard their information by mistake. This will result in accuracy deterioration when the information is related to the target task. For example, lesion color and visual features are critical for skin lesion classification [5], but are also useful for inferring the individual's skin type. To address this issue, we introduce contrastive learning that has shown impressive success in self-supervised pre-training tasks [24] and many other fields [30,41,43] for blindness-based fairness algorithms. Our contrastive loss encourages the representations of samples from the same target class to be proximate regardless of the value of their sensitive attributes, and those from different target classes to be distant. The network is thus enforced to retain discriminative semantic information about the target task.

In this work, we propose FairDisCo, a disentanglement framework with contrastive learning for fairness in dermatology, which not only discourages discrimination against sensitive attributes, i.e. skin-type information, but also preserves encoding the visual characteristics related to classification in the hidden representations through contrastive learning. The network contains a feature extractor and three branches: *target branch*, *sensitive attribute (SA) branch*, and *contrastive branch*. Specifically, the feature extractor encodes the input images into representations. The target branch utilizes these representations to make skin condition predictions. To mitigate the skin-type unfairness, the SA branch enforces the feature extractor to discard the skin-type information by minimizing the likelihood that the model correctly predicts skin types based on the representations. The contrastive branch utilizes supervised contrastive learning to improve the quality of representations and boost classification accuracy. To investigate the fairness in skin type comprehensively and in-depth, we conduct experiments using in-domain and out-domain classification tasks on multiple skin conditions and skin types. In in-domain classification tasks, the training and test sets have the same skin types, whereas in out-domain, they have distinct skin types. We also adapt two fairness metrics from binary classification and modify them for nonbinary classification. We additionally evaluate the performance of three widely used fairness methods (reweighting, resampling, and attribute-aware) to compare our FairDisCo against.

Our contributions could be summarized: (1) We propose a novel framework FairDisCo, featuring disentangled representation learning and contrastive learning, to promote fairness and boost classification accuracy. (2) To the best of our knowledge, we are the first to examine unfairness in skin lesion datasets using a variety of approaches and extensive experiments. (3) We employ three fairness metrics to better compare models, including two that we adapted from the binary sensitive attributes task. (4) FairDisCo achieves the best classification accuracy and fairness scores compared to other fairness-based methods and the baseline. Our code is available at https://github.com/siyi-wind/FairDisCo.

2   Related Works

2.1   Skin Lesion Diagnosis

Skin cancer is mostly diagnosed clinically by experts, starting with a preliminary clinical screening and possibly followed by a dermoscopic evaluation, a biopsy, or histopathological examination [16]. Traditionally, automatic image-based diagnosis is comprised of pre-processing, feature extraction, lesion segmentation, and classification. Dermatologists or machine learning classifiers make a diagnosis based on hand-crafted lesion features [27], following the ABCD rule [35], the CASH algorithm [26], or the seven-point checklist [25].

Deep learning classification techniques, which do not require segmentation or hand-crafted feature extraction stages, are currently the most popular approaches to automate skin disease detection. Kawahara et al. [29] applied a pre-trained convolutional neural network (CNN) as a feature extractor, which yields better results than prior works relying on general engineered features. Esteva et al. [16] then trained an end-to-end CNN outperforming 21 board-certified dermatologists on biopsy-proven clinical images. However, its excellent performance demands a sizable training dataset. Harangi et al. [22] further fused the outputs of several CNNs to reach higher accuracy with limited skin lesion images. One of the state-of-the-art algorithms in this field is the work by Gessert et al. [18]. They used different cropping strategies for input images with different resolutions. The final optimal model is aggregated from eight different CNNs inputting diverse image sizes through an ensemble method. Despite automatic classification methods flourishing in the past several years, the fairness issue in these models has not received much attention [1,10].

2.2   Fairness

Fairness in machine learning is an increasing concern for the public, governments, and scientists [34]. Many works have been proposed to reduce unfairness in deep learning, which can be categorized into 3 groups: pre-processing [6,28], in-processing [3,7,42,49], and post-processing [36,46].

Pre-processing methods aim to transform the data so that the underlying discrimination is removed. A representative work is from Kamiran et al. [28] who introduced and investigated four intuitive pre-processing methods to get classifiers in an optimal trade-off between accuracy and non-discrimination. To train a fairer model, *In-processing* techniques either modified the model architecture or added fairness-related penalties. Wadsworth et al. [42] added an adversarial architecture after a recidivism-predicting neural network, which helps to eliminate racial bias in the prediction. To get a sensitive-attribute agnostic representation, Sarhan et al. [39] disentangled meaningful and sensitive information by enforcing orthogonality constraints as a proxy for independence. Adding a fairness regularizer is a prospective direction. However, the main problem is that common fairness formulas are not differentiable, so they cannot be directly used in the objective function. Bendekgey et al. [7] introduced three new surrogates of fairness constraints for non-convex models, which can be applied to challenging computer vision and natural language processing problems. As for the *post-processing* methods, they aim to utilize the model's outputs and sensitive attributes to calibrate the model's prediction during inference. Petersen et al. [36] cast the individual fairness post-processing problem as a graph smoothing problem corresponding to graph Laplacian regularization that preserves the desired "treat similar individuals similarly" interpretation.

While creating models that are fair to age, sex, or race has become increasingly common, skin-type fairness in skin lesion diagnosis draws little attention. Kinyanjui et al. [32] found no clear relationship between skin type and segmentation performance in datasets. Bevan et al. [9] presented a modified variational autoencoder to uncover skin-type bias and executed a partial experiment, an out-domain classification on two skin-type groups. Wu et al. [48] proposed Fair-Prune, a strategy that pruned parameters during training to reduce the accuracy gap between different skin types, and validated the model on binary sensitive attributes by grouping six-scale Fitzpatrick annotations to two groups (light and dark). In this paper, we directly use the skin-type annotations in the dataset to

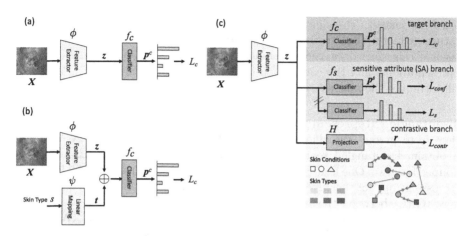

Fig. 1. Diagram of 3 skin disease classifiers: (a) Baseline; (b) Attribute-aware method; (c) Our proposed disentanglement network with contrastive learning (FairDisCo).

study multi-class classification with multiple sensitive attributes and perform in-domain and out-domain classification experiments on two datasets.

3 Methodology

In a multi-class skin lesion classification (M classes), the model is required to output a skin condition prediction y based on an RGB skin image $\boldsymbol{X} \in \mathbb{R}^{H \times W \times 3}$. We treat the skin type as a sensitive attribute s, including N groups with diverse types. Our goal is to model $p(y|\boldsymbol{X})$ without being affected by s. In Sect. 3.1, we outline our proposed framework FairDisCo, in which we incorporate three branches to disentangle skin-type information from the learned latent representations and enhance the feature extraction for higher classification accuracy. In Sect. 3.2, we further explore the existence and effects of the unfair skin-type issue in skin disease datasets. Specifically, we conduct a systematic study by comparing the proposed framework against the baseline and three simple but widely used fairness approaches: two pre-processing approaches (reweighting and resampling) and one in-processing approach (attribute-aware).

In the baseline (BASE) method (Fig. 1(a)), an image is input to a feature extractor ϕ to get a representation $\boldsymbol{z} = \phi(\boldsymbol{X})$ that is then passed to a classifier f_c, which includes one linear layer and a softmax activation function, to get a skin condition prediction $\boldsymbol{p}^c$. We utilize a cross-entropy loss L_c on skin conditions to optimize the whole architecture. Figure 1(b) depicts an attribute-aware method (ATRB) with an additional skin-type branch (further discussed in Sect. 3.2).

3.1 Proposed FairDisCo Model

The intuition behind FairDisCo is to train a model to avoid capturing spurious correlations between skin conditions and types while learning semantic information from images. To achieve that, we incorporate three branches following a feature extractor (Fig. 1(c)): the target branch to predict skin conditions as the baseline's classifier does, the SA branch to decouple skin-type information from representations, and the contrastive branch to enhance feature extraction. We will introduce the sensitive and the contrastive branches in detail.

Disentangled Representation Learning: The SA branch consists of a classifier f_s to predict the skin type p^s based on the representation z. Methods to obtain disentangled representation in *fairness through blindness* could be grouped into two. One is adversarial learning that utilizes an adversarial loss such as cross-entropy for sensitive attributes on the SA branch and a gradient-reversal layer (GRL) between the feature extractor and the classifier, i.e. the feature extractor receives opposite gradients from the SA branch [8,15,44]. When the adversarial loss is minimized, the classifier's ability to predict sensitive attributes is maximized, and the feature extractor's ability to attain sensitive attributes is minimized. Another direction is disentangled representation learning by using two different losses without GRL [3,49]. We follow the latter direction. FairDisCo minimizes a confusion loss given in Eq. 1 to confuse the feature extractor and remove the skin-type information from representations.

$$L_{conf} = -\sum_{i=1}^{N} \frac{1}{N} \log(p_i^s).$$ (1)

This loss is minimized when the classifier outputs equal probability p_i^s for all skin types i, i.e., the representations are free of skin-type information. Notice the classifier f_s might learn a tricky solution like setting all the weights to zero, then p^s is a zero vector, and L_{conf} becomes the smallest even though the representation still contains skin-type information. Thus, we add a skin-type predictive cross-entropy loss L_s only optimizing f_s. These two losses are opposite and serve the same purpose as the adversarial loss and GRL in the first research direction.

Contrastive Feature Extraction Enhancement: Although disentangled representation learning keeps the feature extractor blind to skin-type information, it might hurt the extractor to encode semantic information by discarding important features related to the target task that could indicate skin type. Thus we add a contrastive loss [30] that promotes intra-class cohesion and inter-class diversity to protect target features and improve the representation learning. In the contrastive branch (Fig. 1(c)), we first project representations from the feature extractor into a low dimensional latent space $r = H(z)$. For each embedding in the mini-batch, we split other embeddings with the same disease labels into a positive set P_y and the rest into a negative set N_y (regardless of the skin type). We calculate a contrastive loss as follows.

$$L_{contr} = -\frac{1}{|P_y|} \sum_{p \in P_y} \log \frac{\exp(\Psi(r,p)/\tau)}{\exp(\Psi(r,p)/\tau) + \sum_{n \in N_y} \exp(\Psi(r,n)/\tau)}.$$ (2)

$\Psi(\cdot,\cdot)$ is the cosine similarity between two vectors. $\tau > 0$ is a temperature parameter. We minimize the contrastive loss to optimize the feature extractor and the projection head, thus enforcing the network to keep as many factors to push samples in the same class together (orange lines in Fig. 1(c)) and push away those in other classes without considering differing skin types (blue lines in Fig. 1). Our final loss function for FairDisCo is:

$$L_{total} = L_c(\theta_\phi, \theta_{f_c}) + \alpha L_{conf}(\theta_\phi, \theta_{f_s}) + L_s(\theta_{f_s}) + \beta L_{contr}(\theta_\phi, \theta_H). \quad (3)$$

We use α and β to adjust contributions of confusion loss and contrastive loss. Notice L_s is only used to optimize f_s.

3.2 An Investigation for Three Approaches

To better understand the unfairness issue in skin lesion datasets and enrich our assessment of different fairness models, we study three widely used pre-processing and in-processing fairness algorithms.

Resampling Algorithm (RESM) considers samples in the same skin type and condition are in the same group and then oversamples minorities and undersamples majorities to construct a balanced dataset [28], forcing the model to treat groups equally. We set sampling weights for groups as their inverse frequency.

Reweighting Algorithm (REWT) seeks to make skin types and conditions independent to prevent the model from learning discriminatory features [28]. The expected probability of one group is a multiplication between the probabilities of the skin condition group and the skin-type group. However, we usually observed a lower probability due to the correlation between two attributes. To cut off the connection between them, we assign each sample the weight as follows:

$$w(X) = \frac{P_{\exp}(X(s) = s_i, X(y) = y_j)}{P_{obs}(X(s) = s_i, X(y) = y_j)}. \quad (4)$$

The weight will be multiplied by the cross-entropy loss L_c to adjust the loss contribution for each group.

Attribute-aware Algorithm (ATRB) was described in [49] where it adds skin-type information to the model so that its prediction will not be dominated by major groups. In Fig. 1(b), the feature extractor ϕ outputs a representation vector z. The skin type s is first converted to a one-hot vector and then mapped through one linear layer ψ to a representation vector t that is the same size as z. These two representations are then summed together to get a sensitive attribute-aware representation, which is finally sent into a classifier to get a skin condition prediction. The model is trained through L_c as the baseline.

4 Experiments

Datasets: We study two skin lesion datasets: Fitzpatrick17k dataset [21] and Diverse Dermatology Images (DDI) dataset [12]. In the Fitzpatrick17k dataset,

Groh et al. [21] compiled 16,577 clinical images with skin condition labels and annotated them with Fitzpatrick skin-type labels. There are 114 different skin conditions, and each one has at least 53 images. They further divided these skin conditions into two more advanced categories: 3 (malignant, non-neoplastic, benign) and 9. Fitzpatrick labelling system is a six-point scale initially developed for classifying sun reactivity of skin and adjusting clinical treatment according to skin phenotype [17]. Recently, it has been used in computer vision to evaluate algorithmic fairness [23]. The samples in the dataset are labelled by 6 Fitzpatrick skin types and 1 unknown type. In our experiments, we ignore all samples of the unknown skin type. The DDI dataset contains 656 images with diverse skin types and pathologically confirmed skin condition labels, including 78 detailed disease labels and malignant identification. They grouped 6 Fitzpatrick scales into 3 groups: Fitzpatrick-12, Fitzpatrick-34, and Fitzpatrick-56, where each contains a pair of skin-type classes, i.e., {1,2}, {3,4} and {5,6}, respectively.

Metrics: We use accuracy to measure models' skin condition classification performance. To quantify fairness, we adapt earlier fairness metrics, which were restricted to binary classification or binary sensitive attributes [13], to our task, i.e., >2 disease classes and >2 skin types, resulting in 3 metrics: (i) Predictive Quality Disparity (PQD) measures the prediction quality difference between each sensitive group, which we compute as the ratio between the lowest accuracy to the highest accuracy across different skin-type groups, i.e.,

$$PQD = \frac{\min(acc_j, j \in S)}{\max(acc_j, j \in S)}, \tag{5}$$

where S is the set of skin types. (ii) Demographic Disparity (DP) computes the percentage diversities of positive outcomes for each sensitive group. (iii) Equality of Opportunity (EO) asserts that different sensitive groups should have similar true positive rates. We calculate DPM and EOM across multiple skin conditions, $m \in \{1, 2, \cdots, M\}$, as follows:

$$DPM = \frac{1}{M} \sum_{i=1}^{M} \frac{\min[p(\hat{y} = i | s = j), j \in S]}{\max[p(\hat{y} = i | s = j), j \in S]} \tag{6}$$

$$EOM = \frac{1}{M} \sum_{i=1}^{M} \frac{\min[p(\hat{y} = i | y = i, s = j), j \in S]}{\max[p(\hat{y} = i | y = i, s = j), j \in S]}, \tag{7}$$

where y is the ground-truth skin condition label and $\hat{y}$ is the model prediction. A model is fairer if it has higher values for the above three metrics.

Implementation Details: For the Fitzpatrick17k dataset, we carry out a three-class classification and follow Groh et al. [21] in performing two experimental tasks. The first is an in-domain classification, in which the train and test sets are randomly split in an 8:2 ratio (13261:3318). The second is an out-domain classification, where we train models on samples of two skin types and test on samples of other skin types. We only use samples from those skin conditions present in both the train and the test sets and finally conduct three experimental settings $\mathcal{A} - \mathcal{C}$. $\mathcal{A}$ trains models on skin type 1–2 (7,755 samples) and tests on others (8,257 samples), $\mathcal{B}$ trains models on skin type 3–4 (6,089 samples) and tests on others (10,488 samples), and $\mathcal{C}$ trains models on skin type 5–6 (2,168 samples) and tests on others (14,409 samples). For the DDI dataset, we perform an in-domain binary classification (malignant vs non-malignant) using the same train-test ratio of 8:2 (524:132). For all the models, we use a pre-trained ResNet-18 without the final fully connected layer as the feature extractor. The dimension of the representation vector is 512. A Multi-Layer Perceptron (MLP) with a single hidden layer of size 512 and an output layer of size 128 serves as the projection head of the contrastive branch in FairDisCo. We follow Groh et al. [21] in using a weighted random sampler to ensure equal numbers of samples for each skin condition in one mini-batch, except in RESM. Images are augmented through random cropping, rotation, and flipping to boost data diversity, then resized to $224 \times 224 \times 3$. We use Adam [31] optimizer to train the model with an initial learning rate 1×10^{-4}, which changes through a linear decay scheduler whose step size is 2, and decay factor γ is 0.9. We deploy models on a single TITAN V GPU and train them with a batch size of 64. We set the training epochs for the Fitzpatrick17k dataset to 20 and the DDI dataset to 15.

4.1   Results on the Fitzpatrick17k Dataset

Data Statistic: In Table 1, numerous data biases occur in skin type and skin condition. The non-neoplastic group dominates the skin conditions with a 73% share, and the type-2 group has the most samples of all skin-type groups, with a proportion that is 7.5 times higher than that of the minor type-6 group. Additionally, there are apparent discriminations between target and sensitive attributes. For instance, the group whose condition is malignant and type is 1 has an expected probability of 0.134 yet its observed probability is lower, 0.128, indicating that these two attributes are relevant (calculated as the way in REWT).

In-domain Classification Results and Discussion: We use accuracy and fairness metrics to examine the models in Sect. 3 and the model in [21] (we name it GROH) that contains a fixed VGG-16 and one learnable linear classifier. FairDisCo$^{\oslash}$ is the framework FairDisCo without the contrastive loss. Examining the results (Table 2 and Fig. 2), we make the following observations.

(1) In Table 2, our baseline (BASE) exhibits inconsistent performance (accuracy) that varies across skin types, showing the existence of skin-type unfairness. Nevertheless, by changing the backbone and adjusting the training procedure, the baseline achieves much higher accuracy than GROH (on average 85.11% vs 60.07%). Since the average accuracy for GROH is too low compared with other methods, we exclude it from subsequent analysis.

(2) We observe the phenomena mentioned in [21] where the accuracy of the major skin-type group is lower than the underrepresented skin-type group.

Table 1. Data distributions for skin type and skin condition from the Fitzpatrick17k (Fitz) and DDI datasets.

	Skin Condition	Skin Type						
		T1	T2	T3	T4	T5	T6	Total
Fitz	Benign	444	671	475	367	159	44	2160
	Malignant	453	742	456	301	147	61	2160
	Non-neoplastic	2050	3395	2377	2113	1227	530	11692
	Total	2947	4808	3308	2781	1533	635	16012
DDI		T12		T34		T45		Total
	Malignant	49		74		48		171
	Non-malignant	159		167		159		485
	Total	208		241		207		656

Table 2. In-domain classification result comparison on the Fitzpatrick17k dataset. Accuracy, PQD, DPM, and EOM are expressed in percentages (%).

Model	Accuracy (%) ↑							PQD ↑	DPM ↑	EOM ↑
	Avg	T1	T2	T3	T4	T5	T6			
GROH [21]	60.07	56.19	60.40	57.58	61.51	65.72	70.83	79.32	60.75	77.86
BASE	85.11	82.94	82.21	86.45	87.77	89.75	88.33	91.60	46.99	67.88
RESM	85.45	82.27	82.63	86.45	89.39	90.11	89.17	91.31	46.98	67.56
REWT	85.23	81.77	83.04	85.57	89.39	89.40	89.17	91.47	48.92	70.64
ATRB	85.30	82.78	81.70	85.86	90.65	90.81	85.83	90.00	56.67	67.91
FairDisCo$^{\oslash}$	85.61	83.44	83.17	85.42	89.57	90.81	86.67	91.60	57.64	74.52
FairDisCo	85.76	84.45	82.94	86.60	88.49	90.11	87.50	92.04	55.53	75.62

The ambiguous systematic pattern in accuracy across skin types could be seen in all the models in Table 2, e.g. type 6 has higher accuracy than type 2. We attribute it to the domain distribution diversity, different diagnosis difficulties across skin types, and the possibly untrustworthy results for those groups with few test samples. We analyze the data in Fig. 2(a, b).

The Fitzpatrick17k dataset is collected from two domains: DermaAmin (Derm) and Atlas Dermatologico (Atla) [2,38]. As illustrated in Fig. 2(a), Derm has the most samples in type 2, the fewest samples in type 6, and an average accuracy of 83.05%, whereas Atla has the most samples in type 4, the fewest samples in type 1, and an average accuracy of 91.75%. When we look into accuracy by skin type, type 6 in Atla (black line) has the highest accuracy compared to other skin types, but type 6 in Derm (blue line) is in fourth place. When combining these two domains, its total accuracy (orange line) is in second place, different from that in Derm or Atla. The diversities of distribution and classification difficulty across domains conceal the real pat-

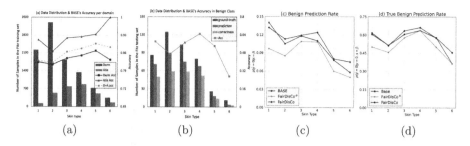

Fig. 2. (a) Training data distributions in Derm and Atla and corresponding accuracy of BASE; (b) Ground-truth, prediction, correctness, and accuracy distributions of the benign class in the Fitzpatrick17k test set; (c) Comparison of BASE, FairDisCo$^{\oslash}$, and FairDisCo in benign prediction rate (the main term for calculating DPM) across skin types; (d) Comparison of BASE, FairDisCo$^{\oslash}$, and FairDisCo in true benign rate (the main term for calculating EOM) across skin types. (Color figure online)

tern when we directly train models on two domains and only report the overall accuracy. In addition, Fig. 2(b) shows that BASE's prediction (gray bars) and ground-truth (blue bars) in benign class have similar distributions, with the most number of samples in type 2 and the least in type 6, indicating the model's prediction is affected by skin type. However, the correctness distribution is quite different. Type 2–4 have a similar number of correctness (green bars) despite type 2 having more benign predictions. Thus type 2 is less accurate than other skin types, i.e. it has more ground-truth and predictions, but less correctness. Then we could hypothesize the images of type 2 might be more challenging to classify than those of type 4 and 5, confusing the relationship between accuracy and statistical data distribution. Moreover, the small group size might make the test results unreliable. In our test case, the group whose skin condition is benign and skin type is 6 only has 11 samples, and the group whose condition is malignant and type is 6 just has 7 samples. A correct or wrong prediction will significantly impact their accuracy, making the results of underrepresented groups untrustworthy.

(3) Compared with BASE, all fairness techniques (RESM, REWT, ATRB, FairDisCo$^{\oslash}$, and FairDisCo) achieve higher average accuracy (Table 2). Additionally, these methods, except RESM, surpass the baseline on more than two fairness metrics. RESM performs worse than BASE on three fairness metrics due to the serious data imbalance, as illustrated in Table 2.

(4) In Table 2, FairDisCo achieves the highest accuracy (outperforms BASE by 0.65% on average accuracy) and fairness (outperforms BASE by 0.44%, 8.54%, and 7.74% on three fairness metrics). The difference between PQD is much smaller than DPM and EOM because the latter two give equal weight to each class when summing them together; thus, they will obviously decrease when one class is extremely unfair. In addition, FairDisCo outperforms FairDisCo$^{\oslash}$ on average accuracy and two fairness metrics, demonstrating the effectiveness of contrastive learning. The gain in fairness might be

because the contrastive loss only uses target label information to cluster the data, resulting in representations invariant across skin types when regarding each skin-type group as a domain.

(5) We further visualize the benign prediction rate $p(\hat{y} = 0|s = j), j \in S$ used to calculate DPM and true benign rate $p(\hat{y} = 0|y = 0, s = j)$ used to calculate EOM in Fig. 2(c, d). BASE has diverse values across skin types, indicating the model is affected by skin-type information. Unlike BASE (black line), the benign prediction rates for both FairDisCo (blue line) and FairDisCo$^{\oslash}$ (green line) present little variation across skin types. In particular, FairDisCo boosts the minor skin-type group's benign prediction rate and true benign rate and performs the best in decreasing the model's diversities across skin types, which exemplifies that our model reduces skin-type unfairness. Finally, we could conclude that skin type in the Fitzpatrick17k dataset influences the models' decisions despite no systematic pattern between accuracy and sensitive group size.

Table 3. Out-domain classification result comparison on the Fitzpatrick17k dataset. The first column (E) presents the experiment index. FairDisCo is abbreviated as FDC.

E	Model	Accuracy (%) ↑							PQD ↑	DPM ↑	EOM ↑
		Avg	T1	T2	T3	T4	T5	T6			
$\mathcal{A}$	BASE	80.33	–	–	80.20	80.65	<u>79.58</u>	<u>81.42</u>	97.74	67.48	69.09
	RESM	79.13	–	–	80.44	79.11	77.76	75.75	94.17	79.49	63.29
	REWT	79.16	–	–	79.05	79.04	79.19	80.16	<u>98.60</u>	69.56	<u>72.30</u>
	ATRB	79.52	–	–	80.17	80.04	77.56	78.58	96.74	65.76	62.60
	FDC$^{\oslash}$	<u>80.50</u>	–	–	80.93	80.80	79.00	80.79	97.80	67.34	71.15
	FDC	80.37	–	–	<u>81.41</u>	<u>81.12</u>	78.02	77.32	94.98	<u>74.34</u>	62.11
$\mathcal{B}$	BASE	78.38	73.91	77.83	–	–	86.82	82.99	85.12	61.33	71.51
	RESM	78.74	74.69	<u>78.62</u>	–	–	85.91	81.10	86.94	71.03	<u>74.96</u>
	REWT	77.60	72.89	77.18	–	–	85.58	83.31	85.17	53.97	64.90
	ATRB	78.23	72.99	77.79	–	–	<u>86.95</u>	<u>84.88</u>	83.94	57.34	73.53
	FDC$^{\oslash}$	<u>78.81</u>	<u>75.20</u>	78.12	–	–	85.52	84.57	<u>87.93</u>	53.80	70.48
	FDC	78.61	74.62	78.56	–	–	85.52	80.79	87.25	<u>71.44</u>	69.53
$\mathcal{C}$	BASE	73.49	69.11	70.98	<u>74.65</u>	81.01	–	–	85.31	61.78	79.33
	RESM	73.70	69.05	71.83	74.56	80.76	–	–	85.50	63.23	77.65
	REWT	69.75	61.70	66.95	72.30	79.93	–	–	77.20	56.24	75.50
	ATRB	72.92	68.18	70.90	73.49	80.65	–	–	84.54	67.80	77.79
	FDC$^{\oslash}$	<u>74.11</u>	70.36	<u>72.08</u>	74.10	<u>81.56</u>	–	–	86.27	67.18	79.02
	FDC	73.64	<u>70.63</u>	71.64	73.61	80.25	–	–	<u>88.01</u>	70.69	<u>83.69</u>

Out-Domain Classification Results and Discussion. We conduct experiments on the baseline and other four fairness methods and show their accuracies and fairness metrics. From Table 3, we could acquire the following findings.

(1) BASE has higher accuracy on those skin types similar to those it is trained on, such as type 4's accuracy in Experiment $\mathcal{C}$ is 11.90% higher than type 1's. This reveals that the model's decision-making is relevant to the skin type, which is consistent with the phenomenon indicated in [21].

(2) For fairness approaches, ATRB performs worse than BASE, since it depends on skin-type information to pay equal attention to each group but cannot access skin type in the test set during training. RESM improves the accuracy and at least two fairness metrics in Experiments $\mathcal{C}$ and $\mathcal{D}$.

Table 4. In-domain classification result comparison on the DDI dataset.

Model	Accuracy (%) ↑				PQD ↑	DPM ↑	EOM ↑	AUC ↑
	Avg	T12	T34	T56				
BASE	82.58	83.78	84.31	79.55	94.34	79.28	80.35	0.81
RESM	82.58	83.78	80.39	84.09	95.60	79.28	93.01	0.80
REWT	82.58	81.08	78.43	88.64	88.49	83.39	82.45	0.82
ATRB	81.82	75.68	82.35	86.36	87.62	85.16	72.42	0.82
FairDisCo$^{\oslash}$	82.58	81.08	80.39	86.37	93.09	77.45	90.54	0.82
FairDisCo	83.33	83.78	84.31	81.82	97.04	83.62	83.46	0.85

(3) FairDisCo$^{\oslash}$ reaches the best results (average accuracy and three fairness metrics) in three experiments. Its increase in minority accuracy serves as further evidence of the unfair skin-type problem. FairDisCo is the second-best among all the algorithms, which raises the average accuracy on three experiments and has better fairness performance on $\mathcal{B}$ and $\mathcal{C}$ in comparison to BASE. We attribute its inferior performance compared with FairDisCo$^{\oslash}$ to that the contrastive loss only helps the model learn better representations of skin-type groups appearing in the training set. However, the out-domain classification task has a large diversity across skin types between the training and test sets.

4.2 Results on the DDI Dataset

Data Statistic: Compared with the Fitzpatrick17k dataset, the DDI dataset contains similar skin-condition data imbalance but less skin-type bias (Table 1). 74% of cases are non-malignant, while the rates for the three skin-type groups are 1:1.16:1. Skin condition and skin type are also dependent in this dataset, e.g. the group whose condition is non-malignant and type is 12 has an expected probability of 0.083 whereas an observed probability of 0.075.

In-domain Classification Results and Discussion: The compared models are the same as those of the in-domain classification in Sect. 4.1 except GROH. We further add another classification performance measurement, area under the ROC curve (AUC). From Table 4, we observe the following.

(1) For BASE, the major skin-type group (T34) is more accurate than the minor groups (T12 and T34), showing it ignores the underrepresented groups.
(2) All the fairness approaches boost the minor group's (T56) accuracy. RESM and REWT improve two fairness metrics while maintaining the average accuracy, but ATRB does not outperform BASE.
(3) FairDisCo not only yields the best classification performance (outperforms BASE by 0.75% on average accuracy and 4% on AUC) but also achieves the best fairness results (outperforms BASE by 2.7%, 4.34%, and 3.11% on three fairness metrics). Compared to FairDisCo$^{\oslash}$, the superior average accuracy, PQD, DPM, and AUC of FairDisCo further demonstrate that the contrastive loss plays an integral role in learning a better semantic representation.

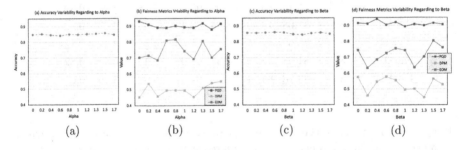

(a)   (b)   (c)   (d)

Fig. 3. (a) and (b) are average accuracy and fairness metrics trends based on confusion loss weight α. (c) and (d) are the trends based on contrastive loss weight β.

4.3   Loss Analysis

To analyze the sensitivity of FairDisCo to the confusion loss weight α and contrastive loss weight β of Eq. 3, we run experiments on the Fitzpatrick17k dataset where we fix α to 1.0 while adjusting β, and vice versa, and report the accuracy and fairness metric values in Fig. 3. When α and β change, even as the accuracy, PQD, DPM, and EOM values vary by 1.84%, 5.19%, 13.1% and 18.2%, respectively, FairDisCo mostly continues to outperform BASE on the 4 metrics, and outperforms RESM, REWT, and ATRB on at least two fairness metrics. We found no obvious relation between the model's performance and the loss composition. Though, unsurprisingly, excessively increasing the contribution of either loss can lead to a decreased model performance.

5 Conclusion

We investigated skin-type fairness in skin lesion classification datasets and models. We improved the fairness and boosted the classification accuracy by proposing a novel deep neural network, FairDisCo. FairDisCo leverages a network branch to disentangle skin-type information from the latent representation and a contrastive branch to avoid accuracy deterioration. To demonstrate the presence of unfairness in datasets and the effectiveness of FairDisCo, we compared the baseline and FairDisCo to 3 widely used fairness techniques (reweighting, resampling, and attribute-aware). Moreover, as there are no standard fairness metrics for evaluating multi-class classification with multiple sensitive attributes, and as using only the ratio of accuracy per skin type ignores the accuracy per skin condition, we proposed three fairness metrics PQD, DPM, and EOM for our task which could also be applied to other fairness tasks. We compared models on in-domain and out-domain classification experiments using the Fitzpatrick17k and DDI, two datasets with diverse skin types and diseases, and clarified the relationship between accuracy and group size in the Fitzpatrick17k in-domain classification. The results indicate that the datasets have skin-type bias, and fairness methods could diminish the discrimination. Our proposed model FairDisCo achieved the best overall accuracy and fairness on in-domain classification tasks and surpassed the baseline on the out-domain classification task.

References

1. Adegun, A., Viriri, S.: Deep learning techniques for skin lesion analysis and melanoma cancer detection: a survey of state-of-the-art. Artif. Intell. Rev. **54**(2), 811–841 (2021)
2. AlKattash, J.A.: Dermaamin. https://www.dermaamin.com/site/ (2022)
3. Alvi, Mohsan, Zisserman, Andrew, Nellåker, Christoffer: Turning a blind eye: explicit removal of biases and variation from deep neural network embeddings. In: Leal-Taixé, Laura, Roth, Stefan (eds.) ECCV 2018. LNCS, vol. 11129, pp. 556–572. Springer, Cham (2019). https://doi.org/10.1007/978-3-030-11009-3_34
4. Balch, C.M., et al.: Final version of 2009 AJCC melanoma staging and classification. J. Clin. Oncol. **27**(36), 6199 (2009)
5. Barata, C., Celebi, M.E., Marques, J.S.: A survey of feature extraction in dermoscopy image analysis of skin cancer. IEEE J. Biomed. Health Inform. **23**(3), 1096–1109 (2018)
6. Bellamy, R.K., et al.: Ai fairness 360: an extensible toolkit for detecting and mitigating algorithmic bias. IBM J. Res. Dev. **63**(4/5), 1–4 (2019)
7. Bendekgey, H., Sudderth, E.: Scalable and stable surrogates for flexible classifiers with fairness constraints. In: Advances in Neural Information Processing Systems, vol. 34 (2021)
8. Beutel, A., Chen, J., Zhao, Z., Chi, E.H.: Data decisions and theoretical implications when adversarially learning fair representations. arXiv preprint arXiv:1707.00075 (2017)
9. Bevan, P.J., Atapour-Abarghouei, A.: Detecting melanoma fairly: skin tone detection and debiasing for skin lesion classification. arXiv preprint arXiv:2202.02832 (2022)

10. Bhardwaj, Aditya, Rege, Priti P..: Skin lesion classification using deep learning. In: Merchant, S.. N.., Warhade, Krishna, Adhikari, Debashis (eds.) Advances in Signal and Data Processing. LNEE, vol. 703, pp. 575–589. Springer, Singapore (2021). https://doi.org/10.1007/978-981-15-8391-9_42

11. Chabi Adjobo, E., Sanda Mahama, A.T., Gouton, P., Tossa, J.: Towards accurate skin lesion classification across all skin categories using a PCNN fusion-based data augmentation approach. Computers 11(3), 44 (2022)

12. Daneshjou, R., et al.: Disparities in dermatology AI: assessments using diverse clinical images. arXiv preprint arXiv:2111.08006 (2021)

13. Du, M., Yang, F., Zou, N., Hu, X.: Fairness in deep learning: a computational perspective. IEEE Intell. Syst. 36(4), 25–34 (2020)

14. El-Khatib, H., Popescu, D., Ichim, L.: Deep learning-based methods for automatic diagnosis of skin lesions. Sensors 20(6), 1753 (2020)

15. Elazar, Y., Goldberg, Y.: Adversarial removal of demographic attributes from text data. In: Proceedings of the 2018 Conference on Empirical Methods in Natural Language Processing (EMNLP), pp. 11–21 (2018)

16. Esteva, A., et al.: Dermatologist-level classification of skin cancer with deep neural networks. Nature 542(7639), 115–118 (2017)

17. Fitzpatrick, T.B.: The validity and practicality of sun-reactive skin types I through VI. Arch. Dermatol. 124(6), 869–871 (1988)

18. Gessert, N., Nielsen, M., Shaikh, M., Werner, R., Schlaefer, A.: Skin lesion classification using ensembles of multi-resolution efficientnets with meta data. MethodsX 7, 100864 (2020)

19. Gessert, N., et al.: Skin lesion classification using CNNs with patch-based attention and diagnosis-guided loss weighting. IEEE Trans. Biomed. Eng. 67(2), 495–503 (2019)

20. Groh, M., Harris, C., Daneshjou, R., Badri, O., Koochek, A.: Towards transparency in dermatology image datasets with skin tone annotations by experts, crowds, and an algorithm. arXiv preprint arXiv:2207.02942 (2022)

21. Groh, M., et al.: Evaluating deep neural networks trained on clinical images in dermatology with the fitzpatrick 17k dataset. In: Proceedings of the IEEE/CVF Conference on Computer Vision and Pattern Recognition (CVPR), pp. 1820–1828 (2021)

22. Harangi, B.: Skin lesion classification with ensembles of deep convolutional neural networks. J. Biomed. Inform. 86, 25–32 (2018)

23. Hazirbas, C., Bitton, J., Dolhansky, B., Pan, J., Gordo, A., Ferrer, C.C.: Casual conversations: A dataset for measuring fairness in AI. In: Proceedings of the IEEE/CVF Conference on Computer Vision and Pattern Recognition (CVPR), pp. 2289–2293 (2021)

24. He, K., Fan, H., Wu, Y., Xie, S., Girshick, R.: Momentum contrast for unsupervised visual representation learning. In: Proceedings of the IEEE/CVF Conference on Computer Vision and Pattern Recognition (CVPR), pp. 9729–9738 (2020)

25. Healsmith, M., Bourke, J., Osborne, J., Graham-Brown, R.: An evaluation of the revised seven-point checklist for the early diagnosis of cutaneous malignant melanoma. Br. J. Dermatol. 130(1), 48–50 (1994)

26. Henning, J.S., et al.: The cash (color, architecture, symmetry, and homogeneity) algorithm for dermoscopy. J. Am. Acad. Dermatol. 56(1), 45–52 (2007)

27. Jamil, U., Khalid, S.: Comparative study of classification techniques used in skin lesion detection systems. In: 17th IEEE International Multi Topic Conference 2014, pp. 266–271. IEEE (2014)

28. Kamiran, F., Calders, T.: Data preprocessing techniques for classification without discrimination. Knowl. Inf. Syst. **33**(1), 1–33 (2012)
29. Kawahara, J., BenTaieb, A., Hamarneh, G.: Deep features to classify skin lesions. In: 2016 IEEE 13th international symposium on biomedical imaging (ISBI), pp. 1397–1400. IEEE (2016)
30. Khosla, P., et al.: Supervised contrastive learning. Adv. Neural. Inf. Process. Syst. **33**, 18661–18673 (2020)
31. Kingma, D.P., Ba, J.: Adam: a method for stochastic optimization. In: ICLR (2015)
32. Kinyanjui, Newton M.., et al.: Fairness of classifiers across skin tones in dermatology. In: Martel, Anne L.., et al. (eds.) MICCAI 2020. LNCS, vol. 12266, pp. 320–329. Springer, Cham (2020). https://doi.org/10.1007/978-3-030-59725-2_31
33. Li, X., Cui, Z., Wu, Y., Gu, L., Harada, T.: Estimating and improving fairness with adversarial learning. arXiv preprint arXiv:2103.04243 (2021)
34. Mehrabi, N., Morstatter, F., Saxena, N., Lerman, K., Galstyan, A.: A survey on bias and fairness in machine learning. ACM Comput Surv. (CSUR) **54**(6), 1–35 (2021)
35. Nachbar, F., et al.: The ABCD rule of dermatoscopy: high prospective value in the diagnosis of doubtful melanocytic skin lesions. J. Am. Acad. Dermatol. **30**(4), 551–559 (1994)
36. Petersen, F., Mukherjee, D., Sun, Y., Yurochkin, M.: Post-processing for individual fairness. In: Advances in Neural Information Processing Systems, vol. 34 (2021)
37. Puyol-Antón, Esther, et al.: Fairness in cardiac MR image analysis: an investigation of bias due to data imbalance in deep learning based segmentation. In: de Bruijne, Marleen, et al. (eds.) MICCAI 2021. LNCS, vol. 12903, pp. 413–423. Springer, Cham (2021). https://doi.org/10.1007/978-3-030-87199-4_39
38. Samuel, F.D.S.: Atlas dermatologico. http://atlasdermatologico.com.br/index.jsf (2022)
39. Sarhan, Mhd Hasan, Navab, Nassir, Eslami, Abouzar, Albarqouni, Shadi: Fairness by learning orthogonal disentangled representations. In: Vedaldi, Andrea, Bischof, Horst, Brox, Thomas, Frahm, Jan-Michael. (eds.) ECCV 2020. LNCS, vol. 12374, pp. 746–761. Springer, Cham (2020). https://doi.org/10.1007/978-3-030-58526-6_44
40. Sung, H., et al.: Global cancer statistics 2020: Globocan estimates of incidence and mortality worldwide for 36 cancers in 185 countries. CA: Can. J. Clin. **71**(3), 209–249 (2021)
41. Thota, M., Leontidis, G.: Contrastive domain adaptation. In: Proceedings of the IEEE/CVF Conference on Computer Vision and Pattern Recognition (CVPR), pp. 2209–2218 (2021)
42. Wadsworth, C., Vera, F., Piech, C.: Achieving fairness through adversarial learning: an application to recidivism prediction (2018)
43. Wang, P., Han, K., Wei, X.S., Zhang, L., Wang, L.: Contrastive learning based hybrid networks for long-tailed image classification. In: Proceedings of the IEEE/CVF Conference on Computer Vision and Pattern Recognition (CVPR), pp. 943–952 (2021)
44. Wang, T., Zhao, J., Yatskar, M., Chang, K.W., Ordonez, V.: Balanced datasets are not enough: estimating and mitigating gender bias in deep image representations. In: Proceedings of the IEEE/CVF International Conference on Computer Vision (ICCV), pp. 5310–5319 (2019)
45. Wang, Z., et al.: Towards fairness in visual recognition: effective strategies for bias mitigation. In: Proceedings of the IEEE/CVF Conference on Computer Vision and Pattern Recognition (CVPR), pp. 8919–8928 (2020)

46. Wang, Z., et al.: Fairness-aware adversarial perturbation towards bias mitigation for deployed deep models. In: Proceedings of the IEEE/CVF Conference on Computer Vision and Pattern Recognition (CVPR), pp. 10379–10388 (2022)
47. WHO: Cancer (2022). https://www.who.int/news-room/fact-sheets/detail/cancer
48. Wu, Y., Zeng, D., Xu, X., Shi, Y., Hu, J.: Fairprune: achieving fairness through pruning for dermatological disease diagnosis. arXiv preprint arXiv:2203.02110 (2022)
49. Xu, Tian, White, Jennifer, Kalkan, Sinan, Gunes, Hatice: Investigating bias and fairness in facial expression recognition. In: Bartoli, Adrien, Fusiello, Andrea (eds.) ECCV 2020. LNCS, vol. 12540, pp. 506–523. Springer, Cham (2020). https://doi.org/10.1007/978-3-030-65414-6_35

CIRCLe: Color Invariant Representation Learning for Unbiased Classification of Skin Lesions

Arezou Pakzad$^{(\boxtimes)}$ ⓘ, Kumar Abhishek ⓘ, and Ghassan Hamarneh ⓘ

School of Computing Science, Simon Fraser University, Burnaby, Canada
{arezou_pakzad,kabhishe,hamarneh}@sfu.ca

Abstract. While deep learning based approaches have demonstrated expert-level performance in dermatological diagnosis tasks, they have also been shown to exhibit biases toward certain demographic attributes, particularly skin types (e.g., light versus dark), a fairness concern that must be addressed. We propose CIRCLe, a skin color invariant deep representation learning method for improving fairness in skin lesion classification. CIRCLe is trained to classify images by utilizing a regularization loss that encourages images with the same diagnosis but different skin types to have similar latent representations. Through extensive evaluation and ablation studies, we demonstrate CIRCLe's superior performance over the state-of-the-art when evaluated on 16k+ images spanning 6 Fitzpatrick skin types and 114 diseases, using classification accuracy, equal opportunity difference (for light versus dark groups), and normalized accuracy range, a new measure we propose to assess fairness on multiple skin type groups. Our code is available at https://github.com/arezou-pakzad/CIRCLe.

Keywords: Fair AI · Skin type bias · Dermatology · Classification · Representation learning

1 Introduction

Owing to the advancements in deep learning (DL)-based data-driven learning paradigm, convolutional neural networks (CNNs) can be helpful decision support tools in healthcare. This is particularly true for dermatological applications where recent research has shown that DL-based models can reach the dermatologist-level classification accuracies for skin diseases [10,20,24] while doing so in a clinically interpretable manner [7,38]. However, this data-driven learning paradigm that allows models to automatically learn meaningful representations from data leads DL models to mimic biases found in the data, i.e., biases in the data can propagate through the learning process and result in an inherently biased model, and consequently in a biased output.

Most public skin disease image datasets are acquired from demographics consisting primarily of fair-skinned people. However, skin conditions exhibit vast visual differences in manifestations across different skin types [56]. Lighter

© The Author(s), under exclusive license to Springer Nature Switzerland AG 2023
L. Karlinsky et al. (Eds.): ECCV 2022 Workshops, LNCS 13804, pp. 203–219, 2023.
https://doi.org/10.1007/978-3-031-25069-9_14

skinned populations suffer from over-diagnosis of melanoma [2] while darker skinned patients get diagnosed at later stages, leading to increased morbidity and mortality [4]. Despite this, darker skin is under-represented in most publicly available data sets [31,36], reported studies [16], and in dermatology textbooks [3]. Kinyanjui et al. [31] performed an analysis on two popular benchmark dermatology datasets: ISIC 2018 Challenge dataset [13] and SD-198 dataset [53], to understand the skin type representations. They measured the individual typology angle (ITA), which measures the constitutive pigementation of skin images [43], to estimate the skin tone on these datasets, and found that the majority of the images in the two datasets ITA values between 34.8° and 48°, which are associated with lighter skin. This is consistent with the under-representation of darker skinned populations in these datasets. It has been shown that CNNs perform best at classifying skin conditions for skin types that are similar to those they were trained on [23]. Thus, the data imbalance across different skin types in the majority of the skin disease image datasets can manifest as racial biases in the DL models' predictions, leading to racial disparities [1]. However, despite these well-documented concerns, very little research has been directed towards evaluating these DL-based skin disease diagnosis models on diverse skin types, and therefore, their utility and reliability as disease screening tools remains untested.

Although research into algorithmic bias and fairness has been an active area of research, interest in fairness of machine learning algorithms in particular is fairly recent. Multiple studies have shown the inherent racial disparities in machine learning algorithms' decisions for a wide range of areas: pre-trial bail decisions [32], recidivism [5], healthcare [42], facial recognition [11], and college admissions [33]. Specific to healthcare applications, previous research has shown the effect of dataset biases on DL models' performance across genders and racial groups in cardiac MR imaging [47], chest X-rays [35,50,51], and skin disease imaging [23]. Recently, Groh et al. [23] showed that CNNs are the most accurate when classifying skin diseases manifesting on skin types similar to those they were trained on.

Learning domain invariant representations, a predominant approach in domain generalization [40], attempts to learn data distributions that are independent of the underlying domains, and therefore addresses the issue of training models on data from a set of source domains that can generalize well to previously unseen test domains. Domain invariant representation learning has been used in medical imaging for histopathology image analysis [34] and for learning domain-invariant shape priors in segmentation of prostrate MR and retinal fundus images [37]. On the other hand, previous works on fair classification and diagnosis of skin diseases have relied on skin type detection and debiasing [9] and classification model pruning [57].

One of the common definitions of algorithmic fairness for classification tasks, based on measuring statistical parity, aims to seek independence between the bias attribute (also known as the protected attribute; i.e., the skin type for our task) and the model's prediction (i.e., the skin disease prediction). Our proposed approach, **C**olor **I**nvariant **R**epresentation learning for unbiased **C**lassification of skin Lesions (**CIRCLe**), employs a color-invariant model that is trained to classify skin conditions independent of the underlying skin type. In this work, we aim to mitigate the skin type bias learnt by the CNNs and reduce the accuracy disparities across skin types. We address this problem by enforcing the feature representation to be invariant across different skin types. We adopt a domain-invariant representation learning method [41] and modify it to transform skin types from clinical skin images and propose a color-invariant skin condition classifier. In particular, we make the following contributions:

- To the best of our knowledge, this is the first work that uses skin type transformations and skin color-invariant disease classification to tackle the problem of skin type bias present in large scale clinical image datasets and how these biases permeate through the prediction models.
- We present a new state-of-the-art classification accuracy over 114 skin conditions and 6 Fitzpatrick skin types (FSTs) from the Fitzpatrick17K dataset. While previous works had either limited their analysis to a subset of diagnoses [9] or less granular FST labels [57], our proposed method achieves superior performance over a much larger set of diagnoses spanning over all the FST labels.
- We provide a comprehensive evaluation of our proposed method, CIRCLe, on 6 different CNN architectures, along with ablation studies to demonstrate the efficacy of the proposed domain regularization loss. Furthermore, we also assess the impact of varying the size and the FST distribution of the training dataset partitions on the generalization performance of the classification models.
- Finally, we propose a new fairness metric called Normalized Accuracy Range that, unlike several existing fairness metrics, works with multiple protected groups (6 different FSTs in our problem).

2 Method

2.1 Problem Definition

Given a dataset $\mathcal{D} = \{X, Y, Z\}$, consider x_i, y_i, z_i to be the input, the label, and the protected attribute for the ith sample respectively, where we have M classes ($|Y| = M$) and N protected groups ($|Z| = N$). Let $\hat{y}_i$ denote the predicted label of sample i. Our goal is to train a classification model $f_\theta(\cdot)$ parametrized by θ that maps the input x_i to the final prediction $\hat{y}_i = f_\theta(x_i)$, such that (1) the prediction $\hat{y}_i$ is *invariant* to the protected attribute z_i and (2) the model's classification loss is minimized.

2.2 Feature Extractor and Classifier

In the representation learning framework, the prediction function $\hat{y}_i = f_\theta(x_i)$ is obtained as a composition $\hat{y}_i = \phi_C \circ \phi_E(x_i)$ of a feature extractor $r_i = \phi_E(x_i)$, where $r_i \in \mathbb{R}^p$ is a learned representation of data x_i, and a classifier $\hat{y}_i = \phi_C(r_i)$,

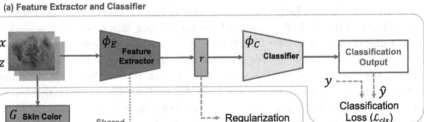

Fig. 1. Overview of CIRCLe. (a) The skin lesion image x with skin type z and diagnosis label y is passed through the feature extractor ϕ_E. The learned representation r goes through the classifier ϕ_C to obtain the predicted label $\hat{y}$. The classification loss enforces the correct classification objective. (b) The skin color transformer (G), transforms x with skin type z into x' with the new skin type z'. The generated image x' is fed into the feature extractor to get the representation r'. The regularization loss enforces r and r' to be similar. (c) The skin color transformer's schematic view with the possible transformed images, where one of the possible transformations is randomly chosen for generating x'.

predicting the label $\hat{y}_i$, given the representation r_i (Fig. 1(a)). Thus, we aim to learn a feature representation r that is invariant to the protected attributes, and hypothesize that this will lead to better generalization for classification.

2.3   Regularization Network

Inspired by the method proposed by Nguyen et al. [41], we use a generative modelling framework to learn a function g that transforms the data distributions between skin types. To this end, we employ a method to synthesize a new image corresponding to a given input image with the subject's skin type in that image changed according to the desired Fitzpatrick skin type (FST) score. We call this model our Skin Color Transformer. After training the Skin Color Transformer model, we introduce an auxiliary loss term to our learning objective, whose aim is to enforce the domain invariance constraint (Fig. 1(b)).

Skin Color Transformer. We learn the function G that performs image-to-image transformations between skin type domains. To this end, we use a Star Generative Adversarial Network (StarGAN) [12]. The goal of the StarGAN is to learn a unified network G (generator) that transforms the data density among multiple domains. In particular, the network $G(x, z, z')$ transforms an image x

from skin type z to skin type z'. The generator's goal is to fool the discriminator D into classifying the transformed image as the destination skin type z'. In other words, the equilibrium state of StarGAN is when G successfully transforms the data density of the original skin type to that of the destination skin type. After training, we use $G(., z, z')$ as the Skin Color Transformer. This model takes the image x_i with skin type z_i as the input, along with a target skin type z_j and synthesizes a new image $z'_i = G(x_i, z_i, z_j)$ similar to x_i, only with the skin type of the image changed in accordance with z_j.

Domain Regularization Loss. In the training process of the disease classifier, for each input image x_i with skin type s_i, we randomly select another skin type $s_j \neq s_i$, and use the Skin Type Transformer to synthesize a new image $x'_i = G(x_i, s_i, s_j)$. After that, we obtain the latent representations $r_i = \phi_E(x_i)$, and $r'_i = \phi_E(x'_i)$ for the original image and the synthetic image respectively. Then we enforce the model to learn similar representations for r_i and r'_i by adding a regularization loss term to the overall loss function of the model:

$$\mathcal{L}_{total} = \mathcal{L}_{cls} + \lambda \mathcal{L}_{reg} \tag{1}$$

where $\mathcal{L}_{cls}$ is the prediction loss of the network that predicts $\hat{y}_i$ given $r_i = \phi_E(x_i)$, and $\mathcal{L}_{reg}$ is the regularization loss. In this equation, $\lambda \in [0, 1]$ is a hyper-parameter controlling the trade-off between the classification and regularization losses. We define $\mathcal{L}_{reg}$ as the distance between the two representations r_i and r'_i to enforce the invariant condition. In our implementation, we use cross entropy as the classification loss $\mathcal{L}_{cls}$:

$$\mathcal{L}_{cls} = -\sum_{j=1}^{M} y_{ij} \log(\hat{y}_{ij}), \tag{2}$$

where y_{ij} is a binary indicator (0 or 1) if class label j is the correct classification for the sample i and $\hat{y}_{ij}$ is the predicted probability the sample i is of class j. The final predicted class $\hat{y}_i$ is calculated as

$$\hat{y}_i = \arg\max_j \hat{y}_{ij}. \tag{3}$$

We use squared error distance for computing the regularization loss $\mathcal{L}_{reg}$:

$$\mathcal{L}_{reg} = ||r_i - r'_i||_2^2. \tag{4}$$

3 Experiments

3.1 Dataset

We evaluate the performance of the proposed method on the Fitzpatrick17K dataset [23]. The Fitzpatrick17K dataset contains 16,577 clinical images with

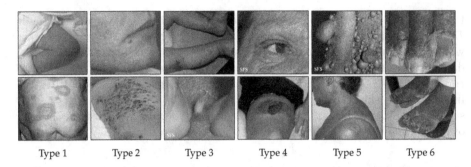

Type 1   Type 2   Type 3   Type 4   Type 5   Type 6

Fig. 2. Sample images of all six Fitzpatrick skin types (FSTs) from the Fitzpatrick17K dataset [23]. Notice the wide varieties in disease appearance, field of view, illumination, presence of imaging artifacts including non-standard background consistent with clinical images in the wild, and watermarks on some images.

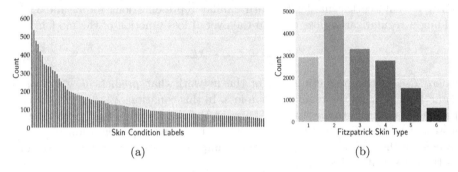

(a)   (b)

Fig. 3. Visualizing the distribution of (a) the skin condition labels and (b) the Fitzpatrick skin type (FST) labels in the Fitzpatrick17K dataset. Notice that the number of images across different skin conditions is not uniformly distributed. Moreover, the number of images is considerably lower for darker skin types.

skin condition labels and skin type labels based on the Fitzpatrick scoring system [21]. The dataset includes 114 conditions with at least 53 images (and a maximum of 653 images) per skin condition, as shown in Fig. 3 (a). The images in this dataset are annotated with (FST) labels by a team of non-dermatologist annotators. Figure 2 shows some sample images from this dataset along with their skin types. The Fitzpatrick labeling system is a six-point scale originally developed for classifying sun reactivity of skin and adjusting clinical medicine according to skin phenotype [21]. In this scale, the skin types are categorized in six levels from 1 to 6, from lightest to darkest skin types. Although Fitzpatrick labels are commonly used for categorizing skin types, we note that not all skin types are represented by the Fitzpatrick scale. [55].

In the Fitzpatrick17K dataset, there are significantly more images of light skin types than dark skin. There are 11,060 images of *light* skin types (FSTs 1, 2, and 3), and 4,949 images of *dark* skin types (FSTs 4, 5, and 6), as shown in Fig. 3 (b).

3.2 Implementation Details

Dataset Construction Details. We randomly select 70%, 10%, and 20% of the images for the train, validation, and test splits, where the random selection is stratified on skin conditions. We repeat the experiments with five different random seeds for splitting the data. A series of transformations are applied to the training images which include: resize to 128×128 resolution, random rotations in $[-15°, 15°]$, and random horizontal flips. We also use ImageNet [18] training partition's mean and standard deviation values to normalize our images for training and evaluation.

Feature Extractor and Classifier's Details. We choose VGG-16 [52] pretrained on ImageNet as our base network. We use the convolutional layers of VGG-16 as the feature extractor ϕ_E. We replace the VGG-16's fully-connected layers with a fully connected 256-to-114 layer as the classifier ϕ_C. We train the network for 100 epochs with plain stochastic gradient descent (SGD) using learning rate 1e-3, momentum 0.9, minibatch size 16, and weight decay 1e-3. We report the results for the epoch with the highest accuracy on the validation set.

StarGAN Details. StarGAN [12] implementation is taken from the authors' original source code with no significant modifications. We train StarGAN on the same train split used for training the classifier. As for the training configurations we use a minibatch size of 16. We train the StarGAN for 200,000 iterations and use the Adam [30] optimizer with a learning rate of 1e−4. For training the StarGAN's discriminator, we use cross entropy loss.

Model Training and Evaluation Setup. We use the PyTorch library [45] to implement our framework and train all our models on a workstation with AMD Ryzen 9 5950X processor, 32 GB of memory, and Nvidia GeForce RTX 3090 GPU with 24 GB of memory.

3.3 Metrics

We aim for an *accurate* and *fair* skin condition classifier. Therefore, we assess our method's performance using metrics for both accuracy and fairness. We use the well-known and commonly-used recall, F1-score, and accuracy metrics for evaluating our model's classification performance. For fairness, we use the equal opportunity difference (EOD) metric [25]. EOD measures the difference in true positive rates (TPR) for the two protected groups. Let TPR_z denote true positive rate of group z and $z \in \{0, 1\}$. Then EOD can be computed as:

$$EOD = |TPR_{z=0} - TPR_{z=1}|. \tag{5}$$

A value of 0 implies both protected groups have equal benefit. Given that the above metric (and other common fairness metrics in the literature [8, 19, 25]) are defined for two groups: privileged and under-privileged, w.r.t the protected attribute, we adopt the light (FSTs 1, 2, and 3) versus dark (FSTs 4, 5, and 6) as the two groups.

Additionally, to measure fairness in the model's accuracy for multiple groups of skin types, we assess the accuracy (ACC) disparities across all the six skin types by proposing the Normalized Accuracy Range (NAR) as follows:

$$NAR = \frac{ACC_{max} - ACC_{min}}{mean(ACC)}, \tag{6}$$

where ACC_{max} and ACC_{min} are the maximum and minimum accuracy achieved across skin types and $mean(ACC)$ is the mean accuracy across skin types, i.e.:

$$\begin{aligned} ACC_{max} &= max\{ACC_i : 1 \leq i \leq N\}, \\ ACC_{min} &= min\{ACC_i : 1 \leq i \leq N\}, \\ mean(ACC) &= \frac{1}{N} \sum_{i=1}^{N} ACC_i \end{aligned} \tag{7}$$

A perfectly fair performance of a model would result in equal accuracy across the different protected groups on a test set, i.e. $ACC_{max} = ACC_{min}$, leading to $NAR = 0$. As the accuracies across protected groups diverge, $ACC_{max} > ACC_{min}$, NAR will change even if the mean accuracy remains the same, thus indicating that the model's fairness is also changed. Moreover, NAR also takes into account the overall mean accuracy: this implies that in cases where the accuracies range $(ACC_{max} - ACC_{min})$ is the same, the model with the overall higher accuracy leads to a lower NAR, which is desirable. In our quantitative results, we report EOD for completeness; however, it is not an ideal measure, given it is restricted to only two protected groups whereas we have six. Therefore, we focus our attention on NAR.

3.4   Models

Baseline. For evaluating our method, we compare our results with the method proposed by Groh et al. [23], which has the current state-of-the-art performance on the Fitzpatrick17K dataset. We call their method the *Baseline*. To obtain a fair comparison, we use the same train and test sets they used.

Table 1. Comparing the model capacities and computational requirements of different backbones evaluated. For all the six backbones, we report the number of parameters and the number of multiply-add operations (**MulAddOps**). All numbers are in millions (**M**). Note how the six backbones encompass several architectural families and a large range of model capacities ($\sim$ 2M to $\sim$ 135M parameters) and computational requirements ($\sim$ 72M MulAddOps to $\sim$ 5136M MulAddOps).

	MobileNetV2	MobileNetV3L	DenseNet-121	ResNet-18	ResNet-50	VGG-16
Parameters (M)	2.55	4.53	7.22	11.31	24.03	135.31
MulAddOps (M)	98.16	72.51	925.45	592.32	1335.15	5136.16

Improved Baseline (Ours). In order to evaluate the effectiveness of the color-invariant representation learning process, we perform an ablation study, in which we remove the regularization loss $\mathcal{L}_{reg}$ from the learning objective of the model and train the classifier with only the classification objective. We call this model the *Improved Baseline*.

CIRCLe (Ours). The proposed model for unbiased skin condition classification, CIRCLe, is composed of two main components: the feature extractor and classifier, and the regularization network (Fig. 1).

Multiple Backbones. To demonstrate the efficacy of our method, we present evaluation with several other backbone architectures in addition to VGG-16 [52] used by Groh et al. [23]. In particular, we use MobileNetV2 [49], MobileNetV3-Large (referred to as MobileNetV3L hereafter) [27], DenseNet-121 [28], ResNet-18 [26], and ResNet-50 [26], thus covering a wide range of CNN architecture families and a considerable variety in model capacities, i.e. from 2.55 million parameters in MobileNetV2 to 135.31 million parameters in VGG-16 (Table 1).

For all the models, we perform an ablation study to evaluate if adding the regularization loss $\mathcal{L}_{reg}$ helps improve the performance.

4   Results and Analysis

4.1   Classification and Fairness Performance

Table 2 shows the accuracy and fairness results for the proposed method in comparison with the baseline. From the table, we can see that our Improved Baseline method recognizably outperforms the baseline method in accuracy and fairness. By using a powerful backbone and a better and longer training process, we more than doubled the classification accuracy on the Fitzpatrick17K dataset for all the skin types. This indicates that the choice of the base classifier and training settings plays a significant role in achieving higher accuracy rates on the Fitzpatrick17K dataset. Moreover, we can see that CIRCLe further improves the performance of our Improved Baseline across all the skin types, as well as the

Table 2. Classification performance and fairness of CIRCLe for classifying 114 skin conditions across skin types as assessed by the mean (std. dev.) of the metrics described in Sect. 3.3. We compute the overall accuracy based on the micro average accuracy across all skin types. Values in bold indicate the best results. CIRCLe yields the best performance while also improving fairness.

| Model | Recall | F1-score | Accuracy | | | | | | | EOD ↓ | NAR ↓ |
			Overall	Type 1	Type 2	Type 3	Type 4	Type 5	Type 6		
Baseline	0.251	0.193	0.202	0.158	0.169	0.222	0.241	0.289	0.155	0.309	0.652
Improved Baseline (Ours)	0.444 (0.007)	0.441 (0.009)	0.471 (0.004)	0.358 (0.026)	0.408 (0.014)	0.506 (0.023)	0.572 (0.022)	0.604 (0.029)	0.507 (0.027)	0.261 (0.028)	0.512 (0.078)
CIRCLe (Ours)	**0.459** (0.003)	**0.459** (0.003)	**0.488** (0.005)	**0.379** (0.019)	**0.423** (0.011)	**0.528** (0.024)	**0.592** (0.022)	**0.617** (0.021)	**0.512** (0.043)	**0.252** (0.031)	**0.474** (0.047)

overall accuracy. This significant improvement demonstrates the effectiveness of the color-invariant representation learning method in increasing the model's generalizability. This observation shows that when the model is constrained to learn similar representations from different skin types that the skin condition appears on, it can learn richer features from the disease information in the image, and its overall performance improves. In addition, CIRCLe shows improved fairness scores (lower EOD and lower NAR), which indicates that the model is less biased. To the best of our knowledge, we set a new state-of-the-art performance on the Fitzpatrick17K dataset for the task of classifying the 114 skin conditions.

Different model architectures may show different disparities across protected groups [46].

We can see in Table 3 that the color-invariant representation learning (i.e. with the regularization loss $\mathcal{L}_{reg}$ activated) significantly improves the accuracy and fairness results in different model architecture choices across skin types, which indicates the effectiveness of the proposed method independently from the backbone choice and its capacity. We can see that while the regularization loss does not necessarily improve the EOD for all the backbones, EOD is not the ideal measure of fairness for our task since as explained in Sect. 3.3, it can only be applied to a lighter-versus-darker skin tone fairness assessment. However, employing the regularization loss does improve the NAR for all the backbone architectures.

4.2 Domain Adaptation Performance

For evaluating the model's performance on adapting to unseen domains, we perform a "two-to-other" experiment, where we train the model on all the images from two FST domains and test it on all the other FST domains. Table 4 shows the performance of our model for this experiment. CIRCLe recognizably improves the domain adaptation performance in comparison with the Baseline and Improved Baseline, demonstrating the effectiveness of the proposed method in learning a color-invariant representation.

Table 3. Evaluating the classification performance improvement contribution of the regularization loss $\mathcal{L}_{reg}$ with multiple different feature extractor backbones. Reported values are the mean of the metrics described in Sect. 3.3. Best values for each backbone are presented in bold. EOD reported (for two groups of light and dark FSTs) for completeness but evaluation over all the 6 FSTs uses NAR (see text for details). Observe that $\mathcal{L}_{reg}$ improves the classification accuracy and the fairness metric NAR for all backbones.

Model	$\mathcal{L}_{reg}$	Recall	F1-score	Accuracy							EOD ↓	NAR ↓
				Overall	Type 1	Type 2	Type 3	Type 4	Type 5	Type 6		
MobileNetV2	✗	0.375	0.365	0.398	0.313	**0.364**	0.409	0.503	0.491	0.333	0.280	0.472
	✓	**0.404**	**0.397**	**0.434**	**0.354**	0.357	**0.471**	**0.559**	**0.544**	**0.421**	0.258	**0.455**
MobileNetV3L	✗	**0.427**	0.403	0.438	0.357	0.388	0.449	0.543	**0.560**	0.413	0.271	0.449
	✓	0.425	**0.412**	**0.451**	**0.369**	**0.400**	**0.464**	**0.565**	0.550	**0.444**	0.275	**0.420**
DenseNet-121	✗	0.425	0.416	0.451	0.393	0.397	0.452	**0.565**	0.522	**0.500**	0.278	0.364
	✓	**0.441**	**0.430**	**0.462**	**0.413**	**0.406**	**0.473**	0.561	**0.550**	0.452	0.294	**0.324**
ResNet-18	✗	0.391	0.381	0.417	0.355	0.353	0.431	0.538	0.516	**0.389**	0.263	0.430
	✓	**0.416**	**0.410**	**0.436**	**0.367**	**0.380**	**0.458**	**0.543**	**0.538**	**0.389**	0.282	**0.395**
ResNet-50	✗	0.390	0.382	0.416	0.337	0.363	0.422	0.549	0.506	0.389	0.257	0.497
	✓	**0.440**	**0.429**	**0.466**	**0.384**	**0.402**	**0.502**	**0.580**	**0.569**	**0.421**	0.283	**0.411**

Table 4. Classification performance measured by micro average accuracy when trained and evaluated on holdout sets composed of different Fitzpatrick skin types (FSTs). For example, "FST3-6" denotes that the model was trained on images only from FSTs 1 and 2 and evaluated on FSTs 3, 4, 5, and 6. CIRCLe achieves higher classification accuracies than Baseline (Groh et al. [23]) and Improved Baseline (also ours) for all holdout partitions and for all skin types.

Holdout Partition	Method	Overall	Type 1	Type 2	Type 3	Type 4	Type 5	Type 6
FST3-6	Baseline	0.138	–	–	0.159	0.142	0.101	0.090
	Improved Baseline	0.249	–	–	0.308	0.246	0.185	0.113
	CIRCLe	**0.260**	–	–	**0.327**	**0.250**	**0.193**	**0.115**
FST12 and FST56	Baseline	0.134	0.100	0.130	–	–	0.211	0.121
	Improved Baseline	0.272	0.181	0.274	–	–	0.453	0.227
	CIRCLe	**0.285**	**0.199**	**0.285**	–	–	**0.469**	**0.233**
FST1-4	Baseline	0.077	0.044	0.055	0.091	0.129	–	–
	Improved Baseline	0.152	0.078	0.111	0.167	0.280	–	–
	CIRCLe	**0.163**	**0.095**	**0.121**	**0.177**	**0.293**	–	–

4.3 Classification Performance Relation with Training Size

As CIRCLe's performance improvement and effectiveness in comparison with the baselines is established in Sect. 4.1, we further analyze the relation of CIRCLe's classification performance with the percentage of images of the FST groups in the training data. To this end, we consider the FST groups of light skin types (FSTs 1 and 2) with 5,549 images, medium skin types (FSTs 3 and 4) with 4,425 images, and dark skin types (FSTs 5 and 6) with 1,539 images in the training set. For each FST group, we gradually increase the number of images of that group in the training set, while the number of training images in other groups remains

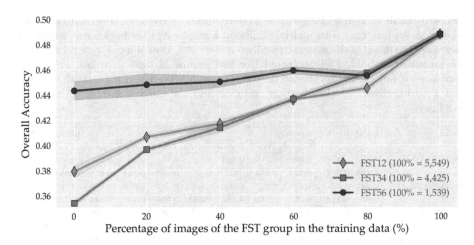

Fig. 4. Classification performance of CIRCLe on the test set as the number of training images of the FST groups increases. Each FST group line plot indicates the series of experiments in which the percentage of number of training images of that FST group changes as the rest of the training images remain idle. The rightmost point in the plot, with 100%, is identical for all the FST groups, which is the overall accuracy achieved by CIRCLe in Table 2. The std. dev. error band, illustrated in the figure, is computed by repetition of experiments with three different random seeds.

Table 5. Total number of training images for each experiment illustrated in Fig. 4. Note that the test set for all these experiments is the original test split with 3,205 images (20% of the Fitzpatrick17K dataset images), and the number of training images for experiments with 100% of each FST group is the same for all three groups, and is equal to the original train split with 11,934 images (70% of the Fitzpatrick17K dataset images).

	0%	20%	40%	60%	80%
FST12	5,964	7,073	8,183	9,293	10,403
FST34	7,088	7,973	8,858	9,743	10,628
FST56	9,974	1,0281	10,589	10,897	11,205

unchanged, and report the model's overall accuracy on the test set. The total number of training images for each of these experiments is provided in Table 5. As we can see in Fig. 4, as the number of training images in a certain FST group increases, the overall performance improves, which is expected since DL-based models generalize better with larger training datasets. However, we can see that for the least populated FST group, i.e., dark skin types (FST56) with 13% of the training data, our method demonstrates a more robust performance across experiments, and even with 0% training data of FST56, it achieves a relatively high classification accuracy of 0.443. In addition, note that in these experiments, FST groups with lower number of images in the dataset, would have a larger number of total training images, since removing a percentage of them from the training images will leave a larger portion of images available for training (see Table 5). This indicates that when the number of training images is large enough, even if images of a certain skin type are not available, or are very limited, our model can perform well overall. This observation signifies our method's ability to effectively utilize the disease-related features in the images from the training set, independently from their skin types, as well as the ability to generalize well to minority groups in the training set.

5   Discussion and Future Work

In order to develop fair and accurate DL-based data-driven diagnosis methods in demotology, we need annotated datasets that include a diversity of skin types and a range of skin conditions. However, only a few publicly available datasets satisfy these criteria. Out of all the datasets identified by the Seventh ISIC Skin Image Analysis Workshop at ECCV 2022 (derm7pt [29], Dermofit Image Library [6], Diverse Dermatology Images (DDI) [17], Fitzpatrick17K [23], ISIC 2018 [13], ISIC 2019 [14,15,54], ISIC 2020 [48], MED-NODE [22], PAD-UFES-20 [44], PH2 [39], SD-128 [53], SD-198 [53], SD-260 [58]), only three datasets contain Fitzpatrick skin type labels: Fitzpatrick17K with 16,577, DDI with 656, and PAD-UFES-20 with 2,298 clinical images. The Fitzpatrick17K dataset is the only dataset out of these three which covers all the 6 different skin types (with over 600 images per skin type) and contains more than 10K images, suitable for training high-capacity DL-based networks and our GAN-based color transformer. It also contains samples from 114 different skin conditions, which is the largest number compared to the other two. For these reasons, in this work, we used the Fitzpatrick17K dataset for training and evaluating our proposed method. However, skin conditions in the Fitzpatrick17K dataset images are not verified by dermotologists and skin types in this dataset are annotated by non-dermatologists. Also, the patient images captured in the clinical settings exhibit various lighting conditions and perspectives. During our experiments, we found many erroneous and wrongly labeled images in the Fitzpatrick17K dataset, which could affect the training process. Figure 5 shows some erroneous images in the Fitzpatrick17K dataset. Therefore, one possible future work can be cleaning the Fitzpatrick17K dataset and verifying its skin conditions and skin types by dermatologists.

Fig. 5. Sample erroneous images from the Fitzpatrick17K dataset that are not clinical images of skin conditions, but are included in the dataset and are wrongly labeled with skin conditions.

In addition, as we can see in Sect. 4.3 and Fig. 4, the number of training images plays a significant role in the model's performance across different skin types. Although in this paper we proposed a method for improving the skin condition classifier's fairness and generalizability, the importance of obtaining large and diverse datasets must not be neglected. Mitigating bias in AI diagnosis tools in the algorithm stage, as we proposed, can be effective and is particularly essential for the currently developed models, however, future research at the intersection of dermatology and computer vision should have specific focus on adding more diverse and annotated images to existing databases.

6   Conclusion

In this work, we proposed CIRCLe, a method based on domain invariant representation learning, for mitigating skin type bias in clinical image classification. Using a domain-invariant representation learning approach and training a color-invariant model, CIRCLe improved the accuracy for skin disease classification across different skin types for the Fitzpatrick17K dataset and set a new state-of-the-art performance on the classification of the 114 skin conditions. We also proposed a new fairness metric Normalized Accuracy Range for assessing fairness of classification in the presence of multiple protected groups, and showed that CIRCLe improves fairness of classification. Additionally, we presented an extensive evaluation over multiple CNN backbones as well as experiments to analyze CIRCLe's domain adaptation performance and the effect of varying the number of training images of different FST groups on its performance.

Acknowledgements. We would like to thank lab members Jeremy Kawahara and Ashish Sinha for their helpful discussions and comments on this work. We would also like to thank the reviewers for their valuable feedback that helped in improving this work. This project was partially funded by the Natural Sciences and Engineering Research Council of Canada (NSERC), and its computational resources were provided by NVIDIA and Compute Canada (computecanada.ca).

References

1. Adamson, A.S., Smith, A.: Machine learning and health care disparities in dermatology. JAMA Dermatol. **154**(11), 1247–1248 (2018)
2. Adamson, A.S., Suarez, E.A., Welch, H.G.: Estimating overdiagnosis of melanoma using trends among black and white patients in the US. JAMA Dermatol. **158**(4), 426–431 (2022)
3. Adelekun, A., Onyekaba, G., Lipoff, J.B.: Skin color in dermatology textbooks: an updated evaluation and analysis. J. Am. Acad. Dermatol. **84**(1), 194–196 (2021)
4. Agbai, O.N., et al.: Skin cancer and photoprotection in people of color: a review and recommendations for physicians and the public. J. Am. Acad. Dermatol. **70**(4), 748–762 (2014)
5. Angwin, J., Larson, J., Mattu, S., Kirchner, L.: Machine bias. In: Ethics of Data and Analytics, pp. 254–264 (2016)

6. Ballerini, L., Fisher, R.B., Aldridge, B., Rees, J.: A color and texture based hierarchical K-NN approach to the classification of non-melanoma skin lesions. In: Color Medical Image Analysis, pp. 63–86 (2013)
7. Barata, C., Marques, J.S., Emre Celebi, M.: Deep attention model for the hierarchical diagnosis of skin lesions. In: Proceedings of the IEEE/CVF Conference on Computer Vision and Pattern Recognition Workshops. pp. 2757–2765 (2019)
8. Bellamy, R.K., et al.: AI fairness 360: an extensible toolkit for detecting and mitigating algorithmic bias. IBM J. Res. Dev. **63**(4/5), 1–4 (2019)
9. Bevan, P.J., Atapour-Abarghouei, A.: Detecting melanoma fairly: skin tone detection and debiasing for skin lesion classification. arXiv preprint arXiv:2202.02832 (2022)
10. Brinker, T.J., et al.: A convolutional neural network trained with dermoscopic images performed on par with 145 dermatologists in a clinical melanoma image classification task. Eur. J. Cancer **111**, 148–154 (2019)
11. Buolamwini, J., Gebru, T.: Gender shades: intersectional accuracy disparities in commercial gender classification. In: Proceedings of the 1st Conference on Fairness, Accountability and Transparency, pp. 77–91 (2018)
12. Choi, Y., Choi, M., Kim, M., Ha, J.W., Kim, S., Choo, J.: StarGAN: unified generative adversarial networks for multi-domain image-to-image translation. In: Proceedings of the IEEE Conference on Computer Vision and Pattern Recognition, pp. 8789–8797 (2018)
13. Codella, N., et al.: Skin lesion analysis toward melanoma detection 2018: A challenge hosted by the international skin imaging collaboration (ISIC). arXiv preprint arXiv:1902.03368 (2019)
14. Codella, N.C., et al.: Skin Lesion Analysis Toward Melanoma Detection: A Challenge at the 2017 International Symposium on Biomedical Imaging (ISBI), Hosted by the International Skin Imaging Collaboration (ISIC). In: 2018 IEEE 15th International Symposium on Biomedical Imaging, pp. 168–172 (2018)
15. Combalia, M., et al.: BCN20000: Dermoscopic lesions in the wild. arXiv preprint arXiv:1908.02288 (2019)
16. Daneshjou, R., Smith, M.P., Sun, M.D., Rotemberg, V., Zou, J.: Lack of transparency and potential bias in artificial intelligence data sets and algorithms: a scoping review. JAMA Dermatol. **157**(11), 1362–1369 (2021)
17. Daneshjou, R., et al.: Disparities in dermatology AI performance on a diverse, curated clinical image set. Sci. Adv.**8**(31), eabq6147 (2022)
18. Deng, J., Dong, W., Socher, R., Li, L.J., Li, K., Fei-Fei, L.: ImageNet: a large-scale hierarchical image database. In: 2009 Proceedings of the IEEE Conference on Computer Vision and Pattern Recognition, pp. 248–255 (2009)
19. Dwork, C., Hardt, M., Pitassi, T., Reingold, O., Zemel, R.: Fairness through awareness. In: Proceedings of the 3rd Innovations in Theoretical Computer Science Conference, pp. 214–226 (2012)
20. Esteva, A., et al.: Dermatologist-level classification of skin cancer with deep neural networks. Nature **542**(7639), 115–118 (2017)
21. Fitzpatrick, T.B.: The validity and practicality of sun-reactive skin types I through VI. Arch. Dermatol. **124**(6), 869–871 (1988)
22. Giotis, I., Molders, N., Land, S., Biehl, M., Jonkman, M.F., Petkov, N.: MED-NODE: a computer-assisted melanoma diagnosis system using non-dermoscopic images. Expert Syst. Appl. **42**(19), 6578–6585 (2015)
23. Groh, M., et al.: Evaluating deep neural networks trained on clinical images in dermatology with the Fitzpatrick 17k dataset. In: Proceedings of the IEEE/CVF Conference on Computer Vision and Pattern Recognition, pp. 1820–1828 (2021)

24. Haenssle, H., et al.: Man against machine: diagnostic performance of a deep learn- ing convolutional neural network for dermoscopic melanoma recognition in com- parison to 58 dermatologists. Ann. Oncol. **29**(8), 1836–1842 (2018)
25. Hardt, M., Price, E., Srebro, N.: Equality of opportunity in supervised learning. Adv. Neural. Inf. Process. Syst. **29**, 3323–3331 (2016)
26. He, K., Zhang, X., Ren, S., Sun, J.: Deep residual learning for image recognition. In: Proceedings of the IEEE Conference on Computer Vision and Pattern Recognition, pp. 770–778 (2016)
27. Howard, A., et al.: Searching for MobileNetV3. In: Proceedings of the IEEE/CVF International Conference on Computer Vision, pp. 1314–1324 (2019)
28. Huang, G., Liu, Z., Van Der Maaten, L., Weinberger, K.Q.: Densely connected convolutional networks. In: Proceedings of the IEEE Conference on Computer Vision and Pattern Recognition, pp. 4700–4708 (2017)
29. Kawahara, J., Daneshvar, S., Argenziano, G., Hamarneh, G.: Seven-point checklist and skin lesion classification using multitask multimodal neural nets. IEEE J. Biomed. Health Inform. **23**(2), 538–546 (2019)
30. Kingma, D.P., Ba, J.: Adam: a method for stochastic optimization. arXiv preprint arXiv:1412.6980 (2014)
31. Kinyanjui, N.M., et al.: Fairness of classifiers across skin tones in dermatology. In: Medical Image Computing and Computer-Assisted Intervention, pp. 320–329 (2020)
32. Kleinberg, J., Lakkaraju, H., Leskovec, J., Ludwig, J., Mullainathan, S.: Human decisions and machine predictions. Q. J. Econ. **133**(1), 237–293 (2018)
33. Kleinberg, J., Ludwig, J., Mullainathan, S., Rambachan, A.: Algorithmic fairness. In: American Economic Association Papers and Proceedings, vol. 108, pp. 22–27 (2018)
34. Lafarge, M.W., Pluim, J.P., Eppenhof, K.A., Veta, M.: Learning domain-invariant representations of histological images. Front. Med. **6**, 162 (2019)
35. Larrazabal, A.J., Nieto, N., Peterson, V., Milone, D.H., Ferrante, E.: Gender imbal- ance in medical imaging datasets produces biased classifiers for computer-aided diagnosis. Proc. Natl. Acad. Sci. **117**(23), 12592–12594 (2020)
36. Lester, J., Jia, J., Zhang, L., Okoye, G., Linos, E.: Absence of images of skin of colour in publications of COVID-19 skin manifestations. Br. J. Dermatol. **183**(3), 593–595 (2020)
37. Liu, Q., Chen, C., Dou, Q., Heng, P.A.: Single-domain generalization in medical image segmentation via test-time adaptation from shape dictionary. In: Proceed- ings of the AAAI Conference on Artificial Intelligence, vol. 36, no. 2, pp. 1756–1764 (2022)
38. Liu, Y., et al.: A deep learning system for differential diagnosis of skin diseases. Nat. Med. **26**(6), 900–908 (2020)
39. Mendonça, T., Ferreira, P.M., Marques, J.S., Marcal, A.R., Rozeira, J.: PH^2–a dermoscopic image database for research and benchmarking. In: 2013 35th Annual International Conference of the IEEE Engineering in Medicine and Biology Society, pp. 5437–5440 (2013)
40. Muandet, K., Balduzzi, D., Schölkopf, B.: Domain generalization via invariant feature representation. In: International Conference on International Conference on Machine Learning, pp. 10–18 (2013)
41. Nguyen, A.T., Tran, T., Gal, Y., Baydin, A.G.: Domain invariant representation learning with domain density transformations. Adv. Neural. Inf. Process. Syst. **34**, 5264–5275 (2021)

42. Obermeyer, Z., Powers, B., Vogeli, C., Mullainathan, S.: Dissecting racial bias in an algorithm used to manage the health of populations. Science **366**(6464), 447–453 (2019)
43. Osto, M., Hamzavi, I.H., Lim, H.W., Kohli, I.: Individual typology angle and Fitzpatrick skin phototypes are not equivalent in photodermatology. Photochem. Photobiol. **98**(1), 127–129 (2022)
44. Pacheco, A.G., et al.: PAD-UFES-20: a skin lesion dataset composed of patient data and clinical images collected from smartphones. Data Brief **32**, 106221 (2020)
45. Paszke, A., et al.: PyTorch: an imperative style, high-performance deep learning library. In: Advances in Neural Information Processing Systems, vol. 32, pp. 8024–8035 (2019)
46. Prince, S.: Tutorial #1: Bias and fairness in AI (2019). https://www.borealisai.com/en/blog/tutorial1-bias-and-fairness-ai/. Accessed 14 Apr 2022
47. Puyol-Antón, E., et al.: Fairness in cardiac MR image analysis: an investigation of bias due to data imbalance in deep learning based segmentation. In: Medical Image Computing and Computer Assisted Intervention, pp. 413–423 (2021)
48. Rotemberg, V., et al.: A patient-centric dataset of images and metadata for identifying melanomas using clinical context. Sci. Data **8**(1), 1–8 (2021)
49. Sandler, M., Howard, A., Zhu, M., Zhmoginov, A., Chen, L.C.: MobileNetV2: inverted residuals and linear bottlenecks. In: Proceedings of the IEEE Conference on Computer Vision and Pattern Recognition, pp. 4510–4520 (2018)
50. Seyyed-Kalantari, L., Liu, G., McDermott, M., Chen, I.Y., Ghassemi, M.: CheXclusion: fairness gaps in deep chest X-ray classifiers. In: Biocomputing 2021: Proceedings of the Pacific Symposium, pp. 232–243 (2020)
51. Seyyed-Kalantari, L., Zhang, H., McDermott, M., Chen, I.Y., Ghassemi, M.: Underdiagnosis bias of artificial intelligence algorithms applied to chest radiographs in under-served patient populations. Nat. Med. **27**(12), 2176–2182 (2021)
52. Simonyan, K., Zisserman, A.: Very deep convolutional networks for large-scale image recognition. arXiv preprint arXiv:1409.1556 (2014)
53. Sun, X., Yang, J., Sun, M., Wang, K.: A benchmark for automatic visual classification of clinical skin disease images. In: Leibe, B., Matas, J., Sebe, N., Welling, M. (eds.) ECCV 2016. LNCS, vol. 9910, pp. 206–222. Springer, Cham (2016). https://doi.org/10.1007/978-3-319-46466-4_13
54. Tschandl, P., Rosendahl, C., Kittler, H.: The HAM10000 dataset, a large collection of multi-source dermatoscopic images of common pigmented skin lesions. Sci. Data **5**(1), 1–9 (2018)
55. Ware, O.R., Dawson, J.E., Shinohara, M.M., Taylor, S.C.: Racial limitations of Fitzpatrick skin type. Cutis **105**(2), 77–80 (2020)
56. Weiss, E.B.: Brown skin matters. https://brownskinmatters.com/. Accessed 23 Jun 2022
57. Wu, Y., Zeng, D., Xu, X., Shi, Y., Hu, J.: FairPrune: achieving fairness through pruning for dermatological disease diagnosis. arXiv preprint arXiv:2203.02110 (2022)
58. Yang, J., et al.: Self-paced balance learning for clinical skin disease recognition. IEEE Trans. Neural Networks Learn. Syst. **31**(8), 2832–2846 (2019)

W12 - Cross-Modal Human-Robot Interaction

W12 - Cross-Modal Human-Robot Interaction

A long-term goal of AI research is to build intelligent agents that can see the rich visual environment around us, interact with humans in multiple modalities, and act in a physical or embodied environment. As one of the most promising directions, cross-modal human-robot interaction has increasingly attracted attention from both academic and industrial fields. The community has developed numerous methods to address the problems in cross-modal human-robot interaction. Visual recognition methods like detection and segmentation enable the robot to understand the semantics in an environment. Large-scale pretraining methods and cross-modal representation learning aim at effective cross-modal alignment. Reinforcement learning methods are applied to learn human-robotic interaction policy. Moreover, the community requires the agent to have other abilities such as life-long/incremental learning or active learning, which broadens the application of real-world human-robot interaction.

Many research works have been devoted to related topics, leading to rapid growth of related publications in the top-tier conferences and journals such as CVPR, ICCV, ECCV, NeurIPS, ACL, EMNLP, T-PAMI, etc. This workshop aimed to further stimulate the progress in the field of human-robot interaction.

October 2022

Fengda Zhu
Yi Zhu
Xiaodan Liang
Liwei Wang
Xiaojun Chang
Nicu Sebe

Distinctive Image Captioning via CLIP Guided Group Optimization

Youyuan Zhang[2], Jiuniu Wang[3], Hao Wu[4], and Wenjia Xu[1(✉)]

[1] State Key Laboratory of Networking and Switching Technology, Beijing University of Posts and Telecommunications, Beijing, China
xuwenjia@bupt.edu.cn
[2] McGill University, Montreal, Canada
youyuan.zhang@mail.mcgill.ca
[3] Department of Computer Science, City University of Hong Kong, Kowloon Tong, Hong Kong
[4] South China University of Technology, Guangzhou, China

Abstract. Image captioning models are usually trained according to human annotated ground-truth captions, which could generate accurate but generic captions. In this paper, we focus on generating distinctive captions that can distinguish the target image from other similar images. To evaluate the distinctiveness of captions, we introduce a series of metrics that use large-scale vision-language pre-training model CLIP to quantify the distinctiveness. To further improve the distinctiveness of captioning models, we propose a simple and effective training strategy that trains the model by comparing target image with similar image group and optimizing the group embedding gap. Extensive experiments are conducted on various baseline models to demonstrate the wide applicability of our strategy and the consistency of metric results with human evaluation. By comparing the performance of our best model with existing state-of-the-art models, we claim that our model achieves new state-of-the-art towards distinctiveness objective.

Keywords: Distinctive image captioning · CLIP · Similar image group · Group embedding gap

1 Introduction

The task of image captioning involves interpreting the semantic information of images into natural language sentences. It is considered as a fundamental task in both computer vision and natural language processing. Image captioning is applied in various scenario such as assisting visually impaired people [23], search engine [12], medical imaging [41], etc.

General requirements of image captioning include two aspects: fluency, as the sentence should be well-formed as natural language; descriptiveness, as the sentence should accurately describe salient information of the image. In this paper, we focus on another important aspect: distinctiveness, which aims to generate captions with sufficient details and distinguish the target image from

ⓒ The Author(s), under exclusive license to Springer Nature Switzerland AG 2023
L. Karlinsky et al. (Eds.): ECCV 2022 Workshops, LNCS 13804, pp. 223–238, 2023.
https://doi.org/10.1007/978-3-031-25069-9_15

other similar ones [37]. Currently, most of the existing models only use human annotated ground-truth captions as labels. These models naturally focus on the similarity of generated captions and ground-truth captions. As a result, the generated captions are usually too generic because they are trained to be close to all ground-truth captions. Some pioneers [21,33] introduce diversity and discriminability objective that operate on semantic embeddings of images to generate more diverse or discriminative captions. However, as discussed in [37], models following these objectives generate grammarly different sentences with similar meanings, which does not follow the requirement of distinctiveness.

Some of the restrictions of traditional training strategies that prevent models from being distinctive include: 1) Only traditional language quality metrics, such as CIDEr [35], are used for training; 2) Ground-truth captions are annotated by human which does not necessarily include much details other than salient objects. Therefore, optimizing towards ground-truth captions through language quality metrics naturally makes captioning models focus on salient information and ignore distinctive details in images. To overcome the intrinsic shortcomings of traditional training strategy, we hereby propose a group based optimization strategy that uses CLIP [28] to guide distinctive training. We propose a series of reference-free metric based on CLIP to evaluate the relevance and distinctiveness of captions and images. Furthermore, we introduce similar group which is formed by the similar images of an target image. The similar group can be seen as hard negatives from which the model explicitly learns to distinguish detailed features. During training, our group based optimization takes the similarity of generated captions with target images as positive reward and takes the similarity with similar images as negative reward. The similarity is obtained from CLIP thus is not affected by the drawbacks of human annotations. Our proposed optimization strategy is "plug and play" which can be applied on any sequence prediction model. Experimental results show that our proposed strategy achieves outstanding performance on a variety of existing models.

To summarize, the contributions in this paper are three fold: 1) We propose a series of reference-free metrics to quantify the distinctiveness of image captioning. 2) We propose a group based optimization strategy that exploits the new metrics as reward to guide distinctive training. 3) We conduct extensive experiments on various state-of-the-art captioning models and demonstrate the effectiveness of our strategy.

2   Related Work

2.1   Image Captioning

Deep learning based captioning models generally employ encoder-decoder framework. Early attempts encode images using CNN [14,34] based feature extractor, and output captions through RNN [7] and LSTM [15] based decoder [16,22,24,36]. Some works use object-level features extracted by Faster-RCNN [30] and improves the performance. Recent works introduce attention mechanism to encourage cross-modal interaction between the two modalities and mitigate the long-term dependency problem in LSTM [6]. For example, [2] applies bottom-up attention at object-level features and map salient visual features to output words via top-down attention. [26] applies X-Linear attention to leverage higher-order intra and inter-modal interactions.

2.2   Objectives for Image Captioning

Sequence prediction models are typically trained with maximum likelihood estimation (MLE) objective by cross-entropy loss. However, as discussed in [29], this method causes the problem of exposure bias, which accumulates the error of caption generation at test time. To address this problem, [4] introduces scheduled sampling to reduce the bias. [29,31] treat sequence generation as reinforcement learning (RL) problem. In specific, [31] proposes self-critical sequence training that optimizes the policy gradient of the model parameters using REINFORCE with baseline. However, as pointed out in [9,10,39] such approaches usually result in over generic captions. Therefore, different objectives, such as diversity and discriminability, are proposed to enrich the expressive capacity of captions. In specific, Div-Cap [39] propose to measure diversity by analyzing latent semantic space of image encoding. G-GAN [9] use conditional generative adversarial network (CGAN) [25] to improve diversity of captions. Disc-Cap [21] trains a retrieval model to guide discriminative captioning. In terms of distinctiveness objective, VisPara-Cap [20] employs a two-stage LSTM model and generates two captions, where the second caption is pharaphrased from the first caption and is more distinctive. CiderBtw-Cap [37] re-weights the ground-truth captions by comparing their distinctiveness and trains the model using a weighted loss. Gdis-Cap [38] designs a group attention module that re-weights the attention features of images to highlight the distinctive features. CLIP-Cap [5] directly use CLIP [28] to compute relevance of image-caption pairs and generates more distinctive captions. Compared with these methods, our proposed method integrates the advantages of the following: 1) We do not rely on a specific model and therefore our method is more widely applicable. 2) We use reference-free reward to avoid the drawbacks of human annotation. 3) We use similar image group to better guide distinctive training via comparing.

2.3   Metrics for Distinctive Image Captioning

Traditional metrics, such as CIDEr [35], SPICE [1], BLEU [27], ROUGE [18] and MENTEOR [3], are used to measure the similarity between generated captions and ground-truth captions. In terms of distinctiveness metrics, [40] designs SPICE-U and introduces a notion of uniqueness over concepts in captions. [37] proposes CIDErBtw metric which reflects the distinctiveness by measuring the similarity of the target captions and captions of similar images. [38] proposes Dis-WordRate metric which reflects the percentage of distinctive words in captions (Fig. 1).

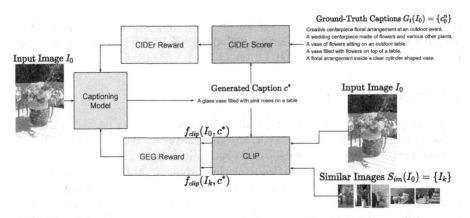

Fig. 1. Pipeline of group optimization using weighted group embedding gap reward.

3 Methodology

The goal of an image captioning model is to generate a caption c_0 for the target image I_0. In this paper, we focus on distinctive image captioning, which could not only describe the target image I_0 correctly but also include as many details as possible to distinguish I_0 from other similar images $\{I_1, ..., I_K\}$, where K denotes the number of similar images.

Following the goal of generating distinctive captions, we first propose the R@K-CLIP metric which uses the language-vision pre-training model CLIP [28] to quantify the distinctiveness of target caption. We then propose a series of metrics called Group Embedding Gap (GEG) including GEG-Avg and GEG-Min that compare the similarity between images and captions, and use these metrics for both training and evaluation. We design a weighted GEG reward which is used in self-critical sequence training (SCST) to guide distinctive training. Note that the GEG training strategy is "plug and play" and can be used on any sequence prediction models.

3.1 Similar Image Group

We start by introducing the construction process of the similar image groups $\{I_1, ..., I_K\}$. In each split of the training, validation and test dataset, we construct the similar image group within the same split. The K similar images are retrieved according to the semantic similarity between their ground truth captions and the target image I_0 by image-to-text retrieval.

Different from [37] which uses VSE++ [11] pre-trained model, we use CLIP (Contrastive Language-Image Pre-Training) [28] trained on larger dataset to perform image-to-text retrieval. CLIP is a neural network trained on a 400 million of image-text pairs collected from a variety of publicly available sources on the Internet. CLIP contains an image encoder $E_i(I)$ and a text encoder $E_t(c)$, which takes I and c as the input image and text, and output their embeddings respectively. In this paper, we use the ViT-B/32 architecture, of which the image encoder $E_i(\cdot)$ is a vision transformer with 12 transformer layers. The architecture of this vision transformer has minor change compared with the original version as an layer normalization is added to the combined patch and positional embeddings before the transformer. The text encoder $E_t(\cdot)$ first encode the raw text using a lower-cased byte paired encoding (BPE) with 49,152 vocab size and then fed them to a Transformer also with 12 transformer layers. Experimental result in Sect. 4.2 demonstrates that CLIP has better image retrieval performance than VSE++. The similarity of image I and caption c is defined as the inner product of their encoding representations:

$$f_{\text{CLIP}}(I, c) = \frac{E_i(I)^T E_t(c)}{\|E_i(I)\| \, \|E_t(c).\|} \tag{1}$$

Note that CLIP, as well as other cross-modal retrieval models are not trained on image-image pairs. Therefore, to obtain the semantic similarity between images, we define the similarity score of target images I_0 and candidate similar image I_i through the ground-truth captions of I_i as following:

$$s(I_0, I_i) = \max_{n \in \{1, 2, \ldots, N\}} f_{\text{CLIP}}(I_0, c_i^n), \tag{2}$$

where N is number of ground-truth captions for each image and $c_i^n \in G_t(I_i)$ is the n-th ground-truth caption of image I_i.

In practice, we first construct the set of similar captions $\{c_1', c_2', \ldots, c_{N'}'\}$, where $N' = N(K + 1)$ and K is the size of similar image group. In this way, at least K different images corresponding to the captions in the similar caption set are included. Then we choose the top K images to form the similar image group $S_{im}(I_0)$.

The similar image groups for all images in each split of the dataset are generated offline. The training process uses similar image groups in both forward and backward propagation. However, even though the evaluation of distinctiveness of captions involves their similar image groups, the model is unaware of the similar image groups and only takes target images as input in the test set at inference time. Therefore, we claim that similar image groups supervise the model to learn distinctive features of images in a contrastive way.

3.2   Metrics

R@K-CLIP. The distinctiveness of image captioning is hard to quantify because it depends on the dataset as well as the retrieval model. A general idea of evaluating distinctiveness is that the distinct caption should be able to retrieve its target image over a bunch of images via a text-image retrieval model. Previous work [37,38] use VSE++ to the retrieve images. In this paper, we choose CLIP as the retrieval model because it is pre-trained on larger dataset thus more reliable. User study in Sect. 4.5 suggests that our metric for distinctivness is highly consistent with human judgement.

In this paper, we fix the retrieval model to be CLIP ViT-B/32 and evaluate the distinctiveness of the generated captions by computing the recall rate of captions that retrieves their corresponding images, which is denoted as Recall at K (R@K). In specific, we define our proposed retrieval metric R@K-CLIP as the percentage of captions which retrieves their corresponding images within the top K images by CLIP ViT-B/32 model.

Group Embedding Gap. We propose the Group Embedding Gap (GEG) metrics that can be used for both training and evaluating a distinctive captioning model. These metrics are group specified, which means they depend on the choice of similar image group for each target image. Intuitively, a distinctive caption C should be close to the target image I while be distant to other similar images $S_{im}(I)$. To this end, the GEG metrics measure the distinctiveness of generated caption c for target image I by subtracting the similarity of c and $S_{im}(I)$ from the similarity of c and I. The difference is GEG-Avg (G_{avg}) computes the average similarity of c and every similar images in $S_{im}(I)$ while GEG-Min (G_{min}) focuses on the most similar one in $S_{im}(I)$:

$$G_{avg}(I,c) = f_{\text{CLIP}}(I,c) - \frac{1}{\|S_{im}(I)\|} \sum_{I' \in S_{im}(I)} f_{\text{CLIP}}(I',c), \tag{3}$$

$$G_{min}(I,c) = f_{\text{CLIP}}(I,c) - \max_{I' \in S_{im}(I)} f_{\text{CLIP}}(I',c), \tag{4}$$

where the CLIP model f_{CLIP} is used to measure the similarity between captions and images.

3.3   Group Embedding Gap Reward

Usually the training schedule for caption models involves both maximum likelihood estimation (MLE) training and self-critical sequence training (SCST) [31]. The purpose of SCST is to optimize a non-differentiable metric, such as CIDEr [35]. SCST treats the generation of captions as a reinforcement learning problem, where the expected reward of non-differentiable metric is optimized by evaluating policy gradient.

Following the sketch of SCST, we compute the policy gradient as follows:

$$\nabla L(\theta) = -\mathbb{E}_{c \sim P(c|I;\theta)}[r(c) - r(c_{greedy}) \nabla_\theta \log P(c|I;\theta)], \tag{5}$$

where $P(c|I;\theta)$ is the probability distribution of the model generating caption c for target image I with model parameter θ. c_{greedy} is the caption generated via greedy decoding, i.e. choosing the word with highest probability at every time step. The expectation of the gradient of loss can be estimated by sampling c using beam search. Therefore, for each sampled caption c_{beam},

$$\nabla L(\theta) = -(r(c_{beam}) - r(c_{greedy})) \nabla_\theta \log P(c_{beam}|I;\theta). \tag{6}$$

$r(c)$ originally take the CIDEr score of caption c as reward. Here we add the group embedding gap metric as a new component and re-weight the original CIDEr score reward and group embedding gap reward to get the final reward for sampled caption c:

$$r(c) = \alpha\mathrm{CIDEr}(c, \hat{c}) + \beta G(I, c), \tag{7}$$

where G represents the choice of group embedding gap metric of either G_{avg} or G_{min}. α and β are hyper-parameters to control weights of the two rewards.

4 Experiments

In this section, we first introduce our experiment settings. Then we validate the reliability of the R@K-CLIP metrics. Next we evaluate the improvement of our group based optimization strategy as for distinctiveness on a variety of captioning models, and compare the best model with existing state-of-the-art. Finally, user study and qualitative results are provided to show the consistency of metric based results and the human evaluation.

4.1 Implementation Details

Dataset. All our experiments are performed on MSCOCO dataset [19] with Karpathy split [16], which consists of a training set with 113,287 images, validation set with 5,000 images, and test set with 5,000 images. Each image is annotated with 5 captions. In this paper, all experiments results are reported on the test set of Karpathy split.

Models. Following [17], we evaluate our optimization strategy on seven different baseline model architectures: LSTM-A3 [44], Attention [42], Up-Down [2], GCN-LSTM [43], Transformer [32], Meshed-Memory [8] and X-LAN [26].

Two types of features, global features and top-down features are used in these models. Global features are extracted from Resnet-101 [14] and top-down features are spatial features extracted from Faster-RCNN [30]. For each image, the global feature is a vector of dimension 2,048 and top-down feature is a matrix of shape $N_{RoI} \times 1,024$ where N_{RoI} is the number of regions of interest. In specific, global features are used by LSTM-A3. Top-down features are used by Attention, Up-down, GCN-LSTM, Transformer, Meshed-Memory and X-LAN.

Each model is trained using three methods: 1) standard MLE objective with cross-entropy loss, denoted as "model"; 2) SCST optimization, which is first trained with MLE objective then optimized by SCST by CIDEr reward, denoted as "model+SCST"; 3) GEG optimization, which uses SCST to optimize the weighted group embedding gap reward based on the SCST model, and is denoted as "model+SCST+GEG".

Table 1. Image Retrieval performance of different models on MS COCO Karpathy test split.

Model	R@1	R@5	R@10
VSE++ [11]	24.1	**52.8**	66.2
CLIP [28]	**30.4**	50.0	**66.9**

Training Details. We apply GEG reward training on each model after SCST optimization. For GEG reward training, we collect $K = 5$ images to form the similar image group and set $\alpha = 0.1$ and $\beta = 10$. Maintaining a non-zero α is crucial because CLIP is not trained with language modeling objective. Ablation study in Sect. 4.4 shows that training using only GEG reward results in grammarly incorrect captions. We use Adam optimizer with base learning rate $r = 5 \times 10^{-5}$, $\beta_1 = 0.9$, $\beta_2 = 0.999$ and $\epsilon = 10^{-8}$. Each model is trained with GEG reward for 10 epochs.

Metrics. We use two groups of metrics. The first group includes language quality metrics, i.e., CIDEr, BLEU@1, BLEU@2, BLEU@3, BLEU@4, METEOR, ROUGE-L and SPICE, to evaluate the similarity of generated captions and ground-truth captions. The second group includes distinctiveness metrics of retrieval metrics, i.e., R@1-CLIP, R@5-CLIP, R@10-CLIP and GEG metrics GEG-Avg, GEG-Min.

We report the image retrieval performance of CLIP ViT-B/32 on MSCOCO test set in Table 1. The results suggest that CLIP ViT-B/32 has generally better performance than VSE++ because of higher R@1 and R@10 and slightly lower R@5.

4.2 Main Results

We report the main results of applying different training strategies on seven baseline captioning models in Table 2. The seven models are ranked by CIDEr score of "model+SCST" from low to high. In specific, we consider CIDEr as the main language quality metric to measure the similarity of generated caption and ground-truth captions. The remaining language quality metrics are positively correlated to CIDEr. We also consider R@K-CLIP metrics as the main distinctiveness metrics, and GEG metrics are positively correlated to R@K-CLIP.

By observing the results in Table 2, we have the following observations. Firstly, by comparing the models trained by MLE or SCST from top to bottom, the results suggest that models with higher language quality metrics tend to have higher distinctiveness metrics. For instance, compared to Up-Down [2] getting CIDEr value 113.1 and R@1-CLIP at 16.08, GCN-LSTM [43] achieves higher CIDEr at 116.3 and R@1-CLIP at 17.46. This correlation implies the consistency of the two types of metrics. When no extra supervision or objective is provided, models which can better mimic human annotations describes more distinctive details of the image. Then, by comparing "model+SCST+GEG" with

Table 2. The main results of applying different training strategy, i.e., the MLE optimization (denoted as "model"), the SCST optimization (denoted as "model + SCST"), and the GEG optimization (denoted as "+ GEG"), on seven baseline models. Here R@K represents R@K-CLIP and B@K represents BLEU@K. M, R and S represent METEOR, ROUGE-L and SPICE respectively.

Model	CIDEr	R@1	R@5	R@10	B@1	B@2	B@3	B@4	M	R	S	GEG-Avg	GEG-Min
LSTM-A3 [44]	107.7	11.52	30.88	42.34	75.3	59.0	45.4	35.0	26.7	55.6	19.7	0.0296	-0.0064
LSTM-A3+SCST	117.0	10.98	29.38	41.34	77.9	61.5	46.7	35.0	27.1	56.3	20.5	0.0288	−0.0073
LSTM-A3+SCST+GEG	114.3	12.98	32.86	45.7	76.9	60.4	45.6	33.9	26.9	55.8	20.6	0.0330	−0.0046
Attention [42]	113.0	14.74	36.4	48.6	76.4	60.6	46.9	36.1	27.6	56.6	20.4	0.0344	−0.0015
Attention+SCST	123.1	14.8	35.5	47.04	79.4	63.5	48.9	37.1	27.9	57.6	21.3	0.0338	−0.0026
Attention+SCST+GEG	116.6	17.68	40.94	52.84	77.1	60.8	45.9	34.3	27.4	56.4	21.1	0.0406	0.0026
Up-Down [2]	113.1	16.08	37.4	49.18	76.3	60.3	46.6	36.0	27.6	56.6	20.7	0.0351	−0.0008
Up-Down+SCST	124.7	15.46	36.2	49.28	80.1	64.3	49.7	37.7	28.0	58.0	21.5	0.0349	−0.0014
Up-Down+SCST+GEG	114.2	19.78	42.7	55.68	76.24	60.15	45.46	33.85	27.41	56.24	21.17	0.0424	0.0048
GCN-LSTM [43]	116.3	17.46	40.02	52.68	76.8	61.1	47.6	36.9	28.2	57.2	21.2	0.0373	0.0011
GCN-LSTM+SCST	127.2	17.22	39.46	52.16	80.2	64.7	50.3	38.5	28.5	58.4	22.1	0.0375	0.0015
GCN-LSTM+SCST+GEG	112.6	20.66	46.28	58.16	76.30	60.04	45.50	33.74	26.92	55.76	20.83	0.0441	0.0063
Transformer [32]	116.6	18.76	42.26	54.16	76.4	60.3	46.5	35.8	28.2	56.7	21.3	0.0391	0.0028
Transformer+SCST	130.0	19.6	42.72	55.62	80.5	65.4	51.1	39.2	29.1	58.7	23.0	0.0403	0.0033
Transformer+SCST+GEG	123.2	25.06	50.8	62.96	77.66	61.83	47.57	35.79	28.58	57.36	22.84	0.0485	0.0102
Meshed-Memory [8]	116.0	18.66	41.12	52.92	76.3	60.2	46.4	35.6	28.1	56.5	21.2	0.0383	0.0024
Meshed-Memory+SCST	**131.1**	19.64	42.94	55.86	**80.7**	**65.5**	**51.4**	39.6	29.2	58.9	22.9	0.0402	0.0034
Meshed-Memory+SCST+GEG	124.4	25.44	50.88	63.7	78.15	62.33	48.17	36.48	28.74	57.48	22.91	0.0484	0.0102
X-LAN [26]	120.7	20.26	43.88	55.4	77.5	61.9	48.3	37.5	28.6	57.6	21.9	0.0402	0.0036
X-LAN+SCST	130.0	21.3	44.74	57.74	80.4	65.2	51.0	39.2	**29.4**	**59.0**	**23.2**	0.0416	0.0045
X-LAN+SCST+GEG	121.7	**28.12**	**50.3**	**67.18**	76.23	60.55	46.37	34.82	28.66	56.96	22.84	**0.0517**	**0.0127**
Stack-Cap [13]	120.4	13.96	34.36	45.78	–	–	47.9	36.1	27.4	56.9	20.9	0.0339	−0.0019
Disc-Cap [21]	120.1	11.24	29.74	41.86	–	–	48.5	36.1	27.7	57.8	21.4	0.0313	−0.0042
CIDErBtw-Cap [37]	127.9	19.16	41.98	54.54	–	–	51.0	38.5	29.1	58.2	23.0	0.0390	0.0025
Gdis-Cap [38]	127.5	17.62	38.94	51.68	–	–	50.0	38.1	–	–	–	–	–

the baseline "model+SCST", the results suggest that models trained by GEG reward generate much more distinctive captions. For instance, compared to X-LAN+SCST getting R@K-CLIP rates at 21.3, 44.74 and 55.74 repectively, X-LAN+SCST+GEG achieves higher R@K-CLIP rates at 28.12, 50.3, 67.18 respectively. This improvement demonstrates the effectiveness of our proposed methods. Although the language metrics get lower, for instance, in the X-LAN group, X-LAN+SCST+GEG has lower CIDEr (121.7) than X-LAN+SCST (130.0), we argue that this is because human annotated ground-truth captions only consider the consistency with target images and are not distinctive in many cases. We provide case study in Sect. 4.6 to demonstrate that the decrease in language metrics does not negatively affect caption quality and on the contrary generates better captions than both baseline captions and ground-truth captions

4.3   Comparison with State-of-the-Art

We compare our model with best performance X-LAN+SCST+GEG with three state-of-the-art captioning models which also focus on distinctiveness object at the bottom of Table 2. We compare distinctiveness of each model based on R@K-CLIP and GEG metrics. Compared with Stack-Cap [13] and Disc-Cap [21], our model shows better performance in both language quality metrics and distinctiveness metrics. For instance, comparing to Stack-Cap [13] which has CIDEr at

120.4 and R@1-CLIP at 13.96, our X-LAN+SCST+GEG has higher CIDEr at 121.7 and higher R@1-CLIP at 28.12. Compared with CIDErBtw-Cap [37] and Gdis-Cap [38], our model has lower CIDEr score but still much better performance in distinctiveness. Recall that decrease in CIDEr score does not imply the reduction of language quality. Therefore, we claim that our model achieves new state-of-the-art towards distinctive captioning.

4.4   Ablation Study

In order to further illustrate the effectiveness of our training strategy. We provide an ablation study including five experiments as shown in Table 3. We compare our proposed standard SCST+GEG strategy with four other strategies on X-LAN model. X-LAN+SCST is the same baseline model trained using SCST as in Table 2. X-LAN+SCST++ is the baseline model trained for 10 more epochs so X-LAN+SCST++ and X-LAN+SCST+GEG are trained for the same number of epochs. Instead of using weighted GEG reward, X-LAN+SCST+ER only takes the CLIP embedding similarity of generated caption and target image as reward. In this case, the embedding reward does not include the negative average similarity of generated caption and similar images. X-LAN+SCST+GEG-sole sets $\alpha = 0$ which removes CIDEr reward and only takes GEG reward.

By analysing Table 3, we have the following observations. Firstly, the results of comparing X-LAN+SCST++ and X-LAN+SCST shows that further training has little improvement on both CIDEr metric and distinctiveness metric. Introducing the possitive similarity of generated caption and the target image as reward (i.e., X-LAN+SCST+ER) already helps the model in improving the distinctivess. For instance, X-LAN+SCST+ER improves the R@1-CLIP score from 21.8 (of X-LAN+SCST) to 25.86. Comparison between X-LAN+SCST+GEG and X-LAN+SCST+ER shows the effectiveness of group based learning. By negatively rewarding the similarity of generated caption and similar images, the X-LAN+SCST+GEG model generates captions that are better in both language quality metric (e.g., improves CIDEr from 120.7 to 121.7) and distinctive metric (e.g., promoting R@1-CLIP from 25.86 to 28.13). X-LAN+SCST+GEG-sole model has unexpectedly high distinctiveness metric without the CIDEr score constraint. However, as discussed in Sect. 4.1, CLIP is not trained with language quality objective. Training with only GEG reward causes frequent grammar mistakes of word disorder and repentance. An example is "a woman eating a white dessert cake with lit candles with lit candles in a restaurant with lit candles with", which is completely unreadable.

4.5   User Study

In order to fairly evaluate the improvement of our model and verify the consistency between our proposed metrics and human perspective, we conducted two user studies.

In the first experiment, each user is given a random group of images and a caption. The caption is generated from either the baseline X-LAN+SCST model or X-LAN+SCST+GEG model. The group of images contains an target image

Table 3. Comparison between different training strategies

Model	R@1-CLIP	R@5-CLIP	R@10-CLIP	CIDEr
X-LAN+SCST	21.3	44.74	57.74	130.0
X-LAN+SCST++	21.8	44.24	57.98	**130.5**
X-LAN+SCST+ER	25.86	52.38	64.88	120.7
X-LAN+SCST+GEG (Full model)	**28.13**	**50.3**	**67.18**	121.7
X-LAN+SCST+GEG-sole	32.82	61.3	72.8	17.4

Table 4. User-evaluated image retrieval

Model	R@1-User	R@5-User	R@10-User
X-LAN+SCST	21.9	46.3	65.4
X-LAN+SCST+GEG	**31.7**	**55.7**	**71.5**

corresponding to the caption and 20 similar images. The user is asked to list in order the 10 most relevant images. In this way, we obtain the R@K-User metric.

In the second experiment, each user is asked to give preference of two captions generated by the above two models following two criteria: accuracy, which caption describes the target image more accurately; distinctiveness, which caption is more informative about the target image. The user can choose win, loss or tie of the X-LAN+SCST+GEG model.

We received 1,200 responses in total. The results are shown in Table 4 and Table 5. Compared with baseline model X-LAN+SCST, our model increases the image retrieval rate R@1-User, R@5-User and R@10-User by 9.8, 9.4 and 6.1, and surpasses the baseline model in both accuracy and distinctiveness. Both observations show that human evaluation is consistent with metric results in Sect. 4.2.

4.6 Qualitative Results

The results shown in Table 6 illustrates the improvement of distinctiveness of our model compared with baseline model. Due to the space limit, we only show five groups of results and analyze the first two groups.

Table 5. Comparison on accuracy and distinctiveness. The rates of win, tie, loss are related to X-LAN+SCST+GEG model.

Criterion	Win (%)	Tie (%)	Loss (%)
Accuracy	16.7	80.3	3.0
Distinctiveness	44.5	47.2	8.3

Table 6. Example captions from the baseline model and our model with similar images. The leftmost images with red borders are the target images. Red words show the distinctive details in the target images which are not mentioned in either baseline model captions or ground-truth captions

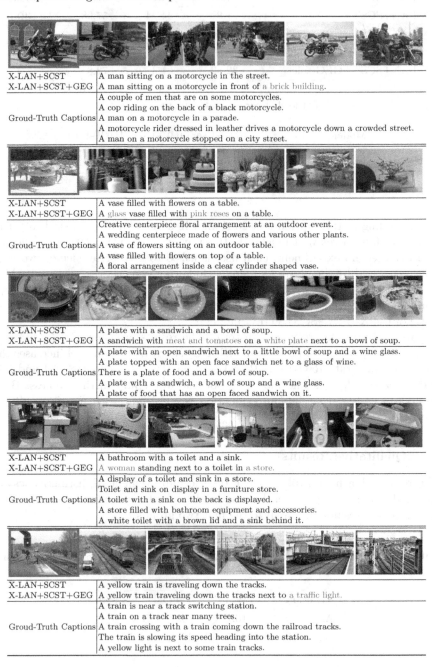

X-LAN+SCST	A man sitting on a motorcycle in the street.
X-LAN+SCST+GEG	A man sitting on a motorcycle in front of a brick building.
	A couple of men that are on some motorcycles.
	A cop riding on the back of a black motorcycle.
Groud-Truth Captions	A man on a motorcycle in a parade.
	A motorcycle rider dressed in leather drives a motorcycle down a crowded street.
	A man on a motorcycle stopped on a city street.

X-LAN+SCST	A vase filled with flowers on a table.
X-LAN+SCST+GEG	A glass vase filled with pink roses on a table.
	Creative centerpiece floral arrangement at an outdoor event.
	A wedding centerpiece made of flowers and various other plants.
Groud-Truth Captions	A vase of flowers sitting on an outdoor table.
	A vase filled with flowers on top of a table.
	A floral arrangement inside a clear cylinder shaped vase.

X-LAN+SCST	A plate with a sandwich and a bowl of soup.
X-LAN+SCST+GEG	A sandwich with meat and tomatoes on a white plate next to a bowl of soup.
	A plate with an open sandwich next to a little bowl of soup and a wine glass.
	A plate topped with an open face sandwich net to a glass of wine.
Groud-Truth Captions	There is a plate of food and a bowl of soup.
	A plate with a sandwich, a bowl of soup and a wine glass.
	A plate of food that has an open faced sandwich on it.

X-LAN+SCST	A bathroom with a toilet and a sink.
X-LAN+SCST+GEG	A woman standing next to a toilet in a store.
	A display of a toilet and sink in a store.
	Toilet and sink on display in a furniture store.
Groud-Truth Captions	A toilet with a sink on the back is displayed.
	A store filled with bathroom equipment and accessories.
	A white toilet with a brown lid and a sink behind it.

X-LAN+SCST	A yellow train is traveling down the tracks.
X-LAN+SCST+GEG	A yellow train traveling down the tracks next to a traffic light.
	A train is near a track switching station.
	A train on a track near many trees.
Groud-Truth Captions	A train crossing with a train coming down the railroad tracks.
	The train is slowing its speed heading into the station.
	A yellow light is next to some train tracks.

In each group of the images, the leftmost image is the target image and followed by five similar images. We compare the generated captions from the baseline model X-LAN+SCST and our model X-LAN+SCST+GEG. In both cases, the baseline model generates accurate but generic captions that describe the salient object in the target image. Contrastively, our model describes more details that distinguish the target image from potential similar images. In the first case, our model not only captures the main object, i.e. "man" and "motorcycle", but also notice the "brick building" in the background. In the second case, our model extends the generic description of "vase" and "flowers" to more specific description "glass vase" and "pink roses". Recall that during inference, our model is unaware of the similar images in test set. In this case, after distinctive training, our model is able to capture the details that distinguish the target image from potential similar images, which demonstrate the effectiveness of our training strategy.

The two cases also explain the decrease in language quality metrics, e.g. CIDEr. In the first case, none of the ground-truth captions mentions "brick building". In the second case, none of the ground-truth captions mentions "glass vase" and "pink roses". However, these ignored details are crucial to distinguish the target images. Training with only language quality objective results in generic captions possibly because the model is stuck in the minimum which is close to every ground-truth captions. After distinctive training, our model can discover more details than human annotated captions, which is considered as the main reason of the decrease in CIDEr metric.

5   Conclusions

In this paper, we focus on the distinctiveness of image captioning. We propose the R@K-CLIP metrics and GEG metrics to evaluate distinctiveness of captioning models. In order to improve distinctiveness, we design a simple and effective strategy which optimize on similar image group to guide distinctive training. In this process, the large-scale vision-language pre-trained model CLIP is used to measure the similarity of image-caption pairs. We conduct extensive experiments on various existing models and demonstrate the wide applicability and effectiveness of the group based optimization strategy in term of improving distinctiveness. Finally, we provide qualitative results and user study to show the consistency of the metric-side improvement with human judgement.

References

1. Anderson, P., Fernando, B., Johnson, M., Gould, S.: SPICE: semantic propositional image caption evaluation. In: Leibe, B., Matas, J., Sebe, N., Welling, M. (eds.) ECCV 2016. LNCS, vol. 9909, pp. 382–398. Springer, Cham (2016). https://doi.org/10.1007/978-3-319-46454-1_24
2. Anderson, P., et al.: Bottom-up and top-down attention for image captioning and visual question answering. In: Proceedings of the IEEE Conference on Computer Vision and Pattern Recognition, pp. 6077–6086 (2018)

3. Banerjee, S., Lavie, A.: Meteor: an automatic metric for MT evaluation with improved correlation with human judgments. In: Proceedings of the ACL Workshop on Intrinsic and Extrinsic Evaluation Measures for Machine Translation and/or Summarization, pp. 65–72 (2005)
4. Bengio, S., Vinyals, O., Jaitly, N., Shazeer, N.: Scheduled sampling for sequence prediction with recurrent neural networks. Advances in neural information processing systems 28 (2015)
5. Cho, J., Yoon, S., Kale, A., Dernoncourt, F., Bui, T., Bansal, M.: Fine-grained image captioning with clip reward. arXiv preprint arXiv:2205.13115 (2022)
6. Cho, K., Courville, A., Bengio, Y.: Describing multimedia content using attention-based encoder-decoder networks. IEEE Trans. Multimedia **17**(11), 1875–1886 (2015)
7. Cho, K., et al.: Learning phrase representations using RNN encoder-decoder for statistical machine translation. arXiv preprint arXiv:1406.1078 (2014)
8. Cornia, M., Stefanini, M., Baraldi, L., Cucchiara, R.: Meshed-memory transformer for image captioning. In: Proceedings of the IEEE/CVF Conference on Computer Vision and Pattern Recognition, pp. 10578–10587 (2020)
9. Dai, B., Fidler, S., Urtasun, R., Lin, D.: Towards diverse and natural image descriptions via a conditional gan. In: Proceedings of the IEEE international conference on computer vision. pp. 2970–2979 (2017)
10. Dai, B., Lin, D.: Contrastive learning for image captioning. Advances in Neural Information Processing Systems 30 (2017)
11. Faghri, F., Fleet, D.J., Kiros, J.R., Fidler, S.: Vse++: improving visual-semantic embeddings with hard negatives. arXiv preprint arXiv:1707.05612 (2017)
12. Frankel, C., Swain, M.J., Athitsos, V.: Webseer: An image search engine for the world wide web. Technical Report 96–14, University of Chicago, Computer Science Department (1996)
13. Gu, J., Cai, J., Wang, G., Chen, T.: Stack-captioning: coarse-to-fine learning for image captioning. In: Proceedings of the AAAI Conference on Artificial Intelligence, vol. 32 (2018)
14. He, K., Zhang, X., Ren, S., Sun, J.: Deep residual learning for image recognition. In: Proceedings of the IEEE Conference on Computer Vision and Pattern Recognition, pp. 770–778 (2016)
15. Hochreiter, S., Schmidhuber, J.: Long short-term memory. Neural Comput. **9**(8), 1735–1780 (1997)
16. Karpathy, A., Fei-Fei, L.: Deep visual-semantic alignments for generating image descriptions. In: Proceedings of the IEEE Conference on Computer Vision and Pattern Recognition, pp. 3128–3137 (2015)
17. Li, Y., Pan, Y., Chen, J., Yao, T., Mei, T.: X-modaler: A versatile and high-performance codebase for cross-modal analytics. In: Proceedings of the 29th ACM International Conference on Multimedia, pp. 3799–3802 (2021)
18. Lin, C.Y.: Rouge: a package for automatic evaluation of summaries. In: Text Summarization Branches Out, pp. 74–81 (2004)
19. Lin, T.-Y., et al.: Microsoft COCO: common objects in context. In: Fleet, D., Pajdla, T., Schiele, B., Tuytelaars, T. (eds.) ECCV 2014. LNCS, vol. 8693, pp. 740–755. Springer, Cham (2014). https://doi.org/10.1007/978-3-319-10602-1_48
20. Liu, L., Tang, J., Wan, X., Guo, Z.: Generating diverse and descriptive image captions using visual paraphrases. In: Proceedings of the IEEE/CVF International Conference on Computer Vision, pp. 4240–4249 (2019)

21. Luo, R., Price, B., Cohen, S., Shakhnarovich, G.: Discriminability objective for training descriptive captions. In: Proceedings of the IEEE Conference on Computer Vision and Pattern Recognition, pp. 6964–6974 (2018)
22. Ma, L., Lu, Z., Shang, L., Li, H.: Multimodal convolutional neural networks for matching image and sentence. In: Proceedings of the IEEE International Conference on Computer Vision, pp. 2623–2631 (2015)
23. Makav, B., Kılıç, V.: A new image captioning approach for visually impaired people. In: 2019 11th International Conference on Electrical and Electronics Engineering (ELECO), pp. 945–949. IEEE (2019)
24. Mao, J., Xu, W., Yang, Y., Wang, J., Huang, Z., Yuille, A.: Deep captioning with multimodal recurrent neural networks (m-rnn). arXiv preprint arXiv:1412.6632 (2014)
25. Mirza, M., Osindero, S.: Conditional generative adversarial nets. arXiv preprint arXiv:1411.1784 (2014)
26. Pan, Y., Yao, T., Li, Y., Mei, T.: X-linear attention networks for image captioning. In: Proceedings of the IEEE/CVF Conference on Computer Vision and Pattern Recognition, pp. 10971–10980 (2020)
27. Papineni, K., Roukos, S., Ward, T., Zhu, W.J.: Bleu: a method for automatic evaluation of machine translation. In: Proceedings of the 40th Annual Meeting of the Association for Computational Linguistics, pp. 311–318 (2002)
28. Radford, A., et al.: Learning transferable visual models from natural language supervision. In: International Conference on Machine Learning, pp. 8748–8763. PMLR (2021)
29. Ranzato, M., Chopra, S., Auli, M., Zaremba, W.: Sequence level training with recurrent neural networks. arXiv preprint arXiv:1511.06732 (2015)
30. Ren, S., He, K., Girshick, R., Sun, J.: Faster r-cnn: towards real-time object detection with region proposal networks. Advances in neural information processing systems 28 (2015)
31. Rennie, S.J., Marcheret, E., Mroueh, Y., Ross, J., Goel, V.: Self-critical sequence training for image captioning. In: Proceedings of the IEEE Conference on Computer Vision and Pattern Recognition, pp. 7008–7024 (2017)
32. Sharma, P., Ding, N., Goodman, S., Soricut, R.: Conceptual captions: a cleaned, hypernymed, image alt-text dataset for automatic image captioning. In: Proceedings of the 56th Annual Meeting of the Association for Computational Linguistics (Volume 1: Long Papers), pp. 2556–2565 (2018)
33. Shetty, R., Rohrbach, M., Anne Hendricks, L., Fritz, M., Schiele, B.: Speaking the same language: Matching machine to human captions by adversarial training. In: Proceedings of the IEEE International Conference on Computer Vision, pp. 4135–4144 (2017)
34. Simonyan, K., Zisserman, A.: Very deep convolutional networks for large-scale image recognition. arXiv preprint arXiv:1409.1556 (2014)
35. Vedantam, R., Lawrence Zitnick, C., Parikh, D.: Cider: consensus-based image description evaluation. In: Proceedings of the IEEE Conference on Computer Vision and Pattern Recognition, pp. 4566–4575 (2015)
36. Vinyals, O., Toshev, A., Bengio, S., Erhan, D.: Show and tell: a neural image caption generator. In: Proceedings of the IEEE Conference on Computer Vision and Pattern Recognition, pp. 3156–3164 (2015)
37. Wang, J., Xu, W., Wang, Q., Chan, A.B.: Compare and reweight: distinctive image captioning using similar images sets. In: Vedaldi, A., Bischof, H., Brox, T., Frahm, J.-M. (eds.) ECCV 2020. LNCS, vol. 12346, pp. 370–386. Springer, Cham (2020). https://doi.org/10.1007/978-3-030-58452-8_22

38. Wang, J., Xu, W., Wang, Q., Chan, A.B.: Group-based distinctive image captioning with memory attention. In: Proceedings of the 29th ACM International Conference on Multimedia, pp. 5020–5028 (2021)
39. Wang, Q., Chan, A.B.: Describing like humans: on diversity in image captioning. In: Proceedings of the IEEE/CVF Conference on Computer Vision and Pattern Recognition, pp. 4195–4203 (2019)
40. Wang, Z., Feng, B., Narasimhan, K., Russakovsky, O.: Towards unique and informative captioning of images. In: Vedaldi, A., Bischof, H., Brox, T., Frahm, J.-M. (eds.) ECCV 2020. LNCS, vol. 12352, pp. 629–644. Springer, Cham (2020). https://doi.org/10.1007/978-3-030-58571-6_37
41. Xiong, Y., Du, B., Yan, P.: Reinforced transformer for medical image captioning. In: Suk, H.-I., Liu, M., Yan, P., Lian, C. (eds.) MLMI 2019. LNCS, vol. 11861, pp. 673–680. Springer, Cham (2019). https://doi.org/10.1007/978-3-030-32692-0_77
42. Xu, K., et al.: Show, attend and tell: Neural image caption generation with visual attention. In: International Conference on Machine Learning, pp. 2048–2057. PMLR (2015)
43. Yao, T., Pan, Y., Li, Y., Mei, T.: Exploring visual relationship for image captioning. In: Ferrari, V., Hebert, M., Sminchisescu, C., Weiss, Y. (eds.) Computer Vision – ECCV 2018. LNCS, vol. 11218, pp. 711–727. Springer, Cham (2018). https://doi.org/10.1007/978-3-030-01264-9_42
44. Yao, T., Pan, Y., Li, Y., Qiu, Z., Mei, T.: Boosting image captioning with attributes. In: Proceedings of the IEEE International Conference on Computer Vision, pp. 4894–4902 (2017)

W13 - Text in Everything

W13 - Text in Everything

Understanding written communication through vision is a key aspect of human civilization and should also be an important capacity of intelligent agents aspiring to function in man-made environments. For example, interpreting written information in natural environments is essential in order to perform most everyday tasks like making a purchase, using public transportation, finding a place in the city, getting an appointment, or checking whether a store is open or not, to mention just a few. As such, the analysis of written communication in images and videos has recently gained an increased interest, as well as significant progress in a variety of text based vision tasks. While in earlier years the main focus of this discipline was on OCR and the ability to read business documents, today this field contains various applications that require going beyond just text recognition and onto additionally reasoning over multiple modalities, such as the structure and layout of documents. Recent advances in this field have been a result of a multi-disciplinary perspective spanning not only computer vision but also natural language processing, document and layout understanding, knowledge representation and reasoning, data mining, information retrieval, and more. The goal of this workshop is to raise awareness about the aforementioned topics in the broader computer vision community, and gather vision, NLP, and other researchers together to drive a new wave of progress by cross-pollinating more ideas between text/documents and non-vision related fields.

October 2022

Ron Litman
Aviad Aberdam
Shai Mazor
Hadar Averbuch-Elor
Dimosthenis Karatzas
R. Manmatha

OCR-IDL: OCR Annotations for Industry Document Library Dataset

Ali Furkan Biten[(✉)], Rubèn Tito, Lluis Gomez, Ernest Valveny,
and Dimosthenis Karatzas

Computer Vision Center, UAB, Barcelona, Spain
{abiten,rperez,lgomez,ernest,dimos}@cvc.uab.cat

Abstract. Pretraining has proven successful in Document Intelligence tasks where deluge of documents are used to pretrain the models only later to be finetuned on downstream tasks. One of the problems of the pretraining approaches is the inconsistent usage of pretraining data with different OCR engines leading to incomparable results between models. In other words, it is not obvious whether the performance gain is coming from diverse usage of amount of data and distinct OCR engines or from the proposed models. To remedy the problem, we make public the OCR annotations for IDL documents using commercial OCR engine given their superior performance over open source OCR models. It is our hope that OCR-IDL can be a starting point for future works on Document Intelligence. All of our data and its collection process with the annotations can be found in https://github.com/furkanbiten/idl_data.

1 Introduction

Analysis of masses of scanned documents is essential in intelligence, law, knowledge management, historical scholarship, and other areas [25]. The documents are often complex and varied in nature that can be digital or scanned born, containing elements such as forms, figures, tables, graphics and photos, while being produced by various printing and handwriting technologies. Some common examples of documents comprise of purchase orders, financial reports, business emails, sales agreements, vendor contracts, letters, invoices, receipts, resumes, and many others [54]. Processing various document types to user's intent is done with manual labor that is time-consuming and expensive, meanwhile requiring manual customization or configuration. In other words, each type of document demands hard-coded changes when there is a slight change in the rules or workflows of documents or even when dealing with multiple formats.

To address these problems, Document Intelligence models and algorithms are created to automatically structure, classify and extract information from documents, improving automated document processing. Particularly, Document Intelligence as a research field aims at creating models for automatically analyzing and understanding documents, reducing the time and the cost associated with it. From a research perspective, what makes Document Intelligence especially challenging is the requirement of combining various disciplines such as

© The Author(s), under exclusive license to Springer Nature Switzerland AG 2023
L. Karlinsky et al. (Eds.): ECCV 2022 Workshops, LNCS 13804, pp. 241–252, 2023.
https://doi.org/10.1007/978-3-031-25069-9_16

optical character recognition (OCR), document structure analysis, named entity recognition, information retrieval, authorship attribution and many more.

Recent methods on Documents Intelligence utilize deep neural networks combining Computer Vision and Natural Language Processing. Hao *et al.* [17] proposed an end-to-end training using Convolutional Neural Networks to detect tables in documents. Several published works [42,46,56] exploit the advances in object detection [19,40] to further improve the accuracy in document layout analysis. Even though these works have advanced the Document Intelligence field, there are two main limitations to be recognized: (i) they rely on a small human annotated dataset and (ii) they use pre-trained networks that have never seen any documents, hence the interaction between text and layout. Inspired by BERT [11], Xu *et al.* [54] identified these problems and propose a pre-training strategy to unlock the potential of large-scale unlabeled documents. More specifically, they obtain OCR annotations from an open source OCR engine Tesseract [45] for 5 Million documents from IIT-CDIP [25] dataset. With the introduction of pre-training strategy and advances in modern OCR engine [1,12,20,28,34], many contemporary approaches [2,7,53] have utilized even more data to advance the Document Intelligence field.

In this work, we make public the OCR annotations for 26 Millions pages using a commercial OCR engine that has the monetary value over 20K US$. Our motivation for releasing a massive scale documents dataset annotated with a commercial OCR engine is two-fold. First of all, the usage of different amount of documents and different OCR engines across the papers makes it impossible to fairly compare their results and hence their architecture. By creating this dataset, we hope that the works in Document Intelligence will become more comparable and have better intuition on what the proposed architecture can actually accomplish.

Secondly, we decide to use a commercial OCR engine, specifically Amazon Textract[1], over Tesseract. It is because the performance of the OCR engines can significantly affect the model's performance which can be seen in fields that use OCR annotations, such as in fine-grained classification [29–31], in scene-text visual question answering [8,9,13,44], in document visual question answering (DocVQA) [33,50]. Apart from improving the annotation quality significantly, we want to level the differences between research groups and companies.

We provide the annotations for publicly available documents from Industry Documents Library (IDL). IDL is a digital archive of documents created by industries which influence public health, hosted by the University of California, San Francisco Library[2]. IDL has already been used in the literature for building datasets: IIT-CDIP [25], RVL-CDIP [18], DocVQA [33,50]. Hence, our OCR annotations can be used to further advance in these tasks.

The rest of the paper is structured as follows. First, we briefly explain all the related works. Next, we will elaborate on our data collection and comparison to

---

[1] https://aws.amazon.com/textract/.

[2] https://www.industrydocuments.ucsf.edu.

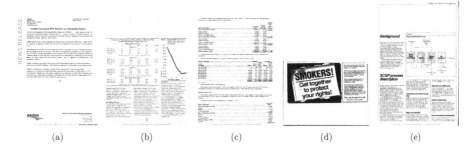

Fig. 1. Document images of OCR-IDL. The dataset includes a wide variety of documents with dense text (a), tables (b), figures (c), and complex layouts that combines different elements (d, e).

other datasets. Finally, we will provide various statistics of the annotations and conclude our paper.

2   Related Work

Document Intelligence can be considered as an umbrella term covering problems of Key Information Extraction [10,54], Table Detection [38,41] and Structure Recognition [39,55], Document Layout Segmentation [4,5] Document Layout Generation [3,6,36,48], Document Visual Question Answering [32,50,51], Document Image Enhancement [22,47,49] which involves the understanding of visually rich semantic information and structure of different layout entities of a whole page.

Early days of Document Intelligence has relied on rule-based handcrafted approaches mainly divided into bottom-up and top-down methods. Bottom-up methods [15,24,35,43] first detect connected components at the pixel level, later to be fused into higher level of structure through various heuristics and name depending on distinct structural features. While top-down methods [16] dissects a page into smaller units such as titles, text blocks, lines, and words.

Lately, the success of large-scale pre-training [11] in Natural Language Processing has been integrated into Document Intelligence, resulting in impressive performance gains. These methods follow a two step procedure where first they pretrain the models on unlabeled documents (OCR annotations are obtained by an off-the-shelf OCR engine), then they finetune it on specific downstream tasks. LayoutLM [54] is one of the first works that pretrain BERT based language model with document layout information, using masked language/vision modeling and multi label classification. BROS [21] is built on top of Span-BERT [23] with spatially aware graph decoder. For the pretraining loss, they use area-masked language model. Self-Doc [27] utilizes two separate transformer [52] encoders for visual and textual features and later to be fed to multi-modal transformers encoder. TILT [37] tries to encode the layout information by integrating pairwise 1D and 2D information into their models. Uni-Doc [14] is designed to do

Table 1. Summary of other Document Intelligence Datasets. *We skipped $145K$ documents that gave xml parsing errors, didn't contain document ID or number of pages. †No traceability between different pages of the same document.

Dataset	# of Docs	# of Pages	Docs source	Docs. description	OCR-Text	OCR-BB	Layout	Doc. type
IIT-CDIP [25]	$6.5M$*	$35.5M$*	UCSF-LTD	Industry documents	Unknown	✗	✗	✓
RVL-CDIP [18]	-†	$400K$	UCSF-LTD	Industry documents	✗	✗	✗	✓
PublayNet [56]	-†	$364K$	PubMedCentral	Journals and articles	✗	✗	✓	✗
DocBank [26]	-†	$500K$	arXiv	Journals and articles	✗	✗	✓	✗
DocVQA [51]	$6K$	$12K$	UCSF-IDL	Industry documents	Microsoft OCR	✓	✗	✓
OCR-IDL	$4.6M$	$26M$	UCSF-IDL	Industry documents	Amazon Textract	✓	✗	✓

most document understanding tasks that takes words and visual features from a semantic region of a document image by combining three self-supervised losses. More recent methods, Doc-Former [2] and LayoutLMv2 [53] combine multiple pretraining losses such as image-text alignment, learning to construct image features and multi-modal masked language modeling together to achieve state-of-the-art results.

Yet, comparing all of these works are cumbersome since each work that performs pretraining uses different amounts of data with diverse OCR engines. Hence, this makes it especially hard to understand where the gain is coming from. In other words, we can not draw clear conclusions to questions such as: "What is the effect of the amount of pretraining data on the performance?", "Is the performance gain coming from a better/stronger OCR engine or from the proposed architecture?", "What is the effect of the pretraining loss on the downstream tasks keeping OCR and the amount of data identical?" To help answer these questions, we collect and annotate the largest public OCR annotated documents dataset (OCR-IDL).

3   OCR-IDL Dataset

In this section, we elaborate on various details regarding OCR-IDL. Firstly, we explain the process we follow on how we get the IDL data and use Amazon-Textract to obtain OCR annotations. Next, we compare OCR-IDL to other datasets that have proven useful for the document intelligence tasks. And finally, we provide in-depth statistics on the documents we use.

3.1   Data Collection

As already mentioned, IDL is an industry documents library hosted by UCSF, its main purpose is to "identify, collect, curate, preserve, and make freely accessible internal documents created by industries and their partners which have an impact on public health, for the benefit and use of researchers, clinicians, educators, students, policymakers, media, and the general public at UCSF and internationally"[3]. IDL in total contains over 70 millions documents. We use the

---

[3] https://en.wikipedia.org/wiki/Industry_Documents_Library.

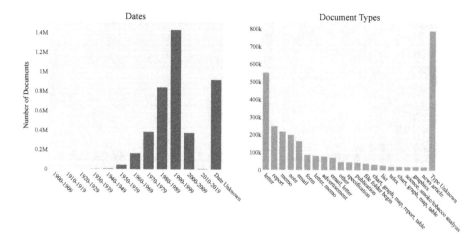

Fig. 2. Distribution of the annotated documents in terms of document types and dates.

publicly available link[4] to downlad the 4.6 Million Documents from IDL dataset which comes in the format of PDFs. Some examples can be viewed in Fig. 1. We appreciate that the documents are quite varied in terms of layout where there are tables, figures, ads and combination of them in a single page. Also, we see that the documents are quite text-rich containing many OCR words. Finally, the pages contain smudges, taints and other discoloration what is found in the in a real use-case scenario.

We choose to annotate only 4.6M documents since annotating 13M documents not only would be much costlier (42K dollars instead of 18K) but also in the literature it is shown that using more data have diminishing returns on the downstream task [7]. After obtaining the data, we preprocess the documents to remove empty, faulty and broken pdfs. This process resulted in the elimination of 6548 documents. Moreover, we remove also documents that have more than 2900 pages which are 71 in total. The necessity of removing such huge documents was because Amazon-Textract OCR only accepts up to 3000 pages. After pre-processing the documents, we feed all the documents to the OCR engine to obtain the annotations. The annotations provided by the Textract engine include transcription of words and lines and their corresponding bounding boxes and polygons with text type that can be printed or handwritten. Processing all 4.6M documents was done by a single machine with 16 parallelized cores and took about 1 month.

---

[4] https://s3-us-west-2.amazonaws.com/edu.ucsf.industrydocuments.artifacts/.

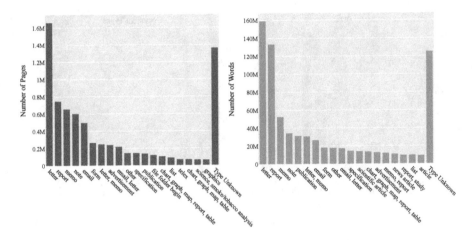

Fig. 3. Distribution of number of pages and number of words per document type.

3.2 Comparison to Existing Datasets

In this section, we compare the statistics of the amount of documents and pages to other datasets that are used in Document Intelligence. To name a few, the Illinois Institute of Technology dataset for Complex Document Information Processing (IIT-CDIP) [25] is the biggest document dataset and it is designed for the task of information retrieval. The Ryerson Vision Lab Complex Document Information Processing (RVL-CDIP) [18] dataset used the IIT-CDIP metadata to create a new dataset for document classification. PublayNet [56] and DocBank [26] are datasets designed for layout analysis tasks and DocVQA [33,51] instead, is designed for Visual Question Answering task over document images.

We summarize all the key information for comparison in Table 1. First, we stress that OCR-IDL is the second biggest dataset in amount for pre-training and biggest dataset with annotations obtained from commercial OCR. This provides unique opportunity for the Document Intelligence research for utilizing the unlabeled documents in their research. Furthermore, even though OCR-IDL uses the documents from the same source as IIT-CDIP and RVL-CDIP, it also contains other type of industrial documents. OCR-IDL contains documents from chemical, medical and drug industries, hence having more variety in terms of content as well as layout information.

3.3 Dataset Statistics

IDL documents come with metadata that is curated by human annotators. They include information about the date, industry, drugs, chemicals, document types and many more. We restrict our analysis on exploring what type of documents we have and the distribution of the dates they are created which can be found in Fig. 2. In IDL metadata, there are 35k various document types from which we show only the most common 20 which include letters, report, email, memo,

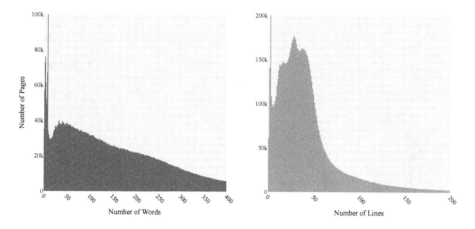

Fig. 4. Distribution of number of words (left) and lines (right) by pages.

note, etc. As can be seen on the left side of Fig. 2, most of the documents' type is unknown. Moreover, even though we have a skewed distribution on letter, report and memo, we also have very distinct distribution such as chart, graphics, news articles. This is especially encouraging because it provides diverse layouts within various contexts. Moreover, the documents are created spanning 100 years where most of the documents are in the range of 1980-2010 as can be seen on the right side of Fig. 2. The date the documents are created contributes to the variability of the documents not only in terms of semantics but more importantly having visual artifacts (smudges, different resolution) with different printing technologies.

On top of the amount of documents, we give more details on the number of pages and words for each document type. The number of pages follows the same distribution as the amount of documents per type, as can be appreciated in Fig. 3. Also, we can see from Fig. 3 that the report type is much richer in terms of text while the rest follows more or less the same distribution.

We turn our attention to OCR annotation statistics. In total, we obtain OCR annotations for 4614232 (4.6M) documents where we have 26621635 (26M) pages, averaging 6 pages per document. Moreover, since documents are known to be a text-rich environment, we provide extra details regarding the amount of words and lines per page and documents. We have 166M words with 46M lines, on average there are 62.5 words and 17.5 lines per page while 360.8 words and 101.25 lines per document. To have a better understanding of the distribution of words and lines per page, we present Fig. 4. As shown in the figure, mean distribution is between 20 to 100 words per page while there is a significant amount of pages that contain more than 200 words per page. The distribution for lines follows a different distribution in which it can be observed that most of the pages contain from 10 to 50 lines. In either case, it is clearly observed that documents at hand are ideal for performing pretraining with their diverse layouts and text-rich settings.

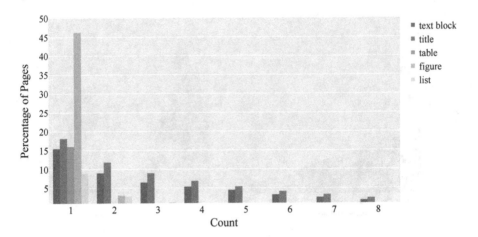

Fig. 5. Distribution of various layout blocks in OCR-IDL. Our documents contain a lot of diversity in terms of title, figure, text block, list and table.

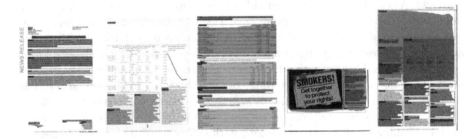

Fig. 6. Qualitative results for segmentation of layout information in OCR-IDL.

Finally, to quantify the diversity of the documents in terms of layout, we run publicly available Faster-RCNN [40] trained on PubLayNet [56] to segment a document into text block, title, figure, list and table. To obtain the segmentation results, we randomly selected 40K pages and some segmentation examples can be found in Fig. 6. It can be appreciated from Fig. 5 that 40% of the documents have at least 1 figure. We also observe that 10-20% of the pages have at least 1 table and list, showing that documents at hand contain very diverse layout information. Moreover, more than 45% of the pages contain more than 1 text block and 1 title, making the documents a text rich environment for pre-training on Document Intelligence tasks.

4 Conclusion

In this paper, we have presented our effort to provide OCR annotations for the large-scale IDL document dataset called OCR-IDL. These annotations have a monetary value over $20,000 and are made publicly available with the aim of advancing the Document Intelligence research field. Our motivation is two-fold, first we make use of a commercial OCR engine to obtain high quality annotations, leading to reduce the noise provided by OCR on pretraining and downstream tasks. Secondly, it is our hope that OCR-IDL can be a starting point for future works on Document Intelligence to be more comparable. Throughout this article we have detailed the process that we have followed to obtain the annotations, we have presented a statistical analysis, and compared them with other datasets in the state of the art. The provided analysis shows that our contribution has a high potential to be used successfully in pre-training strategies for document intelligence models. All the code for data collection process and annotations can be accessed in https://github.com/furkanbiten/idl_data.

Acknowledgments. This work has been supported by projects PDC2021-121512-I00, PLEC2021-00785, PID2020-116298GB-I00, ACE034/21/000084, the CERCA Programme / Generalitat de Catalunya, AGAUR project 2019PROD00090 (BeARS), the Ramon y Cajal RYC2020-030777-I / AEI / 10.13039/501100011033 and PhD scholarship from UAB (B18P0073).

References

1. Aberdam, A., et al.: Sequence-to-sequence contrastive learning for text recognition. In: Proceedings of the IEEE/CVF Conference on Computer Vision and Pattern Recognition, pp. 15302–15312 (2021)
2. Appalaraju, S., Jasani, B., Kota, B.U., Xie, Y., Manmatha, R.: Docformer: end-to-end transformer for document understanding. arXiv preprint arXiv:2106.11539 (2021)
3. Arroyo, D.M., Postels, J., Tombari, F.: Variational transformer networks for layout generation. In: Proceedings of the IEEE/CVF Conference on Computer Vision and Pattern Recognition, pp. 13642–13652 (2021)
4. Biswas, S., Banerjee, A., Lladós, J., Pal, U.: DocEnTr: an end-to-end document image enhancement transformer. arXiv preprint arXiv:2201.11438 (2022)
5. Biswas, S., Riba, P., Lladós, J., Pal, U.: Beyond document object detection: instance-level segmentation of complex layouts. Int. J. Doc. Anal. Recogn. (IJDAR) **24**(3), 269–281 (2021). https://doi.org/10.1007/s10032-021-00380-6
6. Biswas, S., Riba, P., Lladós, J., Pal, U.: DocSynth: a layout guided approach for controllable document image synthesis. In: Lladós, J., Lopresti, D., Uchida, S. (eds.) ICDAR 2021. LNCS, vol. 12823, pp. 555–568. Springer, Cham (2021). https://doi.org/10.1007/978-3-030-86334-0_36
7. Biten, A.F., Litman, R., Xie, Y., Appalaraju, S., Manmatha, R.: LaTr: layout-aware transformer for scene-text VQA. arXiv preprint arXiv:2112.12494 (2021)
8. Biten, A.F., et al.: ICDAR 2019 competition on scene text visual question answering. In: 2019 International Conference on Document Analysis and Recognition (ICDAR), pp. 1563–1570. IEEE (2019)
9. Biten, A.F., et al.: Scene text visual question answering. In: Proceedings of the IEEE/CVF International Conference on Computer Vision, pp. 4291–4301 (2019)

10. Carbonell, M., Riba, P., Villegas, M., Fornés, A., Lladós, J.: Named entity recognition and relation extraction with graph neural networks in semi structured documents. In: 2020 25th International Conference on Pattern Recognition (ICPR), pp. 9622–9627. IEEE (2021)
11. Devlin, J., Chang, M.W., Lee, K., Toutanova, K.: BERT: pre-training of deep bidirectional transformers for language understanding. arXiv preprint arXiv:1810.04805 (2018)
12. Fang, S., Xie, H., Wang, Y., Mao, Z., Zhang, Y.: Read like humans: autonomous, bidirectional and iterative language modeling for scene text recognition. In: Proceedings of the IEEE/CVF Conference on Computer Vision and Pattern Recognition, pp. 7098–7107 (2021)
13. Gómez, L., et al.: Multimodal grid features and cell pointers for scene text visual question answering. Pattern Recogn. Lett. **150**, 242–249 (2021)
14. Gu, J., et al.: UniDoc: unified pretraining framework for document understanding. In: Advances in Neural Information Processing Systems 34 (2021)
15. Ha, J., Haralick, R.M., Phillips, I.T.: Document page decomposition by the bounding-box project. In: Proceedings of 3rd International Conference on Document Analysis and Recognition, vol. 2, pp. 1119–1122. IEEE (1995)
16. Ha, J., Haralick, R.M., Phillips, I.T.: Recursive X-Y cut using bounding boxes of connected components. In: Proceedings of 3rd International Conference on Document Analysis and Recognition, vol. 2, pp. 952–955. IEEE (1995)
17. Hao, L., Gao, L., Yi, X., Tang, Z.: A table detection method for pdf documents based on convolutional neural networks. In: 2016 12th IAPR Workshop on Document Analysis Systems (DAS), pp. 287–292. IEEE (2016)
18. Harley, A.W., Ufkes, A., Derpanis, K.G.: Evaluation of deep convolutional nets for document image classification and retrieval. In: 2015 13th International Conference on Document Analysis and Recognition (ICDAR), pp. 991–995. IEEE (2015)
19. He, K., Gkioxari, G., Dollár, P., Girshick, R.: Mask R-CNN. In: Proceedings of the IEEE International Conference on Computer Vision, pp. 2961–2969 (2017)
20. He, Y., et al.: Visual semantics allow for textual reasoning better in scene text recognition. arXiv preprint arXiv:2112.12916 (2021)
21. Hong, T., Kim, D., Ji, M., Hwang, W., Nam, D., Park, S.: Bros: a pre-trained language model for understanding texts in document (2020)
22. Jemni, S.K., Souibgui, M.A., Kessentini, Y., Fornés, A.: Enhance to read better: a multi-task adversarial network for handwritten document image enhancement. Pattern Recogn. **123**, 108370 (2022)
23. Joshi, M., Chen, D., Liu, Y., Weld, D.S., Zettlemoyer, L., Levy, O.: SpanBERT: improving pre-training by representing and predicting spans. Trans. Assoc. Comput. Linguist. **8**, 64–77 (2020)
24. Lebourgeois, F., Bublinski, Z., Emptoz, H.: A fast and efficient method for extracting text paragraphs and graphics from unconstrained documents. In: 11th IAPR International Conference on Pattern Recognition, Vol. II. Conference B: Pattern Recognition Methodology and Systems, vol. 1, pp. 272–273. IEEE Computer Society (1992)
25. Lewis, D., Agam, G., Argamon, S., Frieder, O., Grossman, D., Heard, J.: Building a test collection for complex document information processing. In: Proceedings of the 29th Annual International ACM SIGIR Conference on Research and Development in Information Retrieval, pp. 665–666 (2006)
26. Li, M., Xu, Y., Cui, L., Huang, S., Wei, F., Li, Z., Zhou, M.: DocBank: a benchmark dataset for document layout analysis. In: Proceedings of the 28th International Conference on Computational Linguistics, pp. 949–960 (2020)

27. Li, P., et al.: SelfDoc: self-supervised document representation learning. In: Proceedings of the IEEE/CVF Conference on Computer Vision and Pattern Recognition, pp. 5652–5660 (2021)
28. Litman, R., Anschel, O., Tsiper, S., Litman, R., Mazor, S., Manmatha, R.: Scatter: selective context attentional scene text recognizer. In: Proceedings of the IEEE/CVF Conference on Computer Vision and Pattern Recognition, pp. 11962–11972 (2020)
29. Mafla, A., Dey, S., Biten, A.F., Gomez, L., Karatzas, D.: Fine-grained image classification and retrieval by combining visual and locally pooled textual features. In: Proceedings of the IEEE/CVF Winter Conference on Applications of Computer Vision, pp. 2950–2959 (2020)
30. Mafla, A., Dey, S., Biten, A.F., Gomez, L., Karatzas, D.: Multi-modal reasoning graph for scene-text based fine-grained image classification and retrieval. In: Proceedings of the IEEE/CVF Winter Conference on Applications of Computer Vision, pp. 4023–4033 (2021)
31. Mafla, A., Rezende, R.S., Gómez, L., Larlus, D., Karatzas, D.: StacMR: scene-text aware cross-modal retrieval. In: Proceedings of the IEEE/CVF Winter Conference on Applications of Computer Vision, pp. 2220–2230 (2021)
32. Mathew, M., Bagal, V., Tito, R., Karatzas, D., Valveny, E., Jawahar, C.: InfographicVQA. In: Proceedings of the IEEE/CVF Winter Conference on Applications of Computer Vision, pp. 1697–1706 (2022)
33. Mathew, M., Karatzas, D., Jawahar, C.: DocVQA: a dataset for VQA on document images. In: Proceedings of the IEEE/CVF Winter Conference on Applications of Computer Vision, pp. 2200–2209 (2021)
34. Na, B., Kim, Y., Park, S.: Multi-modal text recognition networks: Interactive enhancements between visual and semantic features. arXiv preprint arXiv:2111.15263 (2021)
35. O'Gorman, L.: The document spectrum for page layout analysis. IEEE Trans. Pattern Anal. Mach. Intell. 15(11), 1162–1173 (1993)
36. Patil, A.G., Ben-Eliezer, O., Perel, O., Averbuch-Elor, H.: Read: recursive autoencoders for document layout generation. In: Proceedings of the IEEE/CVF Conference on Computer Vision and Pattern Recognition Workshops, pp. 544–545 (2020)
37. Powalski, R., Borchmann, Ł., Jurkiewicz, D., Dwojak, T., Pietruszka, M., Pałka, G.: Going full-TILT boogie on document understanding with text-image-layout transformer. In: Lladós, J., Lopresti, D., Uchida, S. (eds.) ICDAR 2021. LNCS, vol. 12822, pp. 732–747. Springer, Cham (2021). https://doi.org/10.1007/978-3-030-86331-9_47
38. Prasad, D., Gadpal, A., Kapadni, K., Visave, M., Sultanpure, K.: CascadeTabNet: an approach for end to end table detection and structure recognition from image-based documents. In: Proceedings of the IEEE/CVF Conference on Computer Vision and Pattern Recognition Workshops, pp. 572–573 (2020)
39. Raja, S., Mondal, A., Jawahar, C.V.: Table structure recognition using top-down and bottom-up cues. In: Vedaldi, A., Bischof, H., Brox, T., Frahm, J.-M. (eds.) ECCV 2020. LNCS, vol. 12373, pp. 70–86. Springer, Cham (2020). https://doi.org/10.1007/978-3-030-58604-1_5
40. Ren, S., He, K., Girshick, R., Sun, J.: Faster R-CNN: towards real-time object detection with region proposal networks. Adv. Neural. Inf. Process. Syst. 28, 91–99 (2015)
41. Riba, P., Dutta, A., Goldmann, L., Fornés, A., Ramos, O., Lladós, J.: Table detection in invoice documents by graph neural networks. In: 2019 International Conference on Document Analysis and Recognition (ICDAR), pp. 122–127. IEEE (2019)

42. Schreiber, S., Agne, S., Wolf, I., Dengel, A., Ahmed, S.: DeepDeSRT: deep learning for detection and structure recognition of tables in document images. In: 2017 14th IAPR International Conference on Document Analysis and Recognition (ICDAR), vol. 1, pp. 1162–1167. IEEE (2017)
43. Simon, A., Pret, J.C., Johnson, A.P.: A fast algorithm for bottom-up document layout analysis. IEEE Trans. Pattern Anal. Mach. Intell. **19**(3), 273–277 (1997)
44. Singh, A., et al..: Towards VQA models that can read. In: Proceedings of the IEEE/CVF Conference on Computer Vision and Pattern Recognition, pp. 8317–8326 (2019)
45. Smith, R.: An overview of the tesseract OCR engine. In: Ninth International Conference on Document Analysis and Recognition (ICDAR 2007), vol. 2, pp. 629–633. IEEE (2007)
46. Soto, C., Yoo, S.: Visual detection with context for document layout analysis. In: Proceedings of the 2019 Conference on Empirical Methods in Natural Language Processing and the 9th International Joint Conference on Natural Language Processing (EMNLP-IJCNLP), pp. 3464–3470 (2019)
47. Souibgui, M.A., et al.: DocEnTr: an end-to-end document image enhancement transformer. arXiv preprint arXiv:2201.10252 (2022)
48. Souibgui, M.A., et al.: One-shot compositional data generation for low resource handwritten text recognition. In: Proceedings of the IEEE/CVF Winter Conference on Applications of Computer Vision, pp. 935–943 (2022)
49. Souibgui, M.A., Kessentini, Y.: De-GAN: a conditional generative adversarial network for document enhancement. IEEE Transactions on Pattern Analysis and Machine Intelligence (2020)
50. Tito, R., Karatzas, D., Valveny, E.: Document collection visual question answering. arXiv preprint arXiv:2104.14336 (2021)
51. Tito, R., Mathew, M., Jawahar, C.V., Valveny, E., Karatzas, D.: ICDAR 2021 competition on document visual question answering. In: Lladós, J., Lopresti, D., Uchida, S. (eds.) ICDAR 2021. LNCS, vol. 12824, pp. 635–649. Springer, Cham (2021). https://doi.org/10.1007/978-3-030-86337-1_42
52. Vaswani, A., et al.: Attention is all you need. Advances in Neural Information Processing Systems 30 (2017)
53. Xu, Y., et al.: LayoutLMv2: multi-modal pre-training for visually-rich document understanding. arXiv preprint arXiv:2012.14740 (2020)
54. Xu, Y., Li, M., Cui, L., Huang, S., Wei, F., Zhou, M.: LayoutLM: pre-training of text and layout for document image understanding. In: Proceedings of the 26th ACM SIGKDD International Conference on Knowledge Discovery & Data Mining, pp. 1192–1200 (2020)
55. Zhong, X., ShafieiBavani, E., Jimeno Yepes, A.: Image-based table recognition: data, model, and evaluation. In: Vedaldi, A., Bischof, H., Brox, T., Frahm, J.-M. (eds.) ECCV 2020. LNCS, vol. 12366, pp. 564–580. Springer, Cham (2020). https://doi.org/10.1007/978-3-030-58589-1_34
56. Zhong, X., Tang, J., Yepes, A.J.: PubLayNet: largest dataset ever for document layout analysis. In: 2019 International Conference on Document Analysis and Recognition (ICDAR), pp. 1015–1022. IEEE (2019)

Self-paced Learning to Improve Text Row Detection in Historical Documents with Missing Labels

Mihaela Găman, Lida Ghadamiyan, Radu Tudor Ionescu[✉],
and Marius Popescu

Department of Computer Science, University of Bucharest,
14 Academiei, Bucharest, Romania
raducu.ionescu@gmail.com

Abstract. An important preliminary step of optical character recognition systems is the detection of text rows. To address this task in the context of historical data with missing labels, we propose a self-paced learning algorithm capable of improving the row detection performance. We conjecture that pages with more ground-truth bounding boxes are less likely to have missing annotations. Based on this hypothesis, we sort the training examples in descending order with respect to the number of ground-truth bounding boxes, and organize them into k batches. Using our self-paced learning method, we train a row detector over k iterations, progressively adding batches with less ground-truth annotations. At each iteration, we combine the ground-truth bounding boxes with pseudo-bounding boxes (bounding boxes predicted by the model itself) using non-maximum suppression, and we include the resulting annotations at the next training iteration. We demonstrate that our self-paced learning strategy brings significant performance gains on two data sets of historical documents, improving the average precision of YOLOv4 with more than 12% on one data set and 39% on the other.

Keywords: Self-paced learning · Curriculum learning · Text row detection · Neural networks · Training regime

1 Introduction

Automatically processing historical documents to extract and index useful information in digital databases is an important step towards preserving our cultural heritage [10]. This process is commonly based on storing the information as images. However, performing optical character recognition (OCR) on the scanned documents brings major benefits regarding the identification, storage and retrieval of information [11,12,16,17].

Prior to the recognition of handwritten or printed characters, an important step of OCR systems is the detection of text lines (rows) [5,13–15]. Our work is particularly focused on the text row detection task under a difficult setting, where

© The Author(s), under exclusive license to Springer Nature Switzerland AG 2023
L. Karlinsky et al. (Eds.): ECCV 2022 Workshops, LNCS 13804, pp. 253–262, 2023.
https://doi.org/10.1007/978-3-031-25069-9_17

the training data has missing labels (bounding boxes). To improve detection performance on this challenging task, we propose a self-paced learning algorithm that gradually adds pseudo-labels during training. Our algorithm is based on the following hypothesis: pages with more ground-truth bounding boxes are less likely to have missing annotations. We are certain that this hypothesis holds if the number of text rows per page is constant. However, in practice, pages may come from different books having various formats. Moreover, some pages at the end of book chapters may contain only a few rows. Even if such examples are present, we conjecture that our hypothesis holds in a sufficiently large number of cases to be used as a function to schedule the learning process. We thus propose to organize the training examples into k batches, such that the first batch contains the most reliable pages (with the highest number of bounding boxes per page). The subsequent batches have gradually less annotations. The training starts on the first batch and the resulting model is applied on the second batch to enrich it with pseudo-labels (bounding boxes predicted by the model itself). One by one, batches with ground-truth and pseudo-labels are gradually added during training, until the model gets to see all training data. Our self-paced learning process is also illustrated in Fig. 1.

We conduct text row detection experiments on two data sets of historical documents, comparing the state-of-the-art YOLOv4 [1] detector with a version of YOLOv4 trained under our self-paced learning regime. The empirical results show that our self-paced learning algorithm introduces significant performance gains on both benchmarks.

Contribution. In summary, our contribution is twofold:

- We propose a novel self-paced learning algorithm for line detection which considers the training examples in the decreasing order of the number of ground-truth bounding boxes, alleviating the problem of missing labels by introducing pseudo-labels given by the detector.
- We present empirical results demonstrating that our self-paced learning regime brings significant improvements in text row detection on two data sets, under a fair yet challenging evaluation setting that ensures source (document) separation (and, on one data set, even domain shift from the Cyrillic script to the Latin script) across training and test pages.

2   Related Work

Self-paced Learning. Humans hold the ability to guide themselves through the learning process, being capable of learning new concepts at their own pace, without requiring instructions from a teacher. Indeed, learners often choose what, when and how long to study, implicitly using a self-paced curriculum learning process. This learning process inspired researchers to propose and study artificial neural networks based on self-paced learning (SPL) [6,7]. Researchers have proposed a broad range of SPL strategies for various computer vision [6,8,23]

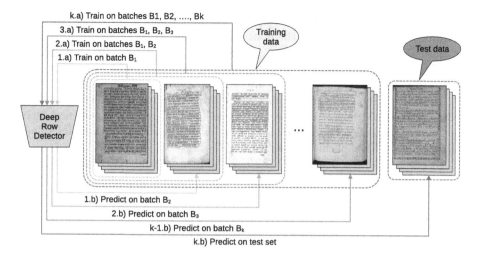

Fig. 1. Our self-paced learning algorithm for row detection in historical documents with missing bounding box annotations. Training examples are sorted into k batches having less and less ground-truth annotations. Best viewed in color. (Color figure online)

and pattern recognition tasks [21, 24, 25], demonstrating performance gains over the standard supervised learning method. Although SPL can be viewed as a particularization of curriculum learning [22], where the complexity of the samples is estimated by the learner, we consider the broader domain of curriculum learning as less relevant to our work and choose not to cover it in this section.

To the best of our knowledge, we are the first to employ self-paced learning for text row detection.

Object Detection. One way to approach text row detection is to employ a state-of-the-art deep object detector. We distinguish two types of deep object detectors in the recent literature. On the one hand, there are two-stage object detectors, e.g. Faster R-CNN [20], that employ a region proposal network to generate object proposals in the first stage. The proposals are subsequently refined in the second detection stage by the final model. On the other hand, there are one-stage detectors, e.g. YOLO [18] and SSD [9], capable of detecting objects in a single run. YOLO is one of the most popular object detectors, being continuously improved by the community with new versions such as YOLOv3 [19] and YOLOv4 [1]. In our self-paced learning algorithm, we employ the recent YOLOv4 as the underlying detector.

Row Detection. Text line extraction is an important initial step in the analysis of document images [5]. Historical writings pose a variety of challenges in the detection of text lines, from multiple different editorial styles to noise such as yellow pages or ink stains, and even to significant physical degradation [13, 14]. Due to such particularities, a reduction in performance is encountered in the processing of ancient text. Thus, even recent works consider the problem of text line detection open and yet to be solved [5,13,14]. In the past few years, researchers continued to make progress in the matter of text line detection in historical documents. We hereby highlight a few recent efforts which successfully employed deep learning to solve this task. Some approaches [13,14] perform text line segmentation, relying on the U-Net architecture or its variations, such as ARU-Net [5]. Hybrid architectures, combining U-Net with traditional image processing techniques worked best in some recent works [5,14]. A trade-off between accuracy and computational efficiency is studied in [15]. Here, the authors introduce a lightweight CNN suitable for mobile devices.

To the best of our knowledge, we are the first to study text line detection under missing labels.

3   Method

Assuming that all pages have about the same number of text rows, we conjecture that pages with more bounding boxes are less likely to have missing annotations. Our self-paced learning algorithm relies on this conjecture to organize and present the training data in a meaningful order, which allows us to learn a more robust detector. The steps of our self-paced learning method are formalized in Algorithm 1.

In the first stage (steps 1–5), we split the training set X into a set of k disjoint (according to step 3) and equally-sized (according to step 4) batches, denoted as $B_1, B_2, ..., B_k$. Before splitting the data, we sort the training images with respect to the number of bounding boxes, in descending order. Hence, for any two images $x \in B_i$ and $z \in B_j$, where $i < j$, the number of bounding boxes associated to x is higher or equal to the number of bounding boxes associated to z (according to step 5).

Once the training data is organized into batches, we proceed with the second stage (steps 6–18). The self-paced learning procedure is carried out for k iterations (step 7), alternating between two loops at each iteration i. In the first loop (steps 10–13), the model f is trained with stochastic gradient descent on batches $B_1, B_2, ..., B_i$, using as target ground-truth bounding boxes as well as pseudo-bounding boxes, which are predicted at the previous iterations $1, 2, ..., i - 1$. In the second loop (steps 15–18), we iterate through the training images in batch B_{i+1} and apply the detector f (step 16) on each sample x to predict the bounding boxes P_x. The predicted bounding boxes are added to the set of ground-truth boxes T_x (step 17). Then, we apply non-maximum suppression (step 18) to eliminate pseudo-bounding boxes that overlap with the ground-truth ones. We hereby note that we assign a confidence score of 1 to all ground-truth bounding boxes, ensuring that these never get eliminated. In other words, the NMS procedure at step 18 can only eliminate pseudo-annotations.

Algorithm 1. Self-paced learning for row detection

Input: $X = \{(x_i, T_{x_i}) \mid x_i \in \mathbb{R}^{h \times w \times c}, T_{x_i} = \{t_{x_i}^j \mid t_{x_i}^j \in \mathbb{R}^5\}, \forall i \in \{1, 2, ..., n\}\}$ - a training set of n images and the associated bounding boxes (each bounding box $t_{x_i}^j$ is represented by its four coordinates and a confidence score); $Y = \{y_i \mid y_i \in \mathbb{R}^{h \times w \times c}, \forall i \in \{1, 2, ..., m\}\}$ - a test set of m images; k - the number of training batches; η - a learning rate; $\mathcal{L}$ - a loss function; p - IoU threshold for the NMS procedure.

Notations: f - a deep row detector; θ - the weights of the detector f; nms - a function that applies non-maximum suppression; $\mathcal{N}(0, \Sigma)$ - the normal distribution of mean 0 and standard deviation Σ; $\mathcal{U}(S)$ - the uniform distribution over the set S.

Initialization: $\theta^{(0)} \sim \mathcal{N}(0, \Sigma)$

Output: $\mathcal{P} = \{P_{y_1}, P_{y_2}, ..., P_{y_m}\}$ - the bounding boxes predicted for $y_i \in Y, \forall i \in \{1, 2, ..., m\}$.

Stage 1: Training data organization

1: sort and split X into $B_1, B_2, ..., B_k$ such that:
2:    $X = \bigcup_{i=1}^{k} B_i$
3:    $B_i \cap B_j = \emptyset, \forall i, j \in \{1, 2, ..., k\}, i \neq j$
4:    $|B_i| - |B_j| \leq 1, \forall i, j \in \{1, 2, ..., k\}$
5:    $\forall x \in B_i, z \in B_j, i < j, |T_x| \geq |T_z|$

Stage 2: Self-paced learning

6: $X' \leftarrow \emptyset$
7: **for** $i \leftarrow 1$ to k **do**
8:    $X' \leftarrow X' \cup B_i$
9:    $t \leftarrow 0$
10:    **while** converge criterion not met **do**
11:       $X^{(t)} \leftarrow$ mini-batch $\sim \mathcal{U}(X')$
12:       $\theta^{(t+1)} \leftarrow \theta^{(t)} - \eta^{(t)} \nabla \mathcal{L}\left(\theta^{(t)}, X^{(t)}\right)$
13:       $t \leftarrow t + 1$
14:    **if** $i + 1 \leq k$ **then**
15:       **for** $x \in B_{i+1}$ **do**
16:          $P_x \leftarrow f(\theta, x)$
17:          $T_x \leftarrow T_x \cup P_x$
18:          $T_x \leftarrow nms(T_x, p)$

Stage 3: Prediction on test set

19: **for** $i \leftarrow 1$ to m **do**
20:    $P_{y_i} \leftarrow f(\theta, y_i)$

In the third stage (steps 19–20), we employ the final model to detect text rows in the test images. The resulting bounding boxes represent the final output of our model.

Aside from the common hyperparameters that configure the standard optimization process based on stochastic gradient descent, our algorithm requires two additional hyperparameters, k and p. The former hyperparameter (k) determines the number of batches, while the second hyperparameter (p) specifies the threshold which decides when to suppress pseudo-bounding boxes.

4   Experiments

4.1   Data Sets

ROCC. The Romanian Old Cyrillic Corpus (ROCC) represents a collection of scanned historical documents, spanning a time frame of more than three centuries [2,3]. The corpus provided by Cristea et al. [2] consists of 367 scanned document pages, with a total of 6418 annotated text lines. We select the training and test pages from different distributions, i.e. distinct books, ensuring a fair and unbiased evaluation. We split the data into a training set consisting of 332 pages (with 5715 annotated bounding boxes) and a test set of 35 pages (with 492 bounding boxes). Both training and test splits have a significant amount of missing bounding boxes (rows that are not annotated). For a correct evaluation, we manually annotate the missing lines in the test set, reaching a total of 703 bounding boxes for the final test set.

cBAD. The cBAD competition [4] for baseline detection featured a corpus formed of 2035 pages of historical documents written in the Latin script. To test the robustness of our self-paced learning method, we randomly select 52 images from the cBAD data set to serve as an out-of-domain test set (the training is conducted on ROCC). We manually annotate the selected pages with 1542 bounding boxes representing text rows.

4.2   Evaluation Setup

Evaluation Metrics. As evaluation measures, we report the average precision (AP) at an Intersection over Union (IoU) threshold of 0.5, as well as the mean IoU.

Baselines. As the first baseline, we consider the YOLOv4 [1] based on the conventional training procedure. We also add a baseline based on self-paced learning with 5 batches, which takes the examples in a random order, in contrast to our approach that sorts the examples based on the number of bounding boxes.

Hyperparameter Choices. We employ the official YOLOv4 implementation from [1], selecting CSPDarknet53 as backbone. We use mini-batches of 4 samples. We employ the Adam optimizer with an initial learning rate of 10^{-3}, leaving the other hyperparameters to their default values. All models are trained for a maximum of 2000 epochs using early stopping. For the self-paced learning strategy, we use $k = 5$ batches and perform 400 epochs with each new batch. The IoU threshold for the NMS procedure is set to $p = 0.5$.

4.3   Results

We present the results on ROCC and cBAD in Table 1. First, we observe that, when the model is trained with either SPL strategy on the first batch only (20% of the training data), the amount of training data seems insufficient for the model

Table 1. Baseline YOLOv4 versus our YOLOv4 based on self-paced learning (SPL). AP and mean IoU scores (in %) are reported on two data sets: ROCC and cBAD. For ROCC, a baseline SPL regime that takes examples in a random order is included. Best scores on each data set are highlighted in bold.

Data Set	Model	Iteration	AP	Mean IoU
ROCC	YOLOv4 (baseline)	-	81.55	70.60
	YOLOv4 + SPL (random)	1	70.06	57.53
		2	74.86	55.53
		3	75.95	54.20
		4	75.16	58.69
		5 (final)	77.56	57.77
	YOLOv4 + SPL (ours)	1	72.43	65.80
		2	87.22	66.97
		3	88.05	67.06
		4	89.86	69.14
		5 (final)	**93.73**	**75.25**
cBAD	YOLOv4 (baseline)	-	35.37	61.30
	YOLOv4 + SPL (ours)	1	22.52	0.00
		2	54.81	56.62
		3	63.14	63.00
		4	63.72	58.96
		5 (final)	**74.57**	**67.52**

to generalize well and compete with the baseline YOLOv4. Both SPL alternatives reach better performance levels once the whole training data is used in the learning process. However, the two SPL strategies improve with each batch at completely different paces. The baseline SPL approach based on randomly ordered examples does not even surpass the YOLOv4 based on conventional training. In contrast, once the model gets to learn at least two training batches with our SPL strategy, our training regime already surpasses the baseline, showing that it is indeed important to sort the examples according to the number of bounding boxes. Moreover, this result indicates that around 60% of the training data is rather harmful for the baseline, due to the missing labels. In general, with each iteration of our SPL strategy, the inclusion of a new batch of images having both ground-truth and pseudo-labels brings consistent performance gains. Our best results are obtained after the last iteration, once the model gets to see the entire training data. We consider that our final improvements (+12.18% on ROCC and +39.20% on cBAD) are remarkable. We conclude that our conjecture is validated by the experiments, and that our self-paced learning method is extremely useful in improving text row detection with missing labels. We also underline that a basic SPL regime is not equally effective.

YOLOv4 (baseline)   YOLOv4+SPL (ours)

Fig. 2. Detections of YOLOv4 based on the standard (baseline) training regime versus our self-paced learning (SPL) regime. Best viewed in color. (Color figure online)

In Fig. 2, we illustrate typical examples of row detections given by the baseline YOLOv4 and the YOLOv4 based on our self-paced learning strategy. We observe that our model detects more rows, while the baseline model sometimes tends to detect two rows in a single bounding box.

5   Conclusion

In this paper, we proposed a self-paced learning algorithm for text row detection in historical documents with missing annotations. Our algorithm is based on the hypothesis that pages with more bounding boxes are more reliable, and thus, can be used sooner during training. Our hypothesis is supported by the empirical results reported on two data sets, which demonstrate significant improvements (between 12% and 40%) in terms of AP. In future work, we aim to extend the applicability of our self-paced learning method to other detection tasks which suffer from the problem of missing labels.

Acknowledgment. This work has been carried out with the financial support of UEFISCDI, within the project PN-III-P2-2.1-PED-2019-3952 entitled "Artificial Intelligence Models (Deep Learning) Applied in the Analysis of Old Romanian Language (DeLORo - Deep Learning for Old Romanian)". This work was also supported by a grant of the Romanian Ministry of Education and Research, CNCS - UEFISCDI, project number PN-III-P1-1.1-TE-2019-0235, within PNCDI III. Authors are alphabetically ordered.

References

1. Bochkovskiy, A., Wang, C.Y., Liao, H.Y.M.: YOLOv4: Optimal speed and accuracy of object detection. arXiv preprint arXiv:2004.10934 (2020)
2. Cristea, D., Pădurariu, C., Rebeja, P., Onofrei, M.: From Scan to Text. Methodology, Solutions and Perspectives of Deciphering Old Cyrillic Romanian Documents into the Latin Script. In: Knowledge, Language, Models, pp. 38–56 (2020)
3. Cristea, D., Rebeja, P., Pădurariu, C., Onofrei, M., Scutelnicu, A.: Data Structure and Acquisition in DeLORo - a Technology for Deciphering Old Cyrillic-Romanian Documents. In: Proceedings of ConsILR (2022)
4. Diem, M., Kleber, F., Fiel, S., Grüning, T., Gatos, B.: cBAD: ICDAR2017 competition on baseline detection. In: Proceedings of ICDAR, pp. 1355–1360 (2017)
5. Grüning, T., Leifert, G., Strauß, T., Michael, J., Labahn, R.: A two-stage method for text line detection in historical documents. Int. J. Document Anal. Recogn. (IJDAR) **22**(3), 285–302 (2019). https://doi.org/10.1007/s10032-019-00332-1
6. Jiang, L., Meng, D., Zhao, Q., Shan, S., Hauptmann, A.: Self-paced curriculum learning. In: Proceedings of AAAI, pp. 2694–2700 (2015)
7. Kumar, M.P., Packer, B., Koller, D.: Self-paced learning for latent variable models. In: Proceedings of NIPS, vol. 23, pp. 1189–1197 (2010)
8. Lin, W., Gao, J., Wang, Q., Li, X.: Pixel-level self-paced learning for super-resolution. In: Proceedings of ICASSP, pp. 2538–2542 (2020)
9. Liu, W., et al.: SSD: single Shot MultiBox Detector. In: Proceedings of ECCV, pp. 21–37 (2016)
10. Lombardi, F., Marinai, S.: Deep learning for historical document analysis and recognition-a survey. J. Imaging **6**(10), 110 (2020)
11. Martínek, J., Lenc, L., Král, P.: Building an efficient OCR system for historical documents with little training data. Neural Comput. Appl. **32**(23), 17209–17227 (2020). https://doi.org/10.1007/s00521-020-04910-x
12. Martínek, J., Lenc, L., Král, P., Nicolaou, A., Christlein, V.: Hybrid Training Data for Historical Text OCR. In: Proceedings of ICDAR, pp. 565–570 (2019)
13. Mechi, O., Mehri, M., Ingold, R., Amara, N.E.B.: Text line segmentation in historical document images using an adaptive U-Net architecture. In: Proceedings of ICDAR, pp. 369–374 (2019)
14. Mechi, O., Mehri, M., Ingold, R., Amara, N.E.B.: Combining deep and ad-hoc solutions to localize text lines in ancient Arabic Document Images. In: Proceedings of ICPR, pp. 7759–7766 (2021)

15. Melnikov, A., Zagaynov, I.: Fast and lightweight text line detection on historical documents. In: Proceedings of DAS, pp. 441–450 (2020)
16. Neudecker, C., et al.: OCR-D: an end-to-end open source OCR framework for historical printed documents. In: Proceedings of DATeCH, pp. 53–58 (2019)
17. Nunamaker, B., Bukhari, S.S., Borth, D., Dengel, A.: A Tesseract-based OCR framework for historical documents lacking ground-truth text. In: Proceedings of ICIP, pp. 3269–3273 (2016)
18. Redmon, J., Divvala, S., Girshick, R., Farhadi, A.: You Only Look Once: Unified, Real-Time Object Detection. In: Proceedings of CVPR, pp. 779–788 (2016)
19. Redmon, J., Farhadi, A.: YOLOv3: An incremental improvement. arXiv preprint arXiv:1804.02767 (2018)
20. Ren, S., He, K., Girshick, R., Sun, J.: Faster R-CNN: towards real-time object detection with region proposal networks. In: Proceedings of NIPS, pp. 91–99 (2015)
21. Ristea, N.C., Ionescu, R.T.: Self-paced ensemble learning for speech and audio classification. In: Proceedings of INTERSPEECH, pp. 2836–2840 (2021)
22. Soviany, P., Ionescu, R.T., Rota, P., Sebe, N.: Curriculum learning: A survey. arXiv preprint arXiv:2101.10382 (2021)
23. Soviany, P., Ionescu, R.T., Rota, P., Sebe, N.: Curriculum self-paced learning for cross-domain object detection. Comput. Vis. Image Underst. **204**, 103166 (2021)
24. Zheng, W., Zhu, X., Wen, G., Zhu, Y., Yu, H., Gan, J.: Unsupervised feature selection by self-paced learning regularization. Pattern Recogn. Lett. **132**, 4–11 (2020)
25. Zhou, P., Du, L., Liu, X., Shen, Y.D., Fan, M., Li, X.: Self-paced clustering ensemble. IEEE Transactions on Neural Networks and Learning Systems (2020)

On Calibration of Scene-Text Recognition Models

Ron Slossberg[1]([✉]), Oron Anschel[2], Amir Markovitz[2], Ron Litman[2],
Aviad Aberdam[1], Shahar Tsiper[2], Shai Mazor[2], Jon Wu[2], and R. Manmatha[2]

[1] Technion Institute, Haifa, Israel
`ronslos@campus.technion.ac.il`, `aaberdam@amazon.com`
[2] Amazon AWS, San Francisco, USA
{`oronans,amirmak,litmanr,tsiper,smazor,jonwu,manmatha`}`@amazon.com`

Abstract. The topics of confidence and trust in modern scene-text recognition (STR) models have been rarely investigated in spite of their prevalent use within critical user-facing applications. We analyze confidence estimation for STR models and find that they tend towards overconfidence thus leading to overestimation of trust in the predicted outcome by users. To overcome this phenomenon we propose a word-level confidence calibration approach. Initially, we adapt existing single-output T-scaling calibration methodologies to suit the case of sequential decoding. Interestingly, extensive experimentation reveals that character-level calibration underperforms word-level calibration and it may even be harmful when employing conditional decoding. In addition, we propose a novel calibration metric better suited for sequential outputs as well as a variant of T-scaling specifically designed for sequential prediction. Finally, we demonstrate that our calibration approach consistently improves prediction accuracy relative to the non-calibrated baseline when employing a beam-search strategy.

1 Introduction

Scene Text Recognition (STR) – the task of extracting text from a cropped word image, has seen an increase in popularity in recent years. While an active research area for almost three decades, STR performance has recently seen a significant boost due to the utilization of deep-learning models [1,3,25,38,41,43]. Some typical applications relying on STR models include assistance to the visually impaired, content moderation in social media and street sign recognition for autonomous vehicles. The above examples, often referred to as user-facing applications entail a high degree of trust. This can be achieved by reliably assessing the prediction confidence, *i.e.*, what the probability for a correct prediction is. Despite the prevalent usage of STR models within critical user-facing applications, confidence estimation for such models has not been thoroughly investigated

---

Supplementary Information The online version contains supplementary material available at https://doi.org/10.1007/978-3-031-25069-9_18.

© The Author(s), under exclusive license to Springer Nature Switzerland AG 2023
L. Karlinsky et al. (Eds.): ECCV 2022 Workshops, LNCS 13804, pp. 263–279, 2023.
https://doi.org/10.1007/978-3-031-25069-9_18

in the past, thus leading to a misjudgment of the risk introduced by the model into the overall application.

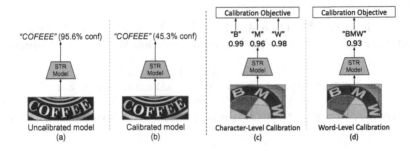

Fig. 1. Overconfidence displayed by uncalibrated STR models. Our analysis exposes the miscalibration of many modern STR models as depicted by (a). The calibrated score in (b) signals downstream tasks that the prediction should not be trusted. **Character vs. Word-Level Calibration.** (c) Individual character score calibration. (d) Word-level calibration *i.e.*the confidence score for the entire word is directly optimized. We demonstrate the importance of adopting the word-level approach.

Confidence calibration is the task of tuning a model's confidence scores to match a successful prediction's underlying probability. For example, within the group of samples producing a confidence score of 0.7, we expect to achieve a prediction success rate of exactly 70%. Confidence calibration and model reliability have been active areas of research for many years [5,7,33]; however, the task of calibrating sequence-level confidence has received little attention and to the best of our knowledge, in STR, it has yet to be explored. In this work we study the confidence characteristics of modern STR models and expose the overconfidence tendency displayed by them. A similar phenomenon in the context of classification was previously observed by [12,14,21,36] (see Fig. 1 (a-b) for illustration). We conduct a comprehensive study encompassing a wide range of recent popular STR methods. Specifically, we examine various encoder choices coupled with conditional as well as non-conditional decoders and propose calibration techniques and practices to improve estimated model confidence scores.

Previous work has examined calibration and confidence intervals for structured prediction in the context of NLP problems [8,20,22,32]. However, these methods focus on the calibration of marginal confidences, analogous to the calibration of each individual decoder output. This methodology resembles single-output model calibration as each classified token is treated individually as depicted in Fig. 1 (c).

In this work, we show that calibration of text recognizers at the character-level is sub-optimal and in fact harmful when employing conditional decoders. Contrarily, we demonstrate that word-level calibration of STR models (see Fig. 1 (d)) are more suitable to sequential decoding tasks independent of decoder choice. This is demonstrated in Fig. 2 where the calibrated sequence length is gradually varied from character-level to word-level calibration. We specifically note the increase in calibration error when applying character-level calibration to conditional decoders. In contrast we notice that error decreases when applying word-level calibration regardless of the decoder type.

Our main objective is to perform confidence calibration within real-time user-facing STR applications. We therefore limit our scope to methodologies that do not increase run-time complexity. Among this class of calibration methods, Temperature-scaling (T-scaling) was shown to be both simple and effective, often more so than other more complex methods [12,37]. While previous works have demonstrated successful calibration of single output (classification) models, we adapt the T-scaling method to the sequence prediction task prescribed by STR. In addition, we extend the Expected Calibration Error (ECE) [31] proposed for binary classification problems to the regime of sequence calibration by incorporating a sequence accuracy measure, namely the edit-distance [24] metric. Furthermore, we present a useful application for confidence calibration by combining calibration with a beam-search decoding scheme, achieving consistent accuracy gains. Finally, we propose a sequence oriented extension to Temperature-scaling named Step Dependent T-Scaling, presenting moderate calibration gains for negligible added computational complexity.

Our key contributions are the following:

- We present the first analysis of confidence estimation and calibration for STR methods, discovering that numerous off-the-shelf STR models are badly calibrated.
- We highlight the importance of directly calibrating for word-level confidence scores and demonstrate that performing character-level optimization often has an adverse effect on calibration error.
- We demonstrate consistent accuracy gains by applying beam-search to calibrated STR models.
- We extend a commonly used calibration metric (ECE) to better suite partial error discovery we term "Edit-Distance ECE", and propose a sequence-based extension to T-scaling termed "Step-dependant T-scaling".

2   Related Work

Scene Text Recognition. Shi *et al.* [41] proposed an end-to-end image to sequence approach without the need for character-level annotations. The authors used a BiLSTM [11] for modeling contextual dependencies, and Connectionist Temporal Classification (CTC) [10] for decoding. Baek *et al.* [1] proposed a four-stage framework unifying several previous techniques [4,27,41,42]. The framework comprises the following building blocks: image transformation, feature extraction, sequence modeling, and decoding. Numerous subsequent methods also conform to this general structure [3,25,38,43]. Currently, SOTA results are often achieved by methods adopting an attention-based [2] decoder scheme. The attention-based decoders usually consist of an RNN cell taking at each step the previously predicted token and a hidden state as inputs and outputting the next token prediction.

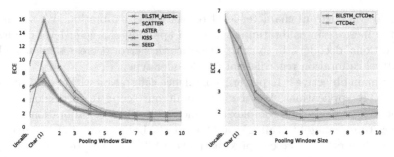

Fig. 2. Calibrated Word-Level ECE Values vs. Pooling Window Sizes (n-grams) for different STR methods. We evaluate our results on 10 saved checkpoints for each model, plotting the mean and standard deviation. **Left:** Conditional decoders. **Right:** Non-conditional decoders. All results are demonstrated on a held-out test-set. "Char" refers to character level calibration corresponding to a window size of 1. We observe that attention based models become uncalibrated when individual character calibration is performed and that longer window sizes are preferred over short ones during calibration.

Confidence Calibration. Model calibration has been a subject of interest within the data modeling and general scientific communities for many decades [5,7,33]. Several recent papers [12,14,16,21,34,36] have studied model calibration in the context of modern neural networks and classifier calibration (scalar or multi-class predictions). Empirically, modern neural networks are poorly calibrated and tend towards overconfidence. A common theme among numerous calibration papers is that Temperature-scaling (T-scaling) [12] is often the most effective calibration method even when compared to complex methods such as Monte-Carlo Dropout, Deep-ensemble, and Bayesian methods [35]. Nixon *et al.* [34] conduct a study on several proposed variations of established calibration metrics and suggest good practices for calibration optimization and evaluation. Similarly, we minimize an ECE calibration objective using a gradient framework.

Confidence Calibration for Sequential Models. Most of the confidence calibration literature is focused on calibrating a single output classifier. Kuleshov and Liang [19] were the first to propose a calibration framework for structured prediction problems. The framework defines the notion of "Events of Interest" coupled with confidence scores allowing event-level calibration. The practical methods laid out by Kuleshov and Liang [19], however, predate the recent advances in DNNs.

Kumar *et al.* [20] address the problem of miscalibration in neural machine translation (NMT) systems. The authors show that NMT models are poorly calibrated and propose a calibration method based on a T-scaling variant where the temperature is predicted at each decoding step. They were also able to improve translation performance by applying beam-search to calibrated models. Our experiments find this to be beneficial for the task of STR as well.

Desai *et al.* [8] suggest the usage of T-Scaling for calibration of pre-trained transformers models (*e.g.*BERT [9]). The authors differentiate between in and out of domain calibration and propose using T-Scaling, and label-smoothing [36] techniques. We point-out that label smoothing is carried out during the training phase and therefore affects the model accuracy. Here, calibration is also conducted at the individual output level.

In another work [22], a proposed extension to T-scaling calibration for sequences is presented. The authors employ a parametric decaying exponential to model the temperature for each decoding step. Again, similarly to [20] calibration is performed for each decoding step and not for entire sequences.

3 Background

Temperature Scaling. T-Scaling [12], a limited version of Platt Scaling [37], is a simple yet effective calibration method. T-scaling utilizes a single parameter $T > 0$. Given a logits vector $\mathbf{z}_i$ the model produces a calibrated score as:

$$\hat{q}_i = \max_k \; \sigma_{\text{SM}} \left(\mathbf{z}_i / T \right)^{(k)}, \tag{1}$$

where $\hat{q}_i$ denotes the estimated confidence of the ith data sample, $\mathbf{z}_i \in \mathbb{R}^K$ is the output logits and K is the number of output classes (number of supported symbols for a STR model). T is a global scaling parameter and σ_{SM} is the softmax function defined as:

$$\sigma_{\text{SM}} \left(\mathbf{z}_i \right)^{(k)} = \frac{\exp \left(\mathbf{z}_i^{(k)} \right)}{\sum_{l=1}^{K} \exp \left(\mathbf{z}_i^{(l)} \right)}. \tag{2}$$

The temperature parameter T scales the logits, either altering the predicted confidence scores as necessary. T-Scaling is a monotonic transform of the confidence values, and therefore *does not affect the classifier model accuracy.*

Reliability Diagrams. Figure 3 presents a visual representation of model calibration [7,33]. Reliability diagrams show the expected accuracy for different confidence bins, where the diagonal represents a perfect calibration. Within each plot, the lower right triangle represents the overconfidence regime, where the estimated sample confidence is higher than its expected accuracy. We observe that the uncalibrated models are overconfident. We note that these plots do not contain the number of bin samples, and therefore, calibration error and accuracy cannot be directly derived from them.

Expected Calibration Error (ECE). Expected Calibration Error (ECE) [31] is perhaps the most commonly used metric for estimating calibration discrepancy. ECE is a discrete empirical approximation of the expected absolute difference between prediction accuracy and confidence estimation. The ECE is formally given by:

$$\mathrm{acc}\,(B_m) = \frac{1}{|B_m|} \sum_{i \in B_m} \mathbb{1}\,(\hat{y}_i = y_i)\,, \tag{3}$$

$$\mathrm{conf}\,(B_m) = \frac{1}{|B_m|} \sum_{i \in B_m} \hat{q}_i, \tag{4}$$

$$\mathrm{ECE} = \sum_{m=1}^{B} \frac{|B_m|}{N} |\,\mathrm{acc}(B_m) - \mathrm{conf}(B_m)|. \tag{5}$$

Here, B_m denotes the set of samples belonging to the mth bin, $|B_m|$ is the number of instances residing in bin b, N is the total number of samples, B is the total number of bins and $\mathbb{1}$ is the indicator function.

Since prediction accuracy cannot be estimated for individual samples but rather by taking the mean accuracy over a group of samples, the ECE score employs a binning scheme aggregating close by confidence values together. There is a resolution-accuracy trade-off between choosing more or fewer bins, and bin boundaries should be chosen carefully. During our experimentation, we choose an adaptive binning strategy proposed by [34], where the boundaries are set such that they split the samples into B even groups of N/B samples each.

This scheme adapts the bins to the natural distribution of confidence scores, thus, trading-off resolution between densely and sparsely populated confidence regions while keeping the accuracy estimation error even among the bins. We refer our readers to [12] for details and experimentation with different variations of the ECE metric.

Negative Log Likelihood. The Negative Log Likelihood (NLL) objective is commonly used for classifier confidence calibration. NLL is defined as:

$$\mathcal{L} = -\sum_{i=1}^{n} \log\left(\hat{\pi}\,(y_i \mid \mathbf{x}_i)\right), \tag{6}$$

where the estimated probability $\hat{\pi}$ for the ground truth label y_i given the sample x_i is formulated as

$$\hat{\pi}\,(y_i \mid \mathbf{x}_i) = \sigma_{\mathrm{SM}}\,(\mathbf{z}_i)^{(y_i)}.$$

Brier Score. Brier score [5] is a scoring method developed in an effort to predict the reliability of weather forecasts and has been subsequently adapted as a proxy for calibration error. Since the number of possible sequential labels is intractable, we treat the problem as a one vs. all classification, enabling the use of the binary Brier formulation. The Brier score as formalized in Eq. 7 is the mean square error between the confidence scores and the binary indicator function over the predicted and ground truth labels.

$$\mathrm{Brier} = \sum_{i=1}^{N} (\mathbb{1}\,(\hat{y}_i = y_i) - \hat{q}_i)^2. \tag{7}$$

As analyzed by [30], the Brier score comprises three components: uncertainty, reliability, and resolution. While confidence calibration is tasked to minimize the reliability term, the other terms carry information regarding the data uncertainty and deviation of the conditional probabilities from the mean. Therefore, while the Brier score contains a calibration error term, it is entangled with two other terms leading to sub-optimal calibration.

The main advantage of minimizing the Brier score is that it is parameter independent as it does not depend on data binning as ECE does. In Sect. 5, we demonstrate that minimization of Brier score leads to reduced ECE on a held-out test-set.

4   Sequence-Level Calibration

We propose to incorporate existing and new calibration methodologies while optimizing for word-level confidence scores, as depicted in Fig. 1 (d). Our optimization scheme consists of a calibration model and a calibration objective applied to word-level scalar confidences.

Starting with a pre-trained STR model, we freeze the model weights and apply the T-scaling calibration method by multiplying the logits for each decoding step by the temperature parameter T. Following Kuleshov and Liang [19] we define the "Event of Interest" as the exact word match between predicted and ground-truth words. For each word prediction, we define a scalar confidence score as the product of individual decoding-step confidences. We assess our performance according to the ECE metric with equal-sized bins as suggested by Nixon et al. [34].

Calibration of Non-IID Predictions. We motivate our choice of word-level calibration from a probabilistic viewpoint. Taking into account inter-sequence dependencies, we assume that the predictions made at each decoding step are non-IID. This is especially evident for RNN-based decoders *e.g.* Attention-based decoders, where each decoded character is provided as an input for the following prediction. In this case, the following inequality holds:

$$\mathbb{P}(\hat{Y} = Y|x) \neq \prod_i \mathbb{P}(\hat{y_i} = y_i|x). \tag{8}$$

Here, $\mathbb{P}(\hat{Y} = Y|x)$ denotes the correct prediction probability for the predicted sequence $\hat{Y}$ for input x, $\mathbb{P}(\hat{y_i} = y_i|x)$ are the marginal probabilities at each decoding step and $\hat{Y} = (\hat{y_1}, ..., \hat{y_L})$ are the predicted sequence tokens.

The calibration process attempts to affect the predicted scores so that they tend towards the prediction probabilities. Therefore, Eq. 8 implies that calibration of the marginals $\mathbb{P}(\hat{y_i} = y_i|x)$, corresponding to character-level calibration (Fig. 1 (c)), will not lead to a calibrated word-level confidence (Fig. 1 (d)). This insight leads us to advocate for direct optimization of the left hand side of Eq. 8 *i.e.*the word-level scalar confidence scores.

Objective Function. In previous work, several calibration objective functions have been proposed. Three of the commonly used functions are ECE, Brier, and NLL. Typically T-scaling is optimized via the NLL objective. Since the proposal of ECE by Naeini *et al.* [31], it has been widely adopted as a standard calibration measure. Adopting the findings of Nixon *et al.* [34], we utilize ECE as the calibration optimization objective. In our experiments, we also examine the Brier and NLL objectives. We find that while Brier can reduce ECE, it does not converge to the same minima. As for the NLL score, we find that it is unsuitable for sequence-level optimization as directly applying it to multiplied character scores achieves the same minima as character-level optimization. This is undesirable due to the inequality from Eq. 8. (see supplementary for more details).

Edit Distance Expected Calibration Error (ED-ECE). A single classification is either correct or incorrect in its prediction. Sequential predictors, however, present a more nuanced sense of correct prediction *e.g.*correctly predicting 4 out of 5 characters is not as bad as predicting only 3. When calibrating in order to minimize the accuracy-confidence gap, the estimated accuracy is obtained by Eq. 3. The indicator function implies a binary classification task, or in the multi-label setting, a one versus all classification. Sequence prediction, however, is more fine-grained and allows for partial errors. We propose to incorporate the error rate into a new calibration metric. By doing so we allow the end user to assess not only the probability for absolute error but also the amount of incorrect predictions within the sequence. To this aim, we propose to manipulate the ECE metric from by replacing Eq. 3 with the following:

$$\mathrm{acc}_n\left(B_m\right) = \frac{1}{|B_m|} \sum_{i \in B_m} \mathbf{1}\left(\mathrm{ED}(\hat{y}_i, y_i) \leq n\right). \qquad (9)$$

Here, ED refers to the Edit Distance function, also know as the Levenshtein Distance [24]. The Edit Distance produces an integer enumerating the minimal number of insertions, deletions, and substitutions performed on one string to produce the other.

We term our modified ECE metric – Edit Distance Expected Calibration Error (ED-ECE). Essentially, ED-ECE is a relaxation of the ECE metric where the error term is relaxed to better suite the sequential nature of textual prediction. ED-ECE provides a fine-grained per-character-centric metric, whereas ECE is a coarse-grained word-centric metric proposed in the context of single-prediction classification tasks. ED-ECE can be minimized for any desired string distance n and produces a confidence scores signifying the likelihood of an erroneous prediction up to an Edit Distance of n. This information is helpful for downstream applications such as dictionary lookup, beam-search or human correction, as each example may be sent down a different correction pathway according to some decision scheme.

Step Dependent T-Scaling (STS). We extend T-Scaling to better suit the case of sequence prediction. As T-Scaling applies a single, global parameter to all model outputs, it does not leverage existing inter-sequence dependencies. This is especially true for context-dependent models such as Attention and Transformer based decoders. Therefore, we propose extending the scalar T-scaling to *Step Dependent T-scaling (STS)* by setting an individual temperature parameter for each character position within the predicted sequence. We replace the scalar temperature T with a vector $\mathbf{T} \in \mathbb{R}^{\tau+1}$, $\mathbf{T} = \{T_0, ..., T_\tau\}$, where τ is a truncation length. This may be formulated as:

$$\hat{q}_{i,j} = \max_k \sigma_{\mathrm{SM}} \left(\mathbf{z}_{i,j} \mathbf{T}_j \right)^{(k)} . \tag{10}$$

Here, $\mathbf{z}_{i,j}$ is the logits vector for the jth character of the ith sample, and $\mathbf{T}_j$ is the temperature value applied to the jth character for all sequences.

Applying this method directly, however, results in sub-optimal results. This is due to the increase in trainable parameters for the same size calibration set. Furthermore, longer words are scarce and present high variability; thus, it may skew the temperature values of time steps above a certain index. We propose a meta-parameter τ that applies to all time steps over a certain value, such that $\mathbf{T}_{j \geq \tau} = \mathbf{T}_\tau$. We establish a value for τ on a held-out subset of the calibration dataset.

5   Experiments

In the following section, we carry out an extensive evaluation and analysis of our proposed optimization framework. We begin by detailing our experimental setup, including datasets, evaluated models, optimization methodology, and implementation details in Sect. 5.1.

In Sect. 5.2 we provide a deeper analysis, including sequence calibration for various aggregation window sizes (n-grams lengths). We further provide detailed results and analysis for the aspects described thus far: ED-ECE metric, calibration by decoding T-scaling step (STS), and the gains obtained through calibration and beam-search based decoding.

5.1   Experimental Setup

Datasets. Following the evaluation protocol set by [44] who also focus on non-accuracy related aspects of STR models, all STR models are retrained on the SynthText [13] dataset. Models are then evaluated using both regular and irregular text datasets. Regular text datasets containing text with nearly horizontally aligned characters include: IIIT5K [29], SVT [45], ICDAR2003 [28], and ICDAR2013 [18]. Irregular text datasets are comprised of arbitrarily shaped text (*e.g.*curved text), and include include: ICDAR2015 [17], SVTP [39], and CUTE 80 [40]. Our calibration set is comprised of the training portions of the aforementioned datasets where available (ICDAR2013, ICDAR2015, IIIT5K and

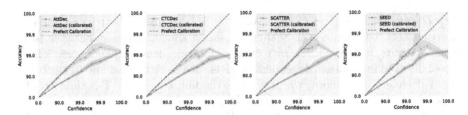

Fig. 3. Reliability Diagrams [7]: (i) AttDec – a variant of [1] with an attention decoder, (ii) CTCDec – a variant of [1] with a CTC decoder, (iii) SCATTER [25] and (iv) SEED [38]. We calibrate using T-scaling coupled with an equal-bin-size ECE objective applied to the word-level scalar confidence scores. The accuracy here is measured w.r.t exact word match. The figure shows accuracy vs. confidence plotted for equally-sized confidence bins, before and after calibration. Over-confidence can be observed for STR models, where the confidence of the model is higher than the expected accuracy.

SVT), while testing was performed on the testing portion of all seven regular and irregular datasets. By following this protocol we keep a strict separation between training, calibration and testing sets.

Text Recognition Models. Our experiments focus on several recent STR models. Baek *et al.* [1] proposed a framework text recognition model comprising four stages: transformation, feature extraction, sequence modeling, and decoding. We consider architectures using four of their proposed variants, including or omitting BiLSTM combined with either a CTC [10] or an attention [23] decoder. In ASTER [43], oriented or curved text is rectified using an STN [15], and a BiLSTM is used for encoding. A GRU [6] with an attention mechanism is used for decoding. SCATTER [25] uses a stacked block architecture that combines both visual and contextual features. They also used a two-step attention decoding mechanism for providing predictions. Bartz *et al.* [3] proposed KISS, combining region of interest prediction and transformer-based recognition. SEED [38] is an attention-based architecture supplemented with a pre-trained language model.

In our work, we evaluate and analyze each of these models' calibration-related behavior, highlighting differences between various decoder types.

Optimization. The task of calibrating confidence scores boils down to minimizing the model parameters w.r.t a given loss function. T-scaling based calibration methods take the predicted logits z_i as input and apply a modified SoftMax operation to arrive at the calibrated confidence score.

Backpropagation is only conducted through the optimized calibration parameter, while the STR model remains unchanged. The model formulations for T-scaling and STS are provided in Eqs. 1 and 10 respectively.

Table 1. Calibrated ECE Scores for Different Objective Functions. Unsurprisingly, ECE values are best optimized w.r.t. the ECE objective. Brier loss is also suitable for reducing calibration error but is less effective. Finally, we observe that NLL is unsuitable for sequence level calibration as detailed in Sect. 4 and in the supplementary.

Method	Uncalib.	ECE	Brier	NLL
CTC [1]	6.9	2.2	5.9	6.8
BiLSTM CTC [1]	6.7	2.0	6.0	6.6
Atten. [1]	5.9	1.8	5.2	5.9
BiLSTM Atten. [1]	5.4	2.0	4.8	5.3
ASTER [43]	1.8	0.8	1.8	1.7
SCATTER [25]	5.8	1.8	4.5	5.7
KISS [3]	9.6	1.4	5.3	9.4
SEED [38]	5.7	2.0	5.7	5.6
Average	5.98	**1.75**	4.9	5.88

We use the L-BFGS optimizer [26] coupled with several calibration objective functions to demonstrate our calibration methodology. Our tested loss functions include ECE, ED-ECE, Brier, and NLL. Table 1 presents ECE achieved by calibrating for NLL, Brier, and the ECE objective functions. We observe that ECE obtains the best calibration error as expected while Brier succeeds to a lower degree. We also demonstrate that, as mentioned, NLL is not suitable for word-level calibration.

5.2   Results and Analysis

Aggregation Window. In order to gain a deeper understanding of the relation between aggregation and calibration performance, we experimented with calibration via partial sequences. To this end, we break up our calibration datasets into all possible sub-sequences of length $\leq n$. We note that when $n = 1$ the calibration is carried out at character-level as depicted by Fig. 1 (c), and for $n = \max \text{length}(w_i)$ we are calibrating on the full sequence (Fig. 1 (d))

Calibration is performed by the T-scaling method coupled with the ECE objective function. All reported results are measured on a held-out test-set. In an attempt to reduce noise, we test the calibration process on 10 training checkpoints of each model and plot the mean and variance measurements in Fig. 2.

We find that for attention decoders (Fig. 2 (Left) $n = 1$ provides worse calibration than the uncalibrated baseline. CTC decoders (Fig. 2 (Right), on the other hand, also exhibit worse ECE scores on per-character calibration; however, the error is still reduced relative to the uncalibrated models. We postulate that this phenomenon relates to the difference between IID and non-IID decoding discussed in Sect. 4. This key observation emphasizes the importance of score aggregation during the calibration process as opposed to individual character calibration.

Beam-Search. Although calibration methods based on T-scaling do not alter prediction accuracy, it is still possible to indirectly affect a model's accuracy rate. This can be achieved through a beam-search methodology, where the space of possible predicted sequences is explored within a tree of possible outcomes. At each leaf, the total score is calculated as the product of all nodes leading up to the leaf.

We note that each predicted character's confidence score does not change ordering with relation to other scores due to the monotonic nature of T-scaling. In contrast, aggregated word-level confidence scores do change ordering in some cases. Score reordering can take place if individual character scores have a non-monotonic dependence on other parameters than the calibrated temperature. This is the case for state-dependant decoders where the scores depend on the internal decoder state, therefore allowing for a reordering of the word-level confidence scores.

We present our calibrated beam-search results in Table 2, showing a consistent gain for each calibrated model relative to the non-calibrated baseline. We also show that this holds for all tested beam widths between two and five. In the supplementary material, we further break down the results according to the individual test datasets. We note that while the overall absolute improvement is below 1%, this is significant when considering the baseline of improvement offered by the uncalibrated beam-search method, in some cases more than doubling beam-search effectiveness. While beam-search on its own is quite intensive, requiring several inference steps for each word, our calibration is performed offline and adds virtually no further computational burden while increasing effectiveness by as much as two fold.

Table 2. Beam-search accuracy gains achieved for calibrated (✓) vs. uncalibrated (X) models. We apply beam widths (bw) between 1 and 5. Displayed results are averaged across test datasets and are reported relative to the baseline of bw = 1, which is equivalent to not using beam-search. CTC based methods are omitted as beam-search requires decoding dependence in order to be effective. We demonstrate consistent improvement across all models and all datasets (see supplementary for breakdown) over the uncalibrated baseline.

Method\Calibration	bw=1 X	bw=2 X	✓	bw=3 X	✓	bw=4 X	✓	bw=5 X	✓
Atten. [1]	86.21	0.18	0.39	0.23	0.43	0.26	**0.47**	0.25	0.45
BiLSTM Atten. [1]	86.2	0.11	0.2	0.18	0.33	0.2	**0.35**	0.21	**0.35**
ASTER [43]	86.03	0.52	0.57	0.63	0.77	0.64	**0.82**	0.64	0.81
SCATTER [25]	87.36	0.16	0.27	0.2	**0.3**	0.19	0.28	0.2	0.29
SEED [38]	81.18	0.2	0.29	0.23	0.35	0.32	**0.42**	0.28	0.4
Average	84.53	0.23	0.34	0.29	0.44	0.32	**0.47**	0.32	0.46

Table 3. ED-ECE Values for uncalibrated (X) and calibrated (✓) models. Calibration was performed by the T-scaling method and ED-ECE objective for $E_d = 0$ (equivalent to ECE) and $E_d \leq 1, 2$. We observe that the optimization process reduces ED-ECE values on the held-out test-set. The calibrated ED-ECE scores may be used to estimate the number of incorrect predictions within the sequence for down-stream applications.

Method\Calibration	$E_d = 0$		$E_d \leq 1$		$E_d \leq 2$	
	X	✓	X	✓	X	✓
CTC [1]	6.9	2.2	4.1	1.8	8.2	1.7
BiLSTM CTC [1]	6.7	2.0	4.2	1.5	7.9	1.9
Atten. [1]	5.9	1.8	2.0	0.9	4.5	0.7
BiLSTM Atten. [1]	5.4	2.0	2.1	1.0	5.0	0.7
ASTER [43]	1.8	0.8	5.3	2.1	9.7	1.6
SCATTER [25]	5.8	1.8	1.6	1.2	3.9	0.9
KISS [3]	9.6	1.4	1.0	0.8	5.2	1.2
SEED [38]	5.7	2.0	2.1	1.2	4.9	0.7
Average	5.98	1.75	2.8	1.31	6.16	1.18

Edit Distance Expected Calibration Error. In Sect. 4 we present a new calibration metric termed Edit Distance Expected Calibration Error (ED-ECE). We calibrate for ED-ECE with $n = [1, 2]$ and present the results in Table 3. As expected, the ED-ECE is reduced significantly due to the optimization. It is worth noting that ED-ECE is often lower than the original ECE score, leading to a more accurate confidence estimation. Once calibrated, three scores corresponding to ECE (ED-ECE for $n = 0$) and ED-ECE for $n = [1, 2]$ are produced for each data sample. This allows us to submit the data to further review according to thresholds on the output scores.

For example given a predicted label "COFEEE" for a ground-truth label of "COFFEE", the absolute and $E_d = 1$ predicted confidences are 75% and 98% respectively. We might recognize such an example and perform a focused search by testing the confidence scores of words that differ by an Edit-Distance of 1 and selecting the most confident prediction within the search space.

Step Dependent T-Scaling (STS). In Sect. 4 we propose *Step Dependent T-Scaling (STS)*. STS extends the previously presented T-scaling by assigning a temperature for each character position in the sequence. Table 4 lists the calibrated ECE values achieved by T-scaling as well as STS calibration schemes coupled with an ECE calibration objective function. We find that a value of $\tau = 5$ is optimal for the held-out validation set and therefore select $\tau = 5$ temperature values while a 6th value is used to calibrate the subsequent sequence positions. Our experimentation demonstrates that the Time-Stamp scaling is beneficial or on par with T-scaling for all but one of the models. Overall, when averaging on

Table 4. ECE Values Comparing T-scaling and STS for uncalibrated (Uncalib.) and calibrated confidence scores obtained on a held-out test-set. We optimize according to our proposed method utilizing the T-scaling (TS) and the proposed STS calibration methods. We demonstrate that STS is slightly advantageous over global TS.

Method	Uncalib.	TS	STS
CTC [1]	6.9	**2.2**	**2.2**
BiLSTM CTC [1]	6.7	2.0	**1.8**
Atten. [1]	5.9	1.8	**1.7**
BiLSTM Atten. [1]	5.4	2.0	**1.7**
ASTER [43]	1.8	**0.8**	1.0
SCATTER [25]	5.8	1.8	**1.6**
KISS [3]	9.6	**1.4**	**1.4**
SEED [38]	5.7	**2.0**	**2.0**
Average	5.98	1.75	**1.67**

all tested models, STS shows a slight benefit over T-scaling. Although relatively small, this benefit is achieved for very low additional complexity to the offline calibration process. We hypothesize that STS is able to improve calibration error due to its finer-grained calibration and exploitation of inter-sequence relations.

6   Conclusion

In this work, we analyze the calibration characteristics of off the shelf STR models finding that they are commonly over confidant in their estimation. We further demonstrate that word-level and, in general, sequence-level calibration should be optimized directly on the per-word scalar confidence outputs. This is motivated by probabilistic reasoning and demonstrated empirically for various STR methods.

To the best of our knowledge, we are the first to conduct an in-depth analysis of the current state of calibration in scene-text recognition models. We perform extensive experimentation with STR model calibration and propose ED-ECE, a text-oriented metric and loss function, extending ECE to calibrate for sequence-specific accuracy measures (e.g. Edit-Distance).

Furthermore, we demonstrate that the calibration of STR models boosts beam-search performance, consistently improving model accuracy for all beam-widths and datasets with relative accuracy improvement of up to double relative to the non-calibrated baseline. Finally we propose to extend the T-scaling calibration method to a sequence-level variant we termed Step dependent T-Scaling (STS), showing moderate gains for very little effort.

References

1. Baek, J., et al.: What is wrong with scene text recognition model comparisons? dataset and model analysis. In: Proceedings of the IEEE International Conference on Computer Vision, pp. 4715–4723 (2019)
2. Bahdanau, D., Cho, K., Bengio, Y.: Neural machine translation by jointly learning to align and translate. In: 3rd International Conference on Learning Representations, ICLR 2015 (2015)
3. Bartz, C., Bethge, J., Yang, H., Meinel, C.: Kiss: Keeping it simple for scene text recognition (2019)
4. Borisyuk, F., Gordo, A., Sivakumar, V.: Rosetta: large scale system for text detection and recognition in images. In: Proceedings of the 24th ACM SIGKDD International Conference on Knowledge Discovery & Data Mining, pp. 71–79 (2018)
5. Brier, G.W.: Verification of forecasts expressed in terms of probability. Mon. Weather Rev. **78**(1), 1–3 (1950)
6. Cho, K., et al.: Learning phrase representations using RNN encoder-decoder for statistical machine translation. In: Proceedings of the 2014 Conference on Empirical Methods in Natural Language Processing (EMNLP), pp. 1724–1734. Association for Computational Linguistics, Doha, Qatar (2014). https://doi.org/10.3115/v1/D14-1179. https://www.aclweb.org/anthology/D14-1179
7. DeGroot, M.H., Fienberg, S.E.: The comparison and evaluation of forecasters. J. Roy. Statist. Soc.: Ser. D (The Statistician) **32**(1–2), 12–22 (1983)
8. Desai, S., Durrett, G.: Calibration of pre-trained transformers. In: Proceedings of the Conference on Empirical Methods in Natural Language Processing (EMNLP) (2020)
9. Devlin, J., Chang, M.W., Lee, K., Toutanova, K.: BERT: pre-training of deep bidirectional transformers for language understanding. In: Proceedings of the 2019 Conference of the North American Chapter of the Association for Computational Linguistics: Human Language Technologies, Volume 1 (Long and Short Papers), pp. 4171–4186. Association for Computational Linguistics, Minneapolis, Minnesota (2019). https://doi.org/10.18653/v1/N19-1423. https://www.aclweb.org/anthology/N19-1423
10. Graves, A., Fernández, S., Gomez, F., Schmidhuber, J.: Connectionist temporal classification: labelling unsegmented sequence data with recurrent neural networks. In: Proceedings of the 23rd international conference on Machine learning, pp. 369–376. ACM (2006)
11. Graves, A., Mohamed, A.R., Hinton, G.: Speech recognition with deep recurrent neural networks. In: 2013 IEEE International Conference on Acoustics, Speech and Signal Processing, pp. 6645–6649. IEEE (2013)
12. Guo, C., Pleiss, G., Sun, Y., Weinberger, K.Q.: On calibration of modern neural networks. In: Proceedings of the 34th International Conference on Machine Learning - Volume 70, pp. 1321–1330. ICML2017, JMLR.org (2017)
13. Gupta, A., Vedaldi, A., Zisserman, A.: Synthetic data for text localisation in natural images. In: Proceedings of the IEEE Conference on Computer Vision and Pattern Recognition, pp. 2315–2324 (2016)
14. Hendrycks, D., Gimpel, K.: A baseline for detecting misclassified and out-of-distribution examples in neural networks (2017)
15. Jaderberg, M., Simonyan, K., Zisserman, A., et al.: Spatial transformer networks. In: Advances in neural information processing systems, pp. 2017–2025 (2015)

16. Ji, B., Jung, H., Yoon, J., Kim, K., et al.: Bin-wise temperature scaling (bts): Improvement in confidence calibration performance through simple scaling techniques. In: 2019 IEEE/CVF International Conference on Computer Vision Workshop (ICCVW), pp. 4190–4196. IEEE (2019)
17. Karatzas, D., et al.: ICDAR 2015 competition on robust reading. In: 2015 13th International Conference on Document Analysis and Recognition (ICDAR), pp. 1156–1160. IEEE (2015)
18. Karatzas, D., et al.: ICDAR 2013 robust reading competition. In: 2013 12th International Conference on Document Analysis and Recognition, pp. 1484–1493. IEEE (2013)
19. Kuleshov, V., Liang, P.S.: Calibrated structured prediction. In: Advances in Neural Information Processing Systems, pp. 3474–3482 (2015)
20. Kumar, A., Sarawagi, S.: Calibration of encoder decoder models for neural machine translation. arXiv preprint arXiv:1903.00802 (2019)
21. Lakshminarayanan, B., Pritzel, A., Blundell, C.: Simple and scalable predictive uncertainty estimation using deep ensembles. In: Advances in Neural Information Processing Systems, pp. 6402–6413 (2017)
22. Leathart, T., Polaczuk, M.: Temporal probability calibration. arXiv preprint arXiv:2002.02644 (2020)
23. Lee, C.Y., Osindero, S.: Recursive recurrent nets with attention modeling for OCR in the wild. In: Proceedings of the IEEE Conference on Computer Vision and Pattern Recognition, pp. 2231–2239 (2016)
24. Levenshtein, V.I.: Binary codes capable of correcting deletions, insertions, and reversals. In: Soviet physics doklady, vol. 10, pp. 707–710 (1966)
25. Litman, R., Anschel, O., Tsiper, S., Litman, R., Mazor, S., Manmatha, R.: Scatter: selective context attentional scene text recognizer. In: Proceedings of the IEEE/CVF Conference on Computer Vision and Pattern Recognition (CVPR) (2020)
26. Liu, D.C., Nocedal, J.: On the limited memory BFGS method for large scale optimization. Math. Program. **45**(1–3), 503–528 (1989)
27. Liu, W., Chen, C., Wong, K., Su, Z., Han, J.: Star-Net: a spatial attention residue network for scene text recognition. In: BMVC (2016)
28. Lucas, S.M., Panaretos, A., Sosa, L., Tang, A., Wong, S., Young, R.: ICDAR 2003 robust reading competitions. In: Seventh International Conference on Document Analysis and Recognition, pp. 682–687 (2003). Proceedings, CiteSeer (2003)
29. Mishra, A., Alahari, K., Jawahar, C.: Scene text recognition using higher order language priors (2012)
30. Murphy, A.H.: A new vector partition of the probability score. J. Appl. Meteorol. **12**(4), 595–600 (1973)
31. Naeini, M.P., Cooper, G.F., Hauskrecht, M.: Obtaining well calibrated probabilities using bayesian binning. In: Proceedings of the AAAI Conference on Artificial Intelligence. AAAI Conference on Artificial Intelligence, vol. 2015, p. 2901. NIH Public Access (2015)
32. Nguyen, K., O'Connor, B.: Posterior calibration and exploratory analysis for natural language processing models. In: EMNLP (2015)
33. Niculescu-Mizil, A., Caruana, R.: Predicting good probabilities with supervised learning. In: Proceedings of the 22nd International Conference on Machine Learning, pp. 625–632 (2005)
34. Nixon, J., Dusenberry, M.W., Zhang, L., Jerfel, G., Tran, D.: Measuring calibration in deep learning. In: Proceedings of the IEEE/CVF Conference on Computer Vision and Pattern Recognition (CVPR) Workshops (2019)

35. Ovadia, Y., et al.: Can you trust your model's uncertainty? evaluating predictive uncertainty under dataset shift. In: Advances in Neural Information Processing Systems, pp. 13991–14002 (2019)
36. Pereyra, G., Tucker, G., Chorowski, J., Kaiser, Ł., Hinton, G.: Regularizing neural networks by penalizing confident output distributions. arXiv preprint arXiv:1701.06548 (2017)
37. Platt, J., et al.: Probabilistic outputs for support vector machines and comparisons to regularized likelihood methods. Adv. Large Margin Class. 10(3), 61–74 (1999)
38. Qiao, Z., Zhou, Y., Yang, D., Zhou, Y., Wang, W.: Seed: semantics enhanced encoder-decoder framework for scene text recognition. In: Proceedings of the IEEE/CVF Conference on Computer Vision and Pattern Recognition, pp. 13528–13537 (2020)
39. Quy Phan, T., Shivakumara, P., Tian, S., Lim Tan, C.: Recognizing text with perspective distortion in natural scenes. In: Proceedings of the IEEE International Conference on Computer Vision, pp. 569–576 (2013)
40. Risnumawan, A., Shivakumara, P., Chan, C.S., Tan, C.L.: A robust arbitrary text detection system for natural scene images. Expert Syst. Appl. 41(18), 8027–8048 (2014)
41. Shi, B., Bai, X., Yao, C.: An end-to-end trainable neural network for image-based sequence recognition and its application to scene text recognition. IEEE Trans. Pattern Anal. Mach. Intell. 39(11), 2298–2304 (2016)
42. Shi, B., Wang, X., Lyu, P., Yao, C., Bai, X.: Robust scene text recognition with automatic rectification. In: Proceedings of the IEEE Conference on Computer Vision and Pattern Recognition, pp. 4168–4176 (2016)
43. Shi, B., Yang, M., Wang, X., Lyu, P., Yao, C., Bai, X.: Aster: an attentional scene text recognizer with flexible rectification. IEEE Transactions on Pattern Analysis and Machine Intelligence (2018)
44. Wan, Z., Zhang, J., Zhang, L., Luo, J., Yao, C.: On vocabulary reliance in scene text recognition. In: Proceedings of the IEEE/CVF Conference on Computer Vision and Pattern Recognition, pp. 11425–11434 (2020)
45. Wang, K., Babenko, B., Belongie, S.: End-to-end scene text recognition. In: 2011 International Conference on Computer Vision, pp. 1457–1464. IEEE (2011)

End-to-End Document Recognition and Understanding with Dessurt

Brian Davis[1,2(✉)], Bryan Morse[2], Brian Price[3], Chris Tensmeyer[3],
Curtis Wigington[3], and Vlad Morariu[3]

[1] AWS AI, Washington, USA
briandvs@amazon.com
[2] Brigham Young University, Provo, USA
morse@byu.edu
[3] Adobe Research, College Park, USA
{bprice,tensmeye,wigingto,morariu}@adobe.com

Abstract. We introduce Dessurt, a relatively simple document understanding transformer capable of being fine-tuned on a greater variety of document tasks than prior methods. It receives a document image and task string as input and generates arbitrary text autoregressively as output. Because Dessurt is an end-to-end architecture that performs text recognition in addition to document understanding, it does not require an external recognition model as prior methods do. Dessurt is a more flexible model than prior methods and is able to handle a variety of document domains and tasks. We show that this model is effective at 9 different dataset-task combinations.

Keywords: Document understanding · End-to-end · Handwriting recognition · Form understanding · OCR

1 Introduction

Document understanding is an area of research attempting to automatically extract information from documents, whether that be specific key information, answers to natural language questions, or other similar elements. While there have been many approaches, the research community has begun to gravitate around pre-trained transformers as general purpose solutions. Beginning with LayoutLM [36], these models began as BERT-like transformers incorporating spatial/layout information and later visual features. In general, we refer to these as the LayoutLM family. The LayoutLM family of models are pre-trained on a large corpus of document images and then fine-tuned to their particular tasks.

B. Davis—Work completed prior to Brian joining AWS.

Supplementary Information The online version contains supplementary material available at https://doi.org/10.1007/978-3-031-25069-9_19.

© The Author(s), under exclusive license to Springer Nature Switzerland AG 2023
L. Karlinsky et al. (Eds.): ECCV 2022 Workshops, LNCS 13804, pp. 280–296, 2023.
https://doi.org/10.1007/978-3-031-25069-9_19

The LayoutLM family consists of encoder-only transformers, meaning predictions are only made for the input tokens. These state-of-the-art models are two-stage models, where text recognition is first performed by an external OCR model to obtain the input text tokens for the transformer. We see two limitations coming from these architecture choices:

1. A limited output space, having predictions only for individual input tokens. While they can classify the input tokens, they cannot produce additional outputs, e.g., arbitrary text or token relationships, without additional submodules.

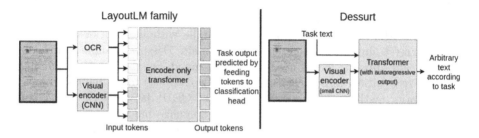

Fig. 1. The LayoutLM family of document transformers require OCR and output is tied to the tokens. Dessurt does not require any separate models and can generate arbitrary text to solve a variety of tasks.

Table 1. Model class capabilities

	Handwriting	Arbitrary output	Apply to different visual domain
LayoutLM family	OCR dependant	✗	Fine-tune two models
Dessurt	✓	✓	Fine-tune single model

2. Dependence on high quality external OCR text segmentation and recognition. Encoder-only transformers are incapable of inserting new tokens if the OCR missed or under-segmented text. A single incorrectly recognized character in an OCR'd word can cause a wrong word embedding to be used or cause the word to be out of vocabulary. Relatedly, discrete input tokens lack the uncertainty the text recognition model may have in its predictions. For clean, modern documents, this generally isn't an issue as the OCR models used are quite robust. However, for handwritten or degraded historical documents, OCR quality can be poor and lead to prediction errors.

To combat these flaws we introduce **Dessurt: D**ocument **e**nd-to-end **s**elf-**s**upervised **u**nderstanding and **r**ecognition **t**ransformer. Dessurt is a novel, general document understanding architecture that can perform a great variety of document tasks. Dessurt operates in an end-to-end manner with a single pass: text segmentation and recognition are learned implicitly. Dessurt takes only the image and task text as input and can auto-regressively produce arbitrary text as output. Figure 1 compares Dessurt to the LayoutLM family at a high level architecturally. The first limitation of the LayoutLM family is easily solved with Dessurt's auto-regressive output. Because text recognition is implicit, rather than provided as explicit OCR results, Dessurt is able to resolve text recognition uncertainty or ambiguity in a task-focused way. Additionally, the auto-regresssive output decouples Dessurt's output from the text recognition. These together address the second limitation. See Table 1 for a comparison of architecture features.

Because Dessurt takes both an input image and text and can output any arbitrary text, it can complete a greater variety of tasks than the LayoutLM family of transformers. Particularly we solve a form parsing task (form image to JSON) that the LayoutLM family cannot handle without additional modules. Also, when retraining for a different visual domain, Dessurt's simple end-to-end design means only one model needs to be fine-tuned. For the LayoutLM family, both the recognition and transformer models would need to be fine-tuned.

Like prior methods, we pre-train on the IIT-CDIP dataset [17], a large collection of document images, with a masked language modeling task. We also introduce three synthetic document datasets to better capture natural language, structured documents, and handwriting recognition. Finally, we introduce new pre-training tasks to teach Dessurt to read and located text, and to parse structured documents.

We validate our claims of Dessurt's flexibility by applying it to six different document datasets across six different tasks: 1) Document question answering, with both DocVQA [23] and HW-SQuAD [22], 2) Form understanding and 3) Form parsing, with both the FUNSD [14] and NAF [5] datasets, 4) Full-page handwriting recognition and 5) Named entity recognition on the IAM handwriting database, and 6) Document classification with the RVL-CDIP dataset. Of particular interest, both NAF and IAM datasets require handwriting recognition, the NAF being comprised of difficult historical documents. These are domains in which the LayoutLM family would need to update its recognition model as well, but Dessurt can fine-tune on without adjustments. We note that Dessurt does not achieve state-of-the-art results for most tasks evaluated, but it is capable of operating on a larger range of tasks than individual state-of-the-art models.

In summary, our primary contributions are

- Dessurt, a novel, general document understanding architecture capable of both performing text recognition and document understanding in an end-to-end manner and producing arbitrary text output,
- A collection of synthetic datasets and tasks for pre-training an end-to-end document understanding model for a variety of possible final tasks,
- An evaluation of Dessurt fine-tuned on 9 dataset-task combinations, and
- Our code, model weights, and datasets which can be found at https://github.com/herobd/dessurt

2   Related Work

2.1   LayoutLM Family

Document understanding has become largely dominated by transformer architectures. Beginning with LayoutLM(v1) [36] the goal was to bring the success of transformers like BERT [8] from the natural language space into the more visual domain of documents. LayoutLM pre-trained in a very similar manner to BERT, but included 2D spatial position information.

BROS [12], TILT [24], and LayoutLMv2 [35] improved the architecture by introducing spatially biased attention, making the spatial information even more influencial. LayoutLMv2 also introduced visual tokens as many layout cues are captured more visually than spatially.

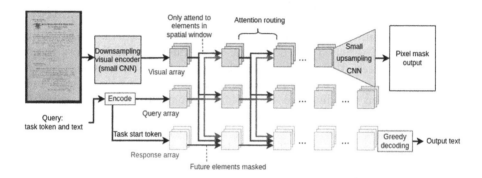

Fig. 2. Dessurt architecture

Visual tokens can be overshadowed by textual tokens. In an effort to make the visual processing more important, DocFormer [1] forced feature updates to be from both textual and visual features.

We note that TILT and DocFormer use only visual features extracted near the text tokens spatially, making them blind to areas of the form without text. LayoutLMv2 extracts visual tokens across the entire document.

2.2   End-to-End Models

Models in the LayoutLM family have been evaluated without taking text recognition into account. Many document understanding datasets come with pre-computed OCR results used by everyone. While this is useful in making comparisons, text recognition is an essential task and for visually difficult documents can become a challenge in itself.

One aim of an end-to-end method can be to accomplish both recognition and understanding in one pass. Another aim might be to learn the output text in a manner that allows arbitrary output predictions. DocReader [16] is an end-to-end method for key information extraction. While it does rely on external OCR, it uses an RNN to predict arbitrary text.

We note that a concurrent pre-print work on end-to-end document understanding, Donut [15], has been introduced, and shares an architecture similar in design to Dessurt. It also utilizes a Swin [19] encoder but uses a BART-like [18] decoder. Donut differs from Dessurt primarily in how the cross attention occurs and in pre-training. Donut shares many of the same advantages of Dessurt.

3 Model

Dessurt is a novel end-to-end framework for handling general document understanding problems. It takes as input an image and query text, and outputs arbitrary text appropriate for the given tasks. It handles character recognition, including handwriting recognition, implicitly as part of the network.

The architecture is shown in Fig. 2. The model processes three streams of tokens: 1) Visual tokens, that encode the visual information, 2) Query tokens that encode the task the model is to perform, and 3) Autoregressive response tokens where the output response is formed. The model progresses through three main stages: Input encoding, cross-attention, and output decoding.

Input Encoding: The input consists of an image and a query token string. Because the Swin [19] layers we use require a fixed size image input, we use an input image size of 1152 × 768. This large size is needed as we process an entire page at once and must ensure small text is legible. The input image is 2-channeled, one being the grayscale document, the other being a highlight mask used in some tasks. The query tokens begin with a special task token indicating the desired task and then potentially have some text providing context for the task (e.g., the question text). The response tokens are initialized with a task specific start token and during training contain the previous ground truth token for teacher-forcing.

The first step of the model is to encode the inputs into feature arrays to initialize the three streams. The input image is tokenized by passing it through a small downsampling CNN and adding learned 2D spatial embeddings. The input query text and response text are tokenized using the same process as BART [18] with standard sinusoidal position encoding. These feature arrays are then passed to their respective token streams.

Note that the model does not require as input any OCR tokens corresponding to the image. The network implicitly recognizes the text.

Cross-attention: The three streams then pass through a series of cross-attention layers to allow them to share information and transfer that information into the response. The visual array is processed by Swin [19] layers modified to not only attend to the other elements in the local window but also the query array. (We note that the biased attention remains for the visual elements.) The query array has standard Transformer [32] attention, but attends to the entire visual array in addition to the query array. The response array has standard autoregressive attention to previous response elements but also attends to the visual and query arrays. The arrays pass through series of eight of their respective cross-attention layers. The last two layers of the model update only the query and response arrays, with both layers attending to the final visual features.

Output Decoding: The final response array is decoded into text using greedy search decoding (where the most likely token is selected at each step), allowing it to predict text not found in the document. Additionally, we also output a pixel mask for use in training. This is produced by a small upsampling network using six transpose convolutions that process the final visual features.

Specific implementation details for the model and its layers can be found in the accompanying Supplementary Materials.

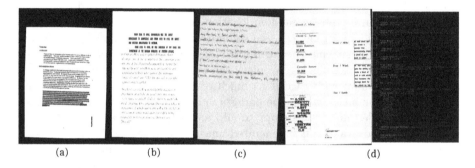

Fig. 3. Examples of data used in pre-training. (a) IIT-CDIP dataset image with Text Infilling task highlighting channel: highlight is magenta (value of 1), removed text is turquoise (value of −1). (b) Synthetic Wikipedia text. (c) Synthetic handwriting. (d) Synthetic form and its parse JSON.

4   Pre-training Procedure

The goal of the pre-training is to teach Dessurt to perform text recognition and document understanding and to have general language model capabilities like BERT. We pretrain several datasets with each dataset having multiple tasks associated with it. An example from each dataset is in Fig. 3.

4.1   IIT-CDIP Dataset

The IIT-CDIP dataset [17] is a pre-training dataset used by several other document understanding transformers [1,35,36]. The OCR method we applied to the IIT-CDIP dataset is in the Supplementary materials.

There are several tasks defined with this dataset all of which are described in the Supplementary Materials. For brevity, we only describe the most important ones here. The primary task (occurring 66% of the time) is a Text Infilling task. It is a masked language modeling task inspired by the text infilling used to train BART [18]; instead of replacing the removed text with a blank token, we delete them from the image. The entire block of text and the deleted areas are marked (with different values) in the input highlight channel, as seen in Fig. 3(a). The model then must predict the text of the entire block, filling in the deleted regions. We also do a variant of this task where a single word is blanked from the image and the model must predict that single word. There are several reading based tasks as well, such as to read on from the text provided in the query.

4.2   Synthetic Wikipedia

We want our pre-training to help the model understand natural language; however, the IIT-CDIP dataset only represents a skewed slice of natural language. Additionally, it represents a limited range of font styles. We choose to create an on-the-fly dataset by selecting random text blocks from Wikipedia[1] [9] and rendering it as paragraphs in random locations with random fonts.

We pick a random article, random column width, random font, random text height, and random spacing (between word and new line). We render the words using the font and text height. We place the words in column/paragraph form, adjusting the column width to fit as much of the article as possible. We find blank space in the image where the paragraph can be added. If one is found the paragraph is added and we attempt to add another paragraph; otherwise, the image is complete. An example generated image is seen in Fig. 3(b).

To obtain our font database, we scrape all the free-for-commercial-use fonts from 1001fonts.com, giving us a set of over 10,000 fonts. The script we used to scrape the fonts will be made available. More details on these fonts and our synthetic dataset creation are found in the Supplementary Materials

This dataset uses the same distribution of tasks as the IIT-CDIP dataset.

4.3   Synthetic Handwriting

Dessurt must be able handle handwriting as several document understanding tasks require this. The IIT-CDIP dataset contains little handwriting and while our font database has "handwritten" fonts, they do not capture the variation present in real handwriting. There is, unfortunately, not a publicly available dataset of handwriting comparable in size to the IIT-CDIP dataset. The IAM handwriting database [21] is frequently used, but with fewer than 800 instances to train on, an autoregressive transformer could overfit during pre-training.

We choose instead to use synthetic handwriting. This allows us to generate a larger breadth of text, but at the cost of realism. We use the full line handwriting synthesis method of Davis et al. [7] to generate 800,000 lines of sequential text from Wikipedia articles, with a randomly sampled style for each line. We compose a document by sampling a random number of consecutive handwriting lines (to maintain language flow), selecting a random text height, random newline height, and random starting location on the page, and then placing the lines in the document in block/paragraph style. We additionally apply warp grid

---

[1] https://huggingface.co/datasets/wikipedia.

augmentation [33] to each line to further add to the visual variation. An example image can be seen in Fig. 3(c).

For the learning task, the model must read the entire page of handwriting.

4.4 Synthetic Forms

We want Dessurt to be capable of extracting the information from highly structured documents, but given the lack of structured information present in our IIT-CDIP annotations, we decided to generate synthetic forms with known structure. The structure is based on the annotations of the FUNSD [14] dataset, which is primarily label-value pairs (or question-answer pairs) which are occasionally grouped under headers. We also include tables.

To come up with label-value pairs, we use GPT-2 [25] to generate synthetic "forms". We give GPT-2 a prompt text (e.g., "This form has been filled out.") followed by an example label-value pair, newline and a label with colon (e.g., "Date: 23 Mar 1999\nName:"). GPT-2 then usually generate a series of label-value pairs separated by colons and newlines, which is easily parsed. All the label-value pairs from one generation are a label-value set in our dataset. We sometimes use Wikipedia article titles as part of the prompt (e.g. "This form should be filled out regarding <u>Marvel Universe</u>") which then become the header for that label-value set. We reuse previously generated labels and label-value pairs as new form prompts. The quality of GPT-2 output is limited, but we hope it reflects at least some of the semantics of label-value relationships.

The data for tables is more random. The row and column headers are random 1 to 3 word snippets from Wikipedia. A cell value is either a random number (with various formatting) or a random word.

A document is composed by randomly placing label-value sets and tables until a placement fails due to there not being enough room. Some cells and values are blanked. More details on the form generation process can be found in the Supplementary Materials.

The primary task on the forms (occurring about half the time) is to parse it into a JSON capturing the text and structure of the form. An example synthetic form and its corresponding JSON are seen in Fig. 3(d). We also have tasks where the query has an entity on the form and the model must predict the class of the entity and then read the entities it is linked to. To ensure an understanding of tables, there are also table-specific tasks such as retrieving a cell based on a query with the row and column header, or listing all the row/column headers for a table. All the tasks used are described in the Supplementary Materials.

4.5 Distillation

Because Dessurt has a unique architecture, we could not use pre-trained transformer weights to initialize our model (like Donut [15] or models in the LayoutLM family). This is clearly a disadvantage, so we attempt to infuse pre-trained language knowledge into Dessurt in a different way: cross-domain distillation. We feed text to a pre-trained transformer teacher, and then render that text in a document image to pass to the student, Dessurt. Then we apply the standard distillation loss introduced by Hinton et al. [11], which guides the logit predictions of the student to match the teachers logits (the "dark knowledge").

Distillation is generally applied with a student and teacher getting the exact same inputs. We are attempting something fairly unique which is to apply distillation across domains, textual to visual.

To ensure architectural similarity, we need the teacher to be an autoregressive model. For this we use BART, an encoder-decoder transformer where the decoder is an autoregressive model with cross attention to the encoder (a vanilla transformer encoder). Both BART and Dessurt will be given the Text Infilling task which BART was pre-trained with. BART gets the masked text as input to its encoder, and Dessurt gets the rendered text with deleted regions as input (and the query token indicating the Test Infilling task) and then they both autoregressively output the input text with the blanks filled in.

The token probabilities Dessurt predicts for a blanked region reflect not only its language modeling, but also the uncertainty it has in reading the other words on the page. For BART the probabilities are only the language modeling; it has no uncertainty about reading. We minimize the reading uncertainty Dessurt has when performing distillation by selecting a subset of "easy" fonts and reducing the variability with which the documents are rendered. More details on this are in the Supplementary Materials.

4.6 Training

We employ a simple curriculum to prioritize certain aspects during early training. This is due to the need for recognition to be learned (to a certain degree) before the understanding tasks can be solved and the difficulty of learning recognition on dense multi-line documents in a semi-supervised fashion.

We first train Dessurt on small images (96×384) of synthetic Wikipedia text with simple reading tasks for 150,000 iterations. Not only is the visual space small, but the output sequence length is short. We then use full-sized synthetic Wikipedia text documents for 200,000 iterations with primarily reading tasks. Finally, the model enters normal pre-training.

The iterations we outline here are what were used for the ablation models. For our primary evaluation we use a model that was pre-trained in total for over 6 million iterations during development (meaning datasets and tasks were added throughout the training), but followed roughly the same curriculum. The ablation models were pre-trained for 1 million iterations with all datasets and tasks being introduced at once.

We used data parallelism over 6 Nvidia Tesla P100s, which can each only hold a batch size of 1. We use gradient accumulation of 128 iterations, leading to an effective batch size of 768, with approximately 7,800 weight update steps for the ablation models. We use the AdamW optimizer [20] and a learning rate of 10^{-4} and weight decay of 0.01.

5   Experiments

To demonstrate the flexibility of Dessurt, we evaluate it on the six document datasets and six tasks listed in Table 2. The RVL-CDIP dataset [10] is a page classification dataset, which requires understanding overall layout and text topics. The DocVQA dataset [23] requires both reading and layout comprehension. HW-SQuAD [22] is more focused on reading comprehension, but has difficult text (synthetic handwriting) to recognize. Both the FUNSD [14] and NAF [5] datasets require form understanding, with a focus on label-value pairs in forms. The FUNSD dataset includes modern business documents, but the NAF dataset is uniquely challenging because it contains historical records with a both printed and handwritten text. We take a task from Tüselmann et al. [31], specifically named entity recognition over the IAM handwriting database (IAM NER), requiring both handwriting recognition and NLP capabilities. We also evaluate full-page handwriting recognition on the IAM database [21]. Each of these, and our experimental protocol for them, are discussed in more detail in the Supplementary Materials. We also present an ablation study at the end of this section.

Table 2. A summary of the end tasks we use to evaluate Dessurt and their attributes. The term "special output" refers to whether the tasks requires more than standard token prediction employed by most in the LayoutLM family

Dataset	Task	Domain	Requires handwriting recognition	Requires special output	Train set size
RVL-CDIP [10]	Classification	Modern printed	No	No	320K
DocVQA [23]	Question answering	Modern	Occasionally	No	39K
HW-SQuAD [22]	Question answering	Synthetic handwriting	Yes (easier)	No	68K
FUNSD [14]	Entity/Relationship detection	Modern printed forms	No	No/Yes	130
FUNSD	Form parsing	Modern printed forms	No	Yes	130
NAF [5]	Line/Relationship detection	Historic forms	Yes	No/Yes	921
NAF	Form parsing	Historic forms	Yes	Yes	921
IAM [21]	Full page recognition	Handwriting	Yes	Yes	747
IAM NER [31]	Named entityrecognition	Handwriting	Yes	No	747

5.1   RVL-CDIP

We compare Dessurt to several other models in Table 3 on document classification with the RVL-CDIP dataset. Dessurt performs slightly below the state-of-the-art, but is comparable to the other models. We note that this problem requires a holistic view of the document and is likely benefiting from a strong vision model. We note that while Dessurt uses a Swin architecture, it is shallower and narrower than the one used by Donut.

5.2   DocVQA and HW-SQuAD

For DocVQA, the model must locate the text that answers a textual question. The results are presented in Table 3 with ANLS, a text edit-distance metric that accounts for multiple correct answers. Unlike RVL-CDIP, understanding the text in DocVQA is critical, likely leading to both Dessurt's and Donut's comparatively limited performance. Other models rely on strong external recognition methods; LayoutLMv2's performance significantly drops when using a weaker OCR. Dessurt outperforms Donut, likely due to its language-focused tasks and real data in pre-training. Dessurt's weakest areas for DocVQA are Figures/Diagrams and Image/Photo. This makes sense because the pre-training datasets are almost exclusively textual.

Table 3. Results on RVL-CDIP and DocVQA datasets

	use OCR	# params	RVL-CDIP accuracy	DocVQA ANLS
BERT$_{BASE}$ [8]	✓	110M +OCR	89.8	63.5
LayoutLM$_{BASE}$ (w/img) [36]	✓	160M + OCR	94.4	–
LayoutLM$_{BASE}$ [36]	✓	113M + OCR	–	69.8
LayoutLMv2$_{BASE}$ [35]	✓	200M + OCR	95.3	78.1
LayoutLMv2$_{BASE}$ w/Tesseract OCR	✓	200M + OCR	–	48.2
DocFormer$_{BASE}$ [1]	✓	183M + OCR	**96.2**	–
TILT$_{BASE}$ [24]	✓	230M + OCR	93.5	**83.9**
Donut [15]		156M	94.5	47.1
Donut +10k trainset images [15]		156M	–	53.1
Dessurt (ours)		127M	93.6	63.2

The HW-SQuAD dataset [22] is the popular question answering benchmark SQuAD [26] rendered with handwritten fonts and noise. We evaluate on the task of machine comprehension, where the single document containing the answer is fed to the model. Unfortunately, the only prior method on this [30] was doing text snippet retrieval, not question answering, and so is incomparable. We use ANLS as our metric as it seems well suited to the task and achieve 55.5%.

5.3   FUNSD and NAF

Form parsing is the most difficult task we tackle, particularly with the NAF dataset, which is comprised of historical forms and has a good deal of handwriting. Our form parsing task is to reproduce the entire contents of the form in a structured manner, including recognition of text. We have the model predict JSON using the same format used in pre-training (Fig. 3(d)). Normalized tree edit-distance (nTED) metric has been introduced by Hwang et al. [13] for comparing document parses. However, nTED is not permutation invariant, which is undesirable as some forms lack a canonical read order. We introduce a modification, Greedily-Aligned nTED (GAnTED), which is robust to permutation. GAnTED is discussed in detail in the Supplementary Materials. We compute GAnTED for FUDGE [6] by running Tesseract [28] on its predicted bounding boxes and using its class and relationship predictions to build the JSON.

We also compare using standard F-measure for entity detection and relationship detection. We do this by aligning Dessurt's predicted strings to the GT strings. This means our results are dependant on the text recognition of Dessurt. This is in contrast to other models that use the GT word boxes for tokens and need only identify the correct box(es) rather than produce the correct text. Thus we end up below what prior methods achieve. Our results on both the FUNSD and NAF datasets are presented in Tables 4 and 5 respectively. On the NAF dataset, no models rely on external recognition models.

The visual domain of NAF is very different from modern documents, meaning two-stage methods require a specialized recognition model. We compare Dessurt's recognition ability to a CNN-LSTM [33] trained on the NAF dataset in the Supplementary Materials. We also report results pre-training Dessurt on images taken from the U.S.A. 1940 Census (visually similar to NAF data) in Table 5. Details for this pre-training are in the Supplementary Materials.

Table 4. Results on FUNSD dataset

	GT OCR used	# params	Entity Fm	Rel Fm	GAnTED
LayoutLM$_{BASE}$ [36]	boxes + text	-	78.7	42.8 [12]	–
BROS$_{BASE}$ [12]	boxes + text	138M + OCR	83.1	**71.5**	–
LayoutLMv2$_{BASE}$ [35]	boxes + text	200M + OCR	82.8	–	–
DocFormer$_{BASE}$ [1]	boxes + text	183M + OCR	**83.3**	–	–
Word-FUDGE [6]	boxes	17M + OCR	72.2	62.6	–
FUDGE [6] (+Tesseract)	none	17M (+OCR)	66.5	56.6	34.8
Dessurt (ours)	none	127M	65.0	42.3	**23.4**

Table 5. Results on NAF dataset

	# params	Line Fm	Rel Fm	GAnTED
Davis et al. [5]	1.8M	**73.8**	49.6	–
FUDGE [6]	17M	73.7	**57.3**	–
Dessurt (ours)	127M	49.3	29.4	42.5
Dessurt w/ census pretraining	127M	50.2	30.3	**38.8**

5.4 IAM Database

There have been several specialized approaches for doing full-page handwriting recognition, where line segmentation is done implicitly or explicitly. Dessurt is trained to do full-page recognition during its pre-training. We compare it to other full-page recognition models in Table 6. The metrics used are character error rate (CER) and word error rate (WER) across an entire page (or paragraph; the IAM dataset has one paragraph per page). Dessurt performs quite favorably compared to these specialized approaches and even achieves the lowest WER. We note that our pre-training includes synthetic handwriting derived from the IAM training set, so Dessurt is uniquely suited to solve this task on the IAM dataset. The fact that Dessurt's WER is relatively better than its CER is unusual and is likely a result of the word-part token prediction (other models use character prediction) and the language modeling capabilities learned in pre-training. We note that the number of parameters in Dessurt is higher than other models.

We also evaluate using the IAM NER task introduced by Tüselmann et al. [31] as part of a set of named entity recognition problems for handwriting datasets. Tüselmann et al. use a two-stage approach constructed specifically for this problem. They use a word level handwriting recognition model, with its outputs fed to a RoBERTa-based NER model (which sees the whole document). We fine-tune Dessurt on both line level NER and document level NER. In both cases Dessurt sees the entire handwriting image but has the lines it is supposed process highlighted. It performs transcription along with the classification with two tasks: (1) first reading a word, and then predicting its class, and (2) the reverse with class predicted first. This ensures we know which word Dessurt is predicting a class for. We randomly replace words in the teacher-forcing with close edit-distance words to decrease reliance on the recognition output. Additionally, we apply warp

Table 6. Results on IAM page/paragraph recognition

	# params	CER	WER
Bluche [2]	–	7.9	24.6
Chung and Delteil [3]	–	8.5	–
Start, Follow, Read [34]	–	6.4	23.2
OrigamiNet [37]	16.4M	4.7	–
Vertical Attention Network [4]	2.7M	**4.5**	14.6
Dessurt (ours)	127M	4.8	**10.2**

Table 7. Results on IAM NER. Reported in macro F-measure

Split task	RWTH 6 classes	Custom 6 classes	RWTH 18 classes	Custom 18 classes
Toledo et al. [29]	34.0	37.4	14.9	18.0
Rowtula et al. [27]	47.4	54.6	32.3	30.3
Tüselmann et al. [31]	**70.7**	**76.4**	**52.0**	**53.6**
Dessurt (ours)	62.0	71.5	40.4	48.5
Dessurt w/ IAM pretraining	59.5	71.1	39.5	45.3

grid augmentation [33] on the lines of the document. We also experimented with adding recognition on IAM words to the pre-training (more details in Supplementary Materials).

Our results for IAM NER are presented in Table 7. While Dessurt is moderately successful, it falls short of the customized two-stage approach presented by Tüselmann et al. They report that the CER of the HWR model they use is 6.8, which is the same CER as Dessurt. We assume this indicates that (unsurprisingly) RoBERTa is a stronger language model than Dessurt and is responsible for this superior performance.

5.5 Ablation

We performed an ablation study of the different sources used in our model's pretraining as well as some of the architectural choices (Table 8). We begin the data ablation with only the IIT-CDIP [17] dataset (I). We then incrementally add the synthetic Wikipedia (W), synthetic handwriting (H), synthetic forms (F), and distillation from BART (D), as well as only using synthetic data (no IIT-CDIP dataset). We ablate out the predicted spatial mask used in pretraining, and change the Swin window size from 12 to 7. We also ablate the 2-way cross attention by instead only having the query and response tokens attend to the visual tokens without the visual tokens attending to the query tokens. This is very similar to Donut, which lacks 2-way cross attention.

Table 8. Ablation results. The top four rows show the pre-training ablation with I=IIT-CDIP dataset, W=synthetic Wikipedia dataset, H=synthetic handwriting dataset, F=synthetic form dataset, D=distillation from BART. The lower three rows show ablations to the model: removing supervision with output mask, removing supervision with output mask and reducing Swin window size to 7, removing cross attention from image to question tokens. Results for DocVQA are evaluated using the validation set. "PT IAM" indicates IAM data added to last 200k iters of pre-training

	DocVQA (valid) ANLS	IAM NER Macro Fm		FUNSD		NAF		RVL CDIP acc.
			IAM PT	Entity Fm	Rel Fm	Entity Fm	Rel Fm	
Max iterations	500k	200k	200k	34k		300k		500k
I	44.0	42.3	43.4	19.7	10.2	28.7	12.6	89.0
W+I	43.2	45.2	49.0	29.5	16.0	31.0	13.7	89.1
H+W+I	44.4	50.1	49.7	29.3	16.5	31.6	14.9	88.9
F+H+W+I	**46.5**	47.6	50.0	44.8	28.2	**36.5**	**17.6**	**89.5**
(Only synth) D+F+H+W	43.1	**52.7**	**53.3**	39.4	22.0	31.5	14.3	88.5
(All) D+F+H+W+I	45.5	50.4	52.5	**47.8**	**29.5**	34.6	15.3	89.0
All, no mask loss	44.9	45.7	49.7	47.3	26.2	33.2	15.1	88.3
All, no mask loss w =7	44.4	44.8	51.3	45.9	28.6	31.8	15.3	88.6
All, 1-way cross attn	44.9	42.9	46.9	41.0	25.2	33.7	15.9	88.8

As can be seen each pre-training data source adds something to the model. The synthetic handwriting and synthetic forms are aimed at particular downstream tasks (IAM NER and form understanding respectively), but we note that their inclusion generally helps other tasks as well. Only the distillation appears selectively helpful and may not contribute significantly. In general, the ablated model components are helpful to the full model, but not necessary. The results with the RVL-CDIP dataset shows that the data a model is pre-trained with appears to be relatively irrelevant to its performance.

6   Conclusion

We have introduced Dessurt, an end-to-end architecture for solving a wide variety of document problems. Dessurt performs recognition within its single pass, removing reliance on an external recognition model, which most document understanding approaches require, making it a much simpler method. Because Dessurt uses arbitrary text as its output, it is also more flexible in the range of problems it can solve. We evaluate Dessurt on a wider range of tasks than any previous single method has done and show results ranging from promising to state-of-the-art.

References

1. Appalaraju, S., Jasani, B., Kota, B.U., Xie, Y., Manmatha, R.: Docformer: end-to-end transformer for document understanding. In: International Conference on Computer Vision (ICCV) (2021)
2. Bluche, T.: Joint line segmentation and transcription for end-to-end handwritten paragraph recognition. Advances in Neural Information Processing Systems (NIPS) (2016)
3. Chung, J., Delteil, T.: A computationally efficient pipeline approach to full page offline handwritten text recognition. In: International Conference on Document Analysis and Recognition Workshops (ICDARW). IEEE (2019)
4. Coquenet, D., Chatelain, C., Paquet, T.: End-to-end handwritten paragraph text recognition using a vertical attention network. IEEE Trans. Pattern Anal. Mach. Intell. (2022)
5. Davis, B., Morse, B., Cohen, S., Price, B., Tensmeyer, C.: Deep visual template-free form parsing. In: International Conference on Document Analysis and Recognition (ICDAR). IEEE (2019)
6. Davis, B., Morse, B., Price, B., Tensmeyer, C., Wiginton, C.: Visual FUDGE: form understanding via dynamic graph editing. In: Lladós, J., Lopresti, D., Uchida, S. (eds.) ICDAR 2021. LNCS, vol. 12821, pp. 416–431. Springer, Cham (2021). https://doi.org/10.1007/978-3-030-86549-8_27
7. Davis, B., Tensmeyer, C., Price, B., Wigington, C., Morse, B., Jain, R.: Text and style conditioned gan for generation of offline handwriting lines (2020)
8. Devlin, J., Chang, M.W., Lee, K., Toutanova, K.: BERT: Pre-training of deep bidirectional transformers for language understanding. In: Conference of the North American Chapter of the Association for Computational Linguistics: Human Language Technologies (NAACL-HLT) (2019)

9. Foundation, W.: Wikimedia downloads. https://dumps.wikimedia.org
10. Harley, A.W., Ufkes, A., Derpanis, K.G.: Evaluation of deep convolutional nets for document image classification and retrieval. In: International Conference on Document Analysis and Recognition (ICDAR)
11. Hinton, G., Vinyals, O., Dean, J.: Distilling the knowledge in a neural network (2015). arXiv preprint arXiv:1503.02531 2 (2015)
12. Hong, T., Kim, D., Ji, M., Hwang, W., Nam, D., Park, S.: Bros: A pre-trained language model focusing on text and layout for better key information extraction from documents. arXiv preprint arXiv:2108.04539 (2021)
13. Hwang, W., Lee, H., Yim, J., Kim, G., Seo, M.: Cost-effective end-to-end information extraction for semi-structured document images. In: Conference on Empirical Methods in Natural Language Processing (EMNLP) (2021)
14. Jaume, G., Ekenel, H.K., Thiran, J.P.: Funsd: A dataset for form understanding in noisy scanned documents. In: International Conference on Document Analysis and Recognition Workshops (ICDARW). IEEE (2019)
15. Kim, G., et al.: Donut: document understanding transformer without ocr. arXiv preprint arXiv:2111.15664 (2021)
16. Klaiman, S., Lehne, M.: Docreader: bounding-box free training of a document information extraction model. In: International Conference on Document Analysis and Recognition (ICDAR) (2021)
17. Lewis, D., Agam, G., Argamon, S., Frieder, O., Grossman, D., Heard, J.: Building a test collection for complex document information processing. In: ACM SIGIR Conference on Research and Development in Information Retrieval (2006)
18. Lewis, M., et al.: Bart: denoising sequence-to-sequence pre-training for natural language generation, translation, and comprehension. In: 58th Annual Meeting of the Association for Computational Linguistics (ACL) (2020)
19. Liu, Z., et al.: Swin transformer: hierarchical vision transformer using shifted windows. In: International Conference on Computer Vision (ICCV) (2021)
20. Loshchilov, I., Hutter, F.: Decoupled weight decay regularization. In: International Conference on Learning Representations (ICLR) (2019)
21. Marti, U.V., Bunke, H.: The iam-database: an English sentence database for offline handwriting recognition. International Journal on Document Analysis and Recognition 5(1) (2002)
22. Mathew, M., Gomez, L., Karatzas, D., Jawahar, C.: Asking questions on handwritten document collections. Int. J. Document Anal. Recogn. (IJDAR) **24**(3) (2021)
23. Mathew, M., Karatzas, D., Jawahar, C.: Docvqa: a dataset for VQA on document images. In: Winter Conference on Applications of Computer Vision (WACV) (2021)
24. Powalski, R., Borchmann, Ł., Jurkiewicz, D., Dwojak, T., Pietruszka, M., Pałka, G.: Going full-tilt boogie on document understanding with text-image-layout transformer. In: International Conference on Document Analysis and Recognition (ICDAR), pp. 732–747 (2021)
25. Radford, A., Wu, J., Child, R., Luan, D., Amodei, D., Sutskever, I., et al.: Language models are unsupervised multitask learners. OpenAI blog (2019)
26. Rajpurkar, P., Zhang, J., Lopyrev, K., Liang, P.: Squad: 100,000+ questions for machine comprehension of text. In: Conference on Empirical Methods in Natural Language Processing (EMNLP) (2016)
27. Rowtula, V., Krishnan, P., Jawahar, C., CVIT, I.: Pos tagging and named entity recognition on handwritten documents. In: International Conference on Natural Language Processing (ICNLP) (2018)

28. Smith, R.: An overview of the tesseract ocr engine. In: Ninth International Conference on Document Analysis and Recognition (ICDAR 2007) (2007). https://doi.org/10.1109/ICDAR.2007.4376991
29. Toledo, J.I., Carbonell, M., Fornés, A., Lladós, J.: Information extraction from historical handwritten document images with a context-aware neural model. Pattern Recogn. **86** (2019)
30. Tüselmann, O., Müller, F., Wolf, F., Fink, G.A.: Recognition-free question answering on handwritten document collections. arXiv preprint arXiv:2202.06080 (2022)
31. Tüselmann, O., Wolf, F., Fink, G.A.: Are end-to-end systems really necessary for ner on handwritten document images? In: Lladós, J., Lopresti, D., Uchida, S. (eds.) International Conference on Document Analysis and Recognition (ICDAR) (2021)
32. Vaswani, A., et al.: Attention is all you need. In: Neural Information Processing Systems (NIPS) (2017)
33. Wigington, C., Stewart, S., Davis, B., Barrett, B., Price, B., Cohen, S.: Data augmentation for recognition of handwritten words and lines using a cnn-lstm network. In: International Conference on Document Analysis and Recognition (ICDAR) (2017)
34. Wigington, C., Tensmeyer, C., Davis, B., Barrett, W., Price, B., Cohen, S.: Start, follow, read: End-to-end full-page handwriting recognition. In: European Conference on Computer Vision (ECCV) (2018)
35. Xu, Y., et al.: LayoutLMv2: multi-modal pre-training for visually-rich document understanding. In: 59th Annual Meeting of the Association for Computational Linguistics (ACL) (2021)
36. Xu, Y., Li, M., Cui, L., Huang, S., Wei, F., Zhou, M.: LayoutLM: pre-training of text and layout for document image understanding. In: International Conference on Knowledge Discovery & Data Mining (KDD) (2020)
37. Yousef, M., Bishop, T.E.: Origaminet: weakly-supervised, segmentation-free, one-step, full page text recognition by learning to unfold. In: Computer Vision and Pattern Recognition (CVPR) (2020)

Task Grouping for Multilingual Text Recognition

Jing Huang$^{(\boxtimes)}$, Kevin J. Liang , Rama Kovvuri , and Tal Hassner

Meta AI, New York City, USA
{jinghuang,kevinjliang,ramakovvuri,thassner}@fb.com

Abstract. Most existing OCR methods focus on alphanumeric characters due to the popularity of English and numbers, as well as their corresponding datasets. On extending the characters to more languages, recent methods have shown that training different scripts with different recognition heads can greatly improve the end-to-end recognition accuracy compared to combining characters from all languages in the same recognition head. However, we postulate that similarities between some languages could allow sharing of model parameters and benefit from joint training. Determining language groupings, however, is not immediately obvious. To this end, we propose an automatic method for multilingual text recognition with a task grouping and assignment module using Gumbel-Softmax, introducing a task grouping loss and weighted recognition loss to allow for simultaneous training of the models and grouping modules. Experiments on MLT19 lend evidence to our hypothesis that there is a middle ground between combining every task together and separating every task that achieves a better configuration of task grouping/separation.

Keywords: OCR · Multilingual text recognition · Task grouping

1 Introduction

Optical Character Recognition (OCR) has long been the fundamental task in computer vision. There are many applications such as automatic content extraction for documents, scene understanding, navigation, and assistance for translation and for the visually impaired users. From the perspective of research, OCR can range from relatively easy and controlled tasks such as digit recognition [21], to difficult scenarios such as scene text with arbitrary orientations and shapes [7,8,18], and has become an important domain for benchmarking new machine learning techniques.

Nevertheless, most existing OCR approaches focus on numbers and the English alphabet due to English's status as a common *lingua franca* and its subsequent wide availability in popular datasets. Thanks to the introduction of multilingual text detection and recognition datasets and benchmarks [29,30], there is now a unified platform to measure the model performance on challenging scenarios containing thousands of distinctive characters. Recent methods

© The Author(s), under exclusive license to Springer Nature Switzerland AG 2023
L. Karlinsky et al. (Eds.): ECCV 2022 Workshops, LNCS 13804, pp. 297–313, 2023.
https://doi.org/10.1007/978-3-031-25069-9_20

have shown that training different languages with different recognition heads can improve end-to-end recognition accuracy compared to combining characters from all languages in the same recognition head [13]. However, it's not clear whether an individual recognition head for each language is optimal. For example, should English and Spanish be separated into two heads? The answer is probably no since they share most of the characters. Moreover, even if separating two languages into two heads does yield the best accuracy, it might not be worth it if the accuracy gain is marginal compared to the increase in number of parameters and/or the inference time. Therefore, one of the questions our work tries to answer is how to decide whether/how languages should be grouped together under the constraint of a limited number of models.

Do perform this grouping, we treat each of the models at initialization as a generalist agent which looks at all tasks. Then, as the scale tips, each agent is encouraged to be increasingly specialized in one or more tasks; each agent becomes a specialist, each model can still try to learn the other tasks to some lesser extent. Due to different transferability, data variation, and similarities among the tasks, each agent can have different progress in both the specialized tasks and non-specialized tasks, and the specialties will be redistributed automatically as the agents evolve. Eventually, as confirmed by our experiments, this multi-agent system will reach an equilibrium where the specialties for each agent do not change any more, and this is when the task grouping result is finalized.

To summarize, our contributions include:

- To our knowledge, this is *the first work* exploring the grouping of languages for multilingual OCR.
- We propose an automatic grouping mechanism that allows dynamic and differentiable shifting of the tasks to be routed to different heads as training goes on.
- We empirically show that the automatic task grouping model outperforms both the one-task-per-head and the all-tasks-in-one-head baselines. We further show that when the models have different capacities, the task assignment can potentially reflect the underlying task complexity and data distribution.

To promote reproduction of our work, we will publicly release our code.

2   Related Work

2.1   Multilingual Text Spotting

Text spotting systems combine text detection and text recognition modules to identify the location and content of text in images. Early works approached both modules independently; first generating the text proposals for regions containing text and then, follow this up with a recognition module to identify the text given a pre-defined character dictionary. For text detection, current state-of-the-art (SotA) methods are mostly based on Region Proposal Networks (RPN) [33] that has proven successful for object detection. Variants of RPN have been proposed to account for varying text orientations [16], arbitrary shapes [24,32] and character masking [2]. For text recognition, models typically use a RNN-style design to be able to predict a sequence of characters [12, 37]. For the decoder, the representative methods include connectionist temporal classification (CTC) [11] and attention-based decoder [4, 22, 34].

While earlier systems treated detection and recognition as independent modules, most of the recent works train these modules in an end-to-end manner. Given the inter-dependability of these modules, this training methodology results in performance improvements for these systems. Some representative works include Mask TextSpotter [23], FOTS [25], CharNet [39], etc. For deeper insights into the text spotting systems, we refer the readers to the thorough review in [26]. Similar to these works, we also employ end-to-end training for our system. For recognition, we mainly use the attention-based decoder to make a fair comparison with the previous works [13, 24].

With the availability of reliable multilingual datasets such as MLT19 [29], text spotting systems have tried to address the problem of multilingual texts. In addition to detection and recognition modules, some multilingual text spotting systems also include a script identification module [5, 13] to identify the language for text recognition. While text spotting systems such as E2E-MLT [5] and CRAFTS [3] present results for multilingual datasets, they do not explicitly incorporate model specific components adapted for multiple languages. Instead, they combine the characters from all languages to form a larger dictionary for the recognition head. Recently, Multiplexed Multilingual Mask TextSpotter (MMMT; [13] proposed to employ different recognition heads for different scripts routed through a script identification module at the word level. Unlike *MMMT*, which employs hard assignment for routing to an appropriate recognition head, we propose to group the languages by training agents to route words to different heads in a data-driven fashion. This automatic grouping mechanism allows for dynamic and differentiable shifting of the tasks to optimize the language combinations for various recognition heads.

2.2 Multitask Learning and Grouping

Multitask learning methods have a long history [6, 9]. As their name implies, they jointly learn solutions for multiple tasks, sharing or transferring information between tasks to improve overall performance. Recent, deep learning–based methods assume that the parameters for the first few layers, which account for low-level feature extraction, are shared among different tasks, while the parameters for the last few layers, which account for high-level integration of visual signals, are task-specific [14, 42]. Hence, information relevant to all tasks is learned by a shared trunk which later splits to multiple, task specific heads.

A natural question that arises when designing multitask systems is: How should the tasks be grouped to maximise a model's accuracy? To answer this question, Kang *et al.* [17] proposed to learn shared feature representation for related tasks by formulating task grouping as a mixed integer programming problem where binary indicator variables are used to assign tasks to groups. Unlike this hard group assignment, Kumar *et al.* [20] propose to allow for parameter

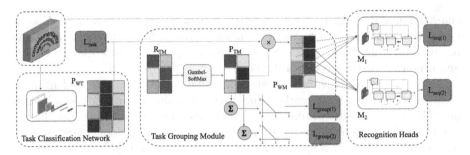

Fig. 1. Our proposed task grouping framework. In this scenario we have a batch of 4 inputs that potentially belong to 3 tasks, with 2 recognition heads. See Sect. 3 for more details.

sharing across groups through soft, latent assignment of task features as a linear combination of a finite number of underlying basis tasks. Zhong *et al.* [43] extend this work by removing constraints on the size of latent basis tasks and adding regularization terms to enforce sparsity in task weights and orthogonality to prohibit commonality among unrelated tasks. Zamir *et al.* [41] proposed a method for modeling the relationship of different visual tasks based on the *transferability* between them. Instead of learning shared representation on a trunk, Strezoski *et al.* [36] introduce a *Task Routing layer* that masks convolutional channels based on the task, effectively creating a sub-network per task.

Our work is similar in spirit to Strezoski *et al.* [36] in that we allow for dynamic routing of tasks to different heads during training. In our approach, however, the routing is done using *Gumbel-Softmax* [15] to ensure probabilistic interpretation of each task and using a novel *grouping loss* for task assignment.

3   Methodology

Given a list of tasks $T = \{T_i\}_{1 \leq i \leq t}$ and a list of models $M = \{M_j\}_{1 \leq j \leq m}$, we can define the task grouping of T over M as a mapping $G : T \rightarrow M$. G is a single-valued function, which means each task will be assigned to exactly one model. On the other hand, G does not need to be an injection, since multiple tasks can be assigned to the same model. G does not need to be a surjection either, in which case some models will not be assigned with any tasks. Our goal is to find out the best assignment G such that the overall performance is maximized.

Figure 1 shows the core architecture of the proposed task grouping framework. Given an input, which could be a batch of already cropped image patches or pre-extracted features, we first pass them through a *task classification network* that predicts the probabilities of each input belonging to each of the tasks.

Under the context of multilingual text recognition, each task T_i can be described as "recognizing the word instance W_k in language set L_i", where W_k is the k-th instance in a batch of w word crops. We can thus define a probability matrix of size $w \times t$ on the likelihood of each word belonging to each task/language set:

$$P_{WT} = \{p(T_i|W_k)\}_{1\leq k\leq w, 1\leq i\leq t} \tag{1}$$

At inference time, P_{WT} can be inferred from a task classification network such as the language prediction network [13]. This is a t-way classification problem and the task classification loss can be computed using a cross entropy loss:

$$L_{task}(W_k) = -\sum_{i=1}^{t} I(T_i = T_{gt}) \log p(T_i|W_k) \tag{2}$$

where $I(T_i = T_{gt})$ is the binary indicator of whether the task matches the ground truth.

At training time, P_{WT} can be inferred from the ground truth, if there is an annotation of which language each word belongs to: $p(T_i|W_k)$ is 1 if W_k belongs to T_i and otherwise becomes 0. When the ground truth annotation for the language information is not directly available but the transcription is available, we can make an educated guess of the probability by calculating the proportion of characters in W_k that are supported by language set L_i.

3.1 Grouping Module

Since task-model mapping G is a discrete function, to be able to learn it we can define the following probability matrix, of size $t \times m$:

$$P_{TM} = \{p(M_j|T_i)\}_{1\leq i\leq t, 1\leq j\leq m}, \tag{3}$$

where $p(M_j|T_i)$ is the probability of an arbitrary word belonging to T_i to be handled by model M_j. Then, we can compute the probability matrix of each word W_k to be handled by model M_j by multiplying P_{WT} and P_{TM}:

$$P_{WM} = P_{WT} \cdot P_{TM} \tag{4}$$

Naive task assignment to a group based on traditional SoftMax is a discrete operation and thus non-differentiable. Backpropagation through only the selected task-group pairing would result in high variance gradients leading to unstable learning. Instead, when computing P_{WM} during training, we apply a soft relaxation of the assignment operation using the Gumbel-Softmax [15]. Gumbel-Softmax if fully differentiable with the reparameterization trick and results in gradient backpropagating through all possible task-group pairings, not just the one with the maximum score. We instantiate learnable parameters for task-model assignment as a real-valued matrix $R_{TM} \in \mathbb{R}^{t \times m}$, initialized with all ones (or any equal numbers) in the beginning, and we set the temperature $\tau = 1.0$ throughout the training. At test time, we can just pick the model corresponding to the maximum, $i.e.$ the hard mode of Gumbel-Softmax.

3.2   Integrated Loss

A key difference of our approach compared to [13] is that in our framework, we do not restrict the capability of each model, or recognition head, to support any specific task, *i.e.*, a certain recognition head can only support certain characters, from the beginning. Instead, we assume each model to be omnipotent in the beginning and has the potential to handle every task. This is necessary since otherwise there is no point in doing the grouping if each model is already designed to do certain tasks.

Therefore, unlike [13], we can directly use the negative log likelihood as the recognition loss $L_{seq(j)}$ for each model M_j without worrying about the unsupported characters:

$$L_{seq} = -\frac{1}{s} \sum_{l=1}^{s} \log p(S_l), \tag{5}$$

where $p(S_l)$ is the predicted probability of character at position l of the sequence, and s is the length of the sequence of character labels.

We can, however, perform the pruning at the output layer to remove any characters that do not belong to the task assigned to certain head, once the grouping is determined. This would reduce the unnecessary weights in the final model.

The integrated loss across all probabilistic instance-model assignments can thus be calculated as the weighted sum of individual losses:

$$L_{integrated0} = \sum_{k=1}^{w} \sum_{j=1}^{m} p(M_j|W_k) \cdot L_{seq(j)}(W_k, M_j), \tag{6}$$

where the probability term is from P_{WM} of Eq. (4), which is essentially the law of total probability:

$$p(M_j|W_k) = \sum_{i=1}^{t} p(T_i|W_k) \cdot p(M_j|T_i) \tag{7}$$

3.3   Integrated Loss with a Base Loss Coefficient

With the integrated loss (Eq. (6)), we can see that in general, a task T_{big} with a bigger probability $p(M_j|T_{big})$ to be assigned to a model M_j will contribute a bigger loss than a task T_{small} with a smaller probability $p(M_j|T_{small})$ to be assigned to the model, encouraging the model to optimize towards a better prediction for T_{big}, which then encourages $p(M_j|T_{big})$ to be bigger until it reaches 1. A similar but opposite process applies to $p(M_j|T_{small})$, which would become smaller until it reaches 0. As a result, the learned task-model assignment P_{TM} will almost certainly be random and fully depending on the first few iterations due to the positive-feedback loop. We resolve this issue by adding a small positive base loss coefficient, ϵ:

$$L_{integrated} = \sum_{k=1}^{w} \sum_{j=1}^{m} (p(M_j|W_k) + \epsilon) \cdot L_{seq(j)}(W_k, M_j). \tag{8}$$

This ensures that the model not only tries to excel at the tasks assigned to it, but also learns the other tasks at a small but positive rate. The effect of ϵ can be quantified from the perspective of training data ratios among different tasks. Assume the original ratio of data from any task is 1, for any model-task pair, the maximum effective data ratio would be $1 + \epsilon$, which is achieved when p reaches 1, and the minimum effective data ratio would be $0 + \epsilon$, which is achieved when p falls to 0. The ratio $\frac{1+\epsilon}{\epsilon}$ can thus be used to measure how biased the model can potentially be trained towards the most vs. least important task. Based on our ablation study (Sect. 4.4), we set $\epsilon = 0.2$ when training from scratch, $\epsilon = 0.1$ when fine-tuning from pretrained models and $\epsilon = 0$ for the final head-wise fine-tuning.

3.4   Grouping Loss

While Eq. (8) makes sure that any model has the potential to learn every task, we also would like to ensure that happens within a certain budget, i.e., given the number of different models (heads) we can support, each model is specialized in at least one task. This ensures we do not waste the modeling capacity of an idle head. Therefore, we introduce the following grouping loss

$$L_{group} = \sum_{j=1}^{m} L_{group(j)} = \sum_{j=1}^{m} \max(\mu_j - \sum_{i=1}^{t} p(M_j|T_i), 0), \tag{9}$$

where μ_j is the least number of tasks model M_j is expected to handle. In most experiments, we set $\mu_j = 1$, meaning that if M_j completely takes over at least one task, the grouping loss for M_j would reach the minimum value 0. Note that the converse does not hold - the grouping loss can reach 0 even when certain model do not excel in any specific task. However, in practice, as long as the number of tasks is larger than or equal to the number of models, the small penalty of the grouping loss could help us achieve the minimum task assignment goal.

4   Experimentals

4.1   Datasets

Our work leverages a number of public datasets. These sets are summarized in Table 1. We next offer a brief description of these sets.

ICDAR 2013 Dataset (IC13).   [19] This is the oldest set used in this work, originally released for the ICDAR 2013 Robust Reading Competition. It offers 229 training and 233 test images of English text. Text locations are given as axis aligned, rectangular bounding boxes with text annotated at a word level.

ICDAR 2015 Dataset (IC15).   [18] This dataset was introduced in ICDAR'15 and offers more images than IC13: 1000 training and 500 test. Images in this set are of scene text in English, appearing at different orientations, where words are annotated using quadrangle bounding boxes.

Total Text Dataset. [7] This collection offers 1255 training and 300 test, English scene text images. The images reflect a wide range of text orientations and shapes, including curved text examples. To accommodate different shapes, text locations are provided as polygons; recognition labels are given at word level.

ICDAR 2017 RCTW Dataset (RCTW17). [35] This set was collected to promote development of OCR methods for in the wild Chinese text. It is partitioned to 8034 and 4229 subsets of training and test images, respectively.

ICDAR 2019 MLT Dataset (MLT19) and SynthTextMLT. [29] was an extension of the ICDAR 2017 MLT dataset (MLT17) [30] for multilingual text detection, recognition and script identification, which contains 10000 training images, 2000 validation images and 10000 test images in 7 different scripts from 10 languages. The dataset contains multi-oriented scene text annotated by quadrangle boxes. A synthetic dataset (SynthTextMLT) [5] containing over 250k synthetic data in 7 scripts was also released along with the MLT19 benchmark. Since MLT19 training and validation sets completely covers the training and validation images in MLT17, though the split is a bit different, we only use MLT19 data for training in this paper.

ICDAR 2019 ArT Dataset (ArT19). [8] Contains 5603 training and 4563 test images in both English and Chinese, sourced from Total Text [7] and SCUT-CTW1500 [40]. Released as part of the ICDAR 2019 Robust Reading Competition, the images in this collection depict texts in challenging shapes. Similarly to Total Text, text locations are encoded as polygons. We remove all Total Text test images from this set, ensuring that any training on this set can be applied to other sets without risk of test images influencing models trained on this set.

ICDAR 2019 LSVT Dataset (LSVT19). [38] This is one of the largest data sets used for developing OCR methods: 30000 training and 20000 test images. LSVT images mostly show street views with about 80% of them showing Chinese text and the rest examples in English.

4.2 Model Training

For fair comparison, we adopt the same segmentation-based detection and ROI mask feature extraction modules as [13], and freeze the pretrained weights of these layers throughout training. For language classification, [13] uses 8 classes including Arabic, Bengali, Chinese, Hindi, Japanese, Korean, Latin and Symbol, but in our experiment we only use 7 classes by discarding the 'Symbol' class, since it doesn't not have any dedicated dataset and the number of the samples is too small to make a difference.

To expedite the training, we first combine every dataset to train a single recognition head with hidden size 256 and embed size of 200 covering all datasets using the ratios specified in Table 1 for 40k iterations. Then, we use this weight as a universal pretrained weights for the second stage of training.

Next, we perform a series of experiments that jointly train the grouping module and the recognition heads, each restricting the number of recognition heads to m ($2 \leq m \leq 7$). For each m, we launch three training jobs with different random seeds. Each of the training jobs runs for 20k iterations on the MLT19 training datasets only to reduce the potential data imbalance when including the other training set. We record and summarize the final grouping results in Table 2, which we will discuss in Sect. 4.3.

Table 1. Datasets used in our experiments. #Train: number of training images. Ratio: the relative sampling ratio when the dataset is used in training. Word / Phrase: Annotations given at a word or phrase level. Box type: horizontal, axis aligned (H-Box), arbitrarily rotated (R-Box), quadrangle (Quad), and Polygon. #Lang: Number of languages provided. Note that the Total Text dataset is fully covered in ArT19, and we removed the testing set of Total Text from ArT19.

Name	#Train	Ratio	Word / Phrase	Box type	#Lang.
ICDAR13 [19]	229	20	Word	H-Box	1
ICDAR15 [18]	1000	20	Word	Quad	1
Total Text [7]	1255	50	Word	Polygon	1
RCTW17 [35]	8034	20	Phrase	R-Box	2
MLT19 [29]	10000	100	Word	Quad	10
SynthTextMLT [5]	252599	1	Word	R-Box	7
ArT19 [8]	5303	50	Word	Polygon	2
LSVT19 [38]	30000	20	Phrase	Polygon	2

Finally, based on the grouping result, we fine-tune each recognition head with only the datasets within the assigned group corresponding to the head. At this stage the grouping is essentially frozen and does not change any more. We can prune the output layer of the decoder so that the characters not belonging to the group are removed, to reduce the parameter number for the final model.

4.3 Task Grouping Results

Table 2 shows the aggregation of grouping results from 18 task grouping experiments with 2 to 7 recognition heads, each repeated for 3 times. All task assignments stabilize after about 10000 iterations.

The top 14 groups are ordered first by the number of occurrences and then by the first occurrence, *i.e.* the minimum number of recognition heads when the group first occurs. All exclusive task-model assignments (one head focusing on one task) occur at least twice, showing the effectiveness of having a dedicated model for each task. Chinese ending up as an individual task occurs in 50% of the cases, which is expected given its high character number and the datasets, except that it's grouped together with Japanese, which shares many characters with it, only once. On the other hand, Hindi seems to be suitable to be grouped with many different languages rather than being trained by itself.

Surprisingly, the most frequent task group that has more than one task is Arabic+Korean, which occurs 5 times. This suggests that there are inherent characteristics shared either by these two scripts, or by the examples in the MLT19 dataset itself, that boost the performance for each other. Another unusual cluster is the combination of 5 tasks, Bengali+Chinese+Hindi+Japanese+Latin, which is the only grouping with more than 2 tasks that occurs more than once.

Table 2. Task grouping result. Task combinations that end up being grouped together. The 2nd to the 5th columns indicate task names in the final grouping, the number of tasks in the group, the number of occurrences in the 18 experiments, and the minimum number of recognition heads when the combination first occurs.

Rank	Group	#Tasks within group	#Occurrences	#Heads at first occurrence
1	Chinese (C)	1	9	4
2	Latin (L)	1	7	5
3	Arabic (A)	1	6	3
3	Korean (K)	1	6	3
5	Arabic+Korean	2	5	2
6	Bengali (B)	1	5	5
7	Japanese (J)	1	4	5
8	B+C+H+J+L	5	2	2
9	Hindi+Japanese	2	3	3
10	Japanese+Latin	2	2	4
11	Arabic+Hindi	2	2	5
11	Bengali+Japanese	2	2	5
11	Hindi+Latin	2	2	5
14	Hindi (H)	1	2	6
15	A+K+L	3	1	2
15	H+J+K	3	1	2
15	A+B+C+L	4	1	2
15	B+C+H+J	4	1	2
15	Chinese+Hindi	2	1	3
15	Korean+Latin	2	1	3
15	A+B+J	3	1	3
15	B+C+L	3	1	3
15	Arabic+Bengali	2	1	4
15	Bengali+Hindi	2	1	4
15	Chinese+Latin	2	1	4
15	A+B+H	3	1	4
15	Japanese+Korean	2	1	5
15	B+C+K	3	1	6
15	Hindi+Korean	2	1	7
15	Chinese+Japanese	2	1	7

We should note that, however, the scattering of the grouping results show that there can be many local optima for this specific scenario of 7 distinctive scripts. We shall expect higher frequencies of the same grouping results if certain tasks share greater similarity, and we will leave that as one of our future work.

Besides that, it's interesting that despite we introduced grouping loss to encourage each head to take on at least one task, when there are 6 or 7 tasks, one or two recognition heads might not be assigned with any task in the end. This means for certain combinations of tasks, training them together could outperform training them separately even if there is spare resource for a new head.

4.4 Ablation Study

Base Integrated Loss Coefficient. We train the task grouping network with different base integrated loss coefficient ϵ defined in Eq. (8) on MLT19 training set. The network contains 5 recognition heads that are initialized with the

Fig. 2. Qualitative results on MLT19 test set [29]. The predicted transcription is rendered with green background, along with the detection confidence, language and the assigned group. The model has 5 heads (groups): group 1 - Arabic (ar) and Hindi (hi), group 2 - Bengali (bn) and Japanese (ja), group 3 - Chinese (zh), group 4 - Latin (la), group 5 - Korean (ko). See Sect. 4.6 for more details.

same pretrained weights. We record the number of task assignment changes in the first 3000 iterations. From Table 3 we can see that, when $\epsilon = 0.0$, there's only 1 assignment change since the model does not have much chance to learn the unassigned tasks; interestingly, when ϵ is too big (0.3/0.4), there are also fewer changes happening, possibly because there is not much diversity across the models and everything moves in the same direction. The maximum number of assignment changes happen when ϵ is 0.2 or 0.1. Therefore, in most of our experiments we use 0.2 for early training and 0.1 for fine-tuning.

Table 3. Ablation study for base integrated loss coefficient ϵ.

Base integrated loss coefficient ϵ	0.0	0.1	0.2	0.3	0.4
Assignment changes within 3k iters	1	3	5	2	2

Table 4. Task assignment result for models with different major hyper-parameters. Each model supports all characters in the beginning so the total number of parameters for each head is high, but they can be pruned when the task assignment stabilizes.

Embed size	Hidden size	Parameter number	Assigned task	Charset size	Final parameters
100	224	4.05 M	Arabic	80	1.15 M
150	224	4.51 M	Bengali	110	1.18 M
200	224	4.98 M	Japanese	2300	2.13 M
100	256	4.59 M	Hindi	110	1.42 M
150	256	5.06 M	Korean	1500	2.00 M
200	256	5.52 M	Chinese	5200	3.78 M
250	256	5.98 M	Latin	250	1.54 M

4.5   Task Assignment on Models with Different Hyper-parameters

In this section, we perform an interesting experiment that showcases how our design can help assign different tasks to models with different hyper-parameters based on the potential difficulty and the available data. We set the number of models (recognition heads) to be equal to the number of tasks, but set the key hyper-parameters of the models, embed size and hidden size, to be different from each other. We train the overall model on the weighted combination of all datasets listed above, and Table 4 shows the assigned task corresponding to each of the models. We can clearly see the correlation between the number of parameters versus the number of characters in the corresponding character set, with the exception of Latin. This illustrated that in general, when the number of characters grow, the heavier models will outperform lighter models in the long term; however, since Latin words are dominating in all the datasets including many difficult cases like curved text, the task grouping framework learns to spend the heaviest model on it to boost the overall performance.

4.6   E2E Text Recognition

Table 5 shows the results on MLT19 [29] end-to-end multilingual recognition benchmark. Figure 2 additionally provides qualitative examples of these multilingual. We find that using varying numbers of grouped heads can perform similarly to (and in some cases, better than) the multiplexed approach of a separate recognition head per language [13]. This is an interesting result, as it means we can significantly cut down on the computational cost and model size with little impact or even some gains to the performance. Notably, we also find that increasing the number of heads from a single shared head (Mask TextSpotter V3 [24]) to even just two grouped heads leads to a significant increase in F1-score.

We provide qualitative failure cases in Fig. 3. While detection errors could be attributed to arbitrary text shape, blurred text, glossy surfaces and rare fonts; recognition errors could be attributed to text ordering, text resolution and challenging scripts.

Table 5. End-to-end recognition results on MLT19. Note that there are two versions results of CRAFTS, one from the official MLT19 website and one from paper [3]. Importantly, CRAFTS has a ResNet-based feature extraction which is much bigger than the one with 5-Convs used in our experiments.

Method	F	P	R
E2E-MLT [5]	26.5	37.4	20.5
RRPN+CLTDR [27]	33.8	38.6	30.1
CRAFTS [3]	51.7	65.7	42.7
CRAFTS (paper) [3]	**58.2**	**72.9**	**48.5**
Mask TextSpotter V3 (1 head) [24]	39.7	**71.8**	27.4
Multiplexed TextSpotter (8 heads) [13]	48.2	68.0	37.3
Grouped (2 heads)	45.5	67.7	34.3
Grouped (3 heads)	47.1	67.0	36.3
Grouped (4 heads)	47.9	66.7	37.4
Grouped (5 heads)	**48.5**	67.7	**37.8**
Grouped (6 heads)	48.3	67.8	37.5
Grouped (7 heads)	48.2	68.0	37.3

Fig. 3. Error analysis on MLT19. Detection errors are represented in red outline and recognition errors in purple. See Sect. 4.6. (Color figure online)

5   Conclusions

Text is one of the most ubiquitous visual object classes in real-world scenes, making understanding it practical and critically important. Processing multiple languages, however, requires substantial resources, to accurately recognize the subtleties of appearances variations of different scripts and different intra-script characters. This ability was, therefore, previously accomplished by specializing separate network heads to specific languages. We, instead, are the first to propose automatically grouping different languages together, in the same recognition heads. Our dynamic, and differentiable task shifting approach automatically routes tasks to different heads while the network trains, optimizing for the best, bottom line accuracy across all languages. Extensive tests show our method to not only achieve StoA accuracy, but to do so with fewer recognition heads and hyperparameters, consequently making is a practical design choice for real-world OCR systems.

Future Work. Our work leaves several natural follow-up directions. One interesting question relates to the scalability of our approach: How many multitask heads, for example, would be required to effectively learn hundreds of languages? Another intriguing direction is extending our multitask learning and task grouping to include neural architecture search as part of its design. Such a solution should allow growing heads with different architectures for different languages to account for, *e.g.*, harder vs. easier languages. Finally, another potential extension could be continual language learning [31]: adding more languages as relevant training data becomes available, without retraining or regrouping existing languages. Alternative grouping approaches based on Bayesian nonparametric approaches like the Chinese Restaurant Process [1] or Indian Buffet Process [10, 28] may be natural ways to perform groupings in such settings.

References

1. Aldous, D.J.: Exchangeability and related topics. In: École d'Été de Probabilités de Saint-Flour XIII — 1983 (1985)
2. Baek, Y., Lee, B., Han, D., Yun, S., Lee, H.: Character region awareness for text detection. In: IEEE Conference on Computer Vision and Pattern Recognition, pp. 9365–9374 (2019)
3. Baek, Y., et al.: Character region attention for text spotting. In: Vedaldi, A., Bischof, H., Brox, T., Frahm, J.-M. (eds.) ECCV 2020. LNCS, vol. 12374, pp. 504–521. Springer, Cham (2020). https://doi.org/10.1007/978-3-030-58526-6_30
4. Bahdanau, D., Cho, K., Bengio, Y.: Neural machine translation by jointly learning to align and translate. arXiv preprint arXiv:1409.0473 (2014)
5. Bušta, M., Patel, Y., Matas, J.: E2E-MLT - an unconstrained end-to-end method for multi-language scene text. In: Carneiro, G., You, S. (eds.) ACCV 2018. LNCS, vol. 11367, pp. 127–143. Springer, Cham (2019). https://doi.org/10.1007/978-3-030-21074-8_11
6. Caruana, R.: Multitask learning. Machine Learn. **28**(1), 41–75 (1997)
7. Ch'ng, C.K., Chan, C.S.: Total-text: a comprehensive dataset for scene text detection and recognition. In: International Conference on Document Analysis and Recognition, vol. 1, pp. 935–942. IEEE (2017)
8. Chng, C.K., et al.: ICDAR 2019 robust reading challenge on arbitrary-shaped text-RRC-ArT. In: International Conference on Document Analysis and Recognition, pp. 1571–1576. IEEE (2019)

9. Evgeniou, T., Pontil, M.: Regularized multi-task learning. In: Proceedings of the International Conference on Knowledge Discovery and Data Mining, pp. 109–117 (2004)
10. Ghahramani, Z., Griffiths, T.L.: Infinite latent feature models and the indian buffet process. In: Neural Information Processing Systems (2006)
11. Graves, A., Fernández, S., Gomez, F., Schmidhuber, J.: Connectionist temporal classification: labelling unsegmented sequence data with recurrent neural networks. In: International Conference on Machine Learning, pp. 369–376 (2006)
12. He, P., Huang, W., Qiao, Y., Loy, C.C., Tang, X.: Reading scene text in deep convolutional sequences. In: AAAI, pp. 3501–3508 (2016)
13. Huang, J., et al.: A multiplexed network for end-to-end, multilingual OCR. In: IEEE Conference on Computer Vision and Pattern Recognition, pp. 4547–4557 (2021)
14. Inkawhich, N., Liang, K., Carin, L., Chen, Y.: Transferable perturbations of deep feature distributions. In: International Conference on Learning Representations (2020)
15. Jang, E., Gu, S., Poole, B.: Categorical reparameterization with gumbel-softmax. arXiv preprint arXiv:1611.01144 (2016)
16. Jiang, Y., et al.: R2CNN: rotational region CNN for orientation robust scene text detection. arXiv preprint arXiv:1706.09579 (2017)
17. Kang, Z., Grauman, K., Sha, F.: Learning with whom to share in multi-task feature learning. In: International Conference on Machine Learning, pp. 521–528 (2011)
18. Karatzas, D., et al.: ICDAR 2015 competition on robust reading. In: International Conference on Document Analysis and Recognition, pp. 1156–1160. IEEE (2015)
19. Karatzas, D., et al.: ICDAR 2013 robust reading competition. In: International Conference on Document Analysis and Recognition, pp. 1484–1493. IEEE (2013)
20. Kumar, A., Daumé III, H.: Learning task grouping and overlap in multi-task learning. In: International Conference on Machine Learning, pp. 1723–1730 (2012)
21. LeCun, Y.: The MNIST database of handwritten digits. http://yann.lecun.com/exdb/mnist/ (1998)
22. Lee, C.Y., Osindero, S.: Recursive recurrent nets with attention modeling for OCR in the wild. In: IEEE Conference on Computer Vision and Pattern Recognition, pp. 2231–2239 (2016)
23. Liao, M., Lyu, P., He, M., Yao, C., Wu, W., Bai, X.: Mask textspotter: an end-to-end trainable neural network for spotting text with arbitrary shapes. IEEE Trans. Pattern Anal. Mach. Intell. **43**(2), 532–548 (2021)
24. Liao, M., Pang, G., Huang, J., Hassner, T., Bai, X.: Mask textspotter v3: segmentation proposal network for robust scene text spotting. arXiv preprint arXiv:2007.09482 (2020)
25. Liu, X., Liang, D., Yan, S., Chen, D., Qiao, Y., Yan, J.: FOTS: fast oriented text spotting with a unified network. In: IEEE Conference on Computer Vision and Pattern Recognition, pp. 5676–5685 (2018)
26. Long, S., He, X., Yao, C.: Scene text detection and recognition: the deep learning era. Int. J. Comput. Vis. **129**(1), 161–184 (2020). https://doi.org/10.1007/s11263-020-01369-0
27. Ma, J., et al.: Arbitrary-oriented scene text detection via rotation proposals. IEEE Trans. Multimedia **20**(11), 3111–3122 (2018)
28. Mehta, N., Liang, K., Verma, V.K., Carin, L.: Continual learning using a Bayesian nonparametric dictionary of weight factors. In: International Conference on Artificial Intelligence and Statistics (2021)

29. Nayef, N., et al.: ICDAR 2019 robust reading challenge on multi-lingual scene text detection and recognition-RRC-MLT-2019. In: International Conference on Document Analysis and Recognition, pp. 1582–1587. IEEE (2019)
30. Nayef, N., et al.: ICDAR 2017 robust reading challenge on multi-lingual scene text detection and script identification-RRC-MLT. In: International Conference on Document Analysis and Recognition, vol. 1, pp. 1454–1459. IEEE (2017)
31. Parisi, G.I., Kemker, R., Part, J.L., Kanan, C., Wermter, S.: Continual Lifelong Learning with Neural Networks: a Review. Neural Netw. **113**, 54–71 (2019)
32. Qin, S., Bissacco, A., Raptis, M., Fujii, Y., Xiao, Y.: Towards unconstrained end-to-end text spotting. In: International Conference on Computer Vision, pp. 4704–4714 (2019)
33. Ren, S., He, K., Girshick, R., Sun, J.: Faster R-CNN: towards real-time object detection with region proposal networks. In: Advances in Neural Information Processing Systems, pp. 91–99 (2015)
34. Shi, B., Wang, X., Lyu, P., Yao, C., Bai, X.: Robust scene text recognition with automatic rectification. In: IEEE Conference on Computer Vision and Pattern Recognition, pp. 4168–4176 (2016)
35. Shi, B., et al.: ICDAR 2017 competition on reading Chinese text in the wild (RCTW-17). In: International Conference on Document Analysis and Recognition (2017)
36. Strezoski, G., van Noord, N., Marcel, W.: Learning task relatedness in multi-task learning for images in context. In: Proceedings of the International Conference on Multimedia Retrieval, pp. 78–86 (2019)
37. Su, B., Lu, S.: Accurate scene text recognition based on recurrent neural network. In: Cremers, D., Reid, I., Saito, H., Yang, M.-H. (eds.) ACCV 2014. LNCS, vol. 9003, pp. 35–48. Springer, Cham (2015). https://doi.org/10.1007/978-3-319-16865-4_3
38. Sun, Y., et al.: ICDAR 2019 competition on large-scale street view text with partial labeling - RRC-LSVT. In: International Conference on Document Analysis and Recognition (2019)
39. Xing, L., Tian, Z., Huang, W., Scott, M.R.: Convolutional character networks. In: International Conference on Computer Vision (2019)
40. Yuliang, L., Lianwen, J., Shuaitao, Z., Sheng, Z.: Detecting curve text in the wild: new dataset and new solution. arXiv preprint arXiv:1712.02170 (2017)
41. Zamir, A.R., Sax, A., Shen, W., Guibas, L.J., Malik, J., Savarese, S.: Taskonomy: disentangling task transfer learning. In: IEEE Conference on Computer Vision and Pattern Recognition, pp. 3712–3722 (2018)
42. Zeiler, M.D., Fergus, R.: Visualizing and understanding convolutional networks. In: Fleet, D., Pajdla, T., Schiele, B., Tuytelaars, T. (eds.) ECCV 2014. LNCS, vol. 8689, pp. 818–833. Springer, Cham (2014). https://doi.org/10.1007/978-3-319-10590-1_53
43. Zhong, S., Pu, J., Jiang, Y.G., Feng, R., Xue, X.: Flexible multi-task learning with latent task grouping. Neurocomputing **189**, 179–188 (2016)

Incorporating Self-attention Mechanism and Multi-task Learning into Scene Text Detection

Ning Ding[1,2], Liangrui Peng[1,2(✉)], Changsong Liu[2], Yuqi Zhang[3], Ruixue Zhang[3], and Jie Li[3]

[1] Beijing National Research Center for Information Science and Technology, Tsinghua University, Beijing, China
[2] Department of Electronic Engineering, Tsinghua University, Beijing, China
dn22@mails.tsinghua.edu.cn, {penglr,lcs}@tsinghua.edu.cn
[3] Shanghai Pudong Development Bank, Shanghai, China
{zhangyq26,zhangrx5,lij131}@spdb.com.cn

Abstract. In recent years, Mask R-CNN based methods have achieved promising performance on scene text detection tasks. This paper proposes to incorporate self-attention mechanism and multi-task learning into Mask R-CNN based scene text detection frameworks. For the backbone, self-attention-based Swin Transformer is adopted to replace the original backbone of ResNet, and a composite network scheme is further utilized to combine two Swin Transformer networks as a backbone. For the detection heads, a multi-task learning method by using cascade refinement structure for text/non-text classification, bounding box regression, mask prediction and text line recognition is proposed. Experiments are carried out on the ICDAR MLT 2017 & 2019 datasets, which show that the proposed method has achieved improved performance.

Keywords: Scene text detection · Attention mechanism · Multi-task learning

1 Introduction

Scene text detection is one of the fundamental tasks in computer vision. In an open environment, it is challenging due to variations in text scripts, contents, font-types, backgrounds, illuminations and perspective distortions, as shown in Fig. 1. To overcome the challenges in scene text detection tasks, it is essential to solve two problems, i.e. (i) how to find distinct feature representations for scene text? (ii) how to formulate the optimization problem for scene text detection?

With the emergence of deep learning technologies, various deep neural network based models have been utilized for scene text detection, including Mask R-CNN, etc. However, Mask R-CNN is originally proposed for general object detection. As the appearances of scene text are different from those of general

© The Author(s), under exclusive license to Springer Nature Switzerland AG 2023
L. Karlinsky et al. (Eds.): ECCV 2022 Workshops, LNCS 13804, pp. 314–328, 2023.
https://doi.org/10.1007/978-3-031-25069-9_21

objects in images, it is feasible to improve feature representation and optimization problem formulation for Mask R-CNN framework by exploiting characteristics of scene text.

Human vision excels in analyzing characteristics of scene text. In human visual cognition, attention mechanism is important to focus on specific information, and the processes of detection and recognition of text are closely coupled. Based on these considerations, we focus on incorporating self-attention mechanism and multi-task learning into Mask R-CNN based scene text detection frameworks.

Fig. 1. Examples of scene text images

The main contributions of this paper are two folds: (i) To improve feature representation ability, we design a backbone by using composite structure with self-attention mechanism; (ii) To improve the optimization problem formulation, we exploit multi-task learning with cascade refinement structure for detection heads of Mask R-CNN framework.

Experiments are mainly carried out on the ICDAR MLT 2017 public scene text dataset. Compared with the Mask R-CNN baseline model, the proposed method can improve detection accuracy. In addition to scene text detection task, experimental results for the end-to-end text detection and recognition task on the ICDAR MLT 2019 dataset are also reported.

2   Related Work

2.1   Mask R-CNN

Mask R-CNN [8] is a two-stage detection framework extended from Faster R-CNN. In addition to the two branches of the original detection network (classification, bounding box regression), a mask prediction head is added for pixel-level semantic image segmentation. Because Mask R-CNN is anchor-based, it is often necessary to adjust the number of anchors, anchor scales and ratios by using statistics information in training samples. Anchor-free methods, such as Center-Net [30], are different from Mask R-CNN. CenterNet detects the center point of an object and predicts the width and height of the bounding box. Center-Net2 [29] further combines the merits of both one-stage and two-stage detector by proposing a probabilistic two-stage detector.

To enhance the performance of the Mask R-CNN based methods, many improved schemes have been proposed to modify the backbone or detection heads. For example, Composite Backbone Network (CBNet) [13] assembles multiple identical backbones by composite connections between the adjacent backbones. CBNetV2 [12] is a further improvement of CBNet. One of the key ideas in CBNetV2 is to use assistant supervision during the training of the CBNet-based models. To deal with the case of dense text, Qin et al. [20] propose to use an MLP decoder in the mask head of Mask R-CNN.

In terms of detection network structure, Hybrid Task Cascade (HTC) [2] leverages the reciprocal relationship between detection and segmentation and interweaves them for a joint multi-stage processing. Unlike the method of using multiple detection heads and cascading, LOMO [28] localizes the text progressively for multiple times.

In our methods, Mask R-CNN is used as baseline model. The composite backbone network and hybrid task cascade detection network are explored. Different detection networks including CenterNet2 and other improved Mask R-CNN frameworks are investigated in our experiments.

2.2 Attention Based Methods

From a technical point of view, attention mechanism can be viewed as a dynamic weight adjustment process for better feature representation. Recently, various attention mechanism based deep learning methods have been proposed [7]. Transformer [24] with self-attention mechanism has been successfully used in both natural language processing (NLP) and computer vision tasks.

For scene text detection, Tang et al. [23] present a simple yet effective Transformer-based architecture to avoid the disturbance by background. Hu et al. [9] propose a Transformer-Convolution Network (TCNet) with a CNN module and a Transformer module for scene text detection task. The CNN module is used to extract local features from the input images, while the Transformer module establishes connections among various local features. Double Supervised Network with Attention Mechanism (DSAN) [6] incorporates one text attention module during feature extraction. Under the supervision of two branches, DSAN can focus on text regions and recognize text in arbitrary length.

For scene text recognition, self-attention mechanism has also been used, such as MEAN [26] with multi-element attention-based encoder-decoder. PREN2D [27] can further alleviate the alignment problem of attention mechanism, by introducing primitive learning based parallel decoding.

Swin Transformer [14] with shifted window-based hierarchical structure can be used as a general backbone in the field of computer vision. By using different channel number and layer numbers, Swin Transformer has different architecture variants including Swin-T, Swin-S, Swin-B and Swin-L. The complexity of Swin-T and Swin-S is similar to ResNet-50 and ResNet-101, respectively.

In our methods, we use Swin Transformer network to replace ResNet in Mask R-CNN, and design a composite network scheme to combine two Swin Transformer networks as a backbone.

2.3   Multi-task Learning Based Methods

As the process of text detection in human visual cognition involves multiple stages, especially for long text and irregular text, it is feasible to incorporate multi-task learning in scene text detection. Previous experimental results [1] on the ICDAR 2015 and ICDAR 2013 datasets show that training both text detection and recognition in a single framework outperforms the method that combines two optimal text detection and text recognition networks. By borrowing the idea from the functional organization of human brain, Dasgupta et al. [4] propose to add a Feature Representation Block (FRB) in a stratified manner to learn task specific features for multi-task learning based scene text spotting. Kittenplon et al. [11] propose a Transformer-based approach named TextTranSpotter for text spotting by jointly optimizing the text detection and recognition components with both fully- and weakly-supervised training settings.

Multi-task learning methods have also been applied to other tasks. Sarshogh et al. [22] propose an end-to-end trainable multi-task network for text extraction in complex documents, which consists of three heads: text localization, classification, and text recognition. A multi-task refinement head called MTR-Head [10] is developed by Huang et al. to solve the problem of feature misalignment between different tasks in object detection. For text-video retrieval, Wu et al. [25] incorporate a multi-task learning approach to optimize the construction of text-video common learning space.

In our methods, we adopt a cascaded network configuration to incorporate multi-task learning into Mask R-CNN frameworks. The text line recognition module is based on Transformer [24].

3   Methodology

The system framework is shown in Fig. 2. The composite backbone module is based on self-attention mechanism to improve feature representation ability. In addition to the original text/non-text classification, bounding box regression and mask prediction heads in Mask R-CNN framework, a text-line recognition head is added to enhance multi-task learning. A cascade refinement scheme for text detection is further utilized for text/non-text classification, bounding box regression, mask prediction and text line recognition.

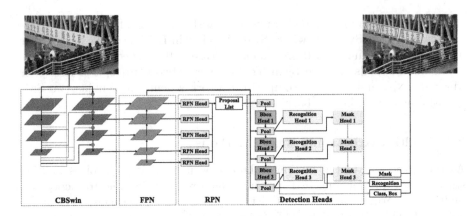

Fig. 2. System framework

3.1   Self-attention Mechanism-based Backbone

In order to obtain a more powerful backbone under limited time and resource conditions, we follow the CBNet method proposed by Liang et al. [12]. Two identical pre-trained backbones, i.e. a lead backbone and an assisting backbone, are integrated to build a backbone for scene text detection, which is named as CBSwin (Composite Backbone Swin Transformer Network). The output of each stage in the assisting backbone becomes the input of the corresponding stage of the lead backbone. The output of the composite backbone combines high-level and low-level features, and can be used as multi-scale features for subsequent FPN (feature pyramid network).

We utilize two Swin Transformer networks to form the composite backbone. Swin Transformer conducts the computing of self-attention mechanism inside local window for an input image. In order to generate information interaction between different windows, the window needs to be shifted. For implementation of fast computing, the effect of shifting window can be fulfilled by shifting the input image.

The calculation of self-attention mechanism is shown in Eq. (1).

$$
\begin{aligned}
O^{C \times n} \\
&= V^{C \times n} \cdot Softmax(\frac{(K^{C \times n})^T \cdot Q^{C \times n}}{\sqrt{C}} + B^{n \times n}) \\
&= W_V^{C \times C} \cdot I^{C \times n} \cdot Softmax(\frac{(I^{C \times n})^T \cdot (W_K^{C \times C})^T \cdot W_Q^{C \times C} \cdot I^{C \times n}}{\sqrt{C}} + B^{n \times n})
\end{aligned}
\tag{1}
$$

where I denotes the input data of the Transformer, which dimension is $C \times n$, O denotes the output data, which dimension is also $C \times n$; W_K, W_Q, W_V represents the weights that can be learned, and their dimensions are all

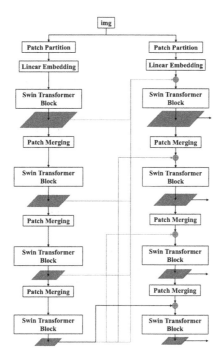

Fig. 3. CBSwin (Composite Backbone Swin Transformer Network)

$C \times C$. B denotes the positional encoding. When using the multi-head attention mechanism, the number of heads is denoted by *num_heads*. For each head, the calculation of self-attention mechanism is shown in Eq. (2).

$$O_i^{d \times n} =$$
$$W_{V_i}^{d \times C} \cdot I^{C \times n} \cdot Softmax(\frac{(I^{C \times n})^T \cdot (W_{K_i}^{d \times C})^T \cdot W_{Q_i}^{d \times C} \cdot I^{C \times n}}{\sqrt{d}} + B^{n \times n}), \quad (2)$$
$$i = 1, 2, \ldots, num\_heads$$

The final output is the multiple $O^{d \times n}$ concatenated as shown in Eq. (3).

$$O^{C \times n} = [O_1^{d \times n}; O_2^{d \times n}; \ldots; O_{num\_heads}^{d \times n}],$$
$$d \times num\_heads = C \tag{3}$$

We use two variants of Swin Transformer networks including Swin-T and Swin-S in the composite backbone, which correspond to CBSwin-T and CBSwin-S respectively in terms of complexity. Similar to CBNetV2, we also add assistant supervision of detection task for the assisting network in CBSwin-T, which corresponds to CBSwin-T-V2. The detailed network structure of CBSwin is shown in Fig. 3.

3.2 Multi-task Cascade Refinement Text Detection

Multi-task Learning. Mask R-CNN has three heads including classification, bounding box regression, and mask prediction. To further adapt to the specific requirements of both scene text detection and recognition tasks, we introduce a Transformer-based text line recognition head into the detection network. The entire network is trained in an end-to-end manner.

The feature maps obtained by the RPN are input into the Transformer for recognition. Text recognition modules generally perform better for horizontal text instances. However, in scene text detection tasks, vertical text instances are often encountered. For the feature maps obtained by the RPN network, the text direction is determined according to the aspect ratio of corresponding region proposals, and then a 90° rotation operation is applied for vertical text instances.

Cascade Refinement Structure. Inspired by the idea of introducing a cascade structure in the detection network [2], we utilize a cascade refinement structure with multiple duplicated stages for multi-task learning, which includes text/non-text classification, bounding box regression, mask prediction and text line recognition. The cascade refinement structure is shown in Fig. 4.

In the cascade network, information from different stages is reused. The bounding box regression results in previous stage is used to update the ROI pooling information in the current stage. The output of the mask prediction head in previous stage is added to the input of the mask prediction head in the current stage.

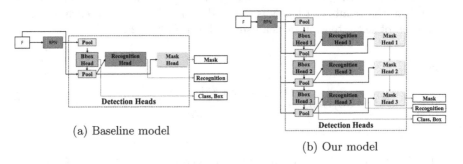

(a) Baseline model

(b) Our model

Fig. 4. Cascade refinement structure of detection heads. F represents the extracted features

4 Experiments

4.1 Experiment Setup

Experiments are carried out on the ICDAR MLT 2017 [16] and ICDAR MLT 2019 [17] datasets.

The ICDAR MLT 2017 [16] dataset was used in the ICDAR 2017 Competition on Multi-lingual scene text detection and script identification, which consists of 9000 training images and 9000 testing images. The text instances in the images are from 9 languages representing 6 different scripts, and text regions in the images can be in arbitrary orientations.

The ICDAR MLT 2019 [17] dataset was used in the ICDAR 2019 Robust Reading Challenge on Multi-lingual scene text detection and recognition, which consists of 10000 training images and 10000 testing images. The text instances in the images are from 10 languages representing 7 different scripts.

The performance of our methods is evaluated for the text detection task on the ICDAR MLT 2017 [16] dataset, and end-to-end scene text detection and recognition task on ICDAR MLT 2019 dataset. We use F_1 score, Precision (P) and Recall (R) as the evaluation metrics. Detailed ablation studies on different network configurations are further reported by exploring the influence of the attention mechanism and multi-task learning in our methods.

Our backbone is pre-trained on the ImageNet [5] dataset, and our model is fine-tuned on the training set of ICDAR MLT 2017 dataset. Data augmentation techniques are utilized, including color space transformations, geometric transformation, etc. Our methods are implemented by using the PyTorch [18] deep learning framework and the open source object detection toolbox MMDetection [3]. The AdamW optimizer [15] with a weight decay of 0.05 is adopted during training. A batch size of 8, and an initial learning rate of 1e-4 are used. Our model is usually fine-tuned for 12 epochs unless otherwise indicated. For multi-scale testing, the long side of an image is set to 1920 and 2560 respectively. The testing environment is Ubuntu 18.04 with Nvidia Tesla P100 GPUs.

4.2 Main Results

Scene Text Detection. The scene text detection results of our method and other SOTA methods on the test set of the ICDAR MLT 2017 dataset are shown in Table 1. Our model can achieve an F_1 score of 79.84% of for scene text detection task on the test set of the ICDAR MLT 2017 dataset.

Table 1. Text detection results on the ICDAR MLT 2017 test set

Method	F_1 score (%)	P (%)	R (%)
Raisi et al. [21]	72.4	84.8	63.2
Zhang et al. [23]	79.6	**87.3**	73.2
Ours	**79.84**	85.20	**75.12**

End-to-end Scene Text Detection and Recognition. Our method is also tested on the end-to-end text detection and recognition task, and has achieved an F_1 score of 59.72% on the test set of the ICDAR MLT 2019 dataset, as Table 2 shows. Examples of the text detection and recognition results are shown in Fig. 5.

Table 2. End-to-end text detection and recognition results on the ICDAR MLT 2019 test set

Method	F_1 score (%)	P (%)	R (%)
Tencent-DPPR Team & USTB-PRIR [17]	59.15	71.26	50.55
CRAFTS [17]	51.74	65.68	42.68
Ours	**59.72**	**72.21**	**50.92**

(a) Detection task

(b) End-to-end detection and recognition task

Fig. 5. Examples of detection and recognition results

4.3 Ablation Studies

Ablation Studies are carried out on scene text detection task of the ICDAR MLT 2017 dataset. Different configurations of backbones and detection heads are compared.

Comparisons of Different Backbones. The baseline model is the Mask R-CNN structure using ResNet-50 as a backbone. We compare different settings for backbones, i.e. ResNet-50, Swin Transformer networks (Swin-T & Swin-S), Swin Transformer networks with composite backbones (CBSwin-T & CBSwin-S), and CBSwin-T with assistant supervision (CBSwin-T-V2). The results are shown in Table 3.

Table 3. Comparison of different backbones

Backbone	F_1 score (%)	P (%)	R (%)
ResNet-50	68.03	76.71	61.11
Swin-T	68.42	74.42	63.32
Swin-S	68.92	74.66	64.00
CBSwin-T	70.48	77.58	64.57
CBSwin-T-V2	70.65	77.68	64.78
CBSwin-S	**71.08**	**78.55**	**64.91**

The results show that Swin Transformer has achieved better performance than ResNet. Using the composite network structure (CBNet) for the backbone can further improve the performance. The detection accuracy of CBSwin-T-V2 is slightly better than that of CBSwin-T, but the time cost for training is increased. Thus the network configuration similar to CBNetV2 is not used in the subsequent experiments due to time limitation.

Table 4. Performance of scene text detection with different detection heads

Backbone	Detector	Recognizer	Multi-scale	Epochs	F_1 score (%)	P (%)	R (%)
CBSwin-S	CenterNet2-mask	–	–	12	70.89	**89.81**	58.55
CBSwin-S	Mask R-CNN	–	–	12	71.08	78.55	64.91
CBSwin-S	Mask R-CNN	–	✓	12	76.86	76.00	77.74
CBSwin-S	Mask R-CNN	–	✓	36	77.13	77.51	76.76
CBSwin-S	Mask R-CNN	PREN2D	✓	36	77.58	75.43	**79.85**
CBSwin-S	Mask R-CNN	Transformer	✓	36	77.92	76.97	78.88
CBSwin-S	HTC	–	–	12	71.17	81.17	63.36
CBSwin-S	HTC	–	✓	12	78.11	83.48	73.39
CBSwin-S	HTC	–	✓	36	78.51	89.00	70.23
CBSwin-S	HTC	PREN2D	✓	36	79.37	84.90	74.51
CBSwin-S	HTC	Transformer	✓	36	**79.84**	85.20	75.12

Comparisons of Different Detection Heads. We also conduct experiments on the ICDAR MLT 2017 dataset to compare different detection heads, including CenterNet2 [29] with additional mask head, Mask R-CNN and the detection heads with cascade refinement structure. The experimental results are shown in Table 4. Compared with CenterNet2 and Mask R-CNN, the detection heads with cascade refinement structure (denoted as HTC in Table 4) can further improve the performance for scene text detection.

For text line recognition modules, we compare Transformer [24] and PREN2D [27], as is also shown in Table 4. By adding the text line recognition module with end-to-end training, the overall performance of the detector is improved. The end-to-end text detection and recognition model with Transformer has achieved the highest F_1 score for scene text detection task in our experiments on the ICDAR MLT 2017 dataset.

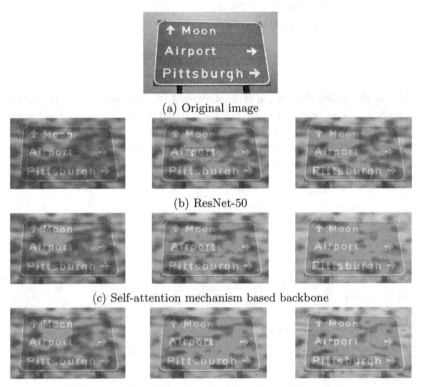

(a) Original image

(b) ResNet-50

(c) Self-attention mechanism based backbone

(d) The model after integrating the recognition module

Fig. 6. Visualization of self-attention mechanism

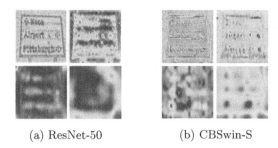

(a) ResNet-50   (b) CBSwin-S

Fig. 7. Averaged feature maps for different channels

4.4 Visualization

Effect of Self-attention Mechanism. We adopt the D-RISE method [19] to visualize the effect of self-attention mechanism. The D-RISE method [19] treats the detector as a black box. A series of masks are added to the input image. The effects of masking randomized regions on the predicted output are measured and used as the weights of masks. The weighted sum of masks corresponding to a text instance is used as a saliency map for visualization [19].

The visualization of self-attention mechanism is shown in Fig. 6. Compared with using ResNet, the detector based on self-attention mechanism can pay more attention to the global feature representation for the input image. After integrating text line recognition module, the model can focus better on textual regions.

Visualization of Feature Maps. The feature maps of different backbones are shown in Fig. 7. Compared with using ResNet, it seems that the composite backbone network based on the self-attention mechanism can extract both global and local textual information for detection tasks.

4.5 Inference Speed

The inference speed of our methods is shown in Table 5. The size of test images is 1280*720, and the batchsize is set to 1.

Table 5. Comparison on inference speed

Method	FPS
ResNet-50	11.4
CBSwin-S	6.8
CBSwin-S with HTC	6.3

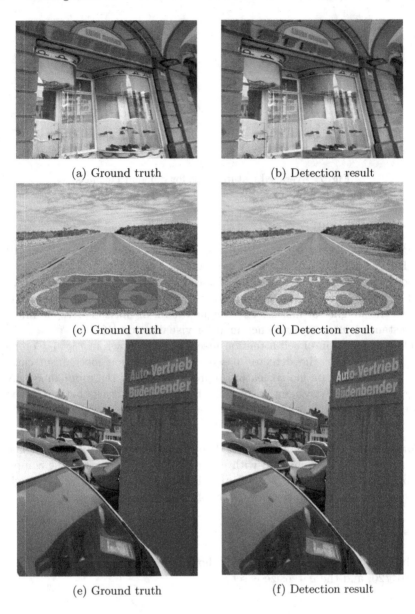

(a) Ground truth   (b) Detection result

(c) Ground truth   (d) Detection result

(e) Ground truth   (f) Detection result

Fig. 8. Examples of missing or inaccurate detection errors

4.6 Error Analysis

As scene text images are complex, there are usually false alarms, missing text instances and inaccurate detection results. Some examples are shown in Fig. 8. An example of false alarms is shown in Fig. 8(b), which is possibly caused by background regions with similar textual appearances. Missing text instances and inaccurate detection results are usually caused by perspective distortions and low quality text images, as Fig. 8(d) and (f) show. Figure 9 shows that our model does not perform well on some vertical or slanted text instances.

Fig. 9. Examples of detection errors of vertical or slant text

5   Conclusion

This paper investigates to incorporate self-attention mechanism and multi-task learning into Mask R-CNN framework for scene text detection. For feature extraction, a composite backbone of two Swin Transformer networks with self-attention mechanism is designed. For detection heads, a cascade refinement structure with multi-task learning is utilized. Experimental results on public scene text datasets show that the proposed method has achieved improved performance. Future work will further explore different forms of attention mechanisms for scene text detection and recognition.

References

1. Bušta, M., Neumann, L., Matas, J.: Deep TextSpotter: an End-to-End Trainable Scene Text Localization and Recognition Framework. In: ICCV, pp. 2223–2231 (2017)
2. Chen, K., Pang, J., Wang, J., et al.: Hybrid task cascade for instance segmentation. In: CVPR, pp. 4974–4983 (2019)
3. Chen, K., Wang, J., Pang, J., et al.: MMDetection: open MMLab detection toolbox and benchmark. arXiv preprint arXiv:1906.07155 (2019)
4. Dasgupta, K., Das, S., Bhattacharya, U.: Stratified multi-task learning for robust spotting of scene texts. In: ICPR, pp. 3130–3137 (2021)
5. Deng, J., Dong, W., Socher, R., et al.: ImageNet: a large-scale hierarchical image database. In: CVPR, pp. 248–255 (2009)
6. Gao, Y., Huang, Z., Dai, Y., et al.: DSAN: double supervised network with attention mechanism for scene text recognition. In: VCIP, pp. 1–4 (2019)
7. Guo, M.H., Xu, T.X., Liu, J.J., et al.: Attention mechanisms in computer vision: a survey. Comput. Vis. Media **8**(3), 331–368 (2022)

8. He, K., Gkioxari, G., Dollár, P., et al.: Mask R-CNN. In: ICCV, pp. 2961–2969 (2017)
9. Hu, Y., Zhang, Y., Yu, W., et al.: Transformer-convolution network for arbitrary shape text detection. In: ICMLSC, pp. 120–126 (2022)
10. Huang, Z., Li, W., Xia, X.G., et al.: A novel nonlocal-aware pyramid and multiscale multitask refinement detector for object detection in remote sensing images. IEEE Trans. Geosci. Remote Sens. **60**, 1–20 (2021)
11. Kittenplon, Y., Lavi, I., Fogel, S., et al.: Towards weakly-supervised text spotting using a multi-task transformer. In: CVPR, pp. 4604–4613 (2022)
12. Liang, T., Chu, X., Liu, Y., et al.: CBNetV2: a composite backbone network architecture for object detection. arXiv preprint arXiv:2107.00420 (2021)
13. Liu, Y., Wang, Y., Wang, S., et al.: CBNet: a novel composite backbone network architecture for object detection. In: AAAI, pp. 11653–11660 (2020)
14. Liu, Z., Lin, Y., Cao, Y., et al.: Swin transformer: hierarchical vision transformer using shifted windows. In: ICCV, pp. 10012–10022 (2021)
15. Loshchilov, I., Hutter, F.: Decoupled weight decay regularization. In: ICLR (2019)
16. Nayef, N., Yin, F., Bizid, I., et al.: ICDAR2017 robust reading challenge on multilingual scene text detection and script identification-RRC-MLT. In: ICDAR, pp. 1454–1459 (2017)
17. Nayef, N., Patel, Y., Busta, M., et al.: ICDAR2019 robust reading challenge on multi-lingual scene text detection and recognition-RRC-MLT-2019. In: ICDAR, pp. 1582–1587 (2019)
18. Paszke, A., Gross, S., Massa, F., et al.: PyTorch: an imperative style, high-performance deep learning library. Adv. Neural. Inf. Process. Syst. **32**, 8024–8035 (2019)
19. Petsiuk, V., Jain, R., Manjunatha, V., et al.: Black-box explanation of object detectors via saliency maps. In: CVPR, pp. 11443–11452 (2021)
20. Qin, X., Zhou, Y., Guo, Y., et al.: Mask is all you need: rethinking mask R-CNN for dense and arbitrary-shaped scene text detection. In: ACM MM, pp. 414–423 (2021)
21. Raisi, Z., Naiel, M.A., Younes, G., et al.: Transformer-based text detection in the wild. In: CVPR Workshop, pp. 3162–3171 (2021)
22. Sarshogh, M.R., Hines, K.: A multi-task network for localization and recognition of text in images. In: ICDAR, pp. 494–501 (2019)
23. Tang, J., Zhang, W., Liu, H., et al.: few could be better than all: feature sampling and grouping for scene text detection. In: CVPR, pp. 4563–4572 (2022)
24. Vaswani, A., Shazeer, N., Parmar, N., et al.: Attention is all you need. In: NIPS (2017)
25. Wu, X., Wang, T., Wang, S.: Cross-modal learning based on semantic correlation and multi-task learning for text-video retrieval. Electronics **9**(12), 2125 (2020)
26. Yan, R., Peng, L., Xiao, S., et al.: MEAN: multi-element attention network for scene text recognition. In: ICPR, pp. 6850–6857 (2021)
27. Yan, R., Peng, L., Xiao, S., et al.: Primitive representation learning for scene text recognition. In: CVPR, pp. 284–293 (2021)
28. Zhang, C., Liang, B., Huang, Z., et al.: Look more than once: an accurate detector for text of arbitrary shapes. In: CVPR, pp. 10552–10561 (2019)
29. Zhou, X., Koltun, V., Krähenbühl, P.: Probabilistic two-stage detection. arXiv preprint arXiv:2103.07461 (2021)
30. Zhou, X., Wang, D., Krähenbühl, P.: Objects as points. arXiv preprint arXiv:1904.07850 (2019)

Doc2Graph: A Task Agnostic Document Understanding Framework Based on Graph Neural Networks

Andrea Gemelli[1]([✉])[ID], Sanket Biswas[2][ID], Enrico Civitelli[1][ID], Josep Lladós[2][ID], and Simone Marinai[1][ID]

[1] Dipartimento di Ingegneria dell'Informazione (DINFO),
Università degli studi di Firenze, Florence, Italy
{andrea.gemelli,enrico.civitelli,simone.marinai}@unifi.it
[2] Computer Vision Center and Computer Science Department,
Universitat Autònoma de Barcelona, Bellaterra, Spain
{sbiswas,josep}@cvc.uab.es

Abstract. Geometric Deep Learning has recently attracted significant interest in a wide range of machine learning fields, including document analysis. The application of Graph Neural Networks (GNNs) has become crucial in various document-related tasks since they can unravel important structural patterns, fundamental in key information extraction processes. Previous works in the literature propose task-driven models and do not take into account the full power of graphs. We propose Doc2Graph, a task-agnostic document understanding framework based on a GNN model, to solve different tasks given different types of documents. We evaluated our approach on two challenging datasets for key information extraction in form understanding, invoice layout analysis and table detection. Our code is freely accessible on https://github.com/andreagemelli/doc2graph.

Keywords: Document analysis and recognition · Graph neural networks · Document understanding · Key information extraction · Table detection

1 Introduction

Document Intelligence deals with the ability to read, understand and interpret documents. Document understanding can be backed by graph representations, that robustly represent objects and relations. Graph reasoning for document parsing involves manipulating structured representations of semantically meaningful document objects (titles, tables, figures) and relations, using compositional rules. Customarily, graphs have been selected as an adequate framework for leveraging structural information from documents, due to their inherent representational power to codify the object components (or semantic entities) and their pairwise relationships. In this context, recently graph neural networks (GNNs)

© The Author(s), under exclusive license to Springer Nature Switzerland AG 2023
L. Karlinsky et al. (Eds.): ECCV 2022 Workshops, LNCS 13804, pp. 329–344, 2023.
https://doi.org/10.1007/978-3-031-25069-9_22

have emerged as a powerful tool to tackle the problems of Key Information Extraction (KIE) [6,35], Document Layout Analysis (DLA) which includes well-studied sub-tasks like table detection [25,26], table structure recognition [20,34] and table extraction [9], Visual Question Answering (VQA) [17,18], synthetic document generation [4] and so on.

Simultaneously, the common state-of-the-art practice in the document understanding community is to utilize the power of huge pre-trained vision-language models [1,32,33] that learn whether the visual, textual and layout cues of the document are correlated. Despite achieving superior performance on most document understanding tasks, large-scale document pre-training comes with a high computational cost both in terms of memory and training time. We present a solution that does not rely on huge vision-language model pre-training modules, but rather recognizes the semantic text entities and their relationships from documents exploiting graphs. The solution has experimented on two challenging benchmarks for forms [15] and invoices [10] with a very small amount of labeled training data.

Inspired by some prior works [8,25,26], we introduce *Doc2Graph*, a novel task-agnostic framework to exploit graph-based representations for document understanding. The proposed model is validated in three different challenges, namely KIE in form understanding, invoice layout analysis and table detection. A graph representation module is proposed to organize the document objects. The graph nodes represent words or the semantic entities while edges the pairwise relationships between them. Finding the optimal set of edges to create the graph is anything but trivial: usually in literature heuristics are applied, e.g. using a visibility graph [25]. In this work, we do not make any assumption a priori on the connectivity: rather we attempt to build a fully connected graph representation over documents and let the network learn by itself what is relevant.

In summary, the primary contributions of this work can be summarized as follows:

- Doc2Graph, the first task-agnostic GNN-based document understanding framework, evaluated on two challenging benchmarks (form and invoice understanding) for three significant tasks, without any requirement of huge pre-training data;
- We propose a general graph representation module for documents, that do not rely on heuristics to build pairwise relationships between words or entities;
- A novel GNN architectural pipeline with node and edge aggregation functions suited for documents, that exploits the relative positioning of document objects through polar coordinates.

The rest of the paper is organized as follows. In Sect. 2 we review the state-of-the-art in graph representation learning and vision-language models for document understanding. Section 3 provides the details of the main methodological contribution. The experimental evaluation is reported in Sect. 4. Finally, the conclusions are drawn in Sect. 5.

2   Related Work

Document understanding has been studied extensively in the last few years, owing to the advent of deep learning, but has been reformulated in a recent survey by Borchmann et al. [5]. The tasks range from KIE performed for understanding forms [15], receipts [14] and invoices [10], to multimodal comprehension of both visual and textual cues in a document for classification [32,33]. It also includes the DLA task where recent works focus on building an end-to-end framework for both detection and classification of page regions [2,3]. Table detection [25,26], structure recognition [20,24] and extraction [9,30] in DLA gathered some special attention in recent years due to the high variability of layouts that make the both necessary to be solved and challenging to be tackled. In addition, question answering [21,29] has emerged as an extension of the KIE task principle, where a natural language question replaces a property name. Current state-of-the-art approaches [1,13,22,32,33] on these document understanding tasks have utilized the power of large pre-trained language models, relying on language more than the visual and geometrical information in a document and also end up using hundreds of millions of parameters in the process. Moreover, most of these models are trained with a huge transformer pipeline, which requires an immense amount of data during pre-training. In this regard, Davis et al. [7] and Sarkar et al. [28] proposed language-agnostic models. In [7] they focused on the entity relationship detection problem in forms [15] using a simple CNN as a text line detector and then detecting key-value relationship pairs using a heuristic based on each relationship candidate score generated from the model. Sarkar et al. [28] rather focused on extracting the form structure by reformulating the problem as a semantic segmentation (pixel labeling) task. They used a U-Net based architectural pipeline, predicting all levels of the document hierarchy in parallel, making it quite efficient.

GNN for document understanding was first introduced for mainly key DLA sub-tasks that include table detection [25] and table structure recognition [23]. The key idea behind its introduction was to utilize the powerful geometrical characteristics of a document using GNN and then to preserve the privacy of confidential textual content (especially for administrative documents) during training, making the model language-independent and more structure-reliant as proposed in [25] for detection of tables in invoices. Carbonell et al. [6] used graph convolutional networks (GCNs) to solve the entity(word) grouping, labeling and entity linking tasks for form analysis. They used the information of the bounding boxes and word embeddings as the principal node features and do not include any visual features, while they used k-nearest neighbours (KNNs) to encode the edge information. The FUDGE [8] framework was then developed for form understanding as an extension of [7] to greatly improve the state-of-the-art on both the semantic entity labeling and entity linking tasks by proposing relationship pairs using the same detection CNN as in [7]. Then a graph convolutional network (GCN) was deployed with plugged visual features from the CNN so that semantic labels for the text entities were predicted jointly with the key-value relationship pairs, as they are quite related tasks.

Inspired by this influential prior work [8], we aim to propose a task-agnostic GNN-based framework called *Doc2Graph* that adapts a similar joint prediction of both the tasks, semantic entity labeling and entity linking using a node classification and edge classification module respectively. Doc2Graph is established to tackle multiple challenges ranging from KIE for form understanding to layout analysis and table detection for invoice understanding, without needing any kind of huge data pre-training and being lightweight and efficient.

3   Method

In this section, we present the proposed approach. First, we describe the preprocessing step that converts document images into graphs. Then, we describe the GNN model designed to tackle different kinds of tasks.

3.1   Documents Graph Structure

A graph is a structure made of nodes and edges. A graph can be seen as a language model representing a document in terms of its segments (text units) and relationships. A preprocessing step is required. Depending on the task, different levels of granularity have to be considered for defining the constituent objects of a document. They can be single words or entities, that is, groups of words that share a certain property (e.g., the name of a company). In our work we try both as the starting point of the pipeline: we apply an OCR to recognize words, while a pre-trained object detection model for detecting entities. The chosen objects, once found, constitute the nodes of the graph.

At this point, nodes need to be connected through edges. Finding the optimal set of edges to create the graph is anything but trivial: usually in literature heuristics are applied, e.g. using a visibility graph [25]. These approaches: (i) do not generalise well on different layouts; (ii) strongly rely on the previous node detection processes, which are often prone to errors; (iii) generate noise in the connections, since bounding box of objects could cut out important relations or allow unwanted ones; (iv) exclude in advance sets of solutions, e.g. answers far from questions. To avoid those behaviours, we do not make any assumption a priori on the connectivity: we build a fully connected graph and we let the network learn by itself what relations are relevant.

3.2   Node and Edge Features

In order to learn, suitable features should be associated to nodes and edges of the graph. In documents, this information can be extracted from sources of different modalities, such as visual, language and layout ones. Different methods can be applied to encode a node (either word or entity) to enrich its representation. In our pipeline, with the aim to possibly keep it lightweight, we include:

- a language model to encode the text. We use the spaCy large English model to get word vector representations of words and entities;

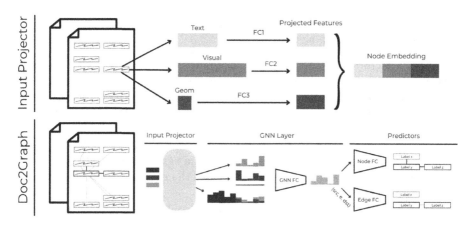

Fig. 1. Our proposed Doc2Graph framework. For visualisation purposes, the architecture shows the perspective of one node (the blue one in Doc2Graph).

- a visual encoder to represent style and formatting. We pretrain a U-Net [27] on FUNSD for entities segmentation. Since U-Net uses feature maps at different encoder's layers to segment the images, we decide to use all these information as visual features. Moreover, it is important to highlight that, for each features map, we used a RoI Alignment layer to extract the features relative to each entities bounding box;
- the absolute normalized positions of objects inside a document; layout and structure are meaningful features to include in industrial documents, e.g. for key-value associations.

As for the edges, to the best of our knowledge, we propose two new sets of features to help both the node and the edge classification tasks:

- a normalized euclidean distance between nodes, by means of the minimum distance between bounding boxes. Since we are using a fully connected graph this is crucial for the aggregation node function in use to keep locality property during the message passing algorithm;
- relative positioning of nodes using polar coordinates. Each source node is considered to be in the center of a Cartesian plane and all its neighbors are encoded by means of distance and angle. We discretize the space into bins (one-hot encoded), which number can be chosen, instead of using normalized angles: a continuous representation of the angle is challenging because, for instance, two points at the same distance with angles 360° and 0° would be encoded differently.

3.3 Architecture

Each node feature vector passes through our proposed architecture (Fig. 1, visualization of GNN layer inspired by "A Gentle Introduction to GNNs"): the connectivity defines the neighborhood for the message passing, while the weight learnable matrices are shared across all nodes and edges, respectively. We make use of four different components:

- Input Projector: this module applies as many fully connected (FC) layers as there are different modalities in use, to project each of their representations into inner spaces of the same dimension; e.g., we found it to be not very informative combine low dimensional geometrical features with high dimensional visual ones, as they are;
- GNN Layer: we make use of a slightly different version of GraphSAGE [11]. Using a fully connected graph, we redefine the aggregation strategy (Eq. 2);
- Node Predictor: this is a FC layer, that maps the representation of each node into the number of target classes;
- Edge Predictor: this is a two-FC layer, that assigns a label to each edge. To do so, we propose a novel aggregation on edges (Eq. 3).

GNN Layer. Our version of GraphSAGE slightly differs in the neighborhood aggregation. At layer l given a node i, h_i its inner representation and $N(i)$ its set of neighbors, the aggregation is defined as:

$$h_{N(i)}^{l+1} = aggregate(\{h_j^l, \forall j \in N(i)\}) \tag{1}$$

where *aggregate* can be any permutation invariant operation, e.g. sum or mean. Usually, in other domains, the graph structure is naturally given by the data itself but, as already stated, in documents this can be challenging (Sect. 3.1). Then, given a document, we redefine the above equation as:

$$h_{N(i)}^{l+1} = \frac{c}{|\Upsilon(i)|} \sum_{j \in \Upsilon(i)} h_j^l \tag{2}$$

where $\Upsilon(i) = \{j \in N(i) : |i - j| < threshold\}$, $|i - j|$ is the Euclidean distance of nodes i and j saved (normalized between 0 and 1) on their connecting edge, and c is a constant scale factor.

Edge Predictor. We consider each edge as a triplet (src, e, dst): e is the edge connecting the source (src) and destination (dst) node. The edge representation h_e to feed into the two-FC classifier is defined as:

$$h_e = h_{src} \parallel h_{dst} \parallel cls_{src} \parallel cls_{dst} \parallel e_{polar} \tag{3}$$

where h_{src} and h_{dst} are the node embeddings output of the last GNN layer, cls_{scr} and cls_{dst} are the softmax of the output logits of the previous node predictor layer, e_{polar} are the polar coordinates described in Sect. 3.2 and $\parallel$ is the concatenation operator. These choices have been made because: (i) relative positioning on edges is stronger compared to absolute positioning on nodes: the local property introduced by means of polar coordinates can be extended to different data, e.g. documents of different sizes or orientations; (ii) if the considered task comprise also the classification of nodes, their classes may help in the classification of edges, e.g. in forms it should not possible to find an answer connected to another answer.

Given the task, graphs can be either undirected or directed: both are represented with two or one directed edge between nodes, respectively. In the first case, the order does not matter and so the above formula can be redefined as:

$$h_e = (h_{src} + h_{dst}) \parallel cls_{src} \parallel cls_{dst} \parallel e_{polar} \tag{4}$$

4   Experiments and Results

In this chapter we present experiments of our method on two different datasets, FUNSD and RVL-CDIP invoices, to tackle three tasks: entity linking, layout analysis and table detection. We also discuss results compared to other methods.

4.1   Proposed Model

We performed ablation studies on our proposed model for entity linking on FUNSD without contribution and classification of nodes (Fig. 1), since we found it to be the most challenging task. In Table 1 we report different combinations of features and hyperparameters. Geometrical and textual features make the largest contribution, while visual features bring almost three points more to the Key-Value F1 score by an important increase in terms of network parameters (2.3 times more). Textual and geometrical features remain crucial for the task at hand, and their combination increase by a large amount both of their scores when used in isolation. This may be due to two facts: (i) our U-Net has not been included during the GNN training time (as done in [8]), unable to adjust the representation for spotting key-value relationship pairs; (ii) the segmentation task used to train the backbone do not yield useful features for that goal (as shown in Table 1). The hyperparameters shown in the table refer to the edge predictor (EP) inner layer input dimension and the input projector fully connected (IP FC) layers (per each modality) output dimension, respectively. A larger EP is much more informative for the classification of links into 'none' (cut edges, meaning no relationship) or 'key-value', while more dimensions for the projected modalities helped the model to better learn the importance of their contributions. These changes bring an improvement of 13 points on the key-value F1 scores, between the third and fourth line of the table where we keep the features fixed. We do not report the score relative to others network settings since their changes only brought a decrease overall metrics. We use a learning rate of 10^{-3} and a weight decay of 10^{-4}, with a dropout of 0.2 over the last FC layer. The threshold over neighbor nodes and their contribution scale factor (Sect. 3.3) are fixed to 0.9 and 0.1, respectively. The bins to discretize the space for angles (Sect. 3.3) are 8. We apply one GNN layer before the node and edge predictors.

Table 1. Ablation studies of Doc2Graph model. EP Inner dim and IP FC dim show edge predictor layer input dimension and the input projector fully connected layers output dimension, respectively. AUC-PR refers to the key-value edge class. The # Params refers to Doc2Graph trainable parameters solely.

Features			EP Inner dim	IP FC dim	F_1 per classes (↑)		AUC-PR (↑)	# Params ×10^6 (↓)
Geometric	Text	Visual			None	Key-Value		
✓	✗	✗	20	100	0.9587	0.1507	0.6301	0.025
✗	✓	✗	20	100	0.9893	0.1981	0.5605	0.054
✓	✓	✗	20	100	0.9941	0.4305	0.7002	0.120
✓	✓	✗	300	300	0.9961	0.5606	0.7733	1.18
✓	✓	✓	300	300	**0.9964**	**0.5895**	**0.7903**	2.68

4.2   FUNSD

Dataset. The dataset [15] comprises 199 real, fully annotated, scanned forms. The documents are selected as a subset of the larger RVL-CDIP [12] dataset, a collection of 400,000 grayscale images of various documents. The authors define the Form Understanding (FoUn) challenge into three different tasks: word grouping, semantic entity labeling and entity linking. A recent work [31] found some inconsistency in the original labeling, which impeded its applicability to the key-value extraction problem. In this work, we are using the revised version of FUNSD (Fig. 2).

Entity Detection. Our focus is on the GNN performances but, for comparison reasons, we used a YOLOv5 small [16] to detect entities (pretrained on COCO [19]). In [15] the word grouping task is evaluated using the ARI metric: since we are not using words, we evaluated the entity detection with F1 score using two

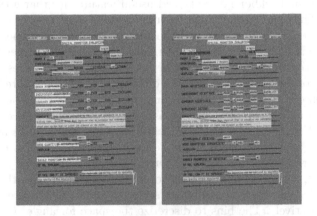

Fig. 2. Image taken from [31]: the document on the right is the revised version of the document on the left, where some answers (green) are mislabeled as question (blue), and some questions (blue) are mislabeled as headers (yellow) (Color figure online)

Table 2. Entity detection results. YOLOv5 [16]-small performances on the entity detection task.

IoU	Metrics (↑)			% Drop rate (↓)	
	Precision	Recall	**F₁**	Entity	Link
0.25	0.8728	0.8712	0.8720	12.72	16.63
0.50	0.8132	0.8109	0.8121	18.67	25.93

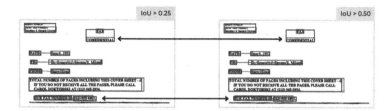

Fig. 3. Blue boxes are FUNSD entities ground truth, green boxes are the correct detected one (with IoU > 0.25/0.50), while red boxes are the false positive ones. (Color figure online)

different IoU thresholds (Table 2). For the semantic entity labeling and entity linking tasks we use IoU > 0.50 as done in [8]: we did not perform any optimization on the detector model, which introduces a high drop rate for both entities and links. We create the graphs on top of YOLO detections, linking the ground truth accordingly (Fig. 3): false positive entities (red boxes) are labeled as class 'other', while false negative entities cause some key-value pairs to be lost (red links). The new connections created as a consequence of wrong detections are considered false positives and labeled as 'none'.

Numerical Results. We trained our architecture (Sect. 3.3) with a 10-fold cross validation. Since we found high variance in the results, we report both

Table 3. Results on FUNSD. The results have been shown for both semantic entity labeling and entity linking tasks with their corresponding metrics.

Method	GNN	F_1 (↑)		# Params ×10⁶ (↓)
		Semantic entity labeling	Entity linking	
BROS [13]	✗	**0.8121**	**0.6696**	138
LayoutLM [13,33]	✗	0.7895	0.4281	343
FUNSD [15]	✓	0.5700	0.0400	–
Carbonell et al. [6]	✓	0.6400	0.3900	201
FUDGE w/o GCN [8]	✗	0.6507	0.5241	12
FUDGE [8]	✓	0.6652	0.5662	17
Doc2Graph + YOLO	✓	0.6581 ± 0.006	0.3882 ± 0.028	13.5
Doc2Graph + GT	✓	<u>0.8225 ± 0.005</u>	0.5336 ± 0.036	**6.2**

mean and variance over the 10 best models chosen over their respective validation sets. The objective function in use (L) is based on both node (L_n) and edge (L_e) classification tasks: $L = L_n + L_e$. In Table 3 we report the performances of our model Doc2Graph compared to other language models [13,33] and graph-based techniques [6,8]. The number of parameters # Params refer to the trainable Doc2Graph pipeline (that includes the U-Net and YOLO backbones); for the spaCy word-embedding details, refer to their documentation. Using YOLO our network outperforms [6] for semantic entity labeling and meets their model on entity linking, using just 13.5 parameters. We could not do better than FUDGE, which still outperforms our scores. Their backbone is trained for both tasks along with the GCN (GCN that adds just minor improvements). The gap, especially on entity linking, is mainly due to the low contributions given by our visual features (Table 1) and the detector in use (Table 3). We also report the results of our model initialized with ground truth (GT) entities, to show how it would perform in the best case scenario. Entity linking remains a harder task compared to semantic entity labeling and only complex language models seem to be able to solve it. Moreover, for the sake of completeness, we highlight that, with good entity representations, our model outperforms all the considered architectures for the Semantic Entity Labeling task. Finally, we want to further stress that the main contribution of a graph-based method is to yield a simpler but more lightweight solution.

Qualitative Results. The order matters for detecting key-value relationship, since the direction of a link induce a property for the destination entity that enrich its meaning. Differently from FUDGE [8] we do make use of directed edges, which led to a better understanding of the document having interpretable results. In Fig. 5 we show our qualitative results using Doc2Graph on groundtruth: green and red dots mean source and destination nodes, respectively. As shown in the different example cases, Fig. 5(a) and 5(b) resemble a simple structured form layout with directed one-to-one key-value association pairs and Doc2Graph manages to extract them. On the contrary, where the layout appears to be more complex as in Fig. 5(d), Doc2Graph fails to generalize the concept of one-to-many key-value relationship pairs. This may be due to the small number of trainable samples we had in our training data and the fact that header-cells usually present different positioning and semantic meaning. In the future we will integrate a table structure recognition path into our pipeline, hoping to improve the extraction of all kinds of key-value relationships in such more complex layout scenarios.

4.3 RVL-CDIP Invoices

Dataset. In the work of Riba et al. [25] another subset of RVL-CDIP has been released. The authors selected 518 documents from the invoices classes, annotating 6 different regions (two examples of annotations are shown in Fig. 4). The task that can be performed are layout analysis, in terms of node classification, and table detection, in terms of bounding box (IoU > 50).

Fig. 4. RVL-CDIP Invoices benchmark in [25]. There are 6 regions: supplier (pink), invoice_info (brown), receiver (green), table (orange), total (light blue), other (gray). (Color figure online)

Numerical Results. As done previously, we perform a k-fold cross validation keeping, for each fold, the same amount of test (104), val (52) and training documents (362). This time we applied an OCR to build the graph. There are two tasks: layout analysis, in terms of accuracy, and table detection, using F1 score and $IoU > 0.50$ for table regions. Our model outperforms [25] in both tasks, as shown in Tables 4 and 5. In particular, for table detection, we extracted the subgraph induced by the edge classified as 'table' (two nodes are linked if they are in the same table) to extract the target region. Riba et al. [25] formulated the problem as a binary classification: we report, for brevity, in Table 5 the threshold on confidence score they use to cut out edges, that in our multi-class setting ('none' or 'table') is implicitly set to 0.50 by the softmax.

Qualitative Results. In Fig. 6 we show the qualitative results. The two documents are duplicated to better visualize the two tasks. For layout analysis, the greater boxes colors indicate the true label that the word inside should have (the colors reflects classes as shown in Fig. 4). For the table detection we use a simple heuristic: we take the enclosing rectangle (green) of the nodes connected by 'table' edges, then we evaluate the IoU with target regions (orange). This heuristic is effective but simple and so error-prone: if a false positive is found

Table 4. Layout analysis results on RVL-CDIP Invoices. Layout analysis accuracy scores depicted in terms of node classification task.

Method	Accuracy (↑)	
	Max	Mean
Riba et al. [25]	62.30	–
Doc2Graph + OCR	**69.80**	**67.80** ± 1.1

Table 5. Table Detection in terms of F1 score. A table is considered correctly detected if its IoU is greater than 0.50. Threshold values refers to the scores an edges has to have in order to do not be cut: in our case is set to 0.50 by the softmax in use.

Method	Threshold	Metrics (↑)		
		Precision	**Recall**	**F₁**
Riba et al. [25]	0.1	0.2520	**0.3960**	0.3080
Riba et al. [25]	0.5	0.1520	0.3650	0.2150
Doc2Graph + OCR	0.5	**0.3786** ± 0.07	0.3723 ± 0.07	**0.3754** ± 0.07

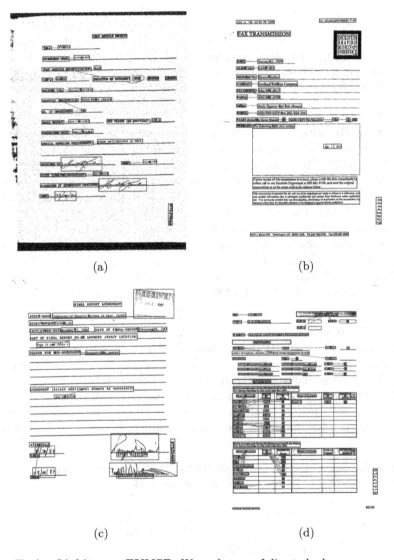

(a)   (b)

(c)   (d)

Fig. 5. Entity Linking on FUNSD. We make use of directed edges: green and red dots mean source and destination nodes, respectively. (Color figure online)

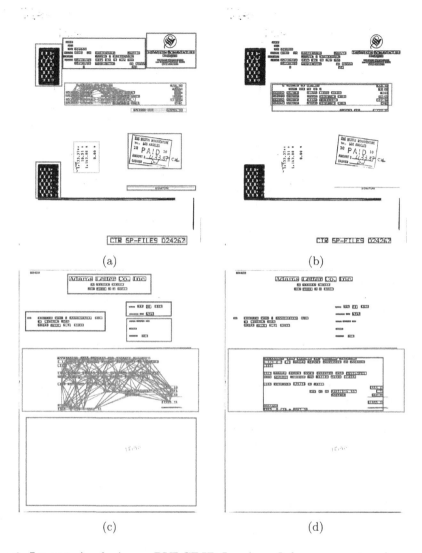

(a)   (b)

(c)   (d)

Fig. 6. Layout Analysis on RVLCDIP Invoices. Inference over two documents from RVL-CDIP Invoices, showing both Layout Analysis (a, c) and Table Detection (b, d) tasks.

outside table regions this could lead to a poor detection result, e.g. a bounding box including also 'sender item' entity or 'receiver item' entity. In addition, as inferred from Figs. 6(a) and 6(b), 'total' regions could be taken out. In the future, we will refine this behaviour by both boosting the node classification task and including 'total' as a table region for the training of edges.

5   Conclusion

In this work, we have presented a task-agnostic document understanding framework based on a Graph Neural Network. We propose a general representation of documents as graphs, exploiting fully connectivity between document objects and letting the network automatically learn meaningful pairwise relationships. Node and edge aggregation functions are defined by taking into account the relative positioning of document objects. We evaluated our model on two challenging benchmarks for three different tasks: entity linking on forms, layout analysis on invoices and table detection. Our preliminary results show that our model can achieve promising results, keeping the network dimensionality considerably low. For future works, we will extend our framework to other documents and tasks, to deeper investigate the generalization property of the GNN. We would like to explore more extensively the contribution of different source features and how to combine them in more meaningful and learnable ways.

Acknowledgment. This work has been partially supported by the Spanish projects MIRANDA RTI2018-095645-B-C21 and GRAIL PID2021-126808OB-I00, the CERCA Program/Generalitat de Catalunya, the FCT-19-15244, and PhD Scholarship from AGAUR (2021FIB-10010).

References

1. Appalaraju, S., Jasani, B., Kota, B.U., Xie, Y., Manmatha, R.: Docformer: end-to-end transformer for document understanding. In: Proceedings of the IEEE/CVF International Conference on Computer Vision, pp. 993–1003 (2021)
2. Biswas, S., Banerjee, A., Lladós, J., Pal, U.: Docsegtr: An instance-level end-to-end document image segmentation transformer. arXiv preprint arXiv:2201.11438 (2022)
3. Biswas, S., Riba, P., Lladós, J., Pal, U.: Beyond document object detection: instance-level segmentation of complex layouts. Int. J. Document Anal. Recogn. (IJDAR) **24**(3), 269–281 (2021). https://doi.org/10.1007/s10032-021-00380-6
4. Biswas, S., Riba, P., Lladós, J., Pal, U.: Graph-based deep generative modelling for document layout generation. In: Barney Smith, E.H., Pal, U. (eds.) ICDAR 2021. LNCS, vol. 12917, pp. 525–537. Springer, Cham (2021). https://doi.org/10.1007/978-3-030-86159-9_38
5. Borchmann, L., Pietruszka, M., Stanislawek, T., Jurkiewicz, D., Turski, M., Szyndler, K., Graliński, F.: Due: end-to-end document understanding benchmark. In: Thirty-fifth Conference on Neural Information Processing Systems Datasets and Benchmarks Track (Round 2) (2021)
6. Carbonell, M., Riba, P., Villegas, M., Fornés, A., Lladós, J.: Named entity recognition and relation extraction with graph neural networks in semi structured documents. In: 2020 25th International Conference on Pattern Recognition (ICPR), pp. 9622–9627. IEEE (2021)
7. Davis, B., Morse, B., Cohen, S., Price, B., Tensmeyer, C.: Deep visual template-free form parsing. In: 2019 International Conference on Document Analysis and Recognition (ICDAR), pp. 134–141. IEEE (2019)

8. Davis, B., Morse, B., Price, B., Tensmeyer, C., Wiginton, C.: Visual FUDGE: form understanding via dynamic graph editing. In: Lladós, J., Lopresti, D., Uchida, S. (eds.) ICDAR 2021. LNCS, vol. 12821, pp. 416–431. Springer, Cham (2021). https://doi.org/10.1007/978-3-030-86549-8_27

9. Gemelli, A., Vivoli, E., Marinai, S.: Graph neural networks and representation embedding for table extraction in PDF documents. In: accepted for publication at ICPR22 (2022)

10. Goldmann, L.: Layout Analysis Groundtruth for the RVL-CDIP Dataset (2019). https://doi.org/10.5281/zenodo.3257319

11. Hamilton, W., Ying, Z., Leskovec, J.: Inductive representation learning on large graphs. Advances in neural information processing systems 30 (2017)

12. Harley, A.W., Ufkes, A., Derpanis, K.G.: Evaluation of deep convolutional nets for document image classification and retrieval. In: International Conference on Document Analysis and Recognition (ICDAR)

13. Hong, T., Kim, D., Ji, M., Hwang, W., Nam, D., Park, S.: Bros: a pre-trained language model for understanding texts in document (2020)

14. Huang, Z., Chen, K., He, J., Bai, X., Karatzas, D., Lu, S., Jawahar, C.: Icdar 2019 competition on scanned receipt OCR and information extraction. In: 2019 International Conference on Document Analysis and Recognition (ICDAR), pp. 1516–1520. IEEE (2019)

15. Jaume, G., Ekenel, H.K., Thiran, J.P.: Funsd: A dataset for form understanding in noisy scanned documents. In: 2019 International Conference on Document Analysis and Recognition Workshops (ICDARW), vol. 2, pp. 1–6. IEEE (2019)

16. Jocher, G., Chaurasia, A., Stoken, A., Borovec, J., NanoCode012, Kwon, Y., TaoXie, Fang, J., imyhxy, Michael, K.: ultralytics/yolov5: v6. 1-tensorrt, tensorflow edge tpu and openvino export and inference. Zenodo, 22 February (2022)

17. Li, X., Wu, B., Song, J., Gao, L., Zeng, P., Gan, C.: Text-instance graph: exploring the relational semantics for text-based visual question answering. Pattern Recogn. **124**, 108455 (2022)

18. Liang, Y., Wang, X., Duan, X., Zhu, W.: Multi-modal contextual graph neural network for text visual question answering. In: 2020 25th International Conference on Pattern Recognition (ICPR), pp. 3491–3498. IEEE (2021)

19. Lin, T.-Y., Maire, M., Belongie, S., Hays, J., Perona, P., Ramanan, D., Dollár, P., Zitnick, C.L.: Microsoft COCO: common objects in context. In: Fleet, D., Pajdla, T., Schiele, B., Tuytelaars, T. (eds.) ECCV 2014. LNCS, vol. 8693, pp. 740–755. Springer, Cham (2014). https://doi.org/10.1007/978-3-319-10602-1_48

20. Liu, H., Li, X., Liu, B., Jiang, D., Liu, Y., Ren, B.: Neural collaborative graph machines for table structure recognition. In: Proceedings of the IEEE/CVF Conference on Computer Vision and Pattern Recognition. pp. 4533–4542 (2022)

21. Mathew, M., Karatzas, D., Jawahar, C.: Docvqa: a dataset for VQA on document images. In: Proceedings of the IEEE/CVF Winter Conference on Applications of Computer Vision, pp. 2200–2209 (2021)

22. Powalski, R., Borchmann, Ł, Jurkiewicz, D., Dwojak, T., Pietruszka, M., Pałka, G.: Going Full-TILT boogie on document understanding with text-image-layout transformer. In: Lladós, J., Lopresti, D., Uchida, S. (eds.) ICDAR 2021. LNCS, vol. 12822, pp. 732–747. Springer, Cham (2021). https://doi.org/10.1007/978-3-030-86331-9_47

23. Qasim, S.R., Mahmood, H., Shafait, F.: Rethinking table recognition using graph neural networks. In: 2019 International Conference on Document Analysis and Recognition (ICDAR), pp. 142–147. IEEE (2019)

24. Raja, S., Mondal, A., Jawahar, C.V.: Table structure recognition using top-down and bottom-up cues. In: Vedaldi, A., Bischof, H., Brox, T., Frahm, J.-M. (eds.) ECCV 2020. LNCS, vol. 12373, pp. 70–86. Springer, Cham (2020). https://doi.org/10.1007/978-3-030-58604-1_5

25. Riba, P., Dutta, A., Goldmann, L., Fornés, A., Ramos, O., Lladós, J.: Table detection in invoice documents by graph neural networks. In: 2019 International Conference on Document Analysis and Recognition (ICDAR), pp. 122–127. IEEE (2019)

26. Riba, P., Goldmann, L., Terrades, O.R., Rusticus, D., Fornés, A., Lladós, J.: Table detection in business document images by message passing networks. Pattern Recogn. **127**, 108641 (2022)

27. Ronneberger, O., Fischer, P., Brox, T.: U-net: convolutional networks for biomedical image segmentation. CoRR abs/1505.04597 (2015). http://arxiv.org/abs/1505.04597

28. Sarkar, M., Aggarwal, M., Jain, A., Gupta, H., Krishnamurthy, B.: Document structure extraction using prior based high resolution hierarchical semantic segmentation. In: Vedaldi, A., Bischof, H., Brox, T., Frahm, J.-M. (eds.) ECCV 2020. LNCS, vol. 12373, pp. 649–666. Springer, Cham (2020). https://doi.org/10.1007/978-3-030-58604-1_39

29. Singh, A., et al.: Towards vqa models that can read. In: Proceedings of the IEEE/CVF Conference on Computer Vision and Pattern Recognition, pp. 8317–8326 (2019)

30. Smock, B., Pesala, R., Abraham, R.: Pubtables-1m: towards comprehensive table extraction from unstructured documents. In: Proceedings of the IEEE/CVF Conference on Computer Vision and Pattern Recognition, pp. 4634–4642 (2022)

31. Vu, H.M., Nguyen, D.T.N.: Revising funsd dataset for key-value detection in document images. arXiv preprint arXiv:2010.05322 (2020)

32. Xu, Y., et al.: Layoutlmv2: Multi-modal pre-training for visually-rich document understanding. arXiv preprint arXiv:2012.14740 (2020)

33. Xu, Y., Li, M., Cui, L., Huang, S., Wei, F., Zhou, M.: Layoutlm: pre-training of text and layout for document image understanding. In: Proceedings of the 26th ACM SIGKDD International Conference on Knowledge Discovery & Data Mining, pp. 1192–1200 (2020)

34. Xue, W., Yu, B., Wang, W., Tao, D., Li, Q.: Tgrnet: a table graph reconstruction network for table structure recognition. In: Proceedings of the IEEE/CVF International Conference on Computer Vision, pp. 1295–1304 (2021)

35. Yu, W., Lu, N., Qi, X., Gong, P., Xiao, R.: Pick: processing key information extraction from documents using improved graph learning-convolutional networks. In: 2020 25th International Conference on Pattern Recognition (ICPR), pp. 4363–4370. IEEE (2021)

MUST-VQA: MUltilingual Scene-Text VQA

Emanuele Vivoli[1,2]([⊠])[iD], Ali Furkan Biten[2][iD], Andres Mafla[2][iD],
Dimosthenis Karatzas[2][iD], and Lluis Gomez[2][iD]

[1] University of Florence, Florence, Italy
emanuele.vivoli@unifi.it
[2] Computer Vision Center, UAB, Barcelona, Spain
{abiten,amafla,dimos,lgomez}@cvc.uab.es

Abstract. In this paper, we present a framework for Multilingual Scene
Text Visual Question Answering that deals with new languages in a zero-
shot fashion. Specifically, we consider the task of Scene Text Visual Question Answering (STVQA) in which the question can be asked in different
languages and it is not necessarily aligned to the scene text language.
Thus, we first introduce a natural step towards a more generalized version
of STVQA: MUST-VQA. Accounting for this, we discuss two evaluation
scenarios in the constrained setting, namely IID and zero-shot and we
demonstrate that the models can perform on a par on a zero-shot setting.
We further provide extensive experimentation and show the effectiveness
of adapting multilingual language models into STVQA tasks.

Keywords: Visual question answering · Scene text · Translation
robustness · Multilingual models · Zero-shot transfer · Power of
language models

1 Introduction

Visual Question Answering is a prominent task that involves two modalities:
vision and language. Language is not only used for expressing the question to the
model, but it's sometimes implicit in the context of text found in the image, such
as in the case of Scene Text Visual Question Answering (STVQA) task [6,33].
The ultimate goal for a holistic STVQA model is to be able to accept questions,
read/analyze the scene text and produce answers in any language or script, this
scenario is referred as *unconstrained setting*. This is especially true considering
the fact that there currently exist more than 7k spoken languages, while more
than 4k have a developed writing system[1], spanning over 100 different scripts.
We believe that the natural extension of the STVQA task in order to benefit
more people while reaching a wider use case, it has to have the capabilities of
dealing with MUltilingual STVQA (MUST-VQA).

---

[1] https://www.ethnologue.com/enterprise-faq/how-many-languages-world-are-
unwritten-0.

© The Author(s), under exclusive license to Springer Nature Switzerland AG 2023
L. Karlinsky et al. (Eds.): ECCV 2022 Workshops, LNCS 13804, pp. 345–358, 2023.
https://doi.org/10.1007/978-3-031-25069-9_23

Evidently, reaching this goal is far from easy as it encapsulates dealing with multiple problems. One of the most important problems is the data scarcity in questions as well as finding images that contain scene text in various languages, being particularly difficult in low-resource languages. Therefore, it is infeasible to collect data for all the languages with all the possible scripts. Moreover, even though STVQA has attracted a lot of research [3,5,13,16,36], the dataset in itself is designed solely for English text. This significantly underpins its use and application in a practical manner considering that roughly 80% of the world population does not speak English [9]. Given the difficulties of obtaining new data and having only English readily available dataset, we define a new practical *constrained setting*. In this setting, we assume that we have questions in multiple languages apart from English. We further divide the constrained setting into IID and zero-shot setting where the models are evaluated with the languages the model is trained with and the languages the models have never seen before, respectively. The zero-shot setting allows models to extend to low-resource languages. Thus, the constrained setting acts as the first step towards the unconstrained, and our aim is to study the behaviour of various models with questions asked in languages other than English.

More specifically, in this work, we take the first steps towards MUltilingual STVQA (MUST-VQA) by automatically translating all the questions in ST-VQA [6] and TextVQA [33] to 5 languages with 3 scripts; namely Spanish, Catalan, Chinese, Italian and Greek, by using automatic translation models and evaluate on IID and zero-shot settings. Furthermore, it is known that neural networks are prone to exploiting shortcuts [12] and thus, we examine our models' robustness to distinct machine translation models. Finally, we study the effect of multiple STVQA models and possible ways to adapt the original architectures to incorporate multilingual inputs.

Our work aims at finding the limitations of the models in MUST-VQA as a preceding step before tackling a full unconstrained multilingual setting. The main contributions of our work are:

- We introduce a natural step towards a more generalized version of STVQA, MUST-VQA and define two settings, unconstrained and constrained.
- We discuss two evaluation scenarios in the constrained setting, namely IID and zero-shot and we demonstrate that our proposed models can perform at the same level in a zero-shot setting.
- We provide extensive experimentation and show the effectiveness of adapting multilingual language models into STVQA tasks.

2   Related Work

The use of scene text in the VQA task is a recent trend in vision and language models' research. Many datasets have been published considering scene text in different domains: natural images [6,33], scanned documents [24] book and movie covers [26], and info-graphics [23]. Additionally, a bilingual (English+Chinese) dataset has been proposed for VQA [14], as well as a captioning dataset with natural images [32].

Alongside with all these datasets, state of the art models have evolved significantly. Singh *et al.* [33] introduced a pointer network to answer either with an answer from the fixed answer vocabulary or by selecting one of the OCR strings. Gómez *et al.* [13] also employed pointer networks directly to the image pixels instead of selecting from a vocabulary. Hu *et al.* [16] as well used pointer networks with a single multi-modal transformer (M4C) to encode all modalities together. Kant *et al.* [17] built on top of M4C with a spatially aware self-attention layer such that each visual entity only looks at neighboring entities defined by a spatial graph. Zhu *et al.* [37] proposes to use an attention mechanism to fuse pairwise modalities.

Recently, following its success in language models [10,20], pre-training has also been successfully used in STVQA. Yang *et al.* [36] performed two stage training where first they do pre-training in a large corpus of images with text to conduct several pretext tasks (OCR token relative position prediction, masked language modelling, and image-text matching) and later fine-tuning for the STVQA task, showing huge performance gains. Finally, Biten *et al.* [3] used layout information via pre-training on IDL [4] data to achieve state-of-the-art performance across multiple benchmarks.

However, the main assumption made until now is that the language of the *question, text in the image* and *answer* is always English. Our belief is that the task of MUST-VQA is still unexplored and lack robust benchmarks. Some recent work has approached the problem of Multilingual Scene-Text VQA (MUST-VQA), but their studies were limited to the use of mono-lingual models (one model per language) [14], or to a single old-fashioned VQA architecture [27].

In this work, we define a customized version of two state of the art transformer-based STVQA models (M4C [16] and LaTr [3]) to incorporate multilingual inputs in a constrained scenario. We employ both approaches as benchmarks for the proposed MUST-VQA task.

3   Method

In this section the main building blocks of our models are introduced and explained. We start by formally defining the task of MUST-VQA in the constrained and unconstrained settings, and then we describe each of these modules.

3.1   Task Definition

Let $v \in I$ be an image of the image space I, and $q \in Q$ a question belonging to the question space Q. The ultimate goal for VQA is to be able to accept questions q and an image v to produce an answer $a \in A_{v,q}$. In our case, we focus on STVQA, task in which the image $v \in \widetilde{I}$ contains scene text, and the question $q \in \widetilde{Q}$ is related to the text in the image. However, the actual state-of-the-art is not able to handle an unconstrained setting, since current models are only trained in English. Therefore, we define additional elements that help towards our goal. First, let $\widetilde{I}_{en} \subset I$ be the subspace of Images containing English scene text, and $\widetilde{Q}_{en} \subset Q$ be the subspace of English questions about text in the image. Let be OCR_{sys} a blackbox which takes as input an image $v \in \widetilde{I}_{en}$ and outputs a set $T = \{(t_v^i, b_v^i)|i = 0, 1, \dots\}$ where t_u^i is a token, and $b_u^i \in [0, 1]^4$ is its normalized position in the image. A common STVQA architecture is able to process all these modalities $v \in \widetilde{I}_{en}$, $q \in \widetilde{Q}_{en}$, $T = OCR_{sys}(v)$ and produce an answer. In order to do that, we need to define some architecture modules.

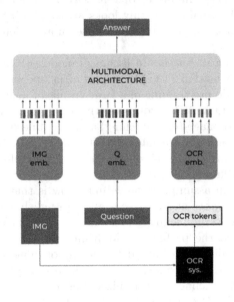

Fig. 1. Proposed model

As we can see from Fig. 1, given an image $v \in \widetilde{I}_{en}$ we obtain a set of M visual features x_{vis}^m and positional information x_b^m through an IMG_{emb} as $\{(x_{vis}^m, x_b^m)|m = 1, \dots M\} = IMG_{emb}(v)$. Additionally, given the question $q \in \widetilde{Q}_{en}$ and a module Q_{emb}, we obtain a set of N textual features y_q^n and positional information y_b^n as $\{(y_q^n, y_b^n)|n = 1, \dots N\} = Q_{emb}(q)$. Lastly, taking into consideration the remaining modality, which is OCR tokens, we can obtain a set of $|T|$ textual features as $z_u^i = OCRemb(t_u^i)$ with $i = 0 \dots |T|$.

MUST-VQA. Until now, the set $\widetilde{I}_{en}$ and $\widetilde{Q}_{en}$ have been defined as set of Images containing scene text and set of Questions about text in the images. However, in the common STVQA task a strong bias is added to the selection of these subsets of I, Q and A: the language. In fact, in STVQA the three elements which are question, scene text and answer all have in common the English language. Thus, we can sample the subspaces $\widetilde{I}_{es} \subset \widetilde{I}$ to get images containing text in Spanish, as well as sample the subspace $\widetilde{Q}_{zh} \subset \widetilde{Q}$ to get Chinese questions about text in the image. The same is true also for the set of answers. With that said, the unconstrained setting of MUST-VQA covers using any sampling, with respect to the language, from $\widetilde{I}$ and $\widetilde{Q}$. However, for most language combinations in the world, data availability is limited, which makes it difficult to obtain for example images with Spanish scene text and original Chinese questions. To this end, we define the constrained MUST-VQA task in which multiple question sets are generated from $\widetilde{Q}_{en}$ by means of an external translator module. The question sets generated with translator g are referred to as $\widetilde{Q}_{ca}^{g}, \widetilde{Q}_{es}^{g}, \widetilde{Q}_{zh}^{g}$, etc. By doing this, we define two experimental settings: IID, in which a subset of question sets is used for training and testing a multimodal architecture $\{\widetilde{Q}_{l}^{g} | l \in (en, ca, es, zh)\}$, and Zero-shot in which we want to test the language transfer capabilities of our models trained under the IID setting to a subset of other languages $\{\widetilde{Q}_{l}^{g} | l \in (it, el)\}$.

3.2 Visual Encoder

In order to obtain the most salient regions of a given image, a Faster R-CNN [29] is used, as proposed initially by [2] and employed in STVQA models as in [33, 36]. The employed Faster R-CNN is pre-trained in ImageNet [30]. Later, Visual Genome [19] is employed to fine-tune the Faster R-CNN to not only predict classes, but also incorporate the prediction of attributes that belong to a specific region that contains an object. The resulting model is employed to extract a set of bounding boxes and visual features enclosed in such regions. In all of our models, the features obtained are then passed through a trainable linear projection layer. The resulting visual features are later fed to each explored model.

3.3 Textual Encoders

In this section, we describe the different textual encoders that have been employed to obtain language features. Specifically, we embed the questions through a given encoder to obtain a set of features to be used as a representation to be later fed into a transformer-based model.

Byte Pair Encoding (BPEmb). The BPEmb [15] is a variable-length encoding that treats text as a symbol sequence. It merges the most common pairs of symbols into a new representation in an iterative manner. This encoding method is trained on Wikipedia in a corpus that employs 275 different languages, thus creating a robust representation that includes most characters found in common human languages. It is shown experimentally that this approach yields rich representations of text that perform on a par compared to other subword embeddings such as Fasttext [7]. BPEmb does not require tokenization and is orders of magnitude smaller than alternative embeddings, allowing for potential applications, specially representing unseen words in different alphabets, thus making it a strong encoder in multilingual scenarios.

Bidirectional Encoder Representations from Transformers (BERT). BERT [10] employs a multi-layer implementation based on the initial Transformer [34]. The work from [10] incorporates two pre-training tasks. The first one, masked-language-modelling (MLM) focuses on predicting a masked tokenized word based on the surrounding words. This pretext task aims to learn semantic representations of words. The second pre-training task is next sentence prediction (NSP) which given a pair of sentence, the model has to predict whether these sentences are consecutive or not. BERT and variations inspired on it are commonly employed as strong semantic descriptors of text in natural language processing. However, the main drawback lies in the lack of sub-word processing to represent out of vocabulary words.

Multilingual-BERT (M-BERT). As in BERT [10], M-BERT is a 12 layer transformer, but rather than relying only on a monolingual English data corpus, it is trained on 104 Wikipedia sites on different languages that share a common vocabulary. It makes no use of a marker to indicate the input language, and there is no explicit mechanism in place to promote translation-equivalent pairings to have comparable representations.

Text-to-Text Transfer Transformer (T5). The T5 [28] is an encoder-decoder transformer. Minor variations are employed from the original [34] implementation. The difference lies in that T5 employs a scaled-down form of layer normalization in which no additive bias is added and the activations are simply rescaled. The T5 architecture is trained on the Colossal Clean Crawled Corpus (C4). The C4 is a text collection that is not only orders of magnitude larger than normal pre-training data sets (about 750 GB), but also comprises clean and curated English material. The model employs a similar query structure describing the task to be performed, such as translation, question answering and classification. The resulting approach can be applied to a variety of tasks, while at the same time similar loss function, model and hyper parameters can be used.

Multilingual-T5. The mT5 [35] model employs a similar set of layers and design as T5. However, they differ in the training corpus. The mT5 model was trained on a 101-language Common Crawl-based multilingual variation. Only English Common Crawl is what T5 has been pre-trained on. Additionally an increase in performance is obtained by the use of GeGLU nonlinearities [31].

3.4   Baselines

In this section we introduce the Scene Text Visual Question Answering models adapted for MUST-VQA. First, we start by introducing the base-model details and then we describe the customized modifications performed on each of them to better adjust to handle multilingual inputs.

M4C. Multimodal Multi-Copy Mesh (M4C) [16] is a multimodal transformer architecture that employs a dynamic pointer network to select among a fixed dictionary or scene text instances. The input comes from two modalities, question and image. However, a scene text recognizer is employed to extract textual instances, which also serve as input to the M4C model. The questions are encoded using BERT [10], while a list of visual object features are obtained by using an off-the-shelf object detector Faster R-CNN [29]. The scene text tokens are obtained by relying on an OCR module, Rosetta-en [8]. The resulting textual transcription is embedded with FastText [7] and a Pyramidal Histogram Of Characters (PHOC) [1]. Such embeddings have shown to be robust representations to encode semantics and morphological information of text [1,22]. The resulting embedded scene text is projected to the same dimension as all other text and visual tokens in order to be used as input to a transformer. The answers are produced in an iterative manner, while the model either selects to output words from a fixed vocabulary or from OCR tokens found in an image by employing a dynamic pointer network.

M5C. The proposed Multilingual-M4C (M5) underwent through a set of custom modifications in order to be able to accept different languages aside from English. To accomplish this goal, we designed a new model: *M5C-mBERT*. The first modification is to substitute the FastText embedding of OCR tokens, since FastText is pre-trained only on English. Next, we replaced the PHOC representation to be able to incorporate different scripts. PHOC encodes only latin-based scripts, therefore it is not suitable for handling unknown languages unless a big descriptor is employed. Therefore we employed a multi-language aligned text embedding method such as BPEmb [15]. In this baseline, we introduce a multilingual Language Model for the question embedding, instead of a pre-trained English based BERT, thus lacking the capability of embedding different languages. By doing that, we designed M5C-mBERT to have multilingual BERT for question embedding.

LaTr (T5). In [3], a Layout-Aware Transformer (LaTr) is proposed, which is based on a T5 [28] encoder-decoder architecture. The pipeline consists of three modules. The first one consists of a Language Model trained specifically on document layouts [3] which contains only text and layout information. The second module is an spatial embedding designed to embed scene text tokens along with positional information. Lastly, a ViT [11] is employed to extract visual tokens. All these three modalities are employed as input to the pre-trained transformer. The encoder learns a suitable representation of the alignment of the 3 modalities to later be used by a decoder to reason and output an answer.

mLaTr (mT5). In this baseline, we replaced the T5 encoder-decoder transformer with the mT5 model in LaTr. Differently from LaTr (which uses layout aware pre-training), we fine-tuned only the text pre-trained multi-lingual Language Model with the multimodal information. Therefore, the input to this mT5 transformers are questions tokens, OCR tokens, and visual features.

4 Experiments

We consider the standard benchmarks of ST-VQA [6] and TextVQA [33]. The proposed MUST-VQA datasets consists of ML-STVQA and ML-TextVQA which are obtained by translating ST-VQA and TextVQA into *Catalan, Spanish, Chinese, Italian*, and *Greek* with Google-Translate-API[2], resulting in a multi-lingual datasets comprised by 6 languages for the constrained setting task of MUST-VQA. In this section, we experimentally examine our baselines in the constrained setting. We further test these baselines for zero-shot multilingual setting, both on ML-STVQA and ML-TextVQA datasets.

4.1   Implementation Details

For all M4C-based methods we used Adam [18] optimizer, with learning rate of 1e-4, and a learning rate decreasing at 14k and 19k iterations. The final model is trained for 24k, while using a 128 batch size.

For all T5-based models we used AdamW [21] optimizer with learning rate of 1e-4, employing a warm up for the first 1000 iterations until reaching 1e-3. Afterwards, we decreased to zero linearly until the end of the training. The batch size employed was 128 and models were trained for 24k iterations for the ML-STVQA dataset. The model trained on ML-TextVQA dataset employed 48k iterations and a batch of 64.

4.2   TextVQA Results

In this section we evaluate results for the dataset ML-TextVQA. We define two evaluation settings. The former is the constrained setting that only uses English, Catalan, Spanish, and Chinese questions for training. On these languages, all the models presented in Sect. 3.4 are trained following the config specifications in Sect. 4.1. Here, either Rosetta-OCR or Microsoft-OCR are used for detection. The latter is the zero-shot transfer setting, in which we measure the performance of the previous models on two new languages that the model has not seen during training (Italian and Greek).

IID Languages (en, ca, es, zh). The first part of Table 1 presents training using Rosetta OCR System, while the bottom part using Microsoft-OCR. Background color is employed to distinguish between monolingual models (white) and our multilingual models (grey). As can be appreciated, our multilingual *M5C-mbert* outperforms *M4C* of about **+1.71%** and **+2.44%** with Rosetta-OCR and Microsoft-OCR respectively, with fewer parameters. These values are the average over the four languages calculated by combining all four subset into a single one. Moreover, as a multilingual model, it is able to perform on Chinese **+5.94%** and **+8.75%** better than its English counterpart (M4C). Increasing model capability to *mLaTr-base* results in a performance gain of **+3.8%**. Furthermore, when training using visual features, performances either recorded a loss of -0.03% and -0.11% for *LaTr-base* Rosetta-OCR and *mLaTr-base* Microsoft-OCR or an increase of +0.58% and +0.16% for *mLaTr-base* Rosetta-OCR and *LaTr-base* Microsoft-OCR. Thus, performances difference is very marginal. Finaly, from

---

[2] cloud.google.com/translate/.

Table 1. Results on the ML-TextVQA dataset. Results refer to multi-lingual training on English, Catalan, Spanish, and Chinese and are reported in term of Accuracy.

Method	OCR	Vis. Feat.	Params	EN	CA	ES	ZH	Avg.
M4C	Ros-en	✔	200M	28.96	29.9	29.60	23.73	28.44
M5C-mbert	Ros-en	✔	162M	28.83	30.26	30.35	29.67	30.15
LaTr-base	Ros-en	✗	226M	41.02	38.35	38.94	20.24	34.64
mLaTr-base	Ros-en	✗	586M	40.35	39.50	39.70	39.49	39.77
LaTr-base	Ros-en	✔	226M	40.92	38.40	38.81	20.34	34.61
mLaTr-base	Ros-en	✔	586M	40.96	40.35	40.35	39.78	40.35
M4C	Ms-OCR	✔	200M	42.16	41.89	41.64	33.60	39.82
M5C-mbert	Ms-OCR	✔	162M	42.36	42.15	42.14	42.35	42.26
LaTr-base	Ms-OCR	✗	226M	46.93	44.32	44.87	23.18	39.83
mLaTr-base	Ms-OCR	✗	586M	46.63	**46.10**	**46.12**	45.38	**46.06**
LaTr-base	Ms-OCR	✔	226M	**47.25**	44.15	44.81	23.79	39.99
mLaTr-base	Ms-OCR	✔	586M	46.65	**46.09**	45.58	45.44	45.95

Table 2. Results on the ML-TextVQA dataset. Results refer to zero-shot transfer on Italian (IT) and Greek (EL) with multi-lingual models trained on English, Catalan, Spanish, and Chinese. Results are reported in term of Accuracy.

Method	OCR	Vis. Feat.	Params	IT	EL	Avg.
M4C	Ros-en	✔	200M	17.45	5.84	28.44
M5C-mbert	Ros-en	✔	162M	24.92	10.88	30.15
LaTr-base	Ros-en	✗	226M	33.35	18.02	34.64
mLaTr-base	Ros-en	✗	586M	3873	37.78	39.77
LaTr-base	Ros-en	✔	226M	33.59	15.01	34.61
mLaTr-base	Ros-en	✔	586M	39.45	38.03	40.35
M4C	Ms-OCR	✔	200M	25.97	14.38	39.83
M5C-mbert	Ms-OCR	✔	162M	33.48	13.11	42.26
LaTr-base	Ms-OCR	✗	226M	36.47	20.25	39.83
mLaTr-base	Ms-OCR	✗	586M	**45**	**44.3**	**46.06**
LaTr-base	Ms-OCR	✔	226M	37.08	21.53	39.99
mLaTr-base	Ms-OCR	✔	586M	**45.01**	**44.25**	45.95

LaTr-base with Microsoft-OCR and visual features we notice that it obtain the best accuracy on Validation set for English Language, which might be due to the distribution of pre-training only-english data samples. In fact, *T5* model has been trained on huge amount of English transcripts (C4), which consist on cleaned English texts from Common Crawl.

Zero-shot Transfer (it, el). A more challenging case for MUST-VQA is the zero-shot cross-lingual setting. Here, a pretrained multilingual model is fine-

Table 3. Results on the ML-STVQA dataset. Results refer to multi-lingual training on English, Catalan, Spanish, and Chinese and are reported in term of Accuracy and ANLS [6]. Microsoft-OCR improve from 5% to 10% over all methods. Visual features do not increase accuracy in general.

Method	OCR	Vis. Feat.	Params	EN		CA		ES		ZH		Avg	
				Acc	ANLS	Acc	ANLS	Acc	ANLS	Acc	ANLS	Acc	ANLS
M4C	Ros-en	✔	200M	35.01	0.439	34.74	0.438	34.36	0.435	30.4	0.384	33.63	0.424
M5C-mbert	Ros-en	✔	162M	35.27	0.438	35.27	0.438	35.81	0.444	35.24	0.438	35.4	0.439
LaTr-base	Ros-en	✗	226M	41.59	0.515	38.78	0.495	38.47	0.497	24.35	0.324	35.8	0.46
mLaTr-base	Ros-en	✗	586M	41.29	0.526	41.29	0.522	41.44	0.528	40.07	0.507	41.03	0.521
LaTr-base	Ros-en	✔	226M	41.67	0.533	39.23	0.51	39	0.5	24.47	0.331	36.09	0.468
mLaTr-base	Ros-en	✔	586M	40.72	0.518	40.68	0.517	40.45	0.514	39.5	0.504	40.33	0.513
M4C	Ms-OCR	✔	200M	41.9	0.507	41.4	0.5	41.51	0.504	36.15	0.44	40.24	0.488
M5C-mbert	Ms-OCR	✔	162M	41.29	0.505	42.39	0.518	42.16	0.514	41.74	0.509	41.4	0.512
LaTr-base	Ms-OCR	✗	226M	47.07	0.559	44.94	0.538	44.86	0.54	28.73	0.352	41.4	0.497
mLaTr-base	Ms-OCR	✗	586M	48.21	0.572	47.72	0.568	47.53	0.566	**47.07**	0.555	47.63	0.565
LaTr-base	Ms-OCR	✔	226M	47.34	0.56	45.4	0.54	45.4	0.542	28.54	0.352	41.67	0.499
mLaTr-base	Ms-OCR	✔	586M	**48.71**	**0.583**	**47.91**	**0.574**	**48.36**	**0.577**	46.84	**0.563**	**47.96**	**0.574**

tuned on TextVQA considering a set of languages but tested on others. In our constrained setting this means testing the models in Sect. 3.4 to generate English answers from Italian or Greek questions, despite having only seen English, Catalan, Spanish and Chinese questions during training. A note for Table 2: last column *Avg.* is the accuracy calculated by combining all four IID subset into a single one. A major observation can be made from Table 2: the best model for IID setting, also perform better on the task of Zero-shot transfer to unseen languages. Moreover, while for Italian the difference is tangible (**+7.92%**), for Greek the gap becomes even wider (**+22.77%**). This behavior might have two main reasons: (1) Italian, Catalan and Spanish are part of the Roman family, descended from Latin, while Greek does not have this common roots with them [25]; (2) Italian share the same script with English, Catalan, and Spanish while Greek has its own script. From these facts we can justify that English-only models trained under constrained settings of EN, CA, ES, ZH languages do have the linguistic and scripting capability of transfer knowledge to Italian setting resulting in **37.08%** accuracy at best, but do not have the same potential for Greek.

4.3   ST-VQA Results

IID Languages (en, ca, es, zh). Table 3 presents the Accuracy and ANLS values in the constrained and unconstrained settings. Similarly to Sect. 4.2, the upper part of the Table refers to Rosetta-OCR, while the bottom to Microsoft-OCR. The grey lines indicates multilingual models, while the white English-only models. However, all these models have been trained on ML-STVQA. One thing to notice is that in this dataset, the best performance is obtained by the *mLaTr-base* model with Microsoft-OCR and visual features. With that said, Chinese is the only exception in which the *mLaTr-base* configuration without visual fea-

Table 4. Results on the ML-STVQA dataset. Results refer to zero-shot transfer on Italian (IT) and Greek (EL) with multi-lingual models trained on English, Catalan, Spanish, and Chinese. Results are reported in term of Accuracy and ANLS (cite ANLS).

Method	OCR	Vis.Feat	Params	IT		EL		Avg	
				acc	ANLS	acc	ANLS	acc	ANLS
M4C	Ros-en	✔	200M	29.15	0.357	21.77	0.288	33.63	0.424
M5C-mbert	Ros-en	✔	162M	30.94	0.389	24.58	0.306	35.4	0.439
LaTr-base	Ros-en	✗	226M	34.78	0.451	23.1	0.307	35.8	0.46
mLaTr-base	Ros-en	✗	586M	39.8	0.505	38.55	0.494	41.03	0.521
LaTr-base	Ros-en	✔	226M	34.89	0.453	24.05	0.324	36.09	0.468
mLaTr-base	Ros-en	✔	586M	39.04	0.501	38.13	0.485	40.33	0.513
M4C	Ms-OCR	✔	200M	34.02	0.413	23.4	0.293	40.24	0.488
M5C-mbert	Ms-OCR	✔	162M	38.58	0.468	30.78	0.384	41.89	0.512
LaTr-base	Ms-OCR	✗	226M	40.6	0.486	27.17	0.329	41.4	0.497
mLaTr-base	Ms-OCR	✗	586M	**46.54**	**0.557**	**45.97**	**0.546**	**47.63**	**0.565**
LaTr-base	Ms-OCR	✔	226M	40.72	0.489	28.16	0.347	41.67	0.498
mLaTr-base	Ms-OCR	✔	586M	46.35	0.554	44.75	0.538	**47.96**	**0.574**

tures actually performs slightly better if considering Accuracy itself. Thus, this empirically confirms, also for this dataset, the fact that visual features might not be relevant to this task. Regarding the comparison of different models in the same ML-STVQA dataset, we can notice once more that *M5C-mbert* obtained +1.77% and (+1.65%) increase in terms of accuracy with respect to M4C English-only baseline. Moreover, from Table 3, we can appreciate three main facts: (1) *mLaTr-base* obtains the best result in overall accuracy, in its variation using Microsoft-OCR and visual features. However, we also observe that visual features don't have considerable impact on the results. (2) When focusing on each language, in the upper part of the table (with Rosetta-OCR) results show that even if *LaTr-base* English-only performs worse than *mLaTr-base* multilingual on almost all the languages with the bigger margin of **-15.72%** for Chinese, it still outperforms the multilingual version for English questions by almost 1 point (**+0.95%**). The last consideration (3) is regarding the pointer network against generative models for languages out of vocabulary. In fact, despite having the lowest score in the overall results, *M4C* obtains higher accuracy in the Chinese questions with both OCR systems, resulting in a margin of **+6.05%** (Rosetta-OCR) and **+7.61%** (Microsoft-OCR) compared to *LaTr-base*.

Zero-shot Transfer (it, el). From Table 4 we can see that the best model for IID setting, also performs better on the task of Zero-shot transfer to unseen languages. Moreover, as saw for ML-TextVQA zero-shot, while for Italian the difference is tangible (+5.82%), for Greek the gap becomes even wider (+17.81%). Possible reasons for that are commented in Sect. 4.2.

5   Analysis

Robustness to Translation Models. In our method, in order to obtain questions in different languages, a translation model is used. Our original translation model is Google-Translate, accessed from its API. To study our approach and how a translation model can influence results, we use three other machine translation models, namely OPUS, M2M_100 and mBART. For all these translation models, we calculate the accuracy of our best model (*mLaTr-base*) for different languages, in term of IID and Zero-shot settings. From Table 5 we can see that accuracy does not drop with other translation models, but instead it has values coherent with the original translation model we use.

Table 5. Results refer to *mLaTr-base* with visual features and Microsoft-OCR. Its average accuracies on Original ML-TextVQA and ML-STVQA questions are reported in the last column *Avg*. Questions have been translated into the 5 languages using OPUS, M2M100 (1.2B), and mBART.

(a) **Results on TextVQA dataset**

	CA	ES	ZH	IT	EL	Avg
OPUS	42.25	45.73	43.82	44.53	43.39	**46.06**
M2M_100	45.73	45.69	44.39	44.91	43.29	**46.06**
mBART	/	45.76	43.53	44.81	/	**46.06**

(b) **Results on STVQA dataset**

	CA	ES	ZH	IT	EL	Avg
OPUS	46.84	47.72	45.74	46.31	46.84	**47.96**
M2M_100	47.22	47.68	45.97	46.16	45.09	**47.96**
mBART	-	46.96	45.89	45.93	-	**47.96**

6   Conclusions and Future Work

In this paper, we present a framework for Multilingual visual question answering that deals with new languages in a zero-shot fashion. Specifically, we defined the task of MUST-VQA and its constrained and unconstrained settings. We defined a multilingual baseline method for MUST-VQA by adopting monolingual architectures. Our results suggest that it is able to operate in a zero-shot fashion, and independent on the translation method used to obtain multilingual questions. In this work, the constrained setting acts as the first step towards the unconstrained, and our aim is to study the behaviour of various models with questions asked in languages other than English. Further work will need to approach also answers in different languages, probably matching the question language.

Acknowledgments. This work has been supported by projects PDC2021-121512-I00, PLEC2021-00785, PID2020-116298GB-I00, ACE034/21/000084, the CERCA Programme/Generalitat de Catalunya, AGAUR project 2019PROD00090 (BeARS), the Ramon y Cajal RYC2020-030777-I/AEI/10.13039/501100011033 and PhD scholarship from UAB (B18P0073).

References

1. Almazán, J., Gordo, A., Fornés, A., Valveny, E.: Word spotting and recognition with embedded attributes. IEEE Trans. Pattern Anal. Mach. Intell. **36**(12), 2552–2566 (2014)

2. Anderson, P., et al.: Bottom-up and top-down attention for image captioning and visual question answering. In: CVPR, pp. 6077–6086 (2018)
3. Biten, A.F., Litman, R., Xie, Y., Appalaraju, S., Manmatha, R.: Latr: layout-aware transformer for scene-text vqa. In: Proceedings of the IEEE/CVF Conference on Computer Vision and Pattern Recognition, pp. 16548–16558 (2022)
4. Biten, A.F., Tito, R., Gomez, L., Valveny, E., Karatzas, D.: Ocr-idl: Ocr annotations for industry document library dataset. arXiv preprint arXiv:2202.12985 (2022)
5. Biten, A.F., et al.: Icdar 2019 competition on scene text visual question answering. In: 2019 International Conference on Document Analysis and Recognition (ICDAR), pp. 1563–1570. IEEE (2019)
6. Biten, A.F., et al.: Scene text visual question answering. In: ICCV, pp. 4291–4301 (2019)
7. Bojanowski, P., Grave, E., Joulin, A., Mikolov, T.: Enriching word vectors with subword information. Trans. Assoc. Comput. Linguistics **5**, 135–146 (2017)
8. Borisyuk, F., Gordo, A., Sivakumar, V.: Rosetta: Large scale system for text detection and recognition in images. In: SIGKDD, pp. 71–79 (2018)
9. Crystal, D.: Two thousand million? English today **24**(1), 3–6 (2008)
10. Devlin, J., Chang, M.W., Lee, K., Toutanova, K.: Bert: pre-training of deep bidirectional transformers for language understanding. arXiv preprint arXiv:1810.04805 (2018)
11. Dosovitskiy, A., et al.: An image is worth 16×16 words: transformers for image recognition at scale. arXiv preprint arXiv:2010.11929 (2020)
12. Geirhos, R., Jacobsen, J.H., Michaelis, C., Zemel, R., Brendel, W., Bethge, M., Wichmann, F.A.: Shortcut learning in deep neural networks. Nature Mach. Intell. **2**(11), 665–673 (2020)
13. Gómez, L., Biten, A.F., Tito, R., Mafla, A., Rusiñol, M., Valveny, E., Karatzas, D.: Multimodal grid features and cell pointers for scene text visual question answering. Pattern Recogn. Lett. **150**, 242–249 (2021)
14. Han, W., Huang, H., Han, T.: Finding the evidence: Localization-aware answer prediction for text visual question answering. arXiv preprint arXiv:2010.02582 (2020)
15. Heinzerling, B., Strube, M.: Bpemb: tokenization-free pre-trained subword embeddings in 275 languages. arXiv preprint arXiv:1710.02187 (2017)
16. Hu, R., Singh, A., Darrell, T., Rohrbach, M.: Iterative answer prediction with pointer-augmented multimodal transformers for textvqa. In: Proceedings of the IEEE/CVF Conference on Computer Vision and Pattern Recognition, pp. 9992–10002 (2020)
17. Kant, Y., Batra, D., Anderson, P., Schwing, A., Parikh, D., Lu, J., Agrawal, H.: Spatially aware multimodal transformers for TextVQA. In: Vedaldi, A., Bischof, H., Brox, T., Frahm, J.-M. (eds.) ECCV 2020. LNCS, vol. 12354, pp. 715–732. Springer, Cham (2020). https://doi.org/10.1007/978-3-030-58545-7_41
18. Kingma, D.P., Ba, J.: Adam: a method for stochastic optimization. arXiv preprint arXiv:1412.6980 (2014)
19. Krishna, R., Zhu, Y., Groth, O., Johnson, J., Hata, K., Kravitz, J., Chen, S., Kalantidis, Y., Li, L.J., Shamma, D.A., et al.: Visual genome: Connecting language and vision using crowdsourced dense image annotations. Int. J. Comput. Vision **123**(1), 32–73 (2017)
20. Liu, Y., et al.: Roberta: a robustly optimized bert pretraining approach. arXiv preprint arXiv:1907.11692 (2019)
21. Loshchilov, I., Hutter, F.: Decoupled weight decay regularization. arXiv preprint arXiv:1711.05101 (2017)

22. Mafla, A., Dey, S., Biten, A.F., Gomez, L., Karatzas, D.: Fine-grained image classification and retrieval by combining visual and locally pooled textual features. In: Proceedings of the IEEE/CVF Winter Conference on Applications of Computer Vision, pp. 2950–2959 (2020)
23. Mathew, M., Bagal, V., Tito, R.P., Karatzas, D., Valveny, E., Jawahar, C.: Infographicvqa. arXiv preprint arXiv:2104.12756 (2021)
24. Mathew, M., Karatzas, D., Jawahar, C.: Docvqa: a dataset for VQA on document images. In: Proceedings of the IEEE/CVF Winter Conference on Applications of Computer Vision, pp. 2200–2209 (2021)
25. Mikulyte, G., Gilbert, D.: An efficient automated data analytics approach to large scale computational comparative linguistics. CoRR (2020)
26. Mishra, A., Shekhar, S., Singh, A.K., Chakraborty, A.: Ocr-vqa: visual question answering by reading text in images. In: 2019 International Conference on Document Analysis and Recognition (ICDAR), pp. 947–952. IEEE (2019)
27. Brugués i Pujolràs, J., Gómez i Bigordà, L., Karatzas, D.: A multilingual approach to scene text visual question answering. In: Uchida, S., Barney, E., Eglin, V. (eds) Document Analysis Systems. DAS 2022. LNCS, vol. 13237, pp. 65–79. Springer, Cham. https://doi.org/10.1007/978-3-031-06555-2_5
28. Raffel, C., Shazeer, N., Roberts, A., Lee, K., Narang, S., Matena, M., Zhou, Y., Li, W., Liu, P.J., et al.: Exploring the limits of transfer learning with a unified text-to-text transformer. J. Mach. Learn. Res. 21(140), 1–67 (2020)
29. Ren, S., He, K., Girshick, R., Sun, J.: Faster r-cnn: towards real-time object detection with region proposal networks. In: Advances in neural information processing systems, pp. 91–99 (2015)
30. Russakovsky, O., Deng, J., Su, H., Krause, J., Satheesh, S., Ma, S., Huang, Z., Karpathy, A., Khosla, A., Bernstein, M., et al.: Imagenet large scale visual recognition challenge. Int. J. Comput. Vision 115(3), 211–252 (2015)
31. Shazeer, N.: Glu variants improve transformer. arXiv preprint arXiv:2002.05202 (2020)
32. Sidorov, O., Hu, R., Rohrbach, M., Singh, A.: Textcaps: a dataset for image captioning with reading comprehension. arXiv preprint arXiv:2003.12462 (2020)
33. Singh, A., et al.: Towards vqa models that can read. In: CVPR, pp. 8317–8326 (2019)
34. Vaswani, A., Shazeer, N., et al.: Attention is all you need. In: Advances in Neural Information Processing Systems, pp. 5998–6008 (2017)
35. Xue, L., et al.: mt5: a massively multilingual pre-trained text-to-text transformer. arXiv preprint arXiv:2010.11934 (2020)
36. Yang, Z., et al.: Tap: text-aware pre-training for text-VQA and text-caption. In: Proceedings of the IEEE/CVF Conference on Computer Vision and Pattern Recognition, pp. 8751–8761 (2021)
37. Zhu, Q., Gao, C., Wang, P., Wu, Q.: Simple is not easy: a simple strong baseline for textvqa and textcaps. arXiv preprint arXiv:2012.05153 (2020)

Out-of-Vocabulary Challenge Report

Sergi Garcia-Bordils[1,3], Andrés Mafla[1], Ali Furkan Biten[1(✉)], Oren Nuriel[2],
Aviad Aberdam[2], Shai Mazor[2], Ron Litman[2], and Dimosthenis Karatzas[1]

[1] Computer Vision Center, Universitat Autonoma de Barcelona, Barcelona, Spain
{sgbordils,amafla,abiten,dimos}@cvc.uab.es
[2] AWS AI Labs, New York, USA
{onuriel,aaberdam,smazor,litmanr}@amazon.com
[3] AllRead MLT, Barcelona, Spain

Abstract. This paper presents final results of the Out-Of-Vocabulary 2022 (OOV) challenge. The OOV contest introduces an important aspect that is not commonly studied by Optical Character Recognition (OCR) models, namely, the recognition of unseen scene text instances at training time. The competition compiles a collection of public scene text datasets comprising of 326,385 images with 4,864,405 scene text instances, thus covering a wide range of data distributions. A new and independent validation and test set is formed with scene text instances that are out of vocabulary at training time. The competition was structured in two tasks, end-to-end and cropped scene text recognition respectively. A thorough analysis of results from baselines and different participants is presented. Interestingly, current state-of-the-art models show a significant performance gap under the newly studied setting. We conclude that the OOV dataset proposed in this challenge will be an essential area to be explored in order to develop scene text models that achieve more robust and generalized predictions.

1 Introduction

Scene-text detection and recognition plays a key role in a multitude of vision and language tasks, such as visual question answering [5,34], image captioning [?] or image retrieval [26]. Performance on classic benchmarks, such as ICDAR13 [13] or ICDAR15 [12] has noticeably increased thanks to the surge of sophisticated deep learning models. Interest in this field has gained traction in the last few years and, as a consequence, multiple new datasets have appeared. Some of them have introduced diverse new challenges, such as irregular text detection and recognition [6,41] or complex layout analysis [24]. At the same time, the scale of new datasets has also noticeably increased, reducing the reliance on synthetic data [16,35].

However, none of the existing benchmarks makes a distinction between out-of-vocabulary (OOV) words and in-vocabulary (IV) words. By OOV word we

---

S. Garcia-Bordils, A. Mafla, A. F. Biten—Equal contribution.
O. Nuriel, A. Aberdam, S. Mazor, R. Litman—Work does not relate to Amazon position.

© The Author(s), under exclusive license to Springer Nature Switzerland AG 2023
L. Karlinsky et al. (Eds.): ECCV 2022 Workshops, LNCS 13804, pp. 359–375, 2023.
https://doi.org/10.1007/978-3-031-25069-9_24

refer to text instances that have never been seen in the training sets of the most common Scene Text understanding datasets to date. Recent research suggests that current OCR systems over-rely on language priors to recognize text [38], by exploiting their explicit or implicit language model. As a consequence, while the performance on IV text is high, recognition performance on unseen vocabulary is lower, showing poor generalization. Since OOV words can convey important high-level information about the scene (such as prices, dates, toponyms, URLs, etc.), performance of OCR systems on unseen vocabulary should also be seen as an important characteristic.

With this motivation in mind we present the Out-of-Vocabulary Challenge, a competition on scene text understanding where the focus is put on unseen vocabulary words. This challenge is formed by two different tasks; an End-to-End Text Recognition task and a Cropped Word Text Recognition task. In the End-To-End task participants were provided with images and were expected to localize and recognize all the text in the image at word granularity. In the Cropped Word task the participants were presented with the cropped word instances of the test set, and were asked to provide a transcription. In order to be able to compare the performance of the submissions on seen and unseen vocabulary, we decided to include both types of instances on the test sets and report the results separately.

The dataset used for this competition is a collection of multiple existing datasets. Some of the featured datasets were collected with text in mind, while others used sources where the text is incidental. We have created our own validation and test splits of the End-to-End challenge to contain at least one OOV word per image. In the validation and test sets of the Cropped Word Recognition task we include the cropped OOV and IV words from the End-to-End dataset.

2   Related Work

The field of scene text recognition can be divided into two main tasks - text detection and text recognition. The OOV challenge addresses methods which either perform end-to-end text recognition [4,9,15,18,23,25,31,32,42] and thus solve both tasks, or methods that tackle just text recognition [1–3,8,22,29,30, 33,36,40], and thus assume the words are already extracted.

The problem of vocabulary reliance in scene text recognition was first revealed by [38]. Through extensive experiments, they found that state-of-the-art methods perform well on previously seen, in vocabulary, word images yet generalize poorly to images with out of vocabulary words, never seen during training. In addition, the authors proposed a mutual learning approach which jointly optimizes two different decoder types, showing that this can alleviate some of the problems of vocabulary reliance. [43] suggested a context-based supervised contrastive learning framework, which pulls together clusters of identical characters within various contexts and pushes apart clusters of different characters in an embedding space. In this way they are able to mitigate some of the gaps between in and out of vocabulary words.

3   Competition Protocol

The OOV Challenge took place from May to July of 2022. A training set was given to participants at the beginning of May, but the images for the test set were only made accessible for a window between June 15 and July 22. Participants were asked to submit results obtained on the public test set images rather than model executables. We rest on the scientific integrity of the participants to adhere to the challenge's specified guidelines.

The Robust Reading Competition (RRC) portal[1] served as the Challenge's host. The RRC site was created in 2011 to host the first rigorous reading contests including text detection and identification from scene photographs and born-digital images, and it has since expanded into a fully-fledged platform for organizing academic competitions. The portal now hosts 19 distinct challenges, with different tasks mostly related to scene text detection and recognition, Scene-Text Visual Question Answering (ST-VQA) and Document VQA. The RRC portal has more than 35,000 registered users from more than 148 countries, and more than 77,000 submitted results have already been evaluated. The findings in this report are an accurate reflection of the submissions' status at the end of the formal challenge period. The RRC portal should be viewed as an archive of results, where any new findings contributed after the compilation of this report will also be included. All submitted findings are automatically analyzed, and the site provides per-task ranking tables and visualization tools to examine the results.

4   The OOV Dataset

The OOV (Out Of Vocabulary) dataset encompasses a collection of images from 7 public datasets, namely: HierText [24], TextOCR [35], ICDAR13 [13], ICDAR15 [12], MLT19 [28], Coco-text [37] and OpenImages [16]. This dataset selection aims to generalize the performance of models overcoming existing biases in datasets [14].

The validation and test splits of the OOV dataset were defined to measure the performance of models on unseen words at training time. To do this we extracted all the words that appear at least once in the training and validation splits of the datasets and, jointly with the 90k word dictionary introduced by Jaderberg et al. [11], we created an in-vocabulary dictionary of words. To create the test set of the OOV dataset we picked those images from the original test sets which contained, at least, one word outside of this vocabulary. Images in the validation dataset were picked from the training and validation splits of the original datasets, and we only kept images that contained words that appear once (and therefore do not appear in the training split). We limited the number of images in the validation set to 5,000 images. The rest of the images were used in the training split.

---

[1] https://rrc.cvc.uab.es/.

Table 1. Dataset size comparison

Dataset	# of Images			# of Cropped Words		
	Train	Validation	Test	Train	Validation	Test
ICDAR13	229	0	233	795	54	0
ICDAR15	1000	0	500	4350	118	0
ICDAR MLT19	10000	0	10000	80937	8499	103297
MSRA-TD500	300	0	200	-	-	-
COCO-Text	43686	10000	10000	80549	6571	11688
TextOCR	24902	0	3232	1155320	47019	96082
HierText	8281	1724	1634	981329	59630	187190
Open Images V5 Text	191059	16731	0	2066451	7788	0
OOV Dataset	312612	5000	8773	4369731	128832	365842

For this first iteration of the competition we focus on text instances in which characters come from a limited alphabet. This alphabet is formed by the Latin alphabet, numbers and a few punctuation signs[2]. Words that contain out of alphabet characters are not considered for the final evaluation (in the End-to-End task they are treated as "don't care").

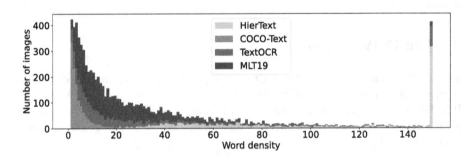

Fig. 1. Number of words per image and dataset of origin. Images that contain more than 150 words have been counted in the last bin.

4.1 Dataset Analysis

Given that this dataset is a collection of datasets from multiple sources, it has the benefit of featuring data that come from different sources and annotation settings. Some of the featured datasets were originally collected with text in mind, while in others the text is more incidental. COCO-Text is an example of a dataset with purely incidental text. Since the source of the images is the MS COCO [21] dataset, the images are not text biased like in other datasets. In

---

[2] See https://rrc.cvc.uab.es/?ch=19&com=tasks for the full alphabet.

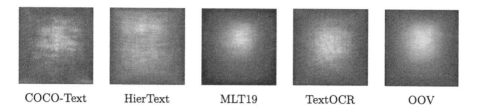

COCO-Text   HierText   MLT19   TextOCR   OOV

Fig. 2. Text spatial distribution of the different datasets featured in the test set.

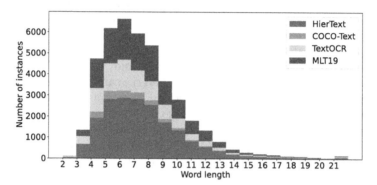

Fig. 3. Character length of OOV words per dataset.

Fig. 1 we can see the distribution of the number of words per image and dataset of origin in the test set. Some datasets contain, in general, a small number of instances per image (such as the aforementioned COCO-Text), while others are more prone to contain images with tens or even hundreds of words (HierText being the most prominent one). Consequently, our test set is less biased towards specific distributions of words.

Table 1 shows the dataset of origin of the cropped words for each one of the splits. The test split features a balanced number of instances coming from 4 different sources, avoiding relying too much on a specific dataset. The spatial distribution of the instances of the test set can be seen in the Fig. 2. Each dataset featured in the test set appears to have different spatial distributions, a consequence of the original source of the images. For example, COCO-Text and TextOCR originally come from datasets that were not collected with text in mind (MS COCO [37] and Open Images V4 [17]), consequently the word instances are incidental and distributed over all the image. On the other hand, the text featured on MLT19 is more focused and more clustered around the center of the images. HierText was collected with text in mind but contains much more text instances per image (as seen in the Fig. 1, which distributes the text uniformly over the images. Combining datasets that contain images from different origins gives us a more varied and rich test set. Finally, Fig. 3 shows the distribution of the lengths of the OOV words of the test set, separated per dataset. The distribution of the lengths appears to be similar for the featured datasets.

5   The OOV Challenge

The Out of Vocabulary challenge aims to evaluate the ability of text extraction models to deal with words that they have never seen before. Our motivation for organizing this challenge is the apparent over-reliance of modern OCR systems on previously seen vocabulary. We argue that generalizing well on in and out of vocabulary text should be considered as important as generalizing well on text with different visual appearances or fonts. OOV text can convey important semantic information about the scene, failing to recognize a proper noun (such as the name of a street in the context of autonomous driving) or a random string of numbers (such as a telephone number) can result in unfortunate consequences.

Wan *et. al.* [38] call the phenomenon of memorizing the words the in the training set *"vocabulary reliance"*. To prove this behaviour, the authors train diverse text recognition models using the same data and the same backbone. The results are reported using the IIIT-5k [27] dataset and they provide results on both in-vocabulary and out-of-vocabulary text. Like us, they consider OOV whichever words are not present in the test set, including the synthetic data. Their results show a gap as small as 15.3% using CA-FCN [20] and as high as 22.5% using CRNN [33], proving how much OCR systems rely on learned vocabulary.

Therefore, in this challenge we evaluate the entries putting special emphasis on words that the models have never seen before. We hope that our curated dataset and our evaluation protocol can be useful to the community to develop more robust and unbiased OCR systems.

5.1   Task 1

The End-to-End task aims to evaluate the performance of the models in both detection and recognition. Unlike some previous competitions in the RRC, we do not provide any vocabulary. For each correct detection we look for a perfect match between the proposed and the ground truth transcriptions. Evaluation is case-sensitive, and punctuation signs are taken into account. The evaluation procedure displays results on both IV and OOV text for each of the methods. Below, we give brief descriptions for some of the submitted methods, as provided by their authors.

CLOVA OCR DEER. An end-to-end scene text spotter based on a CNN backbone, a Deformable Transformer Encoder [44], location decoder and text decoder. The location decoder, based on the segmentation method (Differentiable Binarization [20]), detects text regions, and the text decoder based on the deformable transformer decoder recognizes each instance from image features and detected location information. They use both the training data provided by the challenge, as well as synthetic data.

Detector Free E2E. It is a detection free end-to-end text recognizer, where a CNN with Deformable Encoder & Decoder is used. The models are trained with data from the challenge and additional SynthText data synthesized with MJSynth 90k dictionary.

oCLIP and oCLIP_v2. For detection, they first pre-train their Deformable ResNet-101 by using oCLIP on the provided training set. Then they train TESTR [42], PAN and Mask TextSpotter with different backbones using the pre-trained model. Finally, they combine results from different methods, different backbones, and different scales together while for recognition, they adopt SCATTER [22].

DB_threshold2_TRBA. The detector is based on Differentiable Binarization (DB) [19]. The recognizer is TRBA from WIW [3]. TRBA denotes TPS + ResNet Backbone + BiLSTM + Attention. The models were not jointly trained. Since DB does not output an up-vector, they rotated the detected region according to the aspect ratio. CocoText has label noises (not case sensitive), and thus, they cleaned the dataset using the teacher model. They use synthetic data (ST) as well as challenge-provided data.

E2E_Mask. They only use the OOV dataset to train their model. In the detection stage, they follow TBNet and Mask2Former as the base model with a multi-scale training strategy. To combine the final detection results, they ensemble different detectors with different backbones and different testing sizes. In the recognition stage, they use a vision transformer model that consists of ViT encoder and query-based decoder to generate the recognition results in parallel.

5.2   Task 2

On the Cropped Word Recognition task the participants had to predict the recognition for all the cropped words of the test set. There is a total of 313,751 cropped words in the test set, 271,664 of these words are in-vocabulary and 42,087 are out of vocabulary. Like in Task 1, the evaluation is case-sensitive, and punctuation signs are taken into account. Below, we provide a brief description for some of the submitted methods.

OCRFLY_V2. They design a new text recognition framework for OOV-ST, named Character level Adaptive Mutual Decoder (CAMD), where both multi-arch and multi-direction autoregressive seq2seq heads are jointly used during training and testing. CAMD adopts a CNN-ViT Backbone as encoder, and two different vision-language adaptively balanced decoders: an LSTM and a Transformer decoder, are built upon the aforementioned encoder. Only Syn90k and the training splits given by the challenge are used for training.

OOV3decode. Three models are combined by voting. An Encoder-Decoder Framework with a 12-ViT-based encoder, a CTC Combined Decoder, a CTC-Attention Combined Decoder, and a Mix-CTC-Position-Attn Combined Decoder. They use 300w+ generated images and the challenge training data.

Vision Transformer Based Method (VTBM). They train several models with the same Vision Transformer based backbone and various decoders (CTC and Attention), and they ensemble them based on confidence. They first pre-trained their models on nearly 10 million synthetic images and fine-tuned them on the official training set. Common augmentations such as rotation, blur, etc. are adopted; especially the image concatenate augmentation is used to mine textual context information.

DAT. An Encoder-Decoder transformer-based encoder with 12 layers of VIT-based block and 4×4 patch size is used. An ensemble strategy is used to fuse the results from three decoder types: a CTC-based decoder, an attention-based decoder and a CTC+attention-based decoder.

OCRFLY. This is a simple baseline based on the seq2seq algorithm, where they adopt a CNN-ViT Backbone as encoder and a 6-layer transformer as decoder. Only Syn90k and the training splits are used for training.

5.3   Baselines

For Task 1, we evaluate the recent state of the art methods, TESTR [42], Text-TranSpotter [15], GLASS [32].

TESTR. We provide results for their pretrained model released on their official code package[3] using the default configurations.

TextTranSpotter. TextTranSpotter was trained following the fully-supervised training protocol in the paper with the following datasets: pretrained on Synth-Text, then fine-tuned on a mix of SynthText, ICDAR13, ICDAR15, TotalText and TextOCR. The model weights were then frozen and the mask branch was trained on SynthText and then on the mix of datasets.

GLASS. Training was performed on the following train datasets: SynthText, ICDAR13, ICDAR15, TotalText and TextOCR. The model was pretrained for 250k iterations with a batch size of 24, and then fine tuned for another 100k iterations specifically on TextOCR with a batch size of 8 images. The architecture and parameters chosen for the detection, recognition and fusion branches, are detailed in [32].

For Task 2, we evaluate the models in two types of settings, the first when trained on synthetic data and the second when trained on the real data. The real data consists of the word crops introduced in the OOV dataset.

SCATTER. In both settings SCATTER was trained for 600k iterations from scratch. We employ the exact same training procedure as described in the [22]. We use two selective-contextual refinement blocks and take the output of the last one.

Baek et al. For the synthetic setting, we use the case-sensitive model released in the official repository[4]. For the real data setting the model was trained from scratch using the same training procedure as published in the repository.

---

[3] https://github.com/mlpc-ucsd/testr.
[4] https://github.com/clovaai/deep-text-recognition-benchmark.

ABINET. We used the official codebase[5] and trained a case-sensitive model on the MJ and ST synthetic datasets for the synthetic setting. For the real setting, we also utilized the OOV dataset. In both cases, we only trained the network end-to-end without the pretraining stages.

5.4   Evaluation Metrics

For Task 1 (End-to-End text detection and recognition) we use a modified version of the evaluation method proposed by Wang et. al. [39]. This method considers a correct match when one of the proposed detections overlaps with a ground truth bounding box by more than 50% and their transcriptions match (again, caring about the letter case and punctuation signs). Correctly matched proposals count as true positives, while unmatched proposals count as false positives. Unmatched ground truth annotations count as false negatives. Most of the annotations of the datasets used to form the validation and test sets have annotations with some sort of "unreadable" attribute. These words are treated as "don't care", and do not affect (positively or negatively) the results. As discussed earlier, words that contain characters that are out of alphabet are also considered as "don't care". Jointly with the precision and recall, we also report the harmonic mean (or F-score) of each method:

$$Hmean = \frac{2 * Recall * Precision}{Recall + Precision} \tag{1}$$

Since our evaluation protocol has to distinguish between OOV and IV words, we modified the evaluation procedure to ignore the opposite split during the evaluation. For example, when we are evaluating on OOV words, in-vocabulary words are treated as "don't care". This way, matched ground truth IV words do not count as false positives, and unmatched words do not count as false negatives. The opposite applies when evaluating for IV words. The number of false positives is the same in both cases, since matched ground truth annotations are either true positives or treated as "don't care".

Finally, we report the unweighted average of the OOV and IV Hmean, as a balanced metric for this task. Our motivation behind this metric is to give equal importance to both distributions, as ideally, increasing performance on OOV words should not undermine performance on IV words.

For the Task 2 (Cropped Word Recognition) we report two metrics. The first metric is the total edit distance between each predicted word and its ground truth, considering equal costs for insertions, deletions and substitutions. The second metric is word accuracy, which is calculated as the sum of correctly recognized words divided by the total amount of text instances. We provide both metrics for OOV and IV words. The final, balanced metric reported for the Task 2 is the unweighted average of the OOV and IV accuracy. Similarly to Task 1, we consider performance on both subsets equally important.

---

[5] https://github.com/FangShancheng/ABINet.

6 Results

In this section, we provide the results for the submitted methods on both tasks. We also present baselines to compare to the submitted methods.

Table 2. Harmonic mean, precision and recall for the entire dataset (All), in vocabulary (IV) and out of vocabulary (OOV) across different baseline methods. The average Hmean is the unweighted average of IV and OOV.

Method	Average	All			OOV			IV		
	Hmean	P	R	Hmean	P	R	Hmean	P	R	Hmean
TESTR [42]	15.9	31.4	20.1	25.1	4.4	17.8	7.1	29.2	21.4	24.6
TextTranSpotter [15]	18.6	37.4	25.0	29.9	4.5	16.4	7.0	35.5	26.2	30.1
cre GLASS [32]	34.9	75.8	30.6	43.6	24.9	27.2	26.0	73.7	31.1	43.7
cre CLOVA OCR DEER	**42.39**	**67.17**	**52.04**	**58.64**	18.58	48.72	26.9	**64.51**	**52.49**	**57.88**
cre Detector Free E2E	42.01	66.15	52.44	58.5	17.97	49.35	26.35	63.44	52.86	57.67
cre oCLIP_v2	41.33	67.37	46.82	55.24	20.28	48.42	28.59	64.41	46.6	54.08
cre DB threshold2 TRBA	39.1	64.08	49.93	56.13	15.26	42.29	22.43	61.6	50.96	55.78
cre E2E_Mask	32.13	47.9	54.14	50.83	8.64	46.73	14.58	45.2	55.14	49.68
cre YYDS	28.68	51.53	35.54	42.07	10.63	33.36	16.12	48.57	35.83	41.24
cre Sudokill-9	28.34	51.62	34.08	41.06	11.03	33.22	16.56	48.54	34.2	40.12
cre PAN	28.13	50.5	34.81	41.21	10.5	33.58	16.0	47.45	34.98	40.27
cre oCLIP	24.04	47.72	7.51	12.98	**41.21**	**48.42**	**44.52**	17.46	1.98	3.55
cre DBNetpp	20.34	39.42	27.0	32.05	5.62	20.75	8.85	37.15	27.84	31.83
cre TH-DL	9.32	18.39	13.23	15.39	2.16	10.87	3.6	16.89	13.55	15.04

6.1 Task 1

We provide all the results in Table 2 where the top part represents the baselines while the bottom part is for the submitted methods. We present results on two different subsets, namely in-vocabulary (IV) and out-of-vocabulary (OOV). Moreover, we rank the methods according to the balanced Average Hmean metric.

As can be appreciated from Table 2, CLOVA OCR DEER is the winning method in terms of Average Hmean. The best method in terms of OOV Hmean is oCLIP, surpassing the second best method by **+17.6** points. However, we see that oCLIP's performance on IV set is the lowest between the participated methods. This observation makes us wonder if there is a trade-off between the performance of IV and OOV in the model.

Regarding the architecture choices of the submitted methods, almost all of them, especially the top performing ones make use of the ViT [7] architecture either as backbone for the recognition pipeline or directly for extracting features. Another commonly preferred building block is CTC [10] based encoder or decoder mechanism. Furthermore, the top 2 performing methods, CLOVA OCR DEER and Detector Free E2E, utilize a Deformable DETR architecture [44], showing its effectiveness in the end-to-end text recognition task. Lastly, we observe that almost all the methods make use of synthetic data to either pre-train or finetune together with the real data.

6.2   Task 2

The results can be found in Table 3 where the top part represents the baselines while the bottom part is for the submitted methods. We present results exactly the same way as in Task 1 in two different subsets, namely in-vocabulary (IV) and out-of-vocabulary (OOV) words. Moreover, we rank the methods according to the balanced Total Word Accuracy which is calculated as the average of the Word Accuracy of IV and OOV, giving the same emphasis on both sets. As can be seen from Table 3, our baselines trained with real data clearly outperform the ones trained with synthetic data. We also observe in our baselines that having a boost in IV words also translates to an improvement in OOV performance.

Table 3. Word accuracy and Edit Distance for state of the art recognition models trained on different datasets.

Method	Train Set	Total	IV		OOV	
		Word Acc↑	Word Acc↑	ED↓	Word Acc↑	ED↓
ABINet [8]	Syn	38.01	50.29	342,552	25.73	115536
Baek et al. [3]	Syn	44.47	52.61	365,566	36.34	114,101
SCATTER [22]	Syn	47.79	56.85	321,101	38.74	103,928
ABINet [8]	Real	59.84	71.13	176,126	48.55	67478
Baek et al. [3]	Real	64.97	75.98	138,479	53.96	54,346
SCATTER [22]	Real	66.68	77.98	128,219	55.38	52,535
OCRFLY_v2	Syn + Real	**70.31**	81.02	123,947	**59.61**	46,048
OOV3decode	Syn + Real	70.22	**81.58**	**94,259**	58.86	40,175
VTBM	Syn + Real	70.00	81.36	94,701	58.64	40,187
DAT	-	69.90	80.78	96,513	59.03	**40,082**
OCRFLY	Syn + Real	69.83	80.63	131,232	59.03	53,243
GGUI	-	69.80	80.74	96,597	58.86	40,171
vitE3DCV	Syn + Real	69.74	80.74	96,477	58.74	40,115
DataMatters	Syn + Real	69.68	80.71	96,544	58.65	40,177
MaskOCR	Real	69.63	80.60	108,894	58.65	44,971
SCATTER	Syn + Real	69.58	79.72	113,482	59.45	43,89
Summer	Syn + Real	68.77	79.48	103,211	58.06	42,118
LMSS	Syn + Real	68.46	80.81	116,503	56.11	51,165
UORD	Real	68.28	79.28	118,185	57.27	48,517
PTVIT	Syn + Real	66.29	77.52	120,449	55.06	49,41
GORDON	Syn + Real	65.86	77.25	124,347	54.47	48,907
TRBA_CocoValid	Syn + Real	63.98	77.76	132,781	50.20	60,693
HuiGuan	Real	63.73	74.77	162,87	52.69	68,926
EOCR	-	46.66	55.30	350,166	38.02	113,317
NNRC	-	38.54	45.36	405,603	31.73	136,384
NN	-	37.17	43.38	426,074	30.97	144,032
CCL	Real	31.06	47.40	552,57	14.73	202,087

Regarding the submitted methods, the winner in total word accuracy is OCR-FLY_v2 even though by a slight margin. We also notice that OCRFLY_v2 is the best method in OOV performance in terms of accuracy; however, DAT is the best method in terms of edit distance. On the other hand, OOV3decode achieves state of the art performance in IV. We note that all top 3 methods are trained with both synthetic and real data. As a matter of fact, most of the methods are trained with combined data, confirming the effectiveness of the usage of the combined data. In terms of the favored architectures, we see a similar trend in Task 2 as in Task 1. We observe that the ViT [7] architecture being used in most of the methods combined with CTC and Attention. This demonstrates the clear advantage of the Transformer architecture over LSTMs and RNNs which was state-of-the-art in text recognition literature previously.

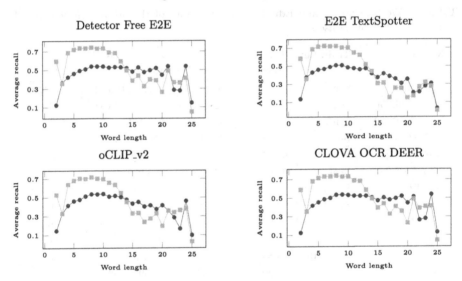

Fig. 4. Average recall for different character lengths in the End-to-End task. The red and blue lines represent the average recall for IV and OOV words, respectively. (Color figure online)

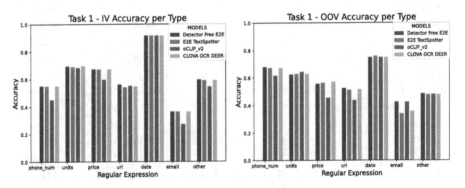

Fig. 5. Model accuracy according to the type of scene text categorized by regular expressions in Task 1. Best viewed in color.

7   Analysis

In this section, we analyze the performance of the top submitted methods in terms of word length and different word categories. Specifically, we study whether the word length has an effect on the performance of a model and how well they perform in terms of the category of the word.

7.1   Task 1

Figure 4 shows the performance of the top 4 methods for words of different character lengths. The metric reported is the average recall of each method for both OOV and IV words. For all methods, the results on IV words shorter than 15 characters are higher than for OOV words. The models seem to have less difficulty dealing with short in-vocabulary words, most likely as a product of vocabulary reliance. For IV words longer than 15 characters, the performance is comparable or sometimes even worse than for OOV words of the same lengths. Interestingly, results on OOV words seem to be consistent regardless of the word length, although we seem to observe fluctuations on the score for words longer than 20 characters. This could be attributed to statistical anomalies due to the low number of OOV words of this length (as seen in Fig. 3, there are fewer words with of than 20 characters).

Additionally, Fig. 5 shows the recognition accuracy by employing an automatic categorization of words via the usage of regular expressions. In both scenarios (IV and OOV), whenever a scene text instance is mostly formed by num-

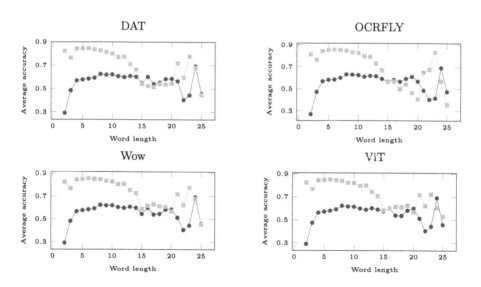

Fig. 6. Average precision for different character lengths in the Cropped Word Recognition task. The red and blue lines represent the average recall for IV and OOV words, respectively. (Color figure online)

bers (units, prices, phone numbers) the accuracy remains uniform. We hypothesize that this outcome is a direct effect of the distribution of the training data that contains numbers. Since numbers do not follow a specific distribution as characters in a given language, scene text models are more flexible at correctly predicting numbers. Subsequently, the direct effect is found in categories where numbers are not common or absent at all, such as in emails, urls and others.

7.2   Task 2

Figure 4 shows the performance of the top 4 methods for words of different lengths of characters. In this case we feature the average precision (correctly recognized words) of the top 4 submitted methods, for both OOV and IV. We observe a similar pattern as in the End-to-End task, the models seem to perform better on IV words of character length of 15 or less. The results on OOV words also seem to remain consistent for different character lengths. Like we have observed in Task 1, performance on OOV words of more than 15 characters is similar or superior to the IV of the same length, which suggest that OCR systems have trouble with longer sequences, regardless of whether they are in vocabulary or not. Similarly to the previous subsection, we show in Fig. 7 the performance of the top 4 models on different word categories. Since in Task 2, no detection is involved, we observe a slightly different behaviour compared to the previous task. Whenever solely numbers are contained in a cropped image, the accuracy in IV and OOV remains similar, as in the case of phone numbers. However, if numbers and characters are expected to be found in a cropped word, the gap in performance is very large in the rest of the categories, except for emails. Even though the performance of all models is very close in task 2 (Table 3), we observe that the winning entry, OCRFLY, gets an edge on prices, emails and other categories in IV and on phone numbers and emails in OOV words.

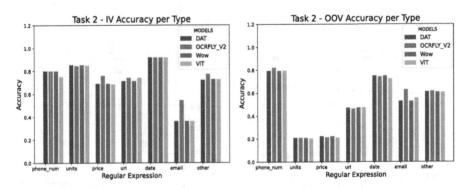

Fig. 7. Model accuracy according to the type of scene text categorized by regular expressions in Task 2. Best viewed in color.

8 Conclusion and Future Work

In this work, we introduce a new task called Out-Of-Vocabulary Challenge, in which end-to-end text recognition and cropped scene text recognition were the two challenges that made up the competition. In order to cover a wide range of data distributions, the competition creates a collection of open scene text datasets that include 326K images and 4.8M scene text instances. Surprisingly, state-of-the-art models exhibit a considerable performance discrepancy on the OOV task. This is especially apparent in the performance gap between in-vocabulary and out-of-vocabulary words. We come to the conclusion that in order to create scene text models that produce more reliable and generalized predictions, the OOV dataset suggested in this challenge would be a crucial area to investigate forward.

References

1. Aberdam, A., Ganz, R., Mazor, S., Litman, R.: Multimodal semi-supervised learning for text recognition. arXiv preprint arXiv:2205.03873 (2022)
2. Aberdam, A., et al.: Sequence-to-sequence contrastive learning for text recognition. In: Proceedings of the IEEE/CVF Conference on Computer Vision and Pattern Recognition, pp. 15302–15312 (2021)
3. Baek, J., et al.: What is wrong with scene text recognition model comparisons? dataset and model analysis. In: Proceedings of the IEEE/CVF International Conference on Computer Vision, pp. 4715–4723 (2019)
4. Baek, Y., et al.: Character region attention for text spotting. In: Vedaldi, A., Bischof, H., Brox, T., Frahm, J.-M. (eds.) ECCV 2020. LNCS, vol. 12374, pp. 504–521. Springer, Cham (2020). https://doi.org/10.1007/978-3-030-58526-6_30
5. Biten, A.F., et al.: Scene text visual question answering. In: Proceedings of the IEEE/CVF International Conference on Computer Vision, pp. 4291–4301 (2019)
6. Ch'ng, C.K., Chan, C.S.: Total-text: a comprehensive dataset for scene text detection and recognition. In: 2017 14th IAPR International Conference on Document Analysis and Recognition (ICDAR), vol. 1, pp. 935–942. IEEE (2017)
7. Dosovitskiy, A., et al.: An image is worth 16 × 16 words: transformers for image recognition at scale. arXiv preprint arXiv:2010.11929 (2020)
8. Fang, S., Xie, H., Wang, Y., Mao, Z., Zhang, Y.: Read like humans: autonomous, bidirectional and iterative language modeling for scene text recognition. In: Proceedings of the IEEE/CVF Conference on Computer Vision and Pattern Recognition, pp. 7098–7107 (2021)
9. Feng, W., He, W., Yin, F., Zhang, X.Y., Liu, C.L.: Textdragon: An end-to-end framework for arbitrary shaped text spotting. In: Proceedings of the IEEE/CVF International Conference on Computer vision, pp. 9076–9085 (2019)
10. Graves, A., Fernández, S., Gomez, F., Schmidhuber, J.: Connectionist temporal classification: labelling unsegmented sequence data with recurrent neural networks. In: Proceedings of the 23rd International Conference on Machine Learning, pp. 369–376 (2006)
11. Jaderberg, M., Simonyan, K., Vedaldi, A., Zisserman, A.: Synthetic data and artificial neural networks for natural scene text recognition. arXiv preprint arXiv:1406.2227 (2014)

12. Karatzas, D., et al.: ICDAR 2015 competition on robust reading. In: 2015 13th International Conference on Document Analysis and Recognition (ICDAR), pp. 1156–1160. IEEE (2015)
13. Karatzas, D., et al.: ICDAR 2013 robust reading competition. In: 2013 12th International Conference on Document Analysis and Recognition, pp. 1484–1493. IEEE (2013)
14. Khosla, A., Zhou, T., Malisiewicz, T., Efros, A.A., Torralba, A.: Undoing the damage of dataset bias. In: Fitzgibbon, A., Lazebnik, S., Perona, P., Sato, Y., Schmid, C. (eds.) ECCV 2012. LNCS, vol. 7572, pp. 158–171. Springer, Heidelberg (2012). https://doi.org/10.1007/978-3-642-33718-5_12
15. Kittenplon, Y., Lavi, I., Fogel, S., Bar, Y., Manmatha, R., Perona, P.: Towards weakly-supervised text spotting using a multi-task transformer. In: Proceedings of the IEEE/CVF Conference on Computer Vision and Pattern Recognition, pp. 4604–4613 (2022)
16. Krylov, I., Nosov, S., Sovrasov, V.: Open images v5 text annotation and yet another mask text spotter. In: Asian Conference on Machine Learning, pp. 379–389. PMLR (2021)
17. Kuznetsova, A., et al.: The open images dataset v4. Int. J. Comput. Vis. **128**(7), 1956–1981 (2020)
18. Li, H., Wang, P., Shen, C.: Towards end-to-end text spotting with convolutional recurrent neural networks. In: Proceedings of the IEEE International Conference on Computer vision, pp. 5238–5246 (2017)
19. Liao, M., Wan, Z., Yao, C., Chen, K., Bai, X.: Real-time scene text detection with differentiable binarization. In: Proceedings of the AAAI Conference on Artificial Intelligence, vol. 34, pp. 11474–11481 (2020)
20. Liao, M., et al.: Scene text recognition from two-dimensional perspective. In: Proceedings of the AAAI Conference on Artificial Intelligence, vol. 33, pp. 8714–8721 (2019)
21. Lin, T.-Y., et al.: Microsoft COCO: common objects in context. In: Fleet, D., Pajdla, T., Schiele, B., Tuytelaars, T. (eds.) ECCV 2014. LNCS, vol. 8693, pp. 740–755. Springer, Cham (2014). https://doi.org/10.1007/978-3-319-10602-1_48
22. Litman, R., Anschel, O., Tsiper, S., Litman, R., Mazor, S., Manmatha, R.: Scatter: selective context attentional scene text recognizer. In: Proceedings of the IEEE/CVF Conference on Computer Vision and Pattern Recognition, pp. 11962–11972 (2020)
23. Liu, Y., Chen, H., Shen, C., He, T., Jin, L., Wang, L.: Abcnet: real-time scene text spotting with adaptive Bezier-curve network. In: Proceedings of the IEEE/CVF Conference on Computer Vision and Pattern Recognition, pp. 9809–9818 (2020)
24. Long, S., Qin, S., Panteleev, D., Bissacco, A., Fujii, Y., Raptis, M.: Towards end-to-end unified scene text detection and layout analysis. In: Proceedings of the IEEE/CVF Conference on Computer Vision and Pattern Recognition, pp. 1049–1059 (2022)
25. Lyu, P., Liao, M., Yao, C., Wu, W., Bai, X.: Mask textSpotter: an end-to-end trainable neural network for spotting text with arbitrary shapes. In: Proceedings of the European Conference on Computer Vision (ECCV), pp. 67–83 (2018)
26. Mafla, A., Dey, S., Biten, A.F., Gomez, L., Karatzas, D.: Multi-modal reasoning graph for scene-text based fine-grained image classification and retrieval. In: Proceedings of the IEEE/CVF Winter Conference on Applications of Computer Vision, pp. 4023–4033 (2021)
27. Mishra, A., Alahari, K., Jawahar, C.: Scene text recognition using higher order language priors. In: BMVC-British Machine Vision Conference. BMVA (2012)

28. Nayef, N., et al.: ICDAR 2019 robust reading challenge on multi-lingual scene text detection and recognition-rrc-mlt-2019. In: 2019 International Conference on Document Analysis and Recognition (ICDAR), pp. 1582–1587. IEEE (2019)

29. Nuriel, O., Fogel, S., Litman, R.: TextadaIN: fine-grained AdaIN for robust text recognition. arXiv preprint arXiv:2105.03906 (2021)

30. Qiao, Z., Zhou, Y., Yang, D., Zhou, Y., Wang, W.: Seed: semantics enhanced encoder-decoder framework for scene text recognition. In: Proceedings of the IEEE/CVF Conference on Computer Vision and Pattern Recognition, pp. 13528–13537 (2020)

31. Qin, S., Bissacco, A., Raptis, M., Fujii, Y., Xiao, Y.: Towards unconstrained end-to-end text spotting. In: Proceedings of the IEEE/CVF International Conference on Computer Vision, pp. 4704–4714 (2019)

32. Ronen, R., Tsiper, S., Anschel, O., Lavi, I., Markovitz, A., Manmatha, R.: Glass: global to local attention for scene-text spotting. arXiv preprint arXiv:2208.03364 (2022)

33. Shi, B., Bai, X., Yao, C.: An end-to-end trainable neural network for image-based sequence recognition and its application to scene text recognition. IEEE Trans. Pattern Anal. Mach. Intell. 39(11), 2298–2304 (2016)

34. Singh, A., Natarajan, V., Shah, M., Jiang, Y., Chen, X., Batra, D., Parikh, D., Rohrbach, M.: Towards VQA models that can read. In: Proceedings of the IEEE/CVF Conference on Computer Vision and Pattern Recognition, pp. 8317–8326 (2019)

35. Singh, A., Pang, G., Toh, M., Huang, J., Galuba, W., Hassner, T.: TextOCR: towards large-scale end-to-end reasoning for arbitrary-shaped scene text. In: Proceedings of the IEEE/CVF Conference on Computer Vision and Pattern Recognition, pp. 8802–8812 (2021)

36. Slossberg, R., et al.: On calibration of scene-text recognition models. arXiv preprint arXiv:2012.12643 (2020)

37. Veit, A., Matera, T., Neumann, L., Matas, J., Belongie, S.: COCO-Text: dataset and benchmark for text detection and recognition in natural images. arXiv preprint arXiv:1601.07140 (2016)

38. Wan, Z., Zhang, J., Zhang, L., Luo, J., Yao, C.: On vocabulary reliance in scene text recognition. In: Proceedings of the IEEE/CVF Conference on Computer Vision and Pattern Recognition, pp. 11425–11434 (2020)

39. Wang, K., Babenko, B., Belongie, S.: End-to-end scene text recognition. In: 2011 International Conference on Computer Vision, pp. 1457–1464. IEEE (2011)

40. Wang, Y., Xie, H., Fang, S., Wang, J., Zhu, S., Zhang, Y.: From two to one: a new scene text recognizer with visual language modeling network. In: Proceedings of the IEEE/CVF International Conference on Computer Vision, pp. 14194–14203 (2021)

41. Yao, C., Bai, X., Liu, W., Ma, Y., Tu, Z.: Detecting texts of arbitrary orientations in natural images. In: 2012 IEEE Conference on Computer Vision and Pattern Recognition, pp. 1083–1090. IEEE (2012)

42. Zhang, X., Su, Y., Tripathi, S., Tu, Z.: Text spotting transformers. In: Proceedings of the IEEE/CVF Conference on Computer Vision and Pattern Recognition, pp. 9519–9528 (2022)

43. Zhang, X., Zhu, B., Yao, X., Sun, Q., Li, R., Yu, B.: Context-based contrastive learning for scene text recognition. AAAI (2022)

44. Zhu, X., Su, W., Lu, L., Li, B., Wang, X., Dai, J.: Deformable DETR: deformable transformers for end-to-end object detection. arXiv preprint arXiv:2010.04159 (2020)

W14 - BioImage Computing

W14 - BioImage Computing

The seventh edition of the BioImage Computing workshop aimed to bring the latest challenges in bio-image computing to the computer vision community. It showcased the specificities of bio-image computing and its current achievements, including issues related to image modeling, denoising, super-resolution, multi-scale instance- and semantic segmentation, motion estimation, image registration, tracking, classification, and event detection.

October 2022

Jan Funke
Alexander Krull
Dagmar Kainmueller
Florian Jug
Anna Kreshuk
Martin Weigert
Virginie Uhlmann
Peter Bajcsy
Erik Meijering

Towards Structured Noise Models for Unsupervised Denoising

Benjamin Salmon⬤ and Alexander Krull$^{(\boxtimes)}$⬤

School of Computer Science, University of Birmingham, Birmingham B15 2TT, UK
brs209@student.bham.ac.uk, a.f.f.krull@bham.ac.uk
https://www.birmingham.ac.uk/schools/computer-science

Abstract. The introduction of unsupervised methods in denoising has shown that unpaired noisy data can be used to train denoising networks, which can not only produce high quality results but also enable us to sample multiple possible diverse denoising solutions. However, these systems rely on a probabilistic description of the imaging noise–a noise model. Until now, imaging noise has been modelled as pixel-independent in this context. While such models often capture shot noise and read-out noise very well, they are unable to describe many of the complex patterns that occur in real life applications. Here, we introduce a novel learning-based autoregressive noise model to describe imaging noise and show how it can enable unsupervised denoising for settings with complex structured noise patterns. We show that our deep autoregressive noise models have the potential to greatly improve denoising quality in structured noise datasets. We showcase the capability of our approach on various simulated datasets and on real photo-acoustic imaging data.

Keywords: Denoising · Deep learning · Autoregressive · Noise · Diverse solutions · VAE · Photoacoustic imaging

1 Introduction

Whenever we attempt to acquire an image **s**, using a microscope or any other recording device, we should generally expect that the result **x** will not perfectly correspond to the signal. Instead, our measurement will be subject to the random inaccuracies of the recording process, resulting in what is referred to as noise. We can define noise **n** = **x**−**s** as the difference between the corrupted observation and the true signal. Noise is especially prevalent in sub-optimal imaging conditions, such as when imaging with only a small amount of light. As a result, noise often becomes the limiting factor in life science imaging, operating right at the boundary of what is possible with current technology. The algorithmic removal of noise (*denoising*) can thus be a vital tool, enabling new previously unfeasible experimental setups [4,16]. Given a noisy image **x**, we can think of the denoising task as finding an estimate ŝ that is close to the true clean image **s**.

© The Author(s), under exclusive license to Springer Nature Switzerland AG 2023
L. Karlinsky et al. (Eds.): ECCV 2022 Workshops, LNCS 13804, pp. 379–394, 2023.
https://doi.org/10.1007/978-3-031-25069-9_25

Noisy input   MMSE output   Ground truth

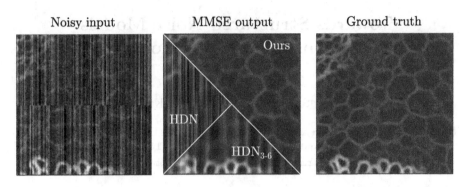

Fig. 1. Comparing HDN with our novel autoregressive noise model to HDN and HDN$_{3-6}$ with the standard pixel-independent noise model. Structured noise can be observed in many imaging modalities. Here, the simulated striped pattern in the noise is designed to mimic noise real noise as it frequently in some sCMOS cameras. HDN with our novel autoregressive noise model is able to remove structured noise, while HDN with the established pixel-independent noise model only removes the pixel-independent component. HDN$_{3-6}$ performs slightly better, but struggles with long range correlation.

Consequently, since the introduction of digital image processing, a plethora of denoising methods have been proposed [7,10,18], to name a few. The last decade however, has seen a revolution of the field, with machine learning (ML) emerging as the technology capable of producing the most accurate results [4,16]. Traditional supervised ML-based methods [28] view denoising as a regression problem, *i.e.*, they attempt to learn a function, mapping noisy images $\mathbf{x}$ to the true clean signal $\mathbf{s}$, based on previously collected training data of noisy-clean-image-pairs.

Despite its success, supervised learning of this form comes with an important caveat, the acquisition of training data can be impractical. Originally, the approach requires us to collect paired clean and noisy images of the same content type we would like to denoise. This is not always possible. Although the problem was partially alleviated by Lehtinen *et al.* [17], showing that pairs of corresponding noisy images are sufficient, the collection of paired data has remained an obstacle for many practical applications.

Only in recent years has this problem been addressed by new self- and unsupervised methods [3,6,14,15,20,23–25], which can be trained on individual (unpaired) noisy images, *e.g.* the very images that are to be denoised. Two of the newest unsupervised techniques [23,24], referred to as DivNoising and HDN [23], provide an additional benefit. They do not produce a single estimate of the true signal, but instead allow us to sample different possible solutions for a noisy input image.

However, to achieve this, these methods require an additional ingredient during training. They rely on a mathematical description of the imaging noise, called *noise model*. The noise model is description of the probability distribution $p(\mathbf{x}|\mathbf{s})$

over the noisy observations **x** we should expect for a given underlying clean image **s**. Noise models can be measured from calibration data [15], bootstrapped (using a self-supervised denoising algorithm) [25], or even co-learned on-the-fly while training the denoiser [24]. Most crucially, noise models are a property of the imaging setup–the camera/detector, amplifier *etc.*, but do not depend on the object that is being imaged. That is, once a noise model has been estimated for an imaging setup it can be reused again and again, opening the door for denoising in many practical applications.

However, previous noise models used in this context are based on a conditional pixel-independence assumption. That is, the model assumes that for an underlying given clean image **s**, noise is generated independently for each pixel in an *unstructured* way, similar to adding the result of separate dice rolls to each pixel without considering its neighbours. This assumption is reasonable for many imaging setups, such as for fluorescence microscopy, where noise is often thought of as a combination of Poisson shot noise and Gaussian readout noise [30]. For simplicity, we will refer to this type of noise simply as *pixel-independent* noise.

Unfortunately, many imaging systems, such as computed tomography (CT) [9] or photo acoustic imaging (PA) [29], do not adhere to this property and can produce structured noise. In practice, even in fluorescence microscopy the *conditional pixel-independence* assumption does not always hold, due to the camera's complex electronics. Many fluorescence microscopy setups suffer from noise that is partially structured. Figure 1 shows an example of simulated structured noise with a pattern close to what is produced by many sCMOS cameras [2].

When DivNoising methods are applied to data containing structured noise which is not accurately represented in their noise model, these methods usually fail to remove it[1]. Even though, Prakash *et al.* [23] show that the effects of this problem can be mitigated by reducing the expressive power of their network, we find that this technique fails to remove noise featuring long range correlations.

Here, we present a new and principled way to address structured noise in the DivNoising framework. We present an autoregressive noise model that is capable of describing structured noise and thus enabling DivNoising to remove it. We evaluate our method quantitatively on various simulated datasets and qualitatively on a PA dataset featuring highly structured noise. We publish our code as well as the our simulated noise datasets.[2]

In summary, our contributions are:

1. We present an autoregressive noise model capable of describing structured noise.
2. We demonstrate that DivNoising together with our noise model can effectively remove simulated structured noise in situations where the previously proposed approach [23] fails.
3. We qualitatively demonstrate structured noise removal on a real PA data.

---

[1] The same is true for self-supervised methods such as [14], which discusses this topic explicitly.

[2] Code and datsets can be found at https://github.com/krulllab/autonoise.

2   Related Work

2.1   Self- And Unsupervised Methods for Removing Structured Noise

Noise2Void [14] is a self-supervised approach to removing pixel-independent noise relying on the assumption that the expected value of a noisy observed pixel, conditioned on the those surrounding it, is the true signal. Using what is known as a *blind spot network*, a model is shown as input a patch of pixels with the one in the centre masked. It is trained to produce an output that is as close as possible to the pixel it did not see for which, under the aforementioned assumption, its best guess is something close to the true signal.

In the case of structured noise, that assumption is broken. Broaddus *et al.* [6] accommodated for this by masking not only the pixel that is to be predicted, but also masking all those for which the conditional expected value of the target pixel is not the true signal. A drawback of this approach is that one must first determine the distance and direction over which noise is correlated. Another is that a considerable amount of valuable information is sacrificed by masking.

As mentioned previously, in [23], Prakash *et al.* demonstrated that tuning the expressive power of a DivNoising based method enables it to remove some cases of structured noise. This method is described in more detail in Sect. 3.1.

2.2   Noise Modelling

In [1], Abdelhamed *et al.* proposed a deep generative noise model known as Noise Flow. It is based on the Glow [12] normalising flow architecture and can be trained for both density estimation and noise generation. In their paper, the authors demonstrated how this noise model could be applied to the problem of denoising by using it to synthesis clean and noisy image pairs. Those pairs could then be used to train a supervised denoising network.

A normalising flow based noise model could be used for the purposes of this paper, but a recent review on deep generative modelling [5] found that auto-regressive models perform slightly better in terms of log-likelihood. As will be seen later, this makes auto-regressive noise models more suitable in a DivNoising framework.

3   Background

Here, we want to give a brief recap of the methods our approach relies on. We will begin with DivNoising Prakash *et al.* [24] and its extension [23] (HDN), which is the framework our method is built upon. We will then discuss the currently used pixel-independent noise models, which are a component in DivNoising and HDN and which we will later compare against our novel autoregressive replacement. Finally, we will have a brief look at deep autoregressive models, which provide the backbone for our noise model.

3.1   DivNoising and HDN

Training DivNoising requires two ingredients, the data that needs to be denoised and a pre-trained or measured noise model, $p_\eta(\mathbf{x}|\mathbf{s})$. We will discuss the noise model in more detail in Sect. 3.2.

Instead of directly providing an estimate $\hat{\mathbf{s}}$ for a noisy image, DivNoising allows us to sample possible solutions $\mathbf{s}^k$ from an approximate *posterior* distribution $p(\mathbf{s}|\mathbf{x})$, *i.e.*, from the distribution of possible clean images given the noisy input. To obtain a sensible single estimate, we can average a large number of these samples to produce the *minimum mean square error* (MMSE) estimate

$$\hat{\mathbf{s}} = \frac{1}{K}\sum_{k=1}^{K}\mathbf{s}^k, \tag{1}$$

which is comparable to the single solution provided by a supervised denoising network.

DivNoising works by training a variational autoencoder (VAE) to approximate the distribution of training images $\mathbf{x}$. VAEs are latent variable models. That is, they can model difficult and high dimensional distributions by introducing an unobserved latent variable $\mathbf{z}$ following a known prior distribution. In DivNoising $p(\mathbf{z})$ is assumed to be a standard normal distribution. DivNoising describes the distribution of noisy images as

$$\log p_{\theta,\eta}(\mathbf{x}) = \log\int p_\eta\left(\mathbf{x}|\mathbf{s} = g_\theta(\mathbf{z})\right)p(\mathbf{z})\,d\mathbf{z}, \tag{2}$$

where $g_\theta : \mathbb{R}^d \rightarrow \mathbb{R}^{D>d}$ is a convolutional neural network (CNN) called the *decoder* that maps from the space of latent variables to the space of signals. We use θ to denote the parameters of the decoder network. Once trained, the decoder warps the simple distribution $p(\mathbf{z})$ to the potentially highly complex distribution of clean images. Even though this is an extremely expressive model, training of the parameters θ is challenging due to the intractable integral Eq. 2. In practice, a VAE can be trained by maximising the variational lower bound

$$\log p_{\theta,\eta}(\mathbf{x}) \geq \mathbb{E}_{q_\phi(\mathbf{z}|\mathbf{x})}[\log p_\eta(\mathbf{x}|\mathbf{s} = g_\theta(\mathbf{z}))] - D_{KL}[q_\phi(\mathbf{z}|\mathbf{x}) \parallel p(\mathbf{z})], \tag{3}$$

where D_{KL} is the Kullback-Liebler divergence, and $q_\phi(\mathbf{z}|\mathbf{x})$ is a parametric distribution in latent space, implemented by a second CNN, called the *encoder*. The encoder network takes a noisy image $\mathbf{x}$ as input and outputs the parameters of the distribution. The encoder, ϕ, and the decoder, θ, are trained in tandem by maximising Eq. 3 based on a set of noisy training images.

Once trained, DivNoising can be used to denoise an image $\mathbf{x}$ by processing it with the encoder, drawing a sample $\mathbf{z}^k$ in latent space from $q_\phi(\mathbf{z}|\mathbf{x})$, and finally decoding the sample $g_\theta(\mathbf{z}^k)$ to obtain sampled solution $\mathbf{s}^k$. The resulting sampled solutions can then be combined to produce an MMSE estimate using Eq. 1.

The original DivNoising shows impressive performance in many cases, but struggles when applied to highly complex datasets, which contain diverse patterns and shapes. In these cases, the results tend to be blurry or contain artifacts.

The reason for this is that DivNoising trains a full model of the image distribution and this is a challenging task for complex datasets. Due to it's architecture, DivNoising performs especially poorly for images that contain a lot of high frequency information.

As a side effect of this, DivNoising was found to at times remove structured noise even when using a pixel-independent noise model [23]. However, this comes at the cost of a blurred denoising result.

In [23], the power of DivNoising was improved with the use of the LadderVAE architecture [26]. This version is known as Hierarchical DivNoising (HDN). The main difference between a LadderVAE and a typical VAE is that the latent variable z is replaced with a hierarchy of latent variables $z = \{z_1, z_2, \ldots z_n\}$ where each z_i is conditionally dependent upon all $z_{i+1}, \ldots, z_n$, so that the prior distribution factorises as:

$$p_\theta(z) = p_\theta(z_n) \prod_{i=1}^{n-1} p_\theta(z_i | z_{i+1}, \ldots, z_n), \tag{4}$$

and the approximate posterior factorises as:

$$q_\phi(z|x) = q_\phi(z_n|x) \prod_{i=1}^{n-1} q_\phi(z_i | z_{i+1}, \ldots, z_n, x). \tag{5}$$

with these changes, the variational lower bound to the log likelihood is now:

$$\begin{aligned}
\log p_\theta(x) \geq\; & \mathbb{E}_{q_\phi(z|x)}[\log p_\eta(x|s = g_\theta(z))] \\
& - D_{KL}[q_\phi(z_n|x) \parallel p(z_n)] \\
& - \sum_{i=1}^{n-1} \mathbb{E}_{q_\phi(z_i|x)}[q_\phi(z_i | z_{i+1}, \ldots, z_n, x) \parallel p_\theta(z_i | z_{i+1}, \ldots, z_n)]
\end{aligned} \tag{6}$$

The authors found that with HDN the denoising capability is greatly improved, especially for complex high detail datasets. However, when HDN is used with a pixel-independent noise model, it will usually also faithfully reconstruct any structured noise instead of removing it. Prakash et al. were able to address this problem in some cases by not conditioning the distribution of the lowest latent variables in the hierarchy on x. They noticed that it was through this conditioning that the model passed information about the structured noise to the output, so by severing the connection, the signal estimate was produced without the structured artifacts.

In their experiments, Prakash et al. mostly used HDN with six latent variables in the hierarchy, and when tackling structured noise they would alter the distribution of the first two. We refer this altered model as HDN_{3-6} for the remainder of this paper.

We find that HDN_{3-6} does not work in all cases (see Fig. 1) and also comes at a cost. By removing some levels of latent variables we also reduce the expressiveness the model. Consequently, when we combine HDN with our autoregressive noise model, we keep all levels of latent variables activated to allow for maximum expressive power.

3.2   Pixel-Independent Noise Models

Noise models, as they have until now been used with DivNoising and HDN, are based on the assumption that when an image is recorded for any underlying signal **s**, noise occurs independently in each pixel i. That is, the distribution factorises over the pixels of the image as

$$p_\eta(\mathbf{x}|\mathbf{s}) = \prod_{i=1}^{N} p_\eta(x_i|s_i), \tag{7}$$

where $p_\eta(x_i|s_i)$ corresponds to the distributions of possible noisy pixel values given an underlying clean pixel value at the same location i. This means that to describe the noise model for an entire image $p_\eta(\mathbf{x}|\mathbf{s})$, we only need to characterise the much simpler 1-dimensional distributions for individual pixel values $p_\eta(x_i|s_i)$. These pixel noise models have been described with the help of 2-dimensional histograms (using one dimension for the clean signal and one for the noisy observation) [15], or parametrically using individual normal distributions [30] or Gaussian mixture models [25] parameterised by the pixel's signal s_i.

3.3   Signal-Independent Noise Models (a Simplification)

Even though the models described in Eq. 7 are unable to capture dependencies on other pixels, importantly, they are able to describe a dependency on the signal at the pixel itself. For many practical applications this is essential. For example, fluorescence microscopy is often heavily influenced by Poisson shot noise [30], following a distribution that depends on the pixel's signal.

However, here in this work, we will consider only a more basic case, in which the noise does not depend on the signal and is purely additive. In this case, we can write

$$p_\eta(\mathbf{x}|\mathbf{s}) \equiv p_\eta(\mathbf{n}), \tag{8}$$

with $\mathbf{n} = \mathbf{x} - \mathbf{s}$, turning Eq. 7 into

$$p_\eta(\mathbf{x}|\mathbf{s}) = \prod_{i=1}^{N} p_\eta(n_i), \tag{9}$$

Allowing us to fully characterise the noise model by defining a single 1-dimensional distribution $p_\eta(n_i)$ describing the noise at the pixel level.

In Sect. 4, we will introduce our novel autoregressive noise model, which will allow us to get rid of the pixel-independence assumption. However, within the scope of this work we are still operating under the assumption of signal-independence (Eq. 8), leaving the more general case of combined signal- and pixel-dependence for future work.

3.4   Deep Autoregressive Models

Generally, the distribution of any high dimensional variable $\mathbf{v} = (v_1, \ldots, v_N)$ can be written as product

$$p(\mathbf{v}) = \prod_{i=1}^{N} p(v_i | v_1, \ldots, v_{i-1}) \tag{10}$$

of 1-dimensional distributions for each element $p(v_i | v_1, \ldots, v_{i-1})$ conditioned on all previous elements.

Oord et $al.$ [27] proposed using a CNN to apply this technique to image data in an algorithm known as PixelCNN. The authors suppose a row-major ordering of the pixels in the image and model the distribution $p(v_i | v_1, \ldots, v_{i-1})$ for each pixel conditioned on all pixels above and to the left of it using a CNN with an adequately shaped receptive field. When applied to the image, the network outputs the parameters of the 1-dimensional conditional distribution for each pixel (Fig. 2).

4   Methods

Considering the signal-independence assumption (Eq. 8), we can see that a structured noise model can be implemented as an image model for the distribution of noise images $\mathbf{n}$. We use the PixelCNN approach to implement this model. To train our autoregressive noise model we require training images containing pure noise. In practice, such noise images might be derived from dark areas of the image, where the signal is close to zero, or could be explicitly recorded for the purpose, e.g. by imaging without a sample. We denote these noise training images as $\mathbf{n}^j$.

To train our noise model based on Eq. 10, we use the following loss function

$$\log p_\eta(\mathbf{n}^j) = \sum_{i=1}^{N} \log p_\eta(n_i | n_1^j, \ldots, n_{i-1}^j), \tag{11}$$

where $p_\eta(n_i | n_1^j, \ldots, n_{i-1}^j)$ are the conditional pixel distributions described by our PixelCNN for pixel i by outputting the parameters of a Gaussian mixture model for each pixel.

Once our noise model is trained, we can proceed to our HDN model for denoising. We follow the training process as described in [23] and use Eq. 6 as training loss. Note that this contains the noise model $\log p_\eta(\mathbf{x} | \mathbf{s} = g_\theta(\mathbf{z}))$.

Considering Eq. 8, we can compute $\hat{\mathbf{n}} = \mathbf{x} - g_\theta(\mathbf{z})$ and insert it into Eq. 6, this time keeping the parameters η fixed.

5   Experiments

We use a total of 5 datasets in our experiments, one is intrinsically noisy PA data and the other four are synthetically corrupted imaging data.

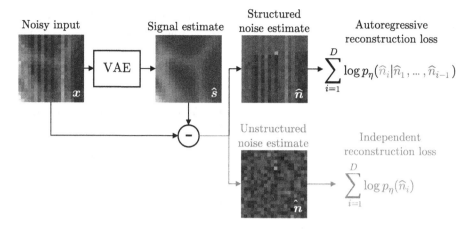

Fig. 2. Our autoregressive noise model as a component in the DivNoising framework. Divnoising trains a VAE to describe the distribution of noisy images **x**. It does so by sampling clean images **ŝ** and using a noise model as part of its loss function, called *reconstruction loss*. The reconstruction loss assess the likelihood of network output **ŝ** giving rise to original noisy training image **x**. It is defined as the logarithm of the noise model. In both cases, for the pixel-independent noise model and our autoregressive noise model, the reconstruction loss can be computed efficiently as a sum over pixels. For the pixel-independent noise model, this is done based on the conditional independence assumption by summing over the pixel noise models $\log p(\hat{n}_i)$, modelled as a Gaussian mixture model. In our autoregressive noise model we sum over the conditional distributions $p(\hat{n}_i|\hat{n}_1, \ldots, \hat{n}_{i-1})$ for the noise in each pixel conditioned on the previous pixels, *i.e.*, the pixels above and left. Our noise models describes these conditional distributions using a modified version of the PixelCNN [21] approach, which is implemented as an efficient fully convolutional network, outputting the parameters of a separate Gaussian mixture model for each pixel.

5.1 Synthetic Noise Datasets

While datasets of paired noisy and clean images are not needed to train our denoiser, they are needed to quantitatively evaluate the denoiser's performance using metrics such as peak signal-to-noise ratio (PSNR). The method proposed here is currently only capable of removing signal-independent noise, with the extension to signal-dependent noise being left for future work. We are not aware of any real datasets of paired noisy and clean images that do not contain signal-dependent noise, and have therefore created synthetic pairs by adding signal-independent noise to clean images for the purpose of quantitative evaluation. The very noise images that were added to the clean images in the simulated datasets were used to train their noise models but this was only for convenience. Any dataset of noise recorded under the same conditions as the signal could be used.

Convallaria sCMOS: Broaddus *et al.* [6] took 1000 images of a stationary section of a *Convallaria* with size 1024×1024. Each image contained signal-dependent noise, but the average of the 1000 images is an estimate of the ground truth. We normalised this ground truth and split it into patches of size 128×128. For each patch, we added the same sample from the standard normal distribution to the upper 64 pixels in a column, taking a different sample for every column, and then did the same for the lower 64 pixels. We then added pixel-independent Gaussian noise with a standard deviation of 0.3. This was an attempt to produce noise similar to the sCMOS noise shown in Fig. 6 of [19].

Brain CT. 2486 clean CT brain scan images were taken from Hssayeni [11] and centre cropped to size 256×256. Independent Gaussian noise was generated with a standard deviation of 110. This noise was smoothed by a Gaussian filter with a standard deviation of 1 vertically and 5 horizontally. More independent Gaussian noise with a standard deviation of 20 was added on top of that. Finally, we subtracted and shifted the noise to have zero mean. This noise was intended to be similar to the CT noise shown in Fig. 3 of [22].

KNIST. The Kuzushiji-MNIST dataset was taken from Clanuwat *et al.* [8]. The data was normalised before adding a value of 1 to diagonal lines to create a stripe pattern. Independent Gaussian noise with a standard deviation of 0.3 was then added on top. This was intended to demonstrate how HDN_{3-6} with a pixel-independent noise model fails on long range, strong correlations while HDN with our noise model is successful.

5.2   Photoacoustic Dataset

PA imaging is the process of detecting ultrasound waves as they are emitted by tissues that are being made to thermoelastically expand and contract by pulses of an infrared laser. The resulting data is a time series, and noise samples can be acquired by taking a recording while the infrared laser is not pulsed.

This particular dataset is afflicted with structured noise (see Fig. 4) that is thought to have been caused by inter-pixel sensitivity variations. It consists of 468 observations of a signal and 200 observations of only noise, with size 128×128.

5.3   Training the Noise Model

The noise model used in experiments uses the architecture in van den Oord *et al.* [21], modified to output the parameters of a Gaussian mixture model. We used the same hyperparameters for each dataset. Those hyperparameters were 5 layers, 128 feature channels and a kernel size of 7. The output of the network was the parameters of a 10 component Gaussian mixture model for each pixel. The Adam optimiser with an initial learning rate of 0.001 was used, and learning rate was reduced by a factor of 0.99 every epoch. Every dataset was trained on for a maximum of 12000 steps, but a patience of 10 on the validation loss was used to avoid overfitting on the training set. Images were randomly cropped to 64×64, except the kanji data which was trained on full images. All experiments used a batch size of 8.

5.4   Training HDN

The HDN architecture was based on that of Prakash *et al.* [23] and was kept the same for all experiments. 6 hierarchical latent variables were used, each with 32 feature channels. There was a dropout probability of 0.2, and to prevent KL vanishing, the free bits approach [13] was used with a lambda of 0.5. The Adamax optimiser was used with a learning rate of 0.0003 and learning rate was reduce by 0.5 when the validation loss plateaued for more than 10 epochs. The same patch and batch size as in the training of the noise model was used.

5.5   Denoising with Autoregressive Noise Models

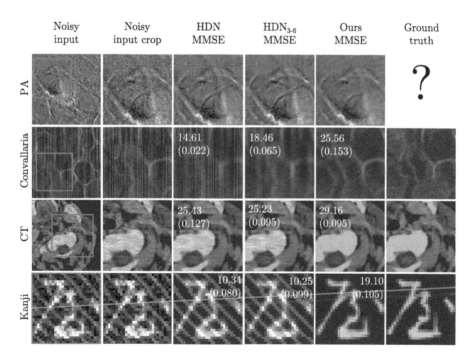

Fig. 3. Denoising results. Here we compare the outputs of different methods on various datasets. The overlaid numbers indicate the mean PSNR values on the dataset after three experiments with the standard deviation in brackets. We find that HDN with a pixel-independent noise model is able to effectively remove some structured artifacts, by removing layers of the latent space space [23], but fails for larger scale structures, spanning over tens of pixels. In contrast, our method reliably removes all small- and large-scale structured noise.

Each of the 4 datasets was denoised using HDN with a pixel-independent Gaussian noise model, HDN_{3-6} with a pixel-independent Gaussian noise model and HDN with our autoregressive noise model. For each test image, 100 samples were generated from each trained model and averaged to produce an MMSE estimate, Each result is shown in Fig. 3, with peak signal-to-noise ratio calculated for the datasets where ground truth is available.

The highest PSNR was achieved by HDN with our noise model. For all of the datasets, HDN with a Gaussian pixel-independent noise model seemed to remove only the pixel-independent component of the noise, while retaining the structured parts. In some cases, HDN_{3-6} manages to partially remove structured noise.

For the KNIST dataset, both HDN and HDN_{3-6} fail to remove the diagonal lines, which are completely removed by our structured noise model.

We believe that HDN_{3-6} is unable to remove these noise structures because they feature long range correlations, which are not only captured by the two lowest latent variables but also by others in the hierarchy, entangled with the signal.

Similarly, for the PA dataset, only our autoregressive noise model is able to remove the structured recording noise. Here, however, we find that our method produces a slightly blurred result. We attribute this to the limited amount of available noise model training data for this dataset. To avoid overfitting, we had to stop noise model training early in this case, which we believe leads to a sub-optimal end result.

5.6   Evaluating the Noise Model

To show how the autoregressive noise model is able to capture dependencies across an image, we calculated the 2dimensional auto-correlation of the real noise from the PA data, samples of noise generated from our autoregressive noise model and samples of noise generated by a pixel-independent noise model. Each of these auto-correlation graphs are shown in Fig. 4, along with an image of each type of noise for visual comparison.

5.7   Choice of Autoregressive Pixel Ordering

Some might be concerned that the choice of autoregressive ordering should take into account the direction of dependencies in the noise, but, fundamentally, this is not the case. Equation 10 generally holds for any distribution of images and also regardless of the used pixel order.

Take, for example, the simulated noise in the Convallaria sCMOS dataset which is designed to be correlated vertically but not horizontally. In the modelling of the noise in this dataset, the distribution over the possible values of one pixel will be more concentrated if it is a function of the other pixels in the same column. However, considering Eq. 10, the autoregressive model must sweep through the whole image one pixel at a time. Therefore, no matter if we choose

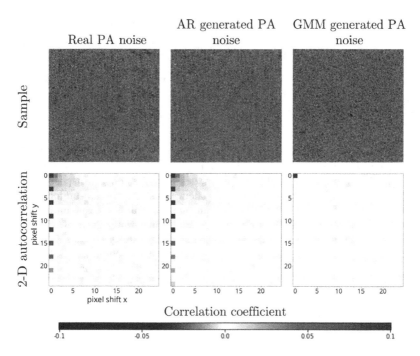

Fig. 4. Comparing the statistics of pixel-independent noise models and our new autoregressive model. Here, we compare generated PA noise samples from our noise model (AR) and a Gaussian mixture pixel-independent noise model (GMM) to real PA noise. The auto-correlation function compares different shifted versions (pixel shift) of the noise images in both directions, characterising the dependencies between pixels values at various distances and directions, *i.e.*, the structure of the noise. As expected, the pixel-independent noise model is unable to capture any such dependencies present in the real noise. In contrast, our autoregressive noise model can faithfully capture and reproduce even longer range dependencies.

a row-major or column-major ordering, at for at least one pixel the distribution has to be computed using without considering relevant correlated pixels. On the other hand, in both cases only one pixel in a column can be a function of all relevant, correlated pixels. Both a row-major and column-major ordering of pixels can achieve this if they have a large enough receptive field.

To demonstrate that there are no practical disadvantages arising from the choice of the pixel order, we ran the experiment on the Convallaria sCMOS dataset with transposed images, which corresponds to changing the pixel order. Figure 5 shows the results of this experiment, where almost no perceptual difference between the MMSE of the two experiments can be detected and only a slight difference in mean PSNR is recorded.

| Noisy input$_1$ | Noisy input$_2$ | MMSE$_1$ | MMSE$_2^T$ | Ground truth |

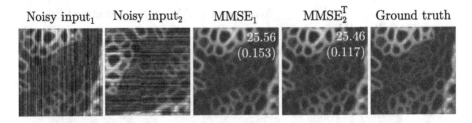

Fig. 5. Our noise model can capture noise patterns regardless of their orientation or the direction of pixel ordering. To demonstrate this, we reran the experiment (including training of the noise model and VAE) on a transposed version of the Convallaria sCMOS dataset. This is equivalent to using a column-major ordering of pixels to train the noise model, while the original experiment used a row-major ordering. We compare denoising results carried out on the original Convallaria sCMOS dataset (Noisy input$_1$, MMSE$_1$) to the transposed version of the dataset (input$_2$, MMSE$_2$). We have transposed the result MMSE$_2^T$ again to allow for easier comparison. The overlayed numbers indicate the average PSNR and its standard deviation (in brackets) over three reruns of the experiment.

6   Conclusion

We have presented a novel type of noise model to be used within the DivNoising framework that addresses, structured noise and outperforms HDN$_{3-6}$ on highly structured, long range noise artefacts. Both the noise model and DivNoising framework can be trained without matched pairs of clean and noisy images. Instead, practitioners require a set of noise samples and the images that are to be denoised. We believe this can potentially have great impact, by enabling applications with structured noise for which no paired data is available.

The key difference between our noise model and those that had been used before [15,24,25] is that ours evaluates the probability of a noise pixel conditioned on other pixels in the image, while previously used noise models evaluate the probability of each pixel independently.

Currently, our method is limited to signal-independent noise, which makes a direct application impossible for many settings, such as fluorescence microscopy, where data is usually affected by signal dependent Poisson shot noise. However, we do believe, that we have made the first step towards widely applied unsupervised removal of structured noise.

In future work, we plan to extend this noise model to learn the distribution of signal-dependent noise, which would vastly increase its utility in the field of life science imaging and beyond.

Acknowledgements. We would like to thank Paul Beard and Nam Huynh for providing us with their photoacoustic imaging dataset, as well as for the insightful discussions we had. Additionally, we want to thank, Ben Cox, James Guggenheim and Dylan Marques for discussing the data and for introducing us to photoacoustic imaging.

References

1. Abdelhamed, A., Brubaker, M.A., Brown, M.S.: Noise flow: noise modeling with conditional normalizing flows. In: Proceedings of the IEEE/CVF International Conference on Computer Vision, pp. 3165–3173 (2019)
2. Babcock, H.P., Huang, F., Speer, C.M.: Correcting artifacts in single molecule localization microscopy analysis arising from pixel quantum efficiency differences in sCMOS cameras. Sci. Rep. **9**(1), 1–10 (2019)
3. Batson, J., Royer, L.: Noise2self: blind denoising by self-supervision (2019)
4. Belthangady, C., Royer, L.A.: Applications, promises, and pitfalls of deep learning for fluorescence image reconstruction. Nat. Methods **16**(12), 1215–1225 (2019)
5. Bond-Taylor, S., Leach, A., Long, Y., Willcocks, C.G.: Deep generative modelling: a comparative review of VAEs, GANs, normalizing flows, energy-based and autoregressive models. arXiv preprint arXiv:2103.04922 (2021)
6. Broaddus, C., Krull, A., Weigert, M., Schmidt, U., Myers, G.: Removing structured noise with self-supervised blind-spot networks. In: 2020 IEEE 17th International Symposium on Biomedical Imaging (ISBI), pp. 159–163 (2020)
7. Buades, A., Coll, B., Morel, J.M.: A non-local algorithm for image denoising. In: 2005 IEEE computer society conference on computer vision and pattern recognition (CVPR2005), vol. 2, pp. 60–65. IEEE (2005)
8. Clanuwat, T., Bober-Irizar, M., Kitamoto, A., Lamb, A., Yamamoto, K., Ha, D.: Deep learning for classical Japanese literature. arXiv preprint arXiv:1812.01718 (2018)
9. Cnudde, V., Boone, M.N.: High-resolution x-ray computed tomography in geosciences: a review of the current technology and applications. Earth Sci. Rev. **123**, 1–17 (2013)
10. Dabov, K., Foi, A., Katkovnik, V., Egiazarian, K.: Image denoising by sparse 3-D transform-domain collaborative filtering. IEEE Trans. Image Process. **16**(8), 2080–2095 (2007)
11. Hssayeni, M., Croock, M., Salman, A., Al-khafaji, H., Yahya, Z., Ghoraani, B.: Computed tomography images for intracranial hemorrhage detection and segmentation. Intracranial Hemorrhage Segmentation Using A Deep Convolutional Model. Data **5**(1), 14 (2020)
12. Kingma, D.P., Dhariwal, P.: Glow: Generative flow with invertible 1x1 convolutions. In: Advances in neural information processing systems 31 (2018)
13. Kingma, D.P., Salimans, T., Jozefowicz, R., Chen, X., Sutskever, I., Welling, M.: Improved variational inference with inverse autoregressive flow. In: Advances in neural information processing systems 29 (2016)
14. Krull, A., Buchholz, T.O., Jug, F.: Noise2void-learning denoising from single noisy images. In: Proceedings of the IEEE Conference on Computer Vision and Pattern Recognition, pp. 2129–2137 (2019)
15. Krull, A., Vicar, T., Prakash, M., Lalit, M., Jug, F.: Probabilistic Noise2Void: unsupervised content-aware denoising. Front. Comput. Sci. **2**, 60 (2020)
16. Laine, R.F., Jacquemet, G., Krull, A.: Imaging in focus: an introduction to denoising bioimages in the era of deep learning. Int. J. Biochemis. Cell Biol. **140**, 106077 (2021)
17. Lehtinen, J., et al.: Noise2noise: Learning image restoration without clean data. In: International Conference on Machine Learning, pp. 2965–2974 (2018)
18. Luisier, F., Vonesch, C., Blu, T., Unser, M.: Fast interscale wavelet denoising of poisson-corrupted images. Signal Process. **90**(2), 415–427 (2010)

19. Mandracchia, B., Hua, X., Guo, C., Son, J., Urner, T., Jia, S.: Fast and accurate sCMOS noise correction for fluorescence microscopy. Nat. Commun. **11**(1), 1–12 (2020)
20. Moran, N., Schmidt, D., Zhong, Y., Coady, P.: Noisier2noise: learning to denoise from unpaired noisy data. In: Proceedings of the IEEE/CVF Conference on Computer Vision and Pattern Recognition, pp. 12064–12072 (2020)
21. Van den Oord, A., Kalchbrenner, N., Espeholt, L., Vinyals, O., Graves, A., et al.: Conditional image generation with pixelCNN decoders. In: Advances in Neural Information Processing Systems 29 (2016)
22. Parakh, A., et al.: Recognizing and minimizing artifacts at dual-energy CT. Radiographics **41**(2), 509 (2021)
23. Prakash, M., Delbracio, M., Milanfar, P., Jug, F.: Interpretable unsupervised diversity denoising and artefact removal. In: International Conference on Learning Representations (2022). https://openreview.net/forum?id=DfMqlB0PXjM
24. Prakash, M., Krull, A., Jug, F.: Fully unsupervised diversity denoising with convolutional variational autoencoders. In: International Conference on Learning Representations (2020)
25. Prakash, M., Lalit, M., Tomancak, P., Krull, A., Jug, F.: Fully unsupervised probabilistic noise2void. arXiv preprint arXiv:1911.12291 (2019)
26. Sønderby, C.K., Raiko, T., Maaløe, L., Sønderby, S.K., Winther, O.: Ladder variational autoencoders. In: Advances in neural information processing systems 29 (2016)
27. Van Oord, A., Kalchbrenner, N., Kavukcuoglu, K.: Pixel recurrent neural networks. In: International Conference on Machine Learning, pp. 1747–1756. PMLR (2016)
28. Weigert, M., et al.: Content-aware image restoration: pushing the limits of fluorescence microscopy. Nat. Methods **15**(12), 1090–1097 (2018)
29. Xu, M., Wang, L.V.: Photoacoustic imaging in biomedicine. Rev. Sci. Instrum. **77**(4), 041101 (2006)
30. Zhang, Y., et al.: A poisson-gaussian denoising dataset with real fluorescence microscopy images. In: CVPR (2019)

Comparison of Semi-supervised Learning Methods for High Content Screening Quality Control

Umar Masud[4], Ethan Cohen[1,2]([✉]), Ihab Bendidi[1,3], Guillaume Bollot[2], and Auguste Genovesio[1]

[1] IBENS, Ecole Normale Supérieure, Paris, France
auguste.genovesio@ens.psl.edu
[2] SYNSIGHT, Evry, France
ecohen@bio.ens.psl.eu
[3] Minos Biosciences, Paris, France
[4] Jamia Millia Islamia, New Delhi, India

Abstract. Progress in automated microscopy and quantitative image analysis has promoted high-content screening (HCS) as an efficient drug discovery and research tool. While HCS offers to quantify complex cellular phenotypes from images at high throughput, this process can be obstructed by image aberrations such as out-of-focus image blur, fluorophore saturation, debris, a high level of noise, unexpected autofluorescence or empty images. While this issue has received moderate attention in the literature, overlooking these artefacts can seriously hamper downstream image processing tasks and hinder detection of subtle phenotypes. It is therefore of primary concern, and a prerequisite, to use quality control in HCS. In this work, we evaluate deep learning options that do not require extensive image annotations to provide a straightforward and easy to use semi-supervised learning solution to this issue. Concretely, we compared the efficacy of recent self-supervised and transfer learning approaches to provide a base encoder to a high throughput artefact image detector. The results of this study suggest that transfer learning methods should be preferred for this task as they not only performed best here but present the advantage of not requiring sensitive hyperparameter settings nor extensive additional training.

Keywords: Cell-based assays · Image analysis · Deep learning · Self-supervised learning

1 Introduction

Image analysis solutions are heavily used in microscopy. They enable the extraction of quantitative information from cells, tissues and organisms. These methods

---

U. Masud and E. Cohen—Equal co-contribution.

© The Author(s), under exclusive license to Springer Nature Switzerland AG 2023
L. Karlinsky et al. (Eds.): ECCV 2022 Workshops, LNCS 13804, pp. 395–405, 2023.
https://doi.org/10.1007/978-3-031-25069-9_26

and tools have proven to be especially useful for high-content screening (HCS), an automated approach that produces a large amount of microscopy image data, to study various mechanisms and identify genetic and chemical modulators in drug discovery and research [19]. However, the success of an HCS screen is often related to the dataset quality obtained at end. In practice, abnormalities in image quality are numerous and can lead to imprecise results at best, and erroneous results or false conclusions at worst. Common abnormalities include noise, out-of-focus, presence of debris, blur or image saturation. Furthermore, in some cases, it can also be convenient to exclude images full of dead or floating cells. More importantly, in HCS, manual inspection of all images in a dataset is intractable, as one such screen typically encompasses hundreds of thousands of images.

Quality control (QC) methods have been investigated for this purpose. Interesting software such as CellProfiler [6] allows end-to-end analysis pipeline with an integrated QC modules. Although powerful, the image quality measures are mainly handcrafted with different computed metrics as described in [2] and therefore hard to generalize. More recently, Yang et al. proposed a method to assess microscope image focus using deep learning [21]. However, this approach is restricted to a specific type of aberration and does not generalize well to other kinds of artefacts. Besides, learning all types of aberrations from scratch in a supervised manner is hardly tractable, given the diversity of both normal and abnormal image types. It would require systematic annotation of all types of aberrations on each new high-throughput assay, and thus would be utterly time-consuming and hardly feasible in practice. For this task, we thus typically seek a semi-supervised solution that would require annotation of a limited amount of data per assay.

Transfer learning typically offers such a solution that relies on little supervision [13]. A network pretrained on a large annotated image set can be reused directly or fine-tuned with a limited set of annotated images to solve a specific task in another domain. Furthermore, recent breakthroughs in self-supervised learning (SSL), which aim to learn representations without any labels data call for new methods [1,4,5,7,9,12,20,24]. For instance, such a framework was successfully used by Perakis et al. [17] to learn single-cell representations for classification of treatments into mechanisms of action. It was shown that SSL performed better than the more established transfer learning (TL) in several applications. However, it is not a strict rule and not systematically the case as assessed by a recent survey [22]. It is still unclear which approach works better on what type of data and tasks.

In this work, we propose to address this question in the context of HCS quality control. To this end we performed a comparative study of a range of SSL and TL approaches to detect abnormal single-cell images in a high-content screening dataset with a low amount of annotated assay specific image data. The paper is organized as follows. In Sect. 2, we briefly describe the various methods we use for transfer and self-supervised representation learning. In Sect. 3, we then detail the setup of this comparative study. We then provide experimental results in Sect. 4, and concluding remarks in Sect. 5.

2 Related Work

We seek a method that would provide a robust base encoder to a quality control downstream task where a low amount of annotated data is available. We thought of several options that could be grouped in two categories: transfer learning and self-supervised learning methods.

2.1 Transfer Learning

Training a deep learning model efficiently necessitates a significant amount of data. In the case of supervised training, it is required that data be annotated with class labels. Transfer learning has become popular to circumvent this issue. It consists in pretraining a network on a large set of annotated images in a given domain, typically a domain where image could be annotated. A variety of tasks in various other domains can then be addressed with decent performance simply by reusing the pretrained network as is or by fine tuning its training on a small available dataset on a specific task in the domain of interest.

In this work, we included three popular networks pretrained with ImageNet for transfer learning. First we used VGG16, a model introduced in 2014 that made a significant improvement over the early AlexNet introduced in 2012, by widening the size of convolutional layer kernels [18]. We also used ResNet18, a network introduced in 2016 that implements residual connections to make possible the stable training of deeper networks [11]. Finally we used ConvNext, one of the most recent convolutional networks introduced in 2022 that competes favorably with most models, including vision transformer, while maintaining the simplicity and efficiency of ConvNets [14].

2.2 Self-supervised Learning

In recent years, self-supervised representation learning has gained popularity thanks to its ability to avoid the need for human annotations. It has provided ways to learn useful and robust representation without labeling any data. Most of these approaches rely on a common and simple principle. Two or more random transformations are applied to the same images to produce a set of images containing different views of the same information content. These images are then passed through an encoder that is trained to somewhat encourage learning of a close and invariant representation through the optimization of a given loss function. The loss function varies depending on the method, but once a self-supervised representation is learned, it can be used to solve downstream tasks that may require little to no annotated data. Various kind of SSL mechanisms have been developed, but a wide range of approaches can be summarized in three classes of methods our study encompasses here, namely contrastive, non contrastive and clustering-based methods:

1. **Contrastive learning methods** aim to group similar samples closer and diverse samples farther from one another. Although powerful, such methods

still need to find some negative examples via a memory bank or to use a large batch size for end to end learning [12].

2. **Non-contrastive learning methods** use only positive sample pairs compared to contrastive methods. These approaches proved to learn good representations without the need for a large batch size or memory bank [1,9,24].

3. **Clustering-based learning methods** ensure that similar samples cluster together but use a clustering algorithm instead of similarity metrics to better generalize by avoiding direct comparison [4,5,20].

In this work, we used a list of methods from the three categories listed above, namely SimCLR [7] for contrastive learning (based on similarity maximisation objective), Barlow Twins [24] and VICReg [1] for non-contrastive techniques (based on redundancy reduction objective) and DeepCluster [4] and SwAV [5] for clustering-based methods.

3   Method

We performed a comparative study that aimed at identifying which of the previously described approaches could be best suited to provide a base encoder, in order to build a classifier for abnormal images from a small annotated image set. In this section we describe the data, the way we perform training for the encoders and the downstream tasks we designed to evaluate and compare them.

3.1   Data

We used the BBBC022 image set, available from the Broad Bioimage Benchmark Collection to conduct our experiments [10,16]. To obtain images at a single-cell level, we cropped a fixed 128×128 pixel square around the center of each nucleus, resulting in a total of $2,122,341 (\approx 2.1M)$ images [15] . Most of these images were used to train the base encoder when needed (i.e. for SSL methods). Separately, we manually annotated 240 abnormal and 240 normal images and split them in a balanced way into training (350) and test (130) sets, with a 50% ratio of normal images in both the training set and the test set, for the downstream tasks. Some annotated images are displayed in Fig. 1. Furthermore, we also used 200 annotated images from the BBBC021 image set, available from the Broad Bioimage Benchmark Collection to test the generalization of our approach [3,16]. This dataset differs from the BBBC022 in the types of cells used, with U20S cells for BBBC022 and MCF7 cells for BBBC021 The difference in cell line results in differences in gene expression which translate in different visual features, even with a similar image acquisition process [23].

3.2   Encoder Training

For all TL methods, we used a model pre-trained on ImageNet as an encoder. For SSL methods, we used two networks - first a ResNet18 as encoder, and a fully

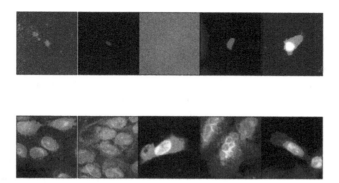

Fig. 1. BBBC022 dataset: the first row displays a few abnormal images, the second row shows a few regular images of cells. U2OS cells with Hoechst 33342 staining for nuclei (blue), WGA + phalloidin staining for actin filaments (Red) and MitoTracker staining for mitochondria (green). (Color figure online)

connected layers (FC layers) as projector. The encoder takes an image as input and outputs a 512 dimensional vector, which goes as input for the projector network, that is in turn made of 2048 dimension FC layers and a temperature of 0.07 for the unsupervised loss. We forward pass the batch of images after producing two different views of them using augmentations. The following augmentations were randomly performed: 90 degree rotation, flip, transpose, shift and scale. We carefully chose these augmentations so as to keep the trained features relevant to our downstream classification tasks. We use a line search strategy to optimize the hyperparameters of the models. The encoder was afterward trained for 5 epochs (about 10 million images) for all the SSL methods, with a batch size of 128, using the SGD optimizer with an initial learning rate of 0.001 and a momentum of 0.9. We also used a warm-up cosine scheduler with warm-up epochs set to 1. For DeepCluster specifically, we set 500 prototypes and 10 K-means iterations. As for SwAV, we also set the number of prototypes to 500 and chose a queue size of 2560. Other projector network parameters were the same as those used in the original papers. We trained models and ran experiments using one Tesla P100 GPU with 16GB vram. All experiments involving SSL methods were done using the solo-learn library [8]. Once the encoder was trained, the projector network was discarded and only the ResNet18 network was used for downstream tasks.

3.3 Downstream Classification Tasks

After training an encoder with each previously described TL or SSL method, we used them to train and test the three following downstream classification tasks, only with the small annotated dataset previously described, after having performed a line search strategy for hyperparameter optimization:

1. **K-nn on a frozen encoder output.** We first aimed to evaluate a simple classification setting that did not necessitate any additional training. To this

end, we performed a K-Nearest Neighbour (KNN) classification (here we chose k=5) on the 512 feature vectors output of the encoder.

2. **Linear classifier trained on a frozen encoder output.** We then evaluated the supervised training of a single dense layer with 2 output classes on top of the pre-trained encoder. In this setting, the weights of the pre-trained encoder were frozen. We trained this layer for 150 epochs and with a batch size of 32. The optimizer was SGD with a learning rate set to 0.001 and momentum to 0.9. We also used a step scheduler with gamma value set to 0.1 at 40, 70 and 120 epochs.

3. **Linear classifier with fine tuned training of the encoder.** We then used a dense linear with 2 output classes as in the previous settings. However, this time we did not freeze the encoder network and allowed it to pursue training. We trained the models for 50 epochs. The learning rate was 0.001 with a momentum of 0.9 with a SGD optimizer. We also used a step scheduler with a gamma value of 0.1 at 25 on 40 epochs.

3.4 Evaluation Criteria

We used Accuracy, F1-Score and the Area Under Curve (AUC) score to assess the classification results. All displayed values are weighted average for the 2 classes. The most important metric is the F1 score because it takes both false positives and false negatives into account. Thus, the higher the F1 score, the better the result. All values mentioned are in percentages.

4 Results

4.1 Evaluations on Downstream Tasks

The results for the Linear Layer classifier and KNN are displayed in Table 1. Among the self-supervised method, we can observe that DeepCluster performs best in both settings and reaches a maximum of 94.57% accuracy. SimCLR also performs best with KNN while VICReg performs poorly, dropping to 76.30% accuracy. However, none of the SSL methods outperforms the three TL encoders with ConvNext culminating at 98.47% accuracy with a Linear Layer classifier.

Furthermore, the results obtained with fine tune trainings of all the encoder weights are displayed in the first 3 columns of Table 2. We can see that with 350 training images (the full annotated training image set), the best results were again obtained with the three TL methods. However, SSL method performed almost as well in this setting with simCLR reaching the best results among the SSL methods with 98.44% accuracy.

4.2 Effect of a Decreasing Amount of Annotated Data

We also performed an ablation study where the number of training images was gradually decreased. We performed training of the third task with 350, 100,

Table 1. Classification with KNN or a single Linear Layer with a frozen encoder using a 350-image training set. 130 images were used for test.

Method	KNN			Linear Layer		
	Acc	F1	AUC	Acc	F1	AUC
VGG16	96.09	95.94	95.38	98.24	98.12	98.09
ResNet18	**97.66**	**97.51**	**97.63**	98.44	98.21	**98.26**
ConvNext	97.65	97.42	97.56	**98.47**	**98.24**	98.17
SimCLR	90.62	89.63	90.30	91.40	90.56	91.20
Barlow Twins	89.84	88.02	88.24	87.28	86.42	87.35
VICReg	76.30	75.28	75.13	87.76	87.36	87.40
DeepClusterV2	90.62	89.50	89.81	94.57	94.02	94.14
SwAV	89.84	88.52	89.46	94.53	94.00	94.08

50, 25, and finally just 10 images. The purpose for decreasing the amount of training images was to evaluate how much supervision the network needs to perform properly.

The results are displayed in Table 2. As the number of training images decrease, VICReg, DeepCluster and SwAV display a drop in performance. With only 25 images, Barlow Twins still produces fairly good results with 94.53% accuracy. With just 10 images, the best result among the SSL methods is sim-CLR with 84.89% accuracy. Overall, semi-supervised training can yield good results even with a few images. However, here again, none of the SSL methods outperforms the transfer learning baselines.

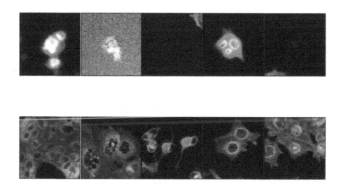

Fig. 2. BBBC021 dataset: the first row displays a few abnormal images, the second row displays a few regular cell images. Fixed human MCF7 cells labeled for DNA (blue), actin (red) and B-tubulin (green) (Color figure online)

4.3   Effect of a Domain Shift

To evaluate how these encoders pretrained on BBBC022 or ImageNet could generalize to a different dataset, we tested them on data taken from BBBC021. For this purpose, we considered our best model, the Linear Layer approach with fine tune training of all the encoder weights on 350 images from the BBBC022 dataset. We then tested it on unseen data taken from BBBC021. We annotated 100 normal and 100 abnormal images from this last dataset for this purpose. Some sample images are displayed in Fig. 2. We made sure to include diverse images in order to thoroughly check the robustness of our trained models.

The results are displayed in Table 3. Among SSL methods, SimCLR and DeepCluster performed best with respectively 73.66% and 72.32% accuracy. These results show that some self-supervised learning methods such as simCLR or DeepCluster trained on a large dataset produce features that could generalize a quality control task to an unseen dataset to some extent. However, in accordance with what was observed in previous sections on BBBC022, none of these approaches outperformed the results obtained with the TL encoders.

Table 2. Effect of a decreasing amount of training images on a Linear Layer classifier with a non-frozen encoder. 130 images were used for test.

Number of Training Images

Method	350			100			50			25			10		
	Acc	F1	AUC	Acc	F1	AUC	Acc	F1	AUC	Acc	F1	AUC	Acc	F1	AUC
VGG16	99.08	98.92	99.01	99.12	98.83	98.89	**99.22**	**99.18**	**99.24**	**98.44**	**98.19**	**98.26**	96.09	95.50	95.81
ResNet18	99.19	99.11	99.02	99.08	98.96	99.07	98.54	98.48	98.47	97.66	97.81	97.62	93.75	93.83	93.54
ConvNext	**99.39**	**99.21**	**99.25**	**99.28**	**99.17**	**99.24**	97.66	97.27	97.53	97.62	97.49	98.02	**96.87**	**96.68**	**96.55**
SimCLR	98.44	98.21	98.34	91.93	92.52	93.00	92.97	93.13	93.15	92.97	93.17	92.80	84.89	84.61	84.32
Barlow Twins	98.44	98.14	98.12	95.05	95.08	94.95	94.53	94.21	93.86	94.53	94.12	93.65	75.00	80.39	83.14
VICReg	98.24	98.00	97.88	89.32	89.03	89.20	71.01	74.33	74.15	72.13	77.37	80.22	66.40	72.12	72.00
DeepClusterV2	96.87	96.18	96.25	81.77	82.47	83.34	83.59	83.00	83.20	86.98	86.53	87.09	83.13	83.31	82.79
SwAV	94.53	94.00	94.10	83.59	83.55	83.70	82.56	82.21	83.09	87.50	87.12	87.25	82.87	82.62	83.39

Table 3. Out of Domain Test. Linear Layer classifier with a non-frozen encoder trained on 350 images from the BBBC022 dataset and tested on the 200 images of the BBBC021 dataset.

Method	Linear Layer		
	Acc	F1	AUC
VGG 16	96.43	97.51	98.02
ResNet18	91.52	93.21	92.87
ConvNext	**98.66**	**98.53**	**98.91**
SimCLR	73.66	78.47	79.00
Barlow Twins	54.91	62.02	58.76
VICReg	37.95	40.00	39.21
DeepClusterV2	72.32	75.99	76.12
SwAV	56.25	57.50	57.30

5 Conclusion

In this work, we conducted a thorough investigation to evaluate transfer and self-supervised representation learning on a large dataset in order to perform a downstream HCS quality control task. The quantitative results we obtained suggest that TL approaches perform better than SSL for this task. Importantly, all SSL methods come with the need to choose crucial hyperparameters that will have significant impact on the learned representation. Among these hyperparameters are the choice of transformations that will define feature invariance in the obtained representation. Furthermore, SSL methods require an additional training on a large set of unannotated images. In contrast, an ImageNet pretrained encoder combined with a KNN downstream can be used out of the box and does not require any training or hyperparameter setting. If training can be performed, then unfreezing the encoder weights and fine tuning the training with a low amount of annotated data will slightly increase the performances, with TL still being a better option than SSL. Altogether this suggests that for the task of identifying abnormal versus normal image, transfer learning should be the preferred choice.

Two reasons could be hypothesized to explain our findings. First, one could argue that our choice of transformations for the SSL approaches may not be the best option to create an optimal representation for our downstream quality control tasks. However, the choices we made were reasonable and relevant, and anyone seeking to solve a task using SSL would face the same issue: choosing hyperparameters and performing an additional training. Importantly, the debate on hyperparameter settings would be sound if transfer learning did not perform so well. Here we show that it is not only performing better than all SSL approaches, but it reaches almost perfect results in several setups, suggesting that even a better choice of SSL augmentations would not necessarily be worth finding. Secondly, this high performance obtained with transfer learning may be related to the specificity of the downstream task. Indeed, the experiments performed in the papers presenting these SSL approaches are often based on ImageNet classification which contains homogeneous semantic classes and therefore represents a different objective than the one presented in this work. Abnormal images do represent a very variable class with, for instance, out-of-focus image of cells being very different than an image containing debris. In this case, the low level features retrieved from the natural images of ImageNet may simply be sufficient and more efficient than higher semantic structure SSL representation typically provides. Although we focused on high-content screening here, we hope our findings will benefit quality control in other imaging modalities.

Acknowledgments. This work was supported by ANR-10-LABX-54 MEMOLIFE and ANR-10 IDEX 0001 -02 PSL* Université Paris and was granted access to the HPC resources of IDRIS under the allocation 2020-AD011011495 made by GENCI.

References

1. Bardes, A., Ponce, J., Lecun, Y.: Vicreg: variance-invariance-covariance regularization for self-supervised learning. In: International Conference on Learning Representations (2022)
2. Bray, M.A., Carpenter, A.E.: Quality control for high-throughput imaging experiments using machine learning in cell profiler. Method. Mol. Biol. **1683**, 89–112 (2018)
3. Caie, P.D., et al.: High-content phenotypic profiling of drug response signatures across distinct cancer cells phenotypic profiling across cancer cell types. Mol. Cancer Ther. **9**(6), 1913–1926 (2010)
4. Caron, M., Bojanowski, P., Joulin, A., Douze, M.: Deep clustering for unsupervised learning of visual features. In: Proceedings of the European Conference on Computer Vision (ECCV), pp. 132–149 (2018)
5. Caron, M., Misra, I., Mairal, J., Goyal, P., Bojanowski, P., Joulin, A.: Unsupervised learning of visual features by contrasting cluster assignments. Adv. Neural. Inf. Process. Syst. **33**, 9912–9924 (2020)
6. Carpenter, A.E., et al.: Cell profiler: image analysis software for identifying and quantifying cell phenotypes. Genome Biol. **7**, R100–R100 (2006)
7. Chen, T., Kornblith, S., Norouzi, M., Hinton, G.: A simple framework for contrastive learning of visual representations. In: International Conference on Machine Learning, pp. 1597–1607. PMLR (2020)
8. da Costa, V.G.T., Fini, E., Nabi, M., Sebe, N., Ricci, E.: solo-learn: a library of self-supervised methods for visual representation learning. J. Mach. Learn. Res. **23**, 56:1-56:6 (2022)
9. Grill, J.B., et al.: Bootstrap your own latent-a new approach to self-supervised learning. Adv. Neural. Inf. Process. Syst. **33**, 21271–21284 (2020)
10. Gustafsdottir, S.M., et al.: Multiplex cytological profiling assay to measure diverse cellular states. PLoS ONE **8**(12), e80999 (2013)
11. He, K., Zhang, X., Ren, S., Sun, J.: Deep residual learning for image recognition. In: Proceedings of the IEEE Conference on Computer Vision and Pattern Recognition, pp. 770–778 (2016)
12. Jaiswal, A., Babu, A.R., Zadeh, M.Z., Banerjee, D., Makedon, F.: A survey on contrastive self-supervised learning. Technologies **9**(1), 2 (2020)
13. Kensert, A., Harrison, P.J., Spjuth, O.: Transfer learning with deep convolutional neural networks for classifying cellular morphological changes. SLAS Discov. **24**(4), 466–475 (2019)
14. Liu, Z., Mao, H., Wu, C.Y., Feichtenhofer, C., Darrell, T., Xie, S.: A convnet for the 2020s. In: Proceedings of the IEEE/CVF Conference on Computer Vision and Pattern Recognition, pp. 11976–11986 (2022)
15. Ljosa, V., et al.: Comparison of methods for image-based profiling of cellular morphological responses to small-molecule treatment. J. Biomol. Screen. **18**, 1321–1329 (2013)
16. Ljosa, V., Sokolnicki, K.L., Carpenter, A.E.: Annotated high-throughput microscopy image sets for validation. Nat. Methods **9**, 637–637 (2012)
17. Perakis, A., Gorji, A., Jain, S., Chaitanya, K., Rizza, S., Konukoglu, E.: Contrastive learning of single-cell phenotypic representations for treatment classification. International Workshop on Machine Learning in Medical Imaging (2021)
18. Simonyan, K., Zisserman, A.: Very deep convolutional networks for large-scale image recognition. arXiv (2014)

19. Singh, S., Carpenter, A.E., Genovesio, A.: Increasing the content of high-content screening. J. Biomol. Screen. **19**, 640–650 (2014)
20. Van Gansbeke, W., Vandenhende, S., Georgoulis, S., Proesmans, M., Van Gool, L.: SCAN: learning to classify images without labels. In: Vedaldi, A., Bischof, H., Brox, T., Frahm, J.-M. (eds.) ECCV 2020. LNCS, vol. 12355, pp. 268–285. Springer, Cham (2020). https://doi.org/10.1007/978-3-030-58607-2_16
21. Yang, S.J., et al.: Assessing microscope image focus quality with deep learning. BMC Bioinformatics 19 (2018)
22. Yang, X., He, X., Liang, Y., Yang, Y., Zhang, S., Xie, P.: Transfer learning or self-supervised learning? a tale of two pretraining paradigms. ArXiv (2020)
23. Yao, K., Rochman, N.D., Sun, S.X.: Cell type classification and unsupervised morphological phenotyping from low-resolution images using deep learning. Sci. Rep. **9**(1), 13467 (2019)
24. Zbontar, J., Jing, L., Misra, I., LeCun, Y., Deny, S.: Barlow twins: Self-supervised learning via redundancy reduction. In: International Conference on Machine Learning, pp. 12310–12320. PMLR (2021)

Discriminative Attribution from Paired Images

Nils Eckstein[1]([envelope]) [iD], Habib Bukhari[1], Alexander S. Bates[2] [iD],
Gregory S. X. E. Jefferis[3] [iD], and Jan Funke[1] [iD]

[1] HHMI Janelia Research Campus, Ashburn, USA
nils.eckstein@googlemail.com
[2] Harvard Medical School, Boston, MA, USA
[3] MRC Laboratory of Molecular Biology, Cambridge CB2 0QH, UK

Abstract. We present a method for deep neural network interpretability by combining feature attribution with counterfactual explanations to generate attribution maps that highlight the most discriminative features between classes. Crucially, this method can be used to quantitatively evaluate the performance of feature attribution methods in an objective manner, thus preventing potential observer bias. We evaluate the proposed method on six diverse datasets, and use it to discover so far unknown morphological features of synapses in *Drosophila melanogaster*. We show quantitatively and qualitatively that the highlighted features are substantially more discriminative than those extracted using conventional attribution methods and improve upon similar approaches for counterfactual explainability. We argue that the extracted explanations are better suited for understanding fine grained class differences as learned by a deep neural network, in particular for image domains where humans have little to no visual priors, such as biomedical datasets.

1 Introduction

Machine Learning—and in particular Deep Learning—continues to see increased adoption in crucial aspects of society such as industry, science, and healthcare. As a result, there is an increasing need for tools that make Deep Neural Networks (DNNs) more interpretable for practitioners. Two popular approaches for explaining DNN classifiers are so called *feature attribution* and *counterfactual* methods. Feature attribution methods highlight the input features that influence the output classification the most, whereas counterfactual methods present the user with a minimally modified input that would have led to a different output classification. In this work, we combine these two complementary approaches and devise a method for DNN interpretability that: (1) is able to highlight the most discriminative features of a given class pair, (2) can be objectively evaluated, and (3) is suitable for knowledge extraction from DNNs in cases where

---

Supplementary Information The online version contains supplementary material available at https://doi.org/10.1007/978-3-031-25069-9_27.

© The Author(s), under exclusive license to Springer Nature Switzerland AG 2023
L. Karlinsky et al. (Eds.): ECCV 2022 Workshops, LNCS 13804, pp. 406–422, 2023.
https://doi.org/10.1007/978-3-031-25069-9_27

humans lack a clear understanding of class differences, a common situation for biomedical image datasets.

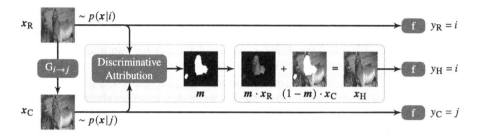

Fig. 1. Overview of the proposed method: An input image x_R of class i is converted through an independently trained cycle-GAN generator $G_{i \to j}$ into a counterfactual image x_C of class j, such that the classifier f we wish to interpret predicts $y_R = i$ and $y_C = j$. A discriminative attribution method then searches for the minimal mask m, such that copying the most discriminative parts of the real image x_R into the counterfactual x_C (resulting in the hybrid x_H) is again classified as $y_H = i$.

Feature attribution methods provide an explanation in terms of a heatmap over input pixels, highlighting and ranking areas of importance. A large number of approaches for feature attribution have been proposed in recent years (for a recent review see [27] and related work below) and they remain a popular choice for practitioners due to their ease of use, availability in popular Deep Learning frameworks, and intuitive outputs. However, the effectiveness, accuracy, and trustworthiness of these approaches is still debated [1,2,11,15], and an objective evaluation of those methods remains a difficult task [14,26]. Of particular concern are explanations that highlight the entire set of features that contributed to the output decision, including features that are shared between two different classes (so-called *distractors* [16]). This is problematic in the setting we are most concerned with, i.e., the extraction of information about class differences from a DNN in order to educate humans. This is particularly relevant in biomedical images, where humans have little to no visual priors (we show one such example in the experimental section of this work).

Counterfactual explanations are a complementary approach for explaining DNN decisions [20,36]. In contrast to feature attribution, counterfactual approaches attempt to explain a DNN output by presenting the user with another input that is close to the original input, but changes the classification decision of the DNN from class i to another class j. Comparing the two inputs then can be used to understand how class i differs from class j, enabling discriminative interpretability. For humans, this approach is arguably more natural and informative than presenting the full set of characteristics of class i, as done by feature attribution maps. However, generating counterfactual explanations typically involves an optimization procedure that needs to be carefully tuned in order to obtain a counterfactual with the desired properties. This process can be computationally expensive and does, in general, not allow for easy computation of attribution maps [35]. Due to these difficulties, counterfactual approaches are comparatively less popular for image data, where feature attribution methods arguably remain the dominant tool for practitioners.

To address these issues, we present a simple method that bridges the gap between counterfactual explainability and feature attribution (DAPI: Discriminative Attribution from Paired Images, see Fig. 1 for a visual summary). We use a cycle-GAN [40] to translate real images x_R of class i to counterfactual images x_C of class $j \neq i$. We then find an attribution map by processing the paired images with a new set of *discriminative attribution methods*, i.e., generalized versions of standard attribution methods. We show that this approach is able to generate sparse, high quality feature attribution maps that highlight the most discriminative features in the real and counterfactual image more precisely than standard attribution methods and prior approaches for discriminative attribution. Crucially, the use of paired images allows quantification of the discriminatory power of attribution maps by performing an intervention, i.e., replacing the highlighted pixels in the counterfactual with the corresponding pixels in the real image. The difference in output classification score of this hybrid image, compared to the real image, then quantifies the importance of the swapped features. We validate DAPI on a set of six diverse datasets, including a challenging artificial dataset, a real world biological dataset (where a DNN solves a task human experts can not), MNIST, and three natural image datasets. For all six datasets, we show quantitatively and qualitatively that DAPI outperforms all other considered methods in identifying key discriminatory features between the classes.

Notably, we use DAPI to perform semi-automated knowledge extraction from a DNN that has been trained to classify electron microscopy images of synapses in *Drosophila melanogaster* by the neurotransmitter they release. Using DAPI, we are able to identify so far unknown morphological differences of synapses that release different neurotransmitters, discovering a crucial structure-function relationship for connectomics. Source code and datasets are publicly available at https://dapi-method.github.io.

2   Related Work

Interpretability methods can be broadly categorized into *local* or *global* methods. Local methods provide an explanation for every input, highlighting the reasons why a particular input is assigned to a certain class by the DNN. Global methods attempt to distill the DNN in a representation that is easier to understand for humans, such as decision trees. One can further distinguish between interpretability methods that are *post-hoc*, i.e., applicable to every DNN after it has been trained, and those methods that require modifications to existing architectures to perform interpretable classification as part of the model. In this work we focus on local post-hoc approaches to DNN interpretability for image classification.

Attribution Methods for Image Classification. Even in this restricted class of approaches there is a large variety of methods [4,5,8,10,16,19,21,24,28–32,34, 38,39,41] They have in common that they aim to highlight the most important features that contributed to the output classification score for a particular class, generating a heatmap indicating the influence of input pixels and features on the output classification. Among those, of particular interest to the work presented here are *baseline* feature attribution methods, which perform feature attribution estimation with reference to a second input. Those methods gained popularity as they assert sensitivity and implementation invariance [30,34]. The baseline is usually chosen to be the zero image as it is assumed to represent a neutral input.

Counterfactual Interpretability. Since the introduction of counterfactual interpretability methods [20], the standard approach for generating counterfactuals broadly follows the procedure proposed by [36]. Concretely, a counterfactual is found as a result of an optimization aiming to maximize output differences while minimizing input differences between the real image x_R and the counterfactual x_C: $x_C = \text{argmin}_x \ L_i(x_R, x) - L_o(f(x_R), f(x))$, with L_i and L_o some loss that measures the distance between inputs and outputs, respectively, and f the classifier in question. Optimizing this objective can be problematic because it contains competing losses and does not guarantee that the generated counterfactual x_C is part of the data distribution $p(x)$. Current approaches try to remedy this by incorporating additional regularizers in the objective [18,35], such as adversarial losses that aim to ensure that the counterfactual x_C is not distinguishable from a sample $x \sim p(x)$ [6,18]. However, this does not address the core problem of competing objectives and will result in a compromise between obtaining in-distribution samples, maximizing class differences, and minimizing input differences. We circumvent this issue by omitting the input similarity loss in the generation of counterfactuals and instead enforce similarity post-hoc, similar to the strategy used by [22]. Similar to ours, the work by [23] uses a cycle-GAN to generate counterfactuals for DNN interpretability. However, this method differs in that the cycle-GAN is applied multiple times to a particular input in order to increase the visual differences in the real and counterfactual images for hypothesis generation. Subsequently, the found features are confirmed by contrasting the original classifiers performance with one that is trained on the discovered features, which does not lead to attribution maps or an objective evaluation of feature importance.

Attribution and Counterfactuals. To the best of our knowledge, two other methods explore the use of counterfactuals for feature attribution: Wang et al. [37] introduces a novel family of so-called *discriminative explanations* that also leverage attribution on a real and counterfactual image in addition to confidence scores from the classifier to derive attributions for the real and counterfactual image that show highly discriminative features. This approach requires calculation of three different attribution maps, which are subsequently combined to produce a discriminative explanation. In addition, this method does not generate new counterfactuals using a generative model, but instead selects a real image from a different class. On one hand this is advantageous because it does not depend on the generator's performance, but on the other hand this does not allow creating hybrid images for the evaluation of attribution maps. Closest to our approach is the method presented by Goyal et al. [12], where counterfactual visual explanations are generated by searching for feature sets in two real images of different classes that, if swapped, influence the classification decision. To this end, an optimization problem is solved to find the best features to swap, utilizing the network's feature representations. The usage of real (instead of generated) counterfactuals can lead to artifacts during the replacement of features. Furthermore, the attributions are limited in resolution to the field of view of the feature layer considered for performing the swap. In addition, depending on the chosen architecture, swapping features between images is not equivalent to swapping the corresponding pixel regions as the field of view of multiple units are overlapping. This makes it difficult to cleanly associate input features with the observed change in classification score, as we show in Sect. 4.

Attribution Evaluation. Being able to objectively evaluate attribution methods is important for method selection and trusting the generated explanations. Prior work evaluated the importance of highlighted features by removing them [26]. However, it has been noted that this strategy is problematic because it is unclear whether any observed performance degradation is due to the removal of relevant features or because the new sample comes from a different distribution. As a result, strategies to remedy this issue have been proposed, for example by retraining classifiers on the modified samples [14]. Instead of removing entire features, in this work we replace them with their corresponding counterfactual features.

3 Method

The method we propose combines counterfactual interpretability with discriminative attribution methods to find and highlight the most important features between images of two distinct classes i and j, given a pretrained classifier f. For that, we first generate for a given input image x_R of class i a counterfactual image x_C of class j. We then use a *discriminative* attribution method to find the attribution map of the classifier for this pair of images. As we will show qualitatively and quantitatively in Sect. 4, using paired images results in attribution maps of higher quality. Furthermore, the use of a counterfactual image gives rise to an objective evaluation procedure for attribution maps. In the next sections we describe (1) our choice for generating counterfactual images, (2) the derivation of discriminative attribution methods from existing baseline attribution methods, and (3) how to use counterfactual images to evaluate attribution maps. We denote with f a pretrained classifier with N output classes, input images $x \in \mathbb{R}^{h \times w}$, and output vector $f(x) = y \in [0, 1]^N$ with $\sum_i y_i = 1$.

3.1 Creation of Counterfactuals

We train a cycle-GAN [40] for each pair of image classes $i \neq j \in \{1, ..., N\}$, which enables translation of images of class i into images of class j and vice versa. We

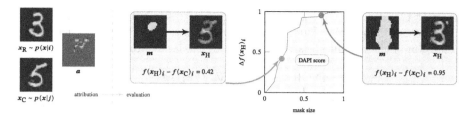

Fig. 2. Evaluation procedure for discriminative attribution methods: Given the real image x_R of class i and its counterfactual x_C of class j, we generate a sequence of binary masks m by applying different thresholds to the attribution map a. Those masks are then used to generate a sequence of hybrid images x_H. The plot shows the change in classifier prediction $\Delta f(x_H)_i = f(x_H)_i - f(x_C)_i$ over the size of the mask m (normalized between 0 and 1). The DAPI score is the area under the curve, i.e., a value between 0 and 1. Higher DAPI scores are better and indicate that a discriminative attribution method found small regions that lead to the starkest change in classification.

perform this translation for each image of class i and each target class $j \neq i$ to obtain datasets of paired images $D_{i \to j} = \{(x_R^k, x_C^k) | k = 1, \ldots, n(i)\}$, where x_R^k denotes the kth real image of class i and x_C^k its counterfactual of class j. We then test for each image in the dataset whether the translation was successful by classifying the counterfactual image x_C and reject a sample pair whenever $f(x_C)_j < \theta$, with θ a threshold parameter (in the rest of this work we set $\theta = 0.8$). This procedure results in a dataset of paired images, where the majority of the differences between an image pair is expected to be relevant for the classifiers decision, i.e., we retain formerly present non-discriminatory distractors such as orientation, lighting, or background. We encourage that the translation makes as little changes as necessary by choosing a ResNet [13] architecture for the cycle-GAN generator, which is able to trivially learn the identity function.

3.2 Discriminative Attribution

The datasets $D_{i \to j}$ are already useful to visualize data-intrinsic class differences (see Fig. 3 for examples). However, we wish to understand which input features the classifier f makes use of. Specifically, we are interested in finding the smallest binary mask m, such that swapping the contents of x_C with x_R within this mask changes the classification under f. To find m, we repurpose existing attribution methods that are amendable to be used with a reference image. The goal of those methods is to produce attribution maps a, which we convert into a binary mask via thresholding. A natural choice for our purposes are so-called *baseline attribution methods*, which derive attribution maps by contrasting an input image with a baseline sample (e.g., a zero image). In the following, we review suitable attribution methods and derive discriminative versions that use the counterfactual image as their baseline. We will denote the discriminative versions with the prefix D.

Input * Gradients. One of the first and simplest attribution methods is *Input * Gradients* (INGRADS) [30,31], which is motivated by the first order Taylor expansion of the output class with respect to the input around the zero point:

$$\text{INGRADS}(\boldsymbol{x}) = |\nabla_{\boldsymbol{x}} f(\boldsymbol{x})_i \cdot \boldsymbol{x}|, \tag{1}$$

where i is the class for which an attribution map is to be generated. We derive an explicit baseline version for the discriminatory attribution of the real $\boldsymbol{x}_R$ and its counterfactual $\boldsymbol{x}_C$ by choosing $\boldsymbol{x}_C$ as the Taylor expansion point:

$$\text{D-INGRADS}(\boldsymbol{x}_R, \boldsymbol{x}_C) = \left| \nabla_{\boldsymbol{x}} f(\boldsymbol{x})_j \right|_{\boldsymbol{x}=\boldsymbol{x}_C} \cdot (\boldsymbol{x}_C - \boldsymbol{x}_R)|, \tag{2}$$

where j is the class of the counterfactual image.

Integrated Gradients. *Integrated Gradients* (IG) is an explicit baseline attribution method, where gradients are accumulated along the straight path from a baseline input $\boldsymbol{x}_0$ to the input image $\boldsymbol{x}$ to generate the attribution map [34]. Integrated gradients along the kth dimension are given by:

$$\text{IG}_k(\boldsymbol{x}) = (\boldsymbol{x} - \boldsymbol{x}_0)_k \cdot \int_{\alpha=0}^{1} \frac{\partial f(\boldsymbol{x}_0 + \alpha(\boldsymbol{x} - \boldsymbol{x}_0))_i}{\partial \boldsymbol{x}_k} d\alpha. \tag{3}$$

We derive a discriminatory version of IG by replacing the baseline as follows:

$$\text{D-IG}_k(\boldsymbol{x}_R, \boldsymbol{x}_C) = (\boldsymbol{x}_C - \boldsymbol{x}_R)_k \cdot \int_{\alpha=0}^{1} \frac{\partial f(\boldsymbol{x}_R + \alpha(\boldsymbol{x}_C - \boldsymbol{x}_R))_j}{\partial \boldsymbol{x}_k} d\alpha. \tag{4}$$

Deep Lift. *Deep Lift* (DL) is also an explicit baseline attribution method which aims to compare individual neuron's activations of an input w.r.t. a reference baseline input [30]. It can be expressed in terms of the gradient in a similar functional form to IG:

$$\text{DL}(\boldsymbol{x}) = (\boldsymbol{x} - \boldsymbol{x}_0) \cdot F_{DL}, \tag{5}$$

where F_{DL} is some function of the gradient of the output (see [3] for the full expression). The discriminative attribution we consider is simply:

$$\text{D-DL}(\boldsymbol{x}_R, \boldsymbol{x}_C) = (\boldsymbol{x}_C - \boldsymbol{x}_R) \cdot F_{DL}. \tag{6}$$

GradCAM. *GradCAM* (GC) is an attribution method that considers the gradient weighted activations of a particular layer, usually the last convolutional layer, and propagates this value back to the input image [28]. We denote the activation of a pixel (u, v) in layer l with size (h, w) and channel k by $C_{k,u,v}^l$ and write the gradient w.r.t. the output $\boldsymbol{y}$ as:

$$\nabla_{C_k^l} \boldsymbol{y} = \left(\frac{d\boldsymbol{y}}{dC_{k,0,0}^l}, \frac{d\boldsymbol{y}}{dC_{k,1,0}^l}, \frac{d\boldsymbol{y}}{dC_{k,2,0}^l}, ..., \frac{d\boldsymbol{y}}{dC_{k,h,w}^l} \right) \tag{7}$$

The original GC is then defined as:

$$GC(\boldsymbol{x}) = \text{ReLU}\left(\sum_k \nabla_{C_k}\boldsymbol{y} \cdot \vec{C}_k\right)$$

$$= \text{ReLU}\left(\sum_k \sum_{u,v} \frac{d\boldsymbol{y}}{dC_{k,u,v}} C_{k,u,v}\right) \quad (8)$$

$$= \text{ReLU}\left(\sum_k \alpha_k C_k\right),$$

where we ommitted the layer index l for brevity. Each term $\frac{d\boldsymbol{y}}{dC_{k,u,v}} C_{k,u,v}$ is the contribution of pixel u,v in channel k to the output classification score $\boldsymbol{y}$ under a linear model. GC utilizes this fact and projects the layer attribution from layer l back to the input image, generating the final attribution map. In contrast to the setting considered by GC, we have access to a matching pair of real and counterfactual images $\boldsymbol{x}_R$ and $\boldsymbol{x}_C$. We extend GC to consider both feature maps $C_k^{\boldsymbol{x}_R}$ and $C_k^{\boldsymbol{x}_C}$ by treating GC as an implicit zero baseline method similar to INGRADS:

$$D\text{-}GC_k(\boldsymbol{x}_R, \boldsymbol{x}_C) = \frac{d\boldsymbol{y}_j}{dC_k}\bigg|_{C=C_k^{\boldsymbol{x}_C}} (C_k(\boldsymbol{x}_C) - C_k(\boldsymbol{x}_R)). \quad (9)$$

Averaging those gradients over feature maps k, and projecting the activations back to image space then highlights pixels that are most discriminative for a particular pair:

$$D\text{-}GC_P(\boldsymbol{x}_R, \boldsymbol{x}_C) = \left|\mathbb{P}\sum_k D\text{-}GC_k(\boldsymbol{x}_R, \boldsymbol{x}_C)\right|, \quad (10)$$

where $\mathbb{P}$ is the projection matrix (in this work we simply rescale to the input size) from feature space C to input space X. Note that in contrast to GC, we use the absolute value of the output attribution, as we do not apply ReLU activations to layer attributions. Because feature maps can be of lower resolution than the input space, GC tends to produce coarse attribution maps [28]. To address this issue it is often combined with *Guided Backpropagation* (GBP), a method that uses the (positive) gradients of the output class w.r.t. the input image as the attribution map [33]. *Guided GradCAM* (GGC) uses this strategy to sharpen the attribution of GC via element-wise multiplication of the attribution maps [28]. For the baseline versions we thus consider multiplication of D-GC with the GBP attribution maps:

$$GBP(\boldsymbol{x}) = \nabla_{\boldsymbol{x}} f(\boldsymbol{x})_i \ , \ \text{with} \nabla \text{ReLU} > 0 \quad (11)$$

$$GGC(\boldsymbol{x}) = GC(\boldsymbol{x}) \cdot GBP(\boldsymbol{x}) \quad (12)$$

$$D\text{-}GGC(\boldsymbol{x}_R, \boldsymbol{x}_C) = D\text{-}GC(\boldsymbol{x}_R, \boldsymbol{x}_C) \cdot GBP(\boldsymbol{x}_R). \quad (13)$$

3.3   Evaluation of Attribution Maps

The discriminative attribution map a obtained for pair of images (x_R, x_C) can be used to quantify the causal effect of the attribution. Specifically, we can copy the area highlighted by a from the real image x_R of class i to the counterfactual image x_C of class j, resulting in a hybrid image x_H. If the attribution accurately captures class-relevant features, we would expect that the classifier f assigns a high probability to x_H being of class i. The ability to create those hybrid images is akin to an intervention, and has two important practical implications: First, it allows us to find a minimal binary mask that captures the most class-relevant areas for a given input image. Second, we can compare the change in classification score for hybrids derived from different attribution maps. This allows us to compare different methods in an objective manner, following the intuition that an attribution map is better, if it changes the classification with less pixels changed. To find a minimal binary mask m_{min}, we search for a threshold of the attribution map a, such that the mask score $\Delta f(x_H) = f(x_H)_i - f(x_C)_i$ (i.e., the change in classification score) is maximized while the size of the mask is minimized, i.e., $m_{min} = \arg\min_m |m| - \Delta f(x_H)$ (where we omitted the dependency of x_H on m for brevity). In order to minimize artifacts in the copying process we also apply a morphological closing operation with a window size of 10 pixels followed by a Gaussian Blur with $\sigma = 11px$. The final masks highlight the relevant class discriminators by showing the user the counterfactual features, the original features they are replaced with, and the corresponding mask score $\Delta f(x_H)$, indicating the quantitative effect of the replacement on the classifier. See Fig. 3 for example pairs and corresponding areas m_{min}. Furthermore, by applying a sequence of thresholds for the attribution map a, we derive an objective evaluation procedure for a given attribution map: For each hybrid image x_H in the sequence of thresholds, we consider the change in classifier prediction relative to the size of the mask that has been used to create the hybrid. We accumulate the change in classifier prediction over all mask sizes to derive our proposed DAPI score. This procedure is explained in detail in Fig. 2 for a single pair of images. When reporting the DAPI score for a particular attribution method, we average the single DAPI scores over all images, and all distinct pairs of classes.

4   Experiments

We evaluate the presented method on six datasets: MNIST [17], SYNAPSES [9][1], two versions of a synthetic dataset that we call DISC-A and DISC-B, and three natural image datasets HORSES, APPLES and SUMMER from [40]. For a more detailed overview of all considered datasets, see Supplementary Sect. 3.

Synapses. A real world biological dataset, consisting of $1 \times 128 \times 128px$ electron microscopy images of synaptic sites in the brain of *Drosophila melanogaster*. Each image is labelled with a functional property of the synapse (six different classes), namely the neurotransmitter it releases (the label was acquired using

---

[1] Dataset kindly provided by the authors of [9].

input image   attribution   evaluation
            S-*   D-* (ours) Counterfactual   hybrid S-*   hybrid D-*   DAPI score

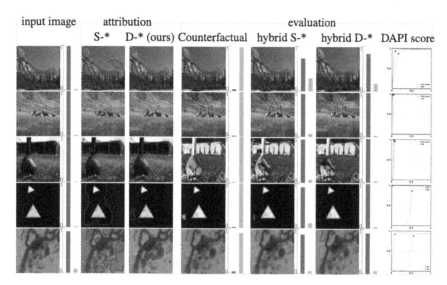

Fig. 3. Samples from the best performing method pairs (S: "single input", D: discriminative) on SUMMER, HORSES, APPLES, DISC-A and SYNAPSES (different rows). Shown in each row is the input image of class i, attribution maps according to S- and D-methods (best performing on the respective dataset), the counterfactual image of class j, and the hybrid images. Bars indicate the classifier's prediction for class i (purple) and j (orange). The last column shows the single image DAPI score. Stars indicate the shown threshold (corresponding to the optimal DAPI score). Additional qualitative results can be found in Supplementary Sect. 2 (Color figure online).

immunohistochemistry labelling, see [9] for details). This dataset is of particular interest for interpretability, since a DNN can recover the neurotransmitter label from the images with high accuracy, but human experts are not able to do so. Interpretability methods like the one presented here can thus be used to gain insights into the relation between structure (from the electron microscopy image) and function (the neurotransmitter released). See Fig. 6 for an example of a discriminatory feature between synapses that release the two different neurotransmitters GABA and Acetylcholine, discovered by applying DAPI on the SYNAPSES dataset. A full description of all discovered feature differences can be found in the supplementary material.

Horses, Apples & Summer. Three natural image datasets corresponding to binary classification tasks. HORSES consists of two sets of images showing horses and zebras respectively, APPLES is a dataset of images depicting apples and oranges, and SUMMER shows landscape pictures in summer and winter. We scale each image to $3 \times 256 \times 256$ for classification and image translation.

Disc-A & Disc-B. Two synthetic datasets with different discriminatory features. Each image is $1 \times 128 \times 128$px in size and contains spheres, triangles or squares. For DISC-A, the goal is to correctly classify images containing an even or odd number of triangles. DISC-B contains images that show exactly two of the three

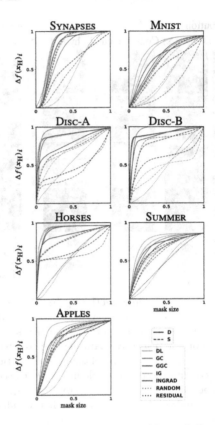

Fig. 4. Quantitative evaluation of discriminative (D - solid) and corresponding original (S for "single input" - dashed) attribution methods over six datasets (SYNAPSES, MNIST, two versions of DISC, HORSES, SUMMER, and APPLES). Attributions for D and S methods are calculated as described in the methods section, with a zero baseline for all S methods that are explicit baseline methods, following standard practice. Corresponding D and S versions of the same method are shown in the same color. For each, we plot the average change of classifier prediction $\Delta f(\boldsymbol{x}_\mathrm{H})_i^k = f(\boldsymbol{x}_\mathrm{H})_i - f(\boldsymbol{x}_\mathrm{C})_i$ as a function of mask size $m \in [0,1]$. In addition we show performance of the two considered baselines: masks derived from random attribution maps (random - red, dotted) and mask derived from the residual of the real and counterfactual image (residual - black, dotted). On all considered datasets all versions of D attribution outperform their S counterparts with the single exception of INGRAD on the APPLES dataset. All experiments shown here are performed with VGG architectures (we observe similar results with ResNet, see Supplementary Sect. 2 (Color figure online)).

available shapes and the goal is to predict which shape is missing (e.g., an image with only triangles and squares is to be classified as "does not contain spheres"). This dataset was deliberately designed to investigate attribution methods in a setting where the discrimination depends on the absence of a feature.

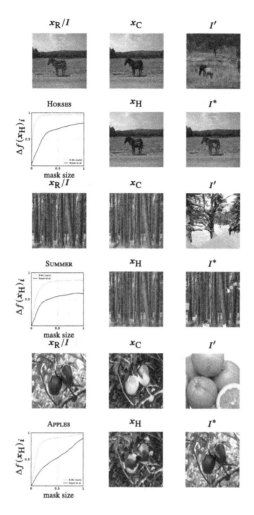

Fig. 5. Comparison of our method to the method presented by Goyal et al. [12] for the three natural image datasets HORSES, SUMMER and APPLES (See Supplementary Sect. 4 for the other datasets and Table 1 for associated DAPI scores). For each dataset we show the real/query image(x_R/I) as well as the counterfactual x_C and hybrid x_H for DAPI. For [12], we show the distractor image I' as well as the hybrid I^* after the minimal amount of swaps needed to change the classifier decision. In addition, we show the DAPI scores for [12] and our best performing method for each dataset averaged over all image pairs. Note that [12] considers replacement of patches from the distractor image I' to the query image I, while we consider the opposite direction $x_R \rightarrow x_C$.

Training. For MNIST, DISC, HORSES, APPLES and SUMMER we train a VGG and ResNet for 100 epochs and select the epoch with highest accuracy on a held out validation dataset. For SYNAPSES we adapt the 3D-VGG architecture from [9] to 2D and train for 500,000 iterations. We select the iteration with the

Table 1. Summary of DAPI scores using VGG architecures for each investigated method on the six datasets MNIST, SYNAPSES, HORSES, SUMMER, APPLES and DISC (two versions) corresponding to 4. Best results are highlighted.

Dataset	D-IG	D-DL	D-INGR	D-GC	D-GGC	RES.	IG	DL	INGR	GC	GGC	RND.	GOYAL
MNIST	0.833289758	0.842277546	0.817678354	0.733100916	0.777029492	0.839020143	0.766036855	0.7850287	0.773991294	0.519173244	0.563839206	0.464561803	0.27
SYNAPSES	0.747423269	0.785551226	0.653958976	0.61527535	0.65284391	0.62920887	0.56225733	0.427523262	0.608311021	0.283937535	0.522137129	0.411464634	0.21
DISC-A	0.89625051	0.899087592	0.882141041	0.948849319	0.787418823	0.896882392	0.693431096	0.701947211	0.719899507	0.427150924	0.483504218	0.544253556	0.41
DISC-B	0.910239637	0.907714689	0.911432742	0.945417258	0.882505759	0.913099126	0.483063519	0.511230496	0.59534397	0.798793785	0.789766349	0.482227646	0.46
HORSES	0.925928342	0.930676525	0.902301773	0.906829632	0.838995055	0.558837054	0.797803627	0.791047188	0.784964749	0.582548534	0.647996254	0.532046845	0.62
SUMMER	0.734584515	0.77116628	0.666113227	0.700197198	0.694408414	0.473990243	0.630162978	0.642697074	0.611374696	0.596378171	0.654900833	0.475837691	0.49
APPLES	0.803571624	0.879397181	0.742535459	0.791326772	0.765578544	0.580580665	0.733091336	0.748294758	0.770055715	0.682398611	0.692690224	0.498499041	0.52

highest validation accuracy for testing. For MNIST, DISC & SYNAPSES we train one cycle-GAN for 200 epochs, on each class pair and on the same training set the respective classifier was trained on. For HORSES, APPLES and SUMMER we use the pretrained cycle-GAN checkpoints provided by [40] (the full network specifications are given in the supplement).

Results. Quantitative results (in terms of the DAPI score, see Sect. 3.3) for each investigated attribution method are shown in Fig. 4 and Table 1. In summary, we find that attribution maps generated from the proposed discriminative attribution methods consistently outperform their original versions in terms of the DAPI score. This observation also holds visually: the generated masks from discriminative attribution methods are smaller and more often highlight the main discriminatory parts of a considered image pair (see Fig. 3). In particular, the proposed method substantially outperforms the considered random baseline, whereas standard attribution methods sometimes fail to do so (e.g., GC on dataset SYNAPSES, IG on dataset DISC-B). Furthermore, on MNIST and DISC-A, the mask derived from the residual of real and counterfactual image is already competitive with the best considered methods and outperforms standard attribution substantially. However, for more complex datasets such as SYNAPSES, HORSES, APPLES and SUMMER, the residual becomes less accurate in highlighting discriminative features. Here, the discriminatory attributions outperform all other considered methods.

In addition to the considered baseline attribution methods we also compare DAPI with the method proposed by Goyal et al. [12]. To this end, we iterated the BESTEDIT procedure until the features of the query image I have been entirely replaced with features from the distractor image I' (see Algorithm 1 in [12]). We then swapped input patches that underlie those features between I and I' accordingly to obtain a sequence of I^* that gradually transforms I into a shuffled version of I'. We measure the impact on the classifier score in the same way as with x_H in our method (see Fig. 5 for examples on HORSES, SUMMER, and APPLES; Table 1 for aggregate results on all datasets; and Supplement Sect. 4 for an extended analysis). We find that DAPI consistently finds smaller regions with larger explanatory power.

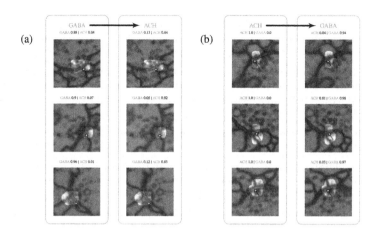

Fig. 6. DAPI reveals so far unnoticed morphological features of synapses. Shown are electron microscopy images and associated masks generated by DAPI for real synapses (blue boxes) that express the neurotransmitters GABA (a) and Acetylcholine (b), next to their respective translations to the other class (orange boxes). The classification score is shown above each image, validating a successful translation. As before, masks highlight regions that change the classification decision back to the real class if swapped from real to fake sample. Masks consistently highlight the synaptic cleft (see yellow arrow), and we observe a brightening of the inside of the cleft when translating from GABA to ACH and a darkening the other way around. (Color figure online)

5   Discussion

This work demonstrates that the combination of counterfactual interpretability with suitable attribution methods is more accurate in extracting key discriminative features between class pairs than standard methods. While the method succeeds in the presented experiments, it comes with a number of limitations. It requires the training of cycle-GANs, one for each pair of output classes. Thus training time and compute cost scale quadratically in the number of output classes and it is therefore not feasible for classification problems with a large number of classes. Furthermore, the translation from the real to the counterfactual image could fail for a large fraction of input images, i.e., $f(x_C) \neq j$. In this work, we only consider those image pairs where translation was successful, as we focus on extracting knowledge about class differences from the classifier. For applications that require an attribution for each input image this approach is not suitable. An additional concern is that focusing only on images that have a successful translation may bias the dataset we consider and with it the results. GANs are known to exhibit so-called mode collapse [7,25], meaning they focus on only a small set of modes of the full distribution. As a consequence, the method described here may miss discriminatory features present in other modes. Furthermore, image classes need to be sufficiently similar in appearance for the cycle-GAN to work, and translating, e.g., an image of a mouse into an image of a tree is unlikely to work and produce meaningful attributions. However, we believe that the generation of masks in combination with the corresponding mask score is superior to classical attribution maps for interpreting DNN decision boundaries, especially for the analysis of fine-grained class differences; a common situation for biomedical image datasets and exemplified here by the SYNAPSES dataset. Although we present this work in the context of understanding DNNs and the features they make use of, an uncritical adaptation of this and other similar interpretability methods can potentially lead to ethical concerns. As an example, results should be critically evaluated when using this method to interpret classifiers that have been trained to predict human behaviour, or demographic and socioeconomic features. As with any data-heavy method, it is important to realize that results will be reflective of data- and model-intrinsic biases. The method presented here should therefore not be used to "prove" that a particular feature leads to a particular outcome.

References

1. Adebayo, J., Gilmer, J., Muelly, M., Goodfellow, I., Hardt, M., Kim, B.: Sanity checks for saliency maps. arXiv preprint arXiv:1810.03292 (2018)
2. Alvarez-Melis, D., Jaakkola, T.S.: On the robustness of interpretability methods. arXiv preprint arXiv:1806.08049 (2018)
3. Ancona, M., Ceolini, E., Öztireli, C., Gross, M.: Towards better understanding of gradient-based attribution methods for deep neural networks. In: International Conference on Learning Representations (2018). https://openreview.net/forum?id=Sy21R9JAW
4. Bach, S., Binder, A., Montavon, G., Klauschen, F., Müller, K.R., Samek, W.: On pixel-wise explanations for non-linear classifier decisions by layer-wise relevance propagation. PLoS ONE **10**(7), e0130140 (2015)
5. Baehrens, D., Schroeter, T., Harmeling, S., Kawanabe, M., Hansen, K., Müller, K.R.: How to explain individual classification decisions. J. Mach. Learn. Res. **11**, 1803–1831 (2010)
6. Barredo-Arrieta, A., Del Ser, J.: Plausible counterfactuals: auditing deep learning classifiers with realistic adversarial examples. In: 2020 International Joint Conference on Neural Networks (IJCNN), pp. 1–7. IEEE (2020)
7. Che, T., Li, Y., Jacob, A.P., Bengio, Y., Li, W.: Mode regularized generative adversarial networks. arXiv preprint arXiv:1612.02136 (2016)
8. Dabkowski, P., Gal, Y.: Real time image saliency for black box classifiers. arXiv preprint arXiv:1705.07857 (2017)
9. Eckstein, N., Bates, A.S., Du, M., Hartenstein, V., Jefferis, G.S., Funke, J.: Neurotransmitter classification from electron microscopy images at synaptic sites in drosophila. BioRxiv (2020)
10. Fong, R.C., Vedaldi, A.: Interpretable explanations of black boxes by meaningful perturbation. In: Proceedings of the IEEE International Conference on Computer Vision, pp. 3429–3437 (2017)
11. Ghorbani, A., Abid, A., Zou, J.: Interpretation of neural networks is fragile. In: Proceedings of the AAAI Conference on Artificial Intelligence, vol. 33, pp. 3681–3688 (2019)

12. Goyal, Y., Wu, Z., Ernst, J., Batra, D., Parikh, D., Lee, S.: Counterfactual visual explanations. In: International Conference on Machine Learning, pp. 2376–2384. PMLR (2019)
13. He, K., Zhang, X., Ren, S., Sun, J.: Deep residual learning for image recognition. In: Proceedings of the IEEE Conference on Computer Vision and Pattern Recognition, pp. 770–778 (2016)
14. Hooker, S., Erhan, D., Kindermans, P.J., Kim, B.: A benchmark for interpretability methods in deep neural networks. arXiv preprint arXiv:1806.10758 (2018)
15. Kindermans, P.-J., et al.: The (Un)reliability of saliency methods. In: Samek, W., Montavon, G., Vedaldi, A., Hansen, L.K., Müller, K.-R. (eds.) Explainable AI: Interpreting, Explaining and Visualizing Deep Learning. LNCS (LNAI), vol. 11700, pp. 267–280. Springer, Cham (2019). https://doi.org/10.1007/978-3-030-28954-6_14
16. Kindermans, P.J., Schütt, K.T., Alber, M., Müller, K.R., Erhan, D., Kim, B., Dähne, S.: Learning how to explain neural networks: Patternnet and patternattribution. arXiv preprint arXiv:1705.05598 (2017)
17. LeCun, Y., Cortes, C.: MNIST handwritten digit database (2010). http://yann.lecun.com/exdb/mnist/
18. Liu, S., Kailkhura, B., Loveland, D., Han, Y.: Generative counterfactual introspection for explainable deep learning. arXiv preprint arXiv:1907.03077 (2019)
19. Lundberg, S., Lee, S.I.: A unified approach to interpreting model predictions. arXiv preprint arXiv:1705.07874 (2017)
20. Martens, D., Provost, F.: Explaining data-driven document classifications. MIS Q. 38(1), 73–100 (2014)
21. Montavon, G., Lapuschkin, S., Binder, A., Samek, W., Müller, K.R.: Explaining nonlinear classification decisions with deep Taylor decomposition. Pattern Recogn. 65, 211–222 (2017)
22. Mothilal, R.K., Sharma, A., Tan, C.: Explaining machine learning classifiers through diverse counterfactual explanations. In: Proceedings of the 2020 Conference on Fairness, Accountability, and Transparency, pp. 607–617 (2020)
23. Narayanaswamy, A., et al.: Scientific discovery by generating counterfactuals using image translation. In: Martel, A.L., et al. (eds.) MICCAI 2020. LNCS, vol. 12261, pp. 273–283. Springer, Cham (2020). https://doi.org/10.1007/978-3-030-59710-8_27
24. Ribeiro, M.T., Singh, S., Guestrin, C.: Model-agnostic interpretability of machine learning. arXiv preprint arXiv:1606.05386 (2016)
25. Salimans, T., Goodfellow, I., Zaremba, W., Cheung, V., Radford, A., Chen, X.: Improved techniques for training GANs. arXiv preprint arXiv:1606.03498 (2016)
26. Samek, W., Binder, A., Montavon, G., Lapuschkin, S., Müller, K.R.: Evaluating the visualization of what a deep neural network has learned. IEEE Trans. Neural Netw. Learn. Syst. 28(11), 2660–2673 (2016)
27. Samek, W., Montavon, G., Lapuschkin, S., Anders, C.J., Müller, K.R.: Explaining deep neural networks and beyond: a review of methods and applications. Proc. IEEE 109(3), 247–278 (2021)
28. Selvaraju, R.R., Cogswell, M., Das, A., Vedantam, R., Parikh, D., Batra, D.: Grad-CAM: visual explanations from deep networks via gradient-based localization. In: Proceedings of the IEEE International Conference on Computer Vision, pp. 618–626 (2017)
29. Shrikumar, A., Greenside, P., Kundaje, A.: Learning important features through propagating activation differences. In: International Conference on Machine Learning, pp. 3145–3153. PMLR (2017)

30. Shrikumar, A., Greenside, P., Shcherbina, A., Kundaje, A.: Not just a black box: learning important features through propagating activation differences. arXiv preprint arXiv:1605.01713 (2016)
31. Simonyan, K., Vedaldi, A., Zisserman, A.: Deep inside convolutional networks: visualising image classification models and saliency maps (2014)
32. Smilkov, D., Thorat, N., Kim, B., Viégas, F., Wattenberg, M.: SmoothGrad: removing noise by adding noise. arXiv preprint arXiv:1706.03825 (2017)
33. Springenberg, J.T., Dosovitskiy, A., Brox, T., Riedmiller, M.: Striving for simplicity: the all convolutional net. arXiv preprint arXiv:1412.6806 (2014)
34. Sundararajan, M., Taly, A., Yan, Q.: Axiomatic attribution for deep networks. In: Proceedings of the 34th International Conference on Machine Learning, Vol. 70, pp. 3319–3328. JMLR.org (2017)
35. Verma, S., Dickerson, J., Hines, K.: Counterfactual explanations for machine learning: a review. arXiv preprint arXiv:2010.10596 (2020)
36. Wachter, S., Mittelstadt, B., Russell, C.: Counterfactual explanations without opening the black box: automated decisions and the GDPR. Harv. JL Tech. **31**, 841 (2017)
37. Wang, P., Vasconcelos, N.: Scout: self-aware discriminant counterfactual explanations. In: Proceedings of the IEEE/CVF Conference on Computer Vision and Pattern Recognition, pp. 8981–8990 (2020)
38. Zeiler, M.D., Fergus, R.: Visualizing and understanding convolutional networks. In: Fleet, D., Pajdla, T., Schiele, B., Tuytelaars, T. (eds.) ECCV 2014. LNCS, vol. 8689, pp. 818–833. Springer, Cham (2014). https://doi.org/10.1007/978-3-319-10590-1_53
39. Zhang, C., Bengio, S., Hardt, M., Recht, B., Vinyals, O.: Understanding deep learning requires rethinking generalization. arXiv preprint arXiv:1611.03530 (2016)
40. Zhu, J.Y., Park, T., Isola, P., Efros, A.A.: Unpaired image-to-image translation using cycle-consistent adversarial networks. In: Proceedings of the IEEE International Conference on Computer Vision, pp. 2223–2232 (2017)
41. Zintgraf, L.M., Cohen, T.S., Adel, T., Welling, M.: Visualizing deep neural network decisions: prediction difference analysis. arXiv preprint arXiv:1702.04595 (2017)

Learning with Minimal Effort: Leveraging in Silico Labeling for Cell and Nucleus Segmentation

Thomas Bonte[1,2,3], Maxence Philbert[1,2,3], Emeline Coleno[4],
Edouard Bertrand[4], Arthur Imbert[1,2,3(✉)], and Thomas Walter[1,2,3(✉)] (iD)

[1] Centre for Computational Biology (CBIO), Mines Paris, PSL University,
75006 Paris, France
{Thomas.Bonte,arthur.imbert,Thomas.Walter}@minesparis.psl.eu
[2] Institut Curie, PSL University, 75248 Paris, Cedex, France
[3] INSERM, U900, 75248 Paris, Cedex, France
[4] IGH, University of Montpellier, CNRS, 34090 Montpellier, France

Abstract. Deep learning provides us with powerful methods to perform nucleus or cell segmentation with unprecedented quality. However, these methods usually require large training sets of manually annotated images, which are tedious — and expensive — to generate. In this paper we propose to use In Silico Labeling (ISL) as a pretraining scheme for segmentation tasks. The strategy is to acquire label-free microscopy images (such as bright-field or phase contrast) along fluorescently labeled images (such as DAPI or CellMask^{TM}). We then train a model to predict the fluorescently labeled images from the label-free microscopy images. By comparing segmentation performance across several training set sizes, we show that such a scheme can dramatically reduce the number of required annotations.

Keywords: Segmentation · Transfer learning · Pretext task · In silico labeling · Fluorescence microscopy

1 Introduction

Detection and segmentation of cells and nuclei, among other cell structures, are essential steps for microscopy image analysis. Deep Learning has provided us with very powerful methods to perform these segmentation tasks. In particular, recently published neural networks, such as NucleAIzer [1], Cellpose [2] or StarDist [3], trained on hundreds of images of different modalities, give excellent results, outperforming by far traditional methods for image segmentation. However, the main drawback of state-of-the-art networks is the need for large amounts of fully annotated ground truth images, which can take a significant amount of time to create. Here, we present an alternative strategy, where we pretrain our segmentation models using In Silico Labeling (ISL) before fine-tuning them on a very small data set to perform nucleus and cell segmentation.

Supplementary Information The online version contains supplementary material available at https://doi.org/10.1007/978-3-031-25069-9_28.

© The Author(s), under exclusive license to Springer Nature Switzerland AG 2023
L. Karlinsky et al. (Eds.): ECCV 2022 Workshops, LNCS 13804, pp. 423–436, 2023.
https://doi.org/10.1007/978-3-031-25069-9_28

ISL was first introduced by [4], aiming to predict fluorescent labels from bright-field inputs. Fluorescence microscopy is the major technique employed in cellular image-based assays, as the use of fluorescence labels allows to highlight particular structures or phenotypic cell states. However, the number of fluorescent labels is limited (typically up to 4). In addition, phototoxicity and photobleaching can also represent serious drawbacks.

To tackle these limitations, several variants have been proposed. In [5], ISL is applied to predict fluorescent labels from transmitted-light images (DIC), or immunofluorescence from electron micrographs. Besides, Generative Adversarial Networks (GAN) are used in [6] to predict different stains: H&E, Jones Silver or Masson's trichrome. They underlie staining standardization as an advantage of ISL. In another paper [7] GANs are also used on different transmitted light images: quantitative phase images (QPI). Moreover, in [8] conditional GANs (cGAN) generate H&E, PSR and Orcein stained images from unstained bright-field inputs. In [9], using the same data set and same tasks as [4], the authors add attention blocks to capture more information than usual convolutions. Finally, stained images of human sperm cells are generated in [10], from quantitative phase images. They use these virtually stained images to recognize normal from abnormal cells. The principle of ISL has also been proposed for experimental ground truth generation for training cell classifiers for the recognition of dead cells [11,12], tumour cells [13] embryo polarization [14] or the cell cycle phase [15].

In this paper we show that models trained to generate fluorescence microscopy images with nuclear or cytoplasmic markers can be used efficiently to pretrain segmentation networks for nuclear and cell segmentation, respectively. To the best of our knowledge, no previous work has used ISL as a pretext task for segmentation of cell structures. This provides us with a powerful strategy to minimize the annotation burden for a given application, and to train models on large data sets, requiring only minimal effort in terms of manual annotation.

2   Materials and Methods

2.1   Image Acquisition

We work on two different data sets. The first dataset has been generated by the Opera Phenix^{TM} Plus High-Content Screening System (Perkin Elmer). It contains 960 images of dimension (2160, 2160). For each position, we acquired bright-field images and DAPI, both at 4 different focal planes. DAPI is a very common fluorescent stain binding to AT-rich regions of the DNA, which can thus be used to locate the nucleus in eukaryotic cells. Additionally we have a phase contrast image, computationally created from the 4 bright-field images by a proprietary algorithm of the Opera system. Images contain on average 15.6±19.6 cells.

Our second data set contains 100 images of dimension (1024, 1024). We used Differential Interference Contrast (DIC) as label-free microscopy technique, and we marked the cytoplasmic membrane with the CellMask^{TM} marker (Life Technologies). Images contain on average 52.4±15.1 cells.

2.2   Nucleus Segmentation

Nucleus segmentation is one of the most important segmentation tasks in biology, as nuclear morphologies are indicative of cellular states, and because they are visually very different from the cytoplasm. Segmentation of the nucleus is usually a comparatively simple segmentation task, and for this reason we assumed that this might be a good first segmentation problem to investigate our ISL-based pretraining.

DAPI Prediction as Pretraining Task. The first step of our strategy for nucleus segmentation is the prediction of DAPI images from bright-field inputs.

We used a data set of 421 images of dimension (2160, 2160), divided into 384 images for training and 37 images for testing. 5 images of dimension (512, 512) were randomly cropped from each initial image (see Fig. 1). Note that we only included images containing at least one nucleus.

Inspired by the work of [4], the model is a U-net-shape model [16] with a densenet121 architecture [17]. It has been previously trained on ImageNet [18], hence it is referred to as 'on steroids' in the following. As input we used 3 channels, 2 being bright-field images of the same field-of-view with different focal planes, and the third the corresponding phase-contrast image. As output we used only one channel, the maximum intensity projection of our DAPI images (z-stack, 4 focal planes) that we have for each field-of-view.

We did not use any data augmentation. All training details are reported in Supplementary Table 1.

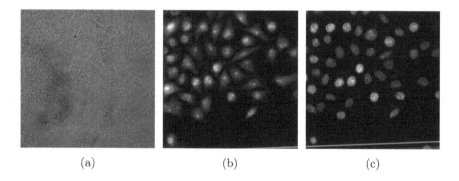

(a)   (b)   (c)

Fig. 1. Images from the same field-of-view, for a given focal plane. (a) Bright-field image. (b) Phase contrast image, computationally generated by the Opera system. (c) Fluorescent DAPI image.

Transfer Learning for Nucleus Segmentation. In a first step, we aimed at investigating how pretraining on fluorescent markers impacts semantic segmentation. For this, we turned to nucleus segmentation.

In order to generate the ground truth, we applied Cellpose [2], a widely used segmentation technique in bioimaging, based on a U-net-shaped network, trained on massive amounts of heterogeneous data. We applied Cellpose to the DAPI channel and corrected the segmentation results manually. As segmentation of nuclei from DAPI images with high resolution is a fairly simple task, as expected the results were overall excellent.

Next, we used training sets with different sizes $N \in \{1, 10, 50, 100, 200, 500\}$, composed of images of dimension (2160, 2160) and evaluated the accuracy for each N. Testing is always performed on the same withheld 190 images. 5 images of dimension (512, 512) were randomly cropped from each initial image.

To investigate whether our pretraining scheme is useful for segmentation, we compare two different models. The first model is composed of the U-net 'on steroids' followed by a sigmoid activation function in order to output, for each pixel, its probability of belonging to a nucleus (Fig. 2a). The second model has the same U-net architecture but is pretrained on DAPI images, and has an activation function displayed in equation (1) that takes a different range into account (Fig. 2b). The reason for this choice is that the model pretrained on DAPI images is likely to output values between 0 and 1, so we centered the following activation function around 0.5.

$$f(x) = \frac{1}{1 + \exp(-(x - 0.5))} \tag{1}$$

We did not use any data augmentation. All training details are reported in Supplementary Table 1.

2.3   Cell Segmentation

We next turned to the application of our pretraining scheme to cell segmentation, a more difficult multiple instance segmentation scenario.

CellMask^{TM} Prediction as Pretraining Task. In our pretraining strategy, the first step of cell segmentation is the prediction of CellMask^{TM} (Fig. 3b) images from DIC microscopy as inputs (Fig. 3a).

We used a data set of 100 images of dimension (1024, 1024), divided into 90 images for training and 10 images for testing. 5 images of dimension (512, 512) were randomly cropped for each initial image.

For comparison, we again used the U-net 'on steroids'. We did not use any data augmentation. All training details are reported in Supplementary Table 2.

Transfer Learning for Cell Segmentation. Segmentation of cells is usually more difficult than nuclear segmentation, because cells tend to touch each other, and the precise detection of the contact line can be challenging. Indeed, we need to turn to multiple instance segmentation, where object properties are predicted together with pixel labels.

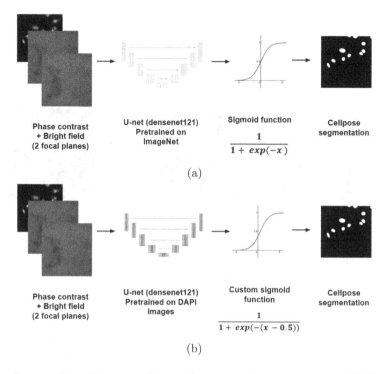

Phase contrast + Bright field (2 focal planes) → U-net (densenet121) Pretrained on ImageNet → Sigmoid function $\frac{1}{1 + exp(-x)}$ → Cellpose segmentation

(a)

Phase contrast + Bright field (2 focal planes) → U-net (densenet121) Pretrained on DAPI images → Custom sigmoid function $\frac{1}{1 + exp(-(x - 0.5))}$ → Cellpose segmentation

(b)

Fig. 2. Compared models to predict nucleus semantic segmentation. (a) U-net 'on steroids' which has not been trained on DAPI images. (b) U-net 'on steroids' pretrained on DAPI images. Note the difference in the activation functions.

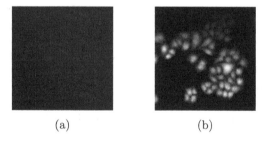

(a)   (b)

Fig. 3. Images from the same field-of-view, for a given focal plane. (a) DIC image. (b) Fluorescent CellMask^{TM}image.

Again, we used Cellpose [2] with manual correction to generate this instance segmentation ground truth images from associated CellMask^{TM}images (Fig. 4a, Fig. 4b).

As for nuclear segmentation, we used training sets of different sizes $N \in \{1, 10, 50, 80\}$ of dimension (1024, 1024) and evaluated the accuracy for each of them. Testing is always performed on the same 17 images. 5 images of dimension (512, 512) were randomly cropped from each initial image.

To tackle the issue of instance segmentation, we implemented a model predicting both a cell semantic segmentation image (Fig. 4c) and a distance map, i.e. an image where pixels values get higher as they are closer to the center of the cell, the background remaining black (Fig. 4d), as proposed in [19,20].

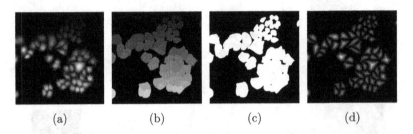

(a)   (b)   (c)   (d)

Fig. 4. (a) Fluorescent CellMask^{TM}image. (b) Corresponding cell instance segmentation image generated by Cellpose. (c) Cell semantic segmentation image generated from Cellpose output. (d) Distance map generated from Cellpose output.

Like in the previous section we compare two models to investigate whether transfer learning from an ISL model can significantly improve the accuracy of our segmentation. The first model is the U-net 'on steroids', outputting 2 channels (Fig. 5a). The second model has the same U-net architecture but is pretrained on CellMask^{TM}images, thus outputting only 1 channel. Hence we add two Conv2d layers at the end to upscale to 2 channels (Fig. 5b).

We did not use any data augmentation. All training details are reported in Supplementary Table 2. Both models use CombinedLoss$_\alpha$, presented in equation (2), as loss function. MSELoss stands for the usual Mean Square Error, while BCEWithLogitsLoss combines a sigmoid layer with the Binary Cross Entropy loss. y represents the output of our model, with the two channels y_d and y_s standing for the distance and semantic segmentation image, respectively. The factor α is used to balance the weights of the different losses during training. It has been set as $\alpha = 2000$, 2000 being the initial ratio between MSELoss and BCEWithLogitsLoss. This has been inspired by the loss function used in Cellpose [2], which also uses a loss function computed as the sum of two loss functions, one for each output channel.

$$\begin{aligned} \text{CombinedLoss}_\alpha(y) &= \text{CombinedLoss}_\alpha((y_d, y_s)) \\ &= \text{MSELoss}(y_d) + \alpha \cdot \text{BCEWithLogitsLoss}(y_s) \end{aligned} \quad (2)$$

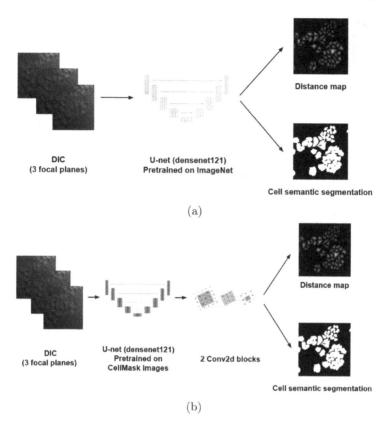

Fig. 5. Models compared to predict cell instance segmentation: (a) U-net 'on steroids' which has not been trained on CellMask^TM images. (b) U-net model pretrained on CellMask^TM images.

Finally, we apply a post-processing step to get the final results. For this, we apply the h-maxima transformation of the predicted distance map, with $h = 10$. The h-maxima transformation is defined as the reconstruction by dilation of $f - h$ under f: $HMAX_h(f) = R_f^\delta(f - h)$, and removes insignificant local maxima. f stands for the initial image, which is in our case the reconstructed distance map displayed in Fig. 6. h stands for the minimum local contrast for a local maximum to be kept; otherwise it will be removed. Each local maximum represents a cell.

The local maxima of $HMAX$ then serve as seed for the watershed algorithm, which splits semantic segmentation result into individual regions, one for each maximum of $HMAX$. This leads to an instance segmentation image, such as the one presented in Fig. 6.

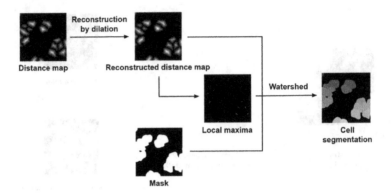

Fig. 6. Pipeline to get instance segmentation image from both distance map and semantic segmentation image. H-maxima transform followed by watershed algorithm enable to segment cells instance-wise.

3   Results

3.1   Evaluation Metrics

The metric used to evaluate DAPI and CellMask^{TM}prediction performance is the Pearson Correlation Coefficient (PCC, equation (3)). PCC is defined as the covariance of two variables divided by the product of their standard deviations. In the equation, x and y are two images to compare, and $\bar{x}$ is the average of x.

$$\text{PCC}(x,y) = \frac{\sum_{i=0}^{n} (x_i - \bar{x})(y_i - \bar{y})}{\sqrt{\sum_{i=0}^{n} (x_i - \bar{x})^2} \sqrt{\sum_{i=0}^{n} (y_i - \bar{y})^2}} \tag{3}$$

To evaluate nucleus semantic segmentation, we use the Jaccard index (equation (4)). The Jaccard index, or Intersection Over Union (IoU), is a very popular metric in segmentation, as it equally penalizes both False Positive and False Negative pixels. A perfect segmentation would lead to IoU of 1, while IoU corresponds to an entirely missed object (no intersection).

$$\text{IoU}(x,y) = \begin{cases} 1 & \text{if } x \cup y = 0 \\ \frac{x \cap y}{x \cup y} & \text{otherwise} \end{cases} \tag{4}$$

While the IoU is perfectly suitable to make pixel-wise comparisons for semantic segmentation, the performance of instance segmentation needs to incorporate an object-wise comparison that does not only penalize wrong pixel decisions, but also fused or split objects. For this, we choose to use the Mean Average Precision (mAP), which is a popular metric for instance segmentation evaluation. For this, a connected component from the ground truth is matched with a connected component from the segmentation result, if the IoU of the two components is above a given threshold. In this case, the object is considered as a TP. Unmatched connected components from the ground truth and the segmentation result are considered as FN and FP, respectively. Thus, given an IoU threshold one can compute the precision as defined in equation (5).

$$\text{Precision} = \frac{\text{TP}}{\text{TP} + \text{FP} + \text{FN}} \tag{5}$$

Precision is computed for all 10 IoU thresholds in $\{0.5 + i \times 0.05, i \in [\![0, 9]\!]\}$. The final result is the mean of these 10 values, hence called mean AP, or mAP.

3.2 Nucleus Segmentation

DAPI prediction yields very good results, with a PCC of 0.95±0.08.

Using the Jaccard index (or IoU) as metric, the U-net 'on steroids' gives 0.64±0.2 after training on 1 single image. In comparison, the model pretrained on DAPI reaches 0.84±0.1, improving the previous score by 31.3% (Fig. 7a). This improvement decreases as the size of the training set increases, being 4.8% (respectively 1.1%, 0.0%, 0.0%, −1.1%) after training on 10 (respectively 50, 100, 200, 500) (Fig. 7b).

Results from both models trained on 1 single image are displayed in Fig. 8.

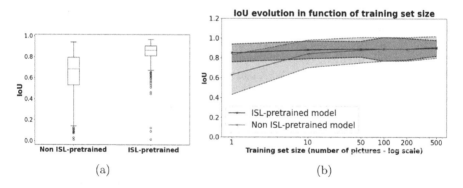

(a)   (b)

Fig. 7. Nucleus segmentation results. (a) Intersection Over Union (IoU) score for non ISL-pretrained and ISL-pretrained models, after training on 1 image. (b) Evolution of IoU average score for both models for different training set sizes.

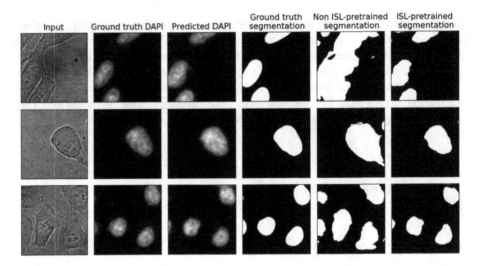

Fig. 8. Input bright-field images, DAPI images, DAPI predictions generated by U-net 'on steroids', ground truth instance segmentation generated by Cellpose, non ISL-pretrained U-net 'on steroids' segmentation prediction, ISL-pretrained U-net 'on steroids' segmentation prediction. Segmentation is performed after training on 1 image for both models.

3.3 Cell Segmentation

CellMask^{TM} prediction also yields very good results, with a PCC of 0.97 ± 0.02.

Using mAP as metric, the U-net 'on steroids' gives 0.17 ± 0.1 after training on 1 single image. In comparison, the model pretrained on CellMask^{TM} reaches 0.33 ± 0.09, improving the previous score by 94.1% (Fig. 9a). As in the previous section this improvement decreases as the size of the training set increases, being 18.5% (respectively -3.0%, -2.9%) after training on 10 (respectively 50, 80) (Fig. 9b).

Results from both models trained on 1 single image are displayed in Fig. 10.

4   Discussion

The results presented in the previous sections show that pretraining with In Silico Labeling as pretext task significantly improves the performance of a segmentation model trained on a very small data set. Indeed, the accuracy raises by 31.3% and 94.1% for nucleus semantic segmentation and cell instance segmentation, respectively, after training on 1 single image, using a model pretrained in an ISL setting.

The fact that pretraining on DAPI images helps to generate a nucleus semantic segmentation was actually expected since the two outputs (DAPI and binary segmentation maps) are very close to each other. On the other hand, cell instance segmentation is a much more complex problem, and our results clearly indicate

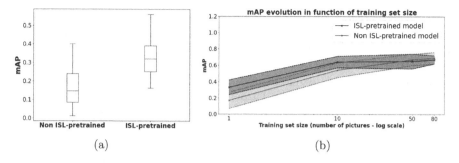

(a)   (b)

Fig. 9. Cell segmentation results. (a) mAP score for non ISL-pretrained and ISL-pretrained models, after training on 1 image. (b) Evolution of mAP average score for both models for different training set sizes.

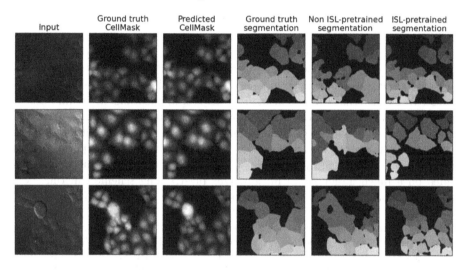

Fig. 10. Input DIC images, CellMask^{TM} images, CellMask^{TM} predictions generated by U-net 'on steroids', ground truth instance segmentation generated by Cellpose, non ISL-pretrained U-net 'on steroids' segmentation prediction, ISL-pretrained U-net 'on steroids' segmentation prediction. Segmentation is performed after training on 1 image for both models.

that also in this situation, pretraining with fluorescent marker prediction as a pretext task significantly improves segmentation accuracy for small datasets. We also observe that transfer learning is useful if we work on a very small data set (1 to 10 images), but that for both nucleus and cytoplasmic segmentation, the accuracy difference disappears if the models are trained on more than 10 images. This being said, if one has access to fluorescent images, it makes sense to use our proposed method to pretrain the network.

From a practical point of view, this idea provides an interesting alternative to manual annotation, in particular in the context of High Content Screening, where it is fairly easy to generate large amounts of data that contain both label-free and fluorescently labeled microscopy images. In this case, we can train efficient models for fluorescence prediction, and use these models in a pre-training scheme to reduce the manual annotation burden. Finally, we showed here that this pre-training scheme is effective for segmentation of nuclei and cells, but we also believe that this could be effective for any other type of cell structures as soon as you can get the associated fluorescent images available. Furthermore, it will be interesting to investigate to which extent the pre-training scheme provides good starting points for generalist networks, applicable to a wide variety of modalities.

5   Conclusion

In this paper, we demonstrated that pretraining on the prediction of relevant fluorescent markers can be very useful to segment nuclei or cells. We showed that a model trained to predict some fluorescent structures from label-free microscopy can learn to segment these structures from a very small data set, down to 1 single image. We believe that this can be of great help for applications where fluorescent data are easily available, if one wants to avoid tedious manual annotation to build large ground truth datasets for the training of neural networks. With only a few images, it is possible to fine-tune a pretrained model achieving performances matching those obtained by ImageNet-pretrained state-of-the-art networks fine-tuned on a much larger set of images. Our pre-training scheme can thus help biologists to save time and money without sacrificing any accuracy.

Acknowledgments. This work has been supported by the French government under management of Agence Nationale de la Recherche (ANR) as part of the "Investissements d'avenir" program, reference ANR-19-P3IA-0001 (PRAIRIE 3IA Institute), the Q-Life funded project CYTODEEP (ANR-17-CONV-0005) and the ANR project TRANSFACT (ANR-19-CE12-0007). Furthermore, we also acknowledge support by France-BioImaging (ANR-10-INBS-04).

Data Availability Statement. Code (pre-processing, training and testing, post-processing pipelines), is available at https://github.com/15bonte/isl_segmentation.

References

1. Hollandi, R., et al.: nucleAIzer: a parameter-free deep learning framework for nucleus segmentation using image style transfer. Cell Syst. **10**(5), 453–458 (2020).e6. ISSN: 24054712. https://doi.org/10.1016/j.cels.2020.04.003
2. Stringer, C., et al.: Cellpose: a generalist algorithm for cellular segmentation. Nature Methods **18**(1), 100–106 (2021). ISSN: 1548–7105. https://doi.org/10.1038/s41592-020-01018-x

3. Schmidt, U., Weigert, M., Broaddus, C., Myers, G.: Cell detection with star-convex polygons. In: Frangi, A.F., Schnabel, J.A., Davatzikos, C., Alberola-López, C., Fichtinger, G. (eds.) MICCAI 2018. LNCS, vol. 11071, pp. 265–273. Springer, Cham (2018). https://doi.org/10.1007/978-3-030-00934-2_30

4. Christiansen, E.M., et al.: In Silico Labeling: predicting fluorescent labels in unlabeled images. Cell **173**(3), 792–803 (2018). e19. ISSN: 0092–8674. https://doi.org/10.1016/j.cell.2018.03.040. https://www.sciencedirect.com/science/article/pii/S0092867418303647

5. Ounkomol, C., et al.: Label-free prediction of three-dimensional fluorescence images from transmitted-light microscopy. Nature Methods **15**(11), 917–920 (2018). ISSN: 1548–7105. https://doi.org/10.1038/s41592-018-0111-2

6. Rivenson, Y., et al.: Virtual histological staining of unlabelled tissueautofluorescence images via deep learning. Nature Biomed. Eng. **3**(6), 466–477 (2019). ISSN: 2157–846X. https://doi.org/10.1038/s41551-019-0362-y

7. Rivenson, Y., et al.: PhaseStain: the digital staining of label-free quantitative phase microscopy images using deep learning. Light Sci. Appl. **8**(1), 23 (2019). ISSN: 2047–7538. https://doi.org/10.1038/s41377-019-0129-y

8. Li, D., et al.: Deep learning for virtual histological staining of bright-field microscopic images of unlabeled carotid artery tissue. Molecular Imaging Biol. **22**(5), 1301–1309, October 2020. ISSN: 1860–2002. https://doi.org/10.1007/s11307-020-01508-6

9. Liu, Y., et al.: Global pixel transformers for virtual staining of microscopy images. IEEE Trans. Med. Imaging, p. 1, January 2020. https://doi.org/10.1109/TMI.2020.2968504

10. Nygate, Y.N., et al.: Holographic virtual staining of individual biological cells. Proc. Natl. Acad. Sci. **117**(17), 9223–9231 (2020). https://www.pnas.org/doi/pdf/10.1073/pnas.1919569117

11. Boyd, J., et al.: Experimentally-generated ground truth for detecting cell types in an image-based immunotherapy screen. In: 2020 IEEE 17th International Symposium on Biomedical Imaging (ISBI), pp. 886–890 (2020). https://doi.org/10.1109/ISBI45749.2020.9098696

12. Hu, C., et al.: Live-dead assay on unlabeled cells using phase imaging with computational specificity. Nature Commun. **13**(1), 713 (2022). ISSN: 2041–1723. https://doi.org/10.1038/s41467-022-28214-x

13. Zhang, J.K., et al.: Automatic colorectal cancer screening using deep learning in spatial light interference microscopy data. en. In: Cells 11.4, February 2022

14. Shen, C., et al.: Stain-free detection of embryo polarization using deep learning. Sci. Rep. **12**(1), 2404 (2022). ISSN: 2045–2322. https://doi.org/10.1038/s41598-022-05990-6

15. He, Y.R., et al.: Cell cycle stage classification using phase imaging with computational specificity. ACS Photonics **9**(4), 1264–1273 (2022). https://doi.org/10.1021/acsphotonics.1c01779

16. Ronneberger, O., Fischer, P., Brox, T.: U-Net: convolutional networks for biomedical image segmentation (2015). https://doi.org/10.48550/ARXIV.1505.04597

17. Yakubovskiy, P.: Segmentation Models Pytorch (2020). https://github.com/qubvel/segmentation_models.pytorch

18. Deng, J., et al.: ImageNet: a large-scale hierarchical image database. In: IEEE Conference on Computer Vision and Pattern Recognition IEEE 2009, pp. 248–255 (2009)

19. Naylor, P., et al.: Nuclei segmentation in histopathology images using deep neural networks. In: IEEE 14th International Symposium on Biomedical Imaging (ISBI 2017) (2017). IEEE, EMB. IEEE Signal Proc Soc. ISSN 19458452 (2017). https://doi.org/10.1109/ISBI.2017.7950669

20. Naylor, P., et al.: Segmentation of nuclei in histopathology images by deep regression of the distance map. IEEE Trans. Med. Imaging **38**(2), 448–459 (2019). https://doi.org/10.1109/TMI.2018.2865709

Towards Better Guided Attention and Human Knowledge Insertion in Deep Convolutional Neural Networks

Ankit Gupta[1(✉)] and Ida-Maria Sintorn[1,2]

[1] Department of Information Technology, Uppsala University,
75236 Uppsala, Sweden
{ankit.gupta,ida.sintorn}@it.uu.se
[2] Vironova AB, 11330 Gävlegatan 22, Stockholm, Sweden

Abstract. Attention Branch Networks (ABNs) have been shown to simultaneously provide visual explanation and improve the performance of deep convolutional neural networks (CNNs). In this work, we introduce Multi-Scale Attention Branch Networks (MSABN), which enhance the resolution of the generated attention maps, and improve the performance. We evaluate MSABN on benchmark image recognition and fine-grained recognition datasets where we observe MSABN outperforms ABN and baseline models. We also introduce a new data augmentation strategy utilizing the attention maps to incorporate human knowledge in the form of bounding box annotations of the objects of interest. We show that even with a limited number of edited samples, a significant performance gain can be achieved with this strategy.

Keywords: Visual explanation · Fine-grained recognition · Attention map · Human-in-the-loop

1 Introduction

CNNs have established themselves as the benchmark approach in image recognition [13,20,30,33]. However, the interpretation of the decision-making process still remains elusive. To be able to visualize and verify that the decision of a CNN is based on correct and meaningful information is important for many computer vision applications like self-driving cars or automated social media content analysis, to increase reliability and trust in the system. For biomedical applications, where instrument settings and small cohorts/datasets easily bias the results this becomes extremely important. This, in addition to the high cost or risk associated with erroneous decisions, limit the reliability and hence also the deployment of CNN-based solutions in clinical diagnostics and biomedical analysis. Tools to explain the decision and be able to correct erroneous conclusions drawn by CNN would improve the trustworthiness of the technology.

Supplementary Information The online version contains supplementary material available at https://doi.org/10.1007/978-3-031-25069-9_29.

© The Author(s), under exclusive license to Springer Nature Switzerland AG 2023
L. Karlinsky et al. (Eds.): ECCV 2022 Workshops, LNCS 13804, pp. 437–453, 2023.
https://doi.org/10.1007/978-3-031-25069-9_29

Visual explanation [35] is used in deep learning to interpret the decisions of the CNNs. Broadly, these methods can be categorized as either requiring additional backpropagation or not. Methods requiring backpropagation [5, 27] can be used out-of-the-box and don't require any network architectural or training method changes. However, they need an extra backpropagation step to find the discriminative regions in images. Response-based methods [10, 36] don't require backpropagation because they generate explanations and predictions simultaneously. In this work, we improve upon [10] and present a response-based visual explanation method that outperforms previous methods.

To fully utilize the benefits of deep learning in practical settings, it would also be beneficial to be able to incorporate human knowledge and interact with the attention maps to correct and improve the networks. One attempt to do this is presented in [22], where the attention map of the images is manually edited and then the network is fine-tuned with this human knowledge. This approach, however, requires manual detailed annotation (drawing) of the attention maps which becomes tedious for large datasets. Here, we propose a softer form of user input in the form of object bounding boxes.

In this paper, we build upon the attention branch network structure and present ways to improve the attention maps and overall network performance even further. In addition, we present a simple and efficient way to incorporate human expertise when training/refining the networks. Figure 1 illustrates the difference in the attention map detail achieved with our proposed multi-scale attention branch network (MSABN) compared to the attention branch network (ABN) and the commonly used Class Activation Mapping (CAM).

Our main contributions are:

- With the introduction of MSABN, we increase the performance of the response-based guided attention models significantly and simultaneously increase the resolution of attention maps by 4x which has not been achieved with the response-based methods yet.
- With the introduction of the puzzle module in the attention branch, the performance is improved further and it also adds better localization performance for fine-grained recognition.
- We provide a human-in-the-loop (HITL) pipeline for inserting human knowledge in the models which require simple annotation (object bounding box) and achieve performance improvement even by annotating/correcting a small portion of the dataset.

2   Related Work

Visual Explanation: Highlighting the regions or details of an image that are important for the decision is commonly used to interpret and verify the decision-making process of CNNs [2, 27]. Overall, the methods for visual explanation fall into two categories: gradient-based and response-based. Gradient-based methods (SmoothGrad [28], Guided Backpropagation [29], LIME [25], grad-CAM

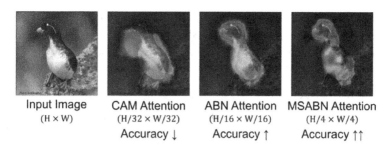

Input Image	CAM Attention	ABN Attention	MSABN Attention
(H × W)	(H/32 × W/32)	(H/16 × W/16)	(H/4 × W/4)
	Accuracy ↓	Accuracy ↑	Accuracy ↑↑

Fig. 1. Comparison of the MSABN attention map with CAM and ABN attention outputs. MSABN provides better accuracy and higher resolution attention maps.

[27], gradCAM++ [5], LRP [2]) rely on the backpropagation of auxiliary data like class index or noise to find the discriminative regions in the image. These methods do not require training or architectural modifications. However, backpropagation adds another step during inference and increases the computational cost. Response-based methods, on the other hand, do not require backpropagation and can provide interpretation at the same time as inference. CAM [36], is a response-based method that provides an attention map for each class by replacing the final fully-connected layer of a CNN with a convolutional layer and global average pooling. The recently proposed attention branch networks [10] extend the response-based model by adding a branch structure to provide the attention to the network in a feed-forward way and show that adding attention in this way improves the performance of the models.

Inserting Human Knowledge: Human-in-the-loop (HITL) refers to approaches in the deep learning framework where a prediction model is trained with integrated human knowledge or user guidance. Several HITL approaches [3,4,23,24,32] have been used to inject human knowledge in different ways into the deep neural networks. In [21], the ClickMe interface is used to get human annotation in the form of mouse clicks to get approximate attention regions in an image. They also modified the model architecture to incorporate this human knowledge and demonstrated improved performance. In [26], the authors proposed a method that leverages the explanation generated by the models such that they should provide the correct explanation for the correct prediction. In [22], Mitsuhara, M. et al. showed that editing the attention maps and sequentially fine-tuning the ABN models improve performance as mentioned in the Introduction. However, their approach involves manually marking the object boundary in a large number (thousands) of images making it unfeasible in practice. In [1,15], the authors attempted to improve model performance by using non-strict attention input and introducing different loss functions and managing to improve performance. In [15], Dharma et al. use different loss functions for the regions inside and outside the object bounding box as a way to incorporate human input. Our method is similar to these methods but focused on augmenting the data rather than changing the loss function.

Copy-paste Augmentation: This refers to an augmentation strategy where a patch from one image is pasted into another image which is then used for training. In [7,8,11] this augmentation strategy is used to improve the performance in instance segmentation and detection tasks. The idea is that by cutting object instances and pasting them into other images a different context for the objects is provided. The CutMix [34] augmentation strategy used this idea more generally and show that simply copying and pasting random patches from one image to another and modifying the output probabilities improves both the classification and localization performance. Finally, PuzzleMix [16] explored CutMix combined with saliency information and image statistics to further improve the performance. In the augmentation method, we propose the way to incorporate human knowledge can be seen as a supervised version of CutMix. We use the object location to copy and replace the objects in the dataset in the images with mismatched attention map and true object location.

3   Methods

3.1   Multi-Scale Attention Branch Network

In [10], a dual attention and perception branch architecture is used to generate the attention map and prediction simultaneously. The output from the third convolutional block (which is referred to as the "feature extractor") is fed into the attention branch which produces the CAM output and the attention map. The attention map is then combined with the feature extractor output and fed into the perception branch which outputs the probability of each class. The CAM output from the attention branch and the probability output of the perception branch are trained on cross-entropy loss simultaneously. The training can then be done in an end-to-end manner in both branches. The loss can be written as,

$$\mathcal{L}(x_i) = \mathcal{L}_{attn}(x_i) + \mathcal{L}_{cls}(x_i) \tag{1}$$

where $\mathcal{L}_{attn}(x_i)$ and $\mathcal{L}_{cls}(x_i)$ are the cross-entropy losses for the attention and perception branch respectively for a training sample x_i.

Unlike [10], where the attention branch is only fed the output of the third convolutional block, the attention branch of our multi-scale attention branch network (MSABN) is fed with input from the first and second convolutional blocks as well. Figure 2 shows the MSABN architecture. The output from the second and third convolutional blocks is upsampled to match the size of the first convolutional block. The output from each of the three convolutional blocks is then passed through 1×1 convolutional blocks such that the total number of channels after concatenating the outputs match the input channels of the attention branch. Our suggested architecture helps the network accumulate hierarchical information at different scales to produce a fine-grained attention map and improve performance. The resolution of the attention map in MSABN is also higher than in ABN as a result of the upsampling.

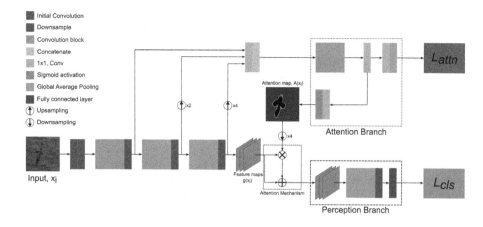

Fig. 2. Overview of our proposed MSABN framework.

3.2   Puzzle Module to Improve Fine-grained Recognition

The puzzle module consists of a tiling and a merging module which aims at minimizing the difference between the merged features from tiled patches in an image and the features from the original image. The tiling module generates non-overlapping tiled patches of the input image. The patches are fed into the network and the merging module merges the patches' outputs into CAMs. The L1 loss between the CAM of the target class from the original image and the reconstruction is then calculated and used for training. Figure 3 illustrates the puzzle module. The overall loss for training MSABN models with the puzzle module becomes:

$$\mathcal{L}(x_i) = \mathcal{L}_{attn}(x_i) + \mathcal{L}_{cls}(x_i) + \mathcal{L}_{re}(x_i), \tag{2}$$

where $\mathcal{L}_{re}(x_i)$ is the L1 loss between the original image CAM and the reconstructed CAM.

The puzzle module was introduced in [14] for weakly-supervised segmentation. It was shown to improve the segmentation performance on the PASCAL VOC 2012 dataset [9] with CAM-based ResNet models. Since the attention branch is trained similarly to CAM models, we decided to apply the puzzle module to the attention branch to observe the effects on attention maps and the performance of the models.

3.3   Embedding Human Knowledge with Copy-Replace Augmentation

We describe a HITL pipeline with the proposed fine-tuning process incorporating human knowledge. First, the MSABN model is trained with images in the training set with corresponding labels. Then, the attention maps of the training samples are obtained from the best-performing model. In a real setting, the

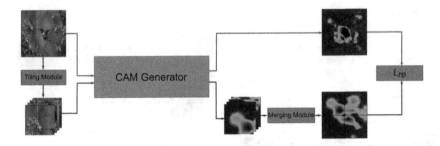

Fig. 3. Overview of the Puzzle Module. The tiling module divides the image into non-overlapping tiles which are then fed into the CAM generator and the merging module merges the tiled CAMs that are then compared with the original CAM.

user can then inspect the attention maps and provide corrective input where the attention is mislocated in the form of a bounding box around the object. This is a softer form of user input compared to object boundary annotation and can be scaled to some degree without it becoming too cumbersome. Then, the training images are divided into two pools, one with bounding box annotations provided by the user and the other without. In the fine-tuning step, we train with all the training images and labels normally, *but* we modify the images where the bounding box annotation was provided. For the images with provided annotations, the patch defined by the bounding box in an image is extracted, resized, and pasted onto the bounding box of another image from the same pool as illustrated in Fig. 4. The resizing to the target bounding box size also acts as scaling augmentation during the training.

4   Experiments

4.1   Image Classification

Datasets: We evaluate MSABN for image classification on the CIFAR100 [19], ImageNet [6] and DiagSet-10x [17] datasets. CIFAR100 contains 60,000 images for training and 10,000 images for testing of 100 classes with an image size of 32×32 pixels. ImageNet consists of 1,281,167 training images and 50,000 validation images of 1000 classes. Finally, the DiagSet-10x dataset is a histopathology dataset for prostate cancer detection with 256×256 pixel images divided into 9 categories with different cancer grading, normal tissue, artifacts, and background. This dataset was included to test the performance of MSABN on texture data rather than object data. The dataset consists of 132,882 images for training, 25,294 for validation, and 30,086 for testing. In the experiments, the background class (empty, non-tissue regions of the slide) was excluded.

Training Details: For the CIFAR100 dataset, ResNet-20,-56,-110 [13] models were used to evaluate the performance. The models were optimized with stochastic gradient descent (SGD) with a momentum of 0.9 and weight decay of 5e-4

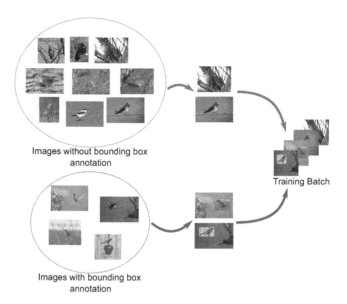

Fig. 4. Overview of the copy-paste training scheme. In our case, objects with misaligned attention are annotated with a bounding box and, cut out, and replaced by other bounding box annotated objects.

for 300 epochs and a batch size of 256. The initial learning rate was set to 0.1 and was divided by 10 at 50% and 75% of the total number of epochs. Random cropping and horizontal flipping were used as augmentations, and model weights were initialized with Kaiming initialization [12]. The experiments were repeated three times with different random seeds and the mean and standard deviation of the accuracy is reported (mainly to assert stability of the models).

For the DiagSet-10x and ImageNet datasets, ResNet-101 and EfficientNet-B1 models were used. The models were optimized with SGD with a momentum of 0.9 and weight decay of 1e-4 for 90 epochs. A batch size of 512 was used for ResNet-101 experiments and 256 for EfficientNet-B1 experiments. The initial learning rate was set to 0.1 and was divided by 10 at 33% and 66% of the total number of epochs. For both datasets, random cropping of 224×224 and horizontal flipping were used as augmentations. For DiagSet-10x vertical flipping and color jitter were also used. Due to a severe imbalance between classes in the DiagSet-10x dataset, the loss was weighted by the number of samples of each class. All model weights were initialized with Kaiming initialization. These experiments were only performed once due to computational constraints. EfficientNet-B1+ABN performance was not reported as the authors in [10] didn't implement ABN on EfficientNets.

Analysis of Attention Mechanism on CIFAR: Following the analysis in [10], two different attention mechanisms (how to combine the attention map with the feature maps), namely $g(x) \cdot A(x)$ and $g(x) \cdot (1 + A(x))$, were compared with the base (no attention branch) and ABN model which uses the $g(x) \cdot (1 + A(x))$ mechanism. As can be seen from Table 1, both mechanisms outperform the ABN for all model versions. $g(x) \cdot (1 + A(x))$ outperforms $g(x) \cdot A(x)$ slightly for the Resnet-20 and 110 models but performs slightly worse for ResNet-56 on average. The standard deviation for $g(x) \cdot A(x)$ mechanism is higher than for $g(x) \cdot (1 + A(x))$ for all models. This might be due to the greater stability provided by the residual connection. We also noticed that the attention maps when using $g(x) \cdot A(x)$ have values close to 0.5 for the background as compared to $g(x) \cdot (1 + A(x))$ where they were close to zero. Hence we decided to use $g(x) \cdot (1 + A(x))$ mechanism as the default for this paper.

Table 1. Comparison of the accuracies (%) on CIFAR100

	ResNet-20	ResNet-56	ResNet-110
BaseModel	68.71 ± 0.202	74.92 ± 0.520	76.83 ± 0.233
BaseModel+ABN	68.97 ± 0.074	76.31 ± 0.029	77.92 ± 0.102
BaseModel+MSABN ($g(x) \cdot A(x)$)	69.47 ± 0.298	$\mathbf{77.16 \pm 0.419}$	78.40 ± 0.254
BaseModel+MSABN ($g(x) \cdot (1 + A(x))$)	$\mathbf{70.06 \pm 0.025}$	76.78 ± 0.149	$\mathbf{78.68 \pm 0.121}$

Accuracy on ImageNet and DiagSet-10x: Table 2 summarizes the results of the experiments on ImageNet. The MSABN model outperforms the ABN model by 0.35% for ResNet-101 and the base model by 0.76% and 1.41% for ResNet-101 and EfficientNet-B1 respectively. The visual comparison of the attention in ABN and MSABN models is shown in Fig. 5. The MSABN attention has a better localization performance and highlights the object boundaries better than ABN attention.

Table 2. Comparison of the accuracies (%) on the ImageNet Dataset

	ResNet-101	EfficientNet-B1
BaseModel	77.33	67.61
BaseModel+ABN	77.74	–
BaseModel+MSABN	**78.09**	**69.08**

Table 3 summarizes the results of the experiments on DiagSet-10x. In addition to accuracy, we also report the balanced accuracy to better estimate the performance on an unbalanced dataset. Here, the MSABN model outperforms the ABN model by 4.23%/2.68% for ResNet-101 and the base model by 2.34%/2.72% and 0.09%/1.91% for ResNet-101 and EfficientNet-B1 respectively. A visual comparison of the attention in ABN and MSABN models is shown in Fig. 6. As compared to ABN, the MSABN attention highlights specific nuclei important for the

Fig. 5. Visualization of high attention regions in the images from the ImageNet dataset.

grading. ABN attention maps also show an activation along the image boundary in the images with non-tissue regions present as an artifact. This artifact was not observed in the MSABN attention maps.

Table 3. Comparison of the accuracies (%) on the DiagSet-10x dataset

	ResNet-101 (Accuracy/Balanced Accuracy)	EfficientNet-B1 (Accuracy/Balanced Accuracy)
BaseModel	54.96/38.57	56.15/40.36
BaseModel+ABN	53.07/38.61	–
BaseModel+MSABN	**57.30/41.29**	**56.24/42.27**

4.2 Fine-grained Recognition

Datasets: The performance of MSABN and Puzzle-MSABN was evaluated in two fine-grained recognition datasets namely CUB-200–2011 [31], and Stanford Cars [18]. The CUB-200–2011 dataset contains 11788 images of 200 categories, 5994 for training and 5794 for testing. The Stanford Cars dataset contains 16185 images of 196 classes and is split into 8144 training images and 8041 testing images.

Training Details: The models used for evaluation of the fine-grained datasets were ResNet-50 [13], ResNext-50 (32×4d) [33], and EfficientNet-B3 [30]. The models showcase different popular architectures and hence were chosen to benchmark the results. The models were optimized with SGD with a momentum of 0.9 and weight decay of 1e-4 for 300 epochs with a batch size of 16. The initial

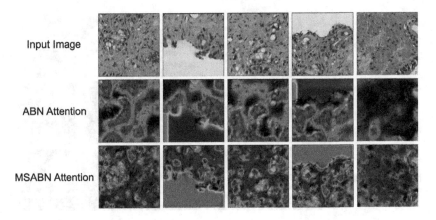

Input Image

ABN Attention

MSABN Attention

Fig. 6. Visualization of high attention regions in the DiagSet-10x dataset.

learning rate was set to 0.1 and was divided by 10 at 50% and 75% of the total number of epochs. The images were resized to 352×352 pixels and the augmentations used were horizontal flip, color jitter, gaussian blur and noise, and solarize. All model weights were initialized with Kaiming initialisation. The experiments were run three times with different random seeds and the mean and standard deviation of the performance are reported.

Table 4 shows the accuracies of the models on the CUB-200–2011 dataset. MSABN improves upon the average performance of ABN models by 4% and 2% for ResNet-50 and ResNext-50 respectively. The performance gain compared with the base model is 14%, 12%, and 12% for ResNet-50, ResNext-50, and EfficientNet-B3 respectively. Furthermore, the introduction of the puzzle module outperforms the ABN models by 6% and 4% respectively. The puzzle module outperforms the base models by 17%, 14%, and 13% for ResNet-50, ResNext-50, and EfficientNet-B3, respectively.

Table 4. Comparison of the accuracies (%) on CUB-200–2011 Dataset

	ResNet-50	ResNext-50	EfficientNet-B3
BaseModel	40.62 ± 1.985	44.68 ± 0.816	53.41 ± 0.792
BaseModel+ABN	50.96 ± 2.145	54.88 ± 0.700	–
BaseModel+MSABN	54.98 ± 0.219	56.86 ± 0.581	65.28 ± 0.397
BaseModel+MSABN+Puzzle	**57.31 ± 0.645**	**58.87 ± 0.592**	**66.21 ± 0.615**

Table 5, shows the performance of the different models on the Stanford Cars dataset. The improvement with MSABN compared to the ABN model is not as significant here but slightly outperforms them on average. The performance gain compared with the base models however is 8%, 3%, and 2% for ResNet-50, ResNext-50, and EfficientNet-B3, respectively. The introduction of the puzzle module outperforms ABN models by 3% and 1% respectively. The puzzle module outperforms the base models by 10%, 4%, and 3% for ResNet-50, ResNext-50, and EfficientNet-B3, respectively.

Table 5. Comparison of the accuracies (%) on Stanford Cars Dataset

	ResNet-50	ResNext-50	EfficientNet-B3
BaseModel	78.70 ± 1.062	84.00 ± 0.817	86.86 ± 0.495
BaseModel+ABN	85.59 ± 0.396	87.34 ± 0.408	–
BaseModel+MSABN	86.81 ± 0.991	87.39 ± 0.526	88.90 ± 0.246
BaseModel+MSABN+Puzzle	**88.25 ± 1.492**	**88.32 ± 1.184**	**89.92 ± 0.189**

Visualization of the attention maps for CUB-200–2011 and Stanford Cars datasets are shown in Fig. 7. Compared to ABN attention maps, MSABN maps are able to delineate the object boundaries better and provide information about the discriminative regions in the image. For CUB-200–2011 dataset, most of the attention in MSABN and MSABN+Puzzle is focused on the key attributes like the bill, wings, or legs of the bird. The puzzle module performs significantly better than the rest of the model configurations here. This can be attributed to the effective regularization puzzle module provided in the case of small datasets and complex images (birds with different angles and actions). For Stanford Cars, the attention is focused on roughly similar regions depending on the car pose and form with different configurations. The performance improvement of MSABN and MSABN+Puzzle is not as significant here as for CUB-200–2011. This can be attributed to the nature of objects, i.e., cars, which have a well-defined shape as compared to birds which will have different shapes depending on the action like, flying, swimming, sitting, or standing.

4.3   Attention Editing Performance

Datasets: We used the previously trained MSABN models on the CUB-200–2011 and Stanford Cars datasets to demonstrate the attention editing process. In both the datasets, bounding boxes of the objects within the image are available and are used here to mimic user input in the HITL pipeline in controlled experiments.

Experimental Setup: We measured the accuracy with respect to the number of annotated samples (bounding boxes) "provided by the user" to estimate the human effort required to achieve performance gain. To decide which samples to "annotate", the attention maps of the training data were saved and a binary map was created by thresholding the intensity values at 0.2. Next, the ratio of attention inside and outside the bounding box of the total attention in the binary image was calculated. The number of samples for which the fraction of attention outside $frac_{attn\_out}$ was above a threshold λ_{out} were put in the bin

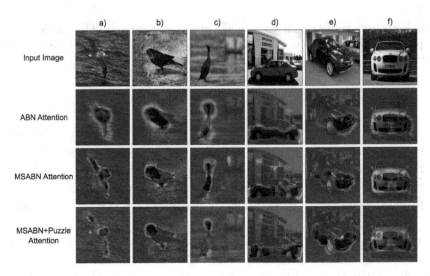

Fig. 7. Visualization of attention maps on the CUB-200–2011 (a, b, and c) and Stanford Cars (c, d, and e) dataset.

of copy-replace augmentation for the fine-tuning training step. For the CUB-200–2011 dataset, the performance at $\lambda_{out} \in \{0.0, 0.1, 0.2, 0.3, 0.4, 0.5, 0.6, 0.7\}$ was obtained and for Stanford Cars, the performance was obtained at $\lambda_{out} \in \{0.0, 0.1, 0.2, 0.3, 0.4, 0.5\}$. We stopped at 0.7 and 0.5 respectively because the number of training samples were too low (less than 30) in some configurations to observe meaningful changes above these thresholds. We chose this way instead of measuring performance at different fractions of the total number of training samples to better infer the trends of attention localization in the datasets and models. If the majority of the attention is focused inside the object box, the number of samples where $frac_{attn\_out} > \lambda_{out}$ would be much lower for higher thresholds. This method puts the images with the worst attention localization first which of course "boosts" the reported results. However, a case can be made that this mimics a potentially real scenario where a user corrects or edits discovered errors for fine-tuning a model: The training samples to show the user can be sorted either on the basis of wrong predictions first or the amount of overall attention (since the models might focus on the background) to be shown first, which will make user input a little less time-consuming. The results were compared with results from vanilla fine-tuning with all the training images to better compare the effects of the copy-replace augmentation.

Training Details: We evaluated all the MSABN models used in the fine-grained recognition. The models were optimized with SGD with a momentum of 0.9 and weight decay of 1e-4 for 50 epochs with a batch size of 16. The initial learning rate was set to 0.1 and was divided by 10 at 50% and 75% of the total number of epochs. No augmentations were applied in the retraining to document the effects of only the copy-replace augmentation. The experiments were repeated three times for each model's (Resnet, ResNext, and EfficientNet) three repetitions in the previous section.

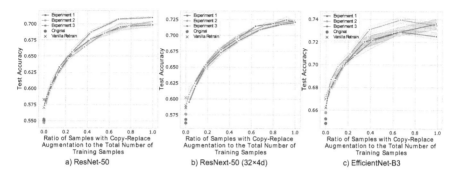

Fig. 8. Fine-tuning performance of different MSABN models on the CUB-200–2011 dataset with different ratios of training samples used for copy-replace augmentation.

Figures 8 and 9 show the results of the fine-tuning with copy-replace augmentation in CUB-200–2011 and Stanford Cars datasets. As can be seen from the figures, fine-tuning improves the performance of the models. The accuracy gain compared to vanilla retraining with only 20% of the training data augmented was approx. 6%, 5%, and 3% for the MSABN models of ResNet-50, ResNext-50, and EfficientNet-B3 respectively for the CUB-200–2011 dataset. For Stanford Cars, the accuracy increased by 2%, 1.5% and 0.5% respectively for the same amount of augmented training samples. The overall gain from augmenting almost all training samples compared with vanilla retraining was approx. 12.5%, 12.5%, and 6% for MSABN models of ResNet-50, ResNext-50, and EfficientNet-B3, respectively for the CUB-200–2011 dataset. For the Stanford dataset, the overall gain was approx. 4.5%, 3.5%, and 1.5%, respectively.

The distribution of data points along the x-axis in Fig. 8 is relatively uniform compared to Fig. 9 where most of the data points are located between zero and 0.2. This means that the number of training samples with $frac_{attn\_out} > \lambda_{out}$ changes proportionally with λ_{out} for the CUB-200–2011 dataset but doesn't change much for the Stanford Cars dataset, which signals that the attention is more focused on the objects in the Stanford Cars dataset than in the CUB-200–2011 dataset. This can be attributed to the nature of objects in the datasets as mentioned earlier.

The experiments showed that a significant performance gain can be achieved with limited and relatively crude human input. As mentioned earlier, the results are biased towards the worst examples, so there is a need to develop a smart sorting system that decides the order of images to be shown to the user.

We also noticed an interesting behaviour with vanilla retraining on the two datasets. While the accuracy after retraining in CUB-200–2011 increased, the accuracy decreased with the Stanford Cars dataset for all the models. The train-

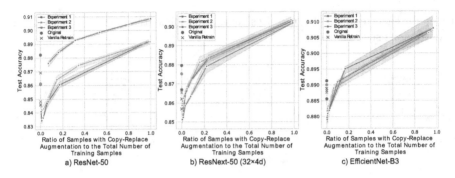

Fig. 9. Fine-tuning performance of different MSABN models on the Stanford Cars dataset with different ratios of training samples used for copy-replace augmentation.

ing loss curves (not shown) indicate that both models were destabilized (training loss increases) at the beginning of the training, however, in CUB-200–2011 dataset the models converged to a minima with lower validation loss than that of the initial training of the models. The observed effect in the Stanford Cars dataset was the opposite. Further experiments to determine the cause were not done, instead, the performance was compared to the vanilla retrain as the benchmark.

Considerations: It's worth noting that due to upscaling of the outputs of intermediate blocks, the attention branch now has to process 16x more input values. This increases the computational cost of the MSABN models. We noticed that it takes roughly twice the time to train MSABN models compared to ABN models. The puzzle module increases the cost further as every image is processed twice, once as original and other as tiled sub-images during training, however, the model behaves like the MSABN model during inference.

5   Conclusion

In this paper, we have presented a multi-scale attention branch network that greatly improves classification performance and also provides more accurate and detailed attention maps. We have evaluated the accuracy of MSABN for image recognition and fine-grained classification on multiple datasets and it was shown to outperform the ABN models. We also showed that using the puzzle module for fine-grained recognition increases the performance of the MSABN models. In addition, we introduced a HITL learning framework that inserts human knowledge in form of object bounding boxes and shows that this is an effective way of improving the performance further. The code is publicly available at https:// github.com/aktgpt/msabn.

Acknowledgments. The work was supported by the Swedish Foundation for Strategic Research (grant BD15-0008SB16-0046) and the European Research Council (grant

ERC-2015-CoG 683810). The computations were enabled by the supercomputing resource Berzelius provided by National Supercomputer Centre at Linköping University and the Knut and Alice Wallenberg foundation.

References

1. Arai, S., Shirakawa, S., Nagao, T.: Non-strict attentional region annotation to improve image classification accuracy. In: 2021 IEEE International Conference on Systems, Man, and Cybernetics (SMC), pp. 2375–2380. IEEE (2021)
2. Bach, S., Binder, A., Montavon, G., Klauschen, F., Müller, K.R., Samek, W.: On pixel-wise explanations for non-linear classifier decisions by layer-wise relevance propagation. PLoS ONE **10**(7), e0130140 (2015)
3. Branson, S., Perona, P., Belongie, S.: Strong supervision from weak annotation: interactive training of deformable part models. In: 2011 International Conference on Computer Vision, pp. 1832–1839. IEEE (2011)
4. Branson, S., et al.: Visual recognition with humans in the loop. In: Daniilidis, K., Maragos, P., Paragios, N. (eds.) ECCV 2010. LNCS, vol. 6314, pp. 438–451. Springer, Heidelberg (2010). https://doi.org/10.1007/978-3-642-15561-1_32
5. Chattopadhay, A., Sarkar, A., Howlader, P., Balasubramanian, V.N.: Gradcam++: generalized gradient-based visual explanations for deep convolutional networks. In: 2018 IEEE Winter Conference on Applications of Computer Vision (WACV), pp. 839–847. IEEE (2018)
6. Deng, J., Dong, W., Socher, R., Li, L.J., Li, K., Fei-Fei, L.: Imagenet: a large-scale hierarchical image database. In: 2009 IEEE Conference on Computer Vision and Pattern Recognition, pp. 248–255. IEEE (2009)
7. Dvornik, N., Mairal, J., Schmid, C.: Modeling visual context is key to augmenting object detection datasets. In: Proceedings of the European Conference on Computer Vision (ECCV), pp. 364–380 (2018)
8. Dwibedi, D., Misra, I., Hebert, M.: Cut, paste and learn: Surprisingly easy synthesis for instance detection. In: Proceedings of the IEEE International Conference on Computer Vision, pp. 1301–1310 (2017)
9. Everingham, M., Van Gool, L., Williams, C.K., Winn, J., Zisserman, A.: The pascal visual object classes (VOC) challenge. Int. J. Comput. Vis. **88**(2), 303–338 (2010)
10. Fukui, H., Hirakawa, T., Yamashita, T., Fujiyoshi, H.: Attention branch network: learning of attention mechanism for visual explanation. In: Proceedings of the IEEE/CVF Conference on Computer Vision and Pattern Recognition, pp. 10705–10714 (2019)
11. Ghiasi, G., Cui, Y., Srinivas, A., Qian, R., Lin, T.Y., Cubuk, E.D., Le, Q.V., Zoph, B.: Simple copy-paste is a strong data augmentation method for instance segmentation. In: Proceedings of the IEEE/CVF Conference on Computer Vision and Pattern Recognition, pp. 2918–2928 (2021)
12. He, K., Zhang, X., Ren, S., Sun, J.: Delving deep into rectifiers: surpassing human-level performance on imagenet classification. In: Proceedings of the IEEE International Conference on Computer Vision, pp. 1026–1034 (2015)
13. He, K., Zhang, X., Ren, S., Sun, J.: Deep residual learning for image recognition. In: Proceedings of the IEEE Conference on Computer Vision and Pattern Recognition, pp. 770–778 (2016)
14. Jo, S., Yu, I.J.: Puzzle-cam: improved localization via matching partial and full features. In: 2021 IEEE International Conference on Image Processing (ICIP), pp. 639–643. IEEE (2021)

15. KC, D., Zhang, C.: Improving the trustworthiness of image classification models by utilizing bounding-box annotations. arXiv preprint arXiv:2108.10131 (2021)
16. Kim, J.H., Choo, W., Song, H.O.: Puzzle mix: exploiting saliency and local statistics for optimal mixup. In: International Conference on Machine Learning, pp. 5275–5285. PMLR (2020)
17. Koziarski, M., et al.: Diagset: a dataset for prostate cancer histopathological image classification. arXiv preprint arXiv:2105.04014 (2021)
18. Krause, J., Stark, M., Deng, J., Fei-Fei, L.: 3D object representations for fine-grained categorization. In: 4th International IEEE Workshop on 3D Representation and Recognition (3dRR-13). Sydney, Australia (2013)
19. Krizhevsky, A., Hinton, G., et al.: Learning multiple layers of features from tiny images (2009)
20. Krizhevsky, A., Sutskever, I., Hinton, G.E.: Imagenet classification with deep convolutional neural networks. In: Advances in Neural Information Processing Systems, vol. 25 (2012)
21. Linsley, D., Shiebler, D., Eberhardt, S., Serre, T.: Learning what and where to attend. In: International Conference on Learning Representations (2019)
22. Mitsuhara, M., et al.: Embedding human knowledge into deep neural network via attention map. arXiv preprint arXiv:1905.03540 (2019)
23. Parikh, D., Grauman, K.: Interactively building a discriminative vocabulary of nameable attributes. In: CVPR 2011, pp. 1681–1688. IEEE (2011)
24. Parkash, A., Parikh, D.: Attributes for classifier feedback. In: Fitzgibbon, A., Lazebnik, S., Perona, P., Sato, Y., Schmid, C. (eds.) ECCV 2012. LNCS, vol. 7574, pp. 354–368. Springer, Heidelberg (2012). https://doi.org/10.1007/978-3-642-33712-3_26
25. Ribeiro, M.T., Singh, S., Guestrin, C.: Why should i trust you? explaining the predictions of any classifier. In: Proceedings of the 22nd ACM SIGKDD International Conference on Knowledge Discovery and Data Mining, pp. 1135–1144 (2016)
26. Rieger, L., Singh, C., Murdoch, W., Yu, B.: Interpretations are useful: penalizing explanations to align neural networks with prior knowledge. In: International Conference on Machine Learning, pp. 8116–8126. PMLR (2020)
27. Selvaraju, R.R., Cogswell, M., Das, A., Vedantam, R., Parikh, D., Batra, D.: Gradcam: visual explanations from deep networks via gradient-based localization. In: Proceedings of the IEEE International Conference on Computer Vision, pp. 618–626 (2017)
28. Smilkov, D., Thorat, N., Kim, B., Viégas, F., Wattenberg, M.: Smoothgrad: removing noise by adding noise. arXiv preprint arXiv:1706.03825 (2017)
29. Springenberg, J.T., Dosovitskiy, A., Brox, T., Riedmiller, M.: Striving for simplicity: the all convolutional net. arXiv preprint arXiv:1412.6806 (2014)
30. Tan, M., Le, Q.: Efficientnet: rethinking model scaling for convolutional neural networks. In: International Conference on Machine Learning, pp. 6105–6114. PMLR (2019)
31. Wah, C., Branson, S., Welinder, P., Perona, P., Belongie, S.: The Caltech-UCSD Birds-200-2011 Dataset. Tech. Rep. CNS-TR-2011-001, California Institute of Technology (2011)
32. Wang, H., Gong, S., Zhu, X., Xiang, T.: Human-in-the-loop person re-identification. In: Leibe, B., Matas, J., Sebe, N., Welling, M. (eds.) ECCV 2016. LNCS, vol. 9908, pp. 405–422. Springer, Cham (2016). https://doi.org/10.1007/978-3-319-46493-0_25

33. Xie, S., Girshick, R., Dollár, P., Tu, Z., He, K.: Aggregated residual transformations for deep neural networks. In: Proceedings of the IEEE Conference on Computer Vision and Pattern Recognition, pp. 1492–1500 (2017)
34. Yun, S., Han, D., Oh, S.J., Chun, S., Choe, J., Yoo, Y.: Cutmix: regularization strategy to train strong classifiers with localizable features. In: Proceedings of the IEEE/CVF International Conference on Computer Vision, pp. 6023–6032 (2019)
35. Zhang, Q.S., Zhu, S.C.: Visual interpretability for deep learning: a survey. Front. Inf. Technol. Electron. Eng. **19**(1), 27–39 (2018)
36. Zhou, B., Khosla, A., Lapedriza, A., Oliva, A., Torralba, A.: Learning deep features for discriminative localization. In: Proceedings of the IEEE Conference on Computer Vision and Pattern Recognition, pp. 2921–2929 (2016)

Characterization of AI Model Configurations for Model Reuse

Peter Bajcsy[1]([⊠])(iD), Michael Majurski[1], Thomas E. Cleveland IV[1], Manuel Carrasco[2], and Walid Keyrouz[1]

[1] National Institute of Standards and Technology, Gaithersburg, MD 20899, USA
peter.bajcsy@nist.gov
[2] George Mason University, Fairfax, VA 22030, USA

Abstract. With the widespread creation of artificial intelligence (AI) models in biosciences, bio-medical researchers are reusing trained AI models from other applications. This work is motivated by the need to characterize trained AI models for reuse based on metrics derived from optimization curves captured during model training. Such AI model characterizations can aid future model accuracy refinement, inform users about model hyper-parameter sensitivity, and assist in model reuse according to multi-purpose objectives. The challenges lie in understanding relationships between trained AI models and optimization curves, defining and validating quantitative AI model metrics, and disseminating metrics with trained AI models. We approach these challenges by analyzing optimization curves generated for image segmentation and classification tasks to assist in a multi-objective reuse of AI models.

Keywords: AI model reuse · Optimization and learning methods · Medical · Biological and cell microscopy

1 Introduction

The problems of reusing trained artificial intelligence (AI) models range from defining a standard AI model file format [5] to sharing an entire software stack [8]. The problems are constrained by the type of shared information that (a) would be useful to a third party reusing trained AI models and (b) would still protect trade secrets of the party providing trained AI models. The optimization curves gathered during AI model training provide such a source of information and the data points in optimization curves could meet both objectives (usefulness and protection). Our focus is on characterizing AI models from optimization curves for the purpose of value added to parties reusing the models.

The need within the scientific imaging community for AI model characterization is driven by several factors. First, the scientific community strives for reproducible research results. Second, domain-specific applications with focus on special objects of interest acquired by unique imaging modalities struggle with insufficient training data (in comparison to typical imaging modalities and objects in computer vision datasets, e.g., ImageNet or Microsoft Common Objects in Context (COCO)) and privacy concerns, especially in the medical imaging field. Finally, the sciences struggle with a general lack of computational resources for AI model training compared to the resources

© The Author(s), under exclusive license to Springer Nature Switzerland AG 2023
L. Karlinsky et al. (Eds.): ECCV 2022 Workshops, LNCS 13804, pp. 454–469, 2023.
https://doi.org/10.1007/978-3-031-25069-9_30

available to large companies. We listed in Table 1 several example tasks that are reusing AI models, utilizing task specific inputs, and benefiting from additional metrics. The terms "optimal configuration" and "explored configurations" in Table 1 refer to the one desirable configuration according to some optimization criteria and to the set of configurations that were evaluated during optimization. The columns labeled as "Reused" and "Task specific" refer to reused data and information generated by other parties, and to hardware and software artifacts that are specific to completing each task. The column "Needed metrics" highlights the key criteria that define a success of completing each task. For instance, the value of 1.7TB of trained AI models for the TrojAI challenge rounds [10] can be increased by metrics about convergence and stability of models (e.g., percentage of AI models that did not converge when trojans were injected), as well as about uniformity of training data defining predicted classes of synthetic traffic signs superimposed on real background. Characterizing AI models improves the input metadata about AI models for reuse and reproducibility. Additionally, model reuse saves computational resources and time while providing higher final model accuracy.

Table 1. Example tasks reusing AI models and benefiting from additional metrics

Task	Reused	Task specific	Needed metrics
1. Inference on new hardware	Trained model	Hardware	GPU utilization and exec. time
2. Reproduce training	Training data Model architecture Optimal configuration	Hardware Training env.	Error GPU utilization and exec. time
3. Model refinement by param. optimization	Training data Trained model Explored configurations	Hyper-param.	Convergence GPU utilization and exec. time
4. Network architecture search	Training datasets	AI graphs Hyper-param.	Data-model representation
5. Transfer learning	Trained models Optimal sets of hyper-param.	Other training datasets	Gain from pretraining Data domain cross compatibility
6. Evaluate robustness to data, architecture and hyper-param. perturbations	Training data Model architecture Optimal configuration	Subsets of training datasets AI graphs Hyper-param.	Data uniformity Error Stability

Our problem space is illustrated in Fig. 1 with a training configuration defined by training dataset, AI architecture, and hyper-parameters. The basic research question is: "what information can be derived from optimization and GPU utilization curves to guide a reuse of a trained AI model?" The objective of this work is to define and implement computable metrics from optimization curves that would be included in model cards [15] and serve as baseline implementations to support decisions about when and how to reuse disseminated trained AI models. Our assumption is that data collected during training sessions are common to all modeling tasks including image classification and segmentation (tasks of our specific interest) and, therefore, the characteristics can be applied to a general set of AI models.

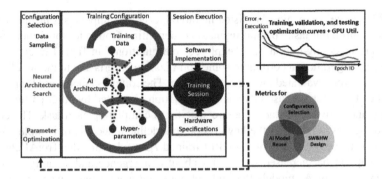

Fig. 1. A space of model training configuration space that is evaluated in each training session by analyzing training, validation, and testing optimization curves, and a set of metrics designed for configuration selection, model reuse, and software+hardware execution design.

Our approach to defining AI model characteristics from train, validation, and test optimization curves consists of three steps:

a) simulate and analyze relationships between AI model coefficients and optimization curves,
b) define mathematical functions used in analytical and statistical analyses that characterize a trained AI model from optimization curves, and
c) design a recommendation system based on extracted and validated quantitative characteristics of trained AI models.

The optimization curves are typically AI model accuracy or error metrics collected over many epochs from train, validation, and test datasets.

The overarching challenges lie in (a) limited information content in optimization curves that combine contributions from model architecture, training hyper-parameters, and training dataset, (b) limited a priori knowledge about relationships among parts of AI solutions that could be used for validation of quantitative AI model characteristics, and (c) computational resources needed to generate a large number of optimization curves for a variety of AI-based modeling tasks. Although the optimization curves contain limited information, they represent an information resource suitable for sharing with AI models to support legitimate reuse while protecting trade secrets (and pirated reuse) in commercial settings. To overcome the challenges, (a) we run simulations for a range of neural network configurations on 2D dot patterns to explore relationships between optimization curves and model configurations, (b) we use apriori information about datasets and modeling task complexity to validate metrics derived from optimization curves, and (c) we leverage optimization curves from a network architecture search database [22] and our model training sessions on five datasets and six architectures.

Relation to Prior Work: The concept of describing AI models has already been discussed in the past (Datasheets for datasets [6], The Dataset Nutrition Labels [9,21], Google AI Model Cards [15]). The published work on Datasheets for datasets and Dataset Nutrition Labels has been focused mainly on training datasets from the perspective of fairness. The fairness aspect is documented via data attributes, motivation for collection, data composition, collection process, and recommended uses in [6], as well as via design of ranking widgets in [9]. In contrast to [6,9,21], our work is focused on documenting lessons-learned from the optimization curves collected during training sessions. While a placeholder for model performance measures has been designated in the AI model cards [15] (i.e., under Metrics heading), the metrics have not been defined yet, which is the gap our work is trying to address. In addition, our work aims at utilizing the information that is not preserved with disseminated AI models for the multi-purpose reuse of the AI models right now, although multiple platforms for optimizing AI model configurations generate such information, such as TensorBoard [1], Optuna [2], Vizier [7], Autotune [11], or Experiment Manager [14]. Finally, our experiments are constrained by computational resources and therefore we leverage optimization values from a neural architecture search (NAS) database [22] with evaluated 5 million models and utilizing over 100 tensor processing unit (TPU) years of computation time.

Our contributions are (a) in exploring relationships between AI model coefficients and optimization curves, (b) in defining, implementing, and validating metrics of AI models from optimization curves for accompanying shared AI models, and (c) generating and leveraging optimization curves for image segmentation and classification tasks in a variety of reuse scenarios. The novelty of this work lies in introducing (a) computable metrics for model cards [15] that can provide cost savings for further reuse of AI models and (b) a recommendation system that can guide scientists in reusing AI models.

2 Methods

This section outlines three key components of using optimization curves for reuse of trained AI models: (1) relationships between AI model coefficients and optimization curves, (2) definitions of AI model metrics, and (3) design of a recommendation system.

Relationship Between AI Model Coefficients and Optimization Curves: To address the basic research question "what information can be derived from optimization and GPU utilization curves to guide a reuse of a particular AI model?", we simulate epoch-dependent AI model losses (train and test) and analyze their relationship to epoch-dependent AI model coefficients. We approach the question by simulating training datasets, architectures, and hyper-parameters at a "playground" scale using the web-based neural network calculator [4]. An example of a simulation is shown in Fig. 2.

Figure 2 (top) illustrates a labeled set with the classification rule described by a rule:

if $x^2 + y^2 > r^2$, then a point is outside of a circle else inside of a circle.

The AI model approximates this mathematical description by a rule:

if $2.5 * (\tanh(-0.2 * x^2 - 0.2 * y^2) + 1.7) < 0$ then a point is outside of a circle else it is inside of a circle.

Figure 2 (bottom) shows the train-test optimization curves (left) and AI model coefficient curves (right) as a function of epochs displayed on log scale. All coefficients go through a relatively large change of values around the epoch 7 with respect to their ranges of values. This change of values in coefficients is not reflected in the change of train-test loss measured via mean-squared error (MSE) since the relationship is a complex mapping from many model coefficients to one test MSE value. However, the absolute values of correlations ρ between MSE_{test} and model coefficients as a function of epochs for epochs ≥ 7 are close to 1 ($\rho(w1, MSE_{test}) = 0.91$, $\rho(w2, MSE_{test}) = 0.84$, $\rho(w3, MSE_{test}) = -0.95$, $\rho(bias, MSE_{test}) = -0.95$). In general, one can partition optimization curves to identify epoch intervals with and without oscillations (or jaggedness of a curve) and with below and above threshold error values. The epoch intervals without oscillations and below error values (e.g., $MSE_{test} < MSE_{test}(7) = 0.1$) can be modeled with curve fits, and the deviations from the curve model as indicators of trained AI model quality.

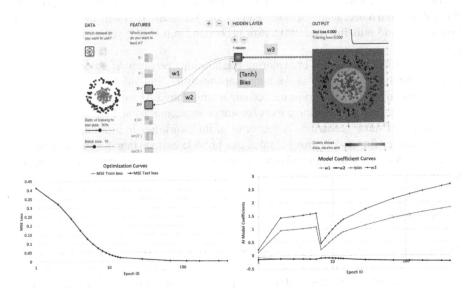

Fig. 2. Top - Simulation of a dot pattern with two clusters of labels separated by a circular boundary, (x^2, y^2) features, a single hidden layer with a single neuron model using $tanh$ activation, and a set of hyper-parameters (batch size = 10, learning rate = 0.3, train-test ratio = 50%) with w1, w2, w3, and bias coefficients. Bottom - The optimization curves (left) and the corresponding AI model coefficient curves (right).

In a similar vein, relationships between test and train curves have been used for assessing data sampling. Using the same web-based neural network calculator [4], one can analyze variable ratios of train: test random sampling for a complex spiral dot pattern. In such simulations, insufficient data sampling for training and insufficient model capacity lead to divergence of train and test curves. Such trends can be quantified, for instance, by a sum of areas under the curves (four-fold difference of the sums in our simulations over the first 14 epochs when comparing sufficient and insufficient data sampling).

These types of simulations suggest that one can derive several indicators (metrics) about convergence, stability, speed, impact of initialization, and uniformity of training and testing data from optimization curves.

Definitions of AI Model Metrics: The AI model characteristics are defined as sums, deltas, correlations, and extreme points as well as least-squared fits of power and exponential models to optimization curves from a varying number of data points. Equations 1–9 denote the index of each epoch as ep, number of epochs as EP, the epoch for which a model achieves the minimum error as $ep^*(M_{er})$, the window around $ep^*(M_{er})$ as $\pm\delta$, initializations as $rand$ (random) or $pretrain$ or a1/a2 graphs, execution time as T, correlation of two curves as ρ, and utilization of memory and processing power of a graphics processing unit as GPU^{mem} and GPU^{util}. In this work, we assume that the optimized error metric M_{er} per AI model is the cross entropy (CE) loss since it is widely used and supported by common AI libraries [1,18].

In Eq. 5, the value of $H^{CE,fit}$ represents a predicted CE values from the first few epochs given power or exponential models for the least-squared fit approximation. In our analyses, we refer to the power model $a*x^b$ as PW and the exponential model $a*b^x$ as EXP for $a,b \in \mathbb{R}$ and $x = H^{CE}(ep)$. The metric $\Delta(fit)$ is a difference between predicted $H^{CE,fit}$ and measured H^{CE} cross entropy loss values. $\Delta(fit)$ is designed as an optimization cost function for finding the most accurate convergence prediction model (min $\Delta(fit)$ constrained by the maximum number of measured epochs over two models $\{Model = PW, EXP\}$ and three sets of AI model optimization curves constrained by maximum of $\{10, 15, 20\}$ measured epochs.

While Eqs. 1 and 6 are commonly used in practice to assess model error and GPU requirements, other metric definitions are either not well-defined (e.g., stability [23]) or not mathematically defined at all. For instance, Eqs. 7, 8, and 9 define D_{re}, D_{unif}, and D_{init} given the train and test curves as (a) representation power of an AI architecture with respect to the non-linear relationship between inputs and outputs defined by the training data D_{re}, (b) uniformity of training and testing data subsets D_{unif}, and (c) compatibility of training data and pretraining data or AI model graphs D_{init}. In this case, D_{re} in Eq. 7 can be interpreted as the overall representation error of encoding training data in an AI architecture represented by a graph with a variety of nodes (modules) and node connectivity.

Table 2. Definition of AI model metrics

Metric name	Math symbol	Eq.
Model error	$M_{er}, ep^*(M_{er})$	1
Model stability	M_{stab}	2
Speed	$T(M_{er})$	3
Initialization gain	G_{init}	4
Predictability	$\Delta(fit)$	5
GPU utilization	$\text{GPU}^{MaxM}, \text{GPU}^{AvgU}$	6
Data-model representation	D_{re}	7
Train-test data uniformity	D_{unif}	8
Data compatibility	D_{init}	9

$$M_{er} = \min_{ep}(H_{test}^{CE}(ep))$$

$$ep^*(M_{er}) = \operatorname{argmin}_{ep} H_{test}^{CE}(ep) \tag{1}$$

$$M_{stab} = \sum_{ep=ep^*(M_{er})-\delta}^{ep^*(M_{er})+\delta} (H_{test}^{CE}(ep) - M_{er}) \tag{2}$$

$$T(M_{er}) = ep^*(M_{er}) * \frac{1}{EP} * \sum_{i=1}^{EP} T_i \tag{3}$$

$$G_{init} = M_{er}^{rand} - M_{er}^{pretrain} \tag{4}$$

$$\Delta(fit) = \sum_{ep=1}^{EP} (H_{test}^{CE}(ep) - H_{test}^{CE,fit}(ep)) \tag{5}$$

$$GPU^{MaxM} = \max_{ep}(GPU^{mem}(ep))$$

$$GPU^{AvgU} = \frac{1}{EP} * \sum_{ep=1}^{EP} GPU^{util}(ep) \tag{6}$$

$$D_{re} = \sum_{ep=1}^{EP} (H_{train}^{CE}(ep) + H_{test}^{CE}(ep)) \tag{7}$$

$$D_{unif} = \rho(H_{train}^{CE}(ep), H_{test}^{CE}(ep)) \tag{8}$$

$$D_{init}(data) = \sum_{ep=1}^{EP} (H_{test}^{CE,rand}(ep) - H_{test}^{CE,pretrain}(ep))$$

$$D_{init}(graph) = \sum_{ep=1}^{EP} (H_{test}^{CE,a1}(ep) - H_{test}^{CE,a2}(ep)) \tag{9}$$

Design of a Recommendation System: Given a set of example tasks shown in Table 1 and a set of derived metrics from optimization curves in Table 2, one needs to rank and recommend AI models, training data, and hyper-parameter configurations to complete a specific task. The optimization curves for deriving characteristics of AI models can be generated by in-house scripts or tools that automate sweeping a range of AI model parameters [2,7,11,14], while following random, grid, simulated annealing, genetic algorithm, or Bayesian search strategies. In our study, we generated pairs of train and test optimization curves using in-house scripts (see Tables 3, 4, and 5) and leveraged NAS-Bench-101 database [22] that contains information about 5 million trained AI model architectures.

Figure 3 (right) shows an example of two pairs of optimization curves with fluctuations due to the complexity of a high-dimensional loss surface traversed during optimization using the training images illustrated in Fig. 3 (left). The optimization curves illustrate the complexity of the cross entropy (CE) loss surface with respect to all contributing variables (i.e., the space is discontinuous and/or frequently flat without expected extrema, evaluations fail due to sparse discrete objective space formed by integer and categorical variables and/or numerical difficulties and hardware failures [19]). Thus, metrics characterizing optimization curves must be presented as a vector and ranked by each vector element interactively. In our recommender design, we chose a parallel coordinate graph to convey ranking and support decisions. We also focus on validating the metrics based on additional information about training datasets.

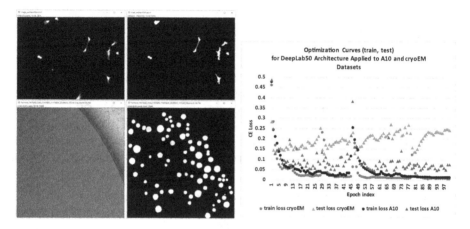

Fig. 3. Left - Examples of training image pairs (intensity, segmentation mask) for A10 dataset (top row), and cryoEM dataset (bottom row). Right - Optimization curves (train, test) for DeepLab50 AI architecture trained on A10 and cryoEM datasets.

3   Experimental Results

We divided the experimental work into (1) generating optimization curves over a range of AI model configurations, (2) validating the designed metrics based on a couple of datasets and their *prior* characterization of segmentation difficulties, and (3) describing recommendations for a reuse of trained AI models (a) driven by use cases listed in Table 1 and (b) applied to image segmentation and classification tasks.

Generation of Optimization Curves: In order to generate optimization curves for varying training datasets, AI model architectures, and hyper-parameters, we gathered five segmentation training images acquired by multiple imaging modalities (see Table 4), implemented six AI segmentation architectures by leveraging the PyTorch library [18] (see Table 5), and varied a couple of hyper-parameters (see Table 3). The model initialization using pre-trained coefficients was based on the COCO dataset [13] for object segmentation (1.5 million object instances).

The training image datasets represent optical florescent, optical bright-field, electron, cryogenic electron, and neutron imaging modalities, and are characterized in terms of the number of predicted classes (#Classes), the number of pixels (#Pixels), and the average coefficient of variation ($\overline{CV}$) over all training images as defined in Eq. 10.

$$\overline{CV} = \frac{1}{N} \sum_{i=1}^{N} \frac{\sigma_i}{\mu_i} \tag{10}$$

where μ_i and σ_i are the mean and standard deviation of each intensity image in the training collection of size N images. The A10 dataset denotes fluorescently labeled optical microscopy images of A10 cells [17]. The concrete dataset came from electron microscopy of concrete samples [3]. The cryoEM dataset was prepared by the authors using cryogenic electron microscopy of lipid nanoparticles. The infer14 dataset was prepared by the authors using data-driven simulations of porous concrete samples from measured neutron images [16]. The rpe2d dataset denotes time-lapse bright-field optical microscopy images of retinal pigment epithelial (RPE) cells published in [20].

For the training runs, we chose to train each model configuration for $EP = 100$ epochs. In general, this value will vary during hyper-parameter optimization runs depending on available computational resources, the definition of model convergence error, or the use of early stopping criterion (an increment observed in CE loss values over consecutive epochs is smaller than ϵ). We also set the value $\delta = 5$ epochs in Eq. 2. All computations were performed on a compute node with a Quadro Ray Tracing Texel eXtreme (RTX) 4000 GPU card and Compute Unified Device Architecture (CUDA) 11.6.

Validation of AI Characteristics: We selected two training image datasets labeled as A10 and cryoEM in Table 4 for validation. Examples of training image pairs are shown in Fig. 3. The A10 dataset has a high contrast ($\overline{CV} = 1.34$) while the cryoEM dataset has a low contrast ($\overline{CV} = 0.06$) and a large heterogeneity in sizes and textures. The datasets were chosen based on the assumption that segmenting images with low contrast is a much harder task than segmenting images with high contrast.

Given the assumption about segmentation difficulty, the complexity of an input-output function for cryoEM dataset is larger than the complexity of such a function for A10 dataset and hence the model utilization or the error must be higher for cryoEM. We observe the worst error over all AI models for A10 ($M_{er} = 0.0568$) to be at least twice smaller than the best error over all AI models for cryoEM ($M_{er} = 0.127$). This implies that any of the explored model capacities could not increase model utilization to accommodate the cryoEM input-output function and hence optimization errors are much higher for cryoEM. Furthermore, the heterogeneity of segments in sizes and textures in cryoEM versus A10 poses challenges on sampling for train-test subsets. Since the sampling is completely random, it is very unlikely that segments with varying sizes and textures from cryoEM will be equally represented in train-test subsets. This implies that $D_{unif} \in [0.226, 0.836]$ for cryoEM is expected to be smaller on average than $D_{unif} \in [-0.359, 0.310]$ for A10 due to the train-test gap. Ideally, the correlation of train and test CE loss curves D_{unif} should be close to one.

One can now validate AI model characteristics against expected inequalities to be satisfied by the values derived from these two datasets. The expected inequalities include model error M_{er}, uniformity of training and testing data D_{unif}, and convergence predictability $\Delta(fit)$ as shown in Eq. 11. These inequalities are validated by comparing the values of M_{er} and D_{unif} in Table 5. Figure 5 shows the values in parallel coordinate plots including the values of $\Delta(fit, A10)$ and $\Delta(fit, CryoEM)$ on the right most vertical line denoted as $\min P(PW\_20)$. The values of $\min P(PW\_20)$ were calculated using the power model fit from the first 20 epochs. The sum of train and test optimization curves D_{re}, as well as the convergence predictability $P(PW\_20)$, quantify the sensitivity of model training to hyper-parameters (i.e., learning rate and initialization). In both A10 and cryoEM datasets, D_{re} and $P(PW\_20)$ values for 2–3 architecture types indicate epoch-specific optimization divergence (as illustrated by the scattered black points in Fig. 4(right) for the A10 dataset).

$$M_{er}(A10) < M_{er}(CryoEM)$$
$$D_{unif}(A10) > D_{unif}(CryoEM) \quad (11)$$
$$\Delta(fit, A10) < \Delta(fit, CryoEM)$$

Table 3. Explored hyper-parameters in AI model configurations

Hyper-parameters	Values
Initialization	Random COCO pre-trained
Learning Rate	$10^{-5}, 10^{-4}, 10^{-3}, 10^{-2}$
Optimizer	Adam
Optimization criterion	Cross entropy loss
Epochs	100
Batch size	2
Class balance method	Weighting by class proportion
Augmentation	None
Train-Test split	80 : 20

Table 4. Training datasets. OF - optical fluorescent, EM - electron microscopy, OB - optical bright field, NI - neutron imaging

Dataset	Modality	#Classes	#Pixels [MPix]	$\overline{CV}$
A10	OF	2	5.79	1.34
concrete	EM	4	71.7	0.31
cryoEM	EM	2	117.44	0.06
infer14	NI	9	125.9	0.24
rpe2d	OB	2	53.22	0.84

Table 5. Summary of AI model characteristics per model architecture and per dataset where the models were optimized over the learning rates and pretraining options listed in Table 3.

Architecture	A10 M_{er}	A10 D_{unif}	cryoEM M_{er}	CryoEM D_{unif}
DeepLab101	0.0528	0.3513	0.1271	−0.0093
DeepLab50	0.0451	0.8356	0.1284	−0.2515
LR-ASPP	0.04	0.7978	0.1435	−0.3585
MobileNetV3	0.0568	0.5794	0.1602	0.3092
ResNet101	0.042	0.2256	0.1379	−0.0867
ResNet50	0.045	0.391	0.1369	−0.2269

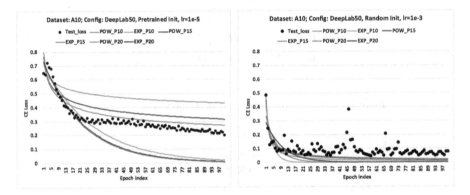

Fig. 4. Predictions of training model convergence from A10 dataset for two configurations. Left configuration: (DeepLab50, COCO pre-trained initialization, learning rate: 1e−5). Right configuration: (DeepLab50, random initialization, learning rate: 1e−3). Black dots are the measured test CE loss values. Color-coded curves are predictions for a set of fitted model parameters.

Reuse of Trained AI Models for Image Segmentation: When the task focus is on convergence predictability $\Delta(fit)$, an AI model configuration in Fig. 4 (left) shows better convergence for power model with 20 epochs than any configuration in Fig. 4 (right). A large divergence from the predicted optimization curves typically indicates that (a) it is not sufficient to predict the model training convergence using a few initial epochs (10, 15, or 20 epochs), (b) training and testing subsets might not have been drawn from the same distribution, and (c) the COCO dataset used for pretraining the AI model might not be compatible with the domain training dataset, and, hence, the test CE loss values vary a lot during the first few epochs. This is undesirable for researchers who would like to predict how many more epochs to run on the existing model while targeting a low test CE loss value. On the other hand, the configuration in Fig. 4(right) achieves lower CE error M_{er} (vertically lowest black point) than the configuration in Fig. 4(left) for the same dataset A10 and the same DeepLab50 AI architecture.

When the task focus is on gain from pretraining G_{init}, the values are less than zero for all datasets except the concrete dataset listed in Table 4. These values indicate that the objects in the COCO dataset are significantly different from the objects annotated

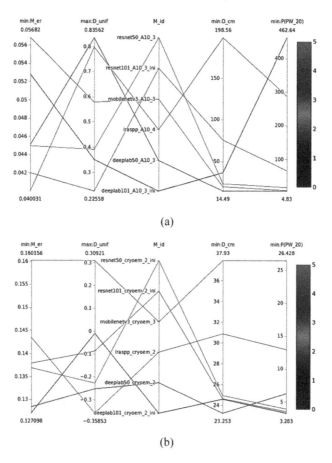

(a)

(b)

Fig. 5. Parallel coordinate plots for A10 (top) and cryoEM (bottom) datasets. The plots are intended to support decisions about which AI model architecture is the most accurate for the image segmentation tasks.

in the five scientific microscopy datasets, and the pre-training on COCO does not yield better model accuracy.

When the task focus is on model stability M_{stab}, Fig. 6 illustrates how stable each optimized AI model is over the configurations listed in Tables 3, 4, and 5. If the test CE loss curve is close to constant within the neighborhood of $\delta = 5$ epochs, then the value of M_{stab}, as defined in Eq. 2, is small indicating model stability. Based on Fig. 6, all model architectures for the rpe2d dataset yielded highly stable, trained models, while the stability of trained models for the infer14 dataset was low and varied depending on a model architecture.

Reuse of Trained AI Models for Image Classification: We analyzed train, validation, and test accuracies obtained using CIFAR-10 training dataset [12] and a network architecture search (NAS) published in NAS-Bench-101 database [22]. NAS-Bench 101

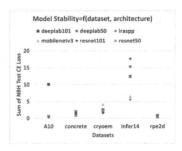

Fig. 6. Model stability of the most accurate AI model per architecture and per dataset

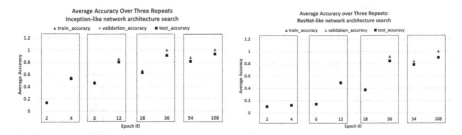

Fig. 7. Accuracies extracted from NAS-Bench101 database for Inception like (left) and ResNet like (right) architectures

contains information about final and halfway accuracies obtained by searching over ResNet and Inception like architectures for 5 million graphs. Accuracies over three repeated training runs were averaged and plotted for the Inception-like and ResNet-like architectures in Fig. 7 at halfway and final epochs and for four epoch budgets $\{4, 12, 36, 108\}$. The standard deviation of average values over all accuracies was 0.038 and 0.042. Since two points in each optimization curve were not very informative, we analyzed all points combined from the epoch budgets per Inception-like and ResNet-like architecture graphs. The correlations of train, validation, and test contours D_{unif} were larger than 0.999 indicating uniformity of train-test splits of CIFAR-10 dataset. Regarding convergence $\Delta(fit)$, a power model is more accurate than an exponential model. We observed the following inequality $\Delta(fit, ResNet) < \Delta(fit, Inception)$ suggesting that one could predict convergence of ResNet-like architecture more accurately than convergence of Inception-like architecture. On the other hand, the Inception-like architecture reaches higher accuracy values faster with epoch numbers than the ResNet-like architecture as it can be documented with a positive sum of deltas $D_{init}(\text{graph}) = 1.45$, where $a1 = $ Inception and $a2 = $ ResNet in Eq. 9.

4 Discussion

The practical value of each metric is task dependent for the use cases listed in Table 1. For instance, if the task is model portability to a new hardware or model-training reproducibility, then knowing maximum required GPU memory and GPU utilization would be very valuable. It is frequent in biology to apply transfer learning and reuse trained AI models due to a limited size of annotated image datasets in scientific experiments. Knowing what architectures, datasets, and hyper-parameters were explored can save computational time when cell type, tissue preparation, or imaging modalities differ between trained AI models and a reuse application.

From the perspective of defining metrics, our goal is not to invent them from scratch but rather to define them in consistent mathematical and computational ways as opposed to the current variations of subsets of presented metrics in practice. Consistent metric definitions enable their use in model cards and reuse of shared AI models. Nonetheless, more metrics will need to be designed and defined to address a broad range of AI model reuses. We shared the code and its documentation in GitHub at https://github.com/usnistgov/ai-model-reuse.

5   Summary

This work presented the problem of learning from a set of optimization curves, defining metrics derived from the curves for model cards [15], and reusing trained AI models based on accompanying metrics. The output quantitative AI model metrics serve multiple purposes: as entries under *Metrics* in the AI model card definition [15] and as inputs to ranking of AI models according to a variety of objectives (e.g., model accuracy refinement, model architecture recommendation).

The impact of sharing AI models with presented metrics is significant for principal investigators limited by their grant budgets and for small research labs limited by their own computational resources or the cost of cloud resources. A higher reuse of shared AI models can save not only cost and time to researchers but also advance their scientific goals more efficiently. The cost of achieving a higher reuse of AI models is the extra summarization of optimization sessions using transparent metrics and sharing them in AI model cards. In the future, we plan to explore how to replace relative with absolute AI model metrics when recommending trained AI models.

6   Disclaimer

Certain commercial equipment, instruments, or materials (or suppliers, or software, ...) are identified in this paper to foster understanding. Such identification does not imply recommendation or endorsement by the National Institute of Standards and Technology, nor does it imply that the materials or equipment identified are necessarily the best available for the purpose.

Acknowledgments. The funding for Peter Bajcsy, Michael Majurski, and Walid Keyrouz was partially provided from the IARPA project: IARPA-20001-D2020-2007180011. We would also like to thank Craig Greenberg, Ivy Liang, and Daniel Gao for providing very insightful feedback and contributing to the code.

References

1. Abadi, M., et al.: TensorFlow: large-scale machine learning on heterogeneous systems (2015). https://doi.org/10.5281/zenodo.5898685. https://www.tensorflow.org/, software available from tensorflow.org
2. Akiba, T., Sano, S., Yanase, T., Ohta, T., Koyama, M.: Optuna: a next-generation hyperparameter optimization framework (2019)
3. Bajcsy, P., Feldman, S., Majurski, M., Snyder, K., Brady, M.: Approaches to training ai-based multi-class semantic image segmentation. J. Microscopy **279**(2), 98–113 (2020). https://doi.org/10.1111/jmi.12906. https://pubmed.ncbi.nlm.nih.gov/32406521/
4. Bajcsy, P., Schaub, N.J., Majurski, M.: Designing trojan detectors in neural networks using interactive simulations. Appl. Sci. **11**(4) (2021). https://doi.org/10.3390/app11041865. https://www.mdpi.com/2076-3417/11/4/1865
5. Community: Open neural network exchange (ONNX) (2022). https://onnx.ai/
6. Gebru, T., Morgenstern, J., Vecchione, B., Vaughan, J.W., Wallach, H., III, H.D., Crawford, K.: Datasheets for datasets. arXiv 1803.09010 (2021)
7. Golovin, D., Solnik, B., Moitra, S., Kochanski, G., Karro, J.E., Sculley, D. (eds.): Google Vizier: A Service for Black-Box Optimization (2017). http://www.kdd.org/kdd2017/papers/view/google-vizier-a-service-for-black-box-optimization
8. Haibe-Kains, B., et al.: Transparency and reproducibility in artificial intelligence. Nature **586**(7829), E14–E16 (2020). https://doi.org/10.1038/s41586-020-2766-y. https://aclanthology.org/Q18-1041
9. Holland, S., Hosny, A., Newman, S., Joseph, J., Chmielinski, K.: The dataset nutrition label: A framework to drive higher data quality standards. arXiv 1805.03677 (2018)
10. IARPA: Trojans in Artificial Intelligence (TrojAI) (2020). https://pages.nist.gov/trojai/, https://www.iarpa.gov/index.php/research-programs/trojai
11. Koch, P., Golovidov, O., Gardner, S., Wujek, B., Griffin, J., Xu, Y.: Autotune. Proceedings of the 24th ACM SIGKDD International Conference on Knowledge Discovery and Data Mining, July 2018. https://doi.org/10.1145/3219819.3219837
12. Krizhevsky, A., Nair, V., Hinton, G.: Cifar-10 (canadian institute for advanced research). http://www.cs.toronto.edu/kriz/cifar.html
13. Lin, T.Y., et al.: Microsoft COCO: common objects in context. arXiv 1405.0312 (2014). http://arxiv.org/abs/1405.0312
14. Long, J., Shelhamer, E., Darrell, T.: Experiment manager (2022). https://www.mathworks.com/help/deeplearning/ref/experimentmanager-app.html
15. Mitchell, M., et al.: Model cards for model reporting. Proceedings of the Conference on Fairness, Accountability, and Transparency, January 2019. https://doi.org/10.1145/3287560.3287596
16. NIST: Data-driven simulations of measured neutron interferometric microscopy images (2022). https://www.nist.gov/programs-projects/interferometry-infer-neutron-interferometric-microscopy-small-forces-and
17. NIST: Fluorescent microscopy images of A-10 rat smooth muscle cells and NIH-3T3 mouse fibro-blasts (2022). https://isg.nist.gov/deepzoomweb/data/dissemination
18. Paszke, A., et al.: PyTorch: an imperative style, high-performance deep learning library. In: Advances in Neural Information Processing Systems 32, pp. 8024–8035. Curran Associates, Inc. (2019). http://papers.neurips.cc/paper/9015-pytorch-an-imperative-style-high-performance-deep-learning-library.pdf

19. Ranjit, M., Ganapathy, G., Sridhar, K., Arumugham, V.: Efficient deep learning hyper-parameter tuning using cloud infrastructure: Intelligent distributed hyperparameter tuning with bayesian optimization in the cloud. In: 2019 IEEE 12th International Conference on Cloud Computing (CLOUD), pp. 520–522. IEEE Computer Society, Los Alamitos, July 2019. https://doi.org/10.1109/CLOUD.2019.00097, https://doi.ieeecomputersociety.org/10.1109/CLOUD.2019.00097

20. Schaub, N., et al.: Deep learning predicts function of live retinal pigment epithelium from quantitative microscopy. J. Clin. Invest. **130**(2), 1010–1023 (2020). https://doi.org/10.1172/JCI131187, https://www.ncbi.nlm.nih.gov/pmc/articles/PMC6994191/

21. Yang, K., Stoyanovich, J., Asudeh, A., Howe, B., Jagadish, H., Miklau, G.: A nutritional label for rankings. Proceedings of the 2018 International Conference on Management of Data, May 2018. https://doi.org/10.1145/3183713.3193568

22. Ying, C., Klein, A., Christiansen, E., Real, E., Murphy, K., Hutter, F.: NAS-bench-101: Towards reproducible neural architecture search. In: Chaudhuri, K., Salakhutdinov, R. (eds.) Proceedings of the 36th International Conference on Machine Learning. Proceedings of Machine Learning Research, vol. 97, pp. 7105–7114. PMLR, Long Beach, California, USA (09–15 Jun 2019), http://proceedings.mlr.press/v97/ying19a.html

23. You, S., et al.: A review on artificial intelligence for grid stability assessment. In: 2020 IEEE International Conference on Communications, Control, and Computing Technologies for Smart Grids (SmartGridComm), pp. 1–6 (2020). https://doi.org/10.1109/SmartGridComm47815.2020.9302990

Empirical Evaluation of Deep Learning Approaches for Landmark Detection in Fish Bioimages

Navdeep Kumar[1(✉)], Claudia Di Biagio[2], Zachary Dellacqua[2,4],
Ratish Raman[1,3], Arianna Martini[5], Clara Boglione[2], Marc Muller[1,3],
Pierre Geurts[1], and Raphaël Marée[1]

[1] University of Liège, Liège, Belgium
nkumar@uliege.be
[2] University of Rome Tor Vergata, Rome, Italy
[3] Laboratory for Organogenesis and Regeneration-GIGA I3, University of Liège,
Liège, Belgium
[4] Marine systems and Sustainable Aquaculture, Ecoaqua Institute, University of Las
Palmas de Gran Canaria. Telde, Gran Canaria, Spain
[5] CREA, Monterotondo, RM, Italy

Abstract. In this paper we perform an empirical evaluation of variants of deep learning methods to automatically localize anatomical landmarks in bioimages of fishes acquired using different imaging modalities (microscopy and radiography). We compare two methodologies namely heatmap based regression and multivariate direct regression, and evaluate them in combination with several Convolutional Neural Network (CNN) architectures. Heatmap based regression approaches employ Gaussian or Exponential heatmap generation functions combined with CNNs to output the heatmaps corresponding to landmark locations whereas direct regression approaches output directly the (x, y) coordinates corresponding to landmark locations. In our experiments, we use two microscopy datasets of Zebrafish and Medaka fish and one radiography dataset of gilthead Seabream. On our three datasets, the heatmap approach with Exponential function and U-Net architecture performs better. Datasets and open-source code for training and prediction are made available to ease future landmark detection research and bioimaging applications.

Keywords: Deep learning · Bioimages · Landmark detection · Heatmap · Multi-variate regression

1 Introduction

In many bioimage studies, detecting anatomical landmarks is a crucial step to perform morphometric analyses and quantify shape, volume, and size parameters of a living entity under study [11]. Landmarks are geometric keypoints

© The Author(s), under exclusive license to Springer Nature Switzerland AG 2023
L. Karlinsky et al. (Eds.): ECCV 2022 Workshops, LNCS 13804, pp. 470–486, 2023.
https://doi.org/10.1007/978-3-031-25069-9_31

localized on an "object" and can be described as coordinate points in a 2D or a 3D space. For example, in human cephalometric study, human cranium is analyzed for diagnosis and treatment of dental disharmonies [23] using X-Ray medical imaging techniques. In biomedical research where fish species such as Zebrafish *(Danio rerio)* and Medaka *(Oryzias letipes)* are used as models, various morphometric analyses are performed to quantify deformities in them and further identify cause and treatment for human related bone disorders [13,37]. Such studies require to analyze and classify deformities in the vertebral column, jaws or caudal fin of the fish, which is addressed by first detecting specific landmark positions in fish images. In aquaculture industry, food fish such as gilthead Seabream suffer from bone related disorders due to the non-natural environment in which they are reared and morphometric studies are carried out to quantify these deformities [7,36,36]. Such studies also require the researchers to select and mark some important landmark locations on fish images in order to perform external shape analyses [19].

Manual annotations of landmarks locations are very labour intensive and require dedicated human expertise. The emergence and heterogeneity of high-throughput image acquisition instruments makes it difficult to continue analyzing these images manually. To address the problem, biomedical researchers began to use automatic landmark localization techniques to speed up the process and analyze large volumes of data. Conventional landmark detection techniques use image processing in order to align two image templates for landmark configurations then applying some Procrustes analysis [4]. Classical machine learning techniques such as random forest based algorithms were also proposed in [17,32,35] to automatically localize landmarks in microscopy images of zebrafish larvae.

Recently, landmark detection or localization has also been extensively studied in the broader computer vision field, especially for real time face recognition systems [8,14,38], hand-gesture recognition [27], and human pose estimation [2,30]. With the advent of more sophisticated techniques such as deep-learning based Convolutional Neural Networks (CNNs), the performance of computerized models for object detection and classification has become comparable to human performance. While deep learning models reach a high level of accuracy in computer vision tasks with natural images (e.g. on ImageNet), there is no guarantee that these methods will give acceptable performance in specific bioimaging applications where the amount of training data is limited. Indeed, learning landmark detection models requires images annotated with precise landmark positions while experts to carry out these annotations are few, the annotation task is tedious and it must be repeated for every new imaging modality and biological entity.

In this paper, we want to evaluate state-of-the-art deep learning based landmark detection techniques to assess if they can simplify and speed up landmark analyses in real-world bioimaging applications, and to derive guidelines for future use. More precisely, we evaluate the two main families of methods in this domain, namely direct multivariate regression and heatmap regression, and we focus our experiments on the identification of anatomical landmarks in 2D images of various fish species. To our knowledge, our work is one of the first few attempts to implement a fully automatic end-to-end deep learning based method for the task of landmark detection in heterogeneous fish bioimages. In Sect. 3, we describe our datasets and image acquisition settings. Methodologies, network architectures and our evaluation protocol are presented in Sect. 4. Then, we present and discuss empirical results in Sect. 5.

2   Related Work

In biomedical image analysis, patch-based deep learning methods are proposed in which local image patches are extracted from the images and fed to the CNN to detect the landmark locations [3,31]. Patch-based methods are usually used to train one landmark model for each landmark location making the whole process computationally very expensive. These models often require plenty of memory storage to operate if the number of landmark points to detect is high. Another drawback of using the patch-based methods is missing global information about all the landmarks combined as local patches represent only limited contextual information about the particular landmark.

Among end-to-end deep learning approaches, the first prominent solution is to output directly the (x, y) coordinates of the landmarks using CNNs regressors [16]. These direct coordinate regression based methods are very simple to design and faster to train. However, to get optimal performances, this approach generally requires large training datasets and deeper networks [10]. Another approach is to output heatmaps corresponding to the landmark locations [6,25,26]. In this scenario, heatmaps are generated from the labelled landmarks locations during training and CNNs are trained to predict these heatmaps. These heatmaps encode per pixel confidence scores for landmark locations rather than numbers or values corresponding to landmark coordinates. The most common heatmap generation methods employ distance (linear) functions or some non-linear gaussian or exponential kernels [39]. In [10] and [20], the authors proposed a method that combines the heatmap based regressors with direct coordinate regressors to automatically localize landmarks in MRI images of spine.

The data scarcity in biomedical image analysis is one of the biggest concerns as it is difficult to train a deep CNN from scratch with limited amount of images and ground truths. To address this issue, the authors of [24,31] explore transfer learning methods such as using a pre-trained CNN as backbone and only training or fine-tuning its last layers for the problem of cephalometric landmark detection. Transfer learning is also used in animal behaviour studies in neuroscience where landmarks are used to aid computer-based tracking systems. [22] devised a transfer learning based landmark detection algorithm that uses pretrained Resnet50 as backbone to automatically track the movements in video recordings of the animals. To tackle the problem of limited data, the authors of [28] proposed a method to train models on thousands of synthetically generated images from other computer vision tasks such as hand recognition systems and evaluate them on MR and CT images.

There are cases in which two landmark points are either very close to each other or one is occluding another landmark. In these cases, a single CNN model is not sufficient to achieve optimal performance in locating the landmarks. To handle these scenarios, authors in [15,34] proposed a combination of CNN regressor and Recurrent Neural Network (RNN) in which RNNs are employed to remember the information for landmark locations to further refine the predictions given by the CNN regressor. Although these methods can lead to very good performance for landmark detection, they are very hard to train on limited image data due to their complex architectural design.

3 Dataset Description

In this work, we use three datasets acquired using different microscopy and radiography imaging protocols. These datasets contain images of three different fish species, namely Zebrafish (*Danio rerio*) and Medaka (*Oryzias latipes*), used in biomedical research as model fishes, and gilthead Seabream (*Sparus aurata*), used for aquaculture research. The Zebrafish microscopy dataset is acquired from GIGA Institute at the University of Liège whereas the Medaka microscopy and gilthead Seabream radiograph datasets are acquired from the department of Biology, University of Rome, Tor Vergata. Table 1 shows the summary of each dataset and detailed dataset descriptions are given below.

3.1 Zebrafish Microscropy Dataset

This dataset is composed of 113 microscropy images of zebrafish (Danio rerio) larvae at 10 dpf (3 mm length). Images were captured using an Olympus SZX10 stereo dissecting microscope coupled with an Olympus XC50 camera with a direct light illumination on a white background. The Olympus XC50 camera allows to acquire 2575 × 1932 pixel resolution images. 25 landmarks are manually annotated by the experts around the head of the zebrafish larvae as follows: 1 and 24: Maxilla; 2 and 23: Branchiostegal ray 2; 3 and 11: Opercle; 4,12,13 and 14: Cleithrum; 5 and 19: Anguloarticular; 6 and 25: Ceratobranchial; 7 and 8: Hyomandibular; 9 and 20: Entopterygoid; 10: Notochord; 21,15 and 18: Parasphenoid; 17 and 22: Dentary; 16: showing anterior end marking. A sample image and its annotations are shown in Fig. 1(A)

3.2 Medaka Microscopy Dataset

This dataset has 470 images of medaka juveniles (40 days after hatching) where each image has size 2560 × 1920. Samples were *in toto* stained with Alizarin red and photographed with the Camera Axiocam 305 color connected to the AxioZoom V.16 (Zeiss) stereomicroscope. A total number of 6 landmarks are manually annotated as follows: 1: rostral tip of the premaxilla (if the head is bent, the landmark was located between the left and right premaxilla); 2: base of the neural arch of the 1st (anteriormost) abdominal vertebra bearing a rib; 3:

base of the neural post-zygapophyses of the first hemal vertebra (*viz.*, vertebra with hemal arch closed by a hemaspine); 4: base of the neural post-zygapophyses of the first preural vertebra; 5: base of the neural post-zygapophyses of the preural-2 vertebra; 6: posteriormost (caudad) ventral extremity of the hypural 1. Figure 1(B) shows a sample image from the dataset with annotated landmarks.

3.3   Seabream Radiography Dataset

In this dataset, the fish species is gilthead Seabream (*Sparus aurata*), sampled at 55 g (average weight). A total of 847 fish were xrayed with a digital DXS Pro X-ray (Bruker) and 19 landmarks are manually annotated on variable image sizes, as follows: A: frontal tip of premaxillary; B: rostral head point in line with the eye center; C: dorsal head point in line with the eye center; D: dorsal extremity of the 1st predorsal bone; E: edge between the dorsal 1st hard ray pterygophore and hard ray; F: edge between the dorsal 1st soft ray pterygophore and soft ray; G: edge between the dorsal last soft ray pterygophore and soft ray; H: dorsal concave inflexion-point of caudal peduncle; I: middle point between the bases of hypurals 2 and 3 (fork); L: ventral concave inflexion-point of caudal peduncle; M: edge between the anal last pterygophore and ray; N: edge between the anal 1st ray pterygophore and ray; O: insertion of the pelvic fin on the body profile; P: preopercle ventral insertion on body profile; Q: frontal tip of dentary; R: neural arch insertion on the 1st abdominal vertebral body; S: neural arch insertion on the 1st hemal vertebral body; T: neural arch insertion on the 6th hemal vertebral body; U: between the pre- and post-zygapophyses of the 1st and 2nd caudal vertebral bodies. Sample images from the dataset with annotated landmarks are shown in Fig. 1(C).

Table 1. Summary of the datasets used in our methodology

Fish species	Number of images	Number of landmarks	Image modality	Research area
Zebrafish	113	25	Microscopy	Bio-medical Science
Medaka	470	6	Microscopy	Bio-medical Science
Gilthead Seabream	847	19	Radiograph	Aquaculture

4   Method Description

We evaluate two types of deep-learning based regression approaches, namely direct regression and heatmap based regression.

4.1   Direct Coordinates Regression

In the direct regression approach, the output is designed to predict $(N \times 2)$ numbers, where the first (resp. last) N numbers correspond to x (resp. y) coordinates of the landmarks.

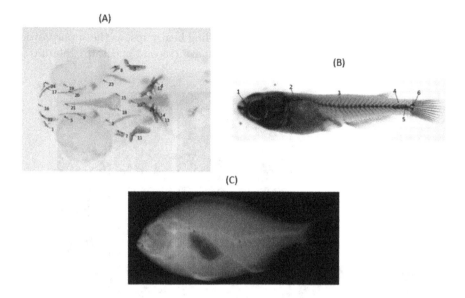

Fig. 1. Sample images with landmarks from three datasets. (A) Zebrafish microscopy (B) Medaka microscopy (C) Seabream radiograph

4.2   Heatmap-Based Regression

The second approach is based on outputting the heatmaps (one per landmark) instead of directly predicting the coordinate points for landmark locations. Each heatmap gives information about the likelihood for each pixel of being the location of a particular landmark. At training, the heatmap is constructed to associate to every pixel a score that takes its highest value (1) at the exact location of the landmark and vanishes towards 0 when moving away from the landmark. The size of the region of influence of a landmark is controlled by a user-defined dispersion parameter σ. More formally, and following [39], we have implemented and compared two probability functions to generate these heatmaps, namely a **Gaussian function** F_G and an **Exponential function** F_E, defined respectively as follows:

$$F_G(x, y) = A \cdot \exp\left(-\frac{1}{2\sigma^2}\left((x - \mu_x)^2 + (y - \mu_y)^2\right)\right),$$

$$F_E(x, y) = A \cdot \exp\left(-\frac{\log(2)}{2\sigma}(|x - \mu_x| + |y - \mu_y|)\right),$$

where x and y are the coordinates of a pixel in the image, μ_x and μ_y are the coordinates of the landmark under consideration, σ is the spread of the distribution, and A is a normalizing constant that gives the amplitude or peak of the curve.

To fix the highest score value as 1 at the exact location of the landmark, we set the normalizing constant A to 1, since it corresponds to the maximum value of the gaussian and exponential functions. Figure 2 shows the original landmarks on the image (first column) and their corresponding heatmaps, as the superposition of the heatmaps corresponding to each landmark (second and third columns).

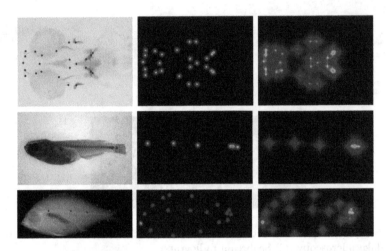

Fig. 2. Original landmarks on the images *(first column)*, their corresponding Gaussian heatmaps *(second column)* and Exponential heatmaps *(third column)*

4.3 Training and Prediction Phases

In the **training phase**, original images are first downscaled to 256×256 to be fed into the network. Since the original images are rectangular, we first downscale the image to a size of 256 along the largest dimension while keeping the aspect ratio unchanged. Padding is then added to the smallest dimension to produce a 256×256 square image. For direct regression, the output of the model consists of $N \times 2$ real numbers, with N the total number of landmark, representing landmark coordinates rescaled between 0 and 1. For heatmap regression, the output is composed of N heatmap slices, each corresponding to one landmark and constructed as described in the previous section.

The **prediction phase** for direct regression based approach is simply predicting the $N \times 2$ numbers and then upscaling them to the original sized image (i.e., multiplying them by the original image width and height after padding is removed). In the case of the heatmap based approach, heatmap slices are first predicted by the network and then, as a post processing step, each heatmap is converted to its corresponding landmark location by taking the *argmax* of the heatmap over all image pixel values. The *argmax* function returns the 2D coordinates of the highest value in a heatmap slice. The corresponding landmark coordinates are then upscaled to the size of the original image to produce the final model predictions.

4.4   Network Architectures

To evaluate our methodology, we implement state-of-art CNNs used in various image recognition, segmentation, and pose estimation tasks. Following are the CNN architectures we implement in both the multivariate and the heatmap regression based output network models. We only give below the main idea of these architectures. Full details and codes are available on GitHub[1].

- **Heatmap based CNN architectures:**
  - *U-Net architecture:* U-Net architecture as described in [29] is a two phase encoder and decoder network in which the encoder module is made up of conventional stack of convolutional layers followed by max-pooling layer and the decoder module consists in a stack of up-sampling layers. The last layer is modified to output the N heatmaps as shown in Fig. 3(A).
  - *FCN8 architecture:* In FCN8 as proposed in [18], the initial layers are made up of stack of convolutional layers followed by maxpooling whereas later layers are upsampling layers that consist in the fusion of intermediate convolutional layers as shown in Fig. 3(B). In the architecture, the last layer is modified to output the probability heatmaps.
  - *ResNet50 backbone:* ResNet50 is a state-of-the-art image recognition CNN model [9] and also successfully used in pose estimation [22]. It is made up of deeper convolutional layers with residual blocks and is capable of solving the vanishing gradient problem in deeper networks by passing the identity information in the subsequent layers. We use the upsampling layers in the decoder part to achieve the same resolution as that of the input size. We use ResNet50 pretrained on ImageNet [5] dataset for our evaluation methodology. Figure 3(C) shows the design of CNN with Resnet50 as backbone. Note that heatmap-based regression with this architecture is very close to the DeepLabCut [12] approach and, thus, can be considered as a reimplementation of this latter method.
  - *HRNet:* The deep High Resolution Network architecture is one of the state-of-the-art architectures for the task of human pose estimation [33]. It maintains the high resolution from end to end and uses other subnetworks in parallel to exchange information between and within the stages. Figure 3(D) shows the HRNet architecture.

- **Multivariate regression based CNN architectures:** To implement multivariate regression that directly regresses coordinate points, we investigate two types of strategies. In the first case, the encoder part of the U-Net architecture shown in Fig. 3(A) is used for learning feature representations. In the second scenario, we explore a transfer learning based approach where a ResNet50 network pretrained on ImageNet is used for learning representations. In both scenarios, a fully connected layer is added at the end of the network to output $N \times 2$ numbers that correspond to (x, y) coordinates of each landmark location, where N is the total number of landmark locations.

---

[1] https://github.com/navdeepkaushish/S_Deep-Fish-Landmark-Prediction.

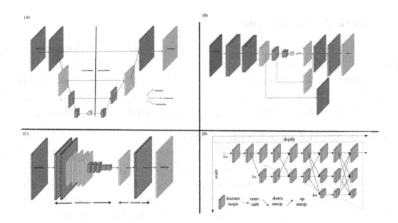

Fig. 3. Illustration of CNN architectures used in our experiments. (A)-U-net, (B)-FCN8, (C)-ResNet50 backbone, (D)-HRNet (reproduced from [33])

4.5 Experimental Protocol and Implementation

To evaluate method variants, we follow a 5-fold cross validation scheme in which each dataset is divided into 5 equal parts. In each iteration, one part is used as test set while the other four parts are merged and shuffled and used as training and validation sets, with a 3:1 ratio. Here the validation set is used for choosing the best model from the number of epochs during training. In each fold, one model is trained for maximum upto 2000 epochs. Mean error is then measured as first upscaling the predictions to the original sized images then taking the Root Mean Square Error (RMSE) (i.e., the Euclidean distance) between original ground-truth landmark locations and upscaled predicted locations for each test image, then calculating the mean over all the test images. The final error is reported by taking the mean error and standard deviation (Std.) over 5-fold cross validation. In all the evaluation protocols, we applied RMSProp optimizer with initial learning rate as 0.001 and Mean Square Error (MSE) as the loss function. To induce variability in the training set, we use data augmentation (scale, shift, rotate, shear, horizontal flip, random brightness, and contrast change) for all methods. We also use some callbacks such as *Early stopping* in which training is stopped when the loss does not improve over 400 epochs and *Learning rate scheduler* in which learning rate is reduced by the factor of 0.2 if validation loss is not improving over 200 epochs. We use *Tensorflow* [1] as the deep learning library and Python as programming language. We have trained the CNNs models on a cluster of roughly 100 NVIDIA's GeForce GTX 1080 GPUs.

5   Results and Discussion

Baseline. We evaluate a first baseline, called *'Mean model'*, that simply predicts for each landmark the mean positions computed for each landmark over original sized images of the training and validation sets. In Table 2, we report the mean error (and standard deviation) of this model across 5-folds for our three datasets. As expected, the errors are very high, showing that landmarks positions are highly variable given the uncontrolled positioning and orientation of the fishes.

Table 2. Mean RMSE for 5-fold cross validation for the baseline *Mean model*

Dataset	Mean error±Std.
Zebrafish microscopy	77.54 ± 8.74
Medaka microscopy	184.96 ± 19.11
Seabream radiography	50.14 ± 1.27

Direct Multivariate Regression. Mean errors and standard deviations over 5-fold cross validation scheme for direct multivariate regression are reported in Table 3. As expected, very significant improvements can be obtained with respect to the Mean model. The only exception is U-Net on the Zebrafish Microscopy dataset that obtains a higher error than the baseline. We hypothesize that this could be due to the significantly lower number of images (113) in this dataset and the fact that U-Net, unlike ResNet50, is not pretrained, which makes this model more difficult to train. U-Net remains however a better model than ResNet50 on the other two, larger, datasets.

Table 3. Mean RMSE for 5-fold cross validation for direct multivariate regression

Dataset	Mean Error±Std.	
	U-Net (31M)	ResNet50 (30M)
Zebrafish Microscopy	121.24 ± 5.38	26.31 ± 6.42
Medaka Microscopy	16.65 ± 2.35	20.44 ± 7.61
Seabream Radiography	7.71 ± 0.2	9.65 ± 2.34

Heatmap Regression. Heatmap regression requires tuning an additional hyper-parameter, the dispersion σ. We carried out some preliminary experiments on the Zebrafish Microscopy Dataset to analyse the impact of this parameter with both heatmap generation strategies. Table 4 shows how the RMSE error, estimated using the validation set of a single dataset split, evolves with σ in the case of the U-Net architecture. The best performance is obtained with $\sigma = 5$ with the Gaussian heatmap and $\sigma = 3$ with the Exponential heatmap. We will therefore set σ to these two values for all subsequent experiments. This will potentially make our results on the Zebrafish Microscopy Dataset a bit positively biased but we expect this bias to be negligible as the errors in Table 4 remain very stable and essentially independent of σ as soon as σ is higher than 3. Note also that better results can be potentially obtained on all problems by tuning σ using some additional internal cross-validation loop (at a higher computational cost).

Table 4. Effect of σ values using Zebrafish microscopy validation data with U-Net

σ	RMSE Error (in pixels)	
	Gaussian	Exponential
1	1202.64	118.87
2	1417.18	1198.1
3	36.38	**19.35**
4	20.66	19.76
5	**19.23**	20.06
6	23.52	19.64
7	20.73	19.68
8	19.58	19.58
9	20.15	20.73
10	20.47	20.11

Table 5. Mean Error (in pixels) from 5-fold cross validation for heatmap regression

Heatmaps	Datasets	Mean Error ± **Std.**			
		U-Net(31M)	FCN8(17M)	RestNet50(51M)	HRNet(6.5M)
Gaussian	Zebrafish Microscopy	13.43±3.14	13.82±2.01	13.77±2.97	**13.16±2.93**
	Medaka Microscopy	10.36±2.45	10.56±1.85	**10.18±1.17**	10.69±2.52
	Seabrean Radiography	**5.69±0.28**	5.74±0.15	6.13±0.31	6.40±0.63
Exponential	Zebrafish Microscopy	**11.29±0.84**	14.28±2.35	13.08±3.24	12.62±2.66
	Medaka Microscopy	**9.34±1.06**	10.12±1.60	9.36±1.05	9.54±1.59
	Seabream Radiography	**5.31±0.13**	5.70±0.16	5.47±0.18	5.90±0.64

Table 5 reports the performance of the different architectures, with both Gaussian and Exponential heatmaps. We observe that CNNs having more parameters tend to perform better in most of the cases (except HRNet with gaussian heatmap) but at the cost of computational efficiency and memory requirements. In particular, **U-Net** is better in terms of accuracy though second largest in size. Pretrained ResNet50 comes next with comparable performance with the largest size among all the models. Exponential heatmap outperforms Gaussian heatmap in almost all situations, although the difference is not very significant.

Comparing Table 5 with Table 3, it can be observed that heatmap based regression clearly outperforms direct multivariate regression on all datasets. From this investigation, we can conclude that, for the problem of landmark detection in Fish bioimages at least, heatmap based regression, with U-Net and Exponential heatmap, is the preferred approach, especially when the dataset is small.

It is interesting to note that because of the downscaling of the input image and the upscaling of the predictions, one can expect that the reported errors will be non zero even if the heatmap is perfectly predicted by the CNN model. We can thus expect that our results could be improved by using higher resolution images/heatmaps, at the price of a higher computational cost.

Hit Rate. To further measure the performance of the model in terms of how many landmarks are correctly predicted, we define a prediction as a **hit** if the predicted landmark location is within some tolerance distance δ from the actual landmark location. The **hit rate** is then the percentage of landmarks in the test images that are having a hit. We choose the best performing method from Table 5 (exponential heatmap based U-Net model) and hit rates with different distance thresholds, estimated by 5-fold cross-validation, are shown in Table 6, with the baseline δ set at the ratio between the original and heatmap resolutions. As expected, there are not many hits at δ, except on the third dataset. At $2 \times \delta$ however, all landmarks are perfectly detected, which suggests that heatmaps are very accurately predicted (2 pixels error in the downscaled resolution) and further supports the idea that better performance could be expected by increasing the resolution of the network input images and heatmaps.

Table 6. Hit rate from the three dataset using best performing models

Dataset	δ (in pixels)	Hit rate (in %)	
		δ	$2 \times \delta$
Zebrafish Microscopy	10	20.0	100
Medaka Microscopy	10	16.66	100
Seabream Radiography	8	94.73	100

Per Landmark Error. To further assess performances hence derive guidelines for practical use in real-world application, we computed mean error per landmark on test sets across 5-folds in order to quantify which landmarks are hard to predict by the models. Figure 4 shows per landmark mean error using the best performing method (exponential heatmap based U-Net model) for all the three datasets. We can observe that in the case of the Zebrafish Microscopy dataset, landmarks 4, 16, and 21 are the most difficult to predict. We hypothesized that these points are largely influenced by their position on the structure which they marked on. These structures exhibit some variability (shape, thickness, overlapping, missing or partially missing). In the case of Seabream Radiography, landmarks G, M, and T are difficult to predict due to their position which is somehow matched with background (see Fig. 1(C)). Lastly, in the case of the Medaka Microscropy dataset, landmark 3 (see Fig. 1(B)) is badly predicted. That might be attributed to the variability of the position it is marked on. As model predictions might vary greatly between landmarks, we believe these approaches should be combined with user interfaces for proofreading to make them effective. In practice, experts would mostly need to focus and proofread badly predicted landmarks, an hybrid human-computer approach which is expected to be much less time consuming than a completely manual approach.

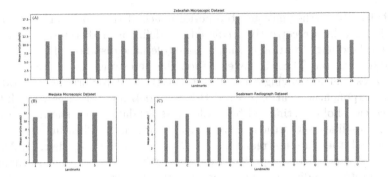

Fig. 4. Mean error per landmark with Exponential heatmap regression based U-Net on Zebrafish (A), Medaka (B), and Seabream (C) datasets

Finally, in Fig. 5, we illustrate the predictions from the best models using one image from the test set of each dataset.

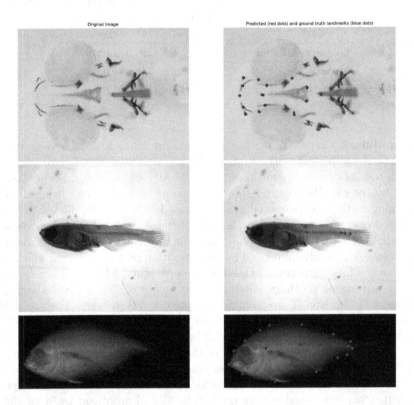

Fig. 5. Sample predictions on one image from each of our three datasets (Zebrafish, Medaka and Seabream) using best performing models (exponential heatmap based U-Net). First column: Original image. Second column: image with predicted landmarks (red dots) and ground truth landmarks (blue dots) (Color figure online)

6   Conclusions

We have evaluated two types of regression based landmark detection strategies combined with four CNN architectures on two microscopy and one radiography imaging datasets of different types of fish species with limited ground truths. The winning strategy (heatmap-based regression with Exponential generation function and U-Net architecture) is a simple end-to-end deep learning methodology where a single model is able to predict all the landmarks in a single run. Datasets and codes are distributed using open licenses and integrated into Cytomine [21][2]. End-users can train models and proofread model predictions, then export all statistics for their morphometric studies. Preliminary experiments have showed that this approach works also well on images of butterfly wings and we expect our work will ease landmark detection in future bioimaging studies.

Acknowledgements. N.K., R.R., C.D.B. and Z.D. are fellows of the BioMedAqu project, M.M. is a "Maître de Recherche" at FNRS. Raphaël Marée is supported by the BigPicture EU Research and Innovation Action (Grant agreement number 945358). This work was partially supported by Service Public de Wallonie Recherche under Grant No. 2010235 - ARIAC by DIGITALWALLONIA4.AI. Computational infrastructure is partially supported by ULiège, Wallonia and Belspo funds. We would like to thank the GIGA zebrafish platform (H. Pendeville-Samain) for taking care of and delivering the zebrafish larvae.

References

1. Abadi, M., et al.: TensorFlow: large-scale machine learning on heterogeneous systems (2015). https://www.tensorflow.org/, software available from tensorflow.org
2. Andriluka, M., Pishchulin, L., Gehler, P., Schiele, B.: 2d human pose estimation: new benchmark and state of the art analysis. In: Proceedings of the IEEE Conference on Computer Vision and Pattern Recognition, pp. 3686–3693 (2014)
3. Aubert, B., Vazquez, C., Cresson, T., Parent, S., De Guise, J.: Automatic spine and pelvis detection in frontal x-rays using deep neural networks for patch displacement learning. In: 2016 IEEE 13th International Symposium on Biomedical Imaging (ISBI), pp. 1426–1429. IEEE (2016)
4. Bookstein, F.L.: Combining the tools of geometric morphometrics. In: Marcus, L.F., Corti, M., Loy, A., Naylor, G.J.P., Slice, D.E. (eds) Advances in Morphometrics. NATO ASI Series, vol 284, pp. 131–151. Springer, Boston (1996). https://doi.org/10.1007/978-1-4757-9083-2_12(1996)
5. Deng, J., Dong, W., Socher, R., Li, L.J., Li, K., Fei-Fei, L.: Imagenet: a large-scale hierarchical image database. In: 2009 IEEE Conference on Computer Vision and Pattern Recognition, pp. 248–255. Ieee (2009)
6. Edwards, C.A., Goyal, A., Rusheen, A.E., Kouzani, A.Z., Lee, K.H.: Deepnavnet: automated landmark localization for neuronavigation. Front. Neurosci. **15**, 730 (2021)

---

[2] Code: https://github.com/cytomine-uliege. Demo server with datasets: http://research.uliege.cytomine.org/ username: eccv2022bic password: deep-fish.

7. Fragkoulis, S., Printzi, A., Geladakis, G., Katribouzas, N., Koumoundouros, G.: Recovery of haemal lordosis in gilthead seabream (sparus aurata l.). Sci. Rep. **9**(1), 1–11 (2019)

8. Guo, Y., Zhang, L., Hu, Y., He, X., Gao, J.: MS-Celeb-1M: a dataset and benchmark for large-scale face recognition. In: Leibe, B., Matas, J., Sebe, N., Welling, M. (eds.) ECCV 2016. LNCS, vol. 9907, pp. 87–102. Springer, Cham (2016). https://doi.org/10.1007/978-3-319-46487-9_6

9. He, K., Zhang, X., Ren, S., Sun, J.: Deep residual learning for image recognition. In: Proceedings of the IEEE Conference on Computer Vision and Pattern Recognition, pp. 770–778 (2016)

10. Huang, W., Yang, C., Hou, T.: Spine landmark localization with combining of heatmap regression and direct coordinate regression. arXiv preprint arXiv:2007.05355 (2020)

11. Ibragimov, B., Vrtovec, T.: Landmark-based statistical shape representations. In: Statistical Shape and Deformation Analysis, pp. 89–113. Elsevier (2017)

12. Insafutdinov, E., Pishchulin, L., Andres, B., Andriluka, M., Schiele, B.: DeeperCut: a deeper, stronger, and faster multi-person pose estimation model. In: Leibe, B., Matas, J., Sebe, N., Welling, M. (eds.) ECCV 2016. LNCS, vol. 9910, pp. 34–50. Springer, Cham (2016). https://doi.org/10.1007/978-3-319-46466-4_3

13. Jarque, S., Rubio-Brotons, M., Ibarra, J., Ordoñez, V., Dyballa, S., Miñana, R., Terriente, J.: Morphometric analysis of developing zebrafish embryos allows predicting teratogenicity modes of action in higher vertebrates. Reprod. Toxicol. **96**, 337–348 (2020)

14. Khabarlak, K., Koriashkina, L.: Fast facial landmark detection and applications: A survey. arXiv preprint arXiv:2101.10808 (2021)

15. Lai, H., Xiao, S., Pan, Y., Cui, Z., Feng, J., Xu, C., Yin, J., Yan, S.: Deep recurrent regression for facial landmark detection. IEEE Trans. Circuits Syst. Video Technol. **28**(5), 1144–1157 (2016)

16. Lee, H., Park, M., Kim, J.: Cephalometric landmark detection in dental x-ray images using convolutional neural networks. In: Medical Imaging 2017: Computer-Aided Diagnosis, vol. 10134, p. 101341W. International Society for Optics and Photonics (2017)

17. Lindner, C., Cootes, T.F.: Fully automatic cephalometric evaluation using random forest regression-voting. In: IEEE International Symposium on Biomedical Imaging (ISBI) 2015-Grand Challenges in Dental X-ray Image Analysis-Automated Detection and Analysis for Diagnosis in Cephalometric X-ray Image (2015)

18. Long, J., Shelhamer, E., Darrell, T.: Fully convolutional networks for semantic segmentation. In: Proceedings of the IEEE Conference on Computer Vision and Pattern Recognition, pp. 3431–3440 (2015)

19. Loy, B.A., Boglione, C., Cataudella, S.: Geometric morphometrics and morphoanatomy: a combined tool in the study of sea bream (sparus aurata, sparidae) shape. J. Appl. Ichthyol. **15**(3), 104–110 (1999)

20. Mahpod, S., Das, R., Maiorana, E., Keller, Y., Campisi, P.: Facial landmarks localization using cascaded neural networks. Comput. Vis. Image Underst. **205**, 103171 (2021)

21. Marée, R., Rollus, L., Stévens, B., Hoyoux, R., Louppe, G., Vandaele, R., Begon, J.M., Kainz, P., Geurts, P., Wehenkel, L.: Collaborative analysis of multi-gigapixel imaging data using cytomine. Bioinformatics **32**(9), 1395–1401 (2016)

22. Mathis, A., Mamidanna, P., Cury, K.M., Abe, T., Murthy, V.N., Mathis, M.W., Bethge, M.: Deeplabcut: markerless pose estimation of user-defined body parts with deep learning. Nat. Neurosci. **21**(9), 1281–1289 (2018)

23. Mohseni, H., Kasaei, S.: Automatic localization of cephalometric landmarks. In: 2007 IEEE International Symposium on Signal Processing and Information Technology, pp. 396–401. IEEE (2007)

24. Park, J.H., Hwang, H.W., Moon, J.H., Yu, Y., Kim, H., Her, S.B., Srinivasan, G., Aljanabi, M.N.A., Donatelli, R.E., Lee, S.J.: Automated identification of cephalometric landmarks: Part 1-comparisons between the latest deep-learning methods yolov3 and ssd. Angle Orthod. **89**(6), 903–909 (2019)

25. Payer, C., Štern, D., Bischof, H., Urschler, M.: Regressing heatmaps for multiple landmark localization using CNNs. In: Ourselin, S., Joskowicz, L., Sabuncu, M.R., Unal, G., Wells, W. (eds.) MICCAI 2016. LNCS, vol. 9901, pp. 230–238. Springer, Cham (2016). https://doi.org/10.1007/978-3-319-46723-8_27

26. Payer, C., Štern, D., Bischof, H., Urschler, M.: Integrating spatial configuration into heatmap regression based CNNs for landmark localization. Med. Image Anal. **54**, 207–219 (2019)

27. Rautaray, S.S., Agrawal, A.: Vision based hand gesture recognition for human computer interaction: a survey. Artif. Intell. Rev. **43**(1), 1–54 (2015)

28. Riegler, G., Urschler, M., Ruther, M., Bischof, H., Stern, D.: Anatomical landmark detection in medical applications driven by synthetic data. In: Proceedings of the IEEE International Conference on Computer Vision Workshops, pp. 12–16 (2015)

29. Ronneberger, O., Fischer, P., Brox, T.: U-Net: convolutional networks for biomedical image segmentation. In: Navab, N., Hornegger, J., Wells, W.M., Frangi, A.F. (eds.) MICCAI 2015. LNCS, vol. 9351, pp. 234–241. Springer, U-net: Convolutional networks for biomedical image segmentation (2015). https://doi.org/10.1007/978-3-319-24574-4_28

30. Samet, N., Akbas, E.: Hprnet: hierarchical point regression for whole-body human pose estimation. Image Vis. Comput. **115**, 104285 (2021)

31. Song, Y., Qiao, X., Iwamoto, Y., Chen, Y.W.: Automatic cephalometric landmark detection on x-ray images using a deep-learning method. Appl. Sci. **10**(7), 2547 (2020)

32. Stern, O., Marée, R., Aceto, J., Jeanray, N., Muller, M., Wehenkel, L., Geurts, P.: Automatic localization of interest points in zebrafish images with tree-based methods. In: Loog, M., Wessels, L., Reinders, M.J.T., de Ridder, D. (eds.) PRIB 2011. LNCS, vol. 7036, pp. 179–190. Springer, Heidelberg (2011). https://doi.org/10.1007/978-3-642-24855-9_16

33. Sun, K., Xiao, B., Liu, D., Wang, J.: Deep high-resolution representation learning for human pose estimation. In: Proceedings of the IEEE/CVF Conference on Computer Vision and Pattern Recognition, pp. 5693–5703 (2019)

34. Torosdagli, N., Liberton, D.K., Verma, P., Sincan, M., Lee, J.S., Bagci, U.: Deep geodesic learning for segmentation and anatomical landmarking. IEEE Trans. Med. Imaging **38**(4), 919–931 (2018)

35. Vandaele, R., Aceto, J., Muller, M., Peronnet, F., Debat, V., Wang, C.W., Huang, C.T., Jodogne, S., Martinive, P., Geurts, P., et al.: Landmark detection in 2d bioimages for geometric morphometrics: a multi-resolution tree-based approach. Sci. Rep. **8**(1), 1–13 (2018)

36. Verhaegen, Y., Adriaens, D., De Wolf, T., Dhert, P., Sorgeloos, P.: Deformities in larval gilthead sea bream (sparus aurata): a qualitative and quantitative analysis using geometric morphometrics. Aquaculture **268**(1–4), 156–168 (2007)

37. Weinhardt, V., Shkarin, R., Wernet, T., Wittbrodt, J., Baumbach, T., Loosli, F.: Quantitative morphometric analysis of adult teleost fish by x-ray computed tomography. Sci. Rep. **8**(1), 1–12 (2018)

38. Xu, Z., Li, B., Yuan, Y., Geng, M.: Anchorface: an anchor-based facial landmark detector across large poses. In: Proceedings of the AAAI Conference on Artificial Intelligence, vol. 35, pp. 3092–3100 (2021)
39. Yeh, Y.C., Weng, C.H., Huang, Y.J., Fu, C.J., Tsai, T.T., Yeh, C.Y.: Deep learning approach for automatic landmark detection and alignment analysis in whole-spine lateral radiographs. Sci. Rep. **11**(1), 1–15 (2021)

PointFISH: Learning Point Cloud Representations for RNA Localization Patterns

Arthur Imbert[1,2,3(✉)], Florian Mueller[4,5], and Thomas Walter[1,2,3(✉)]

[1] Centre for Computational Biology, Mines Paris, PSL University, Paris, France
{Arthur.Imbert,Thomas.Walter}@minesparis.psl.eu
[2] Institut Curie, PSL University, Paris, France
[3] INSERM, U900, Paris, France
[4] Imaging and Modeling Unit, Institut Pasteur and UMR 3691 CNRS, Paris, France
[5] C3BI, USR 3756 IP CNRS, Paris, France

Abstract. Subcellular RNA localization is a critical mechanism for the spatial control of gene expression. Its mechanism and precise functional role is not yet very well understood. Single Molecule Fluorescence in Situ Hybridization (smFISH) images allow for the detection of individual RNA molecules with subcellular accuracy. In return, smFISH requires robust methods to quantify and classify RNA spatial distribution. Here, we present PointFISH, a novel computational approach for the recognition of RNA localization patterns. PointFISH is an attention-based network for computing continuous vector representations of RNA point clouds. Trained on simulations only, it can directly process extracted coordinates from experimental smFISH images. The resulting embedding allows scalable and flexible spatial transcriptomics analysis and matches performance of hand-crafted pipelines.

Keywords: smFISH · RNA localization · Point cloud · Transfer learning · Simulation · Spatial transcriptomics

1 Introduction

Localization of messenger RNAs (mRNAs) are of functional importance for gene expression and in particular its spatial control. RNA localization can be related to RNA metabolism (to store untranslated mRNAs or degrade them) or protein metabolism (to localize translations). RNA localization is not a limited phenomenon but a common mechanism throughout the transcriptome, which might also concern non-coding RNAs [1,2]. Despite the importance of this process, it is still poorly understood, and adequate tools to study this process are still lacking.

The spatial distribution of RNA can be investigated with sequence or image-based techniques. We focus on the latter, since they provide substantially better spatial resolution and are therefore more suitable for the analysis of subcellular RNA localization. The method of choice to visualize RNAs in intact cells is

© The Author(s), under exclusive license to Springer Nature Switzerland AG 2023
L. Karlinsky et al. (Eds.): ECCV 2022 Workshops, LNCS 13804, pp. 487–502, 2023.
https://doi.org/10.1007/978-3-031-25069-9_32

single molecule Fluorescence in Situ Hybridization (smFISH) that comes in many variants, such as a scalable and cost-efficient version [3], which we will use in the following. In smFISH, individual RNA molecules of a given RNA species are targeted with several fluorescently labeled oligonucleotides and appear as bright diffraction-limited spots under a microscope and can thus been detected with traditional image analysis or computer vision methods. Usually, some additional fluorescent markers are used to label relevant cell structures (cytoplasm, nucleus, centrosomes, ...), which allow one to determine the position of each individual RNA with respect to these landmarks. From such a multi-channel microscopy image, we can thus obtain a coordinate representation of each individual cell and its RNAs as illustrated in Fig. 1.

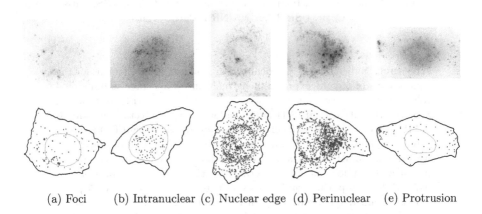

(a) Foci   (b) Intranuclear (c) Nuclear edge (d) Perinuclear   (e) Protrusion

Fig. 1. RNA localization patterns from [4]. (*Top*) Typical smFISH images with different RNA localization patterns. (*Bottom*) Coordinate representations with RNA spots (red), cell membrane (black) and nuclear membrane (blue). Detection and segmentation results are extracted and visualized with FISH-quant [5] (Color figure online)

RNA localization results in several, distinct patterns, which can in general be defined by a local overcrowded subcellular region. In the literature [4], several patterns have been described, even for a simple biological system such as HeLa cells: a random default pattern where RNAs localize uniformly within the cell, RNA clusters (foci), a high density of transcripts along the nuclear membrane (nuclear edge), inside the nucleus itself (intranuclear), in cell extensions (protrusion), or a polarization within the cell, like RNA localizing towards the nucleus (perinuclear).

It is still an open problem how to statistically classify and automatically detect RNA localization patterns and how to represent point cloud distributions. Previous approaches relied essentially on handcrafted features, quantitatively describing RNA spatial distribution within subcellular regions [6–8].

Here, we intend to address the challenge to detect RNA localization patterns with a deep learning approach. The idea is to replace the feature engineering problem by a training design problem. Instead of manually crafting features, we propose a training procedure to learn generic encodings for RNA clouds inside cells allowing us to efficiently address the recognition of RNA localization patterns[1].

2   Related Work

Recognition of RNA Localization Pattern. In previous studies, hand-crafted features to classify RNA localization patterns were developed [4,6–8]. Their design has been inspired by literature on spatial statistics [9] and adapted from analysis pipelines for fluorescence microscopy images [10,11]. Several packages already implement modules to perform smFISH analysis and compute these hand-crafted features [5,12–14]. However, these approaches require to carefully design a set of features corresponding to the concrete biological question under study. For a different study a new set of features might be necessary. Here, we aim to investigate a more general approach to build localization features.

Learning Representations. Neural network learn powerful representations that can often be used for transfer learning. The idea is to pretrain a network and thereby to obtain a generic representation by solving a pretext task on a large annotated dataset, before addressing a more difficult or specific task with sometimes limited data. Often the representation optimized to solve the pretext task can also be useful for the more specific task. Such a model can then be used as a feature extractor by computing features from one of its intermediate layers. The computer vision community progressively replaces hand-crafted features [15, 16] by deep learning features to analyze images. For instance, convolutional neural networks pretrained on large and general classification challenges [17–20] are used as backbone or feature extractor for more complex tasks like face recognition, detection or segmentation. The NLP community follows this trend as well with a heavy use of word embeddings [21,22] or the more recent transformers models. The same strategy has also been applied to graphs: node2vec [23] learns "task-independent representations" for nodes in networks.

Such embeddings can be a continuous and numerical representation of a non-structured data like a text or a graph. In spage2vec [24], the model learned a low dimensional embedding of local spatial gene expression (expressed as graphs). Authors then identified meaningful gene expression signatures by computing this embedding for tissue datasets.

Convolutional Features. Since we analyze imaging data, a first intuition would be to build a convolutional neural network to directly classify localization patterns from these fluorescent images. Such approaches have a long tradition in the classification of subcellular protein localization patterns. Unlike RNAs, proteins are usually difficult to resolve at the single molecule level unless super-resolution microscopy was employed, which is not the case for this kind of studies.

---

[1] Code and data are available in https://github.com/Henley13/PointFISH.

Protein localization patterns is therefore seen as a characteristic texture in the fluorescent image and thus the representation of subcellular protein localization often relies on texture and intensity features. Initial studies [25] computed a set of hand-crafted features from the microscopy image before training a classifier. With the advent of deep learning, protein localization is now tackled with convolutional neural networks, but still framed as a texture classification problem. After crowdsourcing annotations for the Human Protein Atlas dataset [26], researchers trained a machine learning model (Loc-CAT) from hand-crafted features to predict subcellular localization patterns of proteins [27]. More recently, an online challenge [28] was organized, where the majority of top-ranking solutions were based on convolutional neural networks. In summary, for protein localization the shift from hand-crafted features to learned representations allows for more accurate and robust pipelines.

A recent perspective paper [29] suggests the increased use of deep learning models also for RNA localization analysis. The authors emphasize the recent successes and flexibility of neural nets with different types of input, and therefore the possibility to design a multimodal pipeline. However, the fundamental difference to existing protein localization datasets, is that RNA molecules appear as distinguishable spots, and their modeling as a texture seems therefore suboptimal.

Point Cloud Models. We postulate that learning to classify RNA localization patterns directly from detected spot coordinates could be an efficient approach. A point cloud has an unordered and irregular structure. Projecting the point coordinates into images or voxels [30] transforms the problem as an easier vision challenge, but it comes along with some input degradations and dramatically increases the memory needed to process the sample. Also, relevant spatial information can be lost. In case of RNA point clouds, it makes the recognition of 3D localization patterns harder [31].

PointNet [32] is a seminal work that opened the way for innovative models to address shape classification. It directly processes point clouds with shared MLPs and a max pooling layer, making the network invariant to input permutation. However, the pooling step is the only way for the model to share information between close points, which ultimately limits its performance. Yet, recent research dramatically improves point cloud modelling and especially the capture of local information.

PointNet++ [33] learns local geometric structures by recursively applying PointNet to different regions of the point cloud, in a hierarchical manner. This way, local information can be conveyed through the network more efficiently. DGCNN [34] proposes a new EdgeConv layer where edge features are computed between a point and its neighbors. Some models propose to adapt convolutions to point clouds by designing new weighting functions or kernel operations like PointCNN [35], PointConv [36] or KPConv [37]. Another inspiration from the computer vision or NLP literature is the attention-based model. To this end, PointTransformer [38] proposes an attention layer to be applied to local regions within the point cloud. Finally, PointMLP [39] proposes a simple but efficient network with a pure deep hierarchical MLP architecture.

3 Problem Statement

We want to train a model, where we can provide directly the point cloud coordinates as an input and compute a continuous vector representation. This representation can then be used for classification of different RNA localization patterns. Such a deep learning model might require a large volume of annotated data to reach a satisfying performance. To generate such a large data sets, we used simulated data to train our point cloud model and then use it as a trained feature extractor. Eventually we evaluate these learned features on a real dataset.

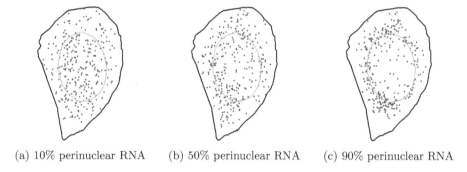

(a) 10% perinuclear RNA   (b) 50% perinuclear RNA   (c) 90% perinuclear RNA

Fig. 2. Perinuclear pattern simulations with increasing pattern strength. Simulated with FISH-quant [5]

Simulated Dataset. Using a Python framework FISH-quant [5], we simulate a dataset with 8 different localization patterns: random, foci, intranuclear, extranuclear, nuclear edge, perinuclear (Fig. 2), cell edge and pericellular. We choose these patterns since they represent a diverse panel of localization patterns in different cellular subregions. We simulate for each pattern 20,000 cells with 50 to 900 RNAs per cell, resulting in a full dataset of

Table 1. Annotated real dataset

Pattern	# of cells
Random	372
Foci	198
Intranuclear	73
Nuclear edge	87
Perinuclear	64
Protrusion	83

160,000 simulated cells. Except for the random pattern, every simulated pattern has a proportion of RNAs with preferential localization ranging from 60% to 100%. In order to test how our trained features generalize to unknown localization patterns, we deliberately omitted one of the patterns (localization in cell protrusions) from the simulation, i.e. the set of real patterns in the experimental dataset in Fig. 1 only partially matches the set of simulated patterns.

We split our dataset into train, validation and test, with 60%, 20% and 20% respectively. FISH-quant simulates point clouds from a limited number of real image templates. To avoid overfitting, we make sure simulations from the same cell template can't be assigned to different splits. Finally, point clouds are augmented with random rotations along the up-axis, centered and divided by their absolute maximum value. This normalization step was initially performed in [32].

Real Dataset. To further validate the learned feature representation on simulated images, we use a previously published experimental dataset [4]. These images are extracted from a smFISH study in HeLa cells targeting 27 different genes. After data cleaning, this dataset consists of 9710 individual cells, with cropped images and coordinates extracted. Cells have on average 346 RNAs and 90% of them have between 39 and 1307 transcripts. Furthermore, 810 cells have manually annotated localization patterns, as detailed in Table 1, providing a ground-truth for validation. Importantly, these patterns are not mutually exclusive since cells can display several patterns at the same time, e.g. foci with a perinuclear distribution.

4   PointFISH

4.1   Input Preparation

For the simulated dataset we directly generate point clouds. In contrast, for any experimental dataset, the starting input is usually a multichannel fluorescent image including a smFISH channel. The first tasks are the detection of RNA molecules and the segmentation of cell and nucleus surfaces. These steps allow to retrieve for each cell a list of RNA coordinates we can format to be used as an input point cloud. For the experimental dataset we reuse the code and the extraction pipeline described in [4].

Besides the original RNA point cloud, we can use an optional second input vector containing additional information as input for our model. Let $X \in \mathbb{R}^{N \times 3}$ be the original input point cloud with N the number of RNAs. We define our second input vector as $\tilde{X} \in \mathbb{R}^{N \times d}$ with $d \in \{1, 2, 3, 4, 5\}$. It is composed of three contextual inputs: morphology input, distance input and cluster input. First, morphological information (i.e. positional information on the plasma and nuclear membrane) is integrated by concatenating the initial RNA point cloud and points uniformly sampled from the 2D polygons outlining the cellular and nuclear membranes. To be consistent with a 3D point cloud input, these 2D coordinates are localized to the average height of the RNA point cloud (0 if it is centered). This morphological input substantially increases the size of the input point cloud X, because we subsample 300 points from the cell membrane and 100 points from the nuclear membrane. To let the network discriminate between

RNA, cell and nucleus points, we define two boolean vectors as contextual inputs to label the points as a cell or a nucleus point. A point sampled from the cell membrane will have a value (*True*, *False*), one from the nucleus membrane (*False*, *True*) and one from the original RNA point cloud (*False*, *False*). We end up with $X \in \mathbb{R}^{\tilde{N} \times 3}$ (with $\tilde{N} = N + 300 + 100$) and $\tilde{X} \in \{0,1\}^{\tilde{N} \times 2}$ as inputs. Second, we compute the distance from cellular and nuclear membrane for each RNA in the point cloud X. This adds an extra input $\tilde{X} \in \mathbb{R}^{N \times 2}$. Third, we leverage the cluster detection algorithm from FISH-quant [5] in order to label each RNA node as clustered or not. It gives us a boolean $\tilde{X} \in \{0,1\}^{N \times 1}$ to indicate if a RNA belongs to a RNA cluster of not. Depending on whether or not we choose to add the morphological information, the distance or the clustering information, we can exploit up to 5 additional dimensions of input.

4.2 Model Architecture

We adopt the generic architecture introduced by PointNet [32]: successive point-wise representations with increasing depth followed by a max pooling operation to keep the network invariant by input permutation. We incorporate state-of-the-art modules to learn efficient local structures within the point cloud. As illustrated in Fig. 3, we also adapt the network to the specificity of RNA point clouds.

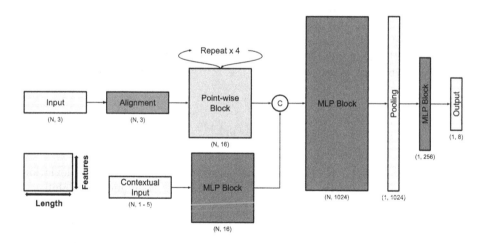

Fig. 3. PointFISH architecture. Width and height of *boxes* represent output length and dimension, respectively. *Tuples* represent output shapes

Point-Wise Block. Instead of shared MLPs like PointNet, we implement a multi-head attention layer based on point transformer layer [38]. First, we assign to each data point x_i its 20 nearest neighbors $X(i) \subset X$, based on the euclidean distance in the feature space. We also compute a position encoding

$\delta_{ij} = \theta(x_i - x_j)$ for every pair within these neighborhoods, with θ a MLP. Three sets of point-wise features are computed for each data point, with shared linear projections ϕ, ψ and α. Relative weights between data points $\gamma(\phi(x_i) - \psi(x_j))$ are computed with the subtraction relation (instead of dot product as in the seminal attention paper [40]) and a MLP γ. These attention weights are then normalized by softmax operation ρ. Eventually, data point's feature y_i is computed as weighted sum of neighbors value $\alpha(x_j)$, weighted by attention. With the position encoding added to both the attention weights and the feature value, the entire layer can be summarized such that:

$$y_i = \sum_{x_j \in X(i)} \rho(\gamma(\phi(x_i) - \psi(x_j) + \delta_{ij})) \odot (\alpha(x_j) + \delta_{ij}) \tag{1}$$

For a multi-head attention layer, the process is repeated in parallel with independent layers, before a last linear projection merge multi-head outputs. A shortcut connection and a layer normalization [41] define the final output of our multi-head attention layer.

Alignment Module. Albeit optional (point clouds can be processed without it), this module dramatically improves performance of the network. Some papers stress the necessity to preprocess the input point cloud by learning a projection to *align* the input coordinates in the right space [32,34]. In addition, density heterogeneity across the point cloud and irregular local geometric structures might require local normalization. To this end, we reuse the geometric affine module described in PointMLP [39] which transforms local data points to a normal distribution. With $\{x_{i,j}\}_{j=1,...,20} \in \mathbb{R}^{20 \times 3}$, the neighborhood's features of x_i, we compute:

$$\{x_{i,j}\} = \alpha \odot \frac{\{x_{i,j}\} - x_i}{\sigma + \epsilon} + \beta \tag{2}$$

where $\alpha \in \mathbb{R}^3$ and $\beta \in \mathbb{R}^3$ are learnable parameters, σ is the feature deviation across all local neighborhoods and ϵ is a small number for numerical stability.

Contextual Inputs. Our RNA point cloud does not include all the necessary information for a localization pattern classification. Especially, information about the morphological properties of the cell and nucleus are lacking. To this end, deep learning architectures allows flexible insertions. Several contextual inputs $\tilde{X}$ can feed the network through a parallel branch, before concatenating RNA and contextual point-wise features. Our best model exploits cluster and distance information in addition to RNA coordinates.

5   Experiment

5.1   Training on Simulated Patterns

We train PointFISH on the simulated dataset. Our implementation is based on TensorFlow [42]. We use ADAM optimizer [43] with a learning rate from 0.001 to 0.00001 and an exponential decay (decay rate of 0.5 every 20,000 steps). Model is trained for a maximum of 150 epochs, with a batch size of 32, but early stopping criterion is implemented if validation loss does not decrease after 10 consecutive epochs. Usually, the model converges after 50 epochs. We apply a 10% dropout for the last layer and classifications are evaluated with a categorical cross entropy loss. Even if localization patterns are not necessarily exclusive, for the simulations we trained the model to predict only one pattern per cell. For this reason, we did not simulate mixed patterns and assume it could help the model to learn disentangled representations. Training takes 6 to 8 h to converge with a Tesla P100 GPU.

A first evaluation can be performed on the simulated test dataset. With our reference PointFISH model, we obtain a general F1-score of 95% over the different patterns. The configurations of this model include the use of distance and cluster contextual inputs, a geometric affine module, an attention layer as a point-wise block and a latent dimension of 256.

5.2   Analysis of the Embeddings Provided by PointFISH

From a trained PointFISH model we can remove the output layer to get a feature extractor that computes a 256-long embedding from a RNA point cloud.

Exploratory Analysis of Experimental Data Embeddings. We compute the embeddings for the entire cell population studied in [4]. All the 9170 cells can be visualized in 2D using a UMAP projection [44]. In Fig. 4 each point represents a cell. Among the 810 annotated cells, those with a unique pattern are colored according to the localization pattern observed in their RNA point cloud. The rest of the dataset is gray. Overall, PointFISH embedding discriminates well the different localization patterns. Intranuclear, nuclear edge and perinuclear cells form distinct clusters, despite their spatial overlap, as well as protrusions. We recall that the protrusion patterns was not used in simulation. Cells with foci can be found in a separated clusters as well, but also mix with nuclear and perinuclear patterns. This confusion is not surprising as a large number of cells in the dataset present a nuclear-related foci pattern (i.e. cells have RNAs clustered in foci, which in turn are close to the nuclear envelope).

Supervised Classification. Because PointFISH already return meaningful embeddings, we can apply a simple classifier on top of these features to learn localization patterns. We use the 810 manually annotated cells from the real

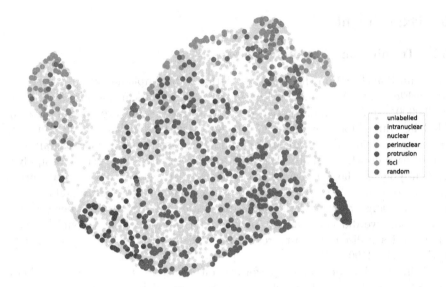

Fig. 4. UMAP embedding with learned features. Each point is a cell from dataset [4]. Manually annotated cells are colored according to their localization pattern (Color figure online)

dataset. We compare the 15 hand-crafted features selected in [4] with our learned embedding. Every set of features (hand-crafted or learned) is rescaled before feeding a classifier "by removing the mean and scaling to unit variance"[2]. Expert features quantify RNA distributions within specific subcellular compartments and compute relevant distances to cell structures. Porportion features excepted, they are normalized by their expected value under a uniform RNA distribution. Hand-crafted features include:

- The number of foci and the proportion of clustered RNA.
- The average foci distance from nucleus and cell.
- The proportion or RNA inside nucleus.
- The average RNA distance from nucleus and cell.
- The number of RNAs detected in cell extensions and a peripheral dispersion index [11].
- The number of RNAs within six subcellular regions (three concentric regions around the nucleus and three others concentric regions around the cell membrane).

We design 5 binary classification tasks, one per localized pattern (random pattern is omitted). The classifier is a SVC model [46]. For evaluation purpose, we apply a nested cross-validation scheme. First, a test dataset is sampled (20%),

---

[2] Features are rescaled with the *StandardScaler* method from scikit-learn [45].

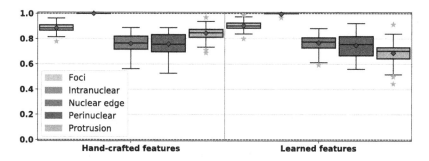

Fig. 5. F1-score distribution with localization pattern classification (SVC model)

then the remaining cells are used for a gridsearch to find an optimal SVC model (with another 20% validation split). Parameters grid includes the choice between a linear or a RBF kernel and the strength of the regularization. The entire process is repeated 50 times, with different test split, and F1-score for each classification task is returned. This full evaluation pipeline is implemented with scikit-learn [45]. F1-score's distribution over 50 splits are summarized in Fig. 5. Learned features match performances of hand-crafted features selected for the tasks. While the recognition of localization in protrusions is slightly worse, it is important to point out that we did not include simulations of this patterns in the training dataset.

5.3 Ablation Studies

We perform ablation studies to evaluate the impact of different components in PointFISH model.

Additional Input. We compare the use of RNA point cloud as unique input or the inclusion of contextual information through a parallel branch. RNA coordinates do not carry any morphological information about the cell. In Table 2, this design logically returns the lowest F1-score. Three additional inputs are available: RNA distance from cell and nucleus (*distance*), RNA clustering flag (*cluster*) and the

Table 2. Impact of contextual inputs. F1-score is averaged over 4 trainings with different random seeds. Best model is in bold. Reference model is labelled with *

Distance	Cluster	Morphology	F1-score
✗	✗	✗	0.42 (± 0.01)
✓	✗	✗	0.74 (± 0.02)
✗	✓	✗	0.45 (± 0.04)
✓	✓	✗	0.81* (± 0.01)
✓	✓	✓	**0.82** (± 0.00)

integration of cell and nucleus membrane coordinates (*morphology*). Both morphology and distance inputs can be added to provide additional information about cell morphology to the network. However, best performances are reached when using at least distance and cluster information. Cell and nucleus coordinates do not increase significantly the classification and dramatically increase the computation time of the model (we need to process a larger point cloud). In particular, cluster information greatly improves the recognition of the foci pattern while morphological distances boost others localization patterns.

Alignment Module and Point-wise Block. To measure the impact of the geometric affine module [39], we compare it with the TNet module implemented in PointNet [32]. We also design a variant of TNetEdge where MLP layers extracting point-wise independent features are replaced with EdgeConv layers [34]. Results are reported in Table 3. An alignment block seems critical at the beginning of the network. In addition, geometric affine module is both more efficient (F1-score of 0.81) and much lighter than TNet and TNetEdge.

Inspired by PointNet and DGCNN, we also compare the use of their respective point-wise blocks with our multi-head attention layer. As expected, Edge-Conv blocks convey a better information than PointNet by exploiting local neighborhood within point cloud (F1-score of 0.78 and 0.75 respectively). Yet, they do not match the performance of multi-head attention layer.

Concerning these layers, we evaluate how the number of parallel heads can influence the performance of PointFISH. By default, we use 3 parallel attention heads to let the model specialize its attentions vectors, but we also test 1, 6 and 9 parallel heads. In Table 3, we only observe a slight benefit between the original point transformer layer [38] (with one attention head) and its augmented implementations.

Latent Dimensions. The second part of PointFISH architecture is standardized: a first MLP block, a max pooling operation, a second MLP block and the output layer. We quantify the impact of additional MLP layers within these blocks. Our reference model returns an embedding with 256 dimensions (before the output layer). In a MLP block, we use ReLU activation and layer normalization, but also increase or decrease the depth by a factor 2 between layers. Before the pooling layer, the first MLP block includes 4 layers with an increasing depth (128, 256, 512 and 1024). After the pooling layer, the second MLP block includes 2 layers with a decreasing depth (512 and 256). Similarly, to return 128, 64 or 32 long embeddings, we implement 6 (128, 256, 512, pooling, 256 and 128), 5 (128, 256, pooling, 128 and 64) or 4 final layers (128, pooling, 64 and 32). We observe in table 3 a reduction in performance for the lowest dimensional embedding (64 and 32). This hyperparameter is also critical to design lighter models, with a division by 4 in terms of trainable parameters between a 256 and a 128 long embedding.

Table 3. Ablation studies on real dataset [4]. F1-score is averaged over 4 trainings with different random seeds. Best models are bold. Reference model is labelled with *

Alignment	Point-wise block	# heads	# dimensions	# parameters	F1-score
-	Attention layer	3	256	1,372,608	0.73 (± 0.00)
TNet	Attention layer	3	256	1,712,521	0.74 (± 0.02)
TNetEdge	Attention layer	3	256	1,589,321	0.74 (± 0.01)
Affine	MLP	–	256	1,374,526	0.75 (± 0.01)
Affine	EdgeConv	–	256	1,387,006	0.78 (± 0.01)
Affine	Attention layer	9	256	1,403,334	**0.82** (± 0.01)
Affine	Attention layer	6	256	1,387,974	**0.82** (± 0.01)
Affine	Attention layer	3	256	1,372,614	0.81* (± 0.01)
Affine	Attention layer	1	256	1,362,374	0.81 (± 0.01)
Affine	Attention layer	3	128	352,966	0.81 (± 0.01)
Affine	Attention layer	3	64	97,094	0.77 (± 0.00)
Affine	Attention layer	3	32	32,646	0.75 (± 0.01)

6   Discussion

We have presented a generic method of quantifying RNA localization patterns operating directly on the extracted point coordinates, without the need to design handcrafted features. For this, we leverage coordinates of simulated localization patterns to train a specifically designed neural network taking as input a list of points and associated features that greatly enhance generalization capabilities. We show that this method is on par with carefully designed, handcrafted feature sets.

Being able to directly process list of points provides the community with a tool to integrate large datasets obtained with very different techniques on different model systems. While the actual image data might look strikingly different between such projects, they can all be summarized by segmentation maps of nuclei and cytoplasm, and a list of coordinates of RNA locations. Having methods that operate directly on point clouds is therefore a strategic advantage.

The idea of training on simulated data provides us the opportunity to query datasets with respect to new localization patterns that have not yet been observed, and for which we do not have real examples so far. In addition, this strategy allows us to control for potential confounders, such as cell morphology, or number of RNAs. Here, we provide a generic method that can leverage these simulations, without the tedious process of handcrafting new features. Of note, it is not necessary that the simulated patterns are optimized as to resemble real data: they rather serve as a pretext task. If a network is capable of distinguishing the simulated patterns, chances are high that the corresponding representation is also informative for slightly or entirely different patterns, in the same way as representations trained on ImageNet can be used for tumor detection in pathology images. We show this by omitting the protrusion pattern from the simulation. We see in Fig. 4 that the protrusion patterns live in a particular region of the feature space, without specific training. Moreover, we see in Fig. 4, that the overall separation between patterns in this exploratory way coincides to a large extent with the figure that has been proposed by the authors of the original paper [4].

7   Conclusion

In this work, we introduce a new approach for the quantification and classification of RNA localization patterns. On the top of existing solutions to extract RNA spots and cell morphology coordinates, we propose to directly process the resulting point clouds. Recent advances in point cloud analysis through deep learning models allows us to build a flexible and scalable pipeline that matches results obtained with specific hand-crafted features.

Overall, with the increasing interest on subcellular RNA localization in the field of spatial transcriptomics, we expect that this approach will be of great use to the scientific community, and that it will contribute to the deciphering of some of the most fundamental processes in life.

Acknowledgments. This work was funded by the ANR (ANR-19-CE12-0007) and by the French government under management of Agence Nationale de la Recherche as part of the "Investissements d'avenir" program, reference ANR-19-P3IA-0001 (PRAIRIE 3IA Institute). Furthermore, we also acknowledge France-BioImaging infrastructure supported by the French National Research Agency (ANR-10-INBS-04).

References

1. Lécuyer, E., et al.: Global analysis of mRNA localization reveals a prominent role in organizing cellular architecture and function. Cell **131**(1), 174–187 (2007)
2. Buxbaum, A.R., Haimovich, G., Singer, R.H.: In the right place at the right time: visualizing and understanding mRNA localization. Nat. Rev. Mol. Cell Biol. **16**(2), 95–109 (2015)
3. Tsanov, N., Samacoits, A., Chouaib, R., Traboulsi, A.M., Gostan, T., Weber, C., Zimmer, C., Zibara, K., Walter, T., Peter, M., Bertrand, E., Mueller, F.: smiFISH and FISH-quant - a flexible single RNA detection approach with super-resolution capability. Nucleic Acids Res. **44**(22), e165–e165 (2016)
4. Chouaib, R., et al.: A dual protein-MRNA localization screen reveals compartmentalized translation and widespread co-translational RNA targeting. Dev. Cell **54**(6), 773-791.e5 (2020)
5. Imbert, A., et al.: FISH-quant v2: a scalable and modular tool for smFISH image analysis. RNA **10**(6), 786–795 (2022)
6. Battich, N., Stoeger, T., Pelkmans, L.: Image-based transcriptomics in thousands of single human cells at single-molecule resolution. Nature Methods **10**(11), 1127–1133 (2013)
7. Stoeger, T., Battich, N., Herrmann, M.D., Yakimovich, Y., Pelkmans, L.: Computer vision for image-based transcriptomics. Methods **85**, 44–53 (2015)
8. Samacoits, A., et al.: A computational framework to study sub-cellular RNA localization. Nat. Commun. **9**(1), 4584 (2018)

9. Ripley, B.: Spatial Statistics. Wiley Series in Probability and Statistics. Wiley (2005)

10. Lagache, T., Sauvonnet, N., Danglot, L., Olivo-Marin, J.C.: Statistical analysis of molecule colocalization in bioimaging. Cytometry A **87**(6), 568–579 (2015)

11. Stueland, M., Wang, T., Park, H.Y., Mili, S.: RDI calculator: an analysis tool to assess RNA distributions in cells. Sci. Rep. **9**(1), 8267 (2019)

12. Mueller, F., et al.: FISH-quant: automatic counting of transcripts in 3D FISH images. Nat. Methods **10**(4), 277–278 (2013)

13. Savulescu, A.F., et al.: DypFISH: dynamic patterned FISH to interrogate RNA and protein spatial and temporal subcellular distribution (2019). https://www.biorxiv.org/content/10.1101/536383v1

14. Mah, C.K., et al.: Bento: a toolkit for subcellular analysis of spatial transcriptomics data (2022). https://www.biorxiv.org/content/10.1101/2022.06.10.495510v1

15. Lowe, D.: Object recognition from local scale-invariant features. In: Proceedings of the Seventh IEEE International Conference on Computer Vision, pp. 1150–1157 (1999)

16. Bay, H., Tuytelaars, T., Van Gool, L.: SURF: speeded up robust features. In: Leonardis, A., Bischof, H., Pinz, A. (eds.) ECCV 2006. LNCS, vol. 3951, pp. 404–417. Springer, Heidelberg (2006). https://doi.org/10.1007/11744023_32

17. He, K., Zhang, X., Ren, S., Sun, J.: Deep residual learning for image recognition. In: Proceedings of the IEEE Conference on Computer Vision and Pattern Recognition (CVPR) (2016)

18. Szegedy, C., Vanhoucke, V., Ioffe, S., Shlens, J., Wojna, Z.: Rethinking the inception architecture for computer vision. In: Proceedings of the IEEE Conference on Computer Vision and Pattern Recognition (CVPR) (2016)

19. Tan, M., Le, Q.: EfficientNet: rethinking model scaling for convolutional neural networks. In: Proceedings of the 36th International Conference on Machine Learning, pp. 6105–6114 (2019)

20. Huang, G., Liu, Z., van der Maaten, L., Weinberger, K.Q.: Densely connected convolutional networks. In: Proceedings of the IEEE Conference on Computer Vision and Pattern Recognition (CVPR) (2017)

21. Mikolov, T., Chen, K., Corrado, G., Dean, J.: Efficient estimation of word representations in vector space (2013). https://arxiv.org/abs/1301.3781

22. Joulin, A., Grave, E., Bojanowski, P., Mikolov, T.: Bag of tricks for efficient text classification (2016). https://arxiv.org/abs/1607.01759

23. Grover, A., Leskovec, J.: Node2vec: scalable feature learning for networks. In: Proceedings of the 22nd ACM SIGKDD International Conference on Knowledge Discovery and Data Mining, pp. 855–864 (2016)

24. Partel, G., Wählby, C.: Spage2vec: unsupervised representation of localized spatial gene expression signatures. FEBS J. **288**(6), 1859–1870 (2021)

25. Boland, M.V., Markey, M.K., Murphy, R.F.: Automated recognition of patterns characteristic of subcellular structures in fluorescence microscopy images. Cytometry **33**(3), 366–375 (1998)

26. Uhlén, M., et al.: Tissue-based map of the human proteome. Science **347**(6220), 1260419 (2015)

27. Sullivan, D.P., Winsnes, C.F., Åkesson, L., Hjelmare, M., Wiking, M., Schutten, R., Campbell, L., Leifsson, H., Rhodes, S., Nordgren, A., Smith, K., Revaz, B., Finnbogason, B., Szantner, A., Lundberg, E.: Deep learning is combined with massive-scale citizen science to improve large-scale image classification. Nat. Biotechnol. **36**(9), 820–828 (2018)

28. Ouyang, W., et al.: Analysis of the human protein atlas image classification competition. Nat. Methods **16**(12), 1254–1261 (2019)
29. Savulescu, A.F., Bouilhol, E., Beaume, N., Nikolski, M.: Prediction of RNA subcellular localization: learning from heterogeneous data sources. iScience **24**(11), 103298 (2021)
30. Maturana, D., Scherer, S.: Voxnet: a 3d convolutional neural network for real-time object recognition. In: 2015 IEEE/RSJ International Conference on Intelligent Robots and Systems (IROS), pp. 922–928 (2015)
31. Dubois, R., et al.: A deep learning approach to identify mRNA localization patterns. In: 2019 IEEE 16th International Symposium on Biomedical Imaging (ISBI), pp. 1386–1390 (2019)
32. Qi, C.R., Su, H., Mo, K., Guibas, L.J.: Pointnet: deep learning on point sets for 3d classification and segmentation. In: Proceedings of the IEEE Conference on Computer Vision and Pattern Recognition (CVPR) (2017)
33. Qi, C.R., Yi, L., Su, H., Guibas, L.J.: Pointnet++: deep hierarchical feature learning on point sets in a metric space. In: Advances in Neural Information Processing Systems, vol. 30 (2017)
34. Wang, Y., Sun, Y., Liu, Z., Sarma, S.E., Bronstein, M.M., Solomon, J.M.: Dynamic graph CNN for learning on point clouds. ACM Trans. Graph. **38**(5) (2019)
35. Li, Y., Bu, R., Sun, M., Wu, W., Di, X., Chen, B.: Pointcnn: convolution on x-transformed points. In: Advances in Neural Information Processing Systems, vol. 31 (2018)
36. Wu, W., Qi, Z., Fuxin, L.: Pointconv: deep convolutional networks on 3d point clouds. In: Proceedings of the IEEE/CVF Conference on Computer Vision and Pattern Recognition (CVPR) (2019)
37. Thomas, H., Qi, C.R., Deschaud, J.E., Marcotegui, B., Goulette, F., Guibas, L.J.: Kpconv: flexible and deformable convolution for point clouds. In: Proceedings of the IEEE/CVF International Conference on Computer Vision (ICCV) (2019)
38. Zhao, H., Jiang, L., Jia, J., Torr, P.H., Koltun, V.: Point transformer. In: Proceedings of the IEEE/CVF International Conference on Computer Vision (ICCV), pp. 16259–16268 (2021)
39. Ma, X., Qin, C., You, H., Ran, H., Fu, Y.: Rethinking network design and local geometry in point cloud: a simple residual MLP framework. In: International Conference on Learning Representations (2022)
40. Vaswani, A., et al.: Attention is all you need. In: Advances in Neural Information Processing Systems, vol. 30 (2017)
41. Ba, J.L., Kiros, J.R., Hinton, G.E.: Layer normalization (2016). http://arxiv.org/abs/1607.06450
42. Abadi, M., et al.: TensorFlow: large-scale machine learning on heterogeneous systems (2015). https://www.tensorflow.org/
43. Kingma, D.P., Ba, J.: Adam: a method for stochastic optimization. CoRR (2015)
44. McInnes, L., Healy, J., Saul, N., Großberger, L.: Umap: uniform manifold approximation and projection. J. Open Source Softw. **3**(29), 861 (2018)
45. Pedregosa, F., et al.: Scikit-learn: machine learning in python. J. Mach. Learn. Res. **12**, 2825–2830 (2011)
46. Chang, C.C., Lin, C.J.: Libsvm: a library for support vector machines. ACM Trans. Intell. Syst. Technol. (TIST) **2**(3), 1–27 (2011)

N2V2 - Fixing Noise2Void Checkerboard Artifacts with Modified Sampling Strategies and a Tweaked Network Architecture

Eva Höck[1] , Tim-Oliver Buchholz[2] , Anselm Brachmann[1], Florian Jug[3] , and Alexander Freytag[1(✉)]

[1] Carl Zeiss AG, Jena, Germany
eva.hoeck@zeiss.com, anselm.brachmann@zeiss.com,
alexander.freytag@zeiss.com
[2] Facility for Advanced Imaging and Microscopy, Friedrich Miescher Institute
for Biomedical Research, Basel, Switzerland
tim-oliver.buchholz@fmi.ch
[3] Jug Group, Fondazione Human Technopole, Milano, Italy
florian.jug@fht.org

Abstract. In recent years, neural network based image denoising approaches have revolutionized the analysis of biomedical microscopy data. Self-supervised methods, such as Noise2Void (N2V), are applicable to virtually all noisy datasets, even without dedicated training data being available. Arguably, this facilitated the fast and widespread adoption of N2V throughout the life sciences. Unfortunately, the blind-spot training underlying N2V can lead to rather visible checkerboard artifacts, thereby reducing the quality of final predictions considerably. In this work, we present two modifications to the vanilla N2V setup that both help to reduce the unwanted artifacts considerably. Firstly, we propose a modified network architecture, *i.e.*, using *BlurPool* instead of *MaxPool* layers throughout the used U-Net, rolling back the residual-U-Net to a non-residual U-Net, and eliminating the skip connections at the uppermost U-Net level. Additionally, we propose new replacement strategies to determine the pixel intensity values that fill in the elected blind-spot pixels. We validate our modifications on a range of microscopy and natural image data. Based on added synthetic noise from multiple noise types and at varying amplitudes, we show that both proposed modifications push the current state-of-the-art for fully self-supervised image denoising.

E. Höck, T.-O. Buchholz, A. Brachmann, F. Jug, A. Freytag—Equal contribution.

Supplementary Information The online version contains supplementary material available at https://doi.org/10.1007/978-3-031-25069-9_33.

© The Author(s), under exclusive license to Springer Nature Switzerland AG 2023
L. Karlinsky et al. (Eds.): ECCV 2022 Workshops, LNCS 13804, pp. 503–518, 2023.
https://doi.org/10.1007/978-3-031-25069-9_33

1   Introduction

Fluorescence microscopy is one of the major drivers for discovery in the life sciences. The quality of possible observations is limited by the optics of the used microscope, the chemistry of used fluorophores, and the maximum light exposure tolerated by the imaged sample. This necessitates trade-offs, frequently leading to rather noisy acquisitions as a consequence of preventing ubiquitous effects such as photo toxicity and/or bleaching. While the light efficiency in fluorescence microscopy can be optimized by specialized hardware, *e.g.*, by using Light Sheet or Lattice Light Sheet microscopes, software solutions that restore noisy or distorted images are a popular additional way to free up some of the limiting photon budget.

Algorithmic image restoration is the reconstruction of clean images from corrupted versions as they were acquired by various optical systems. A plethora of recent work shows that CNNs can be used to build powerful content-aware image restoration (CARE) methods [3,9,17–20]. However, when using supervised CARE approaches, as initially proposed in [18], pairs of clean and distorted images are required for training the method. For many applications in the life-sciences, imaging such clean ground truth data is either impossible or comes at great extra cost, often rendering supervised approaches as being practically infeasible [6].

Hence, self-supervised training methods like Noise2Void (N2V) by Krull *et al.* [6], which operate exclusively on single noisy images, are frequently used in life-science research [1,6–8,14]. Such *blind-spot approaches* are enabled by excluding/masking the center (blind-spot) of a network's receptive field and then training the network to predict the masked intensity. These approaches collectively assume that the noise to be removed is pixel-wise independent (given the signal) and that the true intensity of a pixel can be predicted after learning a content-aware prior of local image structures from a body of noisy data [6].

More recently, methods that can sample the space of diverse interpretations of noisy data were introduced [12,13]. While these approaches show great performance on denoising and even artifact removal tasks, the underlying network

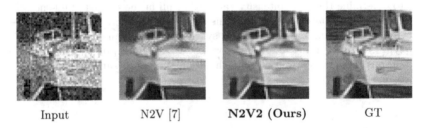

| Input | N2V [7] | **N2V2 (Ours)** | GT |

Fig. 1. Self-supervised denoising of noisy data (left). Results obtained with Noise2Void (N2V) [6] (here shown without residual connection and with sampling without the center point) are subject to clearly visible checkerboard artifacts (2nd column). Our proposed method, Noise2Void v2 (N2V2), visibly reduces these artifacts, leading to improved quality results (3rd column, here shown with median center pixel replacement). The last column shows ground truth (not available to either method)

architectures and training procedures are space and time demanding [12] and can typically not be used on today's typical consumer workstations and laptops. Hence, comparatively small blind-spot networks like N2V are available via consumer solutions such as *Fiji* [4,16], *ZeroCostDL4Mic* [5], or the *BioImage.IO Model Zoo* [11], and are therefore still the most commonly used self-supervised denoising methods.

Still, one decisive problem with blind-spot approaches such as N2V is that checkerboard artifacts can commonly be observed (see Fig. 1 for an illustrative example). Hence, in this work we present Noise2Void v2 (called N2V2), a variation of N2V that addresses the problem with checkerboard artifacts by a series of small but decisive tweaks.

More concretely, the contributions of our work are: (*i*) showcasing and inspecting the short-comings of N2V, (*ii*) proposal of an adapted U-Net architecture that replaces *max-pooling* layers with *max-blur-pooling* [20] layers and omits the top-most skip-connection, (*iii*) introduction of blind-spot pixel replacement strategies, (*iv*) a systematic evaluation of our proposed replacement strategies and architectural changes on the BSD68 dataset [6], the Mouse, Convallaria and Flywing datasets from [13,14] and two newly added salt and pepper (S&P) noise regimes, and (*v*) proposal of a new variation on the Convallaria dataset from [14] that addresses what we believe to be non-ideal setup choices.

2   Related Work

The original CARE work by Weigert *et al.* [18] steered our field away from more established and non-trained denoising methods towards modern data-driven deep denoising methods. With these fully supervised methods deep neural networks are trained on pairs of low-quality and high-quality images that are pixel-perfectly aligned and contain the exact same objects (or 'scene').

Such pairs need to be carefully acquired at the microscope, typically by varying acquisition parameters such as exposure time and illumination intensity. In certain modalities, *e.g.*, cryo transmission electron microscopy (cryo-TEM), acquisition of high-exposure images is impossible and even the acquisition of pairs of noisy images is undesirable [3].

However, if pairs of independently noisy images are available, Noise2Noise (N2N) training [9] can be applied and high quality predictions are still achievable. Later, Buchholz *et al.* [2], extended these ideas to full cryo electron tomography (cryo-ET) workflows [10].

Still, clean ground truth data or a second set of independently noisy images is typically not readily available. This motivated the introduction of self-supervised methods such as Noise2Void [6] and Noise2Self [1]. The simplicity and applicability of these methods makes them, to-date, the de-facto standard approach used by many microscopists on a plethora of imaging modalities and biological samples. All such blind-spot approaches exploit the fact that for noise which is independent per pixel (given the signal), the intensity value of any given pixel can in principle be estimated from examining the pixels image context (surrounding). This is precisely what content-aware image restoration approaches do. Pixel-independent noise, instead, can by definition not be predicted, leading to a situation where the loss minimizing prediction does, in expectation, coincide with the unknown signal at the predicted pixel [1,6,9].

An interesting extension of N2V was introduced by Krull *et al.* [7]. Their method, called *Probabilistic Noise2Void* (PN2V), does not only predict a single (maximum likelihood) intensity value per pixel, but instead an entire distribution of plausible pixel intensity values (prior). Paired with an empirical (measured) noise-model [7,14], *i.e.*, the distributions of noisy observations for any given true signal intensity (likelihood), PN2V computes a posterior distribution of possible predicted pixel intensities and returns, for example, the minimum mean squared error (MMSE) of that posterior.

A slightly different approach to unsupervised image denoising was proposed by Prakash *et al.* [12,13]. Their method is called (Hierarchical) DivNoising and employs a variational auto-encoder (VAE), suitably paired with a noise model of the form described above [7,14], that can be used to sample diverse interpretations of the noisy input data.

In contrast to these probabilistic approaches, we focus on N2V in this work due to its popularity. Hence, we aim at making a popular method, which is at the same time powerful *and* simple, even more powerful.

Particularities of the Publicly Available Convallaria Dataset

Self-supervised denoising methods are built to operate on data for which no high-quality ground truth exists. This makes them notoriously difficult to evaluate quantitatively, unless when applied on data for which ground truth is obtainable.

To enable a fair comparison between existing and newly proposed methods, several benchmark datasets have been made available over the years. One example is the *Convallaria* data, first introduced by Lalit *et al.* [14]. This dataset consists of 100 noisy short exposure fluorescence acquisitions of the same 1024×1024px field of view of the same sample. The corresponding ground truth image used to compare against was created by pixel-wise averaging of these 100 independently noisy observations.

In later work [13,14], the proposed methods were trained on 95 of the individual noisy images, while the remaining 5 images were used for validation purposes. For the peak signal-to-noise ratio (PSNR) values finally reported in these papers, the predictions of the top left 512×512 pixels of all 100 noisy images were compared to the corresponding part of the averaged ground truth image. In this paper we refer to this dataset and associated train/validation/test sets as *Convallaria_95*.

We are convinced that training self-supervised image denoising methods on 95 noisy observations of the exact same field of view is leading to slightly misleading results (that overestimate the performance to be expected from the tested method) in cases where only one noisy image per sample exists. Also note that in cases where already as few as two noisy observations per sample are available, a network can be trained via N2N [9]. With 95 such instances available, one could even average those and use the average as ground truth for fully supervised CARE training [18].

Hence, we propose here to use the Convallaria data differently, namely by selecting one of the 100 images and tiling it into 64 tiles of 128×128px. Of these tiles, 56, 4, and 4 are then used for training, validation, and testing respectively. See the supplementary material for more information. We refer to this data and train/validation/test split as *Convallaria_1* Please see Sect. 4 for a thorough evaluation of achievable denoising results when using *Convallaria_95* versus *Convallaria_1*.

3   Method

As can be seen in Fig. 1, denoising predictions from a vanilla N2V model can exhibit considerable amounts of unwanted checkerboard artifacts. After observing this phenomenon on several datasets, our hypothesis is that these artifacts are caused by two aspects in the vanilla N2V design: (*i*) missing high-frequency suppression techniques to counteract strongly noisy pixel values that really stick out with respect to their close neighbors, and (*ii*) an amplification of this effect due to N2V's self-supervised input replacement scheme (blind-spots). Below, we describe two measures which we introduce in N2V2 to mitigate these problems.

3.1   A Modified Network Architecture for N2V2

The default N2V configuration employs a *residual* U-Net with 2×2 max-pooling layers throughout the encoder [6]. We propose to change this architecture in three decisive ways by (*i*) removing the residual connection and instead use a regular U-Net, (*ii*) removing the top-most skip-connection of the U-Net to further constrain the amount of high-frequency information available for the final decoder layers, and (*iii*) replacing the standard max-pooling layers by max-blur-pool layers [20] to avoid aliasing-related artifacts. In Fig. 2, we highlight all proposed architectural changes which we propose for N2V2.

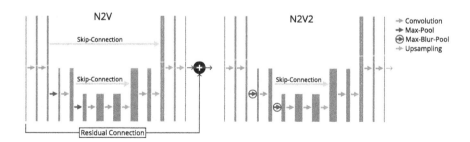

Fig. 2. *Left:* the N2V architecture in [6] is a standard U-Net [15] with a residual connection. *Right:* our N2V2 architecture drops the residual connection, removes the top-most skip-connection and replaces the max-pooling layers with max-blur-pooling layers [20]

(a)   (b)   mean: ▪

Fig. 3. Pixel replacement strategies are key for efficient N2V training. **(a)** the original N2V replacement strategy in [6] chooses a random pixel from the center pixel's local neighborhood, which may lead to artifacts like checkerboard patterns in denoised images (see Fig. 1). **(b)** Our proposed average center pixel replacement strategy calculates the mean or median of the pixel's local neighborhood while excluding the center pixel itself

3.2 New Sampling Strategies to Cover Blind-Spots

As mentioned before, self-supervised denoising methods introduce blind-spots, effectively asking the network to perform content-aware single pixel inpainting [1, 6,7,14].

During training, a self-supervised loss is employed that compares measured (and hence left out) pixel values with the corresponding pixel values predicted by the trained network (to which only the local neighborhood of the respective blind-spot pixels is given).

Let $\mathbf{X} \in \mathfrak{N}^{w \times h}$ be a patch in a given input image with intensity range $\mathfrak{N}$. Without loss of generality, let x_i be a single pixel in $\mathbf{X}$. As loss for a given patch $\mathbf{X}$, N2V [6] starts with proposing

$$\mathfrak{L}_{\text{N2V-naïve}}\big(\mathbf{X}, f\,(\cdot)\big) = \big(f\,(\mathbf{X} \setminus \{x_i\})_i - x_i\big)^2 \quad, \tag{1}$$

where $\mathbf{X} \setminus \{x_i\}$ denotes the exclusion of pixel x_i from $\mathbf{X}$. This exclusion operation would be computational inefficient when implemented naïvely in convolutional networks. Krull *et al.* have therefore proposed not to *exclude* x_i, but rather to *replace* x_i's value and thereby hiding the true intensity of blind-spot pixels:

$$\mathfrak{L}_{\text{N2V}}\big(\mathbf{X}, f\,(\cdot)\big) = \big(f\,(r\,(\mathbf{X}))_i - x_i\big)^2 \quad, \tag{2}$$

where $r\,(\mathbf{X})$ assigns a new value to x_i in $\mathbf{X}$. While Eq. (2) can be evaluated efficiently compared to Eq. (1), it turns out that the choice of $r\,(\mathbf{X})$ is more sensitive than originally believed, with some choices leading to emphasized visual artifacts like the ones shown in Fig. 1.

Default N2V Pixel Sampling Strategies (uwCP and UwoCP). In [6], Krull *et al.* analyze different blind-spot pixel replacement methods and settle for one default method in their public implementation[1]. This default method,

---

[1] https://github.com/juglab/n2v.

called UPS in N2V, is ubiquitously used by virtually all users world-wide and samples a pixel x_j uniformly at random from a small neighborhood $\mathbf{N} \subset \mathbf{X}$ of size $w' \times h'$ around a blind-spot pixel x_i (including x_i itself). We refer to this replacement technique as uwCP, and illustrate it in Fig. 3.

The first obvious observation is that with probability of $1/(w' \cdot h')$, j will be equal to i, $i.e.$, no replacement is happening. In these cases, the best solution to any model $f(\cdot)$ will be the identity, which is clearly not intended for denoising tasks. Therefore, in PN2V [7], the available implementation[2] started using a slightly altered sampling strategy that excludes the center pixel from being sampled, $i.e.$, $i \neq j$, which we refer to as uwoCP.

Blind-Spot Replacement Strategies for N2V2. In contrast to the blind-spot replacement strategies via sampling from $\mathbf{N}$, we propose to compute replacement strategies computed from the entire pixel neighborhood $\mathbf{N}$ (but without x_i). Specifically, we propose $r_{mean}(\mathbf{N}) = mean(\mathbf{N} \setminus \{x_i\})$ and $r_{median}(\mathbf{N}) = median(\mathbf{N} \setminus \{x_i\})$ as replacement strategies, and refer to them as *mean* and *median* replacement strategies, respectively.

Note that the exclusion of the center pixel is important in order to fully remove any residual information about the blind-spot pixels to be masked. Please refer to Fig. 3 for a visual illustration.

4 Evaluation

We evaluate our proposed pixel replacement strategies and the architectural changes on multiple datasets and perform different ablation studies. The covered datasets with their experiment details are described in Sect. 4.1. Evaluation metrics are listed in Sect. 4.2. Results on data with S&P noise are given in Sect. 4.3. Complementary results with other noise types are given in Sect. 4.4. In Sect. 4.5, we finally shed light on aspects of generalization and evaluation in scenarios where only single noisy recordings are available.

4.1 Dataset Descriptions and Training Details

All dataset simulation and method evaluation code, together with the used training configurations, is publicly available on GitHub[3]. The N2V2 code is part of the official Noise2Void implementation[4].

General Settings. All our hyper-parameter choices are based on previous publications [6,7,12] to keep the reported values comparable. Hence, we used an Adam optimizer with a reduce learning rate on plateau scheduler (patience of 10), chose 0.198% random pixels per training patch as blind-spots and the pixel replacement was performed within a neighborhood of size $w' = h' = 5$.

---

[2] https://github.com/juglab/pn2v.
[3] https://github.com/fmi-faim/N2V2_Experiments.
[4] https://github.com/juglab/n2v.

Table 1. Quantitative results on data with simulated S&P noise. Results are given in dB of averaged PSNR on test data. Overall best is <u>underlined</u>. Best fully self-supervised is in **bold**

	Method	Mouse SP3	Mouse SP6	Mouse SP12
	Input	20.03	18.72	17.76
Fully self- supervised	N2V as in [6]	21.32	20.69	20.99
	N2V w/ uwoCP as in [7]	35.17	34.24	33.49
	N2V w/o res, w/uwoCP	35.44	34.89	34.12
	N2V w/o res w/mean	35.29	34.71	33.66
	N2V w/o res w/median	35.23	35.07	33.45
	N2V2 w/uwCP	35.74	35.32	34.19
	N2V2 w/uwoCP	**35.91**	35.47	34.52
	N2V2 w/mean	35.51	35.01	34.17
	N2V2 w/median	35.81	**35.50**	**34.54**
Self- supervised	PN2V [7]	29.67	N/A	N/A
	DivNoising [13]	36.21	N/A	N/A
Supervised	CARE [18]	<u>37.03</u>	N/A	N/A

BSD68. An evaluation on natural images is done with the BSD68 dataset as used in the original N2V paper [6]. For training, we use the same 400 natural gray scale images of size 180×180px from [19]. From those, 396 are used as training data and 4 for validation as described in N2V. BSD68 networks are of depth 2 with 96 initial feature maps and are trained for 200 epochs, with 400 steps per epoch, a batch size of 128, and an initial learning rate of 0.0004.

Convallaria. We evaluate on the fluorescence imaging dataset Convallaria by [14]. Due to its specialities as described in Sect. 2, we call it Convallaria_95. Additionally, we introduce the Convallaria_1 dataset where the input corresponds to only one single noisy observation of 1024×1024px and the corresponding ground truth is the average of the 100 noisy Convallaria observations. This image pair is divided into non-overlapping patches of 128×128px, resulting in 64 patches. These patches are shuffled and 56, 4, and 4 patches are selected as training, validation and test data respectively (see Supplementary Figure S3). We train Convallaria_95 and Convallaria_1 networks with depth 3, with 64 initial feature maps, and for 200 epochs, with 10 steps per epoch, a batch size of 80, and an initial learning rate of 0.001.

Mouse. We further conduct evaluations based on the ground truth Mouse dataset from the DenoiSeg paper [4], showing cell nuclei in the developing mouse skull. The dataset consists of 908 training and 160 validation images of size 128×128px, with another 67 test images of size 256×256px. From this data, we simulate Mouse_G20 by adding Gaussian noise with zero-mean and standard deviation of 20. Furthermore, we simulate Mouse_sp3, Mouse_sp6 and

Table 2. Quantitative results: simulated Gaussian noise. Results are given in dB of averaged PSNR on test data. Overall best is <u>underlined</u>. Best fully self-supervised is in **bold**

	Method	Flywing G70	Mouse G20	BSD68
	Input	17.67	22.52	21.32
Fully self- supervised	N2V as in [6]	25.20	34.12	27.70
	N2V w/ uwoCP as in [7]	25.04	33.94	27.37
	N2V w/o res, w/uwoCP	25.24	34.20	26.95
	N2V w/o res w/mean	25.54	34.49	28.25
	N2V w/o res w/median	**25.57**	34.41	27.49
	N2V w/ bp w/uwCP	25.30	34.17	27.69
	N2V w/o sk w/uwCP	25.49	34.63	27.88
	N2V2 w/uwCP	25.42	34.65	28.04
	N2V2 w/uwoCP	25.49	34.59	27.97
	N2V2 w/mean	25.48	34.61	28.31
	N2V2 w/median	25.46	**34.74**	**28.32**
Self-supervised	PN2V [7]	24.85	34.19	N/A
	DivNoising [13]	25.02	34.13	N/A
Supervised	CARE [18]	<u>25.79</u>	<u>35.11</u>	<u>29.06</u>

Mouse_sp12, three datasets dominated by S&P noise. More specifically, we apply Poisson noise directly to the ground truth intensities, then add Gaussian noise with zero-mean and standard deviation of 10, and clip these noisy observations to the range $[0, 255]$. Then, we randomly select $p\%$ of all pixels ($p \in [3, 6, 12]$) and set them to either 0 or 255 with a probability of 0.5. We train networks on the Mouse dataset with depth 3, with 64 initial feature maps, and for 200 epochs, with 90 steps per epoch, a batch size of 80 and an initial learning rate of 0.001.

Flywing. Finally, we report results on the Flywing dataset from the DenoiSeg [4], showing membrane labeled cells in a flywing. We follow the data generation protocol described in [13], *i.e.*, we add zero-mean Gaussian noise with a standard deviation of 70 to the clean recordings of the dataset. The data consists of 1 428 training and 252 validation patches of size 128 × 128px, with additional 42 images of size 512 × 512px for testing. On the flywing dataset, we train networks with of depth 3, with 64 initial feature maps, and for 200 epochs, with 142 steps per epoch, a batch size of 80 and an initial learning rate of 0.001.

Data Augmentation. All training data is 8-fold augmented by applying three 90 deg rotations and flipping. During training, random 64 × 64 crops are selected from the provided training patches as described in [6].

	N2V	N2V	N2V		
	w/o res	w/o res,	w/o res,	N2V2	
Input	w/ uwoCP	w/ mean	w/ median	w/ median	GT

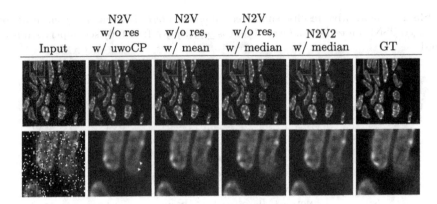

Fig. 4. Qualitative results on Mouse SP12 dataset. This data is dominated by 12% S&P noise, as can be seen in the input image (1st). The results of the N2V method without residual connection and sampling without the center point (2nd, PSNR 33.10) show checkerboard artifacts, see white arrow heads. On BSD68, these artifacts remain when using median replacement (4th, PSNR 32.43), are reduced in the results when using mean replacement (3rd, PSNR 33.01), and are eliminated in the N2V2 results with median replacement (5th, PSNR 33.34)

4.2   Evaluation Metrics

We compute PSNR in all conducted experiments, evaluated with respect to the corresponding high-SNR images. For the BSD68 dataset, the target range of the PSNR computation is set to $[0, 255]$. For all other datasets, the range is obtained by computing the min and max values of each corresponding ground truth image. We finally report PSNR values averaged over the entire test data.

4.3   Results on Mouse SP3, SP6, and SP12 (Salt&Pepper Noise)

The results for the S&P datasets are shown in Table 1. First of all, we see the striking impact of excluding the center pixel from the replacement sampling for S&P noise: while N2V as in [6] can barely increase the PSNR, we see clearly improved results when excluding the center pixel from random sampling for replacement. In addition, a non-residual U-Net further improves the result compared to the residual U-Net that is used by default in the N2V configuration. In a similar line, also our other architecture adaptations yield increased PSNR values. While the proposed replacement strategies mean and median do not result in better quantitative results, we are surprised to see that the mean replacement strategy clearly reduces checkerboard artifacts qualitatively as can be seen in Fig. 4. We finally observe that the best fully self-supervised results in the medium and high noise regime are obtained by combining both architecture and replacement adaptations.

Input	N2V w/o res w/ uwoCP	N2V w/o res, w/ mean	N2V w/o res, w/ median	N2V2 w/ median	GT

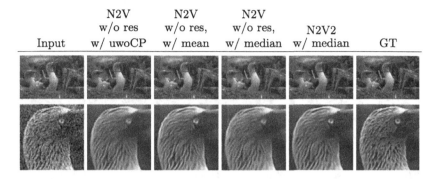

Fig. 5. Qualitative results on the BSD68 dataset. After applying the trained N2V model without residual connection and sampling without the center point to the noisy input (1st), the result shows undesirable checkerboard artifacts (2nd, PSNR 29.01). On BSD68, these artifacts are reduced with mean replacement (3rd, PSNR 29.27), remain when using median replacement (4th, PSNR 28.79), and are eliminated when using N2V2 with median replacement (5th, PSNR 29.24)

4.4   Evaluation Flywing G70, Mouse G20, BSD68

We report results for the datasets with simulated Gaussian noise in Table 2. In contrast to the results for simulated salt and pepper noise, we interestingly see that results do not improve simply by excluding the center pixel from the window for sampling replacement. Also, not using a residual U-Net only yields slight improvements for the microscopy datasets and none for the natural image dataset BSD68, where PSNR even drops. However, the alternative replacement strategies mean and median lead to improved PSNR values, as well as the architecture adaptations bp sk. Combining both adaptations leads to the best self-supervised results for the Mouse G20 and BSD68 datasets.

This is in line with qualitative results shown in Fig. 5 for the BSD68 dataset, where we clearly see checkerboard artifacts in the N2V standard setting, but significantly cleaner predictions with the proposed adaptations. Additional qualitative results are given in the supplementary material in section S.1.

4.5   Evaluation of Real Noisy Data: Convallaria_95 and Convallaria_1

As displayed in Table 3, both the median replacement strategy as well as the N2V2 architecture adaptations improve the results for both Convallaria datasets. This can also be seen in the qualitative example in Fig. 6. N2V2 with median replacement strategy yields the best fully self-supervised results for both cases. Interestingly, according to PSNR values, the mean replacement method does not improve when compared to the baseline N2V performance.

Comparing the two columns in Table 3, a considerable difference in PSNR is apparent, with the denoising results when using the reduced Convallaria_1 dataset being poorer. This leads to two possible interpretations, namely (i) having 95 noisy images of the same field of view allows for better results of the

	N2V	N2V	N2V		
	w/o res	w/o res,	w/o res,	N2V2	
Input	w/ uwoCP	w/ mean	w/ median	w/ median	GT

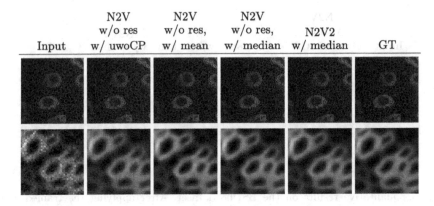

Fig. 6. Qualitative results on the convallaria dataset. After applying the trained N2V model without residual connection and sampling without the center point to the noisy input (1st), the result shows undesirable checkerboard artifacts (2nd, PSNR 35.78). These are eliminated when using mean (3rd, PSNR 35.91) and median replacement (4th, PSNR 36.39) and as well with the N2V2 with median replacement (5th, PSNR 36.37)

self-supervised denoising methods or (ii) results are poorer on the hold-out tiles of the Convallaria_1 test set because they represent parts of the field of view that were not seen during training. However, judging by Table 4, which displays a comparison of the results on the train vs. the test tiles, this seems not to be the case. A similar conclusion is suggested by Fig. 7, showing a qualitative comparison between denoised train and test tiles. Please also refer to the supplementary material section S.2 for additional qualitative results obtained for the whole slide.

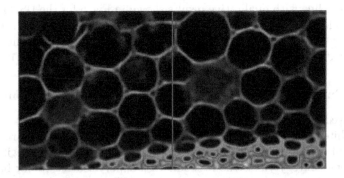

Fig. 7. Does prediction on training data impact N2V quality? Predictions on data which was used for training N2V (left) and hold-out data (right)

Table 3. Quantitative results on real Convallaria data. Results are given in dB of averaged PSNR on test data. Overall best is <u>underlined</u>. Best fully self-supervised is in **bold**

	Method	Convallaria_95	Convallaria_1
	Input	29.40	25.81
Fully self-supervised	N2V as in [6]	35.89	31.43
	N2V w/ uwoCP as in [7]	35.58	31.24
	N2V w/o res, w/uwoCP	35.76	31.27
	N2V w/o res w/mean	35.90	31.34
	N2V w/o res w/median	**36.39**	**31.77**
	N2V2 w/uwCP	36.26	31.45
	N2V2 w/uwoCP	36.31	31.51
	N2V2 w/mean	36.27	31.48
	N2V2 w/median	36.36	31.28
Self-supervised	PN2V [7]	36.47	N/A
	DivNoising [13]	<u>36.90</u>	N/A
Supervised	CARE [18]	36.71	N/A

Table 4. Quantitative results on the Convallaria_1 train and test sets. Results are given in dB of averaged PSNR

	Method	Convallaria_1 train	Convallaria_1 test
	Input	25.21	25.81
Fully self- supervised	N2V w/o res, w/uwoCP	30.52	31.27
	N2V w/o res, w/mean	31.37	31.34
	N2V w/o res, w/median	31.35	31.77
	N2V2 w/mean	31.10	31.48
	N2V2 w/median	31.06	31.28

5   Discussion and Conclusions

In this work, we introduced N2V2, an improved setup for the self-supervised denoising method N2V by Krull *et al.* [6]. N2V2 is build around two complementary contributions: (*i*) a new network architecture, and (*ii*) modified pixel value replacement strategies for blind-spot pixels.

We showed that N2V2 reduces previously observed checkerboard artifacts, which have been responsible for reduced quality of predictions from N2V. While we observed in qualitative examples that the mean replacement strategy is overall more successful than the median replacement strategy, we did not find this trend consistently in all quantitative results. Nonetheless, we have shown that only changing the architecture or only switching to one of our sampling strategies does already lead to improved results. Still, the combination of both yields best overall denoising results (measured by means of PSNR to ground clean truth images).

An interesting observation regards the failed denoising of N2V with uwCP sampling, *i.e.*, the original N2V base line, in the S&P noise setting. The network learns only to remove the Gaussian noise and the pepper noise, but recreates the salt noise pixels in the prediction (see Supplementary Section S.3). We attributed this to the strong contrast of the salt pixels with respect to the rest of the image, and the probability of $\frac{1}{w' \cdot h'}$ that the pixel remains unchanged allowing the network to learn the identity for such pixels. Another commonly used default in N2V denoising might be hurting the performance: When using a residual U-Net, pixels altered by a huge amount of noise appear at times to be strongly biased by the residual input and denoising is therefore negatively effected, as can be seen from Table 1. Without residual connections, on the other hand, this bias is removed and performance therefore improved.

Additionally, we have introduced a modified Convallaria data set (*Convallaria_1*), now featuring (*i*) a clean split between train, validation, and test sets, and offering (*ii*) a more realistic scenario to test self-supervised denoising methods. The newly proposed dataset includes only one noisy input image instead of the previously used 99 noisy acquisitions of the same field of view of the same sample. We strongly urge the community to evaluate future methods on this improved Convallaria setup.

As a final point of discussion, we note that since we decided to train all N2V and N2V2 setups much longer than in previous publications (*e.g.*, [7]), even the baselines we have simply re-run now outperform the corresponding results as reported in the respective original publications. This indicates that original training times were chosen too short and urges all future users of self-supervised denoising methods to ensure that their training runs have indeed converged before stopping them[5].

With N2V2, we presented an improved version of N2V, a self-supervised denoising method leading to denoising results of improved quality on virtually all biomedical microscopy data. At the same time, N2V2 is equally elegant, does not require more or additional training data, and is equally computationally efficient as N2V. Hence, we hope that N2V2 will mark an important update of N2V and will continue the success which N2V has celebrated in the past three years.

Acknowledgements. The authors would like to thank Laurent Gelman of the Facility for Advanced Imaging and Microscopy (FAIM) at the FMI for biomedical research in Basel to provide resources to perform the reported experiments.

---

[5] Note that this is harder to judge for self-supervised compared to supervised methods since loss plots report numbers that are computed between predicted values and *noisy* blind-spot pixel values.

References

1. Batson, J., Royer, L.: Noise2self: Blind denoising by self-supervision. In: International Conference on Machine Learning (ICML), pp. 524–533. PMLR (2019)
2. Buchholz, T.O., Jordan, M., Pigino, G., Jug, F.: Cryo-care: content-aware image restoration for cryo-transmission electron microscopy data. In: 2019 IEEE 16th International Symposium on Biomedical Imaging (ISBI 2019), pp. 502–506. IEEE (2019)
3. Buchholz, T.O., Krull, A., Shahidi, R., Pigino, G., Jékely, G., Jug, F.: Content-aware image restoration for electron microscopy. Methods Cell Biol. **152**, 277–289 (2019)
4. Buchholz, T.-O., Prakash, M., Schmidt, D., Krull, A., Jug, F.: DENOISEG: joint denoising and segmentation. In: Bartoli, A., Fusiello, A. (eds.) ECCV 2020. LNCS, vol. 12535, pp. 324–337. Springer, Cham (2020). https://doi.org/10.1007/978-3-030-66415-2_21
5. von Chamier, L., et al.: Democratising deep learning for microscopy with Zero-CostDL4Mic. Nat. Commun. **12**(1), 2276 (2021)
6. Krull, A., Buchholz, T.O., Jug, F.: Noise2void-learning denoising from single noisy images. In: IEEE/CVF Conference on Computer Vision and Pattern Recognition (CVPR), pp. 2129–2137 (2019)
7. Krull, A., Vičar, T., Prakash, M., Lalit, M., Jug, F.: Probabilistic noise2void: unsupervised content-aware denoising. Frontiers Comput. Sci. **2**, 5 (2020)
8. Laine, S., Karras, T., Lehtinen, J., Aila, T.: High-quality self-supervised deep image denoising. In: Advances in Neural Information Processing Systems (NeurIPS) 32 (2019)
9. Lehtinen, J., et al.: Noise2noise: learning image restoration without clean data. arXiv preprint arXiv:1803.04189 (2018)
10. Jiménez de la Morena, J., et al.: ScipionTomo: towards cryo-electron tomography software integration, reproducibility, and validation. J. Struct. Biol. **214**(3), 107872 (2022)
11. Ouyang, W., et al.: BioImage model zoo: a community-driven resource for accessible deep learning in BioImage analysis, June 2022
12. Prakash, M., Delbracio, M., Milanfar, P., Jug, F.: Interpretable unsupervised diversity denoising and artefact removal. In: International Conference on Learning Representations (ICLR) (2022)
13. Prakash, M., Krull, A., Jug, F.: Fully unsupervised diversity denoising with convolutional variational autoencoders. In: International Conference on Learning Representations (ICLR) (2021)
14. Prakash, M., Lalit, M., Tomancak, P., Krull, A., Jug, F.: Fully unsupervised probabilistic noise2void. In: IEEE International Symposium on Biomedical Imaging (ISBI) (2020)
15. Ronneberger, O., Fischer, P., Brox, T.: U-net: convolutional networks for biomedical image segmentation. In: International Conference on Medical Image Computing and Computer-Assisted Intervention, pp. 234–241. Springer (2015)
16. Schroeder, A.B., Dobson, E.T.A., Rueden, C.T., et al.: The ImageJ ecosystem: Open-source software for image visualization, processing, and analysis. Proteins (2021)

17. Weigert, M., Royer, L., Jug, F., Myers, G.: Isotropic reconstruction of 3D fluorescence microscopy images using convolutional neural networks. In: Descoteaux, M., Maier-Hein, L., Franz, A., Jannin, P., Collins, D.L., Duchesne, S. (eds.) MICCAI 2017. LNCS, vol. 10434, pp. 126–134. Springer, Cham (2017). https://doi.org/10.1007/978-3-319-66185-8_15
18. Weigert, M., et al.: Content-aware image restoration: pushing the limits of fluorescence microscopy. Nat. Methods **15**(12), 1090–1097 (2018)
19. Zhang, K., Zuo, W., Chen, Y., Meng, D., Zhang, L.: Beyond a gaussian denoiser: residual learning of deep CNN for image denoising. IEEE Trans. Image Process. (TIP) **26**(7), 3142–3155 (2017)
20. Zhang, R.: Making convolutional networks shift-invariant again. In: International Conference on Machine Learning (ICML), pp. 7324–7334. PMLR (2019)

W15 - Visual Object-Oriented Learning Meets Interaction: Discovery, Representations, and Applications

W15 - Visual Object-Oriented Learning Meets Interaction: Discovery, Representations, and Applications

Objects, as the most basic and composable units in visual data, exist in specific visual appearances and geometrical forms, carrying rich semantic, functional, dynamic, and relational information. One may discover objects by watching passive videos or actively interacting with the world to find them. Once detected, it is also an open research problem how to interact with the objects to extract and represent such object-oriented semantics. Furthermore, such representations need to be designed to be easily useful for various downstream perception and robotic interaction tasks. In this workshop, we invited experts from related fields but different backgrounds, e.g., vision, robotics, and cognitive psychology, and fostered the community to discuss how to define, discover, and represent objects in visual data from/for interaction for various downstream applications.

October 2022

Kaichun Mo
Yanchao Yang
Jiayuan Gu
Shubham Tulsiani
Hongjing Lu
Leonidas Guibas

Object Detection in Aerial Images with Uncertainty-Aware Graph Network

Jongha Kim[1(✉)], Jinheon Baek[2], and Sung Ju Hwang[2,3]

[1] Korea University, Seoul, South Korea
jonghakim@korea.ac.kr
[2] KAIST, Daejeon, South Korea
jinheon.baek@kaist.ac.kr, sjhwang82@kaist.ac.kr
[3] AITRICS, Seoul, South Korea

Abstract. In this work, we propose a novel uncertainty-aware object detection framework with a structured-graph, where nodes and edges are denoted by objects and their spatial-semantic similarities, respectively. Specifically, we aim to consider relationships among objects for effectively contextualizing them. To achieve this, we first detect objects and then measure their semantic and spatial distances to construct an object graph, which is then represented by a graph neural network (GNN) for refining visual CNN features for objects. However, refining CNN features and detection results of every object are inefficient and may not be necessary, as that include correct predictions with low uncertainties. Therefore, we propose to handle uncertain objects by not only transferring the representation from certain objects (sources) to uncertain objects (targets) over the directed graph, but also improving CNN features only on objects regarded as uncertain with their representational outputs from the GNN. Furthermore, we calculate a training loss by giving larger weights on uncertain objects, to concentrate on improving uncertain object predictions while maintaining high performances on certain objects. We refer to our model as Uncertainty-Aware Graph network for object DETection (UAGDet). We then experimentally validate ours on the challenging large-scale aerial image dataset, namely DOTA, that consists of lots of objects with small to large sizes in an image, on which ours improves the performance of the existing object detection network.

Keywords: Object detection · Graph neural networks · Uncertainty

1 Introduction

Given an input image, the goal of object detection is to find the bounding boxes and their corresponding classes for objects of interests in the image. To tackle this task, various object detection models based on conventional convolutional neural networks (CNNs), including Faster R-CNN [29] and YOLO [28], to recent transformer-based [32] models, such as DETR [3] and Deformable DETR [37], are proposed, showing remarkable performances. In other words, there have been considerable attentions to search

© The Author(s), under exclusive license to Springer Nature Switzerland AG 2023
L. Karlinsky et al. (Eds.): ECCV 2022 Workshops, LNCS 13804, pp. 521–536, 2023.
https://doi.org/10.1007/978-3-031-25069-9_34

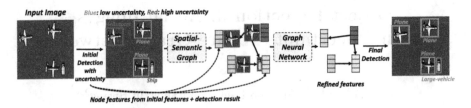

Fig. 1. Concept Figure. Given an input image, initial object detection is done, which generates uncertainties of objects as well. Blue objects with low uncertainties and red objects with high uncertainties are set as *source* and *target* nodes, respectively. A directed graph connecting source nodes to target nodes, where edges are generated based on spatial-semantic distances between nodes, is fed into GNNs to obtain refined representations of objects. (Color figure online)

for new architectures, for improving their performances on various object detection tasks, for example, finding objects in aerial images from Earth Vision [34].

However, despite their huge successes, most existing models are limited in that they do not consider interactions among objects explicitly, which are different from how humans perceive images: considering *how each object is spatially and semantically related to every other objects*, as well as capturing visual features of local regions, for contextualizing the given image. Thus, we suppose that a scheme that does model explicit relationships among objects in the image is necessary for any object detection networks. Note that it becomes more important when handling aerial images [34], which we mainly target, that consist of a large number of objects with varying scales from extremely small to large than images in the conventional datasets (i.e., COCO [26] or Pascal VOC [9]).

The most straightforward approach to consider relationships in the image is to construct an inherent hierarchy, where local-level features for identifying local parts are obtained by outputs of CNNs on sub-regions, whereas global-level features for overall understanding of an entire image are obtained by aggregating locally obtained CNN features [5,25,27]. However, this scheme is highly suboptimal, since it does not explicitly consider semantic and spatial relationships among objects with object representations, but it does implicitly model their relationships with CNN features from different sub-regions. Therefore, in this work, we aim to first detect objects in the given input image with conventional object detection networks (e.g., Faster R-CNN [29] or RoITransformer [7]), and then calculate the edges between them with their distances of semantic and spatial features. After that, over the constructed graph with objects as nodes and their distances as edges, we leverage the graph neural networks (GNNs) [17,23] to obtain graphical features of objects, which are then combined with previously obtained visual CNN features for improving final object prediction performances. Note that, recently, there exist few similar works [20,35] that consider explicit interactions between all objects by constructing either a fully connected object graph [20] or a spatial-aware sparse graph [35], whose object representations are used to improve CNN features for object detection. However, is it necessary to refine CNN features and detection results of every object?

We suppose that, if the prediction results of object detection networks are sufficiently certain, it would not be necessary to replace their features and prediction labels. Therefore, in this work, we aim to improve the prediction results only on uncertain objects that detection models are mostly confused about, by leveraging the *uncertainty* on each object when constructing object graphs and refining object features in the given image. Specifically, when we construct a graph composed by objects (nodes) and their both semantic and spatial similarities (edges), we propose to propagate the information from certain objects to uncertain ones by defining directions on edges represented by a directed graph. This scheme allows the model to improve the features of uncertain objects by contextualizing them with their semantically- and spatially-related certain ones, but also prevents the certain objects to receive noisy information from objects that are uncertain, which is also highly efficient because of ignoring a large number of edges between certain objects compared against existing models [20,35] that consider all object pairs. Moreover, when we refine CNN features by representations of the object graph from GNNs, we propose to manipulate the features for only uncertain objects, rather than changing all object features, which allows the model to focus on improving uncertain objects while maintaining the high prediction performances on certain objects. Lastly, to further make the model focus on uncertain objects, we scale the training loss based on uncertainties, where uncertain predictions get higher weights in their losses.

We refer our novel object detection framework, which explicitly models relationships between objects over the directed graph with their uncertainties, as Uncertainty-Aware Graph network for object DETection (UAGDet), which is illustrated in Fig. 1. We then experimentally validate our UAGDet on the large-scale Dataset for Object deTection in Aerial images (DOTA) [34] with two different versions: DOTA-v1.0 and DOTA-v1.5, containing large numbers of small and uncertain objects with sufficient interaction existing among them. Therefore, we suppose our UAGDet is highly beneficial in such a challenging scenario. Experimentally, we use RoITrans [7] as a backbone object detection network, and the results show that our UAGDet shows 2.32% and 3.82% performance improvements against the backbone RoITrans network on DOTA-v1.0 and -v1.5 datasets, respectively, with mean Average Precision (mAP) as an evaluation metric. Also, further analyses show that our graph construction method considering both semantic and spatial features connects related objects by edges, and also uncertainties are measured high on incorrect predictions which are corrected by refined features from graph representations.

2   Related Works

Object Detection. Object detection is the task of localizing and classifying objects of interest in an image. To mention a few, Faster R-CNN [29] and YOLO [28] are two CNN-based early models, successfully demonstrating deep learning's capability in the object detection task. Recently, following the success of transformers [32] in natural language processing, DETR [3] adopts transformers in the object detection task, which performs self-attention on CNN features from different sub-regions of an input image to consider their relationships. Furthermore, Deformable DETR [37] is proposed to attend only to the relative spatial locations when performing self-attention rather than considering all locations, which makes the transformer-based prediction architecture efficient. However, despite their huge improvements in regard to developing new object detection networks and their performances, they largely overlook the relationships among objects. In particular, CNN-based models rely on constructing hierarchical structures of local and global features for capturing feature-level relationships via multi-level feature map, which is known as a feature pyramid network (FPN) [25], but not capturing the explicit object-level relationships. Meanwhile, transformer-based models only implicitly consider relationships between candidate queries for objects, but also candidate queries often include duplicated objects or meaningless background areas which particularly belong to 'no object'. Thus, unlike these previous works, we propose to explicitly model object-level relationships, by initially detecting objects and then, over the graph structure with objects as nodes, representing them with graph neural networks (GNNs) for obtaining graphical features, which are then combined with CNN features for final object detection.

Graph Neural Networks for Object Detection. Graph Neural Networks (GNNs), which expressively represent graph structured data consisting of nodes and edges by iteratively aggregating features from target node's neighbors, have gained substantial attention in recent years, showing successes in various downstream tasks working on graphs [2,16,17,21,23,33]. As such model architectures explicitly leverage the relationships between connected instances when representing them, there have been recent attempts to use GNNs on the object detection task to capture interactions among objects [4,24,35]. For instance, GAR [24] constructs a context graph, which considers interactions between objects and scenes but also between the objects themselves by forming objects and scenes as nodes and their connections as edges, and then represents the graph with GNNs. Similarly, Relation R-CNN [4] generates semantic and spatial relation networks with objects as nodes, using pre-built co-occurrence matrix and distances between objects, respectively, for forming edges. However, both GAR [24] and Relation R-CNN [4] have an obvious limitation that the edge generation procedure is not end-to-end trainable, but based on simple heuristics using pre-calculated statistics of co-occurrences between instances. On the other hand, SGRN [35] learns a spatial-aware relation network between objects based on visual features and spatial distances of initial proposals, which is then forwarded to GNNs for obtaining representations for objects. Note that our work has key differences against such a relevant work: instead of working on all object proposals, we first largely reduce them with non-maximum suppression (NMS) [6,10], and then aim at improving only the uncertain objects that the model is largely confused about by generating semantic and spatial yet directed edges from certain objects to uncertain ones, which greatly reduces computational costs especially when dealing with lots of objects in a single image (e.g., DOTA [34]).

Uncertainty-Aware Object Detection. One of the main focuses of Bayesian deep learning in computer vision tasks is to accurately capture uncertainty of a model's prediction, which allows the deep learning models to prevent making incorrect predictions when their uncertainties are high. Following the successful applications of uncertainty on semantic segmentation and depth estimation tasks [11,22,30], some works propose to utilize uncertainty in the object detection task as well [13,19]. In particular, Xu et al. [19] propose to merge overlapping bounding boxes based on the weights of their coordinates obtained from uncertainties (i.e., the larger the uncertainties, the lower the weights) when performing soft-NMS [1], which contributes to accurately localizing bounding boxes. Also, CertainNet [13] measures uncertainties using the distances between predicted features and learnable class centroids (i.e., the larger the distances at inference time, the higher the uncertainties), which is based on the work of deterministic uncertainty quantification (DUQ) [31]. However, it does not use uncertainty for improving model's prediction, but only calculates uncertainty values of predicted objects. Note that, in contrast to those previous works [13,19] that use uncertainty either on combining overlapping bounding boxes or on identifying less certain objects, we leverage uncertainties in totally different perspectives. That is we propose to refine representations of uncertain objects by transferring knowledge from certain objects to uncertain ones over the directed graph structure, which may correct initially misclassified objects with high uncertainties.

3   Method

Faster R-CNN Baseline. We first define the notations and terms for objection detection by introducing the object detection pipeline of the Faster R-CNN model [29]. In Faster R-CNN, initial N proposals $prop_i(i = 1, ..., N)$ are generated from an input image by Region Proposal Networks (RPN). Then, we obtain N CNN feature maps $\hat{f}_i^{conv}$, each of which is associated with its proposal $prop_i$, with Region of Interest (RoI) extraction modules (e.g., RoIPool [15] or RoIAlign [18]), to make use of such features for finding bounding boxes and their classes. Specifically, extracted CNN features are fed into two separated fully connected layers, called classification head and regression head, to obtain the class prediction result p_i and the coordinates of detected bounding box $bbox_i$ for each proposal, respectively. The entire process is formally defined as follows:

$$prop_i, conv\_feature = \texttt{RPN}(image),$$
$$\hat{f}_i^{conv} = \texttt{RoIAlign}(prop_i, conv\_feature), \qquad (1)$$
$$\hat{p}_i = \texttt{cls\_head}(\hat{f}_i^{conv}), \quad \widehat{bbox_i} = \texttt{reg\_head}(\hat{f}_i^{conv}),$$

where RPN is a region proposal network for extracting object proposals, RoIAlign is a CNN feature extraction module for each proposal, cls_head is a object classification network, and reg_head is a bounding box regression network. $\hat{p}_i$ and $\widehat{bbox_i}$ denote predicted class label and bounding box coordinates, respectively.

To optimize a model during training, classification loss L_{cls} is calculated from classification results with CrossEntropy loss, and regression loss L_{reg} is calculated from regression results with SmoothL1 loss [14], defined as follows:

$$L_{cls} = \sum_{i}^{N} \texttt{CrossEntropy}(\widehat{p}_i, y_i), \; L_{reg} = \sum_{i}^{N} \texttt{SmoothL1}(\widehat{bbox}_i, bbox_i), \quad (2)$$

where y_i denotes an one-hot vector of a ground-truth label of a proposal i, and $bbox_i$ denotes a ground-truth bounding box coordinates. Then, based on two object detection losses L_{cls} and L_{reg}, overall training loss is formulated as a weighted sum of them: $L_{obj} = L_{cls} + \lambda_1 \times L_{reg}$, where λ is a scaling factor. Note that, in the test time, Non-Maximum Suppression (NMS) is applied so that duplicated bounding boxes with high overlapping areas and non-promising bounding boxes with low confidences are removed.

Overview of Our Uncertainty-Aware Graph Network. In this work, we aim to further improve the object detection network, for example, Faster R-CNN, by leveraging the object representations over the directed graph, which transfers knowledge from certain objects to uncertain objects by reflecting their semantic-spatial distances via Graph Neural Networks (GNNs). To achieve this goal, we first measure the uncertainty of every detected object with minimal costs, while following the initial object detection pipeline represented in Eq. 1, which we specify in Subsect. 3.1. Then, we construct an object graph consisting of objects as nodes and their spatial and semantic relatedness as edges, to contextualize all objects in the input image with a message passing scheme of GNNs. However, since changing features and prediction results for certain objects may not be necessary, we focus on improving uncertain objects by leveraging their contextual knowledge with semantic-spatial relationships to certain objects. We introduce this uncertainty-based graph generation procedure in Subsect. 3.2. After that, based on object representations obtained by GNNs over the object graph: node features from visual CNN features and initial object classes; edge weights from pairwise spatial distances among objects, we refine the features for uncertain objects to improve their prediction performances, which is described in Subsect. 3.3. We finally summarize our overall object detection pipeline in Subsect. 3.4, which is also illustrated in Fig. 1.

3.1 Uncertainty-Aware Initial Object Detection

Using the Faster R-CNN pipeline defined above, we can predict N classification results and their corresponding bounding boxes. Then, in this subsection, we describe how to measure the uncertainty of detected objects along with the Faster R-CNN architecture, to find out the target objects worth refining.

We first define the uncertainty of each object as ϕ_i. Then, to measure this, we use MC Dropout [12]. In particular, MC dropout can approximate an uncertainty of outputs, obtained from M times of repeated forward passes while enabling

---

Algorithm 1. Lightweight MC Dropout

Input: an image I, and the number of dropout iterations M.
Outputs: N numbers of classification results $\{p_1, ..., p_N\}$ with their corresponding bounding boxes $\{bbox_1, ..., bbox_N\}$ and uncertainties $\{\phi_1, ..., \phi_N\}$, where N denotes the number of object proposals.

1: $\{(prop_i, conv\_feature)\}_{i=1}^{N} \leftarrow \mathtt{RPN}(I)$
2: $\{(\hat{f}_i^{conv})\}_{i=1}^{N} \leftarrow \{\mathtt{RoIAlign}(prop_i, conv\_feature)\}_{i=1}^{N}$
3: **for** $j \leftarrow 1, ..., M$ **do**
4:    $\{p_i^j\}_{i=1}^{N} \leftarrow \{\mathtt{cls\_head}(\hat{f}_i^{conv})\}_{i=1}^{N}$
5:    $\{bbox_i^j\}_{i=1}^{N} \leftarrow \{\mathtt{reg\_head}(\hat{f}_i^{conv})\}_{i=1}^{N}$
6: **end for** $//p$ and $bbox$ are stacked to shape : $(N, M, ...)$
7: $p_i \leftarrow \mathtt{mean}(\{p_i^j\}_{j=1}^{M})$
8: $bbox_i \leftarrow \mathtt{mean}(\{bbox_i^j\}_{j=1}^{M})$
9: $\phi_i \leftarrow \mathtt{stdev}(\{p_i^j\}_{j=1}^{M})$
10: **return** $\{p_1, ..., p_n\}, \{bbox_1, ..., bbox_n\}, \{\phi_1, ..., \phi_n\}$

---

dropout, by calculating a variance of them. To alleviate the excessive computational cost of MC Dropout caused by repeated forward passes through every layer, we introduce a slightly modified version of the original MC dropout. This approach is simple – instead of repeating whole forward passes, we only repeat forward passes with dropout in last fully connected layers of Faster R-CNN model's classification head and regression head, that are `cls_head` and `reg_head` in Eq. 1. Therefore, we can measure each object's uncertainty with minimal additional computational costs. Also, the suggested uncertainty measure can be easily implemented to other object detection frameworks, since the only modification required is simply repeating last forward steps M times with dropout. The formal algorithm of the lightweight MC Dropout is shown in Algorithm 1.

After the object detection step above, we apply NMS with 1/10 of the usual threshold value to remove redundant and meaningless objects while maintaining source and targets objects, instead of excessively removing lots of objects except for the most certain ones. Consequently, after the initial detection and NMS phases, N_{nms} objects with their features and uncertainties remain. Note that when any vector existing at this point is used in subsequent layers, their gradients are detached so that the initial detection model is left unaffected by the following modules.

3.2   Uncertainty-Based Spatial-Semantic Graph Generation

In this subsection, we now explain how to construct an object graph based on objects' uncertainties and their spatial-semantic distances. At first, we first sort all N_{nms} objects based on their uncertainties in ascending order. Then, the top half with low uncertainties belongs to the source set $\mathcal{V}_{src}$, and the bottom half with high uncertainties belongs to the target set $\mathcal{V}_{dst}$. After categorizing all objects into source and target, we construct a directed bipartite graph $\mathcal{G}$ that consists of a set of object nodes $\mathcal{V} = \{\mathcal{V}_{src} \cup \mathcal{V}_{dst}\}$ and their edge set $\mathcal{E}$, where source node v_{src} and target node v_{dst} belong to source set $\mathcal{V}_{src}$ and target set

$\mathcal{V}_{dst}$, respectively: $v_{src} \in \mathcal{V}_{src}$ and $v_{dst} \in \mathcal{V}_{dst}$. Note that, in our directed bipartite graph, every edge $e_i \in \mathcal{E}$ connects only the source node to the relevant target node but not in the reverse direction, so that only target nodes (i.e., uncertain nodes) are affected by source nodes (i.e., certain nodes).

Then, to share the knowledge only across relevant nodes, instead of connecting every pair of nodes, we aim at connecting a *related* pair of nodes. To do so, based on the motivation that each object is likely to be affected by its nearby objects, our first criterion of relatedness is defined by a spatial distance between source and target nodes. Specifically, we calculate the Euclidean distance between two nodes' coordinates c_{src} and c_{dst} and consider it as a spatial distance measure, which is formally defined as follows: $d^{spa}_{(v_{src},v_{dst})} = ||c_{src}-c_{dst}||^2$. Here, if $d^{spa}_{(v_{src},v_{dst})}$ is smaller than the certain spatial threshold value thr_{spa}, we add an edge $e_i = (v_{src}, v_{dst})$ to the graph. Therefore, every target node has edges to its nearby source nodes.

On the other hand, we also consider the semantic distance between object nodes to further contextualize them based on their representation-level similarities. To be specific, we similarly calculate the Euclidean distance between CNN features of source and target objects, which is considered as a semantic distance and formally defined as follows: $d^{sem}_{(v_{src},v_{dst})} = ||\hat{f}^{conv}_{src} - \hat{f}^{conv}_{dst}||^2$, where $\hat{f}^{conv}_{src}$ and $\hat{f}^{conv}_{dst}$ denote CNN feature maps for source and target objects, respectively. After that, similarly in the edge addition scheme for spatial distances, if $d^{sem}_{(v_{src},v_{dst})}$ is smaller than the semantic threshold thr_{sem}, we add an edge $e_i = (v_{src}, v_{dst})$.

3.3   Feature Refinement via GNNs with Spatial-Semantic Graph

With the spatial-semantic graph $\mathcal{G} = (\mathcal{V}, \mathcal{E})$ built above, we aim at refining the target (i.e., uncertain) node representations by aggregating features from their source (i.e., certain) node neighbors via GNNs. Before doing so, we have to define node features and edge weights, which we describe in the following paragraphs.

At first, we aim to initialize the node features as features of the CNN outputs and the predicted class. To do so, we first apply 1×1 convolutions to CNN features for each object, to generate a more compact visual representation $\hat{f}^{down}_i$ by downsizing original features $\hat{f}^{conv}_i$ in Eq. 1, therefore having half of the original channel dimension. Also, to explicitly make use of the predicted class information, we regard the index of the maximum value (e.g., argmax) in the initial probability vector $\hat{p}_i$ as its class, and then embed it into the representation space. After that, by concatenating both the down-sized CNN features and the class embedding vector, we initialize the node features $\hat{f}^{node}_i$ for each node. The overall process is formally represented in Eq. 3 below.

On the other hand, edge weight used for neighborhood aggregation in GNNs is defined as a reciprocal of a pairwise spatial distance between source and target nodes divided by the diagonal length of the input image I for normalization, formulated as follows: $w_i = 1/(d^{spa}_{(v_{src},v_{dst})} \times \text{diag\_len}(I))$ for $e_i = (v_{src}, v_{dst})$, where diag_len denotes a function returning the length of a diagonal of the given image. Note that such an edge weighting scheme allows the GNN model to give larger weights on nearby objects during the aggregation of features from neighboring source nodes to the target node. For GNN, we use two Graph Convolutional Network (GCN) [23] layers, and the overall graph feature extraction procedure is as follows:

$$\hat{f}_i^{down} = \texttt{1x1Conv}(\hat{f}_i^{conv}), \quad \hat{c}_i = \texttt{embed}(\texttt{argmax}(\hat{p}_i))$$
$$\hat{f}_i^{node} = \texttt{concat}(\hat{f}_i^{down}, \hat{c}_i), \quad w_i = 1/(d_{(v_{src}, v_{dst})}^{spa} \times \texttt{diag\_len(I)}), \quad (3)$$
$$\hat{f}_i^{gnn} = \texttt{GCN}(\mathcal{E}, \hat{f}_i^{node}, w),$$

where $\texttt{1x1Conv}$ denotes the 1×1 convolutional operation, $\texttt{argmax}$ returns the index of the maximum value, $\texttt{embed}$ denotes the class embedding function, $\texttt{concat}$ denotes the concatenation operation, and we use the GCN for representing nodes in GCN. Based on Eq. 3, the resulting node features for each object $\hat{f}_i^{gnn}$ capture visual and class features based on its relationships to other certain objects as well as itself, but also capture explicit relatedness between source and target nodes via their spatial distances.

3.4 Final Detection Pipeline and Training Losses

In this subsection, we describe the final detection process, which is done with contextualized object representations from GNNs. To be specific, for each object, we first concatenate the visual features $\hat{f}_i^{conv}$ from CNN layers and the graphical features $\hat{f}_i^{gnn}$ from GNN layers in a channel-wise manner. Then, we forward the concatenated features $\hat{f}_i^* = \texttt{concat}(\hat{f}_i^{conv}, \hat{f}_i^{gnn})$ to the fully-connected layers to obtain the final class probability vector $\hat{p}_i^*$, which is similar to the class prediction process in Eq. 1 for Faster R-CNN, while we use the differently parameterized class prediction head $\texttt{cls\_head}^*$. That is defined as follows:

$$\hat{f}_i^* = \texttt{concat}(\hat{f}_i^{conv}, \hat{f}_i^{gnn}),$$
$$\hat{p}_i^* = \texttt{cls\_head}^*(\hat{f}_i^*). \quad (4)$$

Note that one of the ultimate goals of our method is to improve the performance on uncertain objects while maintaining high performance on certain objects. Therefore, we additionally regulate the $\texttt{CrossEntropy}$ loss based on the input prediction's uncertainty value: we give larger weights to objects with high uncertainties so that the model could focus on those objects. To do so, we first use the $\texttt{softmax}$ function to normalize the weight values, and then multiply the number of target objects N_{nms} to keep the scale of the loss: the sum of all w_i from $i = 1$ to N equals to the number of objects N_{nms}, as follows:

$$w_i = \texttt{softmax}(\phi_i) \times N_{nms},$$
$$= \frac{\exp(\phi_i/\tau)}{\sum_{j=1}^{N_{nms}} \exp(\phi_j/\tau)} \times N_{nms}, \quad (5)$$

where τ denotes the temperature scaling value for weights. Then, based on the loss weight for each object proposal, classification loss L_{ref} from refined features consisting of CNN and GNN representations is defined as follows:

$$L_{ref} = \sum_{i}^{N_{nms}} w_i \times \texttt{CrossEntropy}(\hat{p}_i^*, y_i). \tag{6}$$

The overall training loss is then defined by the initial and refined object detection losses, L_{obj} and L_{ref}, as follows: $L_{total} = L_{obj} + \lambda_2 \times L_{ref}$, where λ_2 is the scaling term for the last loss. Note that we refer our overall architecture calculating L_{ref} as Uncertainty-Aware Graph network for object DETection (UAGDet), which is jointly trainable with existing object detection layers by the final objective L_{total}, and easily applicable to any object detection networks. In the test time, we apply NMS to the final object detection results to remain only the most certain bounding boxes and their classes. Also, during evaluation, we replace the initial detection results for uncertain objects associated with target nodes with their final object detection results, while maintaining the initial detection results for certain objects, to prevent the possible performance drop.

4　Experiments

In this section, we validate the proposed Uncertainty-Aware Graph network for object DETection (UAGDet) for its object detection performance on the large-scale Dataset for Object deTection in Aerial images (DOTA) [34].

4.1　Datasets

DOTA [34] is widely known as an object detection dataset in aerial images. There are three versions of DOTA datasets, and we use two of them for evaluation: DOTA-v1.0, and v1.5. Regarding the dataset statistics, DOTA-v1.0 contains 2,806 large-size images from the aerial view, which are then processed to have 188,282 object instances in total within 15 object categories. The 15 object categories for classification are as follows: Plane (PL), Baseball-diamond (BD), Bridge (BR), Ground track field (GTF), Small vehicle (SV), Large vehicle (LV), Ship (SH), Tennis court (TC), Basketball court (BC), Storage tank (ST), Soccer ball field (SBF), Roundabout (RA), Harbor (HB), Swimming pool (SP), and Helicopter (HC). The number of instances per class is provided in Table 1. DOTA-v1.5 uses the same images as in DOTA-v1.0, while extremely small-sized objects and a new category named Container Crane (CC) are additionally annotated. Therefore, DOTA-v1.5 contains 403,318 object instances with 16 classes.

Note that the bounding box regression of the DOTA dataset is different from the conventional datasets, which makes object detection models more challenging. Specifically, in conventional object detection datasets, such as COCO [26] or Pascal VOC [9], the ground truth bounding box of each object is annotated as (x, y, w, h) format. However, since objects have a wide variety of orientations in aerial images, the ground truth bounding box in the DOTA [34] dataset is denoted as (x, y, w, h, θ) format, with an additional *rotation angle* parameter θ.

To reflect such an angle in the object detection architecture, RoITransformer [7] uses additional fully-connected layers to regress additional parameter θ on top of the Faster R-CNN architecture [29], which we follow in our experiments. For model tuning and evaluation on the DOTA dataset, we follow the conventional evaluation setups [7,8,34]. We first tune the hyperparameter of our UAGDet on the *val* dataset while training the model on the *train* set. After the tuning is done, we use both the *train/val* sets to train the model, and infer on the *test* set. Inference results on the *test* set is uploaded to the DOTA evaluation server [8,34], to measure the final performance. For evaluation metrics, we use the Average Precision for each class result, and also the mAP for all results [34].

4.2 Experimental Setups

Baselines and Our Model. We compare our UAGDet with Mask R-CNN [8,18], RoITransformer [7] and GFNet [36] models. Specifically, Mask R-CNN [18] is applied to DOTA [34] by viewing bounding box annotations as coarse pixel-level annotations and finding minimum bounding boxes based on pixel-level segmentation results [8]. RoITransformer (RoITrans) [7] is based on Faster R-CNN [29] with one additional FC layers in order to predict rotation parameter θ for rotated bounding boxes. GFNet [36] uses GNNs to effectively merge dense bounding boxes with a cluster structure, instead of using algorithmic methods such as NMS or Soft-NMS [1]. Our UAGDet uses GNNs for object graphs composed by objects' uncertainties to improve uncertain object predictions with graph representations, where we use RoITrans as a base object detection network.

Implementation Details. Regarding the modules for the architecture including CNN backbone, RPN, and regression/classification heads, we follow the setting of RoITrans [7]. The number of region proposals is set to 1,250. Based on the objects' uncertainties, we set top half nodes as source nodes while bottom half nodes as target nodes. If N_{nms} is larger then 100, we only consider top 100 objects, and exclude the rest of highly uncertain objects that are likely to be noise. We set the dropout ratio as 0.2, M for MC dropout as 50, λ_1, λ_2 for loss weights as 1, thr_{spa} for sparse graph generation as 50, and thr_{sem} for semantic graph generation as 10. The dimensionality of tensors in Eq. 3 is as follows: $\hat{f}_i^{conv} \in \mathbb{R}^{256 \times 7 \times 7}$, $\hat{f}_i^{down} \in \mathbb{R}^{128 \times 7 \times 7}$, and $\hat{c}_i \in \mathbb{R}^{16 \times 7 \times 7}$. For GNN layers, we use two GCN layers [23] with `LeakyReLU` activation between them. Concatenated feature f_i^{node} for each node is first fed into `1x1Conv` for reducing its dimension from $\mathbb{R}^{144 \times 7 \times 7}$ to $\mathbb{R}^{128 \times 7 \times 7}$, and then fed into `GCN`, where its dimension is further reduced by 1/2 and 1/4 in two GCN layers, respectively, i.e., $\hat{f}_i^{gnn} \in \mathbb{R}^{16 \times 7 \times 7}$. We set a temperature scale value τ in `softmax` of Eq. 5 as 0.1. The model is trained for 12 epochs with a batch size of 4, a learning rate of 0.01, and a weight decay of 10^{-4}, and optimized by a SGD. We use 4 TITAN Xp GPUs.

4.3 Quantitative Results

Table 1. Object detection results on DOTA-v1.0 and DOTA-v1.5 test datasets. The CC (Container Crane) category only exists in DOTA-v1.5, thus we remain the performance of it on DOTA-v1.0 as blank. Best results are marked in bold.

	Models	PL	BD	BR	GTF	SV	LV	SH	TC	BC	ST	SBF	RA	HB	SP	HC	CC	mAP
v1.0	# of instances	14,085	1,130	3,760	678	48,891	31,613	52,516	4,654	954	11,794	720	871	12,287	3,507	822	-	
	Mask R-CNN	88.7	74.1	50.8	63.7	73.6	74.0	83.7	89.7	78.9	80.3	47.4	65.1	64.8	66.1	59.8	-	70.7
	GFNet	90.3	83.3	51.9	77.1	65.5	58.2	61.5	90.7	82.1	86.1	65.3	63.9	70.6	69.5	57.7	-	71.6
	RoITrans	87.9	81.1	52.9	68.7	73.9	77.4	86.6	90.2	83.3	78.1	53.5	67.9	76.0	68.7	54.9	-	73.4
	UAGDet (Ours)	89.3	83.3	55.5	73.9	68.4	80.2	87.8	90.8	84.2	81.2	54.3	61.8	76.8	70.9	68.0	-	**75.1**
v1.5	# of instances	14,978	1,127	3,804	689	242,276	39,249	62,258	4,716	988	12,249	727	929	12,377	4,652	833	237	
	Mask R-CNN	76.8	73.5	50.0	57.8	51.3	71.3	79.8	90.5	74.2	66.0	46.2	70.6	63.1	64.5	57.8	9.42	62.7
	RoITrans	71.7	82.7	53.0	71.5	51.3	74.6	80.6	90.4	78.0	68.3	53.1	73.4	73.9	65.6	56.9	3.00	65.5
	UAGDet (Ours)	78.4	82.4	54.4	74.1	50.7	74.2	81.0	90.9	79.3	67.0	52.3	72.8	75.8	72.4	65.3	15.4	**68.0**

Main Results. We report the performances of baseline and our models on both DOTA-v1.0 and DOTA-v1.5 datasets in Table 1. As shown in Table 1, our UAGDet largely outperforms all baselines on both datasets in terms of mAP, obtaining 1.7 and 2.5 point performance gains on DOTA-v1.0 and v1.5 datasets, respectively against RoITrans baseline. The difference between the baseline model and ours is more dramatic in the DOTA-v1.5 dataset, which matches our assumption: our UAGDet is more beneficial in DOTA-v1.5 containing extremely small objects since they get effective representations based on interaction information. Furthermore, our model outperforms GFNet [36], which first builds instance clusters and then applies GNNs to learn comprehensive features of each cluster, since we do not impose a strong assumption as in GFNet [36] that cluster always exists in an image, and also we consider uncertainties of objects in the graph construction and representation learning scheme. Note that our UAGDet only uses 1,250 proposals instead of 2,000 for computational efficiency and no augmentation is applied, thus we believe further performance gains could be easily achieved by additional computation if needed, for real-world applications.

Ablation Study. To see where the performance gain comes from, we conduct an ablation study on the DOTA-v1.0 dataset, and report the results in Table 2. In particular, we ablate two components of our UAGDet: node and edge features in GNNs; uncertainty-scaled loss in Eq. 5. At first, we only use the CNN feature maps as node features, instead of using initially predicted class embedding for nodes and pairwise spatial distances for edges, and we observe the large performance drop of 1.2 point. Furthermore, we do not use the uncertainty-scaled loss for training but rather use the naive `CrossEntropy` loss, and we observe the performance drop of 0.4 point. Those two results suggest that, using complex features in GNN layers, as well as applying uncertainty-aware losses for focusing on uncertain objects help improve the model performances.

Table 2. Results of an ablation study on the DOTA-v1.0 dataset.

Models	mAP
UAGDet (Ours)	75.1
w/o Complex Feature	73.9
w/o Uncertainty-scaled Loss	74.7

Fig. 2. Visualization of Objects with Uncertainties. We visualize detected objects with lowest, intermediate, and highest uncertainties in left, center, and right parts. The larger the uncertainties, more inaccurate detection results are.

4.4 Analyses

Defining Target Objects by Uncertainty. Figure 2 illustrates initially detected objects along with their uncertainties. As shown in Fig. 2, detected objects with the lowest uncertainties are accurate, whereas, objects with the highest uncertainties are completely wrong. Therefore, we only target objects with medium uncertainties, which are neither perfectly detected, nor totally wrong. We can take two advantages when focusing on improving uncertain objects: computational efficiency and performance gain. To analyze this, we compare our UAGDet to SGRN [35], which builds a context graph among objects for contextualizing them. Specifically, in SGRN [35], 50 edges per node are generated for every detected object. However, if SGRN [35] is applied to DOTA [34] with 1,250 initial object proposals, resulting graph contains 50×1250 edges for considering all objects' pair-wise interactions. Compared to such an approach, our UAGDet always generates 50×50 edges as a maximum under its analytical form, and usually generates between 50 to 1,000 edges in most cases. This is because we define the direction of edges only from source to target nodes, which results in much sparser graphs having advantages in terms of time and memory efficiency. Furthermore, we observe that uncertainty-aware graphs with edges from certain nodes to uncertain nodes are more valuable than pair-wise edges between all nodes in terms of performance. In particular, we compare the performances of two models – 73.1% for all pair-wise interactions, whereas 75.1% for our uncertainty-based interactions, on which we observe that ours largely outperforms the baseline. Also, the model with pair-wise edges among all nodes (73.1%) underperforms the RoITrans [7] baseline (73.4%) that does not leverage the relational knowledge between objects. This result confirms that information propagation from uncertain objects to certain objects can harm the original detection results, which we prevent by constructing the bipartite directed graph.

Spatial and Semantic Edges in Graphs. We use both the spatial and semantic similarities to decide if two nodes are related enough to be connected in a graph structure. As depicted in Fig. 3, we observe that spatial and semantic distances play different roles when building a context graph. In particular, spatial distance measure is used to capture relationships within nearby objects. However, it

Table 3. Results of spatial-only and semantic-only graph.

Models	mAP
Ours (Both Distances)	**75.1**
Spatial Distance Only	74.6
Semantic Distance Only	73.9

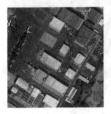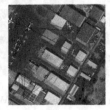

Fig. 3. Visualization of Generated Object Graphs. First and third images show edges generated by a spatial distance, while second and last images show edges generated by a semantic distance.

is suboptimal to only consider geometric distances between objects for contextualizing various objects in the image (e.g., two planes located far away are not connected to each other in Fig. 3 only with spatial distances). Thus, for this case, considering semantic distance helps a model to capture semantically meaningful relationships even though objects are located far from each other, for example, tennis courts and planes in Fig. 3. Experimentally, we observe that the model shows 74.6% and 73.9% performances in terms of mAP, when using spatial and semantic distances independently, which is shown in Table 3. Note that those two results are lower than the performance of 75.1% which considers both spatial and semantic distances for edge generation. Therefore, such results empirically confirm that both spatial and semantic edges contribute to the performance gains, which are in a complementary relationship when contextualizing objects in the image.

5   Conclusion

In this work, we proposed an Uncertainty-Aware Graph network for object DETection (UAGDet), a novel object detection framework focusing on relationships between objects by building a structured graph while considering uncertainties for representing and refining object features. In particular, we first pointed out the importance of object-level relationship which is largely overlooked in existing literature and then proposed to leverage such information by building an object graph and utilizing GNNs. Also, we considered uncertainty as a key factor to define the relationship between objects, where we transferred knowledge from certain objects to uncertain ones, refined only the uncertain object features, and regulated the loss value based on the uncertainty, for improving uncertain object predictions. Experimentally, our method obtained 2.32% and 3.82% performance improvements in the DOTA dataset.

Acknowledgement. This research was supported by the Defense Challengeable Future Technology Program of the Agency for Defense Development, Republic of Korea.

References

1. Bodla, N., Singh, B., Chellappa, R., Davis, L.S.: Soft-NMS- improving object detection with one line of code. In: ICCV, pp. 5562–5570. IEEE Computer Society (2017)
2. Brody, S., Alon, U., Yahav, E.: How attentive are graph attention networks? ArXiv abs/2105.14491 (2021)
3. Carion, N., Massa, F., Synnaeve, G., Usunier, N., Kirillov, A., Zagoruyko, S.: End-to-end object detection with transformers. In: Vedaldi, A., Bischof, H., Brox, T., Frahm, J.-M. (eds.) ECCV 2020. LNCS, vol. 12346, pp. 213–229. Springer, Cham (2020). https://doi.org/10.1007/978-3-030-58452-8_13
4. Chen, S., Li, Z., Tang, Z.: Relation R-CNN: a graph based relation-aware network for object detection. IEEE Signal Process. Lett. **27**, 1680–1684 (2020)
5. Dai, L., et al.: TARDet: two-stage anchor-free rotating object detector in aerial images. In: Proceedings of the IEEE/CVF Conference on Computer Vision and Pattern Recognition, pp. 4267–4275 (2022)
6. Dalal, N., Triggs, B.: Histograms of oriented gradients for human detection. In: 2005 IEEE Computer Society Conference on Computer Vision and Pattern Recognition (CVPR2005), vol. 1, pp. 886–893. IEEE (2005)
7. Ding, J., Xue, N., Long, Y., Xia, G.S., Lu, Q.: Learning ROI transformer for detecting oriented objects in aerial images. ArXiv abs/1812.00155 (2018)
8. Ding, J., et al.: Object detection in aerial images: a large-scale benchmark and challenges. CoRR abs/2102.12219 (2021), https://arxiv.org/abs/2102.12219
9. Everingham, M., Van Gool, L., Williams, C.K.I., Winn, J., Zisserman, A.: The pascal visual object classes challenge 2012 (VOC2012) results. http://www.pascal-network.org/challenges/VOC/voc2012/workshop/index.html
10. Felzenszwalb, P.F., Girshick, R.B., McAllester, D., Ramanan, D.: Object detection with discriminatively trained part-based models. IEEE Trans. Pattern Anal. Mach. Intell. **32**(9), 1627–1645 (2010)
11. Gal, Y.: Uncertainty in Deep Learning, Ph. D. thesis, University of Cambridge (2016)
12. Gal, Y., Ghahramani, Z.: Dropout as a Bayesian approximation: representing model uncertainty in deep learning. In: ICML. JMLR Workshop and Conference Proceedings, vol. 48, pp. 1050–1059. JMLR.org (2016)
13. Gasperini, S., et al.: CertainNet: sampling-free uncertainty estimation for object detection. IEEE Robot. Autom. Lett. **7**, 698–705 (2022)
14. Girshick, R.B.: Fast R-CNN. In: 2015 IEEE International Conference on Computer Vision (ICCV), pp. 1440–1448 (2015)
15. Girshick, R.B., Donahue, J., Darrell, T., Malik, J.: Rich feature hierarchies for accurate object detection and semantic segmentation. In: 2014 IEEE Conference on Computer Vision and Pattern Recognition, pp. 580–587 (2014)
16. Hamilton, W., Ying, Z., Leskovec, J.: Inductive representation learning on large graphs. In: Advances in neural information processing systems 30 (2017)
17. Hamilton, W.L.: Graph representation learning. Synth. Lect. Artif. Intell. Mach. Learn. **14**(3), 1–159 (2020)
18. He, K., Gkioxari, G., Dollár, P., Girshick, R.B.: Mask R-CNN. In: 2017 IEEE International Conference on Computer Vision (ICCV), pp. 2980–2988 (2017)
19. He, Y., Zhu, C., Wang, J., Savvides, M., Zhang, X.: Bounding box regression with uncertainty for accurate object detection. In: CVPR, pp. 2888–2897. Computer Vision Foundation/IEEE (2019)

20. Hu, H., Gu, J., Zhang, Z., Dai, J., Wei, Y.: Relation networks for object detection. In: Proceedings of the IEEE Conference on Computer Vision and Pattern Recognition, pp. 3588–3597 (2018)
21. Jo, J., Baek, J., Lee, S., Kim, D., Kang, M., Hwang, S.J.: Edge representation learning with hypergraphs. Adv. Neural. Inf. Process. Syst. **34**, 7534–7546 (2021)
22. Kendall, A., Gal, Y.: What uncertainties do we need in Bayesian deep learning for computer vision? In: NIPS, pp. 5574–5584 (2017)
23. Kipf, T.N., Welling, M.: Semi-supervised classification with graph convolutional networks. In: ICLR (Poster). OpenReview.net (2017)
24. Li, Z., Du, X., Cao, Y.: GAR: graph assisted reasoning for object detection. In: WACV, pp. 1284–1293. IEEE (2020)
25. Lin, T., Dollár, P., Girshick, R.B., He, K., Hariharan, B., Belongie, S.J.: Feature pyramid networks for object detection. In: CVPR, pp. 936–944. IEEE Computer Society (2017)
26. Lin, T.-Y., et al.: Microsoft COCO: common objects in context. In: Fleet, D., Pajdla, T., Schiele, B., Tuytelaars, T. (eds.) ECCV 2014. LNCS, vol. 8693, pp. 740–755. Springer, Cham (2014). https://doi.org/10.1007/978-3-319-10602-1_48
27. Liu, W., et al.: SSD: single shot multibox detector. In: Leibe, B., Matas, J., Sebe, N., Welling, M. (eds.) ECCV 2016. LNCS, vol. 9905, pp. 21–37. Springer, Cham (2016). https://doi.org/10.1007/978-3-319-46448-0_2
28. Redmon, J., Divvala, S.K., Girshick, R.B., Farhadi, A.: You only look once: Unified, real-time object detection. In: CVPR, pp. 779–788. IEEE Computer Society (2016)
29. Ren, S., He, K., Girshick, R.B., Sun, J.: Faster R-CNN: towards real-time object detection with region proposal networks. IEEE Trans. Pattern Anal. Mach. Intell. **39**, 1137–1149 (2015)
30. Shen, Y., Zhang, Z., Sabuncu, M.R., Sun, L.: Real-time uncertainty estimation in computer vision via uncertainty-aware distribution distillation. In: IEEE Winter Conference on Applications of Computer Vision, WACV 2021, pp. 707–716. Waikoloa, HI, USA, 3–8 January 2021. IEEE (2021). https://doi.org/10.1109/WACV48630.2021.00075
31. Van Amersfoort, J., Smith, L., Teh, Y.W., Gal, Y.: Uncertainty estimation using a single deep deterministic neural network. In: International Conference on Machine Learning, pp. 9690–9700. PMLR (2020)
32. Vaswani, A., et al.: Attention is all you need. ArXiv abs/1706.03762 (2017)
33. Velickovic, P., Cucurull, G., Casanova, A., Romero, A., Liò, P., Bengio, Y.: Graph attention networks. In: ICLR (Poster) (2018)
34. Xia, G.S., et al.: Dota: a large-scale dataset for object detection in aerial images. In: The IEEE Conference on Computer Vision and Pattern Recognition (CVPR) (2018)
35. Xu, H., Jiang, C., Liang, X., Li, Z.: Spatial-aware graph relation network for large-scale object detection. In: CVPR, pp. 9298–9307. Computer Vision Foundation/IEEE (2019)
36. Zhang, S.X., Zhu, X., Hou, J.B., Yin, X.C.: Graph fusion network for multi-oriented object detection. ArXiv abs/2205.03562 (2022)
37. Zhu, X., Su, W., Lu, L., Li, B., Wang, X., Dai, J.: Deformable DETR: deformable transformers for end-to-end object detection. In: 9th International Conference on Learning Representations, ICLR 2021, Virtual Event, Austria, 3–7 May 2021 (2021)

W16 - AI for Creative Video Editing and Understanding

W16 - AI for Creative Video Editing and Understanding

The 2nd installment of the AI for Creative Video Editing and Understanding (CVEU) workshop followed up its previous success at ICCV 2021. The workshop brings together researchers, artists, and entrepreneurs working on computer vision, human-computer interaction, computer graphics, and cognitive research. It aims to bring awareness of recent advances in machine learning technologies that can enable assisted creative-video creation and analysis. This year's workshop included keynote and roundtable sessions and gave the community opportunities to share their work via oral and poster presentations. Topics of interest included but were not limited to Computational Video Editing, Computational Videography, Virtual Cinematography, Story Understanding, Multimodality, Video Description, Search, Retrieval, Shortening and Storyboarding, Diffusion Models, and Neural Rendering.

October 2022

Fabian Caba
Anyi Rao
Alejandro Pardo
Linning Xu
Yu Xiong
Victor A. Escorcia
Ali Thabet
Dong Liu
Dahua Lin
Bernard Ghanem

STC: Spatio-Temporal Contrastive Learning for Video Instance Segmentation

Zhengkai Jiang[1], Zhangxuan Gu[2], Jinlong Peng[1], Hang Zhou[3], Liang Liu[1], Yabiao Wang[1(✉)], Ying Tai[1], Chengjie Wang[1], and Liqing Zhang[2]

[1] Tencent Youtu Lab, Shanghai, China
{zhengkjiang,caseywang}@tencent.com
[2] Shanghai Jiao Tong University, Shanghai, China
[3] The Chinese University of Hong Kong, Sha Tin, Hong Kong

Abstract. Video Instance Segmentation (VIS) is a task that simultaneously requires classification, segmentation, and instance association in a video. Recent VIS approaches rely on sophisticated pipelines to achieve this goal, including RoI-related operations or 3D convolutions. In contrast, we present a simple and efficient single-stage VIS framework based on the instance segmentation method CondInst by adding an extra tracking head. To improve instance association accuracy, a novel bi-directional spatio-temporal contrastive learning strategy for tracking embedding across frames is proposed. Moreover, an instance-wise temporal consistency scheme is utilized to produce temporally coherent results. Experiments conducted on the YouTube-VIS-2019, YouTube-VIS-2021, and OVIS-2021 datasets validate the effectiveness and efficiency of the proposed method. We hope the proposed framework can serve as a simple and strong baseline for other instance-level video association tasks.

Keywords: Video instance segmentation · Spatio-temporal contrastive learning · Temporal consistency

1 Introduction

While significant progress has been made in instance segmentation [4,14,24,28, 30,37,41–43,47] with the development of deep neural networks, less attention has been paid to its challenging variant in the video domain. The video instance segmentation (VIS) [17,44,50,52] task requires not only classifying and segmenting instances but also capturing the instance associations across frames. Such technology can benefit a great variety of scenarios, *e.g.,* video editing, video surveillance, autonomous driving, and augmented reality. As a result, it is in great need of accurate, robust, and fast video instance segmentation approach in practice (Fig. 1).

Previous researchers have developed sophisticated pipelines for tackling this problem [1,2,5,12,39,44,50,51]. Generally speaking, previous studies can be

---

Z. Jiang and Z. Gu—Equal contributions. This work was done while Zhangxuan Gu was interning at Tencent Youtu Lab.

© The Author(s), under exclusive license to Springer Nature Switzerland AG 2023
L. Karlinsky et al. (Eds.): ECCV 2022 Workshops, LNCS 13804, pp. 539–556, 2023.
https://doi.org/10.1007/978-3-031-25069-9_35

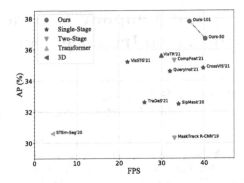

Fig. 1. Speed-Accuracy trade-off curve on the YouTube-VIS-2019 validation set. The baseline results are compared with the same ResNet-50 backbone for fair comparison. We achieve best tradeoff between speed and accuracy. In particular, STC exceeds recent CrossVIS [11] 1.9% mAP with similar running speed.

divided into the categories of two-stage [2,12,39,50,51], feature-aggregation [2, 27] inspired from video object detection domain [19,20,55], 3D convolution-based [1], transformer-based [17,44], and single-stage [5,52] methods. Two-stage methods, *e.g.*, MaskTrack R-CNN [50] and CompFeat [12], usually rely on the RoIAlign operation to crop the feature and obtain the representation of an instance for further binary mask prediction. Such the RoIAlign operation would lead to great computational inefficiency. 3D convolution-based STEm-Seg [1] holds huge complexity and could not achieve good performance. Transformer-based VisTR [17,44] could not handle long videos due to largely increasing memory usage and needs a much longer training time for convergence. Feature-aggregation methods [23,40] enhance features through pixel-wise or instance-wise aggregation from adjacent frames similarly to other video tasks, like video object detection [19,20,45]. Although some attempts [5,46,52] have been made to tackle VIS in a simple single-stage manner, their performances are still not satisfying.

The key difference between video and image instance segmentation lies in the need of capturing robust and accurate instance association across frames. However, most previous works such as MaskTrack R-CNN [50], and CMaskTrack R-CNN [33] formulate instance association as a multi-label classification problem, focusing only on the intrinsic relationship within instances while ignoring the extrinsic constraint between different ones. Thus different instances with similar distributions may be wrongly associated by using previous tracking embeddings only through such multi-label classification loss constraint.

Alternatively, we propose an *efficient* single-stage fully convolutional network for video instance segmentation task, considering that single-stage instance segmentation is simpler and faster. Based on the recent instance segmentation method CondInst [37], an extra tracking head is added to simultaneously learn instance-wise tracking embeddings for instance association besides original classification head, box head, and mask head by dynamic filter. To improve instance association accuracy between adjacent frames, a spatio-temporal contrastive learning strategy is utilized to exploit relations between different instances. Specifically, for a tracking embedding query, we densely sample hundreds of negative and positive embeddings from reference frames based on the label assignment results, acting as a contrastive manner to jointly pull closer to the same instances and push away from different instances. Different from previous metric learning based instance association methods *i.e.*, *Triplet Loss*, the proposed contrastive strategy enables efficient many-to-many relations learning across frames. We believe this contrast mechanism enhances the instance similarity learning, which provides more substantial supervision than using only the labels. Moreover, this contrastive learning scheme is applied in a bi-directional way to better leverage the temporal information from both forward and backward views. At last, we further propose a temporal consistency scheme for instance encoding, which contributes to both the accuracy and smoothness of the video instance segmentation task.

In summary, our main contributions are:

- We propose a single-stage fully convolutional network for video instance segmentation task with an extra tracking head to simultaneously generate instance-specific tracking embeddings for instance association.
- To achieve accurate and robust instance association, we propose a bi-directional spatio-temporal contrastive learning strategy that aims to obtain representative and discriminative tracking embeddings. In addition, we present a novel temporal consistency scheme for instances encoding to achieve temporally coherent results.
- Comprehensive experiments are conducted on the YouTube-VIS-2019, YouTube-VIS-2021, and OVIS-2021 benchmark. Without bells and whistles, we achieve 36.7% AP and 35.5% AP with ResNet-50 backbone on YouTube-VIS-2019 and YouTube-VIS-2021 datasets, which is the best performance among all listed single-model methods with high efficiency. We also achieve best performance on recent proposed OVIS-2021 dataset. In particular, compared to the first VIS method named MaskTrack R-CNN [50], our proposed method (STC) achieves 36.7% AP on YouTube-VIS-2019, outperforming it by **6.4%** AP with the advantage of being much simpler and faster. Compared with recent method CrossVIS [52], STC outperforms it by 1.9% AP with a slightly faster speed.

2   Related Works

2.1   Instance Segmentation

Instance segmentation aims to represent objects at a pixel level, which is a finer-grained representation compared with object detection. There are mainly two kinds of instance segmentation methods, *i.e.*, two-stage [14,16,28], and single-stage [4,6,37,41,42]. Two-stage methods first detect objects, then crop their region features to further classify each pixel into the foreground or background,

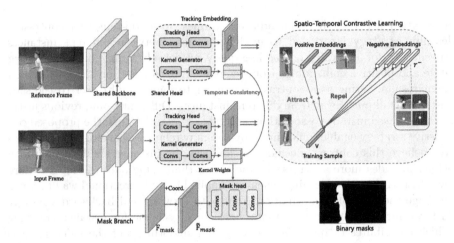

Fig. 2. The overview of our proposed framework. The framework contains the following components: a shared CNN backbone for encoding frames to feature maps, kernel generators with mask heads for instance segmentation, a mask branch to combine multi-scale FPN features, and a shared tracking head with a bi-directional spatio-temporal contrastive learning strategy (the bi-directional learning scheme is omitted here for simplicity) for instance association. A temporal consistency constraint is applied to the kernel weights, as the blue line shows. Best viewed in color. (Color figure online)

while the framework of single-stage instance segmentation is much simpler. For example, YOLACT [4] is proposed to generate a set of prototype masks and predict per-instance mask coefficients. The instance masks are then produced by linearly combining the prototypes with the mask coefficients. SOLO [41,42] reformulates the instance segmentation as two simultaneous category-aware prediction problems, i.e., location prediction, and mask prediction, respectively. Inspired by dynamic filter network [18], CondInst [37] proposes to dynamically predict instance-aware filters for mask generation. SOLOv2 [42] further incorporates dynamic filter scheme to dynamically segments each instance in the image with a novel matrix non-maximum suppression (NMS) technique. Compared to the image instance segmentation, video instance segmentation aims not only to segment object instances in individual frames but also to associate the predicted instances across frames.

2.2 Video Instance Segmentation

Video instance segmentation [50] aims to simultaneously classify, segment, and track instances of the videos. Various complicated pipelines are designed by state-of-the-art methods to solve it. To better introduce the related methods, we separate them into the following groups. (1) The two-stage method Mask-Track R-CNN [50], as the pioneering work for VIS, extends image instance segmentation method Mask R-CNN [14] to video domain by introducing an extra tracking branch for instance association. Another method in the two-stage group is MaskProp [2], which first uses Hybrid Task Cascade (HTC) [7] to generate the predicted masks and propagates them temporally to the other frames in a video. Recently, CompFeat [12] proposed a feature aggregation approach based on MaskTrack R-CNN, which refines features by aggregating multiple adjacent frames features. (2) Relying on 3D convolutions, STEm-Seg [1] models a video clip as a single 3D spatial-temporal volume and separates object instances by clustering. (3) Based on feature-aggregation, STMask [23] proposes a simple spatial feature calibration to detect and segment object masks frame-by-frame, and further introduces a temporal fusion module to track instances across frames. (4) More recently, a transformer-based method VisTR [44] is proposed to reformulate VIS as a parallel sequence decoding problem. (5) There also exist some single-stage VIS methods, *e.g.*, SipMask [5], and TraDeS [46]. SipMask [5] proposes a spatial preservation module to generate spatial coefficients for the mask predictions while recently proposed TraDeS [46] presents a joint detection and tracking model by propagating the previous instance features with the predicted tracking offset. CrossVIS [52] proposes cross-frame instance-wise consistency loss for video instance segmentation. Although current methods have made good progress, their complicated pipelines or unsatisfying performance prohibit practical application. In contrast, the proposed framework acts in a fully convolutional manner with decent performance and efficiency.

2.3 Contrastive Learning

Contrastive learning has lead to considerable progress in many real-world applications [8,9,13,21,31,36,48]. For example, MOCO [13] builds image-level large dictionaries for unsupervised representation learning using contrastive loss. Sim-CLR [8] utilizes the elaborate data augmentation strategies and a large batch, which outperforms MOCO by a large margin on self-supervised learning ImageNet [34] classification task. Different from the above methods, which focus on image-level contrastive learning for unsupervised representation learning, we use modified multiple-positives contrastive learning to learn instance-level tracking embeddings accurately for video instance segmentation tasks.

3   Method

In this section, we first briefly review the instance segmentation method CondInst [37] for mask generation of still-image. Then, we introduce the proposed whole framework for the video instance segmentation task. Next, we present a novel spatio-temporal contrastive learning strategy for tracking embeddings to achieve accurate and robust instance association. In addition, we further propose a bi-directional spatio-temporal contrastive learning strategy. At last, the temporal consistency scheme aiming to achieve temporally coherent results is introduced in detail.

3.1 Mask Generation for Still-Image

For still-image instance segmentation, we use the dynamic conditional convolutions method CondInst [18,37]. Specifically, instance mask at location (x, y) can be generated by convolving an instance-agnostic feature map $\tilde{\mathbf{F}}_{mask}^{x,y}$ from mask branch and instance-specific dynamic filter $\boldsymbol{\theta}_{x,y}$, which is calculated as follows:

$$\mathbf{m}_{x,y} = \mathbf{MaskHead}(\tilde{\mathbf{F}}_{mask}^{x,y}; \boldsymbol{\theta}_{x,y}),\qquad(1)$$

where $\tilde{\mathbf{F}}_{mask}^{x,y}$ is the combination of multi-scale fused feature map $\mathbf{F}_{mask}$ from FPN features $\{P_3, P_4, P_5\}$ and relative coordinates $\mathbf{O}_{x,y}$. The **MaskHead** consists of three 1×1 conv-layers with dynamic filter $\boldsymbol{\theta}_{x,y}$ at location (x, y) as convolution kernels. $\mathbf{m}_{x,y} \in \mathbb{R}^{H \times W}$ is the predicted binary mask at location (x, y) as shown in Fig. 2.

3.2 Proposed Framework for VIS

The overall framework of the proposed method is illustrated in Fig. 2. Based on the instance segmentation method CondInst [37], we add a tracking head for instances association. The whole architecture mainly contains following components: (1) A shared CNN backbone (*e.g.* ResNet-50 [15]) is utilized to extract compact visual feature representations with FPN [25]. (2) Multiple heads including a classification head, a box regression head, a centerness head, a kernel generator head, and a mask head as same as CondInst [37]. Since the architectures of the above classification, box regression, and centerness heads are not our main concerns, we omit them here (please refer to [38] for the details). (3) A tracking head where spatio-temporal contrastive learning strategy is proposed to associate instances across frames with comprehensive relational cues in the tracking embeddings. (4) Temporal consistency scheme on instance-wise kernel weights across frames aims to generate temporally coherent results.

3.3 Spatio-Temporal Contrastive Learning

To associate instances from different frames, an extra lightweight tracking head is added to obtain the tracking embeddings [5,12,50] in parallel with the original kernel generator head. The tracking head consists of several convolutional layers which take multi-scale FPN features $\{P_3, P_4, P_5\}$ as input. And the outputs are fused to obtain the feature map of tracking embedding. As shown in Fig. 2, given an input frame I for training, we randomly select a reference frame I_{ref} from its temporal neighborhood. A location is defined as a positive sample if it falls into any ground-truth box and the class label c of the location is the class label of the ground-truth box. If a location falls into multiple bounding boxes, it is considered as the positive sample of the bounding box with minimal area [38]. Thus, two locations formulate a positive pair if they are associated with the same instance across two frames and a negative pair otherwise.

During training, for a given frame, the model first predicts the object detection results. Then, the tracking embedding of each instance can be extracted from the tracking feature map by the center of the predicted bounding box. For a training sample with extracted tracking embedding q, we can obtain positive embeddings $\mathbf{k}^+$ and negative embeddings $\mathbf{k}^-$ according to label assignment results at reference frame. Note that traditional unsupervised representation learning [8,13] with contrastive learning only uses one positive sample and multiple negative samples as follows:

$$\mathcal{L}_q = -\log \frac{\exp(\mathbf{q} \cdot \mathbf{k}^+)}{\exp(\mathbf{q} \cdot \mathbf{k}^+) + \sum_{\mathbf{k}^-} \exp(\mathbf{q} \cdot \mathbf{k}^-)}. \tag{2}$$

Since there are many positive embeddings at reference frame for each training sample, instead of randomly selecting one positive embedding at reference frames, we optimize the objective loss with multiple positive embeddings and multiple negative embeddings as:

$$
\begin{aligned}
\mathcal{L}_{contra} &= -\sum_{\mathbf{k}^+} \log \frac{\exp(\mathbf{q} \cdot \mathbf{k}^+)}{\exp(\mathbf{q} \cdot \mathbf{k}^+) + \sum_{\mathbf{k}^-} \exp(\mathbf{q} \cdot \mathbf{k}^-)} \\
&= \sum_{\mathbf{k}^+} \log[1 + \sum_{\mathbf{k}^-} \exp(\mathbf{q} \cdot \mathbf{k}^- - \mathbf{q} \cdot \mathbf{k}^+)].
\end{aligned}
\tag{3}
$$

Suppose there are N_{pos} training samples at input frame, the objective track loss with multiple samples is:

$$\mathcal{L}_{track} = \frac{1}{N_{pos}} \sum_{i=1}^{N_{pos}} \mathcal{L}_{contra}^i. \tag{4}$$

Bi-directional Spatio-Temporal Learning. Many video-related tasks have shown the effectiveness of bi-directional modeling [35,54]. To fully exploit such temporal context information, we further propose a bi-directional spatio-temporal learning scheme to learn instance-wise tracking embeddings better. Note that we only utilize this scheme in the training stage, and thus it does not affect the inference speed. Similar to Eq. 4, the objective function of bi-directional spatio-temporal contrastive learning can be denoted as $\hat{\mathcal{L}}_{track}$ by reversing input frame and reference frame. Thus, the final bi-directional spatio-temporal contrastive loss is:

$$\mathcal{L}_{bi-track} = \frac{1}{2}(\mathcal{L}_{track} + \hat{\mathcal{L}}_{track}). \tag{5}$$

3.4   Temporal Consistency

Compared with image data, the coherent property between frames is also crucial to video-related researches. Thus, we add a temporal consistency constraint on the kernel weights, marked as the blue line in Fig. 2, to capture such prior during training so that the predicted masks will be more accurate and robust across frames. Given an instance at location (x, y) appearing at both input and reference frames, we use (x, y) and $(\hat{x}, \hat{y})$ to denote its positive candidate positions from two frames, respectively. Formally, the temporal consistency constraint during training can be formulated as an L2-loss function:

$$\mathcal{L}_{consistency} = ||\boldsymbol{\theta}_{x,y} - \boldsymbol{\theta}_{\hat{x},\hat{y}}^{ref}||^2 + ||\boldsymbol{m}_{x,y} - \boldsymbol{m}_{\hat{x},\hat{y}}^{ref}||^2, \tag{6}$$

where $\boldsymbol{\theta}_{\hat{x},\hat{y}}^{ref}$ is the dynamic filter at reference frame, $\boldsymbol{m}_{\hat{x},\hat{y}}^{ref}$ is the predicted instance mask by reference dynamic filter. With such a simple constraint, our kernel generator can obtain accurate, robust and coherent mask predictions across frames.

3.5   Training and Inference

Training Scheme. Formally, the overall loss function of our model can be formulated as follows:

$$\mathcal{L}_{overall} = \mathcal{L}_{condinst} + \lambda_b \mathcal{L}_{bi-track} + \lambda_c \mathcal{L}_{consistency}, \tag{7}$$

where $\mathcal{L}_{condinst}$ denotes the original loss of CondInst [37] for instance segmentation. We refer readers to [37] for the details of $\mathcal{L}_{condinst}$. λ_b and λ_c are the hyper-parameters.

Inference on Frame. For each frame, we forward it through the model to get the outputs, including classification confidence, centerness scores, box predictions, kernel weights, and tracking embeddings. Then we obtain the box detections by selecting the positive positions whose classification confidence is larger than a threshold (set as 0.03), similar to FCOS [38]. After that, following previous work MaskTrack R-CNN [50], the NMS [4] with the threshold being 0.5 is used to remove duplicated detections. In this step, these boxes are also associated with the kernel weights and tracking embeddings. Supposing that there remain T boxes after the NMS, thus we have T groups of the generated kernel weights. Then T groups of kernel weights are used to produce T mask heads. These instance-specific mask heads are applied to the positions encoded mask feature to predict the instance masks following [37]. T is 10 in default following previous work MaskTrack R-CNN.

Inference on Video. Given a testing video, we first construct an empty memory bank for the predicted instance embeddings. Then our model processes each frame sequentially in an online scheme. Our network generates a set of predicted instance embeddings at each frame. The association with identified instances from previous frames relies on the cues of embedding similarity, box overlap, and category label similar to the MaskTrack R-CNN [50]. All predicted instance embeddings of the first frame are directly regarded as identified instances and saved into the memory bank. After processing all frames, our method produces a set of instances sequence. The majority votes are utilized to decide the unique category label of each instance sequence.

Table 1. Comparisons with some state-of-the-art approaches on **YouTube-VIS-2019 val** set. ✓ indicates using extra data augmentation (*e.g.*, random crop, higher resolution input, multi-scale training) [2] or additional data [1,2,12,44]. [†] indicates the method that reaches higher performance by stacking multiple networks, and we regard it an unfair competitor in general setting. Note that STMask [23] uses deformable convolution network (DCN) [10] as the backbone, which is still inferior to our method at both accuracy and speed, demonstrating the superiority of our proposed framework. [††] means transformer on top of ResNet-50 or ResNet-101.

Method	Publication	Augmentations	Backbone	FPS	AP	AP_{50}	AP_{75}	AR_1	AR_{10}
MaskTrack R-CNN [50]	ICCV'19	✗	ResNet-50	33	30.3	51.1	32.6	31.0	35.5
SipMask [5]	ECCV'20	✗	ResNet-50	34	32.5	53.0	33.3	33.5	38.9
STEm-Seg [1]	ECCV'20	✗	ResNet-50	4.4	30.6	50.7	33.5	31.6	37.1
CompFeat [12]	AAAI'21	✗	ResNet-50	<33	35.3	56.0	38.6	33.1	40.3
TraDeS [46]	CVPR'21	✗	ResNet-50	26	32.6	52.6	32.8	29.1	36.6
QueryInst [11]	ICCV'21	✗	ResNet-50	32	34.6	55.8	36.5	35.4	42.4
CrossVIS [52]	ICCV'21	✗	ResNet-50	39.8	34.8	54.6	37.9	34.0	39.0
VisSTG [40]	ICCV'21	✗	ResNet-50	22	35.2	55.7	38.0	33.6	38.5
PCAN [22]	NeurIPS'21	✗	ResNet-50	–	36.1	54.9	**39.4**	36.3	41.6
Ours (STC)	-	✗	ResNet-50	**40.3**	**36.7**	**57.2**	38.6	**36.9**	**44.5**
STMask [23]	CVPR'21	DCN backbone [10]	ResNet-50	29	33.5	52.1	36.9	31.1	39.2
SG-Net [27]	CVPR'21	multi-scale training	ResNet-50	23	34.8	56.1	36.8	35.8	40.8
VisTR [44]	CVPR'21	random-crop training	ResNet-50	30	35.6	56.8	37.0	35.2	40.2
QueryInst [11]	ICCV'21	multi-scale training	ResNet-50	32	36.2	56.7	39.7	36.1	42.9
CrossVIS [52]	ICCV'21	multi-scale training	ResNet-50	39.8	36.3	56.8	38.9	35.6	40.7
VisSTG [40]	ICCV'21	multi-scale training	ResNet-50	22	36.5	58.6	39.0	35.5	40.8
Ours (STC)	-	multi-scale training	ResNet-50	**40.3**	**37.6**	**58.9**	**39.7**	**38.2**	**46.2**
MaskTrack R-CNN [50]	ICCV'19	✗	ResNet-101	33	30.3	51.1	32.6	31.0	35.5
SRNet [53]	ACMMM'21	✗	ResNet-101	35	32.3	50.2	34.8	32.3	40.1
STEm-Seg [1]	ECCV'20	✗	ResNet-101	2.1	34.6	55.8	37.9	34.4	41.6
PCAN [22]	NeurIPS'21	✗	ResNet-101	–	37.6	57.2	**41.3**	37.2	43.9
Ours (STC)	-	✗	ResNet-101	**36.6**	**37.8**	**58.5**	40.6	**38.5**	**46.3**
SipMask [5]	ECCV'20	multi-scale training	ResNet-101	24	35.8	56.0	39.0	35.4	42.4
STMask [23]	CVPR'21	DCN backbone [10]	ResNet-101	23	36.8	56.8	38.0	34.8	41.8
SG-Net [27]	CVPR'21	multi-scale training	ResNet-101	20	36.3	57.1	39.6	35.9	43.0
VisTR [44]	CVPR'21	random-crop training	ResNet-101	28	38.6	61.3	42.3	37.6	44.2
Ours (STC)	-	multi-scale training	ResNet-101	**36.6**	**39.2**	**61.5**	**42.4**	**39.7**	**47.3**

4   Experiments

4.1   Dataset

To verify the effectiveness of our approach, we evaluate it on recent three video instance segmentation benchmarks, YouTube-VIS-2019 [50], YouTube-VIS-2021 [49] and OVIS-2021 [33] datasets. Following previous works [2,50,52], we evaluate our method on the validation sets of YouTube-VIS-2019, YouTube-VIS-2021 and OVIS-2021.

YouTube-VIS-2019 dataset contains 40 class annotations, including many common objects. The official dataset consists of three subsets: 2238 training videos, 302 validation videos, and 343 test videos.

YouTube-VIS-2021 dataset is an improved version of YouTube-VIS-2019 containing 40 class annotations. It collects more videos and high-quality annotations. This dataset also consists of three subsets: 2985 training videos, 421 validation videos, and 453 test videos.

OVIS-2021 is a new large scale benchmark dataset for video instance segmentation task with 25 common semantic categories. It is designed with object occlusions in videos, which could reveal the complexity of real-world scenes. It consists of 607 training videos, 140 validation videos, and 154 testing videos as the official split.

4.2 Metrics

The evaluation metrics are average precision (AP) and average recall (AR), with the video Intersection over Union (IoU) of the mask sequences as the threshold [50]. Specifically, for a predicted mask $\hat{m}^i$ and a ground-truth mask m^j, we first extend them to the whole video with length T by padding empty mask. Then,

$$\text{IoU}(i, j) = \frac{\sum_{t=1}^{T} \hat{m}_t^i \cap m_t^j}{\sum_{t=1}^{T} \hat{m}_t^i \cup m_t^j}. \tag{8}$$

According to the definition, if the model detects object masks successfully but fails to associate the objects across frames, it still gets a low IoU. Thus, accurate and robust instance association across frames is very crucial for achieving high performance.

4.3 Implementation Details

Model Settings. In our experiments, we choose the ResNet-50 [15] and ResNet-101 with FPN [25] as the backbone in the proposed method. Our model is pretrained on COCO train2017 [26] with 1× schedule following previous works [5,50,52]. We implement the proposed method with PyTorch [32] and the FPS is measured on an RTX 2080 Ti GPU including the pre- and post-processing steps for fair comparison following previous work [52]. The optimizer of the proposed method is SGD, with a learning rate 5e−3 and a weight decay 1e−4. The models are trained with 1× schedule for 12 epoch, and we decay the lr with the ratio 0.1, in the 8-th and 11-th epoch. The input frames are resized to 640 × 360 following previous works [12,50,52].

Hyper-parameters. There exists some hyper-parameters in our proposed framework, *i.e.*, bi-directional contrastive learning loss λ_b, and temporal consistency loss λ_c. In this paper, we set $\lambda_b = 0.2$ and $\lambda_c = 10$ in default.

4.4 Main Results

Here we compare our method with two-stage [2,12,39,50,51], single-stage [5,27, 46], 3D convolution-based [1], feature aggregation-based [23], and transformer-based [44] methods. For some differences in the training settings (*e.g.*, resolution, training epochs) vary from different methods, we strictly follow MaskTrack

Table 2. Comparisons with some recent VIS methods on the **YouTube-VIS-2021** val set. We use ResNet-50 backbone and 1× schedule for all experiments for fair comparison.

Methods	AP	AP_{50}	AP_{75}	AR_1	AR_{10}
SipMask [5]	28.6	48.9	29.6	26.5	33.8
MaskTrack R-CNN [50]	31.7	52.5	34.0	30.8	37.8
STEm-Seg [1]	33.3	53.8	37.0	30.1	37.6
CrossVIS [52]	34.2	54.4	37.9	30.4	38.2
Ours (STC)	**35.5**	**57.4**	**38.0**	**32.8**	**42.2**

Table 3. Comparisons with some recent VIS methods on very challenging **OVIS-2021** val set. We use ResNet-50 backbone and 1× schedule for all experiments for fair comparison.

Methods	AP	AP_{50}	AP_{75}	AR_1	AR_{10}
SipMask [5]	10.3	25.4	7.8	7.9	15.8
MaskTrack R-CNN [50]	10.9	26.0	8.1	8.3	15.2
STEm-Seg [1]	13.8	32.1	11.9	9.1	20.0
CrossVIS [52]	14.9	32.7	12.1	10.3	19.8
Ours (STC)	**15.5**	**33.5**	**13.4**	**11.0**	**20.8**

R-CNN [50], SipMask [5] and CrossVIS [52] with 1× schedule and 640×360 resolution for fair comparison.

YouTube-VIS-2019. Without any bells and whistles, our proposed method achieves the best performance 36.7% AP among the listed single-model methods. More specifically, among the two-stage methods, our model outperforms the original MaskTrack R-CNN [50] by **6.4** % in AP (36.7% vs. 30.3%). As discussed in VisTR [44], we also argue that the performance of MaskProp [2] relies heavily on stacking multiple networks, *e.g.*, Spatio-temporal Sampling Network [3] and Hybrid Task Cascade Network [7], not to mention the larger resolution and more training epochs. Our model also beats the recently proposed CompFeat [12] by 1.4 % in AP with a significant improvement on the performance of speed. Meanwhile, it outperforms STEm-Seg [1] and VisTR [44] with the same backbone on the accuracy, which indicates the superiority of our method. Note that VisTR utilizes multi-scale training and takes a week on 8 NVIDIA Tesla V100 for training. Furthermore, compared with the single-stage methods SipMask [5] and TraDeS [46], our method obtains about 4.2 % and 4.1 % improvement in AP, respectively. Compared with the feature aggregation-based method STMask [23] which uses multi-frames to obtain more robust features, our method surpasses it by 3.2 % in AP for ResNet-50 backbone, and even it uses a stronger ResNet-50-DCN backbone. When compared with recent work CrossVIS [52], our method still shows the superiority of the performance on both performance and speed.

As shown in Table 1, we also compare the FPS (frames per second) with other state-of-the-art methods. Our method achieves 36.7% AP at a 40.3 FPS, which is the best tradeoff for the single model. In addition, our method can run an online mode which is crucial for practical usages.

YouTube-VIS-2021. We evaluate the recently proposed MaskTrack R-CNN [50], SipMask [5] and CrossVIS [52] on YouTube-VIS-2021 using the official implementation for comparison. As shown in Table 2, our method surpasses MaskTrack R-CNN [50] and CrossVIS [52] by 3.8 % and 1.3 % in AP , which verifies the effectiveness of our method.

OVIS-2021. From Table 3 we can observe that all methods meet a large performance degradation due to the complexity and occlusions in OVIS-2021 dataset. Our method achieves the best 15.5% AP, surpassing all methods under the same

Table 4. Ablation studies for each component of the proposed framework on YouTube-VIS-2019 validation set.

Baseline	Consistency	Contrastive	Bi-direction	AP
✓				33.7
✓	✓			34.4
✓	✓	✓		36.3
✓	✓	✓	✓	**36.7**

Table 5. Comparisons among different settings of the track embedding on the YouTube-VIS-2019 validation set.

Contrastive	Bi-direction	Embedding dim	AP
✗	✗	256	34.5
✓	✗	256	36.2
✗	✓	256	35.4
✓	✓	256	**36.7**

Table 6. Comparisons among different settings of the kernel generator head on the YouTube-VIS-2019 validation set.

Consistency	# Conv	AP
✗	1	31.5
✗	2	33.7
✗	3	36.2
✗	4	36.0
✓	3	**36.7**

Table 7. Comparisons among different settings of the mask branch on the YouTube-VIS-2019 validation set.

Coord	# Channel	AP
✗	1	28.7
✗	4	33.6
✗	8	36.2
✗	16	36.1
✓	8	**36.7**

experimental conditions. We hope that our proposed method can serve as a strong baseline for this challenging benchmark.

4.5   Ablation Studies

We conduct experiments on the YouTube-VIS-2019 validation set with the ResNet-50 backbone and 1× schedule for the ablation studies.

Analysis for Each Component. As shown in Table 4, we first use CondInst [37] to obtain the instance masks instead of utilizing RoIAlign and mask head in MaskTrack R-CNN [50], which achieves 3.4 % in AP improvements (33.7% vs. 30.3%). Besides the performance improvement, this component also changes the two-stage model to a simple single-stage and fully convolutional one with faster speed. Note that our temporal consistency constraint for the kernel generator successfully gains 0.7 % in AP by digging deeper into the temporal information in the video sequence. For the instances association across frames, we conduct experiments to verify the effectiveness of two components ("Contrastive" and "Bi-direction"). Specifically, when only using spatio-temporal contrastive learning module, we could achieve 1.9% in AP improvement. When using the bi-directional contrastive learning strategy, we finally obtain 36.7% in AP, surpassing "Contrastive" baseline by 0.4% in AP, demonstrating the effectiveness of the bi-directional learning strategy.

Kernel Generator. Kernel generator from CondInst [37] plays a critical role in our method. Thus, we conduct ablation studies to show the impact of parameters in kernel generator head. As presented in Table 6, with the number of convolutions in kernel generator head increasing, the performance improves steadily and

Table 8. Comparisons among different settings of the negative sampling methods of contrastive learning on the YouTube-VIS-2019 validation set.

Inbox	# Negative	AP
✗	0	34.9
✗	64	36.5
✗	128	**36.7**
✗	256	36.4
✓	128	35.2

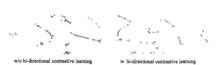

w/o bi-directional contrastive learning   w. bi-directional contrastive learning

Fig. 3. Visualizations of instance embeddings without or with bi-directional contrastive learning module using t-SNE.

achieves the peak 36.7% AP with three stacked convolutions. Temporal consistency obtains 0.5% in AP, which demonstrates the effectiveness.

Mask Branch. To enhance the expressiveness of the mask feature, we further explore the channel number and relative coordinate map ("Coord.") used in the mask branch. As illustrated in Table 7, the 8-channel mask feature achieve 36.2% AP without the coordinate map, and extra channels cannot improve the performance. We set the channel number of the mask feature to 8 by default as a result. Relative coordinates are attached to the mask feature for better performance (about 0.5% in AP improvement).

Tracking Embedding. The tracking embedding is crucial for VIS since AP relies heavily on the accuracy of instance association. We compare with different tracking embedding dimensions. As shown in Table 5, AP improves as the embedding dimension increases. However, we can not afford the complexity cost when the embedding dimension is larger than 256 considering the speed. Thus, we set the embedding dimension as 256 by default.

Negative Sampling. The designed contrastive learning strategy aims to obtain representative and discriminative tracking embeddings. Thus, we further explore different numbers of negative embeddings and how they are selected in Table 8. "Inbox" means we randomly select the negative embeddings within boxes from negative locations according to label assignment results. We find that choosing 128 negative embeddings is a good balance of total training time and accuracy. Moreover, randomly selecting negative embeddings from the whole feature map of the reference frame is much better than "Inbox". This observation verifies that the model can learn more discriminative representations from background stuff or objects.

Fig. 4. Visualization of our proposed method and MaskTrack R-CNN on the YouTube-VIS-2019 val set.

4.6   Visualizations

Instance Embedding. To verify the effectiveness of the proposed method qualitatively, we visualize the instance embeddings of the same video sequence using t-SNE [29], which is shown in Fig. 3. Comparing with Fig. 3(a), the instance embeddings of Fig. 3(b) is more separable, which indicates that our proposed STC module helps to distinguish different instances in the embedding space. Thus, compared with the original multi-class classification loss [50], we could obtain more accurate instance association accuracy for video instance segmentation task.

Video Visualization. The visualization of the proposed method on the YouTube-VIS-2019 validation dataset is shown in Fig. 4. Compared with baseline method MaskTrack R-CNN [50], as shown in the first row and the second row, STC achieve more accurate segmentation results. From the last two rows, STC could achieve more coherent tracking results compared with MaskTrack R-CNN baseline, which demonstrates the effectiveness of the proposed spatio-temporal contrastive learning strategy. In conclusion, our method can segment and associate instances better with more accurate boundary results in challenging situations while MaskTrack R-CNN suffers from the missing instances or identity mistakes.

5   Conclusion

In this work, we introduced a effective architecture for video instance segmentation. Our model is conceptually simple without requiring RoIAlign operation or 3D convolutions. Moreover, it achieves state-of-the-art single-model results (*i.e.*, ResNet-50 backbone) on the YouTube-VIS-2019, YouTube-VIS-2021, and OVIS-2021 datasets in a fully convolutional fashion. We hope our work could serve an strong baseline, which could inspire designing more efficient framework and rethinking the embeddings loss for challenging video instance segmentation task.

References

1. Athar, Ali, Mahadevan, Sabarinath, Ošep, Aljoša, Leal-Taixé, Laura, Leibe, Bastian: STEm-Seg: spatio-temporal embeddings for instance segmentation in videos. In: Vedaldi, Andrea, Bischof, Horst, Brox, Thomas, Frahm, Jan-Michael. (eds.) ECCV 2020. LNCS, vol. 12356, pp. 158–177. Springer, Cham (2020). https://doi.org/10.1007/978-3-030-58621-8_10
2. Bertasius, G., Torresani, L.: Classifying, segmenting, and tracking object instances in video with mask propagation. In: Proceedings of the IEEE Conference on Computer Vision and Pattern Recognition, pp. 9739–9748 (2020)
3. Bertasius, Gedas, Torresani, Lorenzo, Shi, Jianbo: Object detection in video with spatiotemporal sampling networks. In: Ferrari, Vittorio, Hebert, Martial, Sminchisescu, Cristian, Weiss, Yair (eds.) ECCV 2018. LNCS, vol. 11216, pp. 342–357. Springer, Cham (2018). https://doi.org/10.1007/978-3-030-01258-8_21
4. Bolya, D., Zhou, C., Xiao, F., Lee, Y.J.: Yolact: real-time instance segmentation. In: Proceedings of the IEEE International Conference on Computer Vision, pp. 9157–9166 (2019)
5. Cao, Jiale, Anwer, Rao Muhammad, Cholakkal, Hisham, Khan, Fahad Shahbaz, Pang, Yanwei, Shao, Ling: SipMask: spatial information preservation for fast image and video instance segmentation. In: Vedaldi, Andrea, Bischof, Horst, Brox, Thomas, Frahm, Jan-Michael. (eds.) ECCV 2020. LNCS, vol. 12359, pp. 1–18. Springer, Cham (2020). https://doi.org/10.1007/978-3-030-58568-6_1
6. Chen, H., Sun, K., Tian, Z., Shen, C., Huang, Y., Yan, Y.: Blendmask: top-down meets bottom-up for instance segmentation. In: Proceedings of the IEEE Conference on Computer Vision and Pattern Recognition, pp. 8573–8581 (2020)
7. Chen, K., et al.: Hybrid task cascade for instance segmentation. In: Proceedings of the IEEE Conference on Computer Vision and Pattern Recognition, pp. 4974–4983 (2019)
8. Chen, T., Kornblith, S., Norouzi, M., Hinton, G.: A simple framework for contrastive learning of visual representations. In: International Conference on Machine Learning, pp. 1597–1607 (2020)
9. Chen, X., Fan, H., Girshick, R., He, K.: Improved baselines with momentum contrastive learning. arXiv preprint arXiv:2003.04297 (2020)
10. Dai, J., et al.: Deformable convolutional networks. In: Proceedings of the IEEE International Conference on Computer Vision, pp. 764–773 (2017)
11. Fang, Y., et al.: Instances as queries. In: Proceedings of the IEEE/CVF International Conference on Computer Vision, pp. 6910–6919 (2021)
12. Fu, Y., Yang, L., Liu, D., Huang, T.S., Shi, H.: Compfeat: comprehensive feature aggregation for video instance segmentation. arXiv preprint arXiv:2012.03400 (2020)
13. He, K., Fan, H., Wu, Y., Xie, S., Girshick, R.: Momentum contrast for unsupervised visual representation learning. In: Proceedings of the IEEE Conference on Computer Vision and Pattern Recognition, pp. 9729–9738 (2020)
14. He, K., Gkioxari, G., Dollár, P., Girshick, R.: Mask r-cnn. In: Proceedings of the IEEE International Conference on Computer Vision, pp. 2961–2969 (2017)
15. He, K., Zhang, X., Ren, S., Sun, J.: Deep residual learning for image recognition. In: Proceedings of the IEEE Conference on Computer Vision and Pattern Recognition, pp. 770–778 (2016)
16. Huang, Z., Huang, L., Gong, Y., Huang, C., Wang, X.: Mask scoring r-cnn. In: Proceedings of the IEEE Conference on Computer Vision and Pattern Recognition, pp. 6409–6418 (2019)

17. Hwang, S., Heo, M., Oh, S.W., Kim, S.J.: Video instance segmentation using inter-frame communication transformers. Advances in Neural Information Processing Systems 34 (2021)
18. Jia, X., De Brabandere, B., Tuytelaars, T., Gool, L.V.: Dynamic filter networks. In: Advances in Neural Information Processing Systems, pp. 667–675 (2016)
19. Jiang, Z., Gao, P., Guo, C., Zhang, Q., Xiang, S., Pan, C.: Video object detection with locally-weighted deformable neighbors. In: Proceedings of the AAAI Conference on Artificial Intelligence (2019)
20. Jiang, Zhengkai, et al.: Learning where to focus for efficient video object detection. In: Vedaldi, Andrea, Bischof, Horst, Brox, Thomas, Frahm, Jan-Michael. (eds.) ECCV 2020. LNCS, vol. 12361, pp. 18–34. Springer, Cham (2020). https://doi.org/10.1007/978-3-030-58517-4_2
21. Kalantidis, Y., Sariyildiz, M.B., Pion, N., Weinzaepfel, P., Larlus, D.: Hard negative mixing for contrastive learning. arXiv preprint arXiv:2010.01028 (2020)
22. Ke, L., Li, X., Danelljan, M., Tai, Y.W., Tang, C.K., Yu, F.: Prototypical cross-attention networks for multiple object tracking and segmentation. In: Advances in Neural Information Processing Systems 34 (2021)
23. Li, M., Li, S., Li, L., Zhang, L.: Spatial feature calibration and temporal fusion for effective one-stage video instance segmentation. In: Proceedings of the IEEE Conference on Computer Vision and Pattern Recognition, pp. 11215–11224 (2021)
24. Lin, H., Wu, R., Liu, S., Lu, J., Jia, J.: Video instance segmentation with a propose-reduce paradigm. In: Proceedings of the IEEE/CVF International Conference on Computer Vision, pp. 1739–1748 (2021)
25. Lin, T.Y., Dollár, P., Girshick, R., He, K., Hariharan, B., Belongie, S.: Feature pyramid networks for object detection. In: Proceedings of the IEEE Conference on Computer Vision and Pattern Recognition, pp. 2117–2125 (2017)
26. Lin, Tsung-Yi., Maire, Michael, Belongie, Serge, Hays, James, Perona, Pietro, Ramanan, Deva, Dollár, Piotr, Zitnick, C. Lawrence.: Microsoft COCO: common objects in context. In: Fleet, David, Pajdla, Tomas, Schiele, Bernt, Tuytelaars, Tinne (eds.) ECCV 2014. LNCS, vol. 8693, pp. 740–755. Springer, Cham (2014). https://doi.org/10.1007/978-3-319-10602-1_48
27. Liu, D., Cui, Y., Tan, W., Chen, Y.: Sg-net: spatial granularity network for one-stage video instance segmentation. In: Proceedings of the IEEE Conference on Computer Vision and Pattern Recognition, pp. 9816–9825 (2021)
28. Liu, S., Qi, L., Qin, H., Shi, J., Jia, J.: Path aggregation network for instance segmentation. In: Proceedings of the IEEE Conference on Computer Vision and Pattern Recognition, pp. 8759–8768 (2018)
29. Van der Maaten, L., Hinton, G.: Visualizing data using t-sne. J. Mach. Learn. Res. **9**(11) (2008)
30. Oksuz, K., Cam, B.C., Akbas, E., Kalkan, S.: Rank & sort loss for object detection and instance segmentation. In: Proceedings of the IEEE/CVF International Conference on Computer Vision, pp. 3009–3018 (2021)
31. Pang, J., Qiu, L., Li, X., Chen, H., Li, Q., Darrell, T., Yu, F.: Quasi-dense similarity learning for multiple object tracking. In: Proceedings of the IEEE Conference on Computer Vision and Pattern Recognition, pp. 164–173 (2021)
32. Paszke, A., Gross, S., Massa, F., Lerer, A., Bradbury, J., Chanan, G., Killeen, T., Lin, Z., Gimelshein, N., Antiga, L., et al.: Pytorch: an imperative style, high-performance deep learning library. Adv. Neural. Inf. Process. Syst. **32**, 8026–8037 (2019)
33. Qi, J., et al.: Occluded video instance segmentation. arXiv preprint arXiv:2102.01558 (2021)

34. Russakovsky, O., Deng, J., Su, H., Krause, J., Satheesh, S., Ma, S., Huang, Z., Karpathy, A., Khosla, A., Bernstein, M., et al.: Imagenet large scale visual recognition challenge. Int. J. Comput. Vision **115**(3), 211–252 (2015)

35. Sun, C., Baradel, F., Murphy, K., Schmid, C.: Learning video representations using contrastive bidirectional transformer. arXiv preprint arXiv:1906.05743 (2019)

36. Tian, Y., Sun, C., Poole, B., Krishnan, D., Schmid, C., Isola, P.: What makes for good views for contrastive learning. arXiv preprint arXiv:2005.10243 (2020)

37. Tian, Z., Shen, C., Chen, H.: Conditional convolutions for instance segmentation. In: European Conference on Computer Vision, pp. 282–298 (2020)

38. Tian, Z., Shen, C., Chen, H., He, T.: Fcos: fully convolutional one-stage object detection. In: Proceedings of the IEEE International Conference on Computer Vision, pp. 9627–9636 (2019)

39. Voigtlaender, P., Chai, Y., Schroff, F., Adam, H., Leibe, B., Chen, L.C.: Feelvos: fast end-to-end embedding learning for video object segmentation. In: Proceedings of the IEEE Conference on Computer Vision and Pattern Recognition, pp. 9481–9490 (2019)

40. Wang, T., Xu, N., Chen, K., Lin, W.: End-to-end video instance segmentation via spatial-temporal graph neural networks. In: Proceedings of the IEEE/CVF International Conference on Computer Vision, pp. 10797–10806 (2021)

41. Wang, Xinlong, Kong, Tao, Shen, Chunhua, Jiang, Yuning, Li, Lei: SOLO: segmenting objects by locations. In: Vedaldi, Andrea, Bischof, Horst, Brox, Thomas, Frahm, Jan-Michael. (eds.) ECCV 2020. LNCS, vol. 12363, pp. 649–665. Springer, Cham (2020). https://doi.org/10.1007/978-3-030-58523-5_38

42. Wang, X., Zhang, R., Kong, T., Li, L., Shen, C.: Solov2: dynamic and fast instance segmentation. arXiv preprint arXiv:2003.10152 (2020)

43. Wang, Y., Xu, Z., Shen, H., Cheng, B., Yang, L.: Centermask: single shot instance segmentation with point representation. In: Proceedings of the IEEE Conference on Computer Vision and Pattern Recognition. pp. 9313–9321 (2020)

44. Wang, Y., Xu, Z., Wang, X., Shen, C., Cheng, B., Shen, H., Xia, H.: End-to-end video instance segmentation with transformers. In: Proceedings of the IEEE Conference on Computer Vision and Pattern Recognition. pp. 8741–8750 (2021)

45. Wu, H., Chen, Y., Wang, N., Zhang, Z.: Sequence level semantics aggregation for video object detection. In: Proceedings of the IEEE International Conference on Computer Vision. pp. 9217–9225 (2019)

46. Wu, J., Cao, J., Song, L., Wang, Y., Yang, M., Yuan, J.: Track to detect and segment: An online multi-object tracker. In: Proceedings of the IEEE Conference on Computer Vision and Pattern Recognition. pp. 12352–12361 (2021)

47. Xie, E., Sun, P., Song, X., Wang, W., Liu, X., Liang, D., Shen, C., Luo, P.: Polarmask: Single shot instance segmentation with polar representation. In: Proceedings of the IEEE Conference on Computer Vision and Pattern Recognition. pp. 12193–12202 (2020)

48. Xiong, Y., Ren, M., Urtasun, R.: Loco: Local contrastive representation learning. arXiv preprint arXiv:2008.01342 (2020)

49. Xu, N., Yang, L., Yang, J., Yue, D., Fan, Y., Liang, Y., Huang, T.S.: Youtube-vis dataset 2021 version. https://youtube-vos.org/dataset/vis (2021)

50. Yang, L., Fan, Y., Xu, N.: Video instance segmentation. In: Proceedings of the IEEE International Conference on Computer Vision. pp. 5188–5197 (2019)

51. Yang, L., Wang, Y., Xiong, X., Yang, J., Katsaggelos, A.K.: Efficient video object segmentation via network modulation. In: Proceedings of the IEEE Conference on Computer Vision and Pattern Recognition, pp. 6499–6507 (2018)

52. Yang, S., et al.: Crossover learning for fast online video instance segmentation. arXiv preprint arXiv:2104.05970 (2021)
53. Ying, X., Li, X., Chuah, M.C.: Srnet: Spatial relation network for efficient single-stage instance segmentation in videos. In: Proceedings of the 29th ACM International Conference on Multimedia, pp. 347–356 (2021)
54. Zhu, L., Xu, Z., Yang, Y.: Bidirectional multirate reconstruction for temporal modeling in videos. In: Proceedings of the IEEE Conference on Computer Vision and Pattern Recognition, pp. 2653–2662 (2017)
55. Zhu, X., Wang, Y., Dai, J., Yuan, L., Wei, Y.: Flow-guided feature aggregation for video object detection. In: Proceedings of the IEEE International Conference on Computer Vision (2017)

Mitigating Representation Bias in Action Recognition: Algorithms and Benchmarks

Haodong Duan[1,3](✉), Yue Zhao[2], Kai Chen[3], Yuanjun Xiong[4], and Dahua Lin[1]

[1] The Chinese University of Hong Kong, Hong Kong, China
[2] The University of Texas at Austin, Austin, USA
[3] Shanghai AI Lab, Shanghai, China
dhd.efz@gmail.com
[4] Amazon AI, Shanghai, China

Abstract. Deep learning models have achieved excellent recognition results on large-scale video benchmarks. However, they perform poorly when applied to videos with rare scenes or objects, primarily due to the bias of existing video datasets. We tackle this problem from two different angles: algorithm and dataset. From the perspective of algorithms, we propose Spatial-aware Multi-Aspect Debiasing (**SMAD**), which incorporates both *explicit* debiasing with multi-aspect adversarial training and *implicit* debiasing with the spatial actionness reweighting module, to learn a more generic representation invariant to non-action aspects. To neutralize the intrinsic dataset bias, we propose **OmniDebias** to leverage web data for joint training selectively, which can achieve higher performance with far fewer web data. To verify the effectiveness, we establish evaluation protocols and perform extensive experiments on both re-distributed splits of existing datasets and a new evaluation dataset focusing on the action with rare scenes. We also show that the debiased representation can generalize better when transferred to other datasets and tasks.

1 Introduction

Human beings have cognitive bias, and so do the machine learning systems [36]. Human cognitive bias comes from the uniqueness of individual experiences (learning materials) and the tendency of brains to simplify information processing [25]. Machine learning systems are biased for similar reasons. First, the datasets used for training can be intrinsically biased: *e.g.*, sampled from a shifted distribution [44] or collected with a pre-defined ontology [38]. Even if the dataset faithfully represents the real world, there is human bias in the real world which we do not want the machine learning system to exploit, *e.g.*, gender bias [5,6]. Mitigating the bias in machine learning systems has long been a challenging yet valuable research area [3,17,19].

In computer vision, the efforts for building datasets that faithfully represent the real visual world never end. Better data collection and labeling strategies

Supplementary Information The online version contains supplementary material available at https://doi.org/10.1007/978-3-031-25069-9_36.

© The Author(s), under exclusive license to Springer Nature Switzerland AG 2023
L. Karlinsky et al. (Eds.): ECCV 2022 Workshops, LNCS 13804, pp. 557–575, 2023.
https://doi.org/10.1007/978-3-031-25069-9_36

[14,38,43,57] are designed for building less biased datasets from scratch. Besides, various tools can be applied to a built dataset (visual [47] or tabular [4] data) to detect and mitigate unwanted bias. In action recognition, [34] introduces the concept of representation bias and attempts to reduce it throughout dataset construction. However, the dataset they propose is on a small scale and in a narrow domain. We investigate existing large-scale datasets instead and quantify the representation bias by designing different train-test splits and analyzing the performance gaps.

Besides, we propose **OmniDebias**, which uses external web media as auxiliary data to mitigate the dataset bias. On the one hand, the diversity of web data provide us with rich examples that are uncommon in existing datasets, which makes it a suitable data source for debiasing. On the other hand, web data are also severely biased to some factors, *e.g.*, *scene*. OmniDebias adopts a simple yet effective data selection strategy to sample a less biased subset from the entire dataset. Co-training with the selected subset outperforms the vanilla co-training both in performance and data efficiency.

Though effective in debiasing, constructing 'unbiased' datasets can be difficult and may cost lots of human labor, while designing debiasing algorithms is a much cheaper alternative. A series of works [35,53] aim at devising algorithms to mitigate the bias in the learned representation, preventing the algorithms from amplifying the bias in training data. In particular, SDN [11] proposes to mitigate *scene* bias in action recognition with adversarial training and human mask confusion loss. Previous works usually restrict the debiasing algorithm to a specific factor. In the real world, the bias in the dataset can be complex and non-trivial to understand. To deal with more complicated dataset bias, we extend the single-factor adversarial training to a multi-aspect fashion, which shows better generalization capability.

To mitigate the *generic* representation bias in action recognition, we propose a spatial-aware multi-aspect debiasing framework (**SMAD**). A video can have multiple facets besides the action label, such as the background *scene* or the *object* that people interact with. Video datasets collected for different purposes may emphasize different facets. Considering this characteristic, we propose multi-aspect adversarial training (**MAAT**) to enforce the model invariant to these *non-action* facets. We also introduce Spatial-Aware Actionness Reweighting (**SAAR**) to ensure that the model learns where to focus to recognize action without being affected by features related to other facets. The framework **SMAD** proves to be generic for videos with various kinds of bias and does not depend on extra knowledge of specific datasets.

To fairly exhibit the effectiveness of the proposed debiasing algorithm, we devise a series of evaluation protocols. First, for the existing large-scale dataset Kinetics-400 [7], we re-distribute the original train splits by either *scene* or *object* such that the hidden facet does not overlap between the *re-distributed* train and test sets (**facet-based re-distribution**). Second, we collect an additional Action with RAre Scene (ARAS) dataset[1] for evaluation to simulate the **out-of-distribution** setting. Third, we follow the routine of measuring the debiasing

---

[1] Dataset released at https://github.com/kennymckormick/ARAS-Dataset.

effect by transferring the learned model to downstream tasks (**downstream-task transferring**), such as feature classification, few-shot learning, and fine-tuning on other datasets (as is proposed by [11]).

Our contributions are three-fold: 1. We propose SMAD, which considers multiple aspects in adversarial training and achieves better performance when complex bias exists in the training set. 2. We propose OmniDebias, which exploits the richness and diversity of web data effectively and efficiently. 3. We evaluate our method on both conventional evaluation protocols (downstream-task transferring) as well as the new ones (facet-based re-distribution, out-of-distribution testing). The improvements of our methods on all three benchmarks are consistent and remarkable.

2   Related Work

Action Recognition. Action recognition aims at recognizing human activities in videos. Following the success of deep learning in the image domain, two series of deep ConvNets become the mainstream architectures for action recognition, named 2D-CNN and 3D-CNN methods. 2D-CNN methods like Two-Stream [41] and TSN [49] are light-weight while lacking temporal modeling capability to some extent. 3D-CNN methods [7,18,45,46] use 3D convolutions for temporal modeling and achieve the state-of-the-art on large-scale benchmarks like Kinetics-400 [7]. In this paper, we show that both architectures are vulnerable to biases. Our proposed framework can help to mitigate this problem.

Mitigating Dataset Bias. All datasets, more or less, have dataset bias. In computer vision, [44] studies 12 widely used image datasets and finds their data are of different domains and distant from the real visual world. In natural language processing, gender bias occurs in corpus collected from social media and news [22,29]. There are two main approaches to mitigate dataset bias: The first is to design better data collection and labeling strategies [38] or to calibrate the existing dataset with bias detection tools [27,47]. The second is to compensate dataset bias with domain adaptation techniques [20,30,37]. In this paper, following the first approach, we propose to use diversified web media to neutralize the dataset bias.

Mitigating Algorithm Bias. Even if the dataset faithfully represents the real world, bias still exists. Due to human bias, real-world data may bias towards specific factors, while discriminative models even amplify such unwanted bias [54]. In machine learning, it is intuitive to add constraints or regularizations for the pursued fairness metric to the existing optimization objective [1,50,52]. However, most of these approaches are intractable in deep learning. Meanwhile, adversarial training has broader applications both in machine learning and deep learning. [53] use adversarial debiasing for bias mitigation, but the bias factor is required to be known beforehand. [33] propose to use adversarial example reweighting and achieves good performance on debiasing action recognition.

Domain adaptation. Domain adaptation (DA) aims at learning well-performing models on the target domain with training data from the source domain. To that

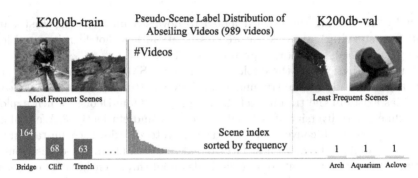

Fig. 1. The long-tailed *scene* distribution of abseiling videos. Most videos belong to several scene categories. In the distribution tail, there are many scene categories that rarely occur in training videos. We sample videos from the distribution head to form K200db-train, from the tail to form K200db-val.

end, many works try to find a common feature space for the source and target domains via adversarial training, both for image tasks [10,21,39] and action recognition [9,12,13]. The setting of debiasing is similar to, but not the same as DA. The main difference is that we have no access to testing videos during training. Besides, the debiasing setting does not assume the amount of testing data, while DA algorithms require a certain number of testing videos to determine the data distribution.

3   Formulation

Following [53], the problem can be formulated as follows. For a video recognition dataset D, we can view each data sample as a tuple (x, y, z) drawn from the joint distribution (X, Y, Z), where x denotes the video, y denotes the *action* label, z denotes one or multiple *non-action* labels, such as *scene*, *object* or other attributes. We consider the supervised learning task, which builds a predictor $\hat{Y} = f(X)$ for Y given X.

Due to the dependence of Y and Z in the training set, the predictors learned via standard supervised learning also yield predictions $\hat{Y}$ dependent on Z given the action label Y. Such behavior will lead to poor generalization capability, severely undermine the testing performance if $P(Z|Y)$ differs a lot between the train and test split.

Our goal is to learn non-discriminatory action recognition models *w.r.t.* Z, which generalize well to testing videos with factors (*scene*, *object*, *e.g.*) that rarely appear in the training set. Non-dicrimination criterias have been of three types in fairness literature [2], *independence* ($Y' \perp Z$), *seperation* ($Y' \perp Z|Y$) and *sufficiency* ($Y \perp Z|Y'$). In the context of video recognition, we pursue EQUALIZED ODDS, similar to *separation*, which is to minimizing the variance of $P(\hat{Y} = y|Y = y, Z = z)$ for different z given y. Besides improving the z-unbiasedness, we also need to maximize $P(\hat{Y} = y|Y = y)$ to secure a good recognition model.

4   Evaluation Benchmark

4.1   Crafting Evaluation Datasets

Most existing datasets assume the joint distribution $P(Y, Z)$ identical between train and validation splits. To find out if an action recognition model is biased towards the *non-action* labels, we design two evaluation protocols based on Kinetics-400 [7]: re-distributing the existing train-val split and constructing a new validation set.

Re-distributing Train-Val Split. We start with the original Kinetics-400 train split (with ~240k videos). We apply a ResNet-50 trained on Places-365 [57] to obtain the pseudo *scene* labels. As shown in Fig. 1, the pseudo *scene* labels have a long-tailed distribution. We take the tail as the validation set and sample a subset from the head to be the training set. To maintain the inter-class sample balance, we select 200 classes with the most training samples and construct a subset that contains 80k videos for training and 10k for validation, (denoted as K200db-train and K200db-val, db for debiasing). We examine the *action-scene* correlation of the two splits by calculating the normalized mutual information (NMI) of *action* and *scene*: for K200db-train, the NMI is 0.466 (0.397 if sampled randomly); for K200db-val, the NMI is 0.374 (0.488 if sampled randomly). Based on the splitting method, we can also tune the overlap of common *scene* labels in K200db-train and K200db-val for varying distribution shift.

Constructing New Validation Set. Beyond being restricted to the original dataset, we can further construct a new dataset for evaluation. This resembles the real-world scenario: the trained model is fixed while the environment changes at deployment. We begin with *action* labels in Kinetics and consider some *rare scenes*. The combinations of *actions* and *rare scenes* are used as queries to obtain

Fig. 2. Samples of ARAS dataset.(ARAS video samples in: https://youtu.be/ j1LA3y-UuEA).

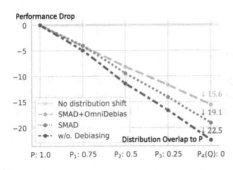

Fig. 3. The Top1-Acc severely drops along with the scene distribution shift.

Table 1. Top1-Acc of TSN and SlowOnly on 3 test sets.

TSN	Top1 Acc
K200-val	76.2
K200-unbias	53.7 ↓ 22.5
ARAS-64	55.8 ↓ 20.4
SlowOnly	Top1 Acc
K200-val	75.1
K200-unbias	51.9 ↓ 23.2
ARAS-64	51.0 ↓ 24.1

web videos from YouTube. We manually examine the web videos and obtain around ten videos for each class in 104 Kinetics classes, denoted as Action with RAre Scenes (ARAS-104). For K200db, there are 64 overlapped classes (ARAS-64). Figure 2 shows several examples. We use ARAS to simulate the out-of-distribution testing for *scene*-debiasing evaluation.

4.2   Evaluation of Existing Methods

We first evaluate existing methods on the new benchmarks, including a 2D-CNN method (TSN-3seg-R50) [49] and a 3D-CNN one (SlowOnly-8 × 8-R18) [18]. From Table 1, we observe that the Top-1 accuracies on both K200db-val and ARAS-64 are significantly lower than the original validation split K200-val. This reflects models learned with vanilla training cannot handle the large discrepancy of the *action-scene* joint distribution between train/val splits. K200db-[train/val] is an extreme case that has disjoint scene labels. We can also vary the overlap of scene labels between K200db-val and K200db-train. Figure 3 demonstrates that the drop of accuracy is positively correlated to the distribution shift. That performance drop can be largely mitigated by **SMAD** and **OmniDebias**, which will be detailed in the following section.

5   Method

We devise Spatial-aware Multi-Aspect Debiasing (**SMAD**) which seeks to learn a representation invariant to multiple aspects of videos, *e.g.*, scene, object, and other attributes, with adversarial training. Besides, we propose OmniDebias to harness the richness and diversity of web data efficiently, to improve the expressive power of the learned representation. We integrate the two complementary aspects into a unified framework, as illustrated in Fig. 4.

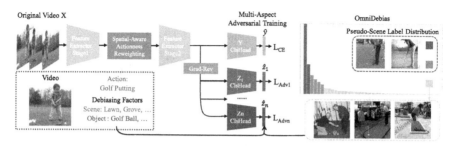

Fig. 4. SMAD & OmniDebias. Left: SMAD framework. Multiple adversarial heads are used in SMAD for Multi-Aspect debiasing. The SAAR module (Fig. 5) is inserted in the backbone to improve the spatial modeling capability. **Right: OmniDebias.** OmniDebias only uses the **unbias** part of web media for joint training, achieving better performance and efficiency.

5.1   Spatial-aware Multi-Aspect Debiasing

SMAD incorporates both **explicit** debiasing using Multiple Aspects as Adversarial Training objectives (MAAT) and **implicit** debiasing with Spatial-Aware Actionness Reweighting (SAAR).

Multi-Aspect Adversarial Training. We denote each input as a tuple $(x, y, z_1, \cdots, z_M) \in \mathcal{X} \times \mathcal{Y} \times \mathcal{Z}_1 \times \cdots \times \mathcal{Z}_M$, where we pre-define M aspects in addition to the set of *action* labels $\mathcal{Y}$. We use a ConvNet f_Θ parameterized by Θ for feature extraction. On top of f_Θ are $(M + 1)$ classification heads: one head $h_{\mathcal{Y}}$ (parameterized by $\theta_{\mathcal{Y}}$) to predict the *action* y and M adversarial heads $h_{\mathcal{Z}_i}$ (parameterized by $\theta_{\mathcal{Z}_i}$) to recognize tags belong to the aspect $\mathcal{Z}_i$. We use the standard cross-entropy loss $L_{\text{ce},y}$ to train $h_{\mathcal{Y}}$, use adversarial losses $L_{\text{adv},\mathcal{Z}_i}$ for the rest *non-action* heads $h_{\mathcal{Z}_i}$.

The optimization can be divided into two parts: classification heads and the backbone. For classification heads, the objective is to minimize $L_{\text{ce},y}$ and $L_{\text{adv},\mathcal{Z}_i}$:

$$\theta_{\mathcal{Y}}, \theta_{\mathcal{Z}_1}, \cdots \theta_{\mathcal{Z}_M} = \underset{\theta_{\mathcal{Y}}, \theta_{\mathcal{Z}_1} \ldots \theta_{\mathcal{Z}_M}}{\text{argmin}} \left(L_{\text{ce},y} + \sum_{i=1}^{i=M} \lambda_i L_{\text{adv},\mathcal{Z}_i} \right). \tag{1}$$

λ_i is the weight of the adversarial loss. For the backbone, since we aim for feature that is both discriminative for $\mathcal{Y}$ and invariant for Z_i, the objective is to minimize $L_{\text{ce},y}$ and *maximize* $L_{\text{adv},\mathcal{Z}_i}$:

$$\Theta = \underset{\Theta}{\text{argmin}} \left(L_{\text{ce},y} - \sum_{i=1}^{i=M} \lambda_i L_{\text{adv},\mathcal{Z}_i} \right). \tag{2}$$

By inserting a gradient reversal layer [21] before $h_{\mathcal{Z}_1}, \cdots, h_{\mathcal{Z}_M}$, we can simultaneously optimize the backbone f_Θ along with all heads efficiently using the standard stochastic gradient descent.

Choice of Adversarial Losses. The type of the adversarial loss depends on the label format of $\mathcal{Z}_i$. For soft-label $\mathcal{Z}_i$, we can use soft cross-entropy loss (**SoftCE**,

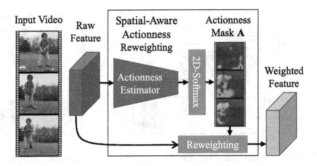

Fig. 5. SAAR module. The Spatial-Aware Actionness Reweighting module learns an actionness mask **A** to reweight features at different locations. The learned mask has low values on irrelated scenes or objects to suppress these features.

Eq. 3) or KL-divergence loss (**KLDiv**, Eq. 4). For multi-label $\mathcal{Z}_i$, we use binary cross-entropy loss (**BCE**).

$$L_{\text{adv},\mathcal{Z}_i} = -\sum_{k=1}^{k=|\mathcal{Z}_i|} z_{i_k} \log[h_{\mathcal{Z}_i}(f(x;\Theta))]_k. \tag{3}$$

$$L_{\text{adv},\mathcal{Z}_i} = \sum_{k=1}^{k=|\mathcal{Z}_i|} z_{i_k} \log \frac{z_{i_k}}{[h_{\mathcal{Z}_i}(f(x;\Theta))]_k}. \tag{4}$$

Source of Non-Action Labels. The labels for $\mathcal{Z}_1, \cdots, \mathcal{Z}_M$ are needed in adversarial training. However, these annotations are usually unavailable for most action recognition datasets. To handle this, we use off-the-shelf ConvNets trained on the specific domain to obtain the pseudo labels. For example, we use ResNet trained on ImageNet [14] and Places365 [57] to assign the pseudo labels for *object* and *scene*, respectively.

Spatial-Aware Actionness Reweighting. The idea of adversarial training is simple but turns out to be fragile, especially when considering multiple aspects. It would be hard for the vanilla algorithm to converge if the adversarial loss weight λ_i is set as a relatively large number. One possible conjecture is that the underlying network pools feature *uniformly* across all positions. Since we can only mitigate the model bias instead of eliminating the dataset bias, the inherent bias from the data would contradict the adversarial objective unless the model selectively attends to the action-related region. To this end, inspired by the idea of actionness estimation [48,55], we propose a Spatial-Aware Actionness Reweighting module (SAAR), illustrated in Fig. 5.

For a feature map $\mathbf{F} \in \mathbf{R}^{C \times T \times H \times W}$, we first estimate an actionness mask $\mathbf{A} \in \mathbf{R}^{T \times H \times W}$, where the scalar for each location represents how much the feature is related to the human action. In experiments, we use a 2D ResNet-Layer with a small bottleneck width for actionness feature extraction, and use another 2D 3×3 convolution as the actionness head, which outputs a 1-channel actionness score map $\mathbf{A}$. On top of the score map, we apply 2D-softmax across the spatial dimensions for normalization: $\mathbf{A}'(t,h,w) = \frac{e^{\mathbf{A}(t,h,w)}}{\sum_{h',w'} e^{\mathbf{A}(t,h',w')}}$. The final modulated feature map is the element-wise multiplication between $\mathbf{F}$ and $\mathbf{A}'$:

$$\mathbf{F}'(c,t,h,w) = (H \times W) \cdot \mathbf{F}(c,t,h,w) \odot \mathbf{A}'(t,h,w) \tag{5}$$

where the coefficient $H \times W$ is used to preserve the magnitude of feature maps after re-weighting. We insert SAAR before the last ResNet-Layer in the backbone. Operating on a small feature map (14×14), the SAAR module adds up to 2% additional computation.

Experiments show that spatial-aware actionness reweighting can not only benefit convergence of training but also lead to better performance. It is worth noting that the benefit of SAAR is much larger when combined with MAAT than used alone, indicating that the adversarial training objective incurs weak supervision implicitly.

5.2   Exploiting Web Media with OmniDebias

Instead of restricting to labeled datasets, we also propose to leverage webly-supervised datasets for bias mitigation via co-training, considering their richness and diversity.

We use GoogleImg (GG) and InsVideo (IG) from the OmniSource dataset [16] as the web data source. Following the same pipeline as the original work, to construct the auxiliary dataset for joint training, we train a teacher network to filter web data and keep high-confidence examples. Joint training with the built auxiliary dataset can lead to much larger improvements on our evaluation benchmarks (K200db-val, ARAS), compared to the improvement on standard validation sets (K200/400-val), mostly because web media contain novel $z \in \mathcal{Z}$ that does not exist in the training set.

However, there is a drawback to the naïve approach. For web data, the distribution over z can be even more imbalanced than the distribution for Kinetics videos. For example, the average entropy of pseudo *scene* distributions of 400 actions is 3.02 for Kinetics, and 2.79 for GoogleImg (larger $\rightarrow$ more diversified). Figure 6 demonstrates pseudo *scene* distributions of 3 action classes. Co-training with such an unbalanced dataset is sub-optimal.

Thus we propose OmniDebias to utilize web media more efficiently. In OmniDebias, we use a simple data selection strategy to select a subset of the entire web dataset for joint training. Specifically, based on the same approach introduced in **Benchmark**, we sort the samples in a same action class by the descending order of z-frequency[2]. Based on the z-frequency, we divide the auxiliary dataset into 3 equal-sized parts, *i.e.* [web]-bias, [web]-mid and [web]-unbias ([web] can be GG, IG, *etc.*), and use [web]-unbias only for joint training. OmniDebias consistently outperforms not only using other parts but also the union, indicating its efficacy and efficiency.

---

[2] For z = scene, if 20 out of 100 samples have the scene label 'cliff', the z-frequency of each of the 20 samples is 0.2 (20 / 100).

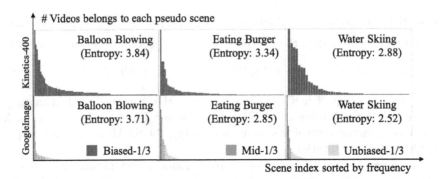

Fig. 6. Pseudo-scene distributions. Visualization of pseudo scene distributions of 3 action categories in Kinetics-400 and GoogleImg. GoogleImg has more imbalanced distributions.

Table 2. The individual and joint effects of SMAD and OmniDebias. We report the Top-1 accuracies on three test sets: ARAS, K200db-val and K200/400-val. ↑ and ↓ denote the improvement or decline to the baseline w/o. debiasing.

Component		SMAD			OmniDebias			Combination		
Model	Train / Test	ARAS-64	K200db-val	K200-val	ARAS-64	K200db-val	K200-val	ARAS-64	K200db-val	K200-val
2D	K200db-train	60.3 ↑ 4.5	55.7 ↑ 2.0	75.5 ↓ 0.7	68.3 ↑ 12.5	60.0 ↑ 6.3	77.4 ↑ 1.2	70.9 ↑ 15.1	62.0 ↑ 8.3	78.1 ↑ 1.9
3D	K200db-train	58.4 ↑ 7.4	55.0 ↑ 3.1	74.6 ↓ 0.5	69.2 ↑ 18.2	60.7 ↑ 8.8	78.7 ↑ 3.6	71.6 ↑ 20.6	62.7 ↑ 10.8	78.4 ↑ 3.3
Model	Train / Test	ARAS-104	-	K400-val	ARAS-104	-	K400-val	ARAS-104	-	K400-val
2D	K400-train	56.2 ↑ 2.1	-	70.0 ↓ 0.6	60.2 ↑ 6.1	-	71.3 ↑ 0.7	61.8 ↑ 7.7	-	71.4 ↑ 0.8
3D	K400-train	55.0 ↑ 3.5	-	68.2 ↓ 0.1	60.2 ↑ 8.7	-	71.0 ↑ 2.7	61.2 ↑ 9.7	-	70.8 ↑ 2.5

6   Experiments

6.1   Experiment Setting

Acquisition of non-action labels. For debiasing, *non-action* labels can be either pseudo labels inferred by a pretrained model or ground-truth labels from a multi-label dataset. To acquire pseudo labels for debiasing, we use ResNet50 [26] pretrained on ImageNet and Places365 as the pseudo label extractor for *scene* and *object*. We also tried ResNet18 and DenseNet161 [28] as the extractor for *scene* labels but observe a subtle difference ($\leq 0.3\%$). For ground-truth labels, we use the HVU dataset [15], which annotates Kinetics videos with three additional tag categories: *context, attribute, event*.

Training and Evaluation. We use TSN-3seg-R50 with ImageNet pretraining as the 2D-CNN baseline, SlowOnly-8 × 8-R18 as the 3D-CNN one. In the choice of adversarial losses, we use a weighted combination of SoftCE and KLDiv for soft-label (better than each individual), use BCE for multi-label. The loss weight is 0.5 for SoftCE and KLDiv losses, 5 for BCE loss. For testing, we uniformly sample 25 frames for TSN or ten clips for SlowOnly with center crop and average

Table 3. SMAD v.s. other debiasing and domain-adaptation algorithms. Access indicates if the algorithm needs to access validation data in training.

Method	Access	K200db-val	ARAS-64
Baseline	✗	51.9	51.0
AdaBN [32]	✗	52.0	52.3
FrameShuffle [8]	✗	52.4	51.3
SDN [11]	✗	54.0	55.3
DANN [21]	✓	52.4	53.3
MMD [23]	✓	52.7	53.1
SAVA [13]	✓	53.4	54.0
SMAD	✗	55.0	58.4

Table 4. Fairness metrics on K200-val. We train the SlowOnly-R18-8 × 8 on K200db-train, I denotes normalized mutual information (lower → more non-discriminative). Red marks denote the *scene*-bias amplified to the oracle independence $I(Y, S) = 0.487$.

Metric	Independence $I(Y', S)$	Separation $I(Y', S\|Y)$	Sufficiency $I(Y, S\|Y')$
Baseline	0.498 ↑ 0.011	0.376	0.356
SMAD	0.490 ↑ 0.003	0.373	0.366
OmniDebias	0.496 ↑ 0.009	0.334	0.321
Combination	0.491 ↑ 0.004	0.321	0.327

the final predictions. In experiments, we report the Top-1 accuracy. Since ARAS is a small evaluation set, we first examine the statistical significance: we train the TSN on K200db-train for 10 times with different random seeds and test it on ARAS-64. We find the standard deviation of accuracy is around 0.3%, which means a difference larger than 0.8% is statistically significant.

6.2 Main Results

Re-distributed Train-Val Splits. When the training and validation subsets have different scene distributions, our method consistently bridges the performance gap between validation sets with *scene* distribution shift and the validation set without *scene* distribution shift, as shown in Fig. 3. The narrower accuracy gaps reflect the improvement made under the fairness metrics EQUALIZED ODDs. When the testing and training scene distributions are completely different, *i.e.* disjoint label sets, the effect is most significant: our methods reduce the accuracy drop by nearly $\frac{1}{3}$: from 22.5% to 15.6%.

Besides EQUALIZED ODDs, we also evaluate three commonly used fairness metrics, namely *independence, separation*, and *sufficiency* in Table 4. SMAD largely mitigates the bias amplified by the algorithm: without SMAD, $I(Y', S)$ increases 0.011 (the *scene*-bias amplified by algorithm) compared to $I(Y, S) = 0.487$, while SMAD reduces it to 0.003. Since Y, S are not statistically independent, the sufficiency and independence cannot both hold. Thus we observe that $I(Y, S|Y')$ increases when we apply SMAD for debiasing. For OmniDebias, since additional web media are used for joint training, all three fairness metrics are improved (lower → better).

We further study the individual and combined effects of SMAD and OmniDebias. Extensive experiments are conducted with both 2D and 3D baselines: models are trained on K200db-train or K400-train and tested on 3 test sets: ARAS-64/104, K200db-val (z-unbiased) and K200/400-val (normal). The

Table 5. Few-shot Learning & Feature Classification. The learned representation achieves good performance on downstream tasks. We report the 3-split average for HMDB51 and UCF101.

Setting	GYM-1shot	GYM-5shot	HMDB51	UCF101	Diving48
w/o. Debiasing	42.2	52.9	49.9	84.3	17.3
+ SMAD	45.5	56.4	50.9	84.9	18.9
+ IG-all	46.6	58.4	54.3	88.4	19.8
+ IG-unbias	47.1	59.5	55.9	88.8	20.9
+ SMAD, IG-unbias	51.7	62.1	57.2	89.9	22.3

results are demonstrated in Table 2. SMAD can improve the performance on z-unbiased test sets by a large margin at the cost of a little accuracy drop on the normal test set. The improvement of OmniDebias is across all 3 test sets since additional web media are used for joint training. while for K200db-val and ARAS the gain is much more noticeable. Combining SMAD and OmniDebias yields the highest accuracy on all z-unbiased test sets, indicates that the two techniques are orthogonal to each other.

A new Debiasing Benchmark. In Table 3, we evaluate multiple debiasing and domain-adaptation algorithms on our new **facet-based re-distribution** and **out-of distribution** benchmarks. The models (backbone: SlowOnly-R18) are trained on K200db-train, tested on K200db-val and ARAS-64. SMAD is a better solution for the debiasing problem compared to the alternatives, considering its superior performance and simple deployment.

Transferring Abilities. The debiased representation is also more useful when transferred to other tasks. We study two cases: few-shot learning and video classification. SlowOnly-8 × 8-R18 trained on K200db-train is used as the feature extractor. For each video, we uniformly sample 10 clips, extract a 512-d feature for each clip, and concatenate them into a 5120-d video-level feature.

We evaluate the few-shot learning performance on FineGYM-99 [40], a fine-grained gymnastic action recognition dataset with less scene bias. We construct 10,000 5-way episodes (1-shot or 5-shot). In each episode, the cosine similarities between the query sample and support samples are used for classification. Table 5 shows that both SMAD and OmniDebias contribute to the few-shot performance on FineGYM-99.

We evelute the performance of video classification on UCF101 [42], HMDB51 [31] and Diving48 [34] with two settings: feature classification and finetuning. For feature classification, we train a linear SVM based on the 5120-d descriptors. Table 5 shows that both SMAD and OmniDebias improve the feature classification performance. Two baselines are used in the finetuning setting. We first use ResNet3D-18 [24] trained on MiniKinetics [51] with input size 112 as the baseline, for a straightforward comparison with SDN [11] (Table 6 upper).

Table 6. Finetuning performance. Our work improves the finetuning performance on 3 datasets significantly under different settings. We report the 3-split average for HMDB51 and UCF101. * denotes using the old version of Diving48 annotations.

Method	Pretrain	HMDB51	UCF101	Diving48*	Diving48
ResNet3D-16 × 1	MiniKinetics	53.6	83.5	18.0	-
+ SDN [11]	MiniKinetics	56.7	84.5	20.5	-
+ SMAD	MiniKinetics	57.2	84.7	20.9	-
+ SMAD, IG-unbias	MiniKinetics	61.2	88.2	22.6	-
SlowOnly-8 × 8	K200db-train	62.6	89.6	25.5	53.9
+ SMAD	K200db-train	64.0	90.4	26.7	55.7
+ SMAD, IG-unbias	K200db-train	67.3	93.3	28.2	59.7

Table 7. Performance of MAAT & SAAR. The baseline is SlowOnly-8 × 8-R18 trained on `K200db-train`.

Test Set	Baseline	MAAT	SAAR	MAAT + SAAR
ARAS-64	51.0	55.8	51.6	58.4
K200db-val	51.9	54.3	52.7	55.0

With pseudo labels for recognition only (much cheaper than pseudo labels for human detection), SMAD can outperform SDN on three downstream tasks. By introducing web data with OmniDebias, the model can obtain much better performance. We further test the finetuning performance on SlowOnly-8 × 8-R18 trained with K200db-train, which is the setting used across this paper (Table 6 lower). The improvement of SMAD and OmniDebias is also steady and distinct upon this much stronger baseline.

6.3 Spatial-aware Multi-Aspect Debiasing

Ablation of SMAD. We first evaluate the efficacy of two components in SMAD, namely Multi-Aspect Adversarial Training (MAAT) and Spatial-Aware Actionness Reweighting (SAAR). Table 7 demonstrates that MAAT itself can largely improve the performance on test videos with novel scenes (`ARAS-64`, `K200db-val`). Upon this decent baseline, SAAR further boost the performance by 0.7% on `K200db-val` and 2.6% on `ARAS-64`. The improvement is non-trivial since SAAR only introduces 2% additional FLOPs and requires no additional explicit supervision. It is also worth noting that without the guidance from MAAT, the gain of SAAR is much reduced. The combination of MAAT and SAAR achieves large improvement on videos with novel scenes.

Table 8. Multi-factor v.s. single-factor debiasing.

Debias Factor	K200db-val
None	51.7
scene	53.2 ↑ 1.5
object	52.9 ↑ 1.2
scene, object	53.5 ↑ 1.8
scene, object, event, attribute, context	53.8 ↑ 2.1

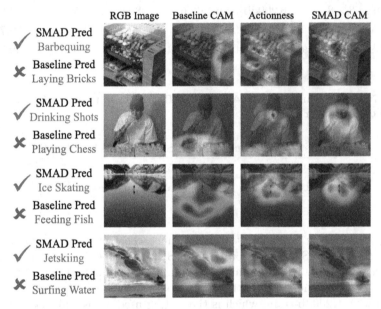

Fig. 7. The visualization of Actionness mask and CAM.[5]

Advantages of Multi-Aspect Debiasing. Multi-aspect debiasing is more generic than the *scene*-debiasing algorithm [11]. To prove that, we design a complex dataset split (K200-both-split, split by both *scene* and *object*) to mimic the real-world debiasing scenario. Specifically, we first create K200-scene-split and K200-obj-split using the introduced re-distributing method, with the factor *scene* and *object* respectively. Then we sample the validation videos from the union of two validation sets and sample the training videos from the remaining videos to form K200-both-split. On that split, we evaluate different debiasing factors. For multi-aspect debiasing with N factors, the weight of each adversarial loss is divided by N. Table 8 shows that multi-aspect debiasing consistently outperforms the single-aspect one under this setting: using both *scene* and *object* as debiasing factors outperforms each individual. Moreover, the best result is

Table 9. OmniDebias. We jointly train `K200db-train` with different web dataset splits. The improvement for z-unbiased test sets (`K200db-val`, `ARAS-64`) is much larger than `K200-val`.

Model	Aux-Data	ARAS-64	K200db-val	K200-val
2D	None	55.8	53.7	76.2
	GG-bias	60.0 ↑ 4.2	53.9 ↑ 0.2	76.0 ↓ 0.2
	GG-mid	60.5 ↑ 4.7	55.1 ↑ 1.4	76.3 ↑ 0.1
	GG-rand	64.1 ↑ 8.3	57.5 ↑ 3.8	77.2 ↑ 1.0
	GG-all	65.5 ↑ 9.7	58.4 ↑ 4.7	77.8 ↑ 1.6
	GG-unbias	68.3 ↑ 12.5	60.0 ↑ 6.3	77.4 ↑ 1.2
3D	None	51.0	51.9	75.1
	IG-rand	62.3 ↑ 11.3	56.8 ↑ 4.9	77.4 ↑ 2.3
	IG-all	63.4 ↑ 12.4	58.5 ↑ 6.6	78.3 ↑ 3.2
	IG-unbias	64.7 ↑ 13.7	59.8 ↑ 7.9	78.1 ↑ 3.0

achieved when using all five factors for debiasing (*event, attribute, context* are not used to create `K200-both-split`).

Qualitative Results. To qualitatively show how SAAR guides feature learning, we visualize the spatial-aware actionness mask predicted by SAAR and the class activation maps (CAM) [56] of models trained with or without SMAD in Fig. 7. Without debiasing, the rare scenes in action videos, *e.g.*, brick grill, many chessman-like shots, transparent ice surface, and huge waves may mislead the model to give out wrong predictions. With SMAD, models can learn to focus on human actions rather than scenes.

6.4 Exploiting Web Media with OmniDebias

Web Data Help in Debiasing. To exploit the richness and diversity of web media, we propose joint training with both labeled datasets and unlabeled web datasets. We first try to use the entire web dataset after teacher filtering for joint training, including both web image dataset `GG-all` and web video dataset `IG-all`. Table 9 shows that the performance improved by web media is considerable for z-unbiased test sets. For both baselines, the gain on `ARAS-64` is around 10%. The improvement on the normal test set `K200-val`, is milder ($1 \sim 4\%$) but also noticeable.

Data Selection Strategy. Although web media contain novel $z \in \mathcal{Z}$ that seldomly or never occurs in the original train set, the per action category z distributions are still highly imbalanced. To study the contribution of each portion of web media, we split each web dataset into 3 equal-sized parts: bias, mid, unbias. We also randomly sample a third from the web dataset (rand) for comparison. Using bias and mid leads to worse performance than rand. Using unbias, however, not only surpasses other subsets, but also outperforms training with all web data. With OmniDebias, the performance gap between z-unbiased test sets and K200-val is largely narrowed: the gap shrinks by around 10% Top1 for ARAS-64 and around 5% Top1 for K200db-val. Besides re-distributed datasets, the improvement of OmniDebias can also be observed when trained on the full Kinetics dataset[6].

7   Conclusion

In this work, we seek to mitigate the *generic* representation bias in action recognition. We propose SMAD and OmniDebias: SMAD integrates multi-head adversarial training and spatial-aware feature reweighting for algorithm debiasing, while OmniDebias exploits the rich diversity of web data efficiently for dataset debiasing. When combined, two components lead to excellent debiasing performance and perform far better on either artificially split test sets or manually collected out-of-distribution ones.

References

1. Agarwal, A., Beygelzimer, A., Dudík, M., Langford, J., Wallach, H.: A reductions approach to fair classification. In: ICML, pp. 60–69. PMLR (2018)
2. Barocas, S., Hardt, M., Narayanan, A.: Fairness and machine learning. Fairmlbook.org (2019). http://www.fairmlbook.org
3. Barocas, S., Selbst, A.D.: Big data's disparate impact. Calif. L. Rev. **104**, 671 (2016)
4. Bellamy, R.K., et al.: AI fairness 360: an extensible toolkit for detecting, understanding, and mitigating unwanted algorithmic bias. arXiv:1810.01943 (2018)
5. Bolukbasi, T., Chang, K.W., Zou, J., Saligrama, V., Kalai, A.: Man is to computer programmer as woman is to homemaker? debiasing word embeddings. arXiv:1607.06520 (2016)
6. Caliskan, A., Bryson, J.J., Narayanan, A.: Semantics derived automatically from language corpora contain human-like biases. Science **356**(6334), 183–186 (2017)
7. Carreira, J., Zisserman, A.: Quo vadis, action recognition? a new model and the kinetics dataset. In: CVPR, pp. 6299–6308 (2017)
8. Carter, K., Shah, M.: An approach for data efficient action recognition and mitigating scene bias (2020)
9. Chen, M.H., Kira, Z., AlRegib, G., Yoo, J., Chen, R., Zheng, J.: Temporal attentive alignment for large-scale video domain adaptation. In: ICCV, pp. 6321–6330 (2019)
10. Chen, Y., Li, W., Sakaridis, C., Dai, D., Van Gool, L.: Domain adaptive faster R-CNN for object detection in the wild. In: CVPR, pp. 3339–3348 (2018)
11. Choi, J., Gao, C., Messou, J.C., Huang, J.B.: Why can't i dance in the mall? learning to mitigate scene bias in action recognition. In: NeurIPS, pp. 853–865 (2019)

---

[6] We list the detailed results in the supplementary material.

12. Choi, J., Sharma, G., Chandraker, M., Huang, J.B.: Unsupervised and semi-supervised domain adaptation for action recognition from drones. In: WACV, pp. 1717–1726 (2020)
13. Choi, J., Sharma, G., Schulter, S., Huang, J.-B.: Shuffle and attend: video domain adaptation. In: Vedaldi, A., Bischof, H., Brox, T., Frahm, J.-M. (eds.) ECCV 2020. LNCS, vol. 12357, pp. 678–695. Springer, Cham (2020). https://doi.org/10.1007/978-3-030-58610-2_40
14. Deng, J., Dong, W., Socher, R., Li, L.J., Li, K., Fei-Fei, L.: Imagenet: a large-scale hierarchical image database. In: CVPR, pp. 248–255. IEEE (2009)
15. Diba, A., et al.: Holistic large scale video understanding. arXiv:1904.11451 38, 39 (2019)
16. Duan, H., Zhao, Y., Xiong, Y., Liu, W., Lin, D.: Omni-sourced webly-supervised learning for video recognition. arXiv:2003.13042 (2020)
17. Dwork, C., Hardt, M., Pitassi, T., Reingold, O., Zemel, R.: Fairness through awareness. In: ITCS, pp. 214–226 (2012)
18. Feichtenhofer, C., Fan, H., Malik, J., He, K.: Slowfast networks for video recognition. In: ICCV, pp. 6202–6211 (2019)
19. Feldman, M., Friedler, S.A., Moeller, J., Scheidegger, C., Venkatasubramanian, S.: Certifying and removing disparate impact. In: KDD, pp. 259–268 (2015)
20. Fernando, B., Habrard, A., Sebban, M., Tuytelaars, T.: Unsupervised visual domain adaptation using subspace alignment. In: ICCV, pp. 2960–2967 (2013)
21. Ganin, Y., Lempitsky, V.: Unsupervised domain adaptation by backpropagation. In: ICML, pp. 1180–1189. PMLR (2015)
22. Garcia, D., Weber, I., Garimella, V.R.K.: Gender asymmetries in reality and fiction: the Bechdel test of social media. arXiv:1404.0163 (2014)
23. Gretton, A., Borgwardt, K.M., Rasch, M.J., Schölkopf, B., Smola, A.: A kernel two-sample test. JMLR 13(1), 723–773 (2012)
24. Hara, K., Kataoka, H., Satoh, Y.: Can spatiotemporal 3D CNNs retrace the history of 2D CNNs and imagenet? In: CVPR, pp. 6546–6555 (2018)
25. Haselton, M.G., Nettle, D., Murray, D.R.: The evolution of cognitive bias. The handbook of evolutionary psychology, pp. 1–20 (2015)
26. He, K., Zhang, X., Ren, S., Sun, J.: Deep residual learning for image recognition. In: CVPR, pp. 770–778 (2016)
27. Holland, S., Hosny, A., Newman, S., Joseph, J., Chmielinski, K.: The dataset nutrition label: a framework to drive higher data quality standards. arXiv:1805.03677 (2018)
28. Huang, G., Liu, Z., Van Der Maaten, L., Weinberger, K.Q.: Densely connected convolutional networks. In: CVPR, pp. 4700–4708 (2017)
29. Jia, S., Lansdall-Welfare, T., Sudhahar, S., Carter, C., Cristianini, N.: Women are seen more than heard in online newspapers. PLoS ONE 11(2), e0148434 (2016)
30. Khosla, A., Zhou, T., Malisiewicz, T., Efros, A.A., Torralba, A.: Undoing the damage of dataset bias. In: Fitzgibbon, A., Lazebnik, S., Perona, P., Sato, Y., Schmid, C. (eds.) ECCV 2012. LNCS, vol. 7572, pp. 158–171. Springer, Heidelberg (2012). https://doi.org/10.1007/978-3-642-33718-5_12
31. Kuehne, H., Jhuang, H., Garrote, E., Poggio, T., Serre, T.: HMDB: a large video database for human motion recognition. In: ICCV, pp. 2556–2563. IEEE (2011)
32. Li, Y., Wang, N., Shi, J., Liu, J., Hou, X.: Revisiting batch normalization for practical domain adaptation (2016)
33. Li, Y., Vasconcelos, N.: Repair: removing representation bias by dataset resampling. In: CVPR, pp. 9572–9581 (2019)

34. Li, Y., Li, Y., Vasconcelos, N.: Resound: towards action recognition without representation bias. In: ECCV, pp. 513–528 (2018)
35. Madras, D., Creager, E., Pitassi, T., Zemel, R.: Learning adversarially fair and transferable representations. arXiv:1802.06309 (2018)
36. Mehrabi, N., Morstatter, F., Saxena, N., Lerman, K., Galstyan, A.: A survey on bias and fairness in machine learning (2019)
37. Oquab, M., Bottou, L., Laptev, I., Sivic, J.: Learning and transferring mid-level image representations using convolutional neural networks. In: CVPR, pp. 1717–1724 (2014)
38. Ray, J., et al.: Scenes-objects-actions: a multi-task, multi-label video dataset. In: ECCV, pp. 635–651 (2018)
39. Saito, K., Watanabe, K., Ushiku, Y., Harada, T.: Maximum classifier discrepancy for unsupervised domain adaptation. In: CVPR, pp. 3723–3732 (2018)
40. Shao, D., Zhao, Y., Dai, B., Lin, D.: FineGym: a hierarchical video dataset for fine-grained action understanding. In: CVPR, pp. 2616–2625 (2020)
41. Simonyan, K., Zisserman, A.: Two-stream convolutional networks for action recognition in videos. In: NeurIPS, pp. 568–576 (2014)
42. Soomro, K., Zamir, A.R., Shah, M.: Ucf101: a dataset of 101 human actions classes from videos in the wild. arXiv:1212.0402 (2012)
43. Thomee, B., et al.: YFCC100M: the new data in multimedia research. Commun. ACM **59**(2), 64–73 (2016)
44. Torralba, A., Efros, A.A.: Unbiased look at dataset bias. In: CVPR, pp. 1521–1528. IEEE (2011)
45. Tran, D., Wang, H., Torresani, L., Feiszli, M.: Video classification with channel-separated convolutional networks. In: ICCV, pp. 5552–5561 (2019)
46. Tran, D., Wang, H., Torresani, L., Ray, J., LeCun, Y., Paluri, M.: A closer look at spatiotemporal convolutions for action recognition. In: CVPR, pp. 6450–6459 (2018)
47. Wang, A., Narayanan, A., Russakovsky, O.: REVISE: a tool for measuring and mitigating bias in visual datasets. In: ECCV (2020)
48. Wang, L., Qiao, Y., Tang, X., Van Gool, L.: Actionness estimation using hybrid fully convolutional networks. In: CVPR, pp. 2708–2717 (2016)
49. Wang, L., et al.: Temporal segment networks: towards good practices for deep action recognition. In: Leibe, B., Matas, J., Sebe, N., Welling, M. (eds.) ECCV 2016. LNCS, vol. 9912, pp. 20–36. Springer, Cham (2016). https://doi.org/10.1007/978-3-319-46484-8_2
50. Woodworth, B., Gunasekar, S., Ohannessian, M.I., Srebro, N.: Learning non-discriminatory predictors. In: COLT, pp. 1920–1953. PMLR (2017)
51. Xie, S., Sun, C., Huang, J., Tu, Z., Murphy, K.: Rethinking spatiotemporal feature learning for video understanding. arXiv:1712.04851 1(2), 5 (2017)
52. Zafar, M.B., Valera, I., Gomez Rodriguez, M., Gummadi, K.P.: Fairness beyond disparate treatment & disparate impact: learning classification without disparate mistreatment. In: WWW, pp. 1171–1180 (2017)
53. Zhang, B.H., Lemoine, B., Mitchell, M.: Mitigating unwanted biases with adversarial learning. In: AIES, pp. 335–340 (2018)
54. Zhao, J., Wang, T., Yatskar, M., Ordonez, V., Chang, K.W.: Men also like shopping: reducing gender bias amplification using corpus-level constraints. arXiv:1707.09457 (2017)
55. Zhao, Y., Xiong, Y., Wang, L., Wu, Z., Tang, X., Lin, D.: Temporal action detection with structured segment networks. In: ICCV, pp. 2914–2923 (2017)

56. Zhou, B., Khosla, A., Lapedriza, A., Oliva, A., Torralba, A.: Learning deep features for discriminative localization. In: CVPR, pp. 2921–2929 (2016)
57. Zhou, B., Lapedriza, A., Khosla, A., Oliva, A., Torralba, A.: Places: A 10 million image database for scene recognition. TPAMI **40**(6), 1452–1464 (2017)

SegTAD: Precise Temporal Action Detection via Semantic Segmentation

Chen Zhao[✉] [iD], Merey Ramazanova[iD], Mengmeng Xu[iD],
and Bernard Ghanem[iD]

King Abdullah University of Science and Technology (KAUST),
Thuwal, Saudi Arabia
{chen.zhao,merey.ramazanova,mengmeng.xu,bernard.ghanem}@kaust.edu.sa

Abstract. Temporal action detection (TAD) is an important yet challenging task in video analysis. Most existing works draw inspiration from image object detection and tend to reformulate it as a proposal generation - classification problem. However, there are two caveats with this paradigm. *First*, proposals are not equipped with annotated labels, which have to be empirically compiled, thus the information in the annotations is not necessarily precisely employed in the model training process. *Second*, there are large variations in the temporal scale of actions, and neglecting this fact may lead to deficient representation in the video features. To address these issues and *precisely* model TAD, we formulate the task in a novel perspective of semantic segmentation. Owing to the 1-dimensional property of TAD, we are able to convert the coarse-grained detection annotations to fine-grained semantic segmentation annotations for free. We take advantage of them to provide precise supervision so as to mitigate the impact induced by the imprecise proposal labels. We propose a unified framework SegTAD composed of a 1D semantic segmentation network (1D-SSN) and a proposal detection network (PDN). We evaluate SegTAD on two important large-scale datasets for action detection and it shows competitive performance on both datasets.

1 Introduction

Nowadays, millions of videos are produced every day, and high demand arises for automatic video processing and analysis. To this end, various tasks have emerged, for example, action recognition [19], active speaker detection [2], video-language grounding [41], temporal action localization [26,42]. Among those tasks, temporal action detection in untrimmed videos, in particular, is one of the fundamental yet challenging tasks. It requires not only to recognize what actions take place in a video but also to localize when they start and end.

Most recent works in the literature regard this task as a temporal version of object detection and tackle it by adapting the 2-dimensional solutions on images (e.g., Faster R-CNN [36]) to the 1-dimensional temporal domain for videos [5,17,23,47]. A conventional pipeline is to first identify candidate action segments (i.e., proposals) by analyzing the entire video sequence and then learn to score each segment with an empirically compiled label for each proposal. This

© The Author(s), under exclusive license to Springer Nature Switzerland AG 2023
L. Karlinsky et al. (Eds.): ECCV 2022 Workshops, LNCS 13804, pp. 576–593, 2023.
https://doi.org/10.1007/978-3-031-25069-9_37

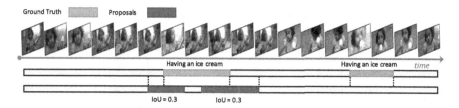

Fig. 1. Proposal Annotations. Proposals different in locations, lengths, and content are assigned the same label if they have the same IoU.

object-detection inspired framework has brought significant improvement on the action detection performance [17], especially with the aid of deep neural network in recent years [5,27,47,48]. However, it lays two caveats that might lead to imprecise action detection modeling.

First, proposals are not accompanied by any annotated labels from the dataset since they are generated on the fly. Their training labels have to be manually compiled based on the ground-truth *action* annotations, i.e., the start/end timestamps of *actions* in each video and their corresponding categories. A common practice is to compare each proposal to each ground-truth action in the video in terms of some metric (e.g., temporal Intersection over Union) and use a preset threshold to determine whether a proposal is positive or negative with respect to each category. However, this is obviously not an optimal approach considering that the mapping between action annotations and proposal annotations are not bijective (as shown in Fig. 1). Noise is inevitably introduced to the compiled proposal labels regardless of what metric or threshold is adopted, resulting in imprecise modeling. Note that even in object detection, it is still an open question on how to identify positive and negative proposals, which is crucial to detection performance [52].

Second, object detection is a relatively coarse-grained problem that does not identify every single pixel but predicts a rectangular box surrounding an entire object. However, videos especially in large-scale datasets, e.g., ActivityNet [10], HACS [55], Ego4D [14] contain actions of dramatically varied temporal duration - from less than a second to minutes. Therefore, shifting from the image domain to the video domain without adapting to the video diversity could lead to deficient feature representation (e.g., burying short actions and under-representing long actions by imprecisely modeling temporal correlations), as well as misalignment between proposals and their receptive fields [5].

To address these issues, in this paper, we propose to formulate the task of temporal action detection (TAD) in a novel perspective with semantic segmentation. In the image 2-dimensional (2D) domain, much more effort is demanded to obtain finer-grained annotations for the tasks such as semantic segmentation, considering that not all pixels in a detection bounding box are contained in the object. In contrast, the task of video TAD requires only 1-dimensional (1D) localization of actions — along the temporal domain. Therefore, all frames within the action boundaries naturally belong to the action category. The detection annotations can be bijectively transformed to segmentation labels without extra effort. We propose a unified TAD framework to take advantage of the fine-grained prediction of semantic segmentation for more precise detection, dubbed as SegTAD. SegTAD contains a 1D semantic segmentation network to learn the category of each single frame using the segmentation labels, which are directly transformed from the detection annotations without introducing any label compilation noise. Regarding the second issue, we design SegTAD modules based on atrous and graph convolutions to precisely represent actions of various temporal duration. **The main contributions are:**

1) We formulate TAD in a novel perspective of semantic segmentation and propose a unified TAD framework SegTAD, which is composed of a 1D semantic segmentation network (1D-SSN) and a proposal detection network (PDN).

2) In 1D-SSN, we design an hourglass architecture with a module of parallel astrous and graph convolutions to effectively aggregate global features and multi-scale local features. In PDN, we incorporate a proposal graph to exploit cross-proposal correlations in our unified framework.

3) The proposed SegTAD achieves competitive performance on two representative large-scale datasets ActivityNet-v1.3 [10] and HACS-v1.1 [55].

2   Related Work

2.1   Temporal Action Detection

Concurrent TAD methods tend to adopt the two-stage framework: 1) generating candidate action segments (i.e., proposals) from the video sequence; 2) cropping each proposal out of the sequence, and classifying each proposal to obtain its confidence score. A large number of these methods focus on improving the first stage to generate proposals with high recall, applying an off-the-shelf classifier (e.g., SVM) to get the detection results [3,9,11,17,29]. Some other methods focus on the second stage, seeking to build more accurate classifiers on proposals produced by other proposal methods (e.g., sliding windows, the above-mentioned first-stage methods) [35,38,39,51,57]. The third category of methods propose unified approaches, where the features of different frames are aggregated and the actions are predicted from the same network [5,16,22,25,30,47,48,50,54]. Our paper belongs to the third category. Among these methods, our SegTAD is related to but essentially different from them in the following aspects.

Snippet-level classification. Multiple methods have identified the coarse granularity and regular distribution issues of anchor-based proposals, such as BSN [25], TAG [57], MGG [29], and CTAP [11]. They have proposed to incorporate snippet-level proposals as a supplement or replacement to the anchor-based ones. They learn a binary classifier for each snippet, either by a 2D convolutional neural network (CNN) on each snippet [11,57] or applying a temporal CNN on the entire sequence [25,29]. By this means, they obtain the probability of being an action/start/end for each snippet, based on which to generate proposals with flexible duration. In addition, the second-stage method CDC [38] also classifies each snippet, but the purpose is to refine an existing proposal instead. In this paper, the proposed SegTAD directly formulates a 1D semantic segmentation problem to classify every single frame into different action categories. It enables the use of large temporal resolution and supports multi-scale feature aggregation with the proposed PAG module. Moreover, it doesn't rely on the actionness/startness/endness scores to generate proposals as in TAG or BSN.

Snippet-and-snippet correlations. The method G-TAD [48] exploits temporal correlations between snippets by adopting a graph convolutional network (GCN). It supports limited temporal resolution due to its lack of multi-scale design and the complexity constraint of GCN when more frames are utilized, consequently sacrificing actions of short duration. Comparatively, our SegTAD adopts an hourglass architecture with an encoder and decoder, and only apply graph convolutions in the intermediate layer with the smallest resolution. In this way, it aggregates global information while preserving the temporal resolution.

Proposal-and-proposal correlations. BMN [22] constructs a boundary map with densely-distributed proposals and apply convolutions on the map to utilize the correlations between proposals, whereas 2D-TAN [53] presents a sparse 2D temporal feature map to represent and correlate proposals. The second-stage method P-GCN [51] uses GCNs [28] on proposals obtained by other methods to improve the proposal scores and boundaries. Our SegTAD incorporates graph and edge convolutions to our *unified* detection framework to exploit cross-proposal correlations. Compared to the standalone P-GCN [51], which is essentially a proposal post-processing method and does not consider correlations between frames, SegTAD jointly learns the graph network with the 1D semantic segmentation network and enhances feature representations via cross-frame and cross-proposal aggregation.

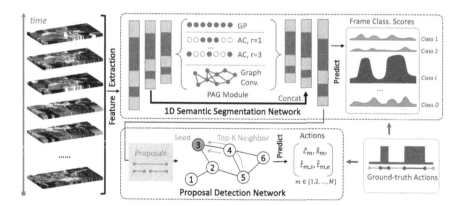

Fig. 2. Illustration of our proposed SegTAD architecture. Input: a sequence of video frames; **Output:** scored candidate actions. **Top:** 1D Semantic segmentation network (1D-SSN) that learns to classify each frame in the sequence. We design a module of parallel atrous and graph convolutions (PAG) to effectively aggregate global features and multi-scale local features. **Bottom:** Proposal detection network (PDN) that scores each candidate action. Graph convolutions are utilized to exploit correlations between proposals.

2.2   Object Detection and Semantic Segmentation

In the image domain, object detection [6,12,13,34,36] is a coarse-grained prediction problem, whose output is a rectangular bounding box that surrounds an object in the image. A widely adopted framework for tackling this task is the two-stage method (e.g., R-CNN [13], Fast R-CNN [12], and Faster R-CNN [36]), which first generates candidate proposals from the original image, and then runs a classifier for each proposal. Recent TAD methods tend to draw inspiration from these object detection methods. But by noticing the 1-dimensional property of TAD problem, we see that besides object detection, TAD has another analogy in the image domain, which is semantic segmentation.

Semantic segmentation [7,18,20,31,37] is a fine-grained prediction task that predicts the class label of every pixel in an image. Thus, annotating for segmentation usually requires extraordinarily more efforts than for object detection. Compared to object detection which resorts to proposals, semantic segmentation usually adopts a different framework, which seeks to preserve the dense grid of the input image while learning its high-level semantic features with a convolutional network. Representative works are U-Net [37] and FCN [31], etc. In videos, temporal semantic annotations and TAD annotations are bijectively transferable, so no extra efforts are required to annotate segmentation. In this work, taking advantage of this discovery, we utilize the semantic segmentation methodology to formulate TAD.

3   Proposed SegTAD

3.1   Problem Formulation and SegTAD Framework

The task of temporal action detection (TAD) is to predict a set of actions $\Phi = \{\phi_m = (t_{m,s}, t_{m,e}, c_m, s_m)\}_{m=1}^M$ given a sequence of T video frames $\{I_t \in \mathbb{R}^{3 \times H \times W}\}_{t=1}^T$, where $t_{m,s}$ and $t_{m,e}$ are action start and end time respectively, c_m is action label, and s_m is prediction confidence. To achieve this, we first transform each frame I_t to a 1-dimensional feature vector $\mathbf{x}_t \in \mathbb{R}^C$ via a feature extraction network (see Sec. 4.1). Using the 1D features $\mathbf{x}_t$ as input, we apply our proposed 1D semantic segmentation network (1D-SSN), followed by a proposal detection network (PDN). The 1D-SSN temporally aggregates features to learn to segment the video sequence in the frame level according to the action annotations, and generates semantic features $\mathbf{y}_t \in \mathbb{R}^{C'}$ for each frame. The PDN learns to score each candidate action and further refines its boundaries. The two components 1D-SSN and PDN are trained in a unified architecture.

We illustrate the entire architecture of SegTAD in Fig. 2. It shows two main components: 1D semantic segmentation network and proposal detection network, which will be described in the following subsections, respectively.

3.2   1D Semantic Segmentation Network

Different from conventional TAD works, which perform prediction in the coarse segment level and compile segment labels from ground-truth action annotations, we use 1D semantic segmentation (1D-SSN) to learn to predict for each single frame. Based on our 1D-SSN, we are able to take advantage of the original true action annotations without introducing any label noise. In the following, we first describe the 1D-SSN architecture that aggregates features from a global temporal range as well as multi-scale local range. Then, we present our segmentation loss that uses the original action annotations to train the segmentation network.

1D-SSN Architecture. Sufficient semantic information from a long temporal range is essential for TAD. This is usually achieved by enlarging the receptive field via strided 1D convolution or pooling. However, aggressively using these operations will dramatically reduce the temporal resolution and severely impair the feature representation of short actions.

To achieve a large receptive field without severely sacrificing temporal resolution, we consider feature aggregation in two aspects: local feature aggregation and global feature aggregation. The former aggregates features in a surrounding temporal window to learn local patterns. We need to make it scale-invariant to represent actions of different duration. The latter associates features in a global range, not constrained in the neighborhood of each frame. This breaks the constraints of the temporal locations of each frame and makes use the correlations between frames in the global context [48].

For the two aspects, we design an hourglass architecture with atrous and graph convolutions in our 1D-SSN. It has the shape of an hourglass, containing an encoder, a parallel module of atrous and graph convolutions (PAG), and a decoder. The encoder temporally downscales the input features by a small ratio, and the decoder is to restore the temporal resolution. The PAG module enables local feature aggregation in multiple scales using atrous convolutions and global feature aggregation via graph convolutions.

1D-SSN Details. The encoder is comprised of a stack of L strided 1D convolution layers Conv1D(k=3,s=2), where k is the kernel size, s is the stride, followed by the non-linear activation function ReLU. We only have L=3 such layers in order not to overly downscale the features. It applies along the temporal dimension on the input video feature sequence $\mathbf{X}=[\mathbf{x}_1, \mathbf{x}_2, \ldots, \mathbf{x}_T]\in\mathbb{R}^{C\times T}$, where C is the input feature dimension, and transforms it into a representation with lower temporal resolution $\mathbf{X}'=[\mathbf{x}'_1, \mathbf{x}'_2, \ldots, \mathbf{x}'_{T/2^L}]\in\mathbb{R}^{C'\times T/2^L}$, where C' is the new feature dimension. The encoder reduces the sequence temporal resolution by a factor of 2^L so as to reduce computation for the subsequent layers, as well as to progressively increase the size of temporal receptive field.

The decoder upscales the temporal resolution of the features $\mathbf{X}'' = $ PAG($\mathbf{X}'$), where PAG stands for operations in the module of parallel atrous and graph convolutions (PAG) (detailed in following paragraphs). It contains a layer of linear interpolation to rescale the features along the temporal dimension to the orignal resolution. In order to complement the details information lost from the encoder, we add a highway connection from the low-level features at the second Conv1D layer in the encoder, which consists of Conv1D($k = 1, s = 1$), batch normalization and ReLU. Then we concatenate the output of this connection with the interpolated features, and apply a Conv1D($k = 3, s = 1$) layer to adaptively fuse them. With this hourglass (encoder-decoder) architecture, we gradually aggregate features from frames further apart while preserving the temporal resolution of the sequences.

The module of parallel atrous and graph convolutions (PAG) (Fig. 2) takes the encoded features $\mathbf{X}'$ as input, and further enlarges the receptive field and empower the features with scale-invariant capability. Considering that a video sequence usually contains actions of various temporal duration, ranging from a couple of seconds to minutes. Excessive pooling or using strided convolutions could impair short actions as mentioned above, whereas long actions require large receptive field to be semantically represented. To adapt to actions of variant temporal scales, we propose this PAG module, which contains atrous convolutions to aggregate features from multi-scale local neighborhood, and graph convolutions to aggregate features from global context. Note that unlike G-TAD [48], which applies graph convolutions in every layer and consequently incurs huge computation cost, we only have them in this intermediate module after the resolution is reduced by the encoder. In this way, it aggregates global information while preserving the temporal resolution.

Atrous convolutions systematically aggregate multi-scale contextual information without losing resolution. They are able to support expansion of the receptive field [49] by filling in empty elements in the convolutional filter. Compared to normal convolutions, they are equipped with a dilation ratio d to specify the number of empty elements in the filter, reflecting the expansion ratio of the receptive field. We use 4 parallel branches of 1D atrous convolutions with different dilation ratios. The choice of dilation ratios will be discussed in Sec. 4.3.

Graph convolutions model the correlations among snippets in a non-local context. We design a graph convolutional network in parallel with the multiple branches of atrous convolutions. Specifically, based on the output features from the encoder $\{\mathbf{x}'_t\}_{t=1}^{T/2^L}$ (we call $\mathbf{x}'_t$ features of a snippet in the following), we build a graph denoted as $\mathcal{G}_s = \{\mathcal{V}_s, \mathcal{E}_s\}$. $\mathcal{V}_s = \{v_t\}_{t=1}^{T/2^L}$ refers to the graph nodes, each corresponding to a snippet, and $\mathcal{E}$ denotes graph edges, which represent the correlations between snippets. To model the correlations of snippets in a global context, we construct the edges dynamically [45] according to the semantic similarity between encoded snippet features rather than their temporal locations, which are computed as minus mean square error (MSE) between two feature vectors. If a snippet is among the top K nearest neighbors of another snippet in terms of their semantic similarity, there is an edge connecting them.

With this graph, we apply one layer of edge convolutions to aggregate features of connected nodes [45], formulated as

$$\mathbf{X_{GC}} = ([\mathbf{X}'^T, \mathbf{A}\mathbf{X}'^T - \mathbf{X}'^T]\mathbf{W})^T, \tag{1}$$

where $\mathbf{A} \in \mathbb{R}^{T/2^L \times T/2^L}$ is the adjacency matrix defined by edge connections between snippets, its $(i, j)^{\text{th}}$ element $\mathbf{a}_{i,j} = 1$ if there is an edge between the ith and the jth snippets, and $\mathbf{a}_{i,j} = 0$, otherwise. For each snippet, $\mathbf{A}\mathbf{X}'^T$ aggregates features from all its connected snippets in the whole sequence. The operation $[\cdot, \cdot]$ concatenates the two feature vectors. $\mathbf{W} \in \mathbb{R}^{C' \times C'}$ denotes trainable parameters.

In order to aggregate the global context information along the temporal dimension, we add a global fast path that first does global average pooling and linearly upsamples back to the original resolution. This mitigates the weight validity issue when large dilation ratios are used [7]. Then we concatenate the output of all atrous convolutional (AC) branches and the graph convolution (GC) network as well as the global fast path (GP), formulated as

$$\text{PAG}(\mathbf{X}') = [\mathbf{X}_{\text{GC}}, \mathbf{X}^1_{\text{AC}}, \dots, \mathbf{X}^B_{\text{AC}}, \mathbf{X}_{\text{GP}}]. \tag{2}$$

Finally, a Conv1D layer followed by ReLU is applied to fuse all branches.

Segmentation Loss. In 1D-SSN, we formulate TAD as a semantic segmentation problem, and predict the category of each single frame to meet their true categories. In the following, we describe how to generate predictions, and formulate the segmentation loss using action annotations.

The output from the decoder in 1D-SSN is a sequence of aggregated feature vectors $\mathbf{Y} = [\mathbf{y}_1, \mathbf{y}_2, \dots, \mathbf{y}_T] \in \mathbb{R}^{C' \times T}$. We use this to predict per-frame classification labels. Suppose we have D action categories, applying one layer of linear transformation and Softmax operation yields

$$\mathbf{p}_t = \text{Softmax}(\mathbf{W}^T_{seg} \mathbf{y}_t), \tag{3}$$

where $\mathbf{p}_t \in \mathbb{R}^D$ refers to the predicted label for t^{th} frame, $\mathbf{W}_{seg} \in \mathbb{R}^{C' \times D}$ contains the parameters in the linear layer.

If we know the ground-truth label $b_t \in \{1, 2, 3, \dots, D\}$ of each frame, then we can compute the segmentation loss using cross-entropy formulated as follows

$$\mathcal{L}_{seg} = -\frac{1}{T} \sum_{1 \leq t \leq T} \sum_{1 \leq d \leq D} \beta_{t,d} \log p_{t,d}, \tag{4}$$

where $\beta_{t,d} = \mathbb{1}\{b_t = d\}$ is the dth one-hot encoding of the label for the tth frame.

Now the question becomes how we obtain the segmentation label b_t. Assume that a video sequence is annotated with N actions $\Psi = \{\psi_n = (t_{n,s}, t_{n,e}, c_n)\}_{n=1}^N$, where $t_{n,s}$ and $t_{n,e}$ denote the start and end time of the nth action instance, respectively, and c_n represents its category. This is a segment-level annotation, without specifying the exact labels of each frame. However, due to the 1D characteristic of TAD, we can easily transform this segment-level annotations to finer-grained frame-level labels. It assigns a frame the label of an action if it falls inside the action boundaries, otherwise, the frame is labeled as background. We see that in the entire process, there is no hyper-parameter and the mapping is bijective, which guarantees precise annotation transformation.

Additionally, considering that action boundaries are important for localize an action, we introduce an auxiliary loss

$$\mathcal{L}_{aux} = \frac{-\sum_{1 \leq t \leq T} \beta^s_t \log p^s_t + (1 - \beta^s_t) \log(1 - p^s_t)}{T}$$

$$= \frac{-\sum_{1 \leq t \leq T} \beta^e_t \log p^e_t + (1 - \beta^e_t) \log(1 - p^e_t)}{T}, \tag{5}$$

where $\beta_t^s, \beta_t^e \in \{0,1\}$ are start and end labels that indicate whether a frame is the first or last frame of an action. p_t^s and p_t^e are predicted confidence scores of a frame being start and end of action, which are generated by

$$\mathbf{p}^s = \text{Sigmoid}(\mathbf{w}_{s2}^T\text{ReLU}(\text{Conv1d}_{k,s=3,1}(\mathbf{Y}; \mathbf{W}_{s1}))) \tag{6}$$

$$\mathbf{p}^e = \text{Sigmoid}(\mathbf{w}_{e2}^T\text{ReLU}(\text{Conv1d}_{k,s=3,1}(\mathbf{Y}; \mathbf{W}_{e1}))) \tag{7}$$

where $\mathbf{W}_{s1}$ and $\mathbf{W}_{e1}$ represent convolutional kernels, and $\mathbf{w}_{s2}, \mathbf{w}_{e2} \in \mathbb{R}^{C' \times 1}$ are parameters of the linear layers.

3.3 Proposal Detection Network

Considering that the actions in the format of $\Phi = \{\phi_m = (t_{m,s}, t_{m,e}, c_m, s_m)\}_{m=1}^M$ are not predicted directly by 1D-SSN, we need an extra detection head, for which we design a proposal detection network (PDN). PDN takes the output features from 1D-SSN along with our designed sparse segment patterns as input, and generate predicted actions. This PDN takes advantage of cross-proposal correlations via a graph network, and further enhances the representation of each frame and each proposal.

PDN Architecture: As shown in Fig. 2, PDN takes the output features $\mathbf{Y}$ from 1D-SSN as well as a sparse pattern of segments. In our framework, in order to precisely detection short actions, we use a high temporal resolution $L = 1000$. Therefore, it is cumbersome to enumerate all possible pairs of frames as proposals as done in [22,48]. Instead, we design a sparse pattern of segments, which covers a large variety of action duration and reduces computation compared to dense segments. Let each element $u_{i,j} \in \{0,1\}$ of the matrix $\mathbf{U} \in \mathbb{R}^{L \times L}$ denote whether the segment starting from i^{th} frame and composed of j frames is selected as a proposal. Its value is determined by the following equation

$$u_{i,j} = \begin{cases} 1, & \text{if } i \% \eta = 0 \text{ and } j \% \eta = 0; \\ 0, & \text{otherwise.} \end{cases} \tag{8}$$

where $\eta=8$ is a step size controlling the sparsity degree. With $\Phi=\{\phi_m=(t_{m,s}, t_{m,e})\}_{m=1}^M$ being all M proposals specified by $\mathbf{U}$, their features $\mathbf{D}=\{\mathbf{d}_m \in \mathbb{R}^{C'}\}_{m=1}^M$ are extracted from video features $\mathbf{Y}$ based on SGAlign [48].

The proposals in the same video are highly correlated and utilizing this property can enhance proposal representations [22]. To model correlations between proposals from any temporal locations, we build a second graph $\mathcal{G}_p = \{\mathcal{V}_p, \mathcal{E}_p\}$ on the proposals and take advantage of graph convolutions in the detection network. Different from the graph $\mathcal{G}_s$ in 1D-SSN, each node in $\mathcal{G}_p$ refers to one proposal, and the edges represent correlations between proposals. Another difference is that the edges $\mathcal{E}_p$ here are constructed based on the temporal intersection over union (tIoU) between proposals, as opposed to the dynamically determined edges in 1D-SSN. We apply the same edge convolutions as shown in Eq. (1), but define each element $a_{i,j}$ in the adjacency matrix $\mathbf{A}$ as an attention value computed as

$a_{i,j} = \mathbf{d}_i^T \mathbf{d}_j / |\mathbf{d}_i| \cdot |\mathbf{d}_j|$ if there is an edge. We stack 3 layers of edge convolutions in PDN.

In order to efficiently train the proposal network as well as to balance the positive and negative samples, we need to sample from our M proposals. Randomly sampling does not guarantee that the sampled proposals have consistent edge connections with each other to form a meaningful graph. So a better strategy is to sample neighborhoods of proposals rather than individual proposals. We adopt the following sampling strategy based on the SAGE method [15]. We first sample a small number of M_0 seed proposals, including $M_0/2$ positive and $M_0/2$ negative samples. Then for each seed proposal, we find its top K neighbors based on its tIoU with other proposals, and put them into the sampling list. For each of the K neighboring proposals, we find its top K neighbors from the remaining proposals, and add these $K \times K$ proposals into the sampling list as well. Hereby, in the sampling list, we totally have $M_0(1 + K + K \times K)$ proposals, all of which have their top K neighbors in the list. $M_0 = 50$ and $K = 4$ by default. In inference, we use all M proposals without sampling.

Detection Loss: The PDN enhances the feature representation of each proposal by aggregating different proposals, formulated as $\mathbf{D}' = \text{PDN}(\mathbf{D})$. We predict proposals' confidence of being actions using the following operation

$$\mathbf{S} = \text{Sigmoid}(\mathbf{W}_{det}^T \mathbf{D}'), \qquad (9)$$

where $\mathbf{W}_{det} \in \mathbb{R}^{C' \times 2}$ contains the parameters in the linear layer to predict the confidence scores. Note that $\mathbf{S} = [\mathbf{s}_1; \mathbf{s}_2]$ contains two different scores for each proposal, each corresponding to one loss function we define in the following

$$\mathcal{L}_{det} = \mathcal{L}_{reg}(\mathbf{h}_{reg}, \mathbf{s}_1) + \mathcal{L}_{cls}(\mathbf{h}_{cls}, \mathbf{s}_2), \qquad (10)$$

where $\mathbf{h}_{reg}$ is the tIoU between each proposals and their closest ground-truth actions, and $\mathcal{L}_{reg}$ is computed using mean square errors. $\mathbf{h}_{cls} = \mathbb{1}(\mathbf{h}_{reg} > \tau)$, where $\tau = 0.5$ is an tIoU threshold determining a proposal's binary label, and $\mathcal{L}_{cls}$ is a binary cross-entropy loss similarly computed as either term in Eq. (5).

3.4 Training and Inference

Training: We train the proposed SegTAD end to end using the segmentation loss and the detection loss, as well as the auxiliary loss as follows

$$\mathcal{L} = \mathcal{L}_{seg} + \lambda_1 \mathcal{L}_{det} + \lambda_2 \mathcal{L}_{aux} + \lambda_3 \mathcal{L}_r, \qquad (11)$$

where $\mathcal{L}_r$ is $\mathcal{L}_2$-norm, λ_1, λ_2, λ_3 denotes the weights for different loss terms, which are all set to 1 by default.

Inference: At inference time, we compute the score of each candidate action using the two scores predicted from the proposal detection network as $s_m = s_{m,1} \times s_{m,2}$. Then we run soft non-maximum suppression (NMS) using the scores and keep the top 100 predicted segments as final output.

Table 1. Action detection result comparisons on validation set of ActivityNet-v1.3, measured by mAP (%) at different tIoU thresholds and the average mAP. G-TAD achieves better performance in average mAP than the other methods, even the latest work of BMN and P-GCN shown in the second-to-last block. (* Re-implemented with the same features as ours.)

Method	0.5	0.75	0.95	Average
Wang et al. [44]	43.65	–	–	–
Singh et al. [40]	34.47	–	–	–
SCC [16]	40.00	17.90	4.70	21.70
Lin et al. [24]	44.39	29.65	7.09	29.17
CDC [38]	45.30	26.00	0.20	23.80
TCN [8]	37.49	23.47	4.47	23.58
R-C3D [47]	26.80	–	–	–
SSN [57]	34.47	–	–	–
BSN [25]	46.45	29.96	8.02	30.03
TAL-Net [5]	38.23	18.30	1.30	20.22
P-GCN+BSN [51]	48.26	33.16	3.27	31.11
BMN [22]	<u>50.07</u>	**34.78**	8.29	<u>33.85</u>
BMN* [22]	48.56	33.66	<u>9.06</u>	33.16
I.C & I.C [56]	43.47	33.91	**9.21**	30.12
SegTAD (top-1 cls.)	49.86	34.37	6.50	33.53
SegTAD (top-2 cls.)	**50.52**	<u>34.76</u>	6.85	**33.99**

Table 2. Action detection results on HACS-v1.1, measured by mAP (%) at different tIoU thresholds and the average mAP.

Method	Validation				Test
	0.5	0.75	0.95	Average	Average
SSN [55]	28.82	18.80	5.32	18.97	16.10
BMN [1]	–	–	–	–	22.10
S-2D-TAN [53]	–	–	–	–	23.49
SegTAD	**43.33**	**29.65**	**6.23**	**29.24**	**28.90**

4   Experimental Results

4.1   Datasets and Implementation Details

Datasets and Evaluation Metric. We conduct our experiments on two large-scale action understanding dataset, **ActivityNet-v1.3** [10] and **HACS-v1.1** [55] for TAD. **ActivityNet-v1.3** contains around $20,000$ temporally annotated untrimmed videos with 200 action categories. Those videos are randomly divided into training, validation and testing sets by the ratio of 2:1:1. **HACS-v1.1** follows the same annotation scheme as ActivityNet-v1.3. It also includes 200 action categories but collects $50,000$ untrimmed videos for TAD. We evaluate SegTAD performance with the average of mean Average Precision (mAP) over 10 different IoU thresholds [0.5:0.05:0.95] on both datasets.

Implementations. For ActivityNet-v1.3, we sample each video at 5 frames per second and adopt the two-stream network by Xiong et. al. [46] pre-trained on Kinetics-400 [4] to extract frame-level features, and rescale each sequence into 1000 snippets as SegTAD input. For HACS, we use the publicly available features extracted using an I3D-50 [4] model pre-trained on Kinetics-400 [4] and temporally rescale them into 400 snippets. We implement and test our framework using PyTorch 1.1, Python 3.7, and CUDA 10.0. In training, the learning rates are $1e-5$ on ActivityNet-1.3 and $2e-3$ on HACS-v1.1 for the first 7 epochs, and are reduced by 10 for the following 8 epochs. In inference, we leverage the global video context and take the top-1 or top-2 video classification scores from the action recognition models of [21,43], respectively for the two datasets, and multiply them by the confidence score c_j for evaluation.

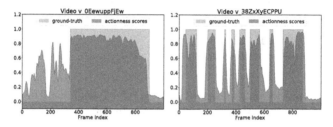

Fig. 3. Predicted per-frame classification scores compared to ground-truth labels. We only plot the scores of the ground-truth category. Green curves represent the ground-truth, and score = 1.0 represents the frames are inside action, and score =0.0 otherwise. Purple curves represent the predicted scores.

4.2   Comparison to State-of-the-Art

In Table 1 and Table 2, we compare SegTAD with representative TAD works in the literature. We report mAP at different tIoU thresholds and average mAP.

On ActivityNet-v1.3, SegTAD achieves competitive average mAP of 33.99%, significantly outperforming the recent works I.C & I.C [56] and BMN [22]. Notably, BMN extracts video features from ActivityNet-finetuned model such that the extracted features are more distinguishable on the target dataset. In contrast, we use more general Kinetics-pretrained features. To achieve fair comparison, we also show the re-produced BMN experimental results with the same features as ours, and our performance gain is even more remarkable. On HACS-v1.1, SegTAD reaches 28.90% average mAP on the test set, surpassing both the challenge winner S-2d-TAN [53] and BMN [22] by large margins. Compared with ActivityNet-v1.3, HACS-v1.1 is more challenging because of its substantial data-scale and precise segment annotations. Therefore, our superior performance on HACS-v1.1 makes SegTAD more remarkable.

4.3   Ablation Study

We provide ablation study to demonstrate the importance of the proposed 1D semantic segmentation network to the detection performance. Also we verify the

Table 3. Effectiveness of our segmentation network.

Segmentation loss	0.5	0.75	0.95	Avg
✗		49.15	33.45	3.81	32.45
✓		**49.86**	**34.37**	**6.50**	**33.53**

Table 4. Different loss functions for segmentation.

Segment. loss types	0.5	0.75	0.95	Avg
Binary	49.35	33.77	4.07	32.79
SegTAD	**49.86**	**34.37**	**6.50**	**33.53**

Table 5. Ablating studies in PAG of 1D-SSN. AC: Atrous convolutions, GC: graph convolutions, GP: global fast path.

PAG Branches			mAP at different tIoUs			
AC	GC	GP	0.5	0.75	0.95	Avg.
✗	✓	✓	48.55	33.04	5.21	32.37
✓	✗	✓	49.48	33.95	**7.50**	33.30
✓	✓	✗	49.75	34.25	5.97	33.29
✓	✓	✓	**49.86**	**34.37**	6.50	**33.53**

Table 6. Different sets of dilation ratios of the AC branches.

Dilation ratios	0.5	0.75	0.95	Avg
1, 2, 4, 6	49.58	34.14	5.58	33.25
1, 6, 12, 18	49.71	34.01	5.90	33.31
1, 10, 20, 30	**49.86**	**34.37**	**6.50**	**33.53**
1, 16, 32, 64	50.00	34.31	6.35	33.45

effectiveness of our design choice for the 1D semantic segmentation network (1D-SSN) and proposal detection network (PDN). In Table 3, we compare SegTAD to its variants of disabling the segmentation loss in the 1D-SSN component. We can see that using the loss leads to obvious improvement compared to not using it. In Table 4, we show the performance of replacing the segmentation loss using a binary classification loss, which learns whether a frame is inside an action or not. Our multi-class segmentation loss in SegTAD is obviously better.

In 1D-SSN, the module of parallel atrous and graph convolutions (PAG) is important to aggregate features from multiple scales. We ablate different branches in PAG to show the performance change in Table 5. It shows that the network with all three kinds of branches produce the best performance.

We also show the results of different sets of dilation ratios for the atrous convolution branches in Table 6 and choose 1, 10, 20, 30 due to its highest mAP. We evaluate the effectiveness of our proposal detection network (PDN) by replacing its each layer of edge convolutions with a layer of Conv1D($k = 1, s = 1$) and apply it on each single proposal independently. In this way, this variant cannot make use of the cross-proposal correlations. We can see from Table 7 that using graph convolutions brings significant improvement compared to independently learning for each proposal. In Table 8, we compare different metrics to determine the similarity between proposals: tIoU and distance between proposal centers. We adopt tIoU in SegTAD due to its better performance.

4.4   Visualization of Segmentation Output

We visualize the output from the 1D semantic segmentation network and compare to ground-truth labels in Fig. 3. Our output tightly matches the ground-truth even for the video that contains many short action instances such as the

Table 7. Ablating the proposal detection network.

PDN layers	0.5	0.75	0.95	Avg
Conv1D($k = 1, s = 1$)	48.31	32.79	5.49	32.12
Graph convolutions	**49.86**	**34.37**	**6.50**	**33.53**

Table 8. Comparing different similarity metrics: temporal intersection over union and center distance between proposals.

Node similarity metric	0.5	0.75	0.95	Avg
Center distance	49.34	33.57	3.62	32.52
Temp. intersection over union	**49.86**	**34.37**	**6.50**	**33.53**

bottom example. Such accurate segmentation is important for learning distinctive features for each frame, and consequently benefits the final detection.

5   Conclusion

In summary, we take a novel perspective to formulate TAD based on 1D semantic segmentation to achieve more accurate label assignment and precise localization. We propose SegTAD, which is composed of a 1D semantic segmentation network (1D-SSN) and a proposal detection network (PDN). To suit the large variety of action temporal duration, in 1D-SSN, we design a module of parallel atrous and graph convolutions (PAG) to aggregate multi-scale local features and global features. In PDN, we design a second graph network to model the cross-proposal correlations. SegTAD is a unified framework that is trained jointly using the segmentation and detection losses from both 1D-SSN and PDN, respectively. As a conclusion, we would like to emphasize the need to focus more on the unique characteristics of videos when dealing with detection problems in video.

Relevance to 'AI for understanding and accelerating video editing'. Given the boom of creative video content on various platforms, such as TikTok, Reels, and YouTube, the tedious and time-consuming editing process urgently needs to be transformed [32]. Our SegTAD is able to localize and recognize actions in long untrimmed videos, which is needed by creative tasks such as cutting for the movie editing [32,33]. More specifically, our well-trained model that predicts the location and the type of human actions can be used to find the places where the transition of the scene should happen.

Acknowledgement. This work was supported by the King Abdullah University of Science and Technology (KAUST) Office of Sponsored Research through the Visual Computing Center (VCC) funding.

References

1. Report of Temporal Action Proposal. http://hacs.csail.mit.edu/challenge/challenge19_report_runnerup.pdf (2020)
2. Alcazar, J.L., Cordes, M., Zhao, C., Ghanem, B.: End-to-end active speaker detection. In: Proceedings of European Conference on Computer Vision (ECCV) (2022)
3. Buch, S., Escorcia, V., Shen, C., Ghanem, B., Niebles, J.C.: SST: single-stream temporal action proposals. In: Proceedings of IEEE Conference on Computer Vision and Pattern Recognition (CVPR) (2017)
4. Carreira, J., Zisserman, A.: Quo vadis, action recognition? A new model and the kinetics dataset. In: Proceedings of the IEEE Conference on Computer Vision and Pattern Recognition (CVPR) (2017)
5. Chao, Y.W., Vijayanarasimhan, S., Seybold, B., Ross, D.A., Deng, J., Sukthankar, R.: Rethinking the faster R-CNN architecture for temporal action localization. In: Proceedings of IEEE Conference on Computer Vision and Pattern Recognition (CVPR) (2018)
6. Chen, C., Ling, Q.: Adaptive convolution for object detection. IEEE Trans. Multimed. (TMM) **21**(12), 3205–3217 (2019). https://doi.org/10.1109/TMM.2019.2916104
7. Chen, L.C., Papandreou, G., Schroff, F., Adam, H.: Rethinking atrous convolution for semantic image segmentation. ArXiv abs/1706.05587 (2017)
8. Dai, X., Singh, B., Zhang, G., Davis, L.S., Qiu Chen, Y.: Temporal context network for activity localization in videos. In: Proceedings of IEEE International Conference on Computer Vision (ICCV) (2017)
9. Escorcia, V., Heilbron, F.C., Niebles, J.C., Ghanem, B.: DAPs: deep action proposals for action understanding. In: Proceedings of European Conference on Computer Vision (ECCV) (2016)
10. Caba Heilbron, F., Victor Escorcia, B.G., Niebles, J.C.: Activitynet: a large-scale video benchmark for human activity understanding. In: Proceedings of IEEE Conference on Computer Vision and Pattern Recognition (CVPR) (2015)
11. Gao, J., Chen, K., Nevatia, R.: CTAP: complementary temporal action proposal generation. In: Proceedings of European Conference on Computer Vision (ECCV) (2018)
12. Girshick, R.B.: Fast R-CNN. In: Proceedings of IEEE International Conference on Computer Vision (ICCV) (2015)
13. Girshick, R.B., Donahue, J., Darrell, T., Malik, J.: Rich feature hierarchies for accurate object detection and semantic segmentation. In: Proceedings of IEEE Conference on Computer Vision and Pattern Recognition (CVPR) (2014)
14. Grauman, K., et al.: Ego4D: around the world in 3,000 hours of egocentric video. In: Proceedings of the IEEE/CVF Conference on Computer Vision and Pattern Recognition, pp. 18995–19012 (2022)
15. Hamilton, W.L., Ying, R., Leskovec, J.: Inductive representation learning on large graphs. In: Proceedings of Neural Information Processing Systems (NeurIPS) (2017)
16. Heilbron, F.C., Barrios, W., Escorcia, V., Ghanem, B.: SCC: semantic context cascade for efficient action detection. In: Proceedings of IEEE Conference on Computer Vision and Pattern Recognition (CVPR) (2017)
17. Heilbron, F.C., Niebles, J.C., Ghanem, B.: Fast temporal activity proposals for efficient detection of human actions in untrimmed videos. In: Proceedings of IEEE Conference on Computer Vision and Pattern Recognition (CVPR) (2016)

18. Kang, B., Lee, Y., Nguyen, T.Q.: Depth-adaptive deep neural network for semantic segmentation. IEEE Trans. Multimed. (TMM) **20**(9), 2478–2490 (2018). https://doi.org/10.1109/TMM.2018.2798282

19. Li, J., Liu, X., Zhang, W., Zhang, M., Song, J., Sebe, N.: Spatio-temporal attention networks for action recognition and detection. IEEE Trans. Multimed. (TMM) **22**(11), 2990–3001 (2020). https://doi.org/10.1109/TMM.2020.2965434

20. Li, Y., Guo, Y., Guo, J., Ma, Z., Kong, X., Liu, Q.: Joint CRF and locality-consistent dictionary learning for semantic segmentation. IEEE Trans. Multimed. (TMM) **21**(4), 875–886 (2019). https://doi.org/10.1109/TMM.2018.2867720

21. Lin, J., Gan, C., Han, S.: TSM: temporal shift module for efficient video understanding. In: Proceedings of the IEEE International Conference on Computer Vision (ICCV) (2019)

22. Lin, T., Liu, X., Li, X., Ding, E., Wen, S.: BMN: boundary-matching network for temporal action proposal generation. In: Proceedings of IEEE International Conference on Computer Vision (ICCV) (2019)

23. Lin, T., Zhao, X., Shou, Z.: Single shot temporal action detection. In: Proceedings of ACM International Conference on Multimedia (ACM MM) (2017)

24. Lin, T., Zhao, X., Shou, Z.: Temporal convolution based action proposal: submission to ActivityNet 2017. ActivityNet Large Scale Activity Recognition Challenge workshop at CVPR (2017)

25. Lin, T., Zhao, X., Su, H., Wang, C., Yang, M.: BSN: boundary sensitive network for temporal action proposal generation. In: Proceedings of European Conference on Computer Vision (ECCV) (2018)

26. Liu, H., Wang, S., Wang, W., Cheng, J.: Multi-scale based context-aware net for action detection. IEEE Trans. Multimed. (TMM) **22**(2), 337–348 (2020). https://doi.org/10.1109/TMM.2019.2929923

27. Liu, H., Wang, S., Wang, W., Cheng, J.: Multi-scale based context-aware net for action detection. IEEE Trans. Multimed. (TMM) **22**(2), 337–348 (2020)

28. Liu, K., Gao, L., Khan, N.M., Qi, L., Guan, L.: A multi-stream graph convolutional networks-hidden conditional random field model for skeleton-based action recognition. IEEE Trans. Multimed. (TMM) **23**, 64–76 (2021). https://doi.org/10.1109/TMM.2020.2974323

29. Liu, Y., Ma, L., Zhang, Y., Liu, W., Chang, S.F.: Multi-granularity generator for temporal action proposal. In: Proceedings of IEEE Conference on Computer Vision and Pattern Recognition (CVPR) (2019)

30. Long, F., Yao, T., Qiu, Z., Tian, X., Luo, J., Mei, T.: Gaussian temporal awareness networks for action localization. In: Proceedings of IEEE Conference on Computer Vision and Pattern Recognition (CVPR) (2019)

31. Long, J., Shelhamer, E., Darrell, T.: Fully convolutional networks for semantic segmentation. In: Proceedings of IEEE Conference on Computer Vision and Pattern Recognition (CVPR) (2015)

32. Pardo, A., Caba, F., Alcázar, J.L., Thabet, A.K., Ghanem, B.: Learning to cut by watching movies. In: Proceedings of the IEEE/CVF International Conference on Computer Vision, pp. 6858–6868 (2021)

33. Pardo, A., Heilbron, F.C., Alcázar, J.L., Thabet, A., Ghanem, B.: Moviecuts: a new dataset and benchmark for cut type recognition. arXiv preprint arXiv:2109.05569 (2021)

34. Qiu, H., et al.: Hierarchical context features embedding for object detection. IEEE Trans. Multimed. (TMM) **22**(12), 3039–3050 (2020). https://doi.org/10.1109/TMM.2020.2971175

35. Ramazanova, M., Escorcia, V., Heilbron, F.C., Zhao, C., Ghanem, B.: Owl (observe, watch, listen): localizing actions in egocentric video via audiovisual temporal context. ArXiv abs/2202.04947 (2022)
36. Ren, S., He, K., Girshick, R., Sun, J.: Faster R-CNN: towards real-time object detection with region proposal networks. IEEE Trans. Pattern Anal. Mach. Intell. **39**(6), 1137–1149 (2016)
37. Ronneberger, O., Fischer, P., Brox, T.: U-net: convolutional networks for biomedical image segmentation. In: Proceedings of Medical Image Computing and Computer-Assisted Intervention (MICCAI) (2015)
38. Shou, Z., Chan, J., Zareian, A., Miyazawa, K., Chang, S.F.: CDC: convolutional-de-convolutional networks for precise temporal action localization in untrimmed videos. In: Proceedings of IEEE Conference on Computer Vision and Pattern Recognition (CVPR) (2017)
39. Shou, Z., Wang, D., Chang, S.F.: Temporal action localization in untrimmed videos via multi-stage CNNs. In: Proceedings of IEEE Conference on Computer Vision and Pattern Recognition (CVPR) (2016)
40. Singh, G., Cuzzolin, F.: Untrimmed video classification for activity detection: submission to ActivityNet Challenge. ActivityNet Large Scale Activity Recognition Challenge workshop at CVPR (2016)
41. Soldan, M., Pardo, A., Alcázar, J.L., Caba, F., Zhao, C., Giancola, S., Ghanem, B.: Mad: a scalable dataset for language grounding in videos from movie audio descriptions. In: Proceedings of the IEEE/CVF Conference on Computer Vision and Pattern Recognition, pp. 5026–5035 (2022)
42. Su, H., Zhao, X., Lin, T., Liu, S., Hu, Z.: Transferable knowledge-based multi-granularity fusion network for weakly supervised temporal action detection. IEEE Trans. Multimed. (TMM) **23**, 1503–1515 (2021). https://doi.org/10.1109/TMM. 2020.2999184
43. Wang, L., Xiong, Y., Lin, D., Van Gool, L.: Untrimmednets for weakly supervised action recognition and detection. In: Proceedings of the IEEE Conference on Computer Vision and Pattern Recognition (CVPR) (2017)
44. Wang, R., Tao, D.: UTS at ActivityNet 2016. ActivityNet Large Scale Activity Recognition Challenge (2016)
45. Wang, Y., Sun, Y., Liu, Z., Sarma, S.E., Bronstein, M.M., Solomon, J.M.: Dynamic graph CNN for learning on point clouds. ACM Trans. Graph. (TOG) **38**(5), 1–12 (2019)
46. Xiong, Y., et al.: CUHK & ETHZ & SIAT submission to ActivityNet Challenge 2016. arXiv:1608.00797 (2016)
47. Xu, H., Das, A., Saenko, K.: R-C3D: region convolutional 3D network for temporal activity detection. In: Proceedings of IEEE International Conference on Computer Vision (ICCV) (2017)
48. Xu, M., Zhao, C., Rojas, D.S., Thabet, A., Ghanem, B.: G-TAD: sub-graph localization for temporal action detection. In: Proceedings of IEEE Conference on Computer Vision and Pattern Recognition (CVPR) (2020)
49. Yu, F., Koltun, V.: Multi-scale context aggregation by dilated convolutions. In: Proceedings of International Conference on Learning Representations (ICLR) (2016)
50. Yuan, Z.H., Stroud, J.C., Lu, T., Deng, J.: Temporal action localization by structured maximal sums. In: Proceedings of IEEE Conference on Computer Vision and Pattern Recognition (CVPR) (2017)

51. Zeng, R., Huang, W., Tan, M., Rong, Y., Zhao, P., Huang, J., Gan, C.: Graph convolutional networks for temporal action localization. In: Proceedings of IEEE International Conference on Computer Vision (ICCV) (2019)
52. Zhang, S., Chi, C., Yao, Y., Lei, Z., Li, S.Z.: Bridging the gap between anchor-based and anchor-free detection via adaptive training sample selection. In: Proceedings of IEEE Conference on Computer Vision and Pattern Recognition (CVPR) (2020)
53. Zhang, S., Peng, H., Yang, L., Fu, J., Luo, J.: Learning sparse 2d temporal adjacent networks for temporal action localization. In: HACS Temporal Action Localization Challenge at IEEE International Conference on Computer Vision (ICCV) (2019)
54. Zhao, C., Thabet, A.K., Ghanem, B.: Video self-stitching graph network for temporal action localization. In: Proceedings of the IEEE/CVF International Conference on Computer Vision (ICCV), pp. 13658–13667 (2021)
55. Zhao, H., Yan, Z., Torresani, L., Torralba, A.: HACS: human action clips and segments dataset for recognition and temporal localization. In: Proceedings of IEEE International Conference on Computer Vision (ICCV) (2019)
56. Zhao, P., Xie, L., Ju, C., Zhang, Y., Wang, Y., Tian, Q.: Bottom-up temporal action localization with mutual regularization. In: Proceedings of European Conference on Computer Vision (ECCV) (2020)
57. Zhao, Y., Xiong, Y., Wang, L., Wu, Z., Tang, X., Lin, D.: Temporal action detection with structured segment networks. In: Proceedings of IEEE International Conference on Computer Vision (ICCV) (2017)

Text-Driven Stylization of Video Objects

Sebastian Loeschcke[1]([✉]), Serge Belongie[2], and Sagie Benaim[2]

[1] Aarhus University, Aarhus, Denmark
201804446@post.au.dk
[2] University of Copenhagen, Copenhagen, Denmark

Abstract. We tackle the task of stylizing video objects in an intuitive and semantic manner following a user-specified text prompt. This is a challenging task as the resulting video must satisfy multiple properties: (1) it has to be temporally consistent and avoid jittering or similar artifacts, (2) the resulting stylization must preserve both the global semantics of the object and its fine-grained details, and (3) it must adhere to the user-specified text prompt. To this end, our method stylizes an object in a video according to two target texts. The first target text prompt describes the global semantics and the second target text prompt describes the local semantics. To modify the style of an object, we harness the representational power of CLIP to get a similarity score between (1) the local target text and a set of local stylized views, and (2) a global target text and a set of stylized global views. We use a pretrained atlas decomposition network to propagate the edits in a temporally consistent manner. We demonstrate that our method can generate consistent style changes over time for a variety of objects and videos, that adhere to the specification of the target texts. We also show how varying the specificity of the target texts and augmenting the texts with a set of prefixes results in stylizations with different levels of detail. Full results are given in the supplementary and in full resolution in the project webpage: https://sloeschcke.github.io/Text-Driven-Stylization-of-Video-Objects/.

Keywords: Video editing · Text-guided stylization · CLIP

1 Introduction

Manipulating semantic object entities in videos using human instructions requires skilled workers with domain knowledge. We seek to eliminate these requirements by specifying a desired edit or stylization through an easy, intuitive, and semantic user instruction in the form of a text-prompt.

However, manipulating video content semantically is a challenging task. One challenge is in generating consistent content or style changes in time, that adhere to the target text specification. Another challenge is to manipulate the content of an object such that it preserves the content of the original video and the global semantics while also adhering to fine-grained details in the target text.

Supplementary Information The online version contains supplementary material available at https://doi.org/10.1007/978-3-031-25069-9_38.

ⓒ The Author(s), under exclusive license to Springer Nature Switzerland AG 2023
L. Karlinsky et al. (Eds.): ECCV 2022 Workshops, LNCS 13804, pp. 594–609, 2023.
https://doi.org/10.1007/978-3-031-25069-9_38

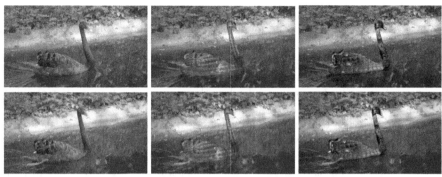

Input video   "Swan made out of cactus" "Swan with crocodile skin"

Fig. 1. Two representative video frames and the edited video frames together with the global target text.

In recent years, advances in computational methods emerged that enable manipulation of appearances and style in images and allow novice users to perform realistic image editing. These methods include manipulation tools that use natural language (text prompts) to express the desired stylization of images or 3D objects [6,18]. The text-driven manipulation is facilitated by recent developments in models for joint embeddings of text and images, e.g. the Contrastive Language Image Pretraining (CLIP [23]) model. Instead of manipulating images or 3D objects, we use CLIP in the context of video manipulation. This is not a straightforward task since simply maximizing the semantic (CLIP-based) similarity between a valid target text and each 2D frame in the video often leads to degenerate solutions. Also, applying methods for image manipulation to each frame in a video results in edits that lack temporal consistency.

The recently introduced Neural Layered Atlases (NLA) work [14], demonstrates the ability to separate a moving object in a video from its background by decomposing the video into a set of 2D atlases. Each atlas provides a unified image representation of an object or background over the video. Edits applied to the image representation are automatically mapped back to the video in a temporally consistent manner. However, editing an image still requires manual effort and editing skills from the user. Another problem with this approach is that the 2D atlas representation can be hard to edit due to local deformations.

We propose a method for performing intuitive and consistent video editing with multiple capabilities by using the representational power of CLIP to express a desired edit through a text-prompt. An example could be to change the style of a swan swimming in a lake according to a target text: "A swan with cactus skin." Text is easily modifiable and allows users to express complex and abstract stylizations intuitively. Using text to express edits reduces the need for manual editing skills and also avoids the problems related to deformation in the 2D atlas representations. An illustration is shown in Fig. 1.

To apply temporally consistent edits to an object in a video, our method uses the atlas decomposition method presented in NLA [14]. We train a generator on a single input video by sampling local and global views of each frame in the video and applying various augmentations to each view. Our method uses a global loss that compares each global view with a global target text and a local loss that compares each local view with a local target text. The global loss then focuses on the global semantics and the local views focus on the fine-grained details. To regularize our learning, we use a sparsity loss that encourages sparse representation and a temporal triplet loss that encourages frames that are close in time to be similar in CLIP's embedding space.

We demonstrate that our method results in natural and consistent stylizations of objects for a diverse set of videos and target texts. We show how varying the specificity of both the local and global target texts varies the stylization and how augmenting the target texts with neutral prefixes can result in more detailed stylizations. We also demonstrate that our global loss focuses on the global semantics and the local losses on the fine-grained details.

2   Related Work

Our work is related to video editing works and to text-based stylization works.

2.1   Video Editing

Unlike images, editing or stylizing objects in videos requires the ability to handle temporal consistency. One natural approach is to propagate the edits from one frame to the next as in Video Propagation Networks [10]. Other approaches use optical flow to propagate edits made on a few key-frames [28]. These approaches work well when there is a clear correspondence between frames, but have difficulties, e.g., when the video contains occlusions.

To address occlusion challenges, recent work has used deep learning approaches, e.g. self-supervised methods for learning visual correspondence from unlabeled videos [9,29] or methods that decompose a video into a 2D representation [14,30]. Our work uses the representation proposed by Neural Layered Atlases (NLA) [14], which decomposes a video into a set of layered 2D atlases. Each atlas provides a unified representation of the appearance of an object or background throughout the video. However, NLA only allows for basic manual editing. We use NLA's atlas separation method for objects in videos, but unlike NLA, we allow for text-driven stylization.

2.2   Text-Based Stylization

Our work bears similarities to recent image and 3D manipulation techniques that edit style and appearances through natural language descriptions. These descriptions are often embedded with the Contrastive Language Image Pretraining (CLIP) [23] model, a multi-modal embedding model that learns an image-text embedding space. Recent work used CLIP together with pretrained generative networks for image editing and stylization [1,3,5,6,8,15,21]. For example, StyleCLIP [21], and StyleGAN-NADA [8] both use a pretrained StyleGAN [12] and CLIP to perform image editing, either by using CLIP to control a latent code or to adapt an image generator to a specific domain [8]. Usually, pretrained generators only work well for the specific domain they are trained on. In contrast, our method does not require a pretrained generator. We train our own generator on the set of video frames we wish to stylize.

Other examples of semantic text-guided manipulation in the context of 3D objects include 3DStyleNet [31], a method for changing the geometric and texture style of 3D objects and ClipMatrix [11] which uses text-prompts to create digital 3D creatures. Another line of recent work uses joint-embedding architectures [19,24–26] for image generation, e.g., DALL-E [25] and its successor, DALL-E 2 [24], which can also be used for stylizing images. DALL-E 2 uses a two-stage model, where a CLIP image embedding is generated using a text prompt and a decoder is then used to generate an image conditioned on the generated image embedding. Training joint embedding architectures requires enormous datasets and many training hours. Instead of training on a large dataset, we train on a set of frames for a single video and use augmentations to extract many different views of the input frames. As opposed to all the abovementioned techniques, we work on videos.

Another line of work [7,17] uses CLIP without relying on a pretrained generator, e.g. Texts2Mesh [18]. In Text2Mesh, the CLIP embedding space is used to enable text-driven editing of 3D meshes. Text2Mesh uses a multi-layer perceptron (MLP) to apply a stylization to (x, y, z)-coordinates of an input mesh. The neural optimization process of the MLP is guided by a semantic loss that computes the similarity between multiple augmented 2D views embedded with CLIP and a target text. Similarly, we do not rely on a pretrained generator.

Our work was developed concurrently to Text2Live [2], which shares many of the same goals and methods as our work. Similarly to our method, Text2Live uses a pretrained Neural Layered Atlases (NLA) model to separate a moving object in a video from its background. Text2Live train a generator to apply text-driven local edits to a single frame and use the NLA model to map the edits back to the input video in a temporally consistent manner. In contrast to our approach, Text2Live does not directly generate the edited output. Instead, it generates an edit layer that is composited with the original input.

3 Method

We wish to apply natural and temporally consistent stylizations to objects in videos using a natural language text prompt as guidance. To change the style of an object to conform with a target text prompt in a temporally consistent manner, we build on top of a Layered Neural Atlas method [14], which separates the appearance of an object in a video from its background. We then use a pretrained text-image multimodal embedding of CLIP [23] in a set of objectives. Minimizing this set of objectives aims at matching the style of a foreground object in a video with that of a target text. The objectives include a global and local objective. The global objective focuses on the global semantics by maximizing the similarity between the global views and a target text that relates to the underlying content. Instead, the local objective focuses on the fine-grained details, by maximizing the similarity between the local views with a target text that relates to local semantics of the stylization. We add a sparsity loss, similar to a L_1-regularization term, that encourages the predicted foreground color values to be minimal. Additionally, we add a temporal triplet loss that encourages the embeddings of frames that are close in time to also be close in CLIP's embedding space. We begin by describing the method of CLIP [23] and that of Neural Layered Atlases (NLA) on which our method is based. We then describe the training and loss formulations used by our method.

3.1   CLIP

CLIP is a multi-modal embedding method that trains an image encoder E_{img} and a text encoder E_{txt} to match between the embeddings of corresponding image-text pairs using a contrastive loss formulation. This loss formulation optimizes the similarity between corresponding image-text pair T and I. More specifically, I and T are first embedded:

$$I_{emb} = E_{img}(I) \in \mathbb{R}^{512}, \quad T_{emb} = E_{txt}(T) \in \mathbb{R}^{512}$$

The similarity between I and T is then measured by $\text{sim}(I_{emb}, T_{emb})$ where $\text{sim}(a, b) = \frac{a \cdot b}{|a||b|}$, is the cosine similarity.

3.2   Neural Layered Atlases (NLA)

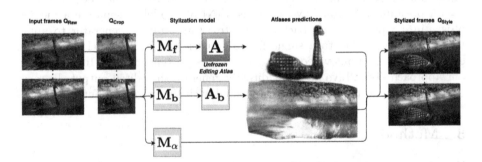

Fig. 2. Our stylization pipeline. In the first step, we train the network using the NLA procedure [14] to reconstruct input video frames. We then finetune the editing atlas A using our approach described in Fig. 3. In our stylization pipeline, we create a set of cropped input video frames Q_{Crop}. This set is passed through a stylization model to create a foreground and background atlas. A set of stylized frames Q_{Style} is produced by α-blending the predicted atlases. All weights of the MLPs are frozen except for the editing atlas MLP A which is finetuned. A closer look at how our editing atlas is trained is given in Fig. 3.

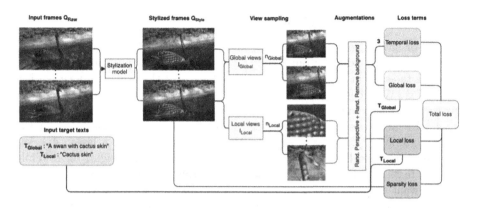

Fig. 3. Finetuning the editing atlas. As described in the stylization pipeline (Fig. 2), we use our stylization pipeline to get a set of stylized frames Q_{Style}. To this end, we finetune our editing atlas, which is part of the stylization model. We sample n_{Global} global views I^{Global} and n_{Local} local views I^{Local}. The set of images I^{Global} are then augmented using random perspectives and random background removal and used together with the global target text T_{Global} to compute a global loss (Eq. 3). Three global augmented images are used to compute the temporal loss (Eq. 4). Similarly, the I^{Local} images are augmented and used together with the local target text T_{Local} to compute the local loss (Eq. 2). Lastly, a sparsity loss (Eq. 5) is computed from the stylized frames Q_{Style}.

Neural Layered Atlases (NLA) [14] decompose a video into a set of layered 2D atlases. Each atlas provides a unified representation of the appearance of an object or the background throughout the video. NLA use two mapping networks M_f and M_b, where each takes a pixel and time location (x, y, t) in the video as input and outputs the corresponding 2D (u, v)-coordinate in each atlas:

$$M_f(p) = (u_f^p, u_f^p), \quad M_b(p) = (u_b^p, u_b^p)$$

An atlas network A takes the predicted 2D (u, v)-coordinate as input and outputs the atlas's RGB color at that location. Additionally, all pixel coordinates are fed into the Alpha MLP network M_α which outputs the opacity of each atlas at that location. The RGB color can then be reconstructed at each pixel location by alpha-blending the predicted atlas points according to the opacity value predicted by M_α.

NLA enables consistent video editing. First, each atlas is discretized into an image. A user can then manually apply edits using an editing program. These edits are mapped back to the input video using the computed (u, v)-mapping. To get the reconstructed color c^p for pixel p, the color of the predicted foreground color c_f^p, background color c_b^p and predicted opacity value α^p of the edited atlas are blended as follows:

$$c^p = (1 - \alpha^p)c_b^p + \alpha^p c_f^p$$

3.3   Our Stylization Pipeline

Instead of having to manually apply edits to the discretized atlases as in [14], we wish to apply edits automatically using a target text prompt. Specifically, we are interested in modifying the RGB values such that they conform with a target text. We focus on the Atlas MLP A, since A makes the RGB predictions. Instead of using a single atlas MLP A for all atlases, we create two copies of the atlas MLP after pre-training the NLA network, one editing atlas MLP for the foreground object we want to stylize (A) and one for all other atlases (A_b). We then freeze the weights of M_b, M_f, M_α, A_b and finetune the weights of A. While the goal in NLA is to optimize the model to produce the best reconstruction, we instead want to create a stylization that conforms with an input target text.

Our stylization pipeline is illustrated in Fig. 2. The input is a set of raw frames Q_{Raw}. Since we know the position of the object in the video from the α-map, we can compute a bounding box containing the whole object. Once we have the bounding box, we crop each frame in Q_{Raw} such that it only contains the content within the bounding box plus a small margin. All the cropped frames $Q_{Cropped}$ are passed through a pre-trained NLA model, where all MLP weights have been frozen, except for the weights of the A MLP. The NLA method produces a set of stylized frames Q_{Style}.

To fine-tune the weights of A, we sample training batches in both time and space. Our sampling method is illustrated in Fig. 3. First, we sample a set Q_{Sample} uniformly at random among all frames in the input video and pass them through the stylization pipeline (Fig. 2) to create a set Q_{Style}.

For each frame in Q_{Style} we sample n_{Global} views I^{Global} and a set of n_{Local} views I^{Local}. Each of the n_{Global} views is produced by sampling a crop with a size in the range $[0.9, 1.0]$ of the original frame size. Each of the n_{Local} views is produced by sampling a crop with a size in the range $[0.1, 0.5]$ of the original frame size. To ensure the local views contain the object we want to stylize, we use the α-map of the frame to determine the position of the object in the frame. We then sample until we get n_{Local} views where at least $\frac{1}{3}$ of the sampled view is part of the object.

Once the local and global views have been sampled, we apply a random perspective transformation and a random background removal which with some probability p removes the background (details in supplementary) Additionally, each augmented frame is normalized with the same mean and standard deviation as used to train the CLIP model [23].

Our objective function is composed of three main losses defined in CLIP's feature space: (1) L_{Local} which focuses on local semantics, (2) L_{Global} which focuses on global semantics, and (3) L_{Temp} which encourages temporal consistency. Additionally, we use the regularization term $L_{sparsity}$ introduced in NLA [14], which encourages sparse representations. In all loss terms, when we compute the similarity between a text and each sampled view, we use the average embedding across all views:

$$I_{emb}^{Local} = \frac{1}{n_{Local}} \sum_{i=1}^{n_{Local}} E_{img}\left(I_i^{Local}\right), I_{emb}^{Global} = \frac{1}{n_{Global}} \sum_{i=1}^{n_{Global}} E_{img}\left(I_i^{Global}\right) \quad (1)$$

Local loss L_{Local} is applied to all views in I^{Local}. The goal is to modify the image, such that the local details conform with a target text T_{Local}:

$$L_{\text{Local}} = 1 - \text{sim}\left(I_{emb}^{Local}, E_{txt}\left(T_{\text{Local}}\right)\right) \tag{2}$$

where $\text{sim}(a, b) = \frac{a \cdot b}{|a||b|}$ is the cosine similarity, and E_{txt} denote CLIP's pre-trained text encoder. The local views have a more zoomed-in view of the object we are stylizing. Additionally, the local target text T_{Local} contains local specific semantics, e.g. *"rough* cactus texture." Hereby, the local loss can focus on the texture and fine-grained details of the stylization we apply to the input video.

Global loss The global loss is applied to views in I^{Global} that all include the entire object being stylized. The intended goal is that the global loss will preserve the overall context. In the target text T_{Global}, we include words that describe the global context of the object we are trying to stylize, e.g. "A *swan* made of cactus." The global loss formulation is then given by:

$$L_{\text{Global}} = 1 - \text{sim}\left(I_{emb}^{Global}, E_{txt}\left(T_{\text{Global}}\right)\right) \tag{3}$$

Temporal Loss. We use a triplet loss to include a temporal aspect and enforce that consecutive frames should be more similar in CLIP's embedding space than frames that are further apart. To compute the temporal loss, we sample three frames t_1, t_2, t_3, where we have that $t_1 < t_2 < t_3$ w.r.t. the order of the frames in the input video. We then enforce that the similarity between the sampled global views of t_1 and t_2 in the CLIP embedding space should be greater than the similarity between t_1 and t_3: Let $I_{emb(t_1)}^{Global}, I_{emb(t_2)}^{Global}, I_{emb(t_3)}^{Global}$ denote the average embedded global views (computed in Eq. 1) for each of the three frames. Then we compute the triplet loss L_{Temp} as follows:

$$\text{Sim}_{t_1 t_3 - t_1 t_2} = \text{sim}\left(I_{emb(t_1)}^{Global}, I_{emb(t_3)}^{Global}\right) - \text{sim}\left(I_{emb(t_1)}^{Global}, I_{emb(t_2)}^{Global}\right)$$
$$L_{Temp} = \lambda_{Temp} \cdot \max\left(0, \text{Sim}_{t_1 t_3 - t_1 t_2}\right) \tag{4}$$

where λ_{Temp} is a weighting of the temporal loss. If the frames are further away from each other, the temporal loss should contribute less to the overall loss. For this reason, we weigh the contribution of the temporal loss by a Gaussian probability density function g with a mean equal to zero and a standard deviation of five. We compute the weight of a triplet by applying g to the difference between t_1 and t_3:

$$\lambda_{Temp} = g(t_3 - t_1)$$

Sparsity Loss. We use the same sparsity loss as in NLA [14]. Its intended function is to encourage points that are mapped to the background atlas to have a zero value in the foreground atlas, e.g. if a point p is mapped to the background atlases it should not contain information about the foreground atlas.

$$L_{\text{sparsity}} = \left\|\left(1 - \alpha^P\right) c_f^P\right\| \tag{5}$$

where c_f^P is the predicted color at p for the foreground layer and α^P is the opacity value at location p.

Full Objective. The full loss term that is minimized is represented as:

$$L = \lambda_{Sparsity}L_{\text{Sparsity}} + \lambda_{Temp}L_{\text{Temp}} + \lambda_{Local}L_{\text{Local}} + \lambda_{Global}L_{\text{Global}} \qquad (6)$$

where $\lambda_{Local}, \lambda_{Global}, \lambda_{Temp}, \lambda_{Sparsity}$ are hyperparameters used to control the weighting of each loss term. As default $\lambda_{Local} = \lambda_{Global} = 1$, while λ_{Temp} and $\lambda_{Sparsity}$ vary depending on the input video.

Table 1. Mean opinion scores (1–5) and standard deviation for Q1 and Q2.

	Q1 (Realism)	Q2 (Matching Text)
Blended Diffusion [1]	2.22 ($\pm$1.08)	2.10 ($\pm$1.00)
Ours	**3.47** ($\pm$1.10)	**3.94** ($\pm$0.99)

4   Experiments

We evaluate our method on a set of videos from the DAVIS dataset [22] across a diverse set of target text prompts. Our goal is to perform consistent and natural video editing. For this purpose, we present both a quantitative and qualitative evaluation of our results and perform a careful ablation study of each loss term in our objective function.

In Sect. 4.1 we demonstrate the capabilities of our method by showing various stylizations for different videos. We also present a quantitative evaluation, where we compare our method to an image baseline method applied to each frame in the video. In Sect. 4.2 we show that the specificity of text prompts influences the details of the results. In Sect. 4.3 we demonstrate how text augmentation affects the results of the stylizations. In Sect. 4.4 we conduct a series of ablations on our loss terms and demonstrate how the local and global losses focus on different semantics. Finally, in Sect. 4.5 we illustrate some limitations of our method.

4.1   Varied Stylizations

In Fig. 4 we illustrate our video stylization method applied to three videos and three texts. All local target texts used for these examples are similar to the global target text, but without containing the information about the underlying content, e.g., for the swan, the local target text is "metal skin." The results show that we can apply temporally consistent stylizations that adhere to the target text specification. The swan example shows fine-grained details of the target texts and also preserves the underlying content of the swan. In the boat example, the stylization captures the details of the target text. The boat has a texture similar to shiny aluminum and also has something that looks like a fishing net at the end of the boat. The dog example shows that our method can apply a realistic and consistent stylization to a video containing occlusions.

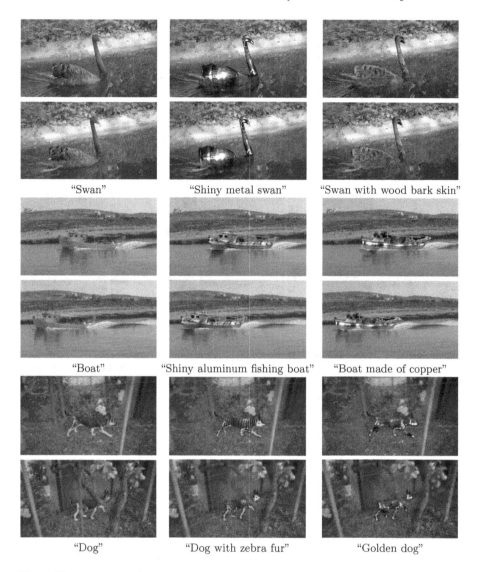

Fig. 4. Example results. Two representative frames from each edited video together with the global target text.

We quantify the effectiveness of our method by comparing it to an image baseline applied to each frame in a video input. As a baseline, we use a pretrained Blended-diffusion (BF) [1] model with standard configurations. The model takes as input an image, a ROI mask, and a target text. BF performs local (region-based) edits based on a target text description and the ROI mask. We conduct a user study to evaluate the perceived quality of the stylized outputs generated by both the BF model and our method, and the degree to which the outputs adhere to the global target text. Our user study comprises 50 users and 20 stylized videos, each with a different target text. For each video and target text combination, the users are asked to assign a score (1–5) to two factors: (Q1) "How realistic is the video?," (Q2) "does the {*object*} in the video adhere to the text {*content*}. For Q1 we make it clear to the user that "realistic" refers to the quality of the video content. The results are shown in Table 1.

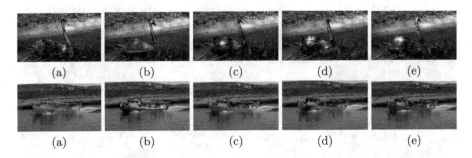

(a)	(b)	(c)	(d)	(e)
(a)	(b)	(c)	(d)	(e)

Fig. 5. Target text specificity. Each example shows a representative frame from the video. The experiment shows the specificity of a target affects the stylization. Global target text prompts, *row one:* (a) "Armor," (b) "Iron armor," (c) "Medieval iron armor," (d) "Suit of shiny medieval iron armor," (e) "Full plate shiny medieval iron armor," *row two:* (a) "Boat made of wood," (b) "Boat made of dark walnut wood (c) "Fishing boat made of wood," (d) "Old fishing boat made of wood," (e) "Fishing boat made of wood planks."

4.2 Prompt Specificity

We demonstrate that varying the specificity of the target text prompt affects the level of detail in the stylization. Our experiment is motivated by recent work on prompt engineering [32] that shows how slight changes in the target text can have a big impact on the CLIP similarity between a text and an image. Figure 5 shows an increasing level of detail for two videos and two target texts. The target text specificity not only influences the level of detail, it also makes it easier for CLIP to navigate in its embedding space. In the swan example in column (a), we have "Swan with an armor," which is a more ambiguous target compared to the other swan examples with more detailed target texts and stylizations. We hypothesize that this is because several stylizations can satisfy the more simple target text, while a more specific target text narrows down the set of possible directions in CLIP's embedding space. The swan examples in columns (d) and (e) indicate that CLIP has some understanding of the different body parts of the swan. The "full plate" target text (d) covers the entire head of the swan while the "suit" of armor in column (e) has a clear cut around the head of the swan.

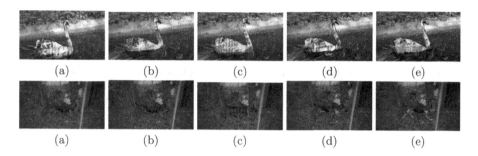

Fig. 6. Prefix augmentations - varying number of prefixes to sample from in each iteration. Each figure shows a representative frame from each video and experiment configuration. The examples in row one use the texts: T_{Global}: "Origami swan with white paper skin," T_{Local}: "Origami white paper skin," and in row two: T_{Global}: "Dog with Bengal tiger fur," T_{Local}: "Bengal tiger fur." (a) no prefixes, (b) 4 local, no global, (c) 4 global, no local (d) 4 global & 4 local, (e) 8 global & 8 local. The prefixes are described in the supplementary.

Fig. 7. Ablation on each loss term - All experiments were run with the same seed and with the texts: T_{Global}: "A swan with crocodile skin," T_{Local}: "Crocodile skin." Each figure shows a representative frame from a video, where we ablate one of our loss terms in Eq. 6. (a) All losses, (b) w/o local loss, (c) w/o global loss, (d) w/o temporal loss, (e) w/o sparsity loss.

4.3 Text Augmentation

We add textual augmentation to our method to address some of the challenges with prompt engineering. Inspired by Zhou et al. [32], we add neutral prefixes to both the local and global target texts, e.g., "a photo of a {}." We then sample a new prefix each iteration for each of the target texts as a form of regularization.

Fig. 8. Global and local semantics. A representative frame from each of the edited video frames. The texts used are: (a) T_{Global}: "Swan with cactus skin," T_{Local}: "Cactus skin," (b) T_{Global}: "Swan with cactus skin," T_{Local}: "Rough cactus skin," (c) T_{Global}: "Swan made out of cactus," T_{Local}: "Catus skin." (a) and (b) have the same global target text while (a) and (c) have the same local target texts. In (b) the local details have changed to a more rough texture. In (c) the global swan semantics are better preserved.

Figure 6 illustrates our text augmentation experiment. We demonstrate that using prefixes increases the quality of the results. In each experiment and each iteration, we sample one prefix among a set of prefixes (details in supplementary) for both the global and local target texts. In this experiment, we vary the number of prefixes to sample from and show that using an equal amount of prefixes for both the local and global target texts produces better quality results than using no prefixes. In both the swan and dog examples for columns (e) and (d) that use multiple prefixes, we see that the stylizations are more detailed than the examples in columns (a-c), e.g., in the dog example (d) and (e) the tiger fur has white pigments around the belly area and also more natural tiger stripes.

4.4   Ablation Study

We validate each term in our objective function (Eq. 6) with an ablation study illustrated in Fig. 7. In (a), we see that all losses combined result in a natural stylization that shows clear characteristics of a swan with crocodile skin. In (b) we see that without the **local loss** the results lack fine-grain details. It is still evident what the intended stylization was but the crocodile texture is not as clear as in (a). The results in (c) show that without the **global loss** the global semantics are less clear, e.g., the swan's neck is not stylized as detailed as the rest of the body. In (d), we see that without the **temporal loss**, we get more edits outside the mask of the object. In (e), we see that without the **sparsity loss**, the stylization is very noisy.

In Fig 8 we illustrate how the local loss affects the fine-grained details and the global loss affects the global semantics. We use (a) as a baseline and then vary the local target texts in (b) and the global target texts in (c). In (b) we see, how changing the local target text to include "rough" affects the details of the cactus texture. In (c) we see how the global semantics of the swan's body become more realistic as an effect of changing the global target text.

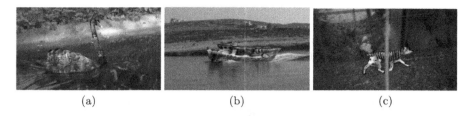

(a) (b) (c)

Fig. 9. Limitations. A representative frame from each of the edited video frames. The texts used are: (a) T_{Global}: "Swan with cactus skin," T_{Local}: "Cactus skin," (b) T_{Global}: "Boat made out of chocolate," T_{Local}: "Chocolate texture," (c) T_{Global}: "Dog with zebra fur," T_{Local}: "Zebra fur." In (a), the cactus texture is applied like photos of cactus. In (b) the chocolate boat has a sailing ship printed at the end of the boat. In (c), the dog has a face of a dog on its haunches.

4.5 Limitations

Figure 9 illustrates some of the limitations of our method. During training our method occasionally starts to overfit and produce unintended stylizations. In (a), we see that the body of the swan contains photo-like cacti instead of natural cactus texture. In (b) and (c), we see how our model has used the contexts of the global target text. In (b), a sailing ship has been added to the end of the boat and in (c), a face of a dog has been added. Our method is limited to text-prompts that do not entail features that cross between the decomposed atlas layers. Some changes are best realized through a shape change, e.g., "Swan with long hair.", which is not currently possible in our framework.

5   Conclusion

We considered the problem of developing intuitive and semantic control for consistent editing and styling of objects in videos. This problem poses a challenge in generating consistent content and style changes over time while being able to produce fine-grained details and preserve the global semantics. We proposed a method that uses CLIP [23] and the video object representation [14] to stylize objects in videos by using both a global target text, to control the global semantics of the stylization, and a local target text, to control the fine-grained details. We demonstrated that the specificity and the prefixes of the target texts can have a significant impact on the details produced by our method's stylization. In future work, it would be interesting to investigate the limitations of CLIP. A model that can generate fine-grained stylizations of videos and images could be leveraged to create data augmentations in other learning settings. Another line of future work is to experiment with background stylizations and to extend our model to be able to generate shape changes or even new objects from scratch.

Acknowledgement. This research was supported by the Pioneer Centre for AI, DNRF grant number P1.

References

1. Avrahami, O., Lischinski, D., Fried, O.: Blended diffusion for text-driven editing of natural images. CoRR abs/2111.14818 (2021). https://arxiv.org/abs/2111.14818
2. Bar-Tal, O., Ofri-Amar, D., Fridman, R., Kasten, Y., Dekel, T.: Text2live: text-driven layered image and video editing (2022). https://doi.org/10.48550/ARXIV.2204.02491. https://arxiv.org/abs/2204.02491
3. Bau, D., et al.: Paint by word. CoRR abs/2103.10951 (2021). https://arxiv.org/abs/2103.10951
4. Brock, A., Donahue, J., Simonyan, K.: Large scale GAN training for high fidelity natural image synthesis. CoRR abs/1809.11096 (2018). http://arxiv.org/abs/1809.11096
5. Chefer, H., Benaim, S., Paiss, R., Wolf, L.: Image-based clip-guided essence transfer. CoRR abs/2110.12427 (2021). https://arxiv.org/abs/2110.12427
6. Crowson, K., et al.: Vqgan-clip: open domain image generation and editing with natural language guidance (2022). https://doi.org/10.48550/ARXIV.2204.08583. https://arxiv.org/abs/2204.08583
7. Frans, K., Soros, L.B., Witkowski, O.: Clipdraw: exploring text-to-drawing synthesis through language-image encoders (2021). https://doi.org/10.48550/ARXIV.2106.14843. https://arxiv.org/abs/2106.14843
8. Gal, R., Patashnik, O., Maron, H., Chechik, G., Cohen-Or, D.: Stylegan-nada: clip-guided domain adaptation of image generators. CoRR abs/2108.00946 (2021). https://arxiv.org/abs/2108.00946
9. Jabri, A., Owens, A., Efros, A.: Space-time correspondence as a contrastive random walk. In: Larochelle, H., Ranzato, M., Hadsell, R., Balcan, M., Lin, H. (eds.) Advances in Neural Information Processing Systems. vol. 33, pp. 19545–19560. Curran Associates, Inc. (2020). https://proceedings.neurips.cc/paper/2020/file/e2ef524fbf3d9fe611d5a8e90fefdc9c-Paper.pdf
10. Jampani, V., Gadde, R., Gehler, P.V.: Video propagation networks. CoRR abs/1612.05478 (2016). http://arxiv.org/abs/1612.05478
11. Jetchev, N.: Clipmatrix: Text-controlled creation of 3d textured meshes. CoRR abs/2109.12922 (2021). https://arxiv.org/abs/2109.12922
12. Karras, T., Laine, S., Aila, T.: A style-based generator architecture for generative adversarial networks. CoRR abs/1812.04948 (2018). http://arxiv.org/abs/1812.04948
13. Karras, T., Laine, S., Aittala, M., Hellsten, J., Lehtinen, J., Aila, T.: Analyzing and improving the image quality of stylegan. CoRR abs/1912.04958 (2019). http://arxiv.org/abs/1912.04958
14. Kasten, Y., Ofri, D., Wang, O., Dekel, T.: Layered neural atlases for consistent video editing. CoRR abs/2109.11418 (2021). https://arxiv.org/abs/2109.11418
15. Kim, G., Ye, J.C.: Diffusionclip: Text-guided image manipulation using diffusion models. CoRR abs/2110.02711 (2021). https://arxiv.org/abs/2110.02711
16. Kingma, D.P., Ba, J.: Adam: A method for stochastic optimization. In: Bengio, Y., LeCun, Y. (eds.) 3rd International Conference on Learning Representations, ICLR 2015, San Diego, CA, USA, 7–9 May, 2015, Conference Track Proceedings (2015). http://arxiv.org/abs/1412.6980
17. Kwon, G., Ye, J.C.: Clipstyler: Image style transfer with a single text condition. CoRR abs/2112.00374 (2021). https://arxiv.org/abs/2112.00374
18. Michel, O., Bar-On, R., Liu, R., Benaim, S., Hanocka, R.: Text2mesh: Text-driven neural stylization for meshes. CoRR abs/2112.03221 (2021). https://arxiv.org/abs/2112.03221

19. Nichol, A., et al.: GLIDE: towards photorealistic image generation and editing with text-guided diffusion models. CoRR abs/2112.10741 (2021). https://arxiv.org/abs/2112.10741

20. Paszke, A., et al.: Pytorch: an imperative style, high-performance deep learning library. In: Advances in Neural Information Processing Systems 32, pp. 8024–8035. Curran Associates, Inc. (2019). http://papers.neurips.cc/paper/9015-pytorch-an-imperative-style-high-performance-deep-learning-library.pdf

21. Patashnik, O., Wu, Z., Shechtman, E., Cohen-Or, D., Lischinski, D.: Styleclip: text-driven manipulation of stylegan imagery. CoRR abs/2103.17249 (2021). https://arxiv.org/abs/2103.17249

22. Pont-Tuset, J., Perazzi, F., Caelles, S., Arbelaez, P., Sorkine-Hornung, A., Gool, L.V.: The 2017 DAVIS challenge on video object segmentation. CoRR abs/1704.00675 (2017). http://arxiv.org/abs/1704.00675

23. Radford, A., et al.: Learning transferable visual models from natural language supervision. CoRR abs/2103.00020 (2021). https://arxiv.org/abs/2103.00020

24. Ramesh, A., Dhariwal, P., Nichol, A., Chu, C., Chen, M.: Hierarchical text-conditional image generation with clip latents (2022). https://doi.org/10.48550/ARXIV.2204.06125. https://arxiv.org/abs/2204.06125

25. Ramesh, A., et al.: Zero-shot text-to-image generation. CoRR abs/2102.12092 (2021). https://arxiv.org/abs/2102.12092

26. Saharia, C., et al.: Photorealistic text-to-image diffusion models with deep language understanding (2022). https://doi.org/10.48550/ARXIV.2205.11487. https://arxiv.org/abs/2205.11487

27. Sanghi, A., Chu, H., Lambourne, J.G., Wang, Y., Cheng, C.Y., Fumero, M., Malekshan, K.R.: Clip-forge: Towards zero-shot text-to-shape generation (2021). https://doi.org/10.48550/ARXIV.2110.02624. https://arxiv.org/abs/2110.02624

28. Texler, O., et al.: Interactive video stylization using few-shot patch-based training. CoRR abs/2004.14489 (2020). https://arxiv.org/abs/2004.14489

29. Wang, X., Jabri, A., Efros, A.A.: Learning correspondence from the cycle-consistency of time. In: Proceedings of the IEEE/CVF Conference on Computer Vision and Pattern Recognition (CVPR), June 2019

30. Ye, V., Li, Z., Tucker, R., Kanazawa, A., Snavely, N.: Deformable sprites for unsupervised video decomposition (2022). https://doi.org/10.48550/ARXIV.2204.07151. https://arxiv.org/abs/2204.07151

31. Yin, K., Gao, J., Shugrina, M., Khamis, S., Fidler, S.: 3dstylenet: creating 3d shapes with geometric and texture style variations. CoRR abs/2108.12958 (2021). https://arxiv.org/abs/2108.12958

32. Zhou, K., Yang, J., Loy, C.C., Liu, Z.: Learning to prompt for vision-language models. CoRR abs/2109.01134 (2021). https://arxiv.org/abs/2109.01134

MND: A New Dataset and Benchmark of Movie Scenes Classified by Their Narrative Function

Chang Liu[✉][iD], Armin Shmilovici[iD], and Mark Last[iD]

Ben-Gurion University of the Negev, P.O.B. 653, Beer-Sheva, Israel
liuc@post.bgu.ac.il, {armin,mlast}@bgu.ac.il

Abstract. The success of Hollywood cinema is partially attributed to the notion that Hollywood film-making constitutes both an art and an industry: an artistic tradition based on a standardized approach to cinematic narration. Film theorists have explored the narrative structure of movies and identified forms and paradigms that are common to many movies - a latent narrative structure. We raise the challenge of understanding and formulating the movie story structure and introduce a novel story-based labeled dataset-the Movie Narrative Dataset (MND). The dataset consists of 6,448 scenes taken from the manual annotation of 45 cinema movies, by 119 distinct annotators. The story-related function of each scene was manually labeled by at least six different human annotators as one of 15 possible key story elements (such as Set-Up, Debate, and Midpoint) defined in screenwriting guidelines. To benchmark the task of scene classification by their narrative function, we trained an XGBoost classifier that uses simple temporal features and character co-occurrence features to classify each movie scene into one of the story beats. With five-fold cross-validation over the movies, the XGBoost classifier produced an F1 measure of 0.31 which is statistically significant above a static baseline classifier. These initial results indicate the ability of machine learning approaches to detect the narrative structure in movies. Hence, the proposed dataset should contribute to the development of story-related video analytics tools, such as automatic video summarization and movie recommendation systems.

Keywords: Computational narrative understanding · Movie understanding · Movie analytics · Plot points detection · Scene classification

1 Introduction

1.1 Background

The success of Hollywood cinema is partially attributed to the notion that Hollywood film-making constitutes both an art and an industry: an artistic tradition based on a standardized approach to cinematic narration [38]. This artistic

Supplementary Information The online version contains supplementary material available at https://doi.org/10.1007/978-3-031-25069-9_39.

© The Author(s), under exclusive license to Springer Nature Switzerland AG 2023
L. Karlinsky et al. (Eds.): ECCV 2022 Workshops, LNCS 13804, pp. 610–626, 2023.
https://doi.org/10.1007/978-3-031-25069-9_39

system influenced other cinemas, creating a sort of international film language. Hollywood has developed some fairly explicit principles for how stories can be told effectively. For example, some scenes will typically contain unresolved issues that demand settling further along [19]. Sometimes a film puzzles or frustrates us when we cannot identify character goals or clear-cut lines of cause and effect [18]. Some manuals of screenwriting have picked up on the principles, turning them into explicit rules [12].

Popular movies present stories - narratives - in an audio-visual manner. Narrative is a core mechanism that human beings use to find meaning that helps them to understand their world [32]. Narratology is the study of stories and story structure and the ways these affect our perception, cognition, and emotion [9]. Most research focuses on the story as it is physically communicated, not on the story, as it is understood. Here, we explore how the story form has been intended by its filmmakers, though implicitly, to engage its spectators.

Narratology is well developed in the "text worlds" (e.g., literature [13] - the earliest known work on Narratology is Aristotle's Poetica). The recent progress in Natural Language Processing has increasingly focused on developing computational models that reason about stories: in Computational Narrative Understanding, theoretical frameworks in narratology are used to develop computational models that can reason about stories [32, 40]. A related problem is narrative scene detection, which attempts to observe the spatial, temporal, and agential boundaries between story segments [10, 28]. At a higher level, narrative plot-line detection is the act of assembling scenes into more general narrative units defined by agents who may range over both time and space [32, 42].

Most works in video understanding are based on computer vision algorithms. Those algorithms perform well on basic, fact-based video understanding tasks, such as recognizing actions in video clips [6, 33], question-answering about the video contents [6, 37], and generating captions for videos [34, 44]. However, most of these algorithms focus on analyzing short video clips (less than 30 s), which makes them very suitable for exploring the detailed (or low-level) information in videos such as "playing soccer" or "running" but very poor at understanding high-level events in those videos (e.g., "enjoying a party" or "going home"), due to the casual and temporal relationships between events, which can be complex and are often implicit.

The huge gap between the state-of-the-art computer vision algorithms and story analytics seems hard to be bridged, and therefore, novel approaches to understanding the video stories are needed. The main contribution of this paper is that we formally raise a novel research task in the field of computational narrative understanding - identifying the narrative function of movie scenes according to screenwriting guidelines. To this end, we collect and assemble a new benchmark dataset of movie scenes labeled by their narrative function - the Movie Narrative dataset (MND[1]). As a complement to the computer vision algorithms, features from the latent story structure can be utilized to enhance applications such as movie summarization [11, 31, 39], and movie recommendation [20].

---

[1] The MND dataset is available at https://github.com/llafcode/The-Movie-Narrative-Dataset-MND.

1.2 Research Objectives and Contributions

Our first objective is to provide a dataset of movie scenes labeled with high-level concepts such as Debate - the internal conflict of the protagonist whether to return to her "comfort zone" after facing a Catalyst event that knocked her out of it. A crowd-sourcing experiment is used for constructing the dataset. The collected labels are analyzed to verify the following two hypotheses: (1) most movies in our dataset adhere fairly well to the latent story structure described by the screenwriting book [35] and (2) even non-experts can identify the scenes' narrative function after reading the annotation guidelines and watching a movie.

Our second objective is to provide a lightweight solution for the challenging task of classifying movie scenes by their narrative function, with the use of relatively simple features and supervised machine learning algorithms. Although a fully automated pipeline is desired, at this stage we use the manually annotated features from the MovieGraphs dataset [41] to avoid the errors prevalent in the current scene splitting, character identification, and other movie annotation tools, which often fail in common cases such as darkly illuminated scenes.

The original contributions of our paper to the domain of computational narrative understanding in movies are two-fold: a) We introduce a new task for movie analytics - learning the latent narrative function of each scene; b) We introduce the first benchmark dataset of movie scenes labeled by their narrative function that will be released to the research community; c) We demonstrate that a movie's latent story structure can be automatically detected using machine learning with some relatively simple scene features, outperforming a strong temporal distribution baseline.

The rest of the paper is organized as follows: Sect. 2 presents some background and some related work; Sect. 3 describes the elements of the latent story model that we use; Sect. 4 describes the construction and the features of the MND dataset; Sect. 5 presents an MND task of movie scenes classification by their narrative function; and finally, Sect. 6 concludes the paper. The appendices in the supplementary document provide some technical details about the data collection and pre-processing.

2 Background and Related Work

2.1 Introduction to Story Models

The architecture of a typical movie at its highest level has four 25–35 minute long acts - *Setup, Complication, Development, and Climax* - with two optional shorter sub-units of *Prolog* and *Epilog* [9,38]. At a middle level, there are typically 40–60 scenes. The scenes develop and connect through short-term chains of cause and effect. Characters formulate specific plans, react to changing circumstances, gain or lose allies, and otherwise take specific steps toward or away from their goals [4]. At the third level of organization audiovisual patterning carries the story along bit by bit. For example, within a scene, we often find patterns of cutting: an establishing shot introduces the setting, reverse camera angles meshed with the developing dialogue, and close-ups cue the relation between the characters [1]. This paper focuses on the middle level.

Aristotle's Poetics is the earliest surviving work on dramatic and literary theories in the West. His three-act form - as applied to movies, has come to mean *Act One*, *Act Two*, and *Act Three*. Each act has its own characteristics: Act One introduces the character(s) and the premise - what the movie is about; Act Two focuses on confrontation and struggle; Act Three resolves the crisis introduced in the premise. The three-act structure was extended by [12] with various plot devices - or story "beats". In film development terms, a "beat" refers to a single story event that transforms the character and story at a critical point in time. Beats such as *Inciting Incident* (typically in the first act), *Disaster*, and *Crisis* (*must appear at least once*), intend to intensify conflict, develop characters, and propel the plot forward. Beats can be also considered as "checkpoints" along the way, which will complete the story and reveal the movie's structure. Turning Points (TP) are moments that direct the plot in a different direction, therefore separating between acts. [30] attempts to identify the 5 TP that separate the acts in feature length screenplays by projecting synopsis level annotations. There are also rules of thumb indicating where to expect each TP. For example, the 1st TP - the *Opportunity*, separates between the Setup act and the New Situation acts and is expected to occur after the initial 10% of the movie duration. A comparison between seven different story models is available [17].

In this paper, we decided to use the recent Save the Cat!® theory [35] which is popular among scriptwriters. It suggests that a good story is like music, which has beats that control the rhythm and flow. The theory defines for the writers 15 story beats we introduce in Sect. 3, that play different roles in the story development. The main advantage of this model is that it can also incorporate two common deviations from the main story: a) *B-Story* - a plot device that carries the theme of the story, but in a different way with different characters. b) *Fun and Games* - scenes that are purely for the enjoyment of the audience (e.g., action scenes such as car chases, funny scenes, romantic scenes - depending on the genre). Thus, this story model can handle more complex movies that have side stories and scenes that do not advance the plot.

2.2   Related Datasets and Research

Only recently, large movie datasets have been constructed for the purpose of movie understanding tasks. However, due to movie copyright issues, some large movie datasets do not contain all the scenes of a movie [7,33], or due to annotation difficulties, only some of the scenes are fully annotated [3]. Most datasets focus on a specific aspect of movies, such as, genre [43,45], Question-Answering [15,37], generating textual descriptions [34,44], or integrating vision and language relations [5]. Obtaining quality human annotation for a full movie is challenging, therefore, some of the annotations are generated from text, e.g. from synchronizing the movie with its script or synopsis. Following is the description of datasets that are sufficiently large, full (scene-wise), and contain some high-level story elements labels, therefore, are most related.

MovieNet - 1,100 movies, 40K scenes, many modalities, (e.g., cinematic styles) quality annotations, and metadata [16]. *SyMoN* - 5,193 video summaries [36]. The most important aspect of story analytics is the alignment between a movie and its script and synopsis.

TRIPOD - 122 movies, 11,320 scenes [30,31], metadata, The most important aspect of story analytics is the annotation of the 5 Turning Points in a movie (via textual analysis of their scripts and synopsis).

MovieGraphs - 51 movies, 7.2K scenes, high-quality manual annotation of the relationships and the interactions between movie characters [21,41]. We used some of those quality annotations for the construction of our dataset.

FSD - 60 episodes of the *Flintstones* cartoons, about 26 min long, 1,569 scenes [26]. Each scene was manually labeled with the 9 labels from the story model of [12]. A classifier was trained to predict the label for each scene. This is the most similar dataset to ours. The main differences are that they use a less elaborate story model than ours [35], for much shorter movies (only about 25% long), inaccurate scene splitting, with the same characters in each episode, while we have built a heterogeneous dataset of 45 movies with more quality annotations and features (e.g., manual scene splitting and character identification).

3   The Story Model, 15 Story Beats

What is a story beat? In film development terms, a "beat" refers to a single story event that transforms the character and story at a critical point in time. For each beat, there is a suggested time for it to arrive in the story. In this section, we present the detailed definitions of the 15 story beats as well as their suggested appearance time within a typical 110 min movie [35]. The recommended position of each beat is proportional to the movie length. In our dataset, the function of each movie scene in the progression of the story is labeled by one of those 15 story beats. In addition, a *"None"* label is used for scenes that do not progress the plot significantly.

1. **Opening Image (minute 1):** It presents the first impression and sets the tone, mood, type, and scope of the movie. It is an opportunity to give the audience a starting point of the hero before the story begins.
2. **Theme Stated (minute 5):** A character (often not the main character) will pose a question or make a statement (usually to the main character) that is the theme of the movie. It will not be obvious. Instead, it will be an off-hand conversational remark that the main character does not get at the time, but it will mean a lot later on.
3. **Set-Up (minute 1–10):** Sets-up the hero, the stakes, and the goal of the story. It is also where every character in the "A" story (the main story) is introduced or at least hinted at. Additionally, this is the time when the screenwriter starts to hint at every character's tic, behavior, and flaws that need to be addressed, showing why the hero will need to change later on. There could be scenes that present the hero in his home, work, and "play"

environments. Typically, the hero is presented in a comfortable state of stagnation or "inner death".

4. **Catalyst (minute 12):** A catalyst moment knocks the hero out of his or her "before" world that was shown in the set-up. The hero loses the safety of its current state.

5. **Debate (minute 12–25):** The debate section must answer some questions about how to deal with the catalyst. Debate shows us that the hero declares, "This is crazy!" and is conflicted by the options to resolve the dilemma: "should I go?" "Dare I go?" "Stay here?" The best action will most likely involve overcoming an obstacle, and therefore will result in the beginning change in the hero's character.

6. **Break into Two (minute 25):** The events cannot draw the hero into Act Two. The hero takes an action because he wants something. The hero MUST pro-actively decide to leave the old world and enter a new world because he wants some- thing. This is the point where we leave "the way things were" and enter into an upside-down version of it. "The Before" and "The After" should be distinct, so the movement into "The After" should also be definite.

7. **B Story (minute 30–55):** It is a different story (such as a love story) where the hero deals with its emotional side, perhaps the hero is even nurtured, energized, and motivated. The B story carries the theme of the story, but in a different way with different characters. The characters are often polar opposites of the characters in Act One, the "upside-down versions" of them.

8. **Fun and Games (minute 30–55):** This is where the hero explores the upside down world he/she has entered into. The "Fun and Games" scenes are purely for the enjoyment of the audience - depending on the genre, it could have action scenes (such as car chase); funny scenes, romantic scenes, etc. During "Fun and Games", we are not as concerned with the plot moving forward and the stakes won't be increased here.

9. **Midpoint (minute 55):** This is where the fun and games are over. The mid-point is where the stakes are raised (no turning back) so that it's either a (false) victory where the hero thinks that everything is fixed and he obtained his goal; or it seems like a (false) defeat for the hero. Sometimes a public display of the hero (such as in a big party). Our hero still has a long way to go before he learn the lessons that really matter.

10. **Bad Guys Close In (minute 55–75):** The (internal or external) forces that are aligned against the hero tighten their grip. As an opposite of the midpoint, If the hero had a false victory in the midpoint and the bad guys seem temporarily defeated, it is during Bad Guys Close In that the bad guys regroup and the hero's overconfidence and jealousy within the good guy team start to undermine all that they accomplished. This is because the hero has not fully learned the lesson he or she is supposed to learn, and the bad guys haven't completely been vanquished. As a result, our hero is headed for a huge fall. If it was a false defeat in the mid-point, now the here is hope.

11. **All Is Lost (minute 75):** The hero losses what he wants and feels the smell of death (or defeat). Most often, it is a false defeat. All aspects of the hero's life are in a mess. If the midpoint was a false victory, then this is the low point for the hero when he or she has no hope.

12. **Dark Night of the Soul (minute 75–85):** This is the darkness night before the dawn, when the hero is forced to admit his or her humility and humanity, yielding control to "fate" or to "the universe". It is just before the hero digs deep down and pulls out that last best idea that will save the hero and everyone else. However, at this very moment this idea is no where in sight.

13. **Break into Three (minute 85):** The hero takes an action because he needs something. At this point both the A story (which is the external, obvious story) and the B story (the internal, emotional story) meet and intertwine. The characters in the B story, the insights gleaned during their conversations discussing the theme, along with the hero striving for a solution to win against bad guys all comes together to reveal the solution to the hero. The hero has passed every test, dug down to find the solution. Now he or she just needs to apply it.

14. **Finale (minute 85–110):** The bad guys are defeated in ascending order, meaning that first the weaker enemy loose, then the middle men, and finally the top enemy. The source of "the problem" must be completely and absolutely defeated for the new world to exist. It is more than just the hero winning; the hero must also change the world.

15. **Final Image (minute 110):** In a happy conclusion, the final image is the opposite of the opening image and acts as proof that change has happened in the hero, and that the change is real. The B-story is resolved. In a sad ending, the hero rejects the change.

For example, in the movie *Silver Lining Playbook*[2], in the *Opening Image* (minutes 1–2) we see the main male character alone in a Psychiatric Facility, talking to his absent wife. In the *Midpoint* scenes (minute 58) we see the main male character dancing for the first time with the main female character, then he runs away and we later see him on his bed, in a turmoil of desire and guilt. In the *Final Image* (minutes 114–116) the whole family is back together in the house. The love of the main male and the female characters has not completed them alone; it has completed the greater family circle.

4   The MND Dataset

4.1   Data Collection - Movies and Scenes

We constructed a labeled dataset of scene categories (story beats) to facilitate the use of supervised machine learning algorithms. We limited our movie selection to the 51 movies used in the MovieGraphs dataset [41]. This dataset was constructed for the purpose of understanding human interactions in movies [21],

---

[2] https://savethecat.com/beat-sheets/the-silver-linings-playbook-beat-sheet.

and therefore, it is heavily biased towards realistic stories (many romantic comedies while almost no horror movies, fantasies or science fiction movies). The reason why we chose to use this dataset is that it offers rich and accurate manual annotations of low-level movie features, such as shot/scene splitting, character identification, character attributes and actions etc., which are difficult to extract automatically and accurately with the current video-processing techniques. Our previous studies that used state-of- the-art video analytics software such as Microsoft's Video Indexer[3], suffered from errors in identifying characters, especially in darkly illuminated scenes, and our previous work [25,26] showed the negative consequences of using fully automated, but inaccurate, scene splitting software for narrative understanding. Therefore, as a preliminary study, we base our work on the MovieGraphs dataset, use its provided scene boundaries and extract features using its high-quality annotations. We hope that in the near future, more accurate video processing tools will be developed, allowing a significant extension of this dataset with a minimal manual effort.

We had to discard 6 movies, either because we could not obtain the same version of the movies used in the MovieGraphs dataset, or because of an irrelevant movie style from the plot point of view (e.g. a biographical movie), and eventually labeled 45 movies contains 6,448 labeled scenes. Nine of the movies have their "gold-standard story beats" summarized by professional writers/scriptwriters who are proficient in the Save the Cat!® theory[4]. The gold standard story beats are used to evaluate the quality of the collected labels.

4.2   Collecting Story Beats Labels for the Scenes

We received the departmental ethics committee approval for using student volunteers as annotators. Each volunteer, which completed its task, received two bonus grade points for a data-science class. We selected 119 human annotators (out of 180 applicants, all senior undergraduate students in the Information Systems Engineering Department), based on their English proficiency level and their level of interest in watching movies (refer to the supplementary information for more detail). During the annotation process, we ensured that: (1) each annotator was assigned at least 3 movies (including one of the 9 movies with "gold-standard" annotation); and (2) each movie was annotated by at least 5 different annotators (in practice, except for one, all movies were assigned to 6 or more annotators). The annotators were provided with the guidelines that described in detail the background concepts, definitions of story beats, required workflows and accepted criteria. They were asked to choose one most appropriate story beat for each scene, out of 16 options described in Sect. 3. The annotation experiment was performed on the *Moodle* teaching platform using the H5P interactive video application[5]. The annotators were allowed to jump

---

[3] https://azure.microsoft.com/en-us/services/media-services/video-indexer/.

[4] The "gold-standard story beats" are provided in the manner of movie deconstruction articles, an example can be found here: https://savethecat.com/beat-sheets/the-silver-linings-playbook-beat-sheet.

[5] https://h5p.org/documentation.

forward and backward without limitation so that they can skip any scene at first and label it later if needed. To evaluate the annotator's attention during the task, we used the knowledge about the characters which participate in each scene to automatically generate ten simple quizzes about the participation of a specific character in a specific scene. The quizzes were inserted at the end of randomly selected scenes in each movie.

For choosing the single best scene label from the annotations, we follow a two steps process: a) Select the scene label which received the most votes; b) If tie, use the time dependent label distribution (Fig. 1 right) to select the label with highest likelihood from the tied ones.

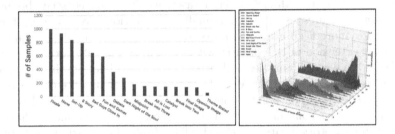

Fig. 1. Label distribution of the MND dataset. **Left:** label histogram. **Right:** the temporal label distribution over all movies. On the class label axis, each tick represents a label, corresponding with that introduced in Sect. 3.

4.3   Dataset Analytics

Evaluation Metrics: Watching and labeling a full length movie is a tough commitment for the crowdsourcing workers. We measure the quality of the collected labels in three ways: (1) *Fleiss' Kappa* [14,22] was used to measure the inter annotator agreement; and (2) Visualization of the labels distribution along the normalized movie duration time were used to verify the compatibility of the collected label with the theory; (3) the similarity between the collected story beats and the gold-standard story beats. A summary of the statistics of the dataset is presented in Table 1. Considering the graphs of the normalized temporal label distributions (Fig. 1, right), the peaks of the distributions correspond fairly well with what is expected from Sect. 3 (e.g., the *Midpoint* is expected at about $55/100 = 0.50$ of the movie).

For each movie, we check if there exists an "outlier" annotator and remove this annotation to increase the agreement. An "outlier" annotator is defined as the annotator whose annotation reduces the agreement the most. For example, the movie *Jerry Maguire* had an initial Kappa of 0.09 (6 annotators) and after removal of the outlier annotator, its Kappa increased to 0.12. The Kappa score presented in Table 1 is the improved score after the outlier removal step.

Table 1. The movie narrative dataset summary

	Min	Max	Avg.	Median
Movie Duration (hh:mm:ss)	01:35:31	02:47:59	02:03:15	01:58:06
# of scenes	51	279	143	143
Scene duration (mm:ss)	00:01	11:10	00:45	00:37
Kappa	0.11	0.67	0.25	0.24

The movies *Dallas Buyers Club* and *Pulp Fiction* obtained the highest Kappa (0.67 and 0.54, respectively), while the movies *Forrest Gump* and *Ocean's Eleven* obtained the lowest Kappa of 0.11. Considering the large number of scenes, categories, movies, and annotators, we consider median Kappa 0.24 as fairly high for such a subjective annotation task. The Kappa score corroborates the hypothesis that most movies adhere fairly well to the latent story structure described by the screenwriting book [35]. A low Kappa score may indicate a movie which is difficult to interpret or may not comply with the story model (e.g., no single main story such as *Crash*, or many *Flash Backs* such as *Forrest Gump*).

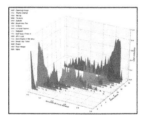 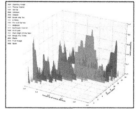

Fig. 2. Visualization of the label distribution for the 9 movies with gold standard story beats. **Left:** visualization of the gold standard story beats; **Right:** visualization of the collected story beats.

By comparing the collected story beats to the 9 movies with gold-standard story beats (i.e., evaluation by experts), we can evaluate how well the annotators understood their task. Figure 2 presents the visualization of label distribution along normalized movie durations for the gold standard labels (right) compared to the collected labels (left). Overall, the two visualizations are similar to each other in the shapes, locations and peaks of the label distribution, therefore we can infer that even non-experts can identify the scenes' narrative function after watching a movie. Therefore, indicating a relatively good quality of the collected labels, and that the entire dataset is reliable and consistent.

The main differences are a) The *None* category: the annotators classified many more scenes as *None*; b) more differences are in the labels that their definition is difficult to understand: *Bad Guys Close In* and *Fun & Games* are selected more in the gold-standard annotations. The distribution difference measures we used to compare the gold-standard annotation with the non-expert annotation (KL-divergence, Bhattacharyya Distance and Earth Movers' distance) also indicate the difference in those labels

5   A MND Task: Movie Scenes Classification by Their Narrative Function

5.1   Data Pre-processing

The MND dataset consists of 6,448 scenes from 45 full-length cinema movies. The scene boundaries are provided by the MovieGraphs dataset [41], and we use the scene boundaries in order to avoid unnecessary noise caused by scene splitting errors. In the dataset, we annotated one label (out of 16 labels introduced in Sect. 3) for each scene. Since there are some infrequent class labels, in order to solve the data imbalance problem, we applied label set reduction by merging 4 labels with other labels based on their definition and their temporal neighborhoods: *Opening Image* and *Theme Stated* are merged with *Set-Up*, *All Is Lost* is replaced by *Dark Night of the Soul*, and *Final Image* is merged with *Finale*. Eventually, each of the 6,448 scenes are labeled by one out of 12 labels. The label histograms before and after reduction are presented in Fig. 3 (left-up and left-down). In the dataset we also all keep the raw annotations, for future research.

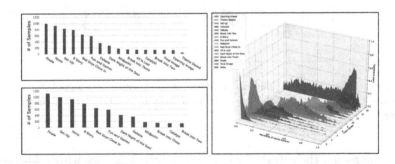

Fig. 3. Label Reduction and temporal distribution of the MND Dataset. **Left-up:** label histogram before reduction; **Left-down:** label histogram after reduction; **Right:** Temporal Distribution of the 16 labels (before reduction).

5.2   Feature Engineering

We use the same feature sets that were introduced in [26]. Specifically, we construct the following two sets of features: The basic feature set is inspired by the localized temporal distribution of most labels (Fig. 3, right), and it contains 5 features that are computed from the position of the scene in the movie, were i is the scene counter. The character network [23,24] feature set is used to capture the protagonist's importance in a scene. We expect the protagonist to participate in most scenes that advance the story - the story beats. The features are described in Table 2. The features' Information Gain (I.G.) column indicates that the most influential feature, r_t_loc, is the normalized position of the start of each scene in the movie. Considering that each story bead in Sect. 3 has a recommended position, this hints that many movies follow that recommendation. Please note that the MND dataset contains potentially many more features than needed to develop a more elaborate and accurate model. For example, we could use the two-clock feature of [25, 27] that attempts to detect turning point scenes. We could extract musical cues from the audio, lighting cue and cinematic cue (e.g., camera motions and close-ups) from the video [3, 16] and possible use metadata information such as the movie genre. We wanted to develop a fairly simple model that may be used to benchmark further improvements in story analytics.

Table 2. Full feature list used for story-based scene classification on full length movies.

Feature set	Name	Type	Description	I.G.
Basic	dur	float	scene duration (seconds)	0.81
	t	float	start time (seconds)	0.76
	$close\_beg\_id$	float	closeness to the beginning, $1/i$	0.83
	r_id_loc	float	i/n, n is number of scenes in the episode	1.75
	r_t_loc	float	t/len, len is the duration of the episode	4.40
Character network	$protagonist\_appear$	bool	1: the episode's protagonist appears in the current scene	0.99
	ava_scores	float	the average character scores	0.95

5.3 Baseline Approach

We present two simple baseline approaches: (1) Majority rule and (2) Maximum likelihood labeling (temporal label distribution baseline). The majority rule baseline is simply labeling all scenes as the most frequent label (e.g., a single label). The maximum likelihood labeling baseline is computed for each given scene based on its normalized location in the movie. In Fig. 3(right), we present the label temporal distribution along the movie time, which is used for the maximum likelihood labeling baseline. Specifically, for a given scene, (e.g., starts at minute 29) we firstly computed the percentile of movie time (t) this scene appeared (e.g., 24%), and from the demonstrated distribution, select the label with maximum probability at t percent.

5.4 Classification Experiment and Baseline Results

We applied five-fold cross validation approach over the Movie Narrative Dataset. In each validation fold, we keep 9 movies for testing and 36 movies for training. We used XGBoost [8] as the classification algorithm and typical evaluation matrices (precision, recall, accuracy and F1 measure) were used for quantitative evaluation. The XGBoost classifier won most of the recent data mining competitions before the introduction of Deep Neural Networks. The implementations of the algorithms used in this work are based on the distributed Python implementation of XGBoost, and scikit-learn, a widely used machine learning library for Python. We used the default parameter settings provided by the implementations. As presented in Table 3, the maximum likelihood labeling baseline significantly outperformed the majority rule baseline, and our classification approach with XGBoost classifier and the features introduced in Table 2 improved the accuracy by 0.03 and the F1 measure by 0.05. The improvements are tested to be statistically significant by t-test. Although the accuracy results on this small dataset are still low, they indicate that (1) The idea of automated scene classification in full-length cinema movies is feasible; (2) The proposed prototype approach and suggested features can be useful for various, more complex story models. We can expect better performance with more advanced feature engineering and larger amounts of annotated data.

Table 3. Baseline classification results on the MND dataset.

Methods	Prec.	Rec.	Acc.	F1.
Majority rule	0.14	0.08	0.17	0.02
Maximum likelihood labeling	0.26	0.27	0.45	0.26
XGBoost	**0.31**	**0.34**	**0.48**	**0.31**

We further experimented with the two-clocks feature of [27] and with more character co-occurrence network features [23]. They did not contribute significantly to the classification performance.

We explored the possibility that different movies in our dataset belong to different story structures, thus reducing the performance of the classification algorithm. Considering the temporal label distribution presented in Fig. 3 (right), we observe that there exist some story elements with a wide temporal distribution (such as *B-Story* and *Bad Guys Close In*). Possible reasons for such wide distributions might be (1) multiple label occurrences per movie (e.g., a story may have more than one *B-Story*); (2) difficult concepts (e.g., the annotators had a hard time understanding the concept of *Theme Stated*), and (3) a movie not conforming to the story model (some movies with very low Kappa scores indicate that the annotators are confused, meaning that the movie itself may not fit our story model well). It might be that the classification performance would improve if the confusing classes were removed from the label set and only the most critical elements were kept (such as *Inciting incident*, and *Debate*), however, label removal or merging might obscure some of the fine elements necessary for understanding the movie.

6   Conclusion and Future Research

In this paper, we defined a novel task for movie analytics: movie scene classification by their narrative function. It is an important step towards understanding the latent story structure within narrative videos such as movies, TV series or animated cartoons. We constructed a novel benchmark labeled scene dataset, the Movie Narrative Dataset (MND). From the manually annotated scene/shot boundaries and character identifications, provided in the MovieGraphs dataset, we constructed two sets of features for 45 movies. The extracted features include the basic information about the movie itself (such as duration, number of scenes etc.), character network features and temporal character appearance features. The features represent the aspects of the movie stories from different angles. The classification and feature selection experiments demonstrated the use of machine learning algorithms for the scene classification task. The evaluated algorithms were able to discover the sequential character of the key elements in the story model we used, which has been further verified by the temporal label distribution baseline and results with the basic feature set. The scene classifiers can serve as a benchmark for future research.

The value of scene classification by their narrative function for movie understanding is in extracting the high level abstract concepts associated with each story type, e.g., *Debate, All is Lost*. These can provide some automatic understanding about the protagonist's character traits and the motivations that drive him in facing his challenges. While the F1 measure of our classification model is relatively small, it might be enhanced with the addition of more elaborate features such as character emotions [29], character interactions [21], dialogue (subtitles) analysis and cinematic mood cues such as shots and camera movements, music and lighting [2]. As a first attempt to learn the story structure of a narrative video, we believe that this work has opened a promising direction for video story understanding.

Future research can naturally follow our results by (1) adding low level story-related features such as automatically detected characters, objects, actions, place, and emotions, etc.; (2) adding cinematic cue features such as shots and camera angle, illumination, music and voice; (3) generating a story-aware summary (in a video or text format) of a given narrative video, using the most important story elements and scenes, (e.g., following the work in [31]); (4) Use the MND in an attempt to detect even higher level narrative concepts such as the 36 dramatic situations of [13]. (5) utilizing story-related scene classification to boost the performance of other story-related tasks, such as movie question-answering; (6) exploring alternative, more elaborate movie story structures; (7) generating additional benchmark collections of story-based video annotations.; (8) constructing a fully automated pipeline that can process a video from start (e.g., scene cutting) to end (the detected story elements) and use it to annotate a large dataset such as [16].

Acknowledgement. This research was partially supported by the Israeli Council for Higher Education (CHE) via the Data Science Research Center, Ben-Gurion University, Israel.

References

1. Arijon, D.: Grammar of the Film Language. Silman-James Press (1991). https://books.google.com.hk/books?id=6bQlAQAAIAAJ
2. Avgerinos, C., Nikolaidis, N., Mygdalis, V., Pitas, I.: Feature extraction and statistical analysis of videos for cinemetric applications. In: 2016 Digital Media Industry Academic Forum (DMIAF), pp. 172–175 (2016). https://doi.org/10.1109/DMIAF.2016.7574926
3. Bain, M., Nagrani, A., Brown, A., Zisserman, A.: Condensed movies: story based retrieval with contextual embeddings. In: Proceedings of the Asian Conference on Computer Vision (ACCV), November 2020
4. Bordwell, D., Staiger, J., Thompson, K.: The Classical Hollywood Cinema: Film Style & Mode of Production to 1960. Basler Studien zur deutschen Sprache und Literatur, Routledge (1985). https://books.google.com.hk/books?id=AkUOAAAAQAAJ
5. Cao, J., Gan, Z., Cheng, Yu., Yu, L., Chen, Y.-C., Liu, J.: Behind the scene: revealing the secrets of pre-trained vision-and-language models. In: Vedaldi, A., Bischof, H., Brox, T., Frahm, J.-M. (eds.) ECCV 2020. LNCS, vol. 12351, pp. 565–580. Springer, Cham (2020). https://doi.org/10.1007/978-3-030-58539-6_34
6. Carreira, J., Zisserman, A.: Quo vadis, action recognition? a new model and the kinetics dataset. In: proceedings of the IEEE Conference on Computer Vision and Pattern Recognition, pp. 6299–6308 (2017)
7. Cascante-Bonilla, P., Sitaraman, K., Luo, M., Ordonez, V.: Moviescope: large-scale analysis of movies using multiple modalities. arXiv preprint arXiv:1908.03180 (2019)
8. Chen, T., Guestrin, C.: XGBoost: a scalable tree boosting system. In: Proceedings of the 22nd ACM SIGKDD International Conference on Knowledge Discovery and Data Mining, KDD 2016, pp. 785–794. ACM, New York (2016). https://doi.org/10.1145/2939672.2939785
9. Cutting, J.E.: Narrative theory and the dynamics of popular movies. Psychonomic Bull. Rev. **23**(6), 1713–1743 (2016). https://doi.org/10.3758/s13423-016-1051-4
10. Delmonte, R., Marchesini, G.: A semantically-based computational approach to narrative structure. In: IWCS 2017–12th International Conference on Computational Semantics-Short papers (2017)
11. Evangelopoulos, G., Zlatintsi, A., Potamianos, A., Maragos, P., Rapantzikos, K., Skoumas, G., Avrithis, Y.: Multimodal saliency and fusion for movie summarization based on aural, visual, and textual attention. IEEE Trans. Multimedia **15**(7), 1553–1568 (2013)
12. Field, S.: Screenplay: The foundations of screenwriting. Delta (2007)
13. Figgis, M.: The Thirty-Six Dramatic Situations. Faber & Faber (2017)
14. Fleiss, J.L.: Measuring nominal scale agreement among many raters. Psychol. Bull. **76**(5), 378 (1971)
15. Garcia, N., Nakashima, Y.: Knowledge-based video question answering with unsupervised scene descriptions. In: Vedaldi, A., Bischof, H., Brox, T., Frahm, J.-M. (eds.) ECCV 2020. LNCS, vol. 12363, pp. 581–598. Springer, Cham (2020). https://doi.org/10.1007/978-3-030-58523-5_34
16. Huang, Q., Xiong, Yu., Rao, A., Wang, J., Lin, D.: MovieNet: a holistic dataset for movie understanding. In: Vedaldi, A., Bischof, H., Brox, T., Frahm, J.-M. (eds.) ECCV 2020. LNCS, vol. 12349, pp. 709–727. Springer, Cham (2020). https://doi.org/10.1007/978-3-030-58548-8_41

17. Huntly, C.: How and Why Dramatica is Different from Six Other Story Paradigms. https://dramatica.com/articles/how-and-why-dramatica-is-different-from-six-other-story-paradigms. Accessed 20 Aug 2022
18. Iglesias, K.: Writing for Emotional Impact. Wingspan (2005)
19. Iglesias, K.: 8 Ways to Hook the Reader, Creative Screenwriting (2006)
20. Ji, J., Krishna, R., Fei-Fei, L., Niebles, J.C.: Action genome: actions as compositions of spatio-temporal scene graphs. In: Proceedings of the IEEE/CVF Conference on Computer Vision and Pattern Recognition (CVPR), June 2020
21. Kukleva, A., Tapaswi, M., Laptev, I.: Learning interactions and relationships between movie characters. In: Proceedings of the IEEE/CVF Conference on Computer Vision and Pattern Recognition, pp. 9849–9858 (2020)
22. Landis, J.R., Koch, G.G.: The measurement of observer agreement for categorical data. Biometrics, pp. 159–174 (1977)
23. Lee, O.J., Jung, J.J.: Modeling affective character network for story analytics. Futur. Gener. Comput. Syst. **92**, 458–478 (2019). https://doi.org/10.1016/j.future.2018.01.030
24. Lee, O.J., You, E.S., Kim, J.T.: Plot structure decomposition in narrative multimedia by analyzing personalities of fictional characters. Appl. Sci. **11**(4) (2021). https://doi.org/10.3390/app11041645
25. Liu, C., Last, M., Shmilovici, A.: Identifying turning points in animated cartoons. Expert Syst. Appl. **123**, 246–255 (2019). https://doi.org/10.1016/j.eswa.2019.01.003
26. Liu, C., Shmilovici, A., Last, M.: Towards story-based classification of movie scenes. PLoS ONE (2020). https://doi.org/10.1371/journal.pone.0228579
27. Lotker, Z.: The tale of two clocks. In: Analyzing Narratives in Social Networks, pp. 223–244. Springer, Cham (2021). https://doi.org/10.1007/978-3-030-68299-6_14
28. Mikhalkova, E., Protasov, T., Sokolova, P., Bashmakova, A., Drozdova, A.: Modelling narrative elements in a short story: a study on annotation schemes and guidelines. In: Proceedings of The 12th Language Resources and Evaluation Conference, pp. 126–132 (2020)
29. Mittal, T., Mathur, P., Bera, A., Manocha, D.: Affect2mm: affective analysis of multimedia content using emotion causality. In: Proceedings of the IEEE/CVF Conference on Computer Vision and Pattern Recognition, pp. 5661–5671 (2021)
30. Papalampidi, P., Keller, F., Lapata, M.: Movie summarization via sparse graph construction (2020)
31. Papalampidi, P., Keller, F., Lapata, M.: Film trailer generation via task decomposition. arXiv preprint arXiv:2111.08774 (2021)
32. Piper, A., So, R.J., Bamman, D.: Narrative theory for computational narrative understanding. In: Proceedings of the 2021 Conference on Empirical Methods in Natural Language Processing, pp. 298–311 (2021)
33. Rohrbach, A., Rohrbach, M., Tandon, N., Schiele, B.: A dataset for movie description. In: Proceedings of the IEEE Conference on Computer Vision and Pattern Recognition, pp. 3202–3212 (2015)
34. Rohrbach, A., Torabi, A., Rohrbach, M., Tandon, N., Pal, C., Larochelle, H., Courville, A., Schiele, B.: Movie description. Int. J. Comput. Vision **123**(1), 94–120 (2017)
35. Snyder, B.: Save the Cat!: The Last Book on Screenwriting You'll Ever Need. M. Wiese Productions, Cinema/Writing (2005)
36. Sun, Y., Chao, Q., Li, B.: Synopses of movie narratives: a video-language dataset for story understanding. arXiv preprint arXiv:2203.05711 (2022)

37. Tapaswi, M., Zhu, Y., Stiefelhagen, R., Torralba, A., Urtasun, R., Fidler, S.: Movieqa: Understanding stories in movies through question-answering. In: Proceedings of the IEEE Conference on Computer Vision and Pattern Recognition, pp. 4631–4640 (2016)

38. Thompson, K.: Storytelling in the new Hollywood: Understanding classical narrative technique. Harvard University Press (1999)

39. Tran, Q.D., Hwang, D., Lee, O., Jung, J.E., et al.: Exploiting character networks for movie summarization. Multimed. Tools Appl. **76**(8), 10357–10369 (2017)

40. Vargas, J.V.: Narrative information extraction with non-linear natural language processing pipelines. Drexel University (2017)

41. Vicol, P., Tapaswi, M., Castrejón, L., Fidler, S.: Moviegraphs: towards understanding human-centric situations from videos. In: Proceedings of the IEEE Conference on Computer Vision and Pattern Recognition (CVPR), June 2018

42. Wallace, B.C.: Multiple narrative disentanglement: unraveling infinite jest. In: Proceedings of the 2012 Conference of the North American Chapter of the Association for Computational Linguistics: Human Language Technologies, pp. 1–10 (2012)

43. Wi, J.A., Jang, S., Kim, Y.: Poster-based multiple movie genre classification using inter-channel features. IEEE Access **8**, 66615–66624 (2020)

44. Zhong, Y., Wang, L., Chen, J., Yu, D., Li, Y.: Comprehensive image captioning via scene graph decomposition. In: Vedaldi, A., Bischof, H., Brox, T., Frahm, J.-M. (eds.) ECCV 2020. LNCS, vol. 12359, pp. 211–229. Springer, Cham (2020). https://doi.org/10.1007/978-3-030-58568-6_13

45. Zhou, H., Hermans, T., Karandikar, A.V., Rehg, J.M.: Movie genre classification via scene categorization. In: Proceedings of the 18th ACM International Conference on Multimedia, pp. 747–750 (2010). https://doi.org/10.1145/1873951.1874068

Are All Combinations Equal? Combining Textual and Visual Features with Multiple Space Learning for Text-Based Video Retrieval

Damianos Galanopoulos$^{(\boxtimes)}$ ⓘ and Vasileios Mezaris ⓘ

CERTH-ITI, 6th Km Charilaou-Thermi Road, P.O. BOX,
60361 Thessaloniki, Greece
{dgalanop,bmezaris}@iti.gr

Abstract. In this paper we tackle the cross-modal video retrieval problem and, more specifically, we focus on text-to-video retrieval. We investigate how to optimally combine multiple diverse textual and visual features into feature pairs that lead to generating multiple joint feature spaces, which encode text-video pairs into comparable representations. To learn these representations our proposed network architecture is trained by following a multiple space learning procedure. Moreover, at the retrieval stage, we introduce additional softmax operations for revising the inferred query-video similarities. Extensive experiments in several setups based on three large-scale datasets (IACC.3, V3C1, and MSR-VTT) lead to conclusions on how to best combine text-visual features and document the performance of the proposed network. (Source code is made publicly available at: https://github.com/bmezaris/TextToVideoRetrieval-TtimesV)

Keywords: Text-based video search · Cross-modal video retrieval · Feature encoders · Multiple space learning

1 Introduction

Cross-modal information retrieval refers to the task where queries from one or more modalities (e.g., text, audio etc.) are used to retrieve items from a different modality (e.g., images or videos). This paper focuses on text-video retrieval, a key sub-task of cross-modal retrieval. The text-video retrieval task aims to retrieve unlabeled videos using only textual descriptions as input. This supports real-life scenarios, such as a human user searching for a video he/she remembers having viewed in the past, e.g., *"I remember a video where a dog and a cat were laying down in front of a fireplace"*, or, searching for a video never seen before, again by expressing their information needs in natural language, e.g., *"I would like to find a video where some kids are playing basketball in an open field"*.

To perform text-video retrieval, typically the videos or video parts, along with the textual queries, need to be embedded into a joint latent feature space.

© The Author(s), under exclusive license to Springer Nature Switzerland AG 2023
L. Karlinsky et al. (Eds.): ECCV 2022 Workshops, LNCS 13804, pp. 627–643, 2023.
https://doi.org/10.1007/978-3-031-25069-9_40

Early approaches to this task [16, 26] tried to annotate both modalities with a set of pre-defined visual concepts, and retrieval was performed by comparing these annotations. With the rise of deep neural networks over the past years, the community turned to them. Although various DNN architectures have been proposed to this end, their general strategy is the same: encode text and video into one or more joint latent feature spaces where text-video similarities can be calculated.

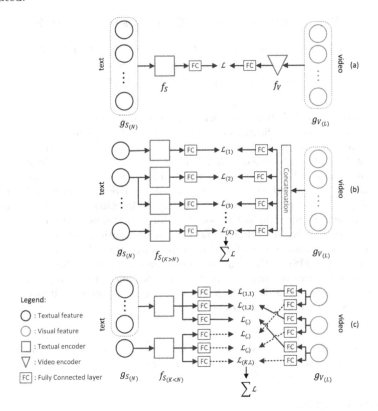

Fig. 1. Illustration of training different network architectures for dealing with the text-video retrieval problem using video-caption pairs. In all illustrations, $\mathcal{L}$ stands for the loss function. (a) All features are fed into one encoder per modality, (b) Every textual feature is used as input to a different encoder (or to more than one encoders), while visual features are simply concatenated, and (c) the proposed T×V approach, where various textual and visual features are selectively combined to create different joint spaces

State-of-the-art cross-modal video retrieval approaches utilize textual information by exploiting several textual features $g_s(\cdot)$ – extracted with the help of already-trained deep networks or non-trainable extractors – and encoding them through one or more trainable textual encoders (which are trained end-to-end as part of the overall cross-modal network training). The simple but widely used Bag-of-Words (bow) feature is often combined with embedding-based features such as Word2vec [28] and Bert [7]. Typically, these features are used as input to a simple or more sophisticated trainable textual encoder $f_s(\cdot)$, e.g., [8,19], that encodes them into one single representation. Similarly to textual information processing, trained image or video networks (e.g. a ResNet-152 trained on ImageNet) are used to extract feature vectors $g_v(\cdot)$ from the video frames. Typically these features are then concatenated and used as input to a trainable video encoder $f_v(\cdot)$. Finally, the outputs of $f_s(\cdot)$ and $f_v(\cdot)$ (after a linear projection and a non-linear transformation) are embedded into a new joint space (Fig. 1a). Methods that follow this general methodology include [9,14,19].

In [20] a new approach of textual encoder assembly was proposed for exploiting diverse textual features. Instead of inputting all these features into a single textual encoder, an architecture where each textual feature is input into a different encoder (or to more than one encoders) was proposed, resulting in multiple joint latent spaces being created. However, when it comes to the video content, its treatment in [20] is much simpler: several video features derived from trained networks are combined via vector concatenation, and individual fully connected layers embed them into a number of joint feature spaces. The cross-modal similarity, which serves as the loss function, is calculated by summing the individual similarities in each latent space. Figure 1b illustrates the above architecture.

In terms of loss function, the majority of the proposed works, e.g. [9,13,19], utilize the improved marginal ranking loss introduced by [12]. This loss utilizes the hard-negative samples within a training batch to separate the positive samples from the samples that are negative but are located near to the positives. In [6] the dual softmax loss, a modification of the symmetric cross-entropy loss, was introduced. It is based on the assumption that the optimal text-video similarity is reached when the diagonal of a constructed similarity matrix contains the highest scores. So, this loss takes into consideration the cross-direction similarities within a training batch and revises the predicted text-video similarities.

In this work, inspired from [20] where multiple textual encoders are introduced, we propose a new cross-modal network architecture to explore the combination of multiple and heterogeneous textual and visual features. We expand the textual information processing strategy of [20], with adaptations, to the visual information processing as well, and we propose a multiple latent space learning approach, as illustrated in Fig. 1c. Moreover, inspired by the dual softmax loss of [6], we examine our network's performance when we introduce a dual softmax operation at the evaluation stage (contrarily to [6] that applies it to the network's training) and use it to revise the inferred text-video similarity scores. The contributions of this work are the following:

- We propose a new network architecture, named T×V, to efficiently combine textual and visual features using multiple loss learning for the text-based video retrieval task.
- We propose introducing a dual softmax operation at the retrieval stage for exploiting prior text-video similarities to revise the ones computed by the network.

2 Related Work

The general idea behind text-video retrieval is to project text and video information into comparable representations. Due to computational resources limitations, early approaches e.g. [17,24,26], dealt with relatively small datasets, and used pre-defined visual concepts as a stepping stone. I.e., videos and text were annotated with concepts, and text-video similarity was calculated by measuring the similarity between these annotations. With the explosion of deep learning, the state-of-the-art moved forward to proposing concept-free methods. The current dominant strategy is to encode both modalities into a joint latent feature space, where the text and video representations are compared.

In [10,14], dual encoding networks were proposed. Two similar sub-networks were introduced, one for the video stream and one for the text, to encode them into a joint feature space. In the dual-task network of [33], a combination of latent space encoding and concept representation, was proposed: the first task encodes text and video into a joint latent space, while the second task encodes video and text as a set of visual concepts. In [19] several textual features were used to create multiple textual encoders, instead of feeding them into a single encoder. In this way, multiple joint text-video latent feature spaces could be learned, leading to more accurate retrieval results. In [31] the problem of understanding textual or visual content with multiple meanings is addressed by combining global and local features through multi-head attention. More recently, inspired by the human reading strategy, [11] proposed a two-branches approach to encode video representations. A preview branch captures the overview information of a video, while the intense-reading branch is designed to extract more in-depth information. Moreover, the two branches interact, and the preview guides the intense-reading branch. As a general trend, the various recent works on text-video retrieval, e.g. [4,11,20], have shown that the utilization of multiple textual features to create more than one video-text joint spaces leads to improved overall performance.

Recent approaches additionally go beyond the standard evaluation protocol (i.e., training the network using the training portion of a dataset and testing it on the testing portion), benefiting from pre-training on further large-scale video-text datasets. This procedure leads to improved performance and learning transferable textual and visual representations. In [27], HowTo100M is introduced: a large-scale dataset of 100M web videos. Using this dataset to pre-train a baseline video retrieval network is shown in [27] to be beneficial. HiT [22] uses a transformer-based architecture to create a hierarchical cross-modal network for creating semantic-level and feature-level encoders. In [22] experimentation with and without pre-training also shows that the network's performance increases with the pre-training step. BridgeFormer [15] introduces a module that is trained to answer textual questions in order to be used as the pre-training step of a dual encoding network. Frozen [2], on the other hand, is based on a transformer architecture and does not use trained image DNNs as feature extractors. It did introduce, though, a large-scale video-text dataset (WebVid-2M) which was used for end-to-end pre-training of their network.

3   Proposed Approach

3.1   Overall Architecture

The text-video retrieval problem is formulated as follows: let $V = \{v_1, v_2, \ldots, v_T\}$ be a large set of T unlabeled video shots and s a free-text query. The goal of the task is, given the query s, to retrieve from V a ranked list with the most relevant video shots.

Our T×V network consists of two key sub-networks, one for the textual and one for the visual stream. The textual sub-network inputs a free-text query and vectorizes it into M textual features $g_S : \{g_s^1(\cdot), g_s^2(\cdot), \ldots, g_s^M(\cdot)\}$. These M features are used as input in a set of carefully-selected K textual encoders $f_S : \{f_s^1(\cdot), f_s^2(\cdot), \ldots, f_s^K(\cdot)\}$ that encode the input sentence. Each of these encoders can be either a trainable network or simply an identity function that just forwards its input. Similarly to the textual one, the visual sub-network inputs a video shot consisting of a sequence of N keyframes $v = \{I_1, I_2, \ldots, I_N\}$. We use L trained DNNs in order to extract the initial frame representations $g_V : \{g_v^1(\cdot), g_v^2(\cdot), \ldots, g_v^L(\cdot)\}$. To obtain video-shot level representations we follow the mean-pooling strategy.

Subsequently, we create all the possible textual encodings-visual feature pairs $(f_s^k(s), g_v^l(v))$ and a joint embedding space is created for each pair, using to this end two fully connected layers. Thus, $K \times L$ different joint spaces are created. The objective of our network is to learn a similarity function $similarity(s, v)$ that will consider every individual similarity in each joint latent space utilizing multi-loss-based training. Figure 1c illustrates our proposed method.

3.2   Multiple Space Learning

To encode the $(f_s^k(\cdot), g_v^l(\cdot))$ pair into its joint feature space, as shown in Fig. 1c, each single part of the pair is linearly transformed by a fully connected layer (FC). A non-linearity is added in the FC output (not illustrated in Fig. 1c for brevity), for which the ReLU activation function is used, as follows:

$$\mathbf{s}_k = ReLU(FC(f_s^k(\cdot))$$
$$\mathbf{v}_l = ReLU(FC(g_v^l(\cdot))$$

This transformation encodes the $(f_s^k(\cdot), g_v^l(\cdot))$ pair into its new joint feature space. The similarity function $sim(\mathbf{s}_k, \mathbf{v}_l)$ calculates the similarity between the output of textual encoder k and video feature l in this joint feature space. The overall similarity between a video-sentence pair is calculated as follows:

$$similarity(s, v) = \sum_{k=1}^{K} \sum_{l=1}^{L} sim(\mathbf{s}_k, \mathbf{v}_l)$$

where $sim(\mathbf{s}_k, \mathbf{v}_l) = cosine\_similarity(\mathbf{s}_k, \mathbf{v}_l)$.

To train our network, similarly to [10,14], we utilize the improved marginal ranking loss introduced in [12]. This emphasizes on the hard-negative samples in order to learn to maximize the similarity between textual and video embeddings. At the training stage, given a sentence-video sample (s, v), for a specific latent feature space (k, l), the improved marginal loss is defined as follows:

$$\mathcal{L}_{(k,l)}(s, v) = max(0, \alpha + sim(\mathbf{s}_k, \mathbf{v}_l^{'}) - sim(\mathbf{s}_k, \mathbf{v}_l))$$
$$+ max(0, \alpha + sim(\mathbf{s}_k^{'}, \mathbf{v}_l) - sim(\mathbf{s}_k, \mathbf{v}_l))$$

where $\mathbf{v}_l^{'}$ and $\mathbf{s}_k^{'}$ are the hardest negatives of $\mathbf{s}_l$ and $\mathbf{v}_k$ respectively and α is a hyperparameter for margin regulation. The overall training loss is calculated as the sum of all $K \times L$ individual loss values:

$$\mathcal{L}(s, v) = \sum_{k=1}^{K} \sum_{l=1}^{L} \mathcal{L}_{(k,l)}(s, v)$$

3.3   Dual Softmax Inference

In [6], a new objective function based on two softmax operations was proposed. According to this, at the training stage the predicted text-video similarities were revised by calculating a so-called *cross-direction* similarity matrix and multiplying it with the predicted one. Specifically, during a training batch, let Q be the number of examined caption-video pairs. By computing the similarities between every caption and all videos, a similarity matrix $\mathbf{X} \in \mathcal{R}^{Q \times Q}$ was generated. Next, by applying two cross-dimension softmax operations (one column-wise and one row-wise) an updated similarity matrix $\mathbf{X}'$ was calculated, and was subsequently used as discussed in [6]. Directly applying this approach at the inference stage, though, would require that all queries to be evaluated are known a priori and are evaluated simultaneously; they would need to be used for calculating matrix $\mathbf{X}$, as illustrated in the left part of Fig. 2. This is not a realistic scenario, especially in real-world retrieval applications.

To deal with this issue and revise the inferred text-video similarities at the retrieval stage, we propose a dual softmax-based inference (DS_{inf}) as illustrated in the right side of Fig. 2. We utilize a fixed set of C pre-defined background textual queries, which are independent of the evaluated dataset, and we calculate their similarities with all D videos of the test set. For the set of background queries, we calculate once the similarity matrix $\mathbf{X}^* \in \mathcal{R}^{C \times D}$. For each individual evaluated query s a similarity vector $\mathbf{y}(s) = [similarity(s, v_1), similarity(s, v_2),$ $\ldots, similarity(s, v_D)]^T$, is calculated. A matrix $\mathbf{Z}(s) = concat(\mathbf{y}(s); \mathbf{X}^*)$ is constructed, and a dual softmax operation revises the similarities as follows:

$$\mathbf{Z}^*(s) = Softmax(\mathbf{Z}(s), \ dim = 0) \odot Softmax(\mathbf{Z}(s), \ dim = 1)$$

where $\odot$ denotes the Hadamard product. Finally, from matrix $\mathbf{Z}^*$ we extract the revised similarity vector $\mathbf{y}^* = [Z_{0,1}^*, Z_{0,2}^*, \cdot, Z_{0,D}^*]$ (Fig. 3). This normalization procedure is meaningful when we expect that there are multiple positive video samples in our dataset for the evaluated query; thus, by normalizing the inferred similarities we can produce a better ranking list.

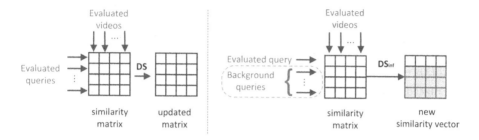

Fig. 2. Different approaches to update the similarity matrix using all the evaluated queries (left subfigure) and pre-defined background queries (right subfigure)

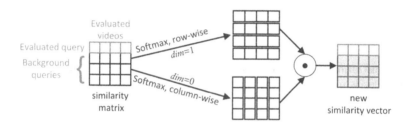

Fig. 3. Illustration of the dual softmax-based inference (DS_{inf}) approach for updating the similarities

3.4 Specifics of Textual Information Processing

In this section, we present every textual feature and every textual encoder we used in order to find the optimal textual encoder combination. Given a sentence s consisting of $\{w_1, w_2, \ldots, w_B\}$ words, we utilize $M = 4$ different textual features. These features are used as input to textual encoders.

Textual features

- Bag-of-Words (bow): We utilize Bag-of-Words to vectorize every sentence into a sparse vector representation expressing the occurrence frequency of every word from a pre-defined vocabulary.
- Word2Vec (w2v): Word2Vec model [28] is an established and well-performing word embedding model. W2v learns to embed words into a word-level representation vectors $\mathbf{W}_{w2v} : \{\mathbf{w}_1^{w2v}, \mathbf{w}_2^{w2v}, \ldots, \mathbf{w}_B^{w2v}\} \in R^{B x D_{w2v}}$. The overall sentence w2v embedding g_s^{w2v} is calculated as the mean pooling of the individual word embeddings.

- Bert: Bert [7] offers contextual embeddings by considering the sequence of all words inside a sentence, which means that the same word may have different embedding when the context of a sentence is different. We utilize the BASE variation of bert consisting of 12 encoders and 12 bidirectional self-attention heads. Similarly to [20], we calculate the bert sentence embedding g_s^{bert} by mean pooling the individual word embedding.
- Clip: The transformer-based trained model of CLIP [30] is used as a textual feature extractor. Sentence s is fed to it as a sequence of words $w_1, ..., w_B$ and token embeddings $\mathbf{W}_{clip} : \{\mathbf{w}_{startoftext}^{clip}, \mathbf{w}_1^{clip}, ..., \mathbf{w}_B^{clip}, \mathbf{w}_{endoftext}^{clip}\}$ are calculated. The last token embedding, $\mathbf{w}_{endoftext}^{clip} \in R^{512}$, is used as our feature vector.

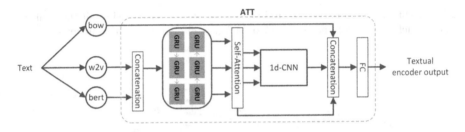

Fig. 4. Illustration of the ATT encoder, inputting several textual features and producing three levels of encodings (bow, self-attention bi-gru and CNN outputs) which are concatenated and contribute to the final output of the encoder

Textual Encoders. The textual encoders input the extracted textual features (one at a time, or a combination of them) and output a new embedding. We experimented with combinations of the followings:

- f_s^{bow}, f_s^{w2v}, f_s^{bert}, f_s^{clip}: these encoders feedforward the corresponding features through an identity layer.
- $f_s^{w2v-bert}$: The concatenation of w2v and bert features is used to feed an identity layer.
- f_s^{bi-gru}: A self-attention bi-gru module, introduced in [14], is trained as part of the complete network architecture; it takes as input the w2v features for each word and their temporal order (i.e., not using the overall sentence w2v embedding, contrarily to f_s^{w2v} and $f_s^{w2v-bert}$).
- Attention-based dual encoding network (ATT): The textual sub-network presented in [14] (illustrated in Fig. 4) is trained (similarly to f_s^{bi-gru}, above), taking as input the bow, w2v and bert features (again, for each individual word rather than the mean-pooled sentence embeddings) and producing a vector in $\mathcal{R}^{2048}$.

Among all the above possible textual encoders, we propose to combine in our network the f_s^{clip} and ATT ones, as experimentally verified in Sect. 4.

3.5 Specifics of Visual Information Processing

Similarly to the textual sub-network, we use several deep networks that have been trained for other visual tasks as frame feature extractors. Considering a video shot v, first we uniformly sample it with a fixed rate of 2 frames per second, resulting in a set of keyframes $\{I_1, I_2, \ldots, I_N\}$. Then, frame-level representations are obtained with the help of the feature extractors listed below, followed by mean pooling of the individual frame representations to get shot-level features.

Visual features

- R_152: The first video feature extractor inputs an image frame into a ResNet-152 [18] network, trained on the ImageNet-11k dataset. The flattened output of the layer before the last fully connected layer of the network is used as a feature representation of every frame in $\mathcal{R}^{2048}$.
- Rx_101: The second feature extractor utilizes a ResNeXt-101 network, pre-trained by weakly supervised learning on web images followed and fine-tuned on ImageNet [25]. Similarly to R_152, Rx_101 inputs frames and the frame representations in $\mathcal{R}^{2048}$, are obtained as the flattened output of the layer before the last fully connected layer.
- Clip: As third video feature extractor we utilise a trained CLIP model (ViT-B/32) [30], to create frame-level representations in $\mathcal{R}^{512}$.

As illustrated in Fig. 1c, we propose using these visual features, without introducing any trainable visual encoder, directly as input to a number of individual FC layers for learning the latent feature spaces. However, in order to examine more design choices, we also tested in our ablation experiments the introduction of a visual encoder, similarly to what we do for textual information processing. To this end, we utilized the visual sub-network of the attention-based dual encoding network of [14] (ATV). Following [14], we input all three aforementioned frame-level features to a single ATV encoder, which (similarly to the ATT one) was trained end-to-end as part of our overall network.

4 Experimental Results

4.1 Datasets and Experimental Setup

We evaluate our approach and report experimental results on three datasets: the two official TRECVID AVS datasets (i.e., IACC.3 and V3C1) [1] and MSR-VTT [34]. The AVS datasets are designed explicitly for text-based video retrieval evaluation; they include the definition of tens of textual queries as well as ground-truth associations of multiple positive samples with each query. The IACC.3 dataset consists of 335.944 test videos, and the V3C1 of 1.082.629 videos for testing (most of which are not associated with any of the textual queries). As evaluation measure we use the mean extended inferred average precision (Mxin-fAP), as proposed in [1] and typically done when working with these datasets. On the other hand, MSR-VTT targets primarily video captioning evaluation, but is also often used for evaluating text-video retrieval methods. It is made of 10.000 videos and each video is annotated with 20 different natural language captions (totaling 200.000 captions, which are generally considered to be unique); for retrieval evaluation experiments, given a caption the goal is to retrieve the single video that is ground-truth-annotated with it. For the MSR-VTT experiments, following the relevant literature, we use as evaluation measures the recall $R@k$, $k = 1, 5, 10$, the median rank ($Medr$) and mean average precision (mAP).

Table 1. Results and comparisons on the IACC.3 and V3C1 datasets. Bold/underline indicates the best-/second-best scores

Model		Datasets (all scores: MxinfAP)						
		IACC.3			V3C1			
	Year	AVS16	AVS17	AVS18	AVS19	AVS20	AVS21	Mean
Dual encoding [9]	2019	0.165	0.228	0.117	0.152	–	–	–
ATT-ATV [14]	2020	0.164	0.243	0.127	–	–	–	–
ATT-ATV [14] (re-training)	–	0.202	0.281	0.146	<u>0.208</u>	0.283	<u>0.289</u>	0.235
Dual-task [33]	2020	0.185	0.241	0.123	0.185	–	–	–
SEA [20]	2021	0.164	0.228	0.125	0.167	–	–	–
SEA [20] (re-training)	–	<u>0.207</u>	0.279	0.148	0.191	0.283	0.283	0.232
SEA-clip [4]	2021	0.203	**0.321**	**0.156**	0.192	**0.329**	–	–
Extended Dual Encoding [10]	2022	0.159	0.244	0.126	–	–	–	–
T×V (proposed)		**0.234**	<u>0.317</u>	<u>0.153</u>	**0.220**	<u>0.316</u>	**0.312**	**0.259**

Regarding the training/testing splits: for the evaluations on the AVS datasets, our cross-modal network (and any network of the literature, i.e. [14,20], that we re-train for comparison) is trained using a combination of four other large-scale video captioning datasets: MSR-VTTTT [34], TGIF [21], ActivityNet [3] and Vatex [32]. For validation purposes, during training, we use the Video-to-Text Description dataset of TRECVID 2016. For testing, all sets of queries specified by NIST for IACC.3 (i.e., AVS16, AVS17 and AVS18) and V3C1 (i.e., AVS19, AVS20 and AVS21) are used. For the evaluations on the MSR-VTT dataset, we experimented with two versions of this dataset: MST-VTT-full [34] and MSR-VTT-1k-A [36]. MST-VTT-full consists of 6.513 videos for training, 497 for validation and 2.990 videos (thus, 2.990 × 20 video-caption pairs) for testing. On the other hand, MSR-VTT-1k-A contains 9.000 videos for training and 1.000 video-caption pairs for testing. For both MSR-VTT versions, we trained our network of the training portion of the dataset and report results on the testing portion, respectively.

Regarding the training conditions and parameters: To train the proposed network, (and again, also for re-training [14,20]) we adopt the setup of [13], where six configurations of the same architecture with different training parameters were combined. Specifically, each model is trained using two optimizers, i.e., Adam and RMSprop, and three learning rates (1×10^4, 5×10^5, 1×10^5). The final results for a given architecture are obtained by combining the six returned ranking lists of the individual configurations in a late fusion scheme, i.e. by averaging the six obtained ranks for each video. For training all configurations, we follow a learning rate decay technique, and we reduce the learning rate 1% per epoch or by 50% if the validation performance does not improve for three epochs. The dropout rate is set to 0.2 to reduce overfitting. Also, following [12] the margin parameter on loss function is set to $\alpha = 0.2$. All experiments were performed on a single computer equipped with Nvidia RTX3090 GPU. Our models were implemented and trained using Pytorch 1.11.

Table 2. Results and comparisons between different encoder strategies when using dual softmax-based inference (DS_{inf}) similarity. Scores in bold/underline indicate the best-/second-best-performing strategy

Model		Datasets (all scores: MxinfAP)						
		IACC.3			V3C1			
		AVS16	AVS17	AVS18	AVS19	AVS20	AVS21	Mean
T×V		0.234	0.317	0.153	0.220	0.316	0.312	0.259
T×V + DS_{inf} on the set of evaluated queries		**0.244**	**0.330**	**0.165**	**0.226**	<u>0.324</u>	**0.324**	**0.269**
T×V + DS_{inf} using as background queries:	60 random captions	0.240	0.323	0.157	0.223	0.318	0.313	0.262
	200 random captions	0.239	0.322	0.158	0.224	0.315	0.315	0.262
	other years' queries of the same dataset	<u>0.243</u>	<u>0.328</u>	<u>0.162</u>	<u>0.226</u>	**0.325**	<u>0.323</u>	<u>0.268</u>

4.2   Results and Comparisons

Table 1 presents the results of the proposed T×V network, i.e. using three visual features (R_152, Rx_101, clip) followed by FC layers and clip and ATT as textual encoders, on IACC.3 and V3C1 datasets and comparisons with state-of-the-art literature approaches. We compare our method with six methods and the presented results are extracted from their original papers. Furthermore, as [4] has shown that the quality of the initial visual features is crucial for the performance of a method, we present results of re-training the ATT-ATV [14] and SEA [20] networks using the same visual features and same training datasets as we did in our experiments, using their publicly available code. Our proposed network outperforms the competitors on AVS16, AVS19 and AVS21. SEA-clip [4] achieves better results on AVS17, AVS18 and AVS20 by exploiting 3D CNN-based visual features. Comparing the mean performance on AVS16-AVS20 (since SEA-clip does not report results on AVS21), our network achieves MxinfAP equal to 0.248 while SEA-clip 0.240.

Table 3. Results and comparisons on the MSR-VTT full and 1k-A datasets. Methods marked with * use an alternative training set of 7.010 video-caption samples for the 1k-A dataset, but still report results on the same test portion of 1k-A as all other methods. Bold/underline indicates the best-/second-best scores

Model	Year	Datasets									
		MSR-VTT full					MSR-VTT 1k-A				
		R@1↑	R@5↑	R@10↑	Medr↓	mAP↑	R@1↑	R@5↑	R@10↑	Medr↓	mAP↑
W2VV [8]	2018	1.1	4.7	8.1	236	3.7	1.9	9.9	15.2	79	6.8
VSE++ [12]	2018	8.7	24.3	34.1	28	16.9	16.0	38.5	50.9	10	27.4
CE [23]	2019	7.9	23.6	34.6	23	16.5	17.2	46.2	58.5	7	30.3
W2VV++ [19]	2019	11.1	29.6	40.5	18	20.6	19.0	45.0	58.7	7	31.8
TCE [35]	2020	9.3	27.3	38.6	19	18.7	17.8	46.0	58.3	7	31.1
HGR [5]	2020	11.1	30.5	42.1	16	20.8	21.7	47.4	61.1	6	34.0
SEA [20] *	2021	12.4	32.1	43.3	15	22.3	23.8	50.3	63.8	5	36.6
HiT [22]	2021	-	-	-	-	–	27.7	59.2	72.0	3	–
HiT (Pre-train. on HT100M) [22]	2021	–	–	–	–	–	30.7	60.9	73.2	2.6	–
CLIP [29]*	2021	**21.4**	41.1	50.4	10	–	31.2	53.7	64.2	4	–
FROZEN [2]	2021	–	–	–	–	–	31.0	59.5	70.5	3	–
Extended Dual Encoding [10]*	2022	11.6	30.3	41.3	17	21.2	21.1	48.7	60.2	6	33.6
RIVRL [11] *	2022	13.0	33.4	44.8	14	23.2	23.3	52.2	63.8	5	36.7
RIVRL with bert [11] *	2022	13.8	35.0	46.5	13	24.3	27.9	59.3	71.2	4	42.0
BridgeFormer [15]	2022	–	–	–	–	–	**37.6**	64.8	73.1	3	–
T×V (proposed) *		–	–	–	–	–	32.3	63.7	74.6	3	46.3
T×V (proposed)		21.2	**46.3**	**58.2**	**7**	**33.1**	36.5	**66.9**	**77.7**	**2**	**50.2**

In Table 2 we experiment with the dual softmax-based (DS_{inf}) inference on IACC.3 and V3C1. We examine the impact of different background query strategies and compare with the proposed network without DS_{inf}. "DS_{inf} on the set of evaluated queries" indicates that the operation was performed using the same year's queries. For example, to retrieve videos for an AVS16 query, the remaining AVS16 queries are used as background queries. This improves the overall performance, but as we already have discussed, is not a realistic application scenario. To overcome this problem, we experiment with using a fixed set of background queries: 60 or 200 randomly selected captions extracted from the training datasets. By using these captions, performance is improved compared to the proposed network in every dataset and every year. The difference between using 60 and 200 captions is marginal. Finally, we try using as background queries all AVS queries defined for the same video dataset but not the same test-year as the examined query. For example, to evaluate each of the AVS16 queries, we use the AVS17 and AVS18 queries as background. This strategy achieves the second-best results among all examined ones, being marginally outperformed by "DS_{inf} on the set of evaluated queries" strategy; and, contrarily to the latter, does not assume knowledge of all the evaluation queries beforehand. For a given query, the retrieval time using DS_{inf} increases (on average) from 0.4 s to 0.6 s on the IACC.3 dataset and from 1.1 s to 1.3 s on the V3C1, respectively.

In Table 3 we present results on the MSR-VTT full and 1k-A datasets and compare with literature methods. Our network outperforms most literature methods, even methods like FROZEN [2] and HiT (pre-trained on HT100M)

Table 4. Comparison of combinations of textual encoders and visual features on the IACC.3 and V3C1 datasets

Model			Datasets (all scores: MxinfAP)						
Visual features combination strategy	Textual encoders (Fig. 1c)	# of feature spaces	IACC.3			V3C1			
			AVS16	AVS17	AVS18	AVS19	AVS20	AVS21	Mean
(feat. concat. + ATV) ↓ FCs (Fig. 1a)	bow, w2v, bert, bi-gru	4	0.169	0.250	0.126	0.170	0.252	0.251	0.203
	bow, w2v, bert, bi-gru, clip	5	0.179	0.265	0.133	0.174	0.260	0.254	0.211
	w2v-bert, clip, ATT	3	0.199	0.265	0.135	0.183	0.271	0.262	0.219
	clip, ATT	2	0.211	0.288	0.143	0.188	0.288	0.271	0.232
(feat. concat.) ↓ FCs (Fig. 1b)	bow, w2v, bert, bi-gru	4	0.184	0.267	0.136	0.202	0.303	0.295	0.231
	bow, w2v, bert, bi-gru, clip	5	0.198	0.275	0.141	0.205	0.305	0.292	0.236
	w2v-bert, clip, ATT	3	0.210	0.281	0.147	0.194	0.290	0.275	0.233
	clip, ATT	2	0.227	0.291	0.149	0.190	0.296	0.292	0.241
Only FCs (Fig. 1c)	**bow, w2v, bert, bi-gru, clip**	**12**	**0.205**	**0.306**	**0.150**	**0.203**	**0.305**	**0.293**	**0.239**
	w2v-bert, clip, ATT	9	0.219	0.312	0.150	0.210	0.307	0.297	0.249
	clip, ATT	6	**0.234**	**0.317**	**0.153**	**0.220**	**0.316**	**0.312**	**0.259**

[22] in which their networks utilize a pre-training step on other large text-video datasets. Moreover, BridgeFormer [15], using a pre-training step on the WebVid-2M [2] dataset, marginally outperforms our network in $R@1$ terms, while our approach achieves better results on the remaining evaluation measures. Finally, we should note that experiments with DS_{inf} on MSR-VTT (not shown in Table 3) lead to only marginal differences in relation to the results of the proposed $T \times V$ network. This is expected because of the nature of MSR-VTT: given a caption the goal is to retrieve the *single* video that is ground-truth-annotated with it; thus, re-ordering the entire ranking list by introducing the DS_{inf} normalization of the caption-video similarities has limited impact.

4.3 Ablation Study

In this section we study the effectiveness of different textual encoders and visual features (or also encoders) combinations. We report results using three visual feature encoding strategies: "*feat. concat. + ATV*" indicates the early fusion of the three visual features that are then fed into the trainable ATV sub-network of [14] (as illustrated in the rightmost part of Fig. 1a), followed by the required FC layers in order to encode the ATV's output into the corresponding joint feature spaces. Similarly, "*feat. concat.*" refers to the early fusion of the three visual features (as illustrated in the rightmost part of Fig. 1b) followed by the required FC layers. Finally, "*only FCs*" refers to the proposed strategy (Fig. 1c), where visual features are individually and directly encoded to the joint spaces using only FC layers.

In Table 4 we report the results on the AVS datasets when using different combinations of textual encoders together with the aforementioned three possible visual encoding strategies. Concerning the visual encoding strategies, the results indicate that the lowest performance is achieved by the "*feat. concat. + ATV*" strategy, regardless of the textual encoders choice. When the early fusion of the trained models "*feat. concat.*" is used instead of ATV, the performance consistently increases. The best results are achieved by forwarding the visual

Table 5. Comparison of combinations of textual encoders and visual features on the full and 1k-A variations of the MSR-VTT dataset

Model		Datasets									
		MSR-VTT full					MSR-VTT 1k-A				
Visual features combination strategy	Textual encoders (Fig. 1c)	R@1↑	R@5↑	R@10↑	Medr↓	mAP↑	R@1↑	R@5↑	R@10↑	Medr↓	mAP↑
(feat. concat. + ATV) ↓ FCs (Fig. 1a)	bow, w2v, bert, bi-gru	17.3	40.9	52.9	9	28.9	32.2	62.9	73.7	3	46.3
	bow, w2v, bert, bi-gru, clip	19.2	43.6	55.6	8	30.1	34.8	63.9	75.8	3	48.6
	w2v-bert, clip, ATT	21.4	46.6	58.5	7	33.4	36.5	67.4	76.4	2	50.3
	clip, ATT	22.3	47.6	59.3	6	34.3	40.1	68.4	78.1	2	52.9
(feat. concat.) ↓ FCs (Fig. 1b)	bow, w2v, bert, bi-gru	17.6	41.3	53.2	9	29.0	32.8	64.4	74.3	3	47.1
	bow, w2v, bert, bi-gru, clip	19.7	44.1	55.9	8	31.4	35.4	65.0	76.0	3	48.5
	w2v-bert, clip, ATT	21.4	46.3	57.7	7	33.3	35.7	67.4	77.4	3	49.8
	clip, ATT	22.0	46.5	57.9	7	33.7	37.1	65.4	75.9	2	50.3
Only FCs (Fig. 1c)	bow, w2v, bert, bi-gru, clip	18.9	40.9	53.0	8	30.6	33.0	63.2	76.3	3	48.9
	w2v-bert, clip, ATT	20.9	45.6	57.7	7	32.6	35.8	66.3	77.2	3	49.6
	clip, ATT	21.2	46.3	58.2	7	33.1	36.5	66.9	77.7	2	50.2

features independently with FC layers, as in the proposed approach. Regarding the combinations of the textual encoders, we can see that the utilization of fewer but more powerful encoders (i.e. clip and ATT) leads to better results than using a multitude of, possibly weak, encoders as in [20], regardless of the employed visual encoding strategy. These evaluations show that *how* the textual and visual features are combined has significant impact on the obtained results.

In Table 5 we presented the same ablation study for the full and 1k-A variations of the MSR-VTT dataset. In these datasets, we can observe a different behavior concerning the visual encoding: while the utilization of a few and powerful textual encoders (i.e. clip and ATT) continues to perform the best, when it comes to the visual modality, the *"feat. concat. + ATV"* strategy consistently performs the best, regardless of the textual encoder choice. This finding, combined with the results of Table 4, shows that for similar yet different problems and datasets there is no universally-optimal way of combining the visual features (and one can reasonably assume that this may also hold for the textual ones).

5   Conclusions

In this work, we presented a new network architecture for efficient text-to-video retrieval. We experimentally examined different combinations of visual and textual features, and concluded that selectively combining the textual features into fewer but more powerful textual encoders leads to improved results. Moreover, we shown how a fixed set of background queries extracted from large-scale captioning datasets can be used together with softmax operations at the inference stage for revising query-video similarities, leading to improved video retrieval. Extensive experiments and comparisons on different datasets document the value of our approach.

Acknowledgements. This work was supported by the EU Horizon 2020 programme under grant agreements H2020-101021866 CRiTERIA and H2020-832921 MIRROR.

References

1. Awad, G., et al.: Evaluating multiple video understanding and retrieval tasks at trecvid 2021. In: Proceedings of TRECVID 2021. NIST, USA (2021)
2. Bain, M., Nagrani, A., Varol, G., Zisserman, A.: Frozen in time: a joint video and image encoder for end-to-end retrieval. In: Proceedings of the IEEE/CVF International Conference on Computer Vision, pp. 1728–1738 (2021)
3. Caba Heilbron, F., Escorcia, V., B., G., Niebles, J.C.: Activitynet: a large-scale video benchmark for human activity understanding. In: Proceedings of the IEEE Conference on Computer Vision and Pattern Recognition, pp. 961–970 (2015)
4. Chen, A., Hu, F., Wang, Z., Zhou, F., Li, X.: What matters for ad-hoc video search? a large-scale evaluation on trecvid. In: Proceedings of the IEEE/CVF International Conference on Computer Vision, pp. 2317–2322 (2021)
5. Chen, S., Zhao, Y., Jin, Q., Wu, Q.: Fine-grained video-text retrieval with hierarchical graph reasoning. In: Proceedings of the IEEE Conference on Computer Vision and Pattern Recognition (CVPR), pp. 10638–10647 (2020)
6. Cheng, X., Lin, H., Wu, X., Yang, F., Shen, D.: Improving video-text retrieval by multi-stream corpus alignment and dual softmax loss. arXiv preprint arXiv:2109.04290 (2021)
7. Devlin, J., Chang, M.W., Lee, K., Toutanova, K.: BERT: pre-training of deep bidirectional transformers for language understanding. In: arXiv preprint arXiv:1810.04805 (2018)
8. Dong, J., Li, X., Snoek, C.G.M.: Predicting visual features from text for image and video caption retrieval. IEEE Trans. Multimedia **20**(12), 3377–3388 (2018)
9. Dong, J., et al.: Dual encoding for zero-example video retrieval. In: Proceedings of the IEEE/CVF Conference on Computer Vision and Pattern Recognition, pp. 9346–9355 (2019)
10. Dong, J., Li, X., Xu, C., Yang, X., Yang, G., Wang, X., Wang, M.: Dual encoding for video retrieval by text. IEEE Trans. Pattern Anal. Mach. Intell. **44**(8), 4065–4080 (2022)
11. Dong, J., Wang, Y., Chen, X., Qu, X., Li, X., He, Y., Wang, X.: Reading-strategy inspired visual representation learning for text-to-video retrieval. IEEE Trans. Circuits Syst. Video Technol. (2022)
12. Faghri, F., Fleet, D.J., Kiros, J.R., Fidler, S.: Vse++: improving visual-semantic embeddings with hard negatives. In: Proceedings of the British Machine Vision Conference (BMVC) (2018)
13. Galanopoulos, D., Mezaris, V.: Hard-negatives or Non-negatives? a hard-negative selection strategy for cross-modal retrieval using the improved marginal ranking loss. In: 2021 IEEE/CVF ICCVW (2021)
14. Galanopoulos, D., Mezaris, V.: Attention mechanisms, signal encodings and fusion strategies for improved ad-hoc video search with dual encoding networks. In: Proceedings of the ACM International Conference on Multimedia Retrieval (ICMR 2020). ACM (2020)
15. Ge, Y., Ge, Y., Liu, X., Li, D., Shan, Y., Qie, X., Luo, P.: Bridging video-text retrieval with multiple choice questions. In: Proceedings of the IEEE/CVF Conference on Computer Vision and Pattern Recognition, pp. 16167–16176 (2022)
16. Habibian, A., Mensink, T., Snoek, C.G.M.: Video2vec embeddings recognize events when examples are scarce. IEEE Trans. Pattern Anal. Mach. Intell. **39**(10), 2089–2103 (2017)

17. Habibian, A., Mensink, T., Snoek, C.G.: Videostory: a new multimedia embedding for few-example recognition and translation of events. In: Proceedings of the 22nd ACM International Conference on Multimedia, MM 2014, pp. 17–26. ACM, New York (2014)
18. He, K., Zhang, X., Ren, S., Sun, J.: Deep residual learning for image recognition. In: Proceedings of the IEEE Conference on Computer Vision and Pattern Recognition, pp. 770–778 (2016)
19. Li, X., Xu, C., Yang, G., Chen, Z., Dong, J.: W2VV++ fully deep learning for ad-hoc video search. In: Proceedings of the 27th ACM International Conference on Multimedia, pp. 1786–1794 (2019)
20. Li, X., Zhou, F., Xu, C., Ji, J., Yang, G.: SEA: sentence encoder assembly for video retrieval by textual queries. IEEE Trans. Multimed. **23**, 4351–4362 (2021)
21. Li, Y., Song, Y., et al.: TGIF: a new dataset and benchmark on animated gif description. In: Proceedings of IEEE CVPR, pp. 4641–4650 (2016)
22. Liu, S., Fan, H., Qian, S., Chen, Y., Ding, W., Wang, Z.: Hit: Hierarchical transformer with momentum contrast for video-text retrieval. In: Proceedings of the IEEE/CVF International Conference on Computer Vision. pp. 11915–11925 (2021)
23. Liu, Y., Albanie, S., Nagrani, A., Zisserman, A.: Use what you have: Video retrieval using representations from collaborative experts. arXiv preprint arXiv:1907.13487 (2019)
24. Lu, Y.J., Zhang, H., de Boer, M., Ngo, C.W.: Event detection with zero example: Select the right and suppress the wrong concepts. In: Proceedings of the 2016 ACM on International Conference on Multimedia Retrieval. p. 127–134. ICMR '16, ACM, New York, NY, USA (2016)
25. Mahajan, D., Girshick, R., Ramanathan, V., He, K., Paluri, M., Li, Y., Bharambe, A., Van Der Maaten, L.: Exploring the limits of weakly supervised pretraining. In: Proceedings of the European conference on computer vision (ECCV). pp. 181–196 (2018)
26. Markatopoulou, F., Galanopoulos, D., Mezaris, V., Patras, I.: Query and keyframe representations for ad-hoc video search. In: Proceedings of the 2017 ACM International Conference on Multimedia Retrieval. pp. 407–411. ICMR '17, ACM (2017)
27. Miech, A., Zhukov, D., Alayrac, J.B., Tapaswi, M., Laptev, I., Sivic, J.: Howto100m: Learning a text-video embedding by watching hundred million narrated video clips. In: Proceedings of the IEEE/CVF International Conference on Computer Vision. pp. 2630–2640 (2019)
28. Mikolov, T., Chen, K., Corrado, G., Dean, J.: Efficient estimation of word representations in vector space. In: 1st International Conference on Learning Representations, Workshop Track Proceedings. ICLR '13 (2013)
29. Portillo-Quintero, J.A., Ortiz-Bayliss, J.C., Terashima-Marín, H.: A straightforward framework for video retrieval using clip. In: Mexican Conference on Pattern Recognition. pp. 3–12. Springer (2021)
30. Radford, A., Kim, J.W., Hallacy, C., Ramesh, A., Goh, G., Agarwal, S., Sastry, G., Askell, A., Mishkin, P., Clark, J., Krueger, G., Sutskever, I.: Learning transferable visual models from natural language supervision. In: Proc. of the 38th Int. Conf. on Machine Learning (ICML) (2021)
31. Song, Y., Soleymani, M.: Polysemous visual-semantic embedding for cross-modal retrieval. In: Proceedings of the IEEE Conference on Computer Vision and Pattern Recognition (CVPR). pp. 1979–1988 (2019)
32. Wang, X., Wu, J., et al.: Vatex: A large-scale, high-quality multilingual dataset for video-and-language research. In: Proc. of the IEEE Int. Conf. on Computer Vision. pp. 4581–4591 (2019)

33. Wu, J., Ngo, C.W.: Interpretable embedding for ad-hoc video search. In: Proceedings of the 28th ACM International Conference on Multimedia. p. 3357–3366. ACM, New York, NY, USA (2020)

34. Xu, J., Mei, T., Yao, T., Rui, Y.: MSR-VTT: A large video description dataset for bridging video and language. In: Proc. of IEEE CVPR. pp. 5288–5296 (2016)

35. Yang, X., Dong, J., Cao, Y., Wang, X., Wang, M., Chua, T.S.: Tree-augmented cross-modal encoding for complex-query video retrieval. In: Proceedings of the 43rd international ACM SIGIR conference on research and development in information retrieval. pp. 1339–1348 (2020)

36. Yu, Y., Kim, J., Kim, G.: A joint sequence fusion model for video question answering and retrieval. In: Proceedings of the European Conference on Computer Vision (ECCV). pp. 471–487 (2018)

Scene-Adaptive Temporal Stabilisation for Video Colourisation Using Deep Video Priors

Marc Gorriz Blanch[1]([✉])([iD]), Noel O'Connor[2]([iD]), and Marta Mrak[1]([iD])

[1] BBC Research and Development, London, UK
{marc.gorrizblanch,marta.mrak}@bbc.co.uk
[2] Dublin City University, Dublin, Ireland
noel.oconnor@dcu.ie

Abstract. Automatic image colourisation methods applied independently to each video frame usually lead to flickering artefacts or propagation of errors because of differences between neighbouring frames. While this can be partially solved using optical flow methods, complex scenarios such as the appearance of new objects in the scene limit the efficiency of such solutions. To address this issue, we propose application of blind temporal consistency, learned during the inference stage, to consistently adapt colourisation to the given frames. However, training at test time is extremely time-consuming and its performance is highly dependent on the content, motion, and length of the input video, requiring a large number of iterations to generalise to complex sequences with multiple shots and scene changes. This paper proposes a generalised framework for colourisation of complex videos with an optimised few-shot training strategy to learn scene-aware video priors. The proposed architecture is jointly trained to stabilise the input video and to cluster its frames with the aim of learning scene-specific modes. Experimental results show performance improvement in complex sequences while requiring less training data and significantly fewer iterations.

Keywords: Video colourisation · Temporal consistency · Deep video prior · Few-shot learning

1 Introduction

Video restoration is in increasing demand in the production industry in order to both deliver historical content in high quality and to support innovation in the creative sector [24]. Video colourisation in particular is still a challenging task due to its ambiguity in the solution space and the requirement of global spatio-temporal consistency. Prior to automatic colourisation methods, producers relied on specialists to perform manual colourisation, resulting in a time consuming and sometimes a prohibitively expensive manual process. Researchers have thus endeavoured to develop computer-assisted methodologies in order to automate the colourisation process and reduce production costs. Early methods relied on

© The Author(s), under exclusive license to Springer Nature Switzerland AG 2023
L. Karlinsky et al. (Eds.): ECCV 2022 Workshops, LNCS 13804, pp. 644–659, 2023.
https://doi.org/10.1007/978-3-031-25069-9_41

frame-to-frame image colourisation techniques propagating colour scribbles [15, 20,33] or reference colours [6,28,31]. The problem that typically occurs when processing is applied on a single frame without consideration of the neighbouring frames is temporal flickering. Similarly, propagation of errors can occur if the temporal dimension is not taken into account when characteristics (e.g. colour) of previous frame are transferred to the current frame. Improved results can be obtained by considering a more robust propagation and imposing refinements with temporal constrains [1,34].

Instead of improving temporal consistency using task-specific solutions, methods that generalise to various tasks can be applied. An example is the work in [5], which proposes a general approach agnostic to a specific image processing algorithm. The method takes the original video (black and white in the case of colourisation) and the per-frame processed counterpart (initially colourised version) and solves a gradient domain optimisation problem to minimise the temporal warping error between consecutive frames. An extension of such an approach takes into account object occlusions by leveraging information from a set of key-frames [32]. Another example was proposed in [17], adopting a perceptual loss to maintain perceptual similarity between output and processed frames. However, most methods rely on a dense correspondence backend (e.g. optical flow or PatchMatch [2]), which quickly becomes impractical in real-world scenarios due to the increased processing time needed. A novel solution proposed the use of Deep Video Prior by training a convolutional network on video content to enforce consistency between pairs of corresponding output patches [19]. The method solves multimodal consistency by means of Iteratively Reweighted Training, which learns to select a main mode among multiple inconsistent ones and discard those outliers leading to flickering artifacts. The main limitation is the requirement to train in test time, which makes the method extremely time-consuming in practice. For instance, training depends on the content, motion and length of the input video, requiring a large number of iterations to generalise to complex sequences with multiple shots and scene changes.

This paper proposes a framework for temporal stabilisation of frame-to-frame colourised videos with an optimised few-shot training strategy to learn scene-aware video priors. The proposed architecture is jointly trained to stabilise the input video and to cluster the input frames with the aim of learning scene-specific modes. Learnt embeddings are posteriorly injected into the decoder process to guide the stabilisation of specific scenes. A clustering algorithm for scene segmentation is used to select meaningful frames and to generate pseudo-labels to supervise the scene-aware training. Experimental results demonstrate the generalisation of the Deep Video Prior baseline [19], obtaining improved performance in complex sequences with small amounts of training data and fewer iterations.

2 Related Work

2.1 Video Colourisation

Although several works attempted to solve video colourisation problem as an end-to-end fully automatic task [18], most rely on single frame colourisation.

This is because image colourisation, compared to video colourisation, achieves higher visual quality and naturalness. Propagation methods are commonly used to stabilise the temporal coherence between frames. For instance, the work in [16] propose Video Propagation Networks (VPN) to process video frames in an adaptive manner. VPN approach applies a neural network for adaptive spatio-temporal filtering. First it connects all the pixels from current and previous frames and propagates associated information across the sequence. Then it uses a spatial network to refine the generated features. Another example is the Switchable Temporal Propagation Network [21], based on a Temporal Propagation Network (TPN), which models the transition-related affinity between a pair of frames in a purely data-driven manner. In this way, a learnable unified framework for propagating a variety of visual properties from video frames, including colour, can be achieved. Aiming at improving the efficiency of deep video processing, colourisation and propagation can be performed at once. An example is the method in [34] that is based on a recurrent video colourisation framework, which combines colourisation and propagation sub-networks to jointly predict and refine results from a previous frame. A direct improvement is the method in [1] that uses masks as temporal correspondences and hence improves the colour leakage between objects by wrapping colours within restricted masked regions over time.

2.2 Deep Video Prior

Methods for temporal stabilisation usually promote blind temporal consistency by means of dense matching (optical flow or PatchMatch [2]) to define a regularisation loss that minimises the distance between correspondences in the stabilised output frames [5]. Such methods are trained with large datasets with pairs of grayscale inputs and colourised frames. Notice that such frameworks are blind to the image processing operator and can be used for multiple tasks such as super-resolution, denoising, dehazing, etc. In contrast, Deep Video Prior (DVP) can implicitly achieve such regularisation by training a convolutional neural network [19]. Such method only requires training on the single test video, and no training dataset is needed. To address the challenging multimodal inconsistency problem, an Iteratively Reweighted Training (IRT) strategy is used in DVP approach. The method selects one mode from multiple possible modes for the processed video to ensure temporal consistency and preserve perceptual quality.

2.3 Few-Shot Learning

Few-shot learning was introduced to learn from a limited number of examples with supervised information [10,11]. For example, although current methods on image classification outperform humans on ImageNet [8], each class needs sufficient amount of labelled images, which can be difficult to obtain. Therefore, few-shot learning can reduce the data gathering effort for data-intensive applications [30]. Many related topics use this methodology, such as meta-learning [12,25], embedding learning [3,27] and generative modelling [9,10]. The method

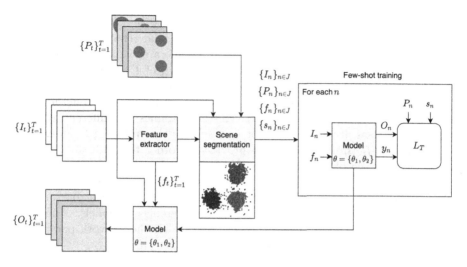

Fig. 1. Proposed framework for temporal stabilisation of frame-to-frame colourised videos. In addition to the DVP baseline, a scene segmentation and a few show training is used to learn scene-aware video priors.

proposed in this paper uses few-shot learning as training strategy to reduce processing time and to generalise to long and complex video sequences.

3   Method

This section describes the proposed extension of DVP baseline for multiple scenes, followed by the optimised few-shot training strategy which enables reduced processing time by removing the time response of DVP conditioned to the number of input frames. Finally, DVP architecture is modified by adding a classification sub-network which clusters the input frames with the objective of learning scene-specific priors.

3.1   Extension of DVP to Multiple Scenes

Given a grayscale input sequence $\{I_t\}_{t=1}^T$ of T frames and its colourised counterpart $\{P_t\}_{t=1}^T$ created using an image colourisation operator F, the goal is to learn the mapping $\hat{G}(\theta) : \{P_t\}_{t=1}^T \longrightarrow \{O_t\}_{t=1}^T$, such that $\{O_t\}_{t=1}^T$ is a temporally stable output without flickering artifacts and θ are the network parameters. Due to the superior performance of image colourisation compared to video methods [5,18], an image operator is applied frame-to-frame and the proposed framework is used to sort out temporal issues. Therefore, from a random initialisation, $\hat{G}(\theta)$ is optimised in each iteration by means of the reconstruction loss L_{data} (e.g. L_1 distance) between $\hat{G}(I_t; \theta)$ and P_t:

$$\arg \min_{\theta} L_{data}(\hat{G}(I_t; \theta), P_t). \tag{1}$$

As shown in Fig. 1, the proposed method extends the DVP framework [19] for video sequence with multiple scenes. In particular, the proposed method defines a scene as a change of content, e.g. a camera shot, appearance of new objects, etc. In particular, the input sequence $\{I_t\}_{t=1}^T$ of T frames is divided into S scenes, where typically $S \ll T$, and $\{s_t\}_{t=1}^T$ is the scene index for each frame. In order to learn scene-specific modes, the proposed network not only learns to stabilise the input sequence, but also to cluster its frames into different scenes by generating a class distribution vector $y_t \in \mathbb{R}^S$. As shown in Fig. 2, an external feature vector f_t (from frame I_t) is provided in order to guide the clustering process. f_t can be obtained from a suitable neural network, e.g. from VGG-16 classification head [26]. Finally, y_t is used to generate scene-specific priors which are posteriorly injected into the different stages of the network decoder. Therefore, the proposed model combines two different sub-models, denoted by $\hat{G}(\theta) = \{\hat{G}_1(\theta_1), \hat{G}_2(\theta_2)\}$, where $\theta = \{\theta_1, \theta_2\}$ are all the network parameters, $\hat{G}_1(\theta_1) : \{P_t\}_{t=1}^T \longrightarrow \{O_t\}_{t=1}^T$ and $\hat{G}_2(\theta_2) : \{f_t\}_{t=1}^T \longrightarrow \{y_t\}_{t=1}^T$.

The neural network is then trained to jointly improve the temporal consistency of the input video frames $\{I_t\}_{t=1}^T$ (enforcing $\{O_t\}_{t=1}^T$ to be close to $\{P_t\}_{t=1}^T$) and classify them into the corresponding scenes $\{s_t\}_{t=1}^T$. Following DVP baseline, an IRT strategy is used to address the problem of averaging when the difference of multiple modes is large (e.g. pixel with more than one possible colourisation solution). In particular, a confidence map C_t is used to enforce the selection of a main mode per pixel from multiple modes, while it ignores the outliers (minor modes leading to flickering artifacts). In practice, DVP doubles the number of output channels (e.g. 6 channels for RGB images) to obtain two output versions: a main frame O_t^{main} and an outlier frame O_t^{minor}. The confidence map $C_{t,i}$ at iteration i is calculated by:

$$C_{t,i} = \begin{cases} 1 & d(O_{t,i}^{main}, P_t) < max\{L_1(O_{t,i}^{minor}, P_t), \delta\} \\ 0 & \text{otherwise} \end{cases}, \tag{2}$$

where d is the function to measure the distance between pixels and δ is a threshold. Therefore, the model parameters at iteration $(i+1)$ can be optimised using $C_{t,i}$ which guides the training loss:

$$\theta^{i+1} = \arg\min_{\theta}\{L_{data}(C_{t,i} \odot O_{t,i}^{main}, C_{t,i} \odot P_t) \\ + L_{data}((1 - C_{t,i}) \odot O_{t,i}^{minor}, (1 - C_{t,i}) \odot P_t)\}. \tag{3}$$

Then, a multi-loss function is proposed combining the IRT loss L_{IRT} between $\hat{G}_1(I_t; \theta_1)$ and P_t, and the cross-entropy loss L_{class} between $\hat{G}_2(f_t; \theta_2)$ and s_t:

$$L_T = L_{IRT}\left(\hat{G}_1(I_t; \theta_1), P_t\right) + L_{class}\left(\hat{G}_2(f_t; \theta_2), s_t\right). \tag{4}$$

3.2   Few-Shot Training Strategy

The main limitation of DVP is the long processing time due to the need for training at inference time. This fact makes the method impractical for long sequences. This paper proposes to speed up the training process reducing the number of iterations by means of a few-shot training strategy. Such strategy selects a reduced set of N frames $\{I_n\}_{n \in J} \subset \{I_t\}_{t=1}^T$, where $J \subset \{1, 2, \dots T\}$ and $N < T$. Notice that for completeness $I_n :\neq I_t$. Selected few-shot samples are then used to train the model for generalisation to the remaining frames during inference time. The proposed model makes this solution feasible thanks to its scene-aware capacity to generalise to variable content. Such approach makes the model more robust for processing of sequences with changes (e.g. with high motion) as it temporally downsamples the input.

The selection of N frames for few-shot training is based on a twofold process: scene segmentation and selection of representative frames per scene. Scene segmentation is performed in an unsupervised way via clustering of deep features $\{f_t\}_{t=1}^T$ with KMeans algorithm [22]. Dimensionality reduction is performed by Principal Components Analysis (PCA) in order to reduce complexity and shorten the clustering time. The number of scenes (e.g. number of clusters) is unknown and variable for each input video. Hence a suitable number of clusters is computed by running KMeans K times and selecting the elbow of the averaged distortion curve, where the distortion of each sample is computed relative the centroid of its cluster. This method allows a fast and effective scene segmentation approach.

Unsupervised clustering of input frames allows the generation of pseudo-labels for training the proposed classification sub-model. Notice that clustering errors will be mitigated thanks to the few-shot training, since the trained classifier will generalise to unseen frames (and potential uncertainties between scenes) during inference time. After segmentation of the input video into the scenes, suitable frames are selected from each scene by sub-clustering frames in that scene to cover a balanced span of different content. KMeans is applied again with a fixed number of clusters and a number of frames is randomly sampled from each sub-cluster. The number of selected frames per cluster and sub-clusters is proportional to the total number of frames in the given sub-cluster.

3.3   Network Architecture

As shown in Fig. 2, the architecture of the model proposed at Sect. 3.1 is composed of two sub-networks (denoted by $\hat{G}_1(\theta_1)$, $\hat{G}_2(\theta_2)$). Its inputs are a frame $I_t \in \mathbb{R}^{1 \times H \times W}$, where $H \times W$ are the input dimensions, and its feature vector $f_t \in \mathbb{R}^{1 \times d}$ (from VGG-16 classification head), where d are the number of its dimensions. The proposed architecture outputs two colour stabilised versions (main and minor frames) $O_t \in \mathbb{R}^{6 \times H \times W}$ of the input frame, and a class distribution vector $y_t \in \mathbb{R}^{1 \times S}$, which is the product of clustering the input to a particular scene.

I_t is processed by 4 encoder blocks which downsample the input by a factor of 2, generating $I_t^b \in \mathbb{R}^{e_b \times H_b \times W_b}$, where $b = 1, \dots 4$ is the block index, e_b is the number of dimensions and $\{H_b, W_b\} = max\left(2^5, \frac{\{H, W\}}{2^b}\right)$. The bottleneck block converts I_t^4 into $O_t^5 \in \mathbb{R}^{o_5 \times H_5 \times W_5}$, where o_5 are the number of

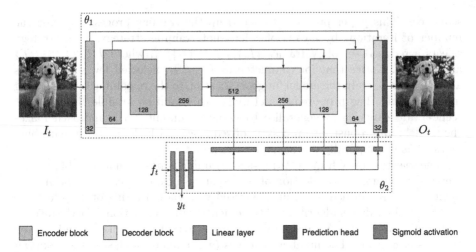

Fig. 2. Proposed architecture for stabilisation of frame-to-frame colourised videos. The model not only learns to stabilise an input sequence, but also to cluster the input frames into different scenes, by generating a class distribution vector y_t.

output dimensions. In parallel, f_t is processed by 2 linear layers to generate deep embeddings $f_t^1 \in \mathbb{R}^{1 \times d}$, $f_t^2 \in \mathbb{R}^{1 \times S}$. f_t^2 is both activated with a softmax operation to generate the class distribution vector y_t and with a sigmoid operation to generate the scene-aware mask a_t that will be injected into the bottleneck and decoder blocks. a_t is processed by a sequence of linear layers which generate 5 scene-aware embeddings $m_t^b \in \mathbb{R}^{1 \times o_b}$, where $b = 1, \dots 5$ and o_b are the dimensions of the bottleneck and decoder outputs. Finally, as shown in Fig. 3, m_t^b are injected into the corresponding blocks as follows: (1) m_t^b is activated with a SoftPlus operation (smooth approximation of ReLU) and spatially repeated to generate a volume $M_t^b \in \mathbb{R}^{e_b \times H_b \times W_b}$, (2) M_t^b is element-wise multiplied to each pre-activation within the corresponding block. 4 decoder blocks with skip connections are then applied to upsample the inputs by factor of 2, generating $O_t^b \in \mathbb{R}^{o_b \times H_b \times W_b}$, where $b = 1, \dots 4$. Finally, a decoder head is applied to map O_t^1 into the output frames O_t.

4   Experiments

4.1   Training Strategy

As shown in DVP, the network needs to be initialised with the main mode in order to guide the main outputs towards a specific mode. DVP selects the first image as reference for the main mode and pre-trains the network for a given number of iterations. However, when the reference image contains outliers, and those are treated as main mode, the performance of such approach is not satisfactory. To address that, this work proposes the use of colour histograms to detect outliers when specific bins present high variance across the sequence. In

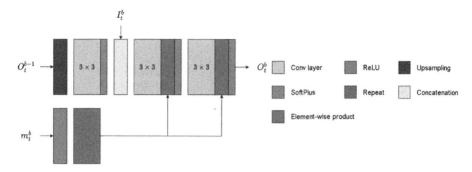

Fig. 3. The proposed decoder block conditioned by the scene-wise embedding m_t^b. Notice that a similar architecture applies to the bottleneck, injecting the embedding vector into the pre-activations.

particular, colourised frames P_t are converted into CIE La*b* colour space [7], and 2D colour histograms $\mathcal{H}_t \in \mathbb{R}^{Q \times Q}$ are obtained by matrix multiplication of individual histograms for a* and b* channels, where Q is the number of bins. Next, a mask $\bar{M} \in \mathbb{R}^{Q \times Q}$ is computed to locate those bins present in all the frames. Hence, bins out of the mask will represent an outlier. $\bar{M} = \prod_{t=1}^{T} M_t$, where M_t masks the bins different than zero. Finally, main mode reference frame P_{t^*} is obtained, where $t^* = \arg\min_t \sum \mathcal{H}_t \odot (M_t - \bar{M})$.

On the other hand, as shown in Fig. 4, few-shot training might lead the network into mode collapse, rapidly converging into a random state. Mode collapse is detected when the Area Under the Curve (AUC) of the generated colour histograms vary below a threshold during a given number of iterations. In this case, the initial pre-training is repeated with random initialisation of the network weights. Due to the significant difference of complexity, classifier and stabiliser (U-Net) sub-networks are optimised using different learning rates. Overall, Adam optimiser is adopted, using a learning rate of 10^{-4} for θ_1 and 10^{-6} for θ_2. All the experiments are performed with a single GPU and using a batch size of 8 samples. Initial pre-training iterations are set to 350, and 150 frames are used for few-shot training.

Following DVP [19], this work uses the test set collected by [5], composed by 20 videos of around 200 frames from Videvo dataset[1], and extended with 8 longer videos from Videvo and Hollywood2 dataset [23], to evaluate the performance for more complex content.

4.2   Evaluation Metrics

Temporal Inconsistency. DVP uses wrapping error to measure temporal inconsistency by means of optical flow. However, the quality of optical flow computation and the corresponding occlusion mask might decrease when dealing with flickering content. To mitigate this issue and to better capture colour

---

[1] https://www.videvo.net/.

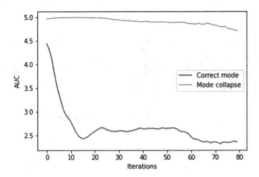

Fig. 4. Example of mode collapse during initial pre-training. The AUC invariance during the initial iterations indicates convergence to a random state, which affects the performance of the posterior IRT training.

artifacts, histogram inconsistency is adopted to measure the temporal similarity in the colour domain. Being $\mathcal{H}_t$ and $\mathcal{H}_{t-1}$ the colour histograms of frames t and $t-1$, respectively, temporal histogram inconsistency E_{hist} is defined as a symmetric χ^2 distance as follows:

$$E_{hist} = 2\sum_{q=1}^{Q^2} \frac{(\mathcal{H}_{t,q} - \mathcal{H}_{t-1,q})^2}{(\mathcal{H}_{t,q} + \mathcal{H}_{t-1,q}) + \epsilon}, \tag{5}$$

where ϵ prevents infinity overflows and Q is the number of bins.

Performance Degradation. Temporal stabilisation has to be achieved without degrading the original colourisation. Since stabilised ground truth is not available, this work uses data fidelity F_{data} between $\{O_t\}_{t=1}^T$ and $\{P_t\}_{t=1}^T$ as follows:

$$F_{data} = \frac{1}{T}\sum_{t=1}^{T} PSNR\,(P_t, O_t). \tag{6}$$

Notice that data fidelity can decrease when frames contain large amount of outliers. Therefore, perceptual quality is also evaluated using Fréchet Inception Distance (FID) [14] with the ground truth.

4.3   Results

Table 1 shows quantitative comparison results between DVP method [19], our method and the proposed ablations in Sect. 4.4. Two image-based fully-automatic colourisation methods are considered: colourful image colourisation (CIC) [35] and ChromaGAN (CGAN) [29]. Reference-based image colourisation method XCNET [4] is also considered. Such methods which colourise frames based on a reference image introduce even larger flickering issues than fully auto-colourisation based networks. References are sampled from Imagenet dataset [8]

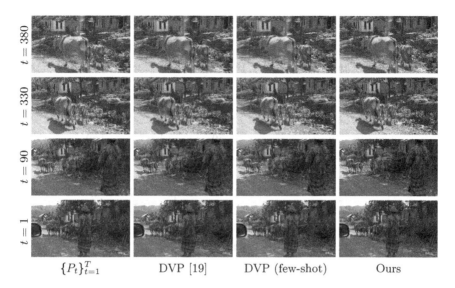

$\{P_t\}_{t=1}^{T}$ DVP [19]   DVP (few-shot)   Ours

Fig. 5. Qualitative comparison with DVP method and the proposed ablations. DVP with few-shot training took around the same processing time as our approach, but failed to generalise to multiple scenes. Moreover, our method achieved better fidelity and perceptual quality than the original DVP.

using the correspondence recommendation pipeline proposed in [4,13]. Finally, quality of the original predictions $\{P_t\}_{t=1}^{T}$ obtained using CIC method is studied to evaluate the effect of the proposed stabilisation. Moreover, Fig. 6 shows the processing time of both DVP and our method in relation to the number of frames.

As can be seen from E_{hist} results, both DVP and our method significantly increase the temporal consistency compared to the original predictions, and although DVP obtains slightly better results, our method significantly reduces the processing time for long scenes. The drop in performance when using XCNET is due to the colourfulness of the corresponding predictions and the higher concentration of flickering artefacts, compared to CIC or CGAN.

As shown in Fig. 5, the frames at different times in the same shot suffer from inconsistent colourisation (notice the same object across various frames with different colour). DVP and DVP with few-shot training temporal both provide more consistent results, but still the main mode is either not correctly chosen or the colours are plain, resulting in less natural appearance. This is reflected in data fidelity results, where our method achieves the best performance. FID also confirm this fact, as DVP lowers the perceptual quality of the original predictions due to its strong stabilisation and degradation of input colours. Finally, as shown in Fig. 6, the few-shot strategy allowed a fix amount of training iterations, resulting into a flat time response independent to the length of the input sequence. Note that the total time may increase proportionally to the number of scenes, due to the individual initial pre-training per scene.

Table 1. E_{hist}, F_{data} and FID comparison for different colourisation methods.

Method	E_{hist} ↓			
	$\{P_t\}_{t=1}^T$	CIC [35]	CGAN [29]	XCNET [4]
DVP [19]		2.30	1.58	3.30
Ours	20.96	3.08	2.54	3.59
DVP (few-shot)		3.75	3.79	3.10
Ours (first frame)		1.39	2.14	2.69

Method	F_{data} ↑ [dB]		
	CIC [35]	CGAN [29]	XCNET [4]
DVP [19]	19.12	19.32	18.94
Ours	28.63	30.31	26.56
DVP (few-shot)	18.14	18.47	18.67
Ours (first frame)	28.46	30.21	26.40

Method	FID ↓			
	$\{P_t\}_{t=1}^T$	CIC [35]	CGAN [29]	XCNET [4]
DVP [19]		126.38	111.16	100.21
Ours	122.74	121.16	105.65	97.96
DVP (few-shot)		129.68	114.98	102.22
Ours (first frame)		119.76	104.03	99.92

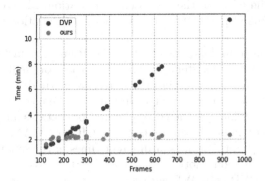

Fig. 6. Comparison of processing time for all test sequences. Notice the significant increase of the processing time for DVP when the number of frames increases.

4.4   Ablations

An ablation study is performed to analyse the importance of the proposed scene-aware architecture. First, DVP is tested with the proposed few-shot training strategy. As shown in Table 1 and Fig. 5 (DVP few-shot), without using a classification sub-network, DVP is unable to generalise to complex sequences and

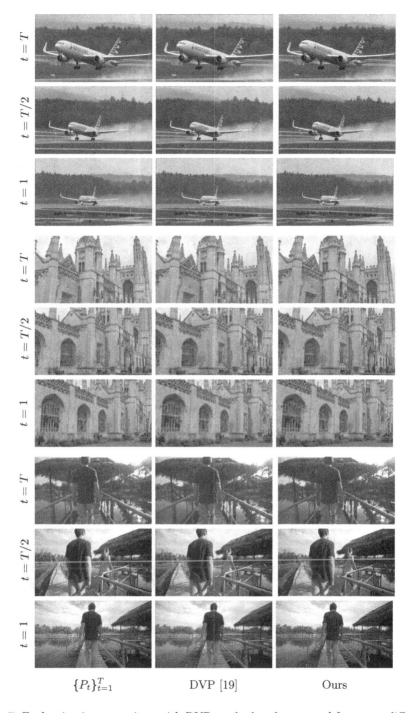

Fig. 7. Evaluation in comparison with DVP method and processed frames at different timestamps.

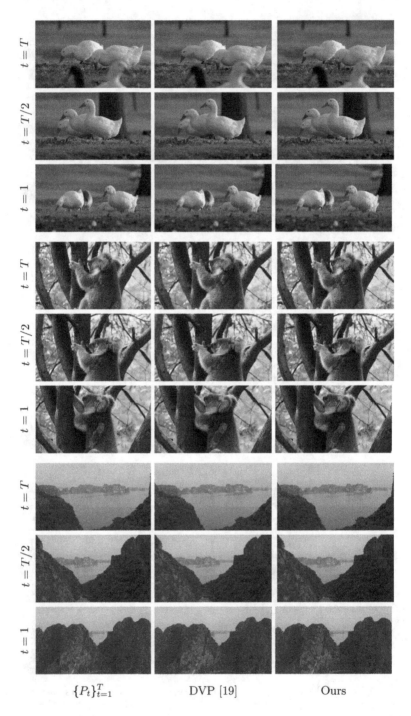

Fig. 8. Evaluation in comparison with DVP method and processed frames at different timestamps.

the input colours are significantly degraded. This drop in performance proves the importance of the classification sub-network to perform effective few-shot training. Finally, a second ablation is performed to evaluate the proposed initialisation mechanism in Sect. 4.1, which proposes the best reference for main mode per scene by means of histogram characteristics. As shown in Table 1 (ours first frame), a drop in performance is observed when using the first frame as main mode reference (as DVP proposes), proving the effectiveness of the proposed methodology. Notice that original DVP performance could be improved by using the same initialisation mechanism.

5   Conclusions

This paper proposed a general framework for temporal stabilisation of frame-to-frame colourised videos using scene-aware deep video priors. The framework includes an optimised few-shot training strategy to reduce the processing time of DVP baseline [19] by removing its time response conditioned on the number of input frames. In order to handle complex sequences with multiple scenes, the DVP architecture is modified by adding a classification sub-network which clusters the input frames with the objective of learning scene-specific priors. Experimental results show that our method improves data fidelity and perceptual quality and achieves similar temporal consistency to DVP while reducing the processing time in long sequences. As future work, model efficiency can be further improved by simplifying the network architecture or by using techniques such as pruning or weights quantisation. Moreover, finer tuning of colourisation could be achieved by improving the scene segmentation process in order to obtain more precise scene priors. Finally, an unified framework for video colourisation can be obtained by integrating the deep video prior methodology into an end-to-end video colourisation pipeline.

References

1. Akimoto, N., Hayakawa, A., Shin, A., Narihira, T.: Reference-based video colorization with spatiotemporal correspondence. arXiv preprint arXiv:2011.12528 (2020)
2. Barnes, C., Shechtman, E., Finkelstein, A., Goldman, D.B.: Patchmatch: A randomized correspondence algorithm for structural image editing. ACM Trans. Graph. **28**(3), 24 (2009)
3. Bertinetto, L., Henriques, J.F., Valmadre, J., Torr, P., Vedaldi, A.: Learning feedforward one-shot learners. Advances in neural information processing systems 29 (2016)
4. Blanch, M.G., Khalifeh, I., O'Connor, N.E., Mrak, M.: Attention-based stylisation for exemplar image colourisation. In: 2021 IEEE 23rd International Workshop on Multimedia Signal Processing (MMSP), pp. 1–6 (2021). https://doi.org/10.1109/MMSP53017.2021.9733506
5. Bonneel, N., Tompkin, J., Sunkavalli, K., Sun, D., Paris, S., Pfister, H.: Blind video temporal consistency. ACM Transactions on Graphics (TOG) **34**(6), 1–9 (2015)

6. Bugeau, A., Ta, V.T., Papadakis, N.: Variational exemplar-based image colorization. IEEE Transactions on Image Processing 23(1), 298–307 (2013)

7. Connolly, C., Fleiss, T.: A study of efficiency and accuracy in the transformation from rgb to cielab color space. IEEE transactions on image processing 6(7), 1046–1048 (1997)

8. Deng, J., Dong, W., Socher, R., Li, L.J., Li, K., Fei-Fei, L.: Imagenet: a large-scale hierarchical image database. In: 2009 IEEE Conference on Computer Vision and Pattern Recognition, pp. 248–255 (2009). https://doi.org/10.1109/CVPR.2009.5206848

9. Edwards, H., Storkey, A.: Towards a neural statistician. arXiv preprint arXiv:1606.02185 (2016)

10. Fei-Fei, L., Fergus, R., Perona, P.: One-shot learning of object categories. IEEE transactions on pattern analysis and machine intelligence 28(4), 594–611 (2006)

11. Fink, M.: Object classification from a single example utilizing class relevance metrics. Advances in neural information processing systems 17 (2004)

12. Finn, C., Abbeel, P., Levine, S.: Model-agnostic meta-learning for fast adaptation of deep networks. In: International Conference on Machine Learning, pp. 1126–1135. PMLR (2017)

13. He, M., Chen, D., Liao, J., Sander, P.V., Yuan, L.: Deep exemplar-based colorization. ACM Transactions on Graphics (TOG) 37(4), 1–16 (2018)

14. Heusel, M., Ramsauer, H., Unterthiner, T., Nessler, B., Hochreiter, S.: Gans trained by a two time-scale update rule converge to a local nash equilibrium. Advances in neural information processing systems 30 (2017)

15. Huang, Y.C., Tung, Y.S., Chen, J.C., Wang, S.W., Wu, J.L.: An adaptive edge detection based colorization algorithm and its applications. In: Proceedings of the 13th Annual ACM International Conference on Multimedia, pp. 351–354 (2005)

16. Jampani, V., Gadde, R., Gehler, P.V.: Video propagation networks. In: Proceedings of the IEEE Conference on Computer Vision and Pattern Recognition, pp. 451–461 (2017)

17. Lai, W.S., Huang, J.B., Wang, O., Shechtman, E., Yumer, E., Yang, M.H.: Learning blind video temporal consistency. In: Proceedings of the European conference on computer vision (ECCV). pp. 170–185 (2018)

18. Lei, C., Chen, Q.: Fully automatic video colorization with self-regularization and diversity. In: Proceedings of the IEEE/CVF Conference on Computer Vision and Pattern Recognition, pp. 3753–3761 (2019)

19. Lei, C., Xing, Y., Chen, Q.: Blind video temporal consistency via deep video prior. Advances in Neural Information Processing Systems 33, 1083–1093 (2020)

20. Levin, A., Lischinski, D., Weiss, Y.: Colorization using optimization. In: ACM SIGGRAPH 2004 Papers, pp. 689–694 (2004)

21. Liu, S., Zhong, G., De Mello, S., Gu, J., Jampani, V., Yang, M.H., Kautz, J.: Switchable temporal propagation network. In: Proceedings of the European Conference on Computer Vision (ECCV), pp. 87–102 (2018)

22. MacQueen, J., et al.: Some methods for classification and analysis of multivariate observations. In: Proceedings of the Fifth Berkeley Symposium on Mathematical Statistics and Probability, vol. 1, pp. 281–297. Oakland, CA, USA (1967)

23. Marszalek, M., Laptev, I., Schmid, C.: Actions in context. In: 2009 IEEE Conference on Computer Vision and Pattern Recognition, pp. 2929–2936. IEEE (2009)

24. Mrak, M.: Ai gets creative. In: Proceedings of the 1st International Workshop on AI for Smart TV Content Production, Access and Delivery, AI4TV 2019, pp. 1–2. Association for Computing Machinery, New York (2019). https://doi.org/10.1145/3347449.3357490

25. Ravi, S., Larochelle, H.: Optimization as a model for few-shot learning (2016)
26. Simonyan, K., Zisserman, A.: Very deep convolutional networks for large-scale image recognition. arXiv preprint arXiv:1409.1556 (2014)
27. Sung, F., Yang, Y., Zhang, L., Xiang, T., Torr, P.H., Hospedales, T.M.: Learning to compare: relation network for few-shot learning. In: Proceedings of the IEEE Conference on Computer Vision and Pattern Recognition, pp. 1199–1208 (2018)
28. Tai, Y.W., Jia, J., Tang, C.K.: Local color transfer via probabilistic segmentation by expectation-maximization. In: 2005 IEEE Computer Society Conference on Computer Vision and Pattern Recognition (CVPR 2005), vol. 1, pp. 747–754. IEEE (2005)
29. Vitoria, P., Raad, L., Ballester, C.: Chromagan: adversarial picture colorization with semantic class distribution. In: Proceedings of the IEEE/CVF Winter Conference on Applications of Computer Vision, pp. 2445–2454 (2020)
30. Wang, Y., Yao, Q., Kwok, J.T., Ni, L.M.: Generalizing from a few examples: A survey on few-shot learning. ACM computing surveys (csur) **53**(3), 1–34 (2020)
31. Welsh, T., Ashikhmin, M., Mueller, K.: Transferring color to greyscale images. In: Proceedings of the 29th Annual Conference on Computer Graphics and Interactive Techniques, pp. 277–280 (2002)
32. Yao, C.H., Chang, C.Y., Chien, S.Y.: Occlusion-aware video temporal consistency. In: Proceedings of the 25th ACM international conference on Multimedia, pp. 777–785 (2017)
33. Yatziv, L., Sapiro, G.: Fast image and video colorization using chrominance blending. IEEE transactions on image processing 15(5), 1120–1129 (2006)
34. Zhang, B., et al.: Deep exemplar-based video colorization. In: Proceedings of the IEEE/CVF Conference on Computer Vision and Pattern Recognition, pp. 8052–8061 (2019)
35. Zhang, R., Isola, P., Efros, A.A.: Colorful image colorization. In: European conference on computer vision. pp. 649–666. Springer (2016)

MOVIE LENS: Discovering and Characterizing Editing Patterns in the Analysis of Short Movie Sequences

Bartolomeo Vacchetti[(✉)] [ID] and Tania Cerquitelli [ID]

Polytechnic of Turin, Torino, Corso Duca degli Abruzzi 24, 10129 Turin, Italy
bart.vacchetti@gmail.com

Abstract. Video is the most widely used media format. Automating the editing process would impact many areas, from the film industry to social media content. The editing process defines the structure of a video. In this paper, we present a new method to analyze and characterize the structure of 30-second videos. Specifically, we study the video structure in terms of sequences of shots. We investigate what type of relation there is between what is shown in the video and the sequence of shots used to represent it and if it is possible to define editing classes. To this aim, labeled data are needed, but unfortunately they are not available. Hence, it is necessary to develop new data-driven methodologies to address this issue. In this paper we present MOVIE LENS, a data driven approach to discover and characterize editing patterns in the analysis of short movie sequences. Its approach relies on the exploitation of the Levenshtein distance, the K-Means algorithm, and a Multilayer Perceptron (MLP). Through the Levenshtein distance and the K-Means algorithm we indirectly label 30 s long movie shot sequences. Then, we train a Multilayer Perceptron to assess the validity of our approach. Additionally the MLP helps domain experts to assess the semantic concepts encapsulated by the identified clusters. We have taken out data from the Cinescale dataset. We have gathered 23 887 shot sequences from 120 different movies. Each sequence is 30 s long. The performance of MOVIE LENS in terms of accuracy varies (93% - 77%) in relation to the number of classes considered (4-32). We also present a preliminary characterization concerning the identified classes and their relative editing patterns in 16 classes scenario, reaching an overall accuracy of 81%.

Keywords: Sequence analysis · Movie editing · K-Means · Multilayer perceptron · Levenshtein distance

1 Introduction

Videos have been deeply studied in the past using machine learning techniques. These studies mainly focus on image, video, or text data. However, a little attention has been devoted to studying the editing process of a video that defines its narrative structure.

© The Author(s), under exclusive license to Springer Nature Switzerland AG 2023
L. Karlinsky et al. (Eds.): ECCV 2022 Workshops, LNCS 13804, pp. 660–675, 2023.
https://doi.org/10.1007/978-3-031-25069-9_42

Editing consists of joining individual shots together to form a scene. How long each shot is, which shot comes before and which after, are all elements that strongly influence the viewer's perception. In other words, editing helps define the narrative and mood of a film, along with other elements such as the setting, the soundtrack, and so on. However, while these elements are perceived by our senses, the audience only notices editing when it is poorly done. For this reason, it has been called invisible art. According to Walter Murch [16], editing a film is like telling a story. If there is a great story, but the narrator tells it in the wrong rhythm and focuses on the wrong parts, it has no impact. If the film is poorly edited, the viewer will be less engaged in the story. Worse, there is a possibility that the viewer will not understand the story at all. This is because editing, among other elements, determines the structure of a film, which consequently affects the mood of the narrative style. Practically speaking, editing is a process of cutting and chaining together individual clips that have no meaning on their own to create a meaningful video.

In the past, some studies, such as [28], have analyzed the challenges of automating the entire video editing process. Some successful efforts have been made to solve some tasks of the process. One example is the autoEdit library, which is based on the research presented in [5]. AutoEdit receives as input a video file and gives as output the text extracted from the people speaking in the video. Then, the user has to select a part of the text, and the algorithm selects the corresponding clip from the video and cuts it. Automating video processing could bring interesting and useful benefits, especially considering that it is the most commonly used media format [35]. For example, a deeper understanding of the narrative structures implemented in videos could lead to the automatic creation of more semantically meaningful videos on social media. It could also speed up the post-editing phase of a video production. Another practical implementation could be an integration with text-image translation, such as DALLE [21]. This type of model is able to generate photo-realistic images from a textual description. This type of network could be used to transform scripts into storyboards. However, to do this, they need to understand the classification of shots and editing patterns.

In this paper we propose a novel data-driven methodology, named MOVIE LENS . It allows identifying editing patterns within video structures and characterizing their main properties. To this aim, we focus on sequences of shot scale classes. The shot scale is defined in relation to what is shown in the camera field of view (see Sect. 3 for further details). We are interested in investigating what kind of correlation exists between sequences of shot classes and what is actually shown in the corresponding video. We selected shot class sequences from the Cinescale dataset [24]. However, the sequences obtained were not labeled. To label them, we have used a novel method based on a joint approach that relies on K-Means and the Levenshtein distance. As a result, we assign a label for each sequence without directly analyzing the content of the sequences. Then we then train and test a classifier using the original sequences and the newly obtained labels.

Specifically, MOVIE LENS introduces the following contributions: (i) a data-driven methodology to cluster short video sequences in homogeneous and well-separated groups; (ii) a machine learning algorithm capable of classifying video sequences assessing the robustness of the clustering analysis. Furthermore, it helps domain experts to easily identify a semantic label for each group of short video sequences. (iii) Additionally we present a preliminary characterization conducted on a real set of data characterized by 23 887 sequences from 120 different movies.

This paper is organized as follows. In Sect. 2 we analyze the existing state-of-the-art on the subject. In Sect. 3 we focus on the data used, while in Sect. 4 we present the MOVIE LENS methodology. Section 5 discusses some preliminary results obtained by MOVIE LENS on a large set of real-data. Section 6 concludes the paper with an in-depth discussion on the weaknesses of the proposed approach and how to address them in our future research studies.

2   Related Works

In recent years, videos and films have been analyzed using different techniques and for different purposes. Most of these studies are related to computer vision and image classification, but other branches of research focus on other aspects. One task that has been addressed in the past using different approaches is the classification of movie images [3,9,25,30,31]. It is a classification that assigns a shot type to an image based on what is in the camera's field of view. Most of these studies rely on convolutional neural networks (CNN) [27] to predict the shot scale of an image. This type of classification can be more or less fine-grained. Some studies also incorporate camera motion classes [22]. In computer vision, there have been significant studies on translating text into images. These algorithms, such as [21], are able to reproduce a more or less accurate image based on a text description. A rather different, but still interesting approach is the work proposed in [20]. Here the authors use natural language processing techniques on the IMDB dataset to perform movie sentiment analysis.

In terms of studies dealing with the editing structure of videos and films, there have been some developments. The first study on this topic dates back to 2002 ([14]). Here, the authors attempted to identify editing rules and patterns. They emphasized the central importance of editing in videos. Their focus was on editing speed and on a preliminary version next shot type prediction with three classes. Recently, in [4], the authors have analyzed how the concatenation of shot types affects viewer attention. In [1] the authors present a new, manually labeled, dataset and a novel methodology to assist the video editing process. Another interesting study is proposed in [11]. Here, the authors present an automatic censoring method that detects inappropriate visual content and classifies it as "Universal," "Universal Adult," or "Adult". In [37], the authors perform video retrieval. They use recording sequences in combination with temporal information and visual features to find a specific video in a large collection. Also in [2], video retrieval is performed using textual data along with other movie metadata such as script, editing speed, release year, and genre of the movie. Also in [33] the authors propose a tool for video editing that relies mostly on textual input. In [26], the authors instead perform movie genre classification by implementing a methodology based on convolutional neural networks. [36] is also about movie genre classification, but the approach here is different. Representative key frames are extracted from trailer clips and genre classification analysis is performed using these key frames. In [17], the authors focus on film structure by analyzing the time interval from one cut to the next. They examine how film genre affects shot duration. In [29], the authors analyze what kind of relationship exists between the director of a film and the shots used in the film. In automated video editing, there have been advances in various aspects of the entire process. In [5], the authors develop a method for cutting a video depending on the part of the dialog that the transcript text receives as input. In [23], the focus is not on classifying frames or videos into recording types, but on video segmentation, i.e., splitting a video into individual clips. In [18] the authors propose a methodology to estimate the cut plausibility based on audiovisual patterns. In [19], the authors present a preliminary methodology that mimics the editing structure of movies. In this study, three shot typologies are considered: close up, medium shot, and long shot. Another interesting study dealing with automatic video editing is presented in [34]. Here, the focus is on the typology of videos to be processed in multi camera environments. For a more comprehensive overview, we recommend the following surveys [15] [28,35]. All of these studies could yield even more surprising results if they integrated even rudimentary narrative editing patterns.

Editing structure influences how the story is perceived. Some studies address specific aspects of editing structure or use automated editing in controlled environments. Differently from all cited works, MOVIE LENS focuses on discovering and characterizing editing patterns to better understand how the sequence of shots changes in relation to what is shown on screen. Only a few research studies, address our research issue, such as [4,19,34], however with a different methodology and a different problem characterization. Differently than [4,19,34], MOVIE LENS considers more shot classes and integrates a novel strategy to automatically discover editing patterns and how to classify them easily.

3 Data

For our experiment, we have used the Cinescale dataset [24]. To be more precise, only its labels were used. Cinescale is a dataset containing 120 different movies realized by six different directors. It is a dataset used to perform cinematographic shot classification. From each movie a frame has been sampled and labeled at every second. The label assigned to each sampled frame corresponds to its shot class. The shot classes considered in this study are the following: (i) Class 0) Foreground Shot (FS): a shot that contains elements of different shot classes. For instance camera movements fall under this category (128 335 frames); (ii) Class 1) Extreme Close Up (ECU): a shot that focuses on details, such as the

Fig. 1. The 8 types of shot considered in this study.

eyes of the subject, what the character is holding and so on (3 367 frames); (iii) Class 2) Close Up (CU): a shot focused on the subject's face, it shows the actor from the shoulder up. It can also be used to focus the viewer's attention on some detail like objects or hands (83 682 frames); (iv) Class 3) Medium Close Up (MCU): the subject figure is shown from the upper half of its torso (252 639 frames); (v) Class 4) Medium Shot (MS): only the upper half of a human subject is shown (78 053 frames); (vi) Class 5) Medium Long Shot (MLS): the human figure is shown from the knee up (89 450 frames); (vii) Class 6) Long Shot (LS): the human figure occupies the totality of the frame height or 2/3 (49 788 frames); (viii) Class 7) Extreme Long Shot (ELS): the human figure is absent or occupies less than a third of the screen height (7 118 frames). Figure 1 shows the different types of shot classes considered in this study.

The extracted video sequences were 30 frames long, which corresponds to 30 s, since in the Cinescale dataset one frame is sampled every second. We focused on 30-second sequences for the following reasons. Not only do scenes in movies vary in length, but even when they are of similar length, the number of shots used varies. Therefore, to simplify the problem, we decided to use 30-second sequences. Usually scenes do not last that long, but it is rare for a scene to be shorter than 30 s. In this time interval, there is also a chance to capture some patterns, while if smaller sequences are chosen, there is a risk that the resulting sequences cannot describe anything in particular. On the other hand, if a larger time interval is chosen, there is a chance that more scenes will be compressed into a single one. Cinescale has two other classes besides the shot type classes. One contains opening and closing titles, the other undefined frames. After removing the sequences that contained these shots, the resulting dataset contained 23 887 sequences. The next section will explain the methodology in more detail.

4   Methodology

MOVIE LENS performs two main analytic building blocks: the *Label Estimation Phase* and the *Editing Patterns Analysis*. The aim of the first task is to label each short movie sequence in the selected portion of data. In order to do that, first we run an analysis based on a distance metric on every sequence. The analysis focuses on how much every sequence is similar to some fixed reference sequences. At the end of this operation we use those distances as points' coordinates. Every point models a sequence. The points are then grouped through a clustering algorithm. Each group should theoretically model a type of sequence. As a result of the first step, Movie Lens defines a cluster identifier for each short movie sequence. To help the domain expert to define a semantic label to each group, we have introduced a second phase named Editing Pattern Analysis. The aim of the second step is to evaluate the results of the Label Estimation Phase through the analysis of the performance yielded by a classifier on a portion of the dataset. After training the model, we test it on the remaining part of the dataset. The obtained results need to be assessed by domain experts. Specifically, in this final step Movie Lens allows the domain expert to analyze which patterns characterize each class and which ones are the most common misclassification errors. The domain experts are asked to assess the quality of the classes identified by Movie Lens . This analysis is performed manually on a subset of short movie sequences for each class along with the label defined by the classifier. Selected sequences include sequences that are classified correctly or misclassified.

Figure 2 shows the main analytic building blocks of Movie Lens . Further details of each task are provided in the next subsections.

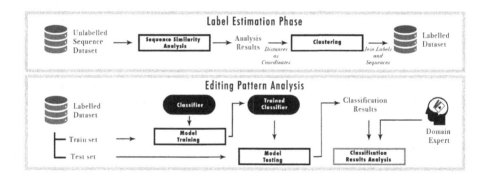

Fig. 2. Overview of the Movie Lens 's methodology.

4.1   Label Estimation Phase

This analytic building block performs the similarity analysis and models its results through a clustering algorithm. Specifically, we have used the Levenshtein distance [8] to compute the sequences' similarities and differences, and the K-Means algorithm [12] to indirect model editing patterns. The Levenshtein distance measures the similarity between sequences by counting the minimum number of changes (substitution, insertions, and deletions) needed to convert one sequence to another. The selection of the distance measure has been guided by the data semantic. Specifically, the shot type classes on Cinescale are defined through numerical values. From label 1 to 7 the higher the class number the wider shot we are considering. However, the foreground shot class, identified by the number 0, is not defined by the shot scale. Since the numerical magnitude of a class is not reliable to compute more traditional distances, such as the Euclidean distance [13], we have chosen a different approach. We have considered our shot sequences as strings and used the Levenshtein distance. However, also the Levenshtein distance has some limitations, because it does not characterize the changes made. However in our scenario it is important to know also to what class the frame has been converted to in order to match the sequence. Hence, instead of using the Levenshtein distance to directly measure the differences between sequences, we took a slightly different approach. First, we have defined 8 reference sequences artificially, one for each shot type (e.g., close up). All reference sequences were 30 frames long, with each frame belonging to the same shot type. Then, we have measured the Levenshtein distance among the short video sequences extracted from the Cinescale dataset and each reference sequence. After computing the distances between each sequence and the 8 reference ones, we have used those distances as coordinates. In this way each sequence corresponds to an 8-dimensional point in the multi-dimensional space. Afterwards we have clustered all the points and labeled them according to which region of the 8-dimensional space they occupy. Among the different clustering techniques we have decided to rely on the K-Means algorithm. We have decided to exploit this specific clustering technique because it converges quickly while providing good results [7]. The only parameter it requires is the number of clusters. To find a reliable value for the input parameter, a joint strategy based on the elbow graph method and Ward's method - i.e., a hierarchical clustering metric - has been adopted. Both these strategies focus on identifying the optimal number of clusters by minimizing the within-cluster variance. We have run different simulations with a varying number of groups. For each 8-dimensional point, the cluster analysis identifies the group of similar sequences to which the corresponding short movie sequence belongs to.

4.2   Editing Patterns Analysis

This phase has a double purpose. On the one hand it assesses the robustness of the clustering analysis with a supervised approach. On the other hand it helps the domain experts to analyze emerging patterns from the clustering groups. Additionally the resulting supervised model can be used to label sequences for which the class is not known. In the editing pattern analysis we split our dataset into train and test sets. The classifier is trained on the shot sequences with the labels defined through the previous step. Then, the performance of the classifier has been evaluated through the most common classification quantitative metrics (see Sect. 4.3 for further detail).

For the classifier choice we have decided to rely on a Multilayer Perceptron (MLP), by adapting the model[1] proposed for sentence classification to short movie sequences classification. After training the MLP we evaluate its performance by analyzing the classification results on the test set. This evaluation step has a double purpose. This evaluation step on the one hand allows understanding if the results obtained from the labeling phase are robust. On the other hand it allows domain experts to easily evaluate the main characteristics of each defined class. In fact, after training and evaluating the model, we analyze the patterns characterizing the different classes. To this aim the domain experts are asked to verify the correctness of the classification results and to extract the main characteristics of each sequence type. For each class, we selected 5% of short movie sequences stratified with respect to the class cardinality and the distribution of misclassified sequences. The domain expert verifies what type of correlation there is among the identified patterns and what is shown in the video. Section 5 offers more insight on this aspect.

4.3   Technical Details

We used a traditional K-Means procedure to cluster the 8-dimensional points representative of the acquisition sequences. After a large set of experiments we have defined the following Movie Lens configuration. The number of initial centroids of the clusters is 20. The MLP has an input layer, a hidden layer, and an output layer. The number of units of the input layer and the hidden layer are 128 and 64 respectively. The number of layers of the output layer depends on the number of classes considered. The optimizer used is the adaptive Nesterov moment estimation (Nadam). The activation function for both the input and hidden layer is the Hyperbolic tangent activation function (tanh), while for the output layer the activation function is the softmax function. The loss function used is the categorical cross entropy. The experiments were conducted on a HPC system with 4 CPU allocated. The code was implemented in Python with the support of the following libraries: Numpy, Keras, Tensorflow, Scikit-learn and Pandas. The metrics considered in this study are the following: (i) accuracy: number of correct predictions over the total amount of predictions; (ii) recall: the ratio of true positives identified over the sum of true positives and false negatives; (iii) precision: the ratio of true positives identified over the total of positives, both true and false; (iv) f1-score: the weighted harmonic mean of precision and recall.

5   Preliminary Results

The following is a description of the initial experiments we conducted to evaluate the quality of Movie Lens. The experiments were performed to show: (i) the

---

[1] the Multilayer Perceptron for sentence classification can be retrieved from a GitHub repository [32].

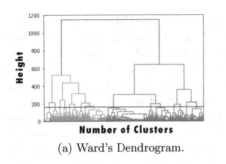
(a) Ward's Dendrogram.

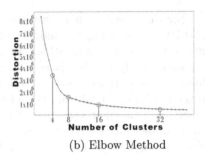
(b) Elbow Method

Fig. 3. Different strategies to identify the optimal number of clusters.

quality of the identified partition by the K-Means algorithm; (ii) if the identified patterns hold any insight in relation to what is happening in the video. We tested different configurations to perform its evaluation and configuration. Thanks to Ward's hierarchical clustering technique and the Elbow method, we were able to correctly configure the K-Means algorithm. Figure 3a 3b shows the obtained results. The height parameter in Ward's dendrogram grows with the intra-cluster variance. Table 1 shows the performance of the MLP with 4, 8, 16 and 32 classes, in terms of accuracy, f1-score macro average and f1-score weighted average.

Table 1. Performance of the MLP classifier with a different amount of classes considered.

Number of classes	Overall accuracy	Macro average	Weighted average
4	93	93	93
8	88	88	88
16	81	79	81
32	77	72	77

With 4 classes our classifier achieves a good performance, however if we take a look at the elbow graph we can see that the number of classes can be incremented. With 8 classes the classifier reaches a slightly lower performance but still satisfying. However, since our sequences are composed of eight elements and our labels are defined in relation to how much every sequence is distant from the 8 artificial sequences, we wanted to see if by increasing the number of classes considered we were able to find more fine-grained editing patterns. Hence we tested MOVIE LENS with also 16 and 32 classes. Figures 4a and b show the average pattern per class, i.e. centroid identified in these last two scenarios. By analyzing the patterns in Fig. 4a we can see that centroids modeling editing pattern groups are well-separated with respect to the ones in Fig. 4b. Thus, fine-grained partitions obtained with 32 classes are too detailed and present overlapped centroids (i.e., some types of frames in the sequences are very similar).

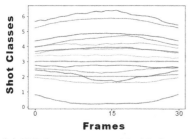

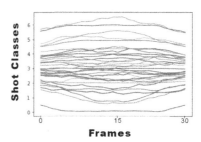

(a) Editing patterns with 16 classes.   (b) Editing patterns with 32 classes.

Fig. 4. These patterns, one for each class, are obtained by averaging all the sequences belonging to the same group.

Hence, in the following we detail the result obtained with 16 classes. The experiments presented here were performed using a stratified K-fold cross-validation with a K value of 10. With 16 classes, the MLP achieves a training accuracy of 88% after 50 epochs and a validation accuracy between 78% and 83%, with an average value of 81.2%. Table 2 shows precision, recall, and f-1 score of the average model.

From the classification report shown in Table 2 we can see that the proposed model performs particularly well in some classes. In a first analysis, with the help of a domain expert we found some insightful patterns allowing us to define three semantic labels (as shown in Table 2): (1) *Character-Environment Relationship (CER)*, (2) *Environment Descriptions (ED)*, and (3) *Character-Character Interaction (CCI)* discussed in the following subsections. Furthermore, an additional label, named *undefined (N)* has been defined to group all ill-defined classes.

5.1   Character-Environment Relationship

The classes in this first semantic group are 0, 1, 3 and 4, as shown in Table 2. Class 0 contains mainly medium close ups and medium long shots. On class 0 the MLP was able to reach an f1-score of 63%. It is one of the lowest scores in terms of performance and it is mainly due to the presence of two types of sequences. The first one consists of dialogues where there is not too much involvement with the characters or in their reaction. The focus of the viewer in these sequences is divided between the characters and the environment in which the scene is set. The other main sequence branch is also centered around characters and the environment, but represents it from a different perspective. In these sequences there is the camera that follows the subject exploring or moving through the scene environment. Some sequences show a mixture of both patterns. For instance two

Table 2. Precision, recall and f1-score of the MLP with 16 labels.

Labels	Semantic Label	Precision	Recall	f1-Score	Support
0	CER	66%	61%	63%	130
1	CER	73%	79%	76%	109
2	N	**99%**	**97%**	**98%**	177
3	CER	76%	72%	74%	101
4	CER	88%	86%	87%	140
5	ED	96%	64%	77%	77
6	CCI	85%	82%	83%	197
7	CCI	**97%**	**95%**	**96%**	369
8	N	67%	61%	64%	127
9	N	69%	83%	75%	237
10	CCI	**99%**	**95%**	**97%**	155
11	ED	79%	71%	75%	21
12	ED	87%	92%	89%	118
13	N	56%	77%	65%	119
14	N	67%	57%	62%	136
15	N	77%	75%	76%	175
Accuracy				81%	2388
Macro avg		80%	78%	79%	2388
Weighted avg		82%	81%	81%	2388

characters may be talking about a friend of theirs in jail and then we see a flashback in which we follow the third character while he is being arrested.

Also class 1 (f1-score = 76%) encapsulates sequences focused on the relationship between characters and the environment. These sequences contain mainly close ups mixed with wider shots. We have found one relevant pattern. The analyzed clips showed mainly how the characters react to changes in the environment. The close ups are used to see the character facial expression, while the wider shots are used to show what is happening. Whether the character interacts with the environment producing a change to which he reacts to or the environment changes by itself they describe action and reaction between the character and the environment. Class 3 (f1-score = 74%) is conceptually similar. The typologies of shot used are similar, although in different proportions. Also here what is shown is the relationship between character and environment. The difference from the previous class is that here the close ups do not show only the character reactions but also objects of the surrounding environment. By using close up to show also details of the environment changes the viewer's perception of the story. Yet another class that focuses on the relationship between character and environment is class 4 (f1-score = 87%). However, here the mood is different. Usually they represent a unique shot, with a fixed shot or a camera movement. Here the characters are presented with the environment itself, whether they are doing nothing or talking to each other. Most of the analyzed sequences depicted scenes in an internal setting, as shown in Fig. 5.

Fig. 5. Character and Environment Introduction. For visualization purposes we show 1 frame every 5.

5.2 Environment Description

Fig. 6. Environment Sequence. For visualization purposes we show 1 frame every 3.

The classes in this semantic group are 5, 11 and 12 (see Table 2). These groups focus on the environment description, what changes mainly is the shot scale used. For instance, class 12 (f1-score = 89%) encapsulates sequences that are mainly composed of medium long shots. This class contains mainly sequence shots. Sequence shot, which is not to not be confused with sequence of shots or shots sequence, is a sequence composed of a unique clip. In other words there are no cuts. These sequence shots are either stationary, or follow at a fixed distance a character that moves around the environment. Sequences belonging to class 11 (f1-score = 75%) contain mainly long shots or extreme long shots. Here the focus is completely on the Environment, if there are subjects on the screen, the viewer's attention dedicated to them is minimal. Figure 6 shows an example of this type of sequence taken from "Bande à Part". Class 5 (f1-score = 77%) uses more narrow shots. compared to class 11. Nonetheless also these are environment descriptive sequences. They tend to be sequence shots that describe the environment. If there are characters framed, the viewer gives a little more attention to them. However there is no emphasis on what the characters are doing and the main focus remains on the environment.

5.3 Character-Character Interaction

This main category, including groups 6, 7, and 10, as shown in Table 2, focuses on dialogue among characters. Class 7 (f1-score = 96%) contains mainly medium close up sequences. This is the class with the highest amount of sequences. All of

Fig. 7. Dialogue Sequence. For visualization purposes we show 1 frame every 3.

these sequences are dialogues. Unfortunately all of these sequences look similar so it is difficult to extract more meaningful patterns (more on this in Sect. 6). Figure 7 shows a sequence belonging to this category.

Class 10 (f1-score = 97%) instead identifies a more specific pattern. In this class there are a lot of close ups. Usually there are two, three at most characters involved in this type of sequence. The interaction between characters is one to one. Instead class 15 (f1-score = 76%) contains dialogues among more characters. Usually there is the main character talking while the others listen or vice versa. The interactions among characters are one to many. In fact this class contains wider shots compared to class 10.

Class 6 (f1-score = 83%) is a class that contains a lot of foreground shots and narrow shots. Even if the sequences are a little noisy in this class there are dialogues, not too centered on the character's facial emotional reaction but also his physical reaction. To follow the character's physical reaction we follow him at a fixed distance using camera movements, labeled on Cinescale as foreground shots.

5.4   Undefined Classes

Classes 8-9-13-14-15 (see Table 2) are ill-defined since no specific and no common meaningful patterns can be identified. For classes 8 (f1-score = 64%), 13 (f1-score = 65%) and 14 (f1-score = 62%) we were not able to extract any type of knowledge. This is probably related to the fact that these classes share similar shot types.

For class 2 (f1-score = 98%) the issue is different. This class contains mainly foreground shots. Foreground shots usually describe camera movements in which the shot scale changes. However there is no difference if the shot scale changes from a close up to a medium shot or from a medium shot to a long shot. However this type of information impacts the viewer's perception. Generally speaking this class encapsulates sequences that contain only camera movements, but for now it is not possible to have a more fine-grained characterization. For this reason we have chosen to include this group of sequences in the undefined classes.

5.5   Misclassified Sequences

We have conducted a preliminary characterization also on the wrongly classified sequences. Here we have reported the analysis conducted on the 5 main misclassification errors, identified through the confusion matrix (not reported here to lack of space). Specifically, misclassified patterns belong to classes 5-0-6-8-14 predicted as 13-9-15-13-9, respectively. These errors mainly arise since the wrong predicted classes are ill-defined, thus the classifier is not so robust to clearly distinguish some editing patterns. A large and comprehensive data set should be considered so that the classifier can better model each specific group and perform the classification easily and correctly.

6   Discussion

This paper proposes a data-driven approach to effectively address editing pattern characterization. From our preliminary characterization, we were able to discern some interesting fine-grained editing patterns. Although these preliminary results are promising, there is room for improvements. Some research directions to be considered are discussed in the following: (1) A *more fine-grained shot type classification* modeling more cinematographic shot categories. This would mean having classes for camera movements, rather than the general class for foreground shots. Also, a distinction between close ups of a character's face and objects would allow for more insightful analysis. (2) A *characterization of single clip versus multiple clips*. It should categorize sequences obtained from a single clip from those composed of different video files. It could improve the Movie Lens overall performance. (3) *Dataset enrichment with video metadata*. Additional movie metadata, such as the starting and ending point of single scenes, could help the editing pattern characterization. (4) *Extension of the* Movie Lens *methodology with different ML algorithms*. The proposed approach could be enriched with other deep learning methods. For example the LSTM [10] networks or graph neural networks [6] could be integrated to take into account the temporal dimension in the analytics pipeline.

References

1. Argaw, D.M., Heilbron, F.C., Lee, J.Y., Woodson, M., Kweon, I.: The anatomy of video editing: a dataset and benchmark suite for AI-assisted video editing. ArXiv abs/2207.09812 (2022)
2. Bain, M., Nagrani, A., Brown, A., Zisserman, A.: Condensed movies: story based retrieval with contextual embeddings. CoRR abs/2005.04208 (2020). https://arxiv.org/abs/2005.04208
3. Bak, H.Y., Park, S.B.: Comparative study of movie shot classification based on semantic segmentation. Applied Sci. **10**, 3390 (2020). https://doi.org/10.3390/app10103390
4. Benini, S., Savardi, M., Balint, K., Kovacs, A., Signoroni, A.: On the influence of shot scale on film mood and narrative engagement in film viewers. IEEE Trans. Affect. Comput. **13**(2), 592–603 (2022). https://doi.org/10.1109/taffc.2019.2939251
5. Berthouzoz, F., Li, W., Agrawala, M.: Tools for placing cuts and transitions in interview video. ACM Trans. Graph. **31**, 1–8 (2012). https://doi.org/10.1145/2185520.2335418
6. Bloemheuvel, S., van den Hoogen, J., Jozinovic, D., Michelini, A., Atzmueller, M.: Multivariate time series regression with graph neural networks. CoRR abs/2201.00818 (2022). https://arxiv.org/abs/2201.00818

7. Chakraborty, S., Nagwani, N., Dey, L.: Performance comparison of incremental k-means and incremental dbscan algorithms. Int. J. Comput. Appl. **27**, 975–8887 (2011)
8. Haldar, R., Mukhopadhyay, D.: Levenshtein distance technique in dictionary lookup methods: an improved approach. Computing Research Repository - CORR (2011)
9. Hasan, M.A., Xu, M., He, X., Xu, C.: CAMHID: camera motion histogram descriptor and its application to cinematographic shot classification. IEEE Trans. Circuits Syst. Video Technol. **24**(10), 1682–1695 (2014). https://doi.org/10.1109/TCSVT. 2014.2345933
10. He, Z., Gao, S., Xiao, L., Liu, D., He, H., Barber, D.: Wider and deeper, cheaper and faster: tensorized LSTMS for sequence learning (2017)
11. Jani, K., Chaudhuri, M., Patel, H., Shah, M.: Machine learning in films: an approach towards automation in film censoring. J. Data Inf. Manage. **2**(1), 55–64 (2019). https://doi.org/10.1007/s42488-019-00016-9
12. Juang, B.H., Rabiner, L.: The segmental k-means algorithm for estimating parameters of hidden Markov models. IEEE Trans. Acoust. Speech Signal Process. **38**(9), 1639–1641 (1990)
13. Liberti, L., Lavor, C., Maculan, N., Mucherino, A.: Euclidean distance geometry and applications. SIAM Rev. **56**, 120875909 (2012). https://doi.org/10.1137/ 120875909
14. Matsuo, Y., Amano, M., Uehara, K.: Mining video editing rules in video streams, pp. 255–258 (2002). https://doi.org/10.1145/641007.641058
15. Mogadala, A., Kalimuthu, M., Klakow, D.: Trends in integration of vision and language research: a survey of tasks, datasets, and methods. J. Artif. Int. Res. **71**, 1183–1317 (2021). https://doi.org/10.1613/jair.1.11688
16. Murch, W.: In the Blink of an Eye. Silman-James Press (2001)
17. Nothelfer, C., DeLong, J., Cutting, J.E.: Shot structure in Hollywood film (2009)
18. Pardo, A., Heilbron, F.C., Alcázar, J.L., Thabet, A.K., Ghanem, B.: Learning to cut by watching movies. CoRR abs/2108.04294 (2021). https://arxiv.org/abs/ 2108.04294
19. Podlesnyy, S.: Towards data-driven automatic video editing (2019)
20. Qaisar, S.: Sentiment analysis of IMDB movie reviews using long short-term memory (2020). https://doi.org/10.1109/ICCIS49240.2020.9257657
21. Ramesh, A., et al.: Zero-shot text-to-image generation (2021). https://doi.org/10. 48550/ARXIV.2102.12092. https://arxiv.org/abs/2102.12092
22. Rao, A., Wang, J., Xu, L., Jiang, X., Huang, Q., Zhou, B., Lin, D.: A unified framework for shot type classification based on subject centric lens. CoRR abs/2008.03548 (2020). https://arxiv.org/abs/2008.03548
23. Ren, J., Shen, X., Lin, Z., Měch, R.: Best frame selection in a short video. In: 2020 IEEE Winter Conference on Applications of Computer Vision (WACV), pp. 3201–3210 (2020). https://doi.org/10.1109/WACV45572.2020.9093615
24. Savardi, M., Kovács, A.B., Signoroni, A., Benini, S.: Cinescale: A dataset of cinematic shot scale in movies. Data Brief **36**, 107002 (2021)
25. Savardi, M., Signoroni, A., Migliorati, P., Benini, S.: Shot scale analysis in movies by convolutional neural networks, pp. 2620–2624 (2018). https://doi.org/10.1109/ ICIP.2018.8451474
26. Simões, G., Wehrmann, J., Barros, R., Ruiz, D.: Movie genre classification with convolutional neural networks, pp. 259–266 (2016). https://doi.org/10.1109/ IJCNN.2016.7727207

27. Simonyan, K., Zisserman, A.: Very deep convolutional networks for large-scale image recognition. CoRR abs/1409.1556 (2014). http://arxiv.org/abs/1409.1556

28. Soe, T.H.: Automation in video editing: assisted workflows in video editing. In: AutomationXP@CHI (2021)

29. Svanera, M., Savardi, M., Signoroni, A., Kovács, A.B., Benini, S.: Who is the film's director? authorship recognition based on shot features. IEEE Multimedia **26**(4), 43–54 (2019). https://doi.org/10.1109/MMUL.2019.2940004

30. Vacchetti, B., Cerquitelli, T.: Cinematographic shot classification with deep ensemble learning. Electronics **11**(10), 1570 (2022)

31. Vacchetti, B., Cerquitelli, T., Antonino, R.: Cinematographic shot classification through deep learning. In: 2020 IEEE 44th Annual Computers, Software, and Applications Conference (COMPSAC), pp. 345–350 (2020). https://doi.org/10.1109/COMPSAC48688.2020.0-222

32. Walters, A.: Sentence classification. https://github.com/lettergram/sentence-classification

33. Wang, M., Yang, G.W., Hu, S.M., Yau, S.T., Shamir, A.: Write-a-video: computational video montage from themed text. ACM Trans. Graph. **38**(6) 1–13 (2019). https://doi.org/10.1145/3355089.3356520

34. Wu, H.Y., Santarra, T., Leece, M., Vargas, R., Jhala, A.: Joint attention for automated video editing. In: ACM International Conference on Interactive Media Experiences, pp. 55–64. IMX 2020, Association for Computing Machinery, New York, NY, USA (2020). https://doi.org/10.1145/3391614.3393656

35. Zhang, X., Li, Y., Han, Y., Wen, J.: AI video editing: a survey (2021). https://doi.org/10.20944/preprints202201.0016.v1

36. Zhou, H., Hermans, T., Karandikar, A., Rehg, J.: Movie genre classification via scene categorization, pp. 747–750 (2010). https://doi.org/10.1145/1873951.1874068

37. Zhou, J., Zhang, X.P.: Automatic identification of digital video based on shot-level sequence matching. In: Proceedings of the 13th Annual ACM International Conference on Multimedia, pp. 515–518. MULTIMEDIA 2005, Association for Computing Machinery, New York, NY, USA (2005). https://doi.org/10.1145/1101149.1101265

W17 - Visual Inductive Priors for Data-Efficient Deep Learning

W17 - Visual Inductive Priors for Data-Efficient Deep Learning

Save data by adding visual knowledge priors to Deep Learning! Data is fueling deep learning, yet it is costly to gather and to annotate. Training on massive datasets has a huge energy consumption adding to our carbon footprint. In addition, there are only a select few deep learning behemoths which have billions of data points and thousands of expensive deep learning hardware GPUs at their disposal. This workshop focuses on how to pre-wire deep networks with generic visual inductive innate knowledge structures, which allows us to incorporate hard won existing generic knowledge. Visual inductive priors are data efficient: what is built-in no longer has to be learned, saving valuable training data.

Excellent recent research has investigated data efficiency in deep networks by exploiting other data sources through unsupervised learning, re-using existing datasets, or synthesizing artificial training data. However, not enough attention has been given to overcoming the data dependency by adding prior knowledge to deep nets. As a consequence, all knowledge has to be (re-)learned implicitly from data, making deep networks hard to understand black boxes which are susceptible to dataset bias requiring huge datasets and compute resources. This workshop aims to remedy this gap by investigating how to flexibly pre-wire deep networks with generic visual innate knowledge structures, which allows us to incorporate hard won existing knowledge from physics such as light reflection or geometry.

October 2022

Jan C. van Gemert
Nergis Tömen
Ekin D. Cubuk
Robert-Jan Bruintjes
Attila Lengyel
Osman Semih Kayhan
Marcos Baptista Rios
Lorenzo Brigato

SKDCGN: Source-free Knowledge Distillation of Counterfactual Generative Networks Using cGANs

Sameer Ambekar$^{(\boxtimes)}$ ⓘ, Matteo Tafuro ⓘ, Ankit Ankit ⓘ,
Diego van der Mast ⓘ, Mark Alence ⓘ, and Christos Athanasiadis ⓘ

University of Amsterdam, Amsterdam, The Netherlands
ambekarsameer@gmail.com, tafuromatteo00@gmail.com,
ankitnitt1721@gmail.com, diego.vandermast@student.uva.nl,
mark.alence@gmail.com, c.athanasiadis@uva.nl

Abstract. With the usage of appropriate inductive biases, Counterfactual Generative Networks (CGNs) can generate novel images from random combinations of shape, texture, and background manifolds. These images can be utilized to train an invariant classifier, avoiding the wide spread problem of deep architectures learning spurious correlations rather than meaningful ones. As a consequence, out-of-domain robustness is improved. However, the CGN architecture comprises multiple over parameterized networks, namely BigGAN and U2-Net. Training these networks requires appropriate background knowledge and extensive computation. Since one does not always have access to the precise training details, nor do they always possess the necessary knowledge of counterfactuals, our work addresses the following question: Can we use the knowledge embedded in pre-trained CGNs to train a lower-capacity model, assuming black-box access (i.e., only access to the pretrained CGN model) to the components of the architecture? In this direction, we propose a novel work named SKDCGN that attempts knowledge transfer using Knowledge Distillation (KD). In our proposed architecture, each independent mechanism (shape, texture, background) is represented by a student 'TinyGAN' that learns from the pretrained teacher 'BigGAN'. We demonstrate the efficacy of the proposed method using state-of-the-art datasets such as ImageNet, and MNIST by using KD and appropriate loss functions. Moreover, as an additional contribution, our paper conducts a thorough study on the composition mechanism of the CGNs, to gain a better understanding of how each mechanism influences the classification accuracy of an invariant classifier. Code available at: https://github.com/ambekarsameer96/SKDCGN.

1 Introduction

Deep neural networks are prone to learning simple functions that fail to capture intricacies of data in higher-dimensional manifolds [1], which causes networks

---

S. Ambekar, M. Tafuro, A. Ankit, D. van der Mast, M. Alence—Equal contribution.

Supplementary Information The online version contains supplementary material available at https://doi.org/10.1007/978-3-031-25069-9_43.

ⓒ The Author(s), under exclusive license to Springer Nature Switzerland AG 2023
L. Karlinsky et al. (Eds.): ECCV 2022 Workshops, LNCS 13804, pp. 679–693, 2023.
https://doi.org/10.1007/978-3-031-25069-9_43

to struggle in generalizing to unseen data. In addition to spectral bias [1] and shortcut learning, which are properties inherent to neural networks [2], spurious learned correlations are also caused by biased datasets. To this end, Counterfactual Generative Networks (CGNs), proposed by Sauer and Geiger [3], have been shown to generate novel images that mitigate this effect. The authors expose the causal structure of image generation and split it into three Independent Mechanisms (IMs) (object shape, texture, and background), to generate synthetic and *counterfactual* images whereon an invariant classifier ensemble can be trained.

The CGN architecture comprises multiple over-parameterized networks, namely BigGANs [4] and U2-Nets [5], and its training procedure generally requires appropriate domain-specific expertise. Moreover, one does not always have access to the precise training details, nor do they necessarily possess the required knowledge of counterfactuals. Motivated by these observations, we propose *Source-free Knowledge Distillation of Counterfactual Generative Networks* (SKDCGN), which aims to use the knowledge embedded in a pre-trained CGN to train a lower capacity model, assuming black-box access (i.e., only inputs and outputs) to the components of the source model. More specifically, we harness the idea of Knowledge Distillation (KD) [6] to train a network comprising three (small) generative models, i.e. TinyGANs [7], each being responsible for a single independent mechanism. SKDCGN carries both practical and theoretical implications, and it is intended to:

1. Obtain a lightweight version of the CGN, reducing its computational cost and memory footprint. This is meant to (i) ease the generation of counterfactual datasets and hence encourage the development of robust and invariant classifiers, as well as (ii) potentially allowing the deployment of the model on resource-constrained devices.
2. Explore whether we can *learn* from a fully trained CGN and distill it to a less parameterized network, assuming that we do not have access to the training process of the model.

Along the lines of the original paper, we demonstrate the ability of our model to generate counterfactual images on ImageNet-1k [8] and Double-Colored MNIST [3]. Furthermore, we compare our outputs to [3] and a simple baseline in terms of out-of-distribution robustness on the original classification task. As an additional contribution, we conduct a study on the shape IM of the CGN.

The paper is organized as follows: firstly, we present a brief literature survey in Sect. 2; next in Sect. 3 the SKDCGN is dissected; Sect. 4 presents the experimental setup and the empirical results, which are finally discussed in Sect. 5.

2   Related Work

This section introduces the fundamental concepts and the related works that we use as a base for our SKDCGN.

Counterfactual Generative Networks. The main idea of CGNs [3] has already been introduced in Sect. 1. Nonetheless, to aid the understanding of our method to readers that are not familiar with the CGN architecture, we summarize its salient components in this paragraph and also provide the network diagram in Appendix Section A.1 Fig. 1. The CGN consists of 4 backbones: (i) the part of the network responsible for the shape mechanism, those responsible for (ii) texture and (iii) background, and a (iv) composition mechanism that combines the previous three using a deterministic function. Given a noise vector **u** (sampled from a spherical Gaussian) and a label y (drawn uniformly from the set of possible labels y) as input, (i) the shape is obtained from a BigGAN-deep-256 [4], whose output is subsequently passed through a U2-Net [5] to obtain a binary mask of the object shape. The (ii) texture and (iii) background are obtained similarly, but the BigGAN's output does not require to be segmented by the U2-Net. Finally, the (iv) composition mechanism outputs the final counterfactual image $\mathbf{x}_{gen}$ using the following analytical function:

$$\mathbf{x}_{gen} = C(\mathbf{m}, \mathbf{f}, \mathbf{b}) = \mathbf{m} \odot \mathbf{f} + (1 - \mathbf{m}) \odot \mathbf{b}, \tag{1}$$

where **m** is the shape mask, **f** is the foreground (or texture), **b** is the background and $\odot$ denotes element-wise multiplication.

More recently, [9] devises an approach that learns a latent transformation that generates visual CFs automatically by steering in the latent space of generative models. Additionally, [10] uses a deep model inversion approach that provides counterfactual explanations by examining the area of an image.

Knowledge Distillation. [11] firstly proposed to transfer the knowledge of a pre-trained cumbersome network (referred to as the *teacher*) to a smaller model (the *student*). This is possible because networks frequently learn low-frequency functions among other things, indicating that the learning capacity of the big network is not being utilized fully [1,2]. Traditional KD approaches (often referred to as *black-box*) simply use the outputs of the large deep model as the teacher knowledge, but other variants have made use of activation, neurons or features of intermediate layers as the knowledge to guide the learning process [12,13]. Existing methods like [7] are also making use of Knowledge distillation for the task of image generation. Our work is similar to this, however, they transfer the knowledge of BigGAN trained on ImageNet dataset to a TinyGAN. In contrast, in our work, we transfer not just the knowledge of image generation but also the task of counterfactual generation from a BigGAN to a TinyGAN.

Distilling GANs using KD. Given its high effectiveness for model compression, KD has been widely used in different fields, including visual recognition and classification, speech recognition, natural language processing (NLP), and recommendation systems [14]. However, it is less studied for image generation. [15] firstly applied KD to GANs. However, our project differs from theirs as they use *unconditional* image generation, less general (DCGAN [16]) architectures and they do not assume a black-box generator. Our setting is much more similar to that of [7], where a BigGAN is distilled to a network with 16× fewer parameters, assuming no access to the teacher's training procedure or parameters. Considering its competitive performance, we use the proposed architecture (TinyGAN) as the student model and use a modified version of their loss function (further details in Sect. 3.1) to optimize our network.

Source-free: We term our method as Source-free since we do not have access to the source data, source training details, procedure, and any knowledge about the counterfactuals, etc., but only have access to trained source models. This method is similar to methods such as [17,18]. With large diffusion models like Imagen [19] and DALL·E 2 [20] where the training process is usually extremely expensive in terms of computation, lack precise details about training them and often not reproducible by academic groups, we often have access to pretrained models. These can be used to transfer knowledge to a smaller network, and perform the same task with model of lower capacity.

3   Approach

This section dives into the details of the SKDCGN architecture, focusing on the training and inference phases separately for ImageNet-1k and MNIST. In addition, we discuss the loss functions that were employed for Knowledge Distillation.

3.1   SKDCGN

Although transferring the knowledge of an entire CGN into a single generative model could drastically reduce the number of parameters, this strategy would compromise the whole purpose of CGNs, i.e. disentangling the three mechanisms and having control over each of them. Therefore, we opt to train a generative model for each individual component. As shown in the architecture diagram (Fig. 1), we treat each IM backbone as a black-box teacher and aim to mimic its output by training a corresponding TinyGAN student. Note that this implies that in the case of the shape mechanism, a single generative model learns to mimic both the BigGAN and the U2-Net. We believe a TinyGAN should be capable of learning binary masks directly, removing the need for the U2-Net and reducing the model size even further. During inference, the outputs of the three students are combined into a final counterfactual image using the composition function defined in Eq. 1.

Training: Distilling the Knowledge of IMs. To train SKDCGN, we utilize each IM backbone from the CGN architecture as a black-box teacher for the student network, as visualized in the training section of Fig. 1 (the backbones are BigGAN + U2-Net for *shape*, BigGAN for *texture*, and BigGAN for *background*). As introduced in the Related work section, [7] proposed an effective KD framework for compressing BigGANs. As the IMs in CGNs rely on BigGANs, we utilize their proposed student architecture. For completeness, the details of the student architecture are reported in Appendix Sect. 1.2 Fig. 2.

We base our training objective on the loss function proposed by [7]. Our full objective comprises multiple terms: (i) a pixel-wise distillation loss, (ii) an adversarial distillation loss, (iii) a feature-level distillation loss, and (iv) KL

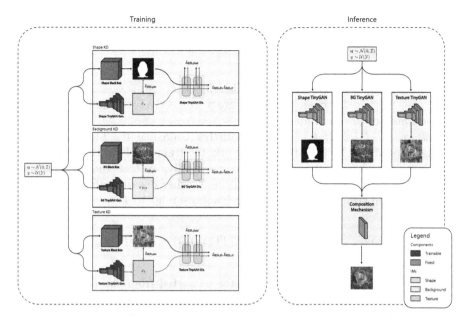

Fig. 1. *Architecture of the SKDCGN.* During training, each independent mechanism serves as a black-box teacher model to train a corresponding student model. During inference, the outputs of the three trained TinyGANs are combined using a Composition Mechanism that returns the final counterfactual image.

Divergence. In addition to introducing KL Divergence, we deviate from the original TinyGAN training objective by omitting the term that allows the model to learn from real images of the ImageNet dataset. This would inevitably compromise the quality of the generated counterfactuals. KL Divergence leads to entropy minimization between the teacher and student, which is why we propose its usage.

The individual loss terms are dissected below as from [7]:

1. *Pixel-wise Distillation Loss*: To imitate the functionality of BigGAN for scaling generation to high-resolution, high-fidelity images, we minimize the pixel-level distance (L1) between the images generated by BigGAN and TinyGAN given the same input:

$$\mathcal{L}_{\mathrm{KD\_pix}} = \mathbb{E}_{z \sim p(z), y \sim q(y)}[\|T(z, y) - S(z, y)\|_1] \qquad (2)$$

where T represents the Teacher network, S represents the Student network, z is a latent variable drawn from the truncated normal distribution $p(z)$, and y is the class label sampled from some categorical distribution $q(y)$.

2. *Adversarial Distillation Loss*: To promote sharper outputs, an adversarial loss is incorporated to make the outputs of S indistinguishable from those of T. It includes a loss for the generator (Eq. 3) and one for the discriminator (Eq. 4):

$$\mathcal{L}_{\text{KD\_G}} = - \mathbb{E}_{z,y}[D(S(z,y),y)] \tag{3}$$

$$\mathcal{L}_{\text{KD\_D}} = - \mathbb{E}_{z,y}[max(0, 1 - D(T(z,y),y)) + max(0, 1 - D(S(z,y),y))], \tag{4}$$

where z is the noise vector, y is the class label, $T(z,y)$ is the image generated by the Teacher T, while G and D are – respectively – the generator and discriminator of the Student S.

3. *Feature Level Distillation Loss*: To further overcome the blurriness in the images produced by the Student network, the training objective also includes a feature-level distillation loss. More specifically, we take the features computed at each convolutional layer in the Teacher discriminator, and with a loss function stimulate S to generate images similar to T:

$$\mathcal{L}_{\text{KD\_feat}} = \mathbb{E}_{z,y}\left[\sum_i \alpha_i \|D_i(T(z,y),y) - D_i(S(z,y),y)\|_1\right] \tag{5}$$

where D_i represents the feature vector extracted from the ith layer of the discriminator and the corresponding weights are given by α_i.

4. *KL Divergence*: L1 alone cannot reduce the entropy between the teacher and target. To improve the proposed method, we use KL Divergence in a similar manner to [21] for the task of knowledge distillation between real images drawn from source $P(x)$ and target images $Q(x)$. The

$$\mathcal{D}_{\text{KL}}(P\|Q) = \sum_{x\in\mathcal{X}} P(x) \log\left(\frac{P(x)}{Q(x)}\right) \tag{6}$$

$$\mathcal{L}_{\text{KL}} = \sum_{x\in X} -p_x^t \log p_x^s + p_x^t \log p_x^t \tag{7}$$

where x is the class label and p contains the output softmax probabilities of the Generator G divided by the temperature t. The latter was experimentally set to 1.

To sum up, the student's generator (G) and discriminator (D) are respectively optimized using the following objectives:

$$\mathcal{L}_{\text{G}} = \mathcal{L}_{\text{KD\_feat}} + \lambda_1 \mathcal{L}_{\text{KD\_pix}} + \lambda_2 \mathcal{L}_{\text{KD\_G}} (+ \mathcal{L}_{\text{KL}}) \tag{8}$$

$$\mathcal{L}_{\text{D}} = \mathcal{L}_{\text{KD\_D}} \tag{9}$$

where λ_1 and λ_2 are the regularization terms mentioned in [7], and the KL divergence term $(\mathcal{L}_{\text{KL}})$ is only used in the enhanced version of SKDCGN.

Implementing the SKDCGN architecture requires training a TinyGAN for each Independent Mechanism of the CGN (see Fig. 1). The KD training procedure, however, requires training data. Hence prior to training, 1000 images per class (totalling 1 million samples) are generated using the IM backbones extracted from the pre-trained CGN (as provided by Sauer and Geiger [3]).

Finally, note that the original CGN architecture (illustrated in Appendix Sect. 1.1, Fig. 1) comprises another BigGAN trained on ImageNet-1k. It is unrelated to the three Independent Mechanisms and provides primary training supervision via reconstruction loss. We discard this component of the architecture for two main reasons: we do not have a dataset of counterfactuals whereon a GAN can be trained; we argue that this additional knowledge is already embedded in the backbones of a pre-trained CGN.

Inference: Generating Counterfactuals. Once the three student networks are trained, their outputs are combined during inference akin to [3] using the analytical function of Eq. 1. Since the composition function is deterministic, we devise inference as a separate task to training.

4   Experiments and Results

This section defines our experimental setup, then proceeds to present the results. First, we test SKDCGN – as defined in the Approach section – on both ImageNet-1k and MNIST (Sect. 4.3), and based on the observed findings we make some changes to the proposed architecture to improve the quality of the results (Sect. 4.4). Due to computational constraints we test these improvements on a smaller dataset, namely the double-colored variant of MNIST [22]. Finally, as an additional contribution, we conduct a thorough study on the composition mechanism, to gain a better understanding of how each mechanism influences the classification accuracy of an invariant classifier. We present the results of such a study in Sect. 4.5.

4.1   Datasets

ImageNet-1k. The ImageNet-1k ILSVRC dataset [8] contains 1,000 classes, with each class consisting of 1.2 million training images, 50,000 validation and 100,000 test images. Images were resized to 256×256 to maintain consistent experiments and to allow direct comparisons with the original results of [3].

Double-colored MNIST. We use the *double-colored* MNIST dataset proposed by Sauer and Geiger in the original CGN paper [3]. This is a variant of the MNIST dataset where both the digits and the background are independently colored. It consists of 60,000 28×28 images of the 10 digits, along with a test set of 10,000 images.

4.2   Baseline Model: CGN with Generator Replaced by TinyGAN Generator

The SKDCGN is compared with a modified version of the original CGN architecture, where each BigGAN has been replaced by the generator model of a

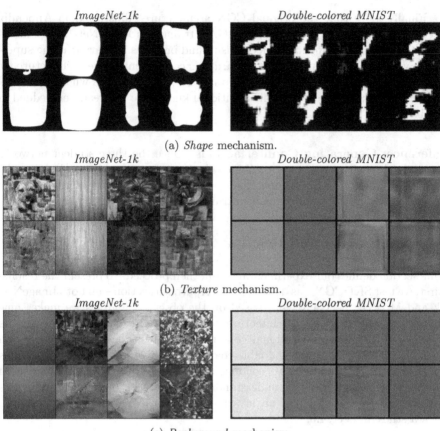

(a) *Shape* mechanism.

(b) *Texture* mechanism.

(c) *Background* mechanism.

Fig. 2. A comparison of images (on both ImageNet-1k and double-colored MNIST) generated by the CGN backbones and those generated by the corresponding SKDCGN's TinyGAN (given the same input), for each independent mechanism.

TinyGAN. Training this baseline using the procedure described by [7], omitting KD, allows for rigorous comparisons that emphasize the effectiveness of the knowledge distillation process. Further training details are provided in Sect. 3.1 of the Appendix.

4.3   Results of SKDCGN

The proposed model was firstly trained and tested on ImageNet-1k. To further validate our method, we repeated the training procedure on MNIST.

The qualitative results are collected in Fig. 2 and demonstrate that Tiny-GANs can closely approximate the output of each IM. While this is true for both datasets, the effectiveness of our method is especially visible in the case of MNIST. It is likely the case that the reduced capacity of the TinyGANs (compared to the original CGN backbones) is sufficient to decently model the under-lying data distribution. ImageNet-1k, on the other hand, reveals more apparent (though still acceptable) discrepancies between the images, especially for the *texture* IM.

However, careful and extensive experiments revealed that the three Tiny-GANs could not generalize when random noise was given to the generator, i.e., they could not produce results beyond the test set. This might be due to a number of reasons. First, the compromised generalization capabilities of each IM's TinyGAN could be caused by their reduced network capacity. Furthermore, each TinyGAN was trained on all 1000 classes of ImageNet-1K, as opposed to Chang and Lu's choice of limiting the training data to the 398 animal labels [7]. Finally, we generate the test samples using the test noise instead of random noise, since we hypothesize that the student networks only learn the manifolds that the teacher networks have been trained on. Additional experiments are required to analyze whether samples generated using random noise are found along the same manifold; unfortunately, we were hindered by the limited time frame allocated for this project, hence we leave this question open for future works.

4.4   Improving the SKDCGN Model

The results presented in the previous section reveal that the outputs are noisy and ambiguous in nature when knowledge distillation is performed using the pre-trained models provided by Sauer and Geiger [3] (note the artifacts in the SKDCGN's outputs of Fig. 2, especially those trained on ImageNet-1k). This statement was supported by an interesting yet unexpected result of the study on the composition mechanism (refer to Sect. 4.5): it was observed that modifying Eq. 1 such that the shape mask **m** is multiplied with a weight factor of 0.75 (i.e., setting the transparency of the shape mask to 75%), yielded an accuracy increase of the CGN's invariant classifier. The findings of this experiment – conducted on the double-colored MNIST dataset – suggest that the mask component is noisy in nature, leading to ambiguities in the decision boundaries during the classification of several digits.

In light of this new hypothesis, we attempt to use the *Kullback-Leibler* (KL) divergence to improve the visual quality of the outputs[1]. Since KL leads to entropy minimization between the teacher and student networks, we deem such a technique adequate for the task at hand. Moreover, the choice of using KL was encouraged by the work of Asano and Saeed [21], which proved the suitability of the measure in this context. Concretely, the KL Divergence loss (as defined in Eq. 7) was included in the overall generator loss $\mathcal{L}_G$ as seen in Eq. 3.

First, the modified SKDCGN was tested on the double-colored MNIST dataset. As depicted in Fig. 3, the introduction of KL divergence improves SKD-CGN's visual fidelity of both *background* and *texture* IMs, while the quality of

---

[1] It is noteworthy that other techniques were tested in the attempt to improve the visual quality of the results. Although they did not prove to be as beneficial, they are described in Sect. 4 of the Appendix (Supplementary Material).

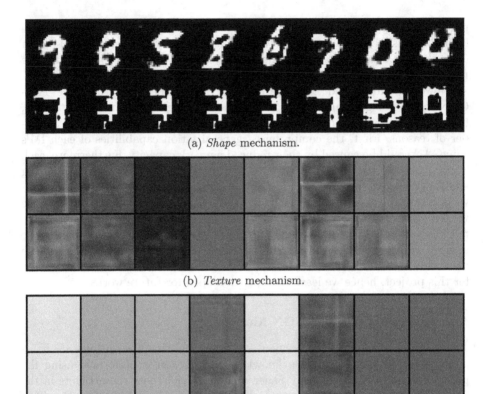

(a) *Shape* mechanism.

(b) *Texture* mechanism.

(c) *Background* mechanism.

Fig. 3. A comparison of double-colored MNIST images generated by the CGN backbones and those generated by the corresponding SKDCGN's TinyGAN (given the same input) for each IM. Here, SKDCGN was tuned such that KL divergence is minimized between the teacher and student networks, and the L1 loss is multiplied with the activation of every layer.

the *shape* masks seems to diminish after a few epochs. Contrarily, this approach appeared to be beneficial for the shape mechanism too, in the context of ImageNet-1k. The shape masks resulted more natural and consistent since the first epoch, whereas the absence of KL yielded noisy masks even at a later stage of training (refer to Fig. 4).

4.5   Additional Results: Study of the Shape IM

As an additional contribution, we conduct a thorough study on the composition mechanism, to gain a better understanding of how the mechanisms influence the classification accuracy of an invariant classifier (i.e., a classifier that predicts whether an image is CGN-generated or real). Due to the limited time at our disposal, we focused on the mechanism that we deem most important in the

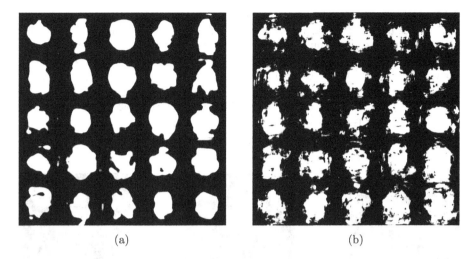

(a)   (b)

Fig. 4. (a) Shape masks obtained after the *first* epoch of SKDCGN training on ImageNet-1k, using KL divergence. (b) Shape masks obtained after the 23rd epoch of SKDCGN training on ImageNet-1k, *without* KL divergence. Evidently, KL enhances the quality of the masks from the first epoch, whereas its absence compromises the results even at a later stage of training.

decision-making of such a classifier, namely the *shape*. To evaluate the effects of the shape IM we trained several (original) CGN models on the double-colored MNIST dataset; we tuned the resulting shape masks prior to the counterfactual image generation (governed by the composition mechanism of Eq. 1) and used the generated images to train an invariant classifier. More specifically, we experimented with (i) the addition of Gaussian noise in the shape mask, (ii) random rotation of the mask, and (iii) multiplying the mask $\mathbf{m}$ in the composition mechanism (Eq. 1) with a factor smaller than 1 (or in other words, lowering the opacity of the shape mask). A transparency of 75% (hence a weight factor of 0.75) was experimentally found to be most beneficial for the accuracy of the classifier.

The influence of the three transformations on the invariant classifier is quantified – in terms of accuracy – in Table 1; sample shape masks generated from each transformation are displayed in Fig. 5. It is apparent from the test accuracy values that Gaussian noise and random rotations do not lead to any remarkable performance of the classifier but, contrarily, degrade its accuracy to values below 15%. This is most likely the result of overfitting on the training set, as supported by the *train* accuracy values. On the other hand, lowering the opacity of the mask substantially boosts the test accuracy, improving the previous results by a factor of 4× (circa). It is noteworthy that the masks obtained using the

Table 1. Results of the invariant classifier for the analysis of the shape IM. The classifier has been trained to predict whether images are CGN-generated or real. The training examples contain counterfactuals whose shape mechanism has been tuned with one of the three transformations indicated in the table (noise, rotation, transparency – refer to Sec.4.5 for further details).

	Noise	Rotation	Transparency
Train Accuracy	99.9	99.1	94.7
Test Accuracy	14.96	13.51	**58.86**

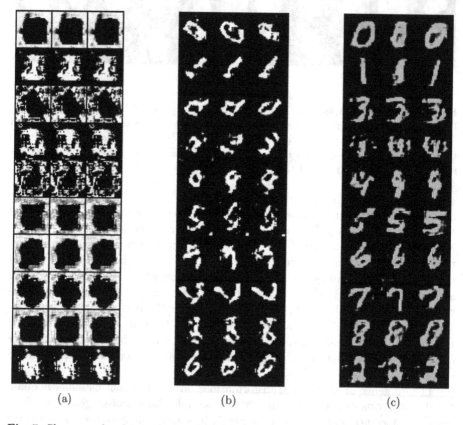

(a)   (b)   (c)

Fig. 5. Shape masks obtained after (a) addition of Gaussian random noise, (b) application of random rotation and (c) decrease of the mask opacity (i.e., lowering its transparency to 75%).

transparency adjustment are more akin to those achieved using regular CGNs (see Fig. 5). The other transformations, instead, result in mask shapes that are particularly different. As such, they can potentially be used to make classifiers more robust when mixed with regular data during training. Because this is an extensive topic, we believe it warrants further research.

5   Discussion and Conclusion

With the prevalence of heavily parameterized architectures such as BigGANs, and with the advent of limited-access models like the trending DALL·E 2, source-free compression becomes a growing necessity. In this paper we explored the possibility to obtain a lightweight version of the CGN network, assuming that we do not have access to the training process of the model. More specifically, we treat the backbone of each independent mechanism (shape, texture and background) as a black-box, then use KD to transfer the knowledge of the pre-trained cumbersome networks to simple TinyGANs.

SKDCGN achieves a remarkable compression of the overall network: it models the shape mechanism – initially controlled by a BigGAN (55.9M parameters) and a U2-Net (44M parameters) – using a single TinyGAN (6.4M parameters); similarly, it replaces the BigGANs responsible for the texture and background IMs with TinyGANs, and discards the forth BigGAN of the original CGN network that provides primary training supervision via reconstruction loss. This translates into four BigGANs and one U2-net (55.9M×4 + 44M parameters, totalling 267.6M) being replaced with three simple TinyGANs (6.4M parameters each, meaning 19.2M parameters in total).

Despite the significant compression, we demonstrate the ability of our model to generate counterfactual images on ImageNet-1k and double-colored MNIST datasets (see Fig. 2). When trained on the latter, SKDCGN's network capacity was proven to be sufficient to model the simple data distribution. If trained on the former, the proposed method exhibited remarkable ability in mimicking the original shape and background generations, while the texture mechanism suffered more from the reduction of size. This finding reveals great potential for future works that would attempt to tune the distillation (and hence enhance the synthesis) of the texture images, for instance by including data augmentation in the training procedure.

Given the obtained results, we attemptedly limit the presence of noisy and ambiguous artifacts by minimizing the entropy between the teacher and student networks. We introduce a new measure in the knowledge distillation loss, i.e. KL divergence, which we find to enhance the visual quality results of some IMs for both Imagenet-1k and MNIST.

Finally, we conduct a study on the composition mechanism to gain a better understanding of how the *shape* IM influences the classification accuracy of an invariant classifier. Though other adjustments were tested, giving a lower weight to the shape mask **m** seemingly boosts the classifier performance.

6   Future Work

To conclude, the experimental findings of SKDCGN prove that, upon the usage of Knowledge Distillation, one can transfer the capacity/ability of a cumbersome network to a lower-capacity model while still maintaining competitive performances. Although this paper unveils its potential, SKDCGN requires further research that we encourage other researchers to undertake. In addition to the suggestions offered throughout the sections, possible avenues of research include and are not limited to: improving the image generation process by using higher-order activation functions, since the utilized datasets consist of rich image data; improving the teacher-student architecture by introducing additional loss functions; using a learnable, neural network-based composition function instead of an analytical expression.

Acknowledgments. We would like to express our sincere gratitude to Prof. dr. Efstratios Gavves and Prof. Wilker Aziz for effectively organizing the *Deep Learning II* course at the University of Amsterdam, which is the main reason this paper exists. We are thankful to our supervisor, Christos Athanasiadis, for his precious guidance throughout the project. Finally, we also thank the former Program Director of the MSc. Artificial Intelligence, Prof. dr. Cees G.M. Snoek, and the current Program Manager, Prof. dr. Evangelos Kanoulas, for effectively conducting the Master's program in Artificial Intelligence at the University of Amsterdam.

References

1. Fridovich-Keil, S., Lopes, R.G., Roelofs, R.: Spectral bias in practice: the role of function frequency in generalization. CoRR, abs/2110.02424 (2021)
2. Geirhos, R., et al.: Shortcut learning in deep neural networks. CoRR, abs/2004.07780 (2020)
3. Sauer, A., Geiger, A.: Counterfactual generative networks. CoRR, abs/2101.06046 (2021)
4. Brock, A., Donahue, J., Simonyan, K.: Large scale GAN training for high fidelity natural image synthesis (2019)
5. Qin, X., Zhang, Z.V., Huang, C., Dehghan, M., Zaïane, O.R., Jägersand, M.: U^2-net: going deeper with nested u-structure for salient object detection. CoRR, abs/2005.09007 (2020)
6. Beyer, L., Zhai, X., Royer, A., Markeeva, L., Anil, R., Kolesnikov, A.: Knowledge distillation: a good teacher is patient and consistent. CoRR, abs/2106.05237 (2021)
7. Chang, T.-Y., Lu, C.-J.: Tinygan: distilling biggan for conditional image generation. CoRR, abs/2009.13829 (2020)
8. Deng, J., Dong, W., Socher, R., Li, L.-J., Li, K., Fei-Fei, L.: Imagenet: a large-scale hierarchical image database. In: 2009 IEEE Conference on Computer Vision and Pattern Recognition, pp. 248–255 (2009)
9. Khorram, S., Fuxin, L.: Cycle-consistent counterfactuals by latent transformations (2022)
10. Thiagarajan, J.J., Narayanaswamy, V.S., Rajan, D., Liang, J., Chaudhari, A., Spanias, A.: Designing counterfactual generators using deep model inversion. CoRR, abs/2109.14274 (2021)
11. Hinton, G., Vinyals, O., Dean, J.: Distilling the knowledge in a neural network. In: NIPS Deep Learning and Representation Learning Workshop (2015)
12. Romero, A., Ballas, N., Kahou, S.E., Chassang, A., Gatta, C., Bengio, Y.: Hints for thin deep nets, Fitnets (2014)
13. Ahn, S., Hu, S.X., Damianou, A., Lawrence, N.D., Dai, Z.: Variational information distillation for knowledge transfer. In: 2019 IEEE/CVF Conference on Computer Vision and Pattern Recognition (CVPR), pp. 9155–9163 (2019)

14. Gou, J., Yu, B., Maybank, S.J., Tao, D.: Knowledge distillation: a survey. CoRR, abs/2006.05525 (2020)
15. Aguinaldo, A., Chiang, P.-Y., Gain, A., Patil, A., Pearson, K., Feizi, S.: Compressing GANs using knowledge distillation. CoRR, abs/1902.00159 (2019)
16. Radford, A., Metz, L., Chintala, S.: Unsupervised representation learning with deep convolutional generative adversarial networks (2015)
17. Yang, S., Wang, Y., van de Weijer, J., Herranz, L., Jui, S.: Generalized source-free domain adaptation. In: Proceedings of the IEEE/CVF International Conference on Computer Vision, pp. 8978–8987 (2021)
18. Ding, N., Xu, Y., Tang, Y., Xu, C., Wang, Y., Tao, D.: Source-free domain adaptation via distribution estimation. arXiv preprint arXiv:2204.11257 (2022)
19. Saharia, C., et al.: Photorealistic text-to-image diffusion models with deep language understanding. arXiv preprint arXiv:2205.11487 (2022)
20. Marcus, G., Davis, E., Aaronson, S.: A very preliminary analysis of dall-e 2 (2022)
21. Asano, Y.M., Saeed, A.: Extrapolating from a single image to a thousand classes using distillation. arXiv preprint arXiv:2112.00725 (2021)
22. Lecun, Y., Bottou, L., Bengio, Y., Haffner, P.: Gradient-based learning applied to document recognition. Proc. IEEE **86**(11), 2278–2324 (1998)

C-3PO: Towards Rotation Equivariant Feature Detection and Description

Piyush Bagad[✉], Floor Eijkelboom, Mark Fokkema, Danilo de Goede,
Paul Hilders, and Miltiadis Kofinas

University of Amsterdam, Amsterdam, The Netherlands
piyush.bagad@student.uva.nl

Abstract. Despite the recent advances in local feature matching, dealing with affine distortions remains a major challenge. While state-of-the-art methods have shown to perform well in the absence of rotation perturbations, some computer vision applications, such as object tracking and image stitching, require keypoint extraction methods that maintain high performance regardless of the image orientation. Current approaches perform extensive data augmentation to artificially acquire a degree of rotation equivariance. However, this does not only induce redundancy in the learned feature representations, but also does not provide any geometric guarantees. To address this issue, this work explores an alternative approach that instead instills rotation equivariance inside the model itself. Leveraging recent advances in group equivariant deep learning, we propose C-3PO, a family of feature detection-and-description models based on steerable group convolutions. We evaluate our method against prior work, and find that it outperforms its non-equivariant counterparts for most rotation perturbations. However, presumably due to the task's inherent sensitivity to interpolation artifacts, extending a discrete rotation equivariant model to a continuous variant provides only marginal performance gains.

Keywords: Feature detection and description · Local feature matching · Rotation equivariance · Steerable CNNs

1 Introduction

Image correspondence, *i.e.* determining which parts of one image correspond to which parts of another image, is a fundamental problem in computer vision[1]. Typically, correspondence is framed as detecting and describing similar points

---

P. Bagad, F. Eijkelboom, M. Fokkema, D. de Goede and P. Hilders—Roughly equal contribution.

[1] As once said by Takeo Kanade when asked about the three most fundamental problems of computer vision: 'Correspondence, correspondence, and correspondence!'.

Supplementary Information The online version contains supplementary material available at https://doi.org/10.1007/978-3-031-25069-9_44.

© The Author(s), under exclusive license to Springer Nature Switzerland AG 2023
L. Karlinsky et al. (Eds.): ECCV 2022 Workshops, LNCS 13804, pp. 694–705, 2023.
https://doi.org/10.1007/978-3-031-25069-9_44

of interest (*keypoints*) in the given pair of images. This forms a foundation of several computer vision applications such as 3D reconstruction [11], structure-from-motion [24], and visual localization [23].

Classical methods, such as Scale-Invariant Feature Transform (SIFT) [14], address this problem by incorporating handcrafted heuristics within the model to identify robust keypoints. While these methods have shown to work well, they have been increasingly outperformed by recent deep learning based approaches that instead *learn* to recognise suitable keypoints [9,19,22]. However, these approaches perform extensive data augmentation to address affine distortions such as rotations, which does not provide any geometric guarantees such as rotation equivariance.

In line with the recent developments in the field of geometric deep learning (as popularised by [4]), we aim to improve local feature matching approaches by introducing geometric priors to a deep feature detection-and-description model. Specifically, we introduce rotation equivariance to the model using steerable group CNNs, where we consider a discrete rotation group C_n and continuous rotation group SO(2) using steerable CNNs. In summary, we make the following contributions:

- We propose C-3PO (Correspondence, Correspondence, Correspondence between Points in different Orientations), a family of rotation equivariant deep feature detection-and-description models, and show the implications of using group convolutions on architectural design and computational costs.
- We perform a comprehensive empirical analysis to highlight the viability of group convolutions for image correspondence. We evaluate these results through both a quantitative and qualitative analysis to study the effect of using equivariant layers compared to a non-equivariant baseline.
- We investigate the impact of both discrete and continuous group convolutions on local feature matching tasks to study the matching performance across several flavours of rotation equivariance.

2   Background

2.1   Theory

Group Theory. A group $(G, \cdot)$ is a pair of a set G and binary operation $\cdot : G \times G \to G$ under which the set is closed, that satisfies the group axioms (associativity, identity, and inverse, see Appendix A.1. Groups are used to describe the symmetries of objects, such as the changes we can apply to an image without changing its semantics. A set X is called a G-space if it is equipped with a group action based on G (Appendix A.1). Two important groups in this research are the cyclic group C_n of all rotations of $\frac{1}{n} \cdot 360°$, and the 2-dimensional special orthogonal group SO(2) of all planar rotations.

For two G-spaces X, Y, a function $f : X \to Y$ is called *equivariant* to G if applying a symmetry transformation $g \in G$ and then computing the function f produces the same result as computing the function f and then applying the transformation g. Similarly, such a function is called *invariant* to G if transforming an element does not change its function value. Formally, the two can be written as

$$\text{Equivariance: } f(g.x) = g.(f(x)), \qquad (1)$$

$$\text{Invariance: } f(g.x) = f(x). \qquad (2)$$

Group CNNs. Standard convolutional networks are equivariant with respect to translations, but not to other transformations such as rotations [7]. Group convolutional networks (G-CNNs) [20] are a common tool to introduce such equivariances by not only determining a response for each translated pixel, but also for each rotated pixel as described by G:

$$[k \star_G f](g) = \int_{x \in \Omega} k(g^{-1}x) f(x) dx, \qquad (3)$$

for some kernel k and function f defined on domain Ω (*i.e.* the image). In the case of a 32×32 image, introducing 4 rotations gives a tensor response of shape $32 \times 32 \times 4$, thereby 'lifting' the image to a higher domain. After a series of such convolutions, where each next convolution integrates over the entire previous group, it is common to perform a rotation-invariant pooling. In the case of rotations, the average over all rotation maps is taken to find a final representation of our image invariant to the initial rotation.

Steerable Group CNNs. When using G-CNNs, the number of responses is expanded by convolving over a larger domain, for example a roto-translation group, rather than just a translation group. An alternative to obtaining more responses on the image is extending the codomain, *i.e.* assigning higher dimensional responses to each input feature corresponding to different rotations. To ensure that our model is still equivariant to rotations, these feature vectors need to be transformed with the image. This property is referred to as *steerability*, and the resulting feature vector is referred to as a *fiber*. To assure that a convolution is steerable, the kernel k needs to satisfy the *steerability constraint* (see Appendix A.2). Similar to the standard group convolution, the arising fiber for some pixel describes the response for each rotation in the rotation group. Using this, a steerable CNN is obtained by connecting a series of steerable group convolutions together.

Continuous Steerability. Using a feature for each rotation comes with the immediate disadvantage of only being able to consider a finite number of filters. By realising that assigning a feature value to each possible rotation implicitly defines a signal on the circle S^1, we can leverage circular harmonics and write that signal as an infinite series of complex exponents. Using a finite subset of our Fourier coefficients, the signal $s(x)$ can be approximated as

$$s(x) \approx \sum_{n=-N}^{N} a_n e^{inx}, \qquad (4)$$

for some $N \in \mathbb{N}$. This allows to indirectly describe the responses for each input pixel for all rotations using a finite set of Fourier coefficients per point, solving the initial problem.

2.2 Related Work

Deep Feature Matching. Feature matching based on robust interest point detection and local feature description is at the core of several computer vision algorithms such as image stitching [1], visual localisation [18], structure-from-motion [25,28], and 3D reconstruction [30]. Classical approaches such as SIFT [14], ORB [21], and SURF [3] devise handcrafted features to achieve invariance to local geometric and photometric transformations in a two-stage manner: detection and description. Modern CNN-powered approaches are based on jointly learning keypoint locations and their descriptions. For instance, R2D2 [19] predicts reliability and repeatability maps to extract not only repeating keypoints, but also those that are reliable for downstream matching. Several recent approaches also leverage transformer-based attention models for feature matching [12,22,26]. However, these deep neural methods are not robust to rotations. One solution for ensuring rotation equivariance is using data augmentation, as is done in methods such as RoRD [15].

Rotation Equivariant Feature Matching. Unfortunately, data augmentation leads to redundancy, since the network has to learn to be robust to each augmented transform. Alternatively, we could introduce geometric priors to the model architecture itself as is done in rotation equivariant CNNs [6,8]. While rotation equivariant feature extraction has not yet received a lot of attention within the field of deep feature matching, some papers have very recently started to pick up on this concept. Notably, [5] replaces the CNN backbone of LoFTR [26] with steerable CNNs based on discrete groups. While their work is primarily focused on extending a transformer-based architecture to be rotation equivariant, our work instead looks into the benefits of applying steerable group convolutions to a *fully-convolutional* network architecture. Closest to our work, [17] extends upon the R2D2 architecture by replacing its fully-convolutional backbone by a C_8 equivariant one, and mainly looks into combining their architecture with others to create an ensemble with high coverage of different rotation angles. In contrast, our work focuses on pure equivariant models, and studies the effects of instilling different levels of rotation equivariance, from the discrete case to the continuous case.

3   Methodology

In order to make local feature matching robust to rotations, we introduce geometric priors to the model directly. For this purpose, we propose C-3PO, a family of novel deep feature detection-and-description models based on steerable group convolutional networks. As illustrated in Fig. 1, each network takes an RGB image $\mathbf{I} \in \mathbb{R}^{H \times W \times 3}$ as input, and produces (i) a set of dense D-dimensional feature descriptors $\mathbf{X} \in \mathbb{R}^{H \times W \times D}$, (ii) a repeatability map $\mathbf{S} \in [0,1]^{H \times W}$, and (iii) an associated reliability map $\mathbf{R} \in [0,1]^{H \times W}$. In the remainder of this section, we introduce these models along with the baseline models we used in our experiments.

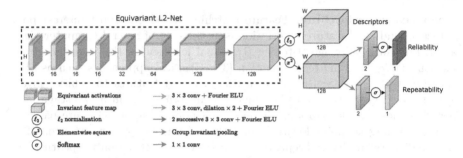

Fig. 1. C-3PO Network Architecture. The network architecture of the SO(2) variant of C-3PO. The initial layers comprise an equivariant variant of L2-Net. In line with [19], the remaining part of the network consists of three heads outputting the feature descriptors, repeatability map, and reliability map.

R2D2 Baseline. As our primary baseline, we consider the R2D2 detection-and-description model for our experimental evaluations [19]. Our formulation is generic for CNN-based feature detection methods. However, for evaluating rotation equivariance, we choose the R2D2 model for its high performance and well-documented code. In addition, the R2D2 model consists of a fully-convolutional network design that enables end-to-end optimisation of both feature extraction and description. This fully-convolutional structure of R2D2 allows for direct substitution with equivariant convolutional layers. R2D2 uses a modified L2-Net [27] backbone, consisting of 7 convolutional layers with monotonically increasing channel-sizes (see [19] for details).

C-3PO. The network architecture of C-3PO, shown in Fig. 1, largely follows the R2D2 model. Analogously to [5], we alter the L2-Net backbone of R2D2 by substituting the convolutional blocks with steerable ones, which makes the backbone equivariant to rotations. Each block applies an equivariant layer, batch normalisation, and an activation function that is applied either to the signal directly, or in the Fourier domain, depending on whether the group is finite.

To prevent the basis for the block expansion of the steerable filter from being empty, we replace the three successive 2×2 convolutional blocks at the end of R2D2 by two successive 3×3 equivariant convolutional blocks. After this sequence of equivariant convolutional blocks, we apply group invariant pooling by performing a max-pooling operation within each regular field to ensure that the final keypoint descriptors of the input pixels are invariant to rotations.

We distinguish between three variants of C-3PO that are each based on a different group. Each variant takes an RGB image as input, and correspondingly, the input types of the first layer are three independent scalar fields in all cases. In contrast, the intermediate signals transform according to the regular representations of their respective group. The first two variants are based on the finite group C_n for $n \in \{4, 8\}$, and the last variant on the infinite group SO(2).

For the discrete variants, we can use normal ReLU pointwise activation functions to ensure equivariance, since the underlying group is finite. The SO(2) variant instead uses a Fourier ELU, which uses the inverse Fourier transform to sample the function, applies the ELU nonlinearity, and finally recovers the Fourier coefficients by performing a Fourier transform. As argued in Sect. 2.1, we can approximate the Fourier transform using a finite subset of Fourier coefficients. While increasing the number of coefficients provides a better approximation of the underlying signal, it also considerably increases the computational cost to train the network. We empirically found that using 4 coefficients provides a good balance between approximation precision and computational cost.

The equivariant L2-Net backbone maintains the number of layers, but reduces the number of channels per layer. The motivation for this is twofold. First, because the equivariant layers inherently capture rotated copies of the same feature, our equivariant backbone theoretically requires less channels than the base R2D2 model. Second, the addition of the group convolutional layers drastically increases the number of parameters. Whereas the base R2D2 model required roughly 0.5M parameters, the $C4$-variant required 12M and the SO(2)-variant required roughly 60M parameters.

In line with [29], we reduce the number of channels for most of the layers such that all models have a similar number of parameters. This does not only make the increased computational cost tractable, but also allows for a fair comparison. Figure 2 shows a comparison of the various models with respect to their corresponding number of trainable parameters and average inference time per input image after reducing the number of channels.

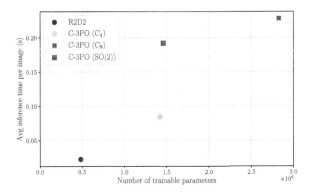

Fig. 2. Model Efficiency. We show the computational efficiency of different network architectures, both in terms of inference time and number of trainable parameters.

Implementation. For our experiments, we use the original PyTorch [16] implementation of the R2D2 model[2], and modify this framework to adapt to our C-3PO model architecture. To this end, we employ the `escnn` E(n)-Equivariant Steerable CNNs library[3] to develop our rotation equivariant C-3PO variants [6,29]. In order to enable fair comparison between our equivariant models and the R2D2 baseline, we keep the R2D2 training pipeline intact and use the same hyperparameter configuration as the original R2D2 implementation.

4   Experiments

Training Configuration. In accordance with the default settings for R2D2, we train for a maximum of 25 epochs using the Adam optimizer [13] with a learning rate of $1 \cdot 10^{-4}$, a weight decay of $5 \cdot 10^{-4}$, and a batch size of 8. Additionally, we adopt an early stopping mechanism where we do not continue training if the evaluation loss does not decrease for at least 3 consecutive epochs. As a result, training all the different models, including the R2D2 baseline model, required approximately 120 h using a single Nvidia Titan RTX GPU. Performing our experiments required 16 GPU hours.

Evaluation Setup. Following [19], we evaluate local feature matching using the Homography-Patches (HPatches) dataset [2]. This dataset consists of 116 image sequences with each image sequence comprising a source image I_S and $M = 5$ target images I_{T_j}, for $j \in \{1 \cdots M\}$. Each pair (I_S, I_{T_j}) is related via a homography $H_j := \mathbb{H}(I_S, I_{T_j})$, where $\mathbb{H}$ is a planar affine transformation between the two images. For a given pair of images, we first resize both to 300×300 following [17], then we use a trained model (*e.g.* R2D2) to detect and describe keypoints in each of them independently. These are then matched using a nearest neighbor search in the feature descriptor space. Next, we apply RANSAC [10] filtering to remove outlier matches. We then transform the keypoints in the first image using $\mathbb{H}$, project them onto the second image and compare with those matched in the second image to compute the metrics. A match is considered correct if its reprojection error, *i.e.*, the distance between a projected point and the corresponding point in second image, is within a certain pixel threshold. For evaluating rotation-robustness, we fix the threshold $\tau = 3$px following [19]. Our primary metric is mean matching accuracy (MMA), which computes the fraction of correct matches in a given pair, averaged across the dataset. To study the benefit of using rotation equivariant CNNs instead of standard CNNs, we compare performance in terms of MMA for input images from the HPatches dataset across rotations from $0°$ to $360°$ with an interval of $15°$.

Quantitative Results. The key results of our experiments are summarized in Fig. 3. First, we show that R2D2 [19] is indeed not robust to rotations in the target image. As rotations are applied incrementally, the MMA drops steeply to zero

---

[2] https://github.com/naver/r2d2.
[3] https://pypi.org/project/escnn/.

at about 60°. Second, we note that the C_4 variant significantly improves upon R2D2 at rotations in and around $\{90°, 180°, 270°\}$. This confirms the effectiveness of rotation equivariance baked into this variant at these special rotations. Third, to our surprise, both C_8 and steerable SO(2) variants follow a similar performance trend as C_4. One reason for this could be the small kernel sizes that were used, as rotating a 3×3 kernel implies only the centre pixel is considered for the response. Among these rotation equivariant models, SO(2) does outperform C_8 which in turn outperforms C_4, but the performance gains are marginal. Further, the dip in performance across models in between the 90° rotations that are odd-multiples of 45° could be a result of rotation artifacts such as artificial edges introduced due to filling of the unoccupied regions in the rotated image.

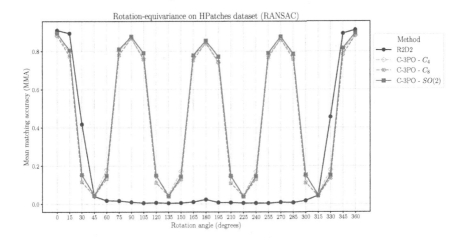

Fig. 3. Evaluation Rotation-Equivariance. We compare vanilla R2D2 model with our model variants that endow R2D2 with rotation-equivariance capabilities. At and around special rotations, SO(2) outperforms other variants but only marginally.

Qualitative Analysis. As a means to provide a more holistic understanding and intuition behind the quantitative results, we show feature matching results on a sample image pair from the HPatches dataset in Fig. 4. All models find robust matches without any rotation of the target image, but with rotation of 90°, the vanilla R2D2 model struggles while the equivariant models are still able to find robust matches.

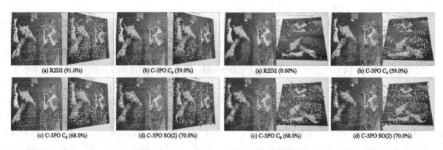

(a) Feature matching with no rotation.   (b) Feature matching with 90° rotation.

Fig. 4. Qualitative Matching Results. Matches found for various models for a pair of images. The percentage in parenthesis shows the fraction of correct matches for each of the models for this particular image-pair. Blue points denote keypoint detected by the model. Yellow points on the target image denote the points in source image transformed by the ground truth homography $\mathbb{H}$. Correct matches are shown in green. (Color figure online)

Furthermore, recall that the R2D2 model jointly detects and describes salient keypoints in the image. We qualitatively analyse its ability to achieve rotation-equivariance for plain keypoint detection. To that end, we observe how detected keypoints vary when applying a 90° rotation on a sample set of images in Fig. 5. Upon visual inspection, it seems keypoint detection may be equivariant to 90° rotations for all models, including vanilla R2D2. Interestingly, vanilla R2D2 seems to detect a large number of keypoints most of which are likely to be harmful for matching. This relates back to the notion of *reliability* in the R2D2 model formulation. Our model variants are not only rotation equivariant to keypoint detection but also seem to only detect *reliable* keypoints even with rotations. We observe this phenomenon in multiple samples and report more examples in Appendix B.

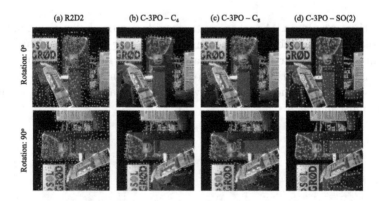

Fig. 5. Keypoint Detection Robustness. Qualitative evaluation of robustness of keypoint-detection against 90° rotation across all models. Our model variants detect *reliable* keypoints while being equivariant to rotations.

5 Conclusion

We have looked into the viability of using steerable group convolutions in feature detection and description in order to introduce rotation invariance to a high-performance feature detector and descriptor. To this end, we introduced C-3PO, a family of rotation equivariant feature detection-and-description models, and studied its robustness to rotations and homographies. Our results indicate that using the rotation equivariant models can provide additional robustness, especially for rotation angles that are multiples of $90°$. However, we also found that extending the simple discrete equivariant C_4 model to C_8 and SO(2) architectures provided only marginal gains.

Limitations and Future Research. While our proposed C-3PO model is able to achieve considerable performance gains when compared to our R2D2 baseline, the experiments and model design have revealed some limitations. Firstly, converting the R2D2 model to a rotation equivariant network using steerable CNN layers came at a substantially higher computational cost than we initially anticipated, due to the large increase in required network parameters. Nonetheless, our experiments also revealed that reducing the channel sizes of the steerable layers is an effective solution to mitigate the computational cost without suffering a performance drop when compared to the baseline. Secondly, our experiments showed that, whereas using the C_4 variant provides a considerable performance boost for 90 degree rotations, extending the rotation equivariance to the C_8 and SO(2) variants provided minimal additional performance gains.

Future research should investigate why the performance difference between the discrete and continuous equivariant models is only very minimal, and how these models can actually be improved to provide reasonable robustness across all rotation angles. Some interesting directions of improvement could be (i) using larger kernel sizes for each layer to incorporate a larger context when rotating the filters and (ii) exploring different interpolation techniques when rotating.

Acknowledgements. We would like to thank the Master AI program at the University of Amsterdam for providing financial support for conference registration. In addition, we would also like to thank Gabriele Cesa and Rob Hesselink for their excellent suggestions for possible future research topics.

References

1. Adel, E., Elmogy, M., Elbakry, H.: Image stitching based on feature extraction techniques: a survey. Int. J. Comput. Appl. **99**(6), 1–8 (2014)
2. Balntas, V., Lenc, K., Vedaldi, A., Mikolajczyk, K.: HPatches: a benchmark and evaluation of handcrafted and learned local descriptors. In: CVPR (2017)
3. Bay, H., Ess, A., Tuytelaars, T., Van Gool, L.: Speeded-up robust features (surf). Comput. Vis. Image Underst. **110**(3), 346–359 (2008)
4. Bronstein, M.M., Bruna, J., LeCun, Y., Szlam, A., Vandergheynst, P.: Geometric deep learning: going beyond Euclidean data. IEEE Signal Process. Mag. **34**(4), 18–42 (2017)

5. Bökman, G., Kahl, F.: A case for using rotation invariant features in state of the art feature matchers (2022). https://doi.org/10.48550/ARXIV.2204.10144. https://arxiv.org/abs/2204.10144

6. Cesa, G., Lang, L., Weiler, M.: A program to build E(N)-equivariant steerable CNNs. In: International Conference on Learning Representations (2022). https://openreview.net/forum?id=WE4qe9xlnQw

7. Cohen, T., Welling, M.: Group equivariant convolutional networks. In: Balcan, M.F., Weinberger, K.Q. (eds.) Proceedings of The 33rd International Conference on Machine Learning. Proceedings of Machine Learning Research, vol. 48, pp. 2990–2999. PMLR, New York, New York, USA (20–22 Jun 2016). https://proceedings.mlr.press/v48/cohenc16.html

8. Cohen, T.S., Welling, M.: Steerable CNNs. arXiv preprint arXiv:1612.08498 (2016)

9. DeTone, D., Malisiewicz, T., Rabinovich, A.: Superpoint: self-supervised interest point detection and description. In: Proceedings of the IEEE Conference on Computer Vision and Pattern Recognition workshops, pp. 224–236 (2018)

10. Fischler, M.A., Bolles, R.C.: Random sample consensus: a paradigm for model fitting with applications to image analysis and automated cartography. Commun. ACM **24**(6), 381–395 (1981)

11. Heinly, J., Schonberger, J.L., Dunn, E., Frahm, J.M.: Reconstructing the world* in six days *(as captured by the yahoo 100 million image dataset). In: Proceedings of the IEEE Conference on Computer Vision and Pattern Recognition (CVPR) (2015)

12. Jiang, W., Trulls, E., Hosang, J., Tagliasacchi, A., Yi, K.M.: COTR: correspondence transformer for matching across images. In: Proceedings of the IEEE/CVF International Conference on Computer Vision, pp. 6207–6217 (2021)

13. Kingma, D.P., Ba, J.: Adam: A method for stochastic optimization. arXiv preprint arXiv:1412.6980 (2014)

14. Lowe, D.G.: Distinctive image features from scale-invariant keypoints. Int. J. Comput. Vision **60**(2), 91–110 (2004)

15. Parihar, U.S., et al.: RORD: rotation-robust descriptors and orthographic views for local feature matching. In: 2021 IEEE/RSJ International Conference on Intelligent Robots and Systems (IROS), pp. 1593–1600. IEEE (2021)

16. Paszke, A., et al.: PyTorch: an imperative style, high-performance deep learning library. In: Wallach, H., Larochelle, H., Beygelzimer, A., d' Alché-Buc, F., Fox, E., Garnett, R. (eds.) Advances in Neural Information Processing Systems 32, pp. 8024–8035. Curran Associates, Inc. (2019). http://papers.neurips.cc/paper/9015-pytorch-an-imperative-style-high-performance-deep-learning-library.pdf

17. Peri, A., Mehta, K., Mishra, A., Milford, M., Garg, S., Krishna, K.M.: Ref-rotation equivariant features for local feature matching. arXiv preprint arXiv:2203.05206 (2022)

18. Piasco, N., Sidibé, D., Demonceaux, C., Gouet-Brunet, V.: A survey on visual-based localization: on the benefit of heterogeneous data. Pattern Recogn. **74**, 90–109 (2018)

19. Revaud, J., Weinzaepfel, P., De Souza, C., Pion, N., Csurka, G., Cabon, Y., Humenberger, M.: R2d2: repeatable and reliable detector and descriptor. arXiv preprint arXiv:1906.06195 (2019)

20. Romero, D., Bekkers, E., Tomczak, J., Hoogendoorn, M.: Attentive group equivariant convolutional networks. In: International Conference on Machine Learning, pp. 8188–8199. PMLR (2020)

21. Rublee, E., Rabaud, V., Konolige, K., Bradski, G.: ORB: an efficient alternative to sift or surf. In: 2011 International Conference on Computer Vision, pp. 2564–2571. IEEE (2011)

22. Sarlin, P.E., DeTone, D., Malisiewicz, T., Rabinovich, A.: Superglue: learning feature matching with graph neural networks. In: Proceedings of the IEEE/CVF Conference on Computer Vision and Pattern Recognition, pp. 4938–4947 (2020)

23. Sattler, T., et al.: Benchmarking 6dof outdoor visual localization in changing conditions. In: Proceedings of the IEEE Conference on Computer Vision and Pattern Recognition (CVPR) (2018)

24. Schonberger, J.L., Frahm, J.M.: Structure-from-motion revisited. In: Proceedings of the IEEE Conference on Computer Vision and Pattern Recognition (CVPR) (2016)

25. Schonberger, J.L., Frahm, J.M.: Structure-from-motion revisited. In: Proceedings of the IEEE Conference on Computer Vision and Pattern Recognition, pp. 4104–4113 (2016)

26. Sun, J., Shen, Z., Wang, Y., Bao, H., Zhou, X.: LoFTR: detector-free local feature matching with transformers. In: Proceedings of the IEEE/CVF Conference on Computer Vision and Pattern Recognition, pp. 8922–8931 (2021)

27. Tian, Y., Fan, B., Wu, F.: L2-Net: deep learning of discriminative patch descriptor in euclidean space. In: Proceedings of the IEEE Conference on Computer Vision and Pattern Recognition, pp. 661–669 (2017)

28. Ullman, S.: The interpretation of structure from motion. Proceed. Royal Soc. London. Series B. Biolog. Sci. **203**(1153), 405–426 (1979)

29. Weiler, M., Cesa, G.: General E(2)-Equivariant Steerable CNNs. In: Conference on Neural Information Processing Systems (NeurIPS) (2019)

30. Yamada, K., Kimura, A.: A performance evaluation of keypoints detection methods SIFT and AKAZE for 3D reconstruction. In: 2018 International Workshop on Advanced Image Technology (IWAIT), pp. 1–4. IEEE (2018)

Towards Flexible Inductive Bias via Progressive Reparameterization Scheduling

Yunsung Lee[1], Gyuseong Lee[2], Kwangrok Ryoo[2], Hyojun Go[1], Jihye Park[2], and Seungryong Kim[2(✉)]

[1] Riiid AI Research, Seoul, South Korea
[2] Korea University, Seoul, South Korea
yunsung.lee@riiid.co

Abstract. There are two *de facto* standard architectures in recent computer vision: Convolutional Neural Networks (CNNs) and Vision Transformers (ViTs). Strong inductive biases of convolutions help the model learn sample effectively, but such strong biases also limit the upper bound of CNNs when sufficient data are available. On the contrary, ViT is inferior to CNNs for small data but superior for sufficient data. Recent approaches attempt to combine the strengths of these two architectures. However, we show these approaches overlook that the optimal inductive bias also changes according to the target data scale changes by comparing various models' accuracy on subsets of sampled ImageNet at different ratios. In addition, through Fourier analysis of feature maps, the model's response patterns according to signal frequency changes, we observe which inductive bias is advantageous for each data scale. The more convolution-like inductive bias is included in the model, the smaller the data scale is required where the ViT-like model outperforms the ResNet performance. To obtain a model with flexible inductive bias on the data scale, we show reparameterization can interpolate inductive bias between convolution and self-attention. By adjusting the number of epochs the model stays in the convolution, we show that reparameterization from convolution to self-attention interpolates the Fourier analysis pattern between CNNs and ViTs. Adapting these findings, we propose Progressive Reparameterization Scheduling (PRS), in which reparameterization adjusts the required amount of convolution-like or self-attention-like inductive bias per layer. For small-scale datasets, our PRS performs reparameterization from convolution to self-attention linearly faster at the late stage layer. PRS outperformed previous studies on the small-scale dataset, e.g., CIFAR-100.

Keywords: Flexible architecture · Vision transformer · Convolution · Self-attention · Inductive bias

---

Y. Lee, G. Lee, K. Ryoo, H. Go, J. Park—Equal contributions.

© The Author(s), under exclusive license to Springer Nature Switzerland AG 2023
L. Karlinsky et al. (Eds.): ECCV 2022 Workshops, LNCS 13804, pp. 706–720, 2023.
https://doi.org/10.1007/978-3-031-25069-9_45

1 Introduction

Architecture advances have enhanced the performance of various tasks in computer vision by improving backbone networks [3,15,16,27,28]. From the success of Transformers in natural language processing [2,10,31], Vision Transformers (ViTs) show that it can outperform Convolutional Neural Networks (CNNs) and its variants have led to architectural advances [22,30,36]. ViTs lack inductive bias such as translation equivariance and locality compared to CNNs. Therefore, ViTs with sufficient training data can outperform CNNs, but ViTs with small data perform worse than CNNs.

To deal with the data-hungry problem, several works try to inject convolution-like inductive bias into ViTs. The straightforward approaches use convolutions to aid tokenization of an input image [14,32–34] or design the modules [6,12,20,35] for improving ViTs with the inductive bias of CNNs. Other approaches use the local attention mechanisms for introducing locality to ViTs [13,22], which attend to the neighbor elements and improve the local extraction ability of global attention mechanisms. These approaches can design architectures that leverage the strength of CNNs and ViTs and can alleviate the data-hungry problem at some data scale that their work target.

However, we show these approaches overlook that the optimal inductive bias also changes according to the target data scale by comparing various models' accuracy on subsets of sampled ImageNet at different ratios. If trained on the excessively tiny dataset, recent ViT variants still show lower accuracy than ResNet, and on the full ImageNet scale, all ViT variants outperform ResNet. Inspired by Park *et al.* [24], we perform Fourier analysis on these models to further analyze inductive biases in the architecture. We observe that ViTs injected convolution-like inductive bias show frequency characteristics between it of ResNet and ViT. In this experiment, the more convolution-like inductive bias is included, the smaller the data scale is required where the model outperforms the ResNet performance. Specifically, their frequency characteristics tend to serve as the high-pass filter in early layers and as more low-pass filter closer to the last layer. Nevertheless, such a fixed architecture in previous approaches has a fixed inductive bias between CNNs and ViTs, making it difficult to design an architecture that performs well on various data scales. Therefore, each time a new target dataset is given, the optimal inductive bias required changes, so each time the model's architectural design needs to be renewed. For example, a CNN-like architecture should be used for small-scale dataset such as CIFAR [17], and a ViT-like architecture should be designed for large-scale datasets such as JFT [26]. Also, this design process requires multiple training for tuning the inductive bias of the model, which is time-consuming.

In this paper, we confirm the possibility of reparameterization technique [5, 19] from convolution to self-attention towards flexible inductive bias between convolution and self-attention during a single training trial. The reparameterization technique can change the learned convolution layer to self-attention, which identically operates like learned convolution. Performing Fourier analysis, we show that reparameterization can interpolate the inductive biases between

Table 1. Comparison of various architectures ✓ means that the model has the corresponding characteristics, and ✗ does not. ✓* indicates that ConViT's convolutional operation is given only in the initial training stage and then learned in the form of gated self-attention.

	DeiT [29]	ResNet [16]	ConViT [12]	ResT [35]	Swin [22]
Hierarchical Structure	✗	✓	✗	✓	✓
Relative Positional Encoding	✗	✗	✓	✗	✓
Local Attention	✗	✗	✗	✗	✓
Convolutional Operation	✗	✓	✓*	✓	✗

convolution and self-attention by adjusting the moment of reparameterization during training. We observe that more training with convolutions than with self-attention makes the model have a similar frequency characteristic to CNN and vice versa. This observation shows that adjusting the schedule of reparameterization can interpolate between the inductive bias of CNNs and ViTs.

From these observations, we propose the Progressive Reparameterization Scheduling (PRS). PRS is to sequentially reparameterize from the last layer to the first layer. Layers closer to the last layers are more trained with self-attention than convolution, making them closer to self-attention. Therefore, we can make the model have a suitable inductive bias for small-scale data with our schedule. We validate the effectiveness of PRS with experiments on the CIFAR-100 dataset.

Our contributions are summarized as follows:

- We observe that architecture with a more convolutional inductive bias in the early stage layers is advantageous on a small data scale. However, if the data scale is large, it is advantageous to have a self-attentional inductive bias.
- We show that adjusting the remaining period as convolution before reparameterization can interpolate the inductive bias between convolution and self-attention.
- Based on observations of favorable conditions in small-scale datasets, we propose the Progressive Reparameterization Scheduling (PRS) which sequentially changes convolution to self-attention from the last layer to the first layer. PRS outperformed previous approaches on the small-scale dataset, e.g., CIFAR-100.

2   Related Work

2.1   Convolution Neural Networks

CNNs, the most representative models in computer vision, have evolved over decades from LeNeT [18] to ResNet [16] in a way that is faster and more accurate. CNNs can effectively capture low-level features of images through inductive biases which are locality and translation invariance. However, CNNs have a weakness in capturing global information due to their limited receptive field.

2.2   Vision Transformers

Despite the great success of vision transformer [11] in computer vision, ViT has several fatal limitations that it requires high cost and is difficult to extract the low-level features which contain fundamental structures, so that it shows inferior performance than CNNs in small data scales. There are several attempts to overcome the limitations of ViT and improve its performance by injecting a convolution inductive bias into the Transformer.

DeiT [29] allows ViT to take the knowledge of convolution through distillation token. They can converge a model, which fails in ViT. On the other hand, The straightforward approaches [4,20,34,35] employ inductive bias to augment ViT by adding depthwise convolution to the FFN of the Transformer. ConViT [12] presents a new form of self-attention(SA) called Gated positional self-attention (GPSA) that can be initialized as a convolution layer. After being initialized as convolution only at the start of learning, ConViT learns only in the form of self-attention. Thus, it does not give sufficient inductive bias on small resources. Swin Transformer [22] imposes a bias for the locality to ViT in a way that limits the receptive field by local attention mechanisms. A brief comparison of these methods is shown in Table 1.

2.3   Vision Transformers and Convolutions

There have been several studies analyzing the difference between CNNs and ViTs [24,25]. Park et al. [24] and Raghu et al. [25] prove that CNN and Transformer extract entirely different visual representations. In particular, Park et al. [24] present the several analysis of self-attention and convolution that self-attention acts as a low-pass filter while convolution acts as a high pass filter. Furthermore, several approaches [5,8,19] have reparameterized convolution to self-attention by proving that their operations can be substituted for each other. Cordonnier et al. [5] demonstrates that self-attention and convolution can have the same operation when relative positional encoding and the particular settings are applied. T-CNN [8] presents the model using GPSA proposed by ConViT, which reparameterizes convolution layer as GPSA layers. C-MHSA [19] prove that reparameterization between two models is also possible even when the input was patch unit, and propose a two-phase training model, which initializes ViT from a well-trained CNN utilizing the construction in above theoretical proof.

3   Preliminaries

Here, we recall the mathematical definitions of multi-head self-attention and convolution to help understand the next section. Then, we briefly introduce the background of reparameterization from convolution layer to self-attention layer. We follow the notation in [5].

Convolution Layer. The convolution layer has locality and translation equivariance characteristics, which are useful inductive biases in many vision tasks. Those inductive biases are encoded in the model through parameter sharing and local information aggregation. Thanks to the inductive biases, better performance can be obtained with a low data regime compared to a transformer that has a global receptive field. The output of the convolution layer can be roughly formulated as follows:

$$\text{Conv}(\boldsymbol{X}) = \sum_{\Delta} \boldsymbol{X}\,\boldsymbol{W}^C, \tag{1}$$

where $\boldsymbol{X} \in \mathbb{R}^{H \times W \times C}$ is an image tensor, H,W,C is the image height, width and channel, $\boldsymbol{W}^C$ is convolution filter weight and the set

$$\Delta = \left[-\left\lfloor \frac{K}{2} \right\rfloor, \cdots, \left\lfloor \frac{K}{2} \right\rfloor \right] \times \left[-\left\lfloor \frac{K}{2} \right\rfloor, \cdots, \left\lfloor \frac{K}{2} \right\rfloor \right] \tag{2}$$

is the receptive field with $K \times K$ kernel.

Multi-head Self-Attention Mechanism. Multi-head self-attention(MHSA) mechanism [31] trains the model to find semantic meaning by finding associations among a total of N elements using query $\boldsymbol{Q} \in \mathbb{R}^{N \times d_H}$, key $\boldsymbol{K} \in \mathbb{R}^{N \times d_H}$, and value $\boldsymbol{V} \in \mathbb{R}^{N \times d_H}$, where d_H is the size of each head. After embedding the sequence $\boldsymbol{X} \in \mathbb{R}^{N \times d}$ as a query and key using $\boldsymbol{W}^Q \in \mathbb{R}^{d \times d_H}$ and $\boldsymbol{W}^K \in \mathbb{R}^{d \times d_H}$, an attention score $\boldsymbol{A} \in \mathbb{R}^{N \times N}$ can be obtained by applying softmax to the value obtained by inner producting $\boldsymbol{Q}$ and $\boldsymbol{K}$, where d is the size of an input token. Self-attention(SA) is obtained through matrix multiplication of $\boldsymbol{V}$ embedded by $\boldsymbol{W}^V \in \mathbb{R}^{N \times d_H}$ and $\boldsymbol{A}$:

$$\text{SA}(\boldsymbol{X}) = \boldsymbol{A}(\boldsymbol{X}\boldsymbol{W}^Q, \boldsymbol{X}\boldsymbol{W}^K)\boldsymbol{X}\boldsymbol{W}^V,$$

$$\boldsymbol{A}(\mathbf{Q}, \mathbf{K}) = \text{softmax}\left(\frac{\boldsymbol{Q}\boldsymbol{K}^\top}{\sqrt{d}} + \boldsymbol{B} \right), \tag{3}$$

where $\boldsymbol{B}$ is a relative position suggested in [7]. By properly setting the relative positional embedding $\boldsymbol{B}$, we can force the query pixel to focus on only one key pixel. MHSA allows the model to attend information from different representation subspaces by performing an attention function in parallel using multiple heads. MHSA with a total of N_H heads can be formulated as follows:

$$\text{MHSA}(\boldsymbol{X}) = \sum_{k=1}^{N_H} \text{SA}_k(\boldsymbol{X})\,\boldsymbol{W}_k^O, \tag{4}$$

where $\boldsymbol{W}^O$ is learnable projection and k is the index of the head.

Reparameterizing MHSA into Convolution Layer. [19] showed that $K \times K$ kernels can be performed through K^2 heads, where K is the size of the kernel. Since the convolution layer is agnostic to the context of the input, it is necessary to set $\boldsymbol{W}^Q$ and $\boldsymbol{W}^K$ as $\boldsymbol{0}$ to convert the convolution to MHSA. Using equations (3) and (4) together, MHSA can be formulated as follows:

$$\mathrm{MHSA}(\boldsymbol{X}) = \sum_{k=1}^{N_H} \boldsymbol{A}_k \boldsymbol{X} \boldsymbol{W}_k^V \boldsymbol{W}_k^O. \tag{5}$$

As $\boldsymbol{A}_k \boldsymbol{X}$ is used to select the desired pixel, the knowledge of the convolution layer can be completely transferred to the MHSA by setting $\boldsymbol{W}^V$ to $\boldsymbol{I}$ and initializing $\boldsymbol{W}^O$ to $\boldsymbol{W}^C$.

4   Inductive Bias Analysis of Various Architectures

In this section, we analyze various architectures through Fourier analysis and accuracy tendency according to data scale. Previous works designing the modules by mixing convolution-like inductive bias to ViTs overlook that a fixed architecture has a fixed inductive bias and optimal inductive bias can change according to data scale. To confirm it, we conduct experiments that measure the accuracy of various architectures by changing the data scale of ImageNet [9]. In these experiments, we observe that the required data scale for outperforming ResNet is different for each architecture.

Then, we link frequency characteristics of the recent ViT variants and the tendency of their accuracy with data scale by expanding observations of Park et al. [24]. In [23,24], they analyze feature maps in Fourier space and demonstrate that self-attention is a low-pass filter, and convolution is a high-pass filter. This phenomenon of filtering noise of different frequencies is caused by different inductive biases of self-attention and convolution. With Fourier analysis of Park et al. [24], we observe that architecture having more CNN-like frequency characteristics shows CNN-like efficiency and accuracy tendency in the small-scale datasets. Park et al. [24] conducted Fourier analysis only for ViT and ResNet, but we analyzed several models with various attempts to inject convolutional induction biases into ViT architecture. In this section, the Fourier characteristics vary for each injected inductive bias, and we can see which model among the ViT variables was more convolution-like or self-attention-like. Section 5 will show that we can interpolate from these convolution-like Fourier features to self-attention-like Fourier features with reparameterization.

4.1   Our Hypothesis

We hypothesize that 1) the more convolution-like inductive bias is included, the smaller the data scale is required where the ViT-like model outperforms CNNs,

Table 2. Data scale experiment of various model architectures. For a fair comparison, the data augmentation and regulation techniques during the learning process of all experimental models followed those of DeiT [29].

Model	ImgNet Ratio	Acc@1	Acc@5	Flops	# params
DeiT-Ti [29]	0.01	6.43	16.37	1.25G	5M
	0.05	24.82	46.40		
	0.1	38.61	63.26		
	0.5	67.03	88.11		
	1	**72.2**	**91.1**		
ConViT-Ti [12]	0.01	6.08	15.82	1G	6M
	0.05	26.93	49.86		
	0.1	42.92	67.78		
	0.5	68.21	88.93		
	1	**73.1**	**91.7**		
ResTv1-Lite [35]	0.01	11.19	26.542	1.4G	11M
	0.05	**42.92**	**67.91**		
	0.1	**52.88**	**76.62**		
	0.5	**73.03**	**91.39**		
	1	**77.0**	**93.6**		
ResNet-18 [16]	**0.01**	**13.93**	**30.85**	1.8G	11.6M
	0.05	42.04	67.58		
	0.1	52.24	76.38		
	0.5	66.38	87.30		
	1	69.53	89.08		
Swin-T [22]	0.01	13.20	27.39	4.5G	28M
	0.05	38.69	61.88		
	0.1	53.46	75.57		
	0.5	**76.21**	**92.86**		
	1	**81.2**	**95.5**		
ResNet-50 [16]	**0.01**	**13.67**	**30.19**	3.6G	23.9M
	0.05	**46.82**	**70.85**		
	0.1	**58.14**	**80.53**		
	0.5	75.23	92.42		
	1	80.15	94.49		

and 2) frequency characteristics can explain whether the inductive bias of model is closer to CNNs or ViTs. Specifically, the incapacity to which the layer amplifies the high-frequency signal tends to dramatically increase from the first layer to the last layer in CNN, whereas ViT does not increase well. ViTs injected with the inductive bias of convolutions tend to increase it, but not as drastic as CNN.

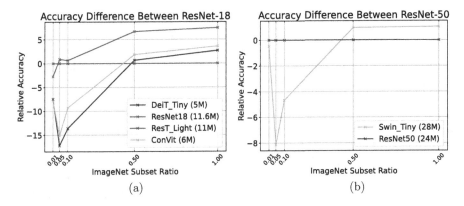

Fig. 1. Comparisons of accuracy between ResNet and various ViT-like architectures. Each model is trained on the subsets of imagenet, specifically 1%, 5%, 10%, 50%, and 100%. We plot the accuracy difference between ResNet and other architectures with the increasing subset ratio. The numbers in parentheses mean the number of parameters of each model.

Here, we observe that ViTs increasing this incapacity more dramatically perform well on smaller scale data like CNNs.

4.2 Data Scale Experiment

CNNs have inductive biases such as locality and translation invariance and ViTs do not. Because of the difference in inductive bias that architecture has, the data scale determines their superiority. In small-scale data, CNNs outperform ViTs, and at some point, ViTs outperform CNNs as the data scale grows. ViT variants injected with the convolution-like inductive bias have stronger inductive bias compared to naïve ViT, and the amount of data required to outperform ResNet will be less than it. In this subsection, we identify accuracy trends and the amount of data required to outperform ResNet for various architectures by changing the data scale.

As shown in Table 2 and Fig. 1, we make subsets with the ratio of 0.01, 0.05, 0.1, and 0.5 respectively in ImageNet for experiments in various settings with the same data distribution and different data scales. By utilizing the taxonomy of vision transformer proposed in [21], We choose the representatives in each category as ViT variants to compare together. ResT [35] injects inductive bias directly by adding convolution layers, whereas Swin [22] and ConViT [12] add locality in a new way. Swin uses a method that constrains global attention, while ConViT proposes a new self-attention layer that can act as a convolution layer in the initial stage of training. Therefore, we select ResNet-18 and ResNet-50 as the basic architecture of CNN, DeiT-Ti as Vanilla ViT and ResT-Light, ConViT-Ti, and Swin-T as the variations of the ViT to be tested. Since the number of parameters also significantly affects the performance, we compare the tiny version of Swin (Swin-T) [22] with ResNet-50 [16] and the remaining ViT variants with ResNet-18 [16]. Swin-T has more parameters than other models since the dimension is doubled every time it passes through one layer.

At 0.01, the smallest data scale, the ResNet series consisting of only CNNs shows better performance, and between them, ResNet-18 with smaller parameters has the highest accuracy. However, as the data scale increase, the accuracy of other ViT models increase more rapidly than ResNet. In particular, ResTv1-Light [35] and Swin-T [22], which have hierarchical structures, show superior performance among ViT variants and ResTv1-Light even records the highest accuracy of all models when the data scale is 0.05 or more.

As illustrated in Fig. 1, DeiT-Ti [29] shows better performance than ResNet when the data scale is close to 1, while ConViT-Ti [12] and Swin-T [22] outperform it at 0.5 or more. meanwhile, the accuracy of ResT is higher than ResNet-18 from quite a small data scale of 0.05. Therefore, we argue that the inductive bias is strong in the order of ResTv1-Light, Swin-T, ConViT-Ti, and DeiT-Ti. Through these experiments, we can prove that inductive bias and hierarchical structure have a great influence on accuracy improvement.

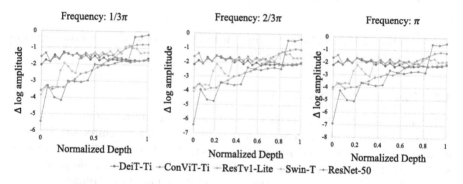

Fig. 2. Frequency characteristics of ViTs and ResNet. In ResNet-50, ResTv1-Lite, and Swin-T, the difference in log amplitude sharply increases as the normalized depth increase. On the other side, DeiT and ConViT which softly inject inductive biases into models do not have this tendency.

4.3   Fourier Analysis

As shown in Sect. 4.2, the required data scale for outperforming ResNet is different for each architecture. Inspired by the analysis of Park *et al.* [24], we show that the architectures with frequency characteristics more similar to ResNet tend to outperform ResNet at smaller data scales through Fourier analysis.

As in [23,24], the feature maps of each layer can be converted to a two-dimensional frequency domain with Fourier transform. Transformed feature maps can be represented on normalized frequency, which frequency is normalized to $[-\pi, \pi]$. The high-frequency components are represented at $-\pi$ and π

and the lowest frequency components are represented at 0. Then, we use the difference in log amplitude to report the amplitude ratio of high-frequency to low-frequency components. For better visualization, differences in log amplitude between 0 and $1/3\pi$, 0 and $2/3\pi$, and 0 and π are used to capture the overall frequency characteristics well.

Figure 2 shows frequency characteristics through Fourier analysis. In the ResNet results, the difference in log amplitude sharply increases as the normalized depth increases. This shows that early layers tend to amplify the high-frequency signal, and the tendency to amplify the high-frequency signal decreases sharply as closer to the last layer. However, DeiT and ConViT which softly inject inductive biases into models do not have this tendency and their frequency characteristics are similar through the layers. The results of Swin and ResT that strongly inject inductive biases into models with the local attention mechanism or convolution illustrate that the increase of the difference in log amplitude shows an intermediate level between it ResNet and DeiT.

By combining the results of Fig. 2 and Table 2, we can see that the model performs well for small-scale data if the increase in the difference in log amplitude through layers is sharp. It becomes smoother in the order of ResNet, ResT, Swin, ConViT, and DeiT, the accuracy is higher in the low-data regime in this order. These results are consistent with the observations of previous work that the inductive bias of CNNs helps the model to learn on small-scale data. From these, we address that the difference in log amplitude through the layers can measure the CNN-like inductive bias of the model. If it increases sharply similar to CNNs, the model has strong inductive biases and performs well in a low-data regime.

5   Reparameterization Can Interpolate Inductive Biases

As shown on Sect. 4, a fixed architecture does not have flexible inductive bias, causing them to have be tuned for each data. Since modifying the architecture to have a suitable inductive bias for each data is too time-consuming, the method which can flexibly adjust the inductive bias during the training process is needed.

We observe that the model trained more with CNN than self-attention have more CNN-like frequency characteristics through reparameterization. With these results, we show that reparameterization can interpolate the inductive bias between CNNs and ViT by adjusting the moment of reparameterization during training.

5.1   Experimental Settings

Because reparameterization can change convolution to self-attention, we can adjust the ratio of epochs that each layer is trained with convolution and self-attention. In a 10% subset of the ImageNet data, we adjust this ratio by four settings: model trained with 1) convolution for 300 epochs and self-attention for

0 epochs, 2) convolution for 250 epochs and self-attention for 50 epochs 3) convolution for 150 epochs and self-attention for 150 epochs and 4) convolution for 50 epochs and self-attention for 250 epochs. We note that the model is more trained with convolution from 1) to 4). We follow the setting for reparameterization as in CMHSA-3 [19] and Fourier analysis as in Sect. 4.3.

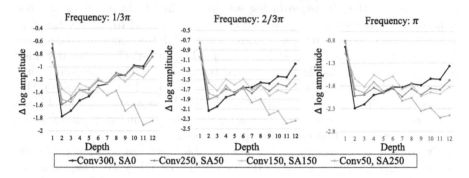

Fig. 3. Visualization of Interpolation. As the ratio trained with self-attention increases, the difference in log amplitude of early stage layers tends to increase, and the difference in log amplitude of late stage layers tends to decrease. Conv x, SA y denotes that the model is trained with convolution for x epochs and self-attention for y epochs.

5.2   Interpolation of Convolutional Inductive Bias

Figure 3 shows the results of Fourier analysis according to the ratio of trained epoch with convolution and self-attention. When comparing 1) to 4), we can see that the degree of increase become smaller from 1) to 4). As the ratio trained with self-attention increases, the difference in log amplitude of early stage layers tends to increase, and the difference in log amplitude of late stage layers tends to decrease. These results show that the more training with convolution make the degree of increase sharper. As we observed in the Sect. 4.3, the more sharply increasing the difference of log amplitude through normalized depth represents that the model have more CNN-like inductive biases. By combining the results of Fig. 3 and this observation, we can see that the more trained with convolution make the model have more CNN-like inductive biases.

6   Progressive Reparameterization Scheduling

We now propose Progressive Reparameterization Scheduling (PRS) which adjusts the inductive bias of ViT for learning on small-scale data. PRS is based on our findings as:

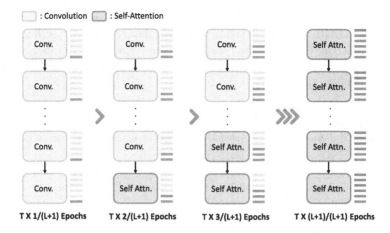

□ : Convolution ▨ : Self-Attention

Fig. 4. Illustration of PRS. Conv. is a block with a convolutional layer, and Self Attn. is a block with a self-attention layer. Each block is progressively transformed from a convolution block to a self-attention block as the training progresses.

- As shown in Sect. 4, the more convolution-like inductive bias is included, the smaller the data scale is required where the ViT-like model outperforms CNNs. In more detail, we can see that the model performs well for small-scale data if the increase in the difference of log amplitude through layers is sharp.
- Furthermore, in the interpolation experiment in Sect. 5, if the layer is trained in a convolution state for longer epochs, the layer has more convolution-like characteristics. If the layer is trained in a self-attention state for longer epochs, the layer has more self-attention-like characteristics. That is, by adjusting the schedule, it is possible to interpolate how much inductive bias the model will have between self-attention and convolution.

From these findings, PRS makes the early layer have a small difference in log amplitude as a high-pass filter and the last layer has a large difference in log amplitude as a low-pass filter. Because convolution and self-attention serve as high-pass filter and low-pass filter respectively as in Park $et\ al.$ [24], PRS wants the rear layer to play the role of self-attention and the front layer to play the role of convolution. In order to force the rear layers to focus more on the role of self-attention than the front layers, PRS reparameterizes according to linear time scheduling from convolution to self-attention, starting from the rear part. PRS is depicted in Fig. 4 and can be expressed as a formula as follows:

$$z_0 = \text{PE}(X), \tag{6}$$

$$z_l' = \begin{cases} \text{Conv}(\text{LN}(z_{l-1})) + z_{l-1}, & (t \leq T \cdot (1 - \frac{l}{L+1})) \\ \text{MHSA}(\text{LN}(z_{l-1})) + z_{l-1}, & (t > T \cdot (1 - \frac{l}{L+1})) \end{cases} \tag{7}$$

$$z_l = \text{MLP}(\text{LN}(z_l')) + z_l', \tag{8}$$

$$\mathbf{y} = \text{Linear}(\text{GAP}(z_L)), \tag{9}$$

where $PE(\cdot)$ is the patch embedding function that follows [19], $LN(\cdot)$ is LayerNorm [1], $GAP(\cdot)$ is global average pooling layer, $Linear(\cdot)$ is linear layer, t denotes current epoch at training, L denotes the total number of layers, $l = 1, 2, \cdots, L$ denotes the layer index and T denotes the total number of training epochs, $\mathbf{y}$ denotes the output of the model.

Table 3. Training results of PRS. We train the model for 400 epochs on the CIFAR-100 [17] dataset with our method, Progressive Reparameterization Scheduling.

Model	Acc@1	Acc@5	Layers	#Heads	$\dim_{\text{emb}}$
ViT-base [11]	60.90	86.66	12	12	768
DeiT-small [29]	71.83	90.99	12	6	384
DeiT-base [29]	69.98	88.91	12	12	768
CMHSA-3 [19]	76.72	93.74	6	9	768
CMHSA-5 [19]	78.74	94.40	6	9	768
Ours w/CMHSA-3	**79.09**	**94.86**	6	9	768

Table 3 shows the effectiveness of PRS in CIFAR-100 dataset. PRS outperforms the baseline with a top-1 accuracy score of +2.37p on the CIFAR-100 dataset, showing that the performance can be boosted by a simple scheduling. We note that our PRS achieves better 1023 performance than the previous two-stage reparameterization strategy [19]. These results show that PRS can dynamically apply an appropriate inductive bias for each layer. Through the successful result of PRS, we conjecture that flexibly inducing inductive bias with reparameterization has the potential for designing the model on various scale data.

7 Conclusion

From the analysis of existing ViT-variant models, we have the following conclusion: the more convolution-like inductive bias is included in the model, the smaller the data scale is required where the ViT-like model outperforms CNNs. Furthermore, we empirically show that reparameterization can interpolate inductive biases between convolution and self-attention by adjusting the moment of reparameterization during training. Through this empirical observation, we propose PRS, Progressive Reparameterization Scheduling, a flexible method that embeds the required amount of inductive bias for each layer. PRS outperforms existing approaches on the small-scale dataset, e.g., CIFAR-100.

Limitations and Future Works. Although linear scheduling is performed in this paper, there is no guarantee that linear scheduling is optimal. Therefore, through subsequent experiments on scheduling, PRS can be improved by changing it to learnable rather than linearly. In this paper, we only covered datasets with scales below ImageNet, but we will also proceed with an analysis of larger data scales than ImageNet. We also find that the hierarchical architectures tend to have more CNNs-like characteristics than the non-hierarchical architectures. This finding about hierarchy can further improve our inductive bias analysis and PRS.

References

1. Ba, J.L., Kiros, J.R., Hinton, G.E.: Layer normalization. arXiv preprint arXiv:1607.06450 (2016)
2. Brown, T., et al.: Language models are few-shot learners. Adv. Neural. Inf. Process. Syst. **33**, 1877–1901 (2020)
3. Carion, N., Massa, F., Synnaeve, G., Usunier, N., Kirillov, A., Zagoruyko, S.: End-to-end object detection with transformers. In: Vedaldi, A., Bischof, H., Brox, T., Frahm, J.-M. (eds.) ECCV 2020. LNCS, vol. 12346, pp. 213–229. Springer, Cham (2020). https://doi.org/10.1007/978-3-030-58452-8_13
4. Chu, X., et al.: Conditional positional encodings for vision transformers. arXiv preprint arXiv:2102.10882 (2021)
5. Cordonnier, J.B., Loukas, A., Jaggi, M.: On the relationship between self-attention and convolutional layers. arXiv preprint arXiv:1911.03584 (2019)
6. Dai, Z., Liu, H., Le, Q.V., Tan, M.: CoAtNet: marrying convolution and attention for all data sizes. Adv. Neural. Inf. Process. Syst. **34**, 3965–3977 (2021)
7. Dai, Z., Yang, Z., Yang, Y., Carbonell, J., Le, Q.V., Salakhutdinov, R.: Transformer-xl: attentive language models beyond a fixed-length context. arXiv preprint arXiv:1901.02860 (2019)
8. d'Ascoli, S., Sagun, L., Biroli, G., Morcos, A.: Transformed CNNs: recasting pre-trained convolutional layers with self-attention. arXiv preprint arXiv:2106.05795 (2021)
9. Deng, J., Dong, W., Socher, R., Li, L.J., Li, K., Fei-Fei, L.: Imagenet: a large-scale hierarchical image database. In: CVPR (2009)
10. Devlin, J., Chang, M.W., Lee, K., Toutanova, K.: BERT: pre-training of deep bidirectional transformers for language understanding. In: Proceedings of the 2019 Conference of the North American Chapter of the Association for Computational Linguistics: Human Language Technologies, Volume 1 (Long and Short Papers), pp. 4171–4186 (2019)
11. Dosovitskiy, A., et al.: An image is worth 16 × 16 words: transformers for image recognition at scale. arXiv preprint arXiv:2010.11929 (2020)
12. d'Ascoli, S., Touvron, H., Leavitt, M.L., Morcos, A.S., Biroli, G., Sagun, L.: Convit: improving vision transformers with soft convolutional inductive biases. In: International Conference on Machine Learning, pp. 2286–2296. PMLR (2021)
13. Han, K., Xiao, A., Wu, E., Guo, J., Xu, C., Wang, Y.: Transformer in transformer. Adv. Neural. Inf. Process. Syst. **34**, 15908–15919 (2021)
14. Hassani, A., Walton, S., Shah, N., Abuduweili, A., Li, J., Shi, H.: Escaping the big data paradigm with compact transformers. arXiv preprint arXiv:2104.05704 (2021)
15. He, K., Gkioxari, G., Dollár, P., Girshick, R.: Mask R-CNN. In: Proceedings of the IEEE International Conference on Computer Vision, pp. 2961–2969 (2017)
16. He, K., Zhang, X., Ren, S., Sun, J.: Deep residual learning for image recognition. In: Proceedings of the IEEE Conference on Computer Vision and Pattern Recognition, pp. 770–778 (2016)

17. Krizhevsky, A.: Learning multiple layers of features from tiny images. Master's thesis, University of Tront (2009)
18. LeCun, Y., Bottou, L., Bengio, Y., Haffner, P.: Gradient-based learning applied to document recognition. Proc. IEEE **86**(11), 2278–2324 (1998)
19. Li, S., Chen, X., He, D., Hsieh, C.J.: Can vision transformers perform convolution? arXiv preprint arXiv:2111.01353 (2021)
20. Li, Y., Zhang, K., Cao, J., Timofte, R., Van Gool, L.: Localvit: bringing locality to vision transformers. arXiv preprint arXiv:2104.05707 (2021)
21. Liu, Y., et al.: A survey of visual transformers. arXiv preprint arXiv:2111.06091 (2021)
22. Liu, Z., et al.: Swin transformer: hierarchical vision transformer using shifted windows. In: Proceedings of the IEEE/CVF International Conference on Computer Vision, pp. 10012–10022 (2021)
23. Park, N., Kim, S.: Blurs behave like ensembles: spatial smoothings to improve accuracy, uncertainty, and robustness. In: International Conference on Machine Learning, pp. 17390–17419. PMLR (2022)
24. Park, N., Kim, S.: How do vision transformers work? arXiv preprint arXiv:2202.06709 (2022)
25. Raghu, M., Unterthiner, T., Kornblith, S., Zhang, C., Dosovitskiy, A.: Do vision transformers see like convolutional neural networks? Adv. Neural. Inf. Process. Syst. **34**, 12116–12128 (2021)
26. Sun, C., Shrivastava, A., Singh, S., Gupta, A.: Revisiting unreasonable effectiveness of data in deep learning era. In: Proceedings of the IEEE International Conference on Computer Vision, pp. 843–852 (2017)
27. Tian, Z., Shen, C., Chen, H.: Conditional convolutions for instance segmentation. In: Vedaldi, A., Bischof, H., Brox, T., Frahm, J.-M. (eds.) ECCV 2020. LNCS, vol. 12346, pp. 282–298. Springer, Cham (2020). https://doi.org/10.1007/978-3-030-58452-8_17
28. Tian, Z., Shen, C., Chen, H., He, T.: FCOS: A simple and strong anchor-free object detector. IEEE Transactions on Pattern Analysis and Machine Intelligence (2020)
29. Touvron, H., Cord, M., Douze, M., Massa, F., Sablayrolles, A., Jégou, H.: Training data-efficient image transformers & distillation through attention. In: International Conference on Machine Learning, pp. 10347–10357. PMLR (2021)
30. Touvron, H., Cord, M., Sablayrolles, A., Synnaeve, G., Jégou, H.: Going deeper with image transformers. In: Proceedings of the IEEE/CVF International Conference on Computer Vision, pp. 32–42 (2021)
31. Vaswani, A., et al.: Attention is all you need. In: Advances in Neural Information Processing Systems, vol. 30 (2017)
32. Wu, H., et al.: CvT: introducing convolutions to vision transformers. In: Proceedings of the IEEE/CVF International Conference on Computer Vision, pp. 22–31 (2021)
33. Xiao, T., Singh, M., Mintun, E., Darrell, T., Dollár, P., Girshick, R.: Early convolutions help transformers see better. Adv. Neural. Inf. Process. Syst. **34**, 30392–30400 (2021)
34. Yuan, K., Guo, S., Liu, Z., Zhou, A., Yu, F., Wu, W.: Incorporating convolution designs into visual transformers. In: Proceedings of the IEEE/CVF International Conference on Computer Vision, pp. 579–588 (2021)
35. Zhang, Q., Yang, Y.B.: Rest: an efficient transformer for visual recognition. Adv. Neural. Inf. Process. Syst. **34**, 15475–15485 (2021)
36. Zhou, D., et al.: Deepvit: towards deeper vision transformer. arXiv preprint arXiv:2103.11886 (2021)

Zero-Shot Image Enhancement
with Renovated Laplacian Pyramid

Shunsuke Takao$^{(\boxtimes)}$ ⓘ

Port and Airport Research Institute, Yokosuka, Japan
`takao.s.work@gmail.com`

Abstract. In this research, we tackle image enhancement task both in
the traditional and Zero-Shot learning scheme with renovated Laplacian
pyramid. Recent image enhancement fields experience power of Zero-
Shot learning, estimating output from information of an input image
itself without additional ground truth data, aiming for avoiding collection of training dataset and domain shift. As requiring "zero" training
data, introducing effective visual prior is particularly important in Zero-
Shot image enhancement. Previous studies mainly focus on designing
task specific loss function to capture its internal physical process. On
the other, though incorporating signal processing methods into enhancement model is efficaciously performed in supervised learning, is less common in Zero-Shot learning. Aiming for further improvement and adding
promising leaps to Zero-Shot learning, this research proposes to incorporate Laplacian pyramid to network process. First, Multiscale Laplacian
Enhancement (MLE) is formulated, simply enhancing an input image
in the hierarchical Laplacian pyramid representation, resulting in detail
enhancement, image sharpening, and contrast improvement depending
on its hyper parameters. By combining MLE and introducing visual prior
specific to underwater images, Zero-Shot underwater image enhancement
model with only seven convolutional layers is proposed. Without prior
training and any training data, proposed model attains comparative performance compared with previous state-of-the-art models.

Keywords: Zero-shot learning · Underwater image enhancement ·
Laplacian pyramid · Image restoration

1 Introduction

Image enhancement is essential in measuring physical environment and progressively has been improved with deep learning. Previous deep learning models
mainly focus on supervised approaches by mapping degraded images to clear
images employing large scale image dataset in various image enhancement fields
[25,26,39]. However, constructing real large scale image pairs requires tremendous costs, as well as obtaining massive clear images is inherently difficult [25]

Supplementary Information The online version contains supplementary material
available at https://doi.org/10.1007/978-3-031-25069-9_46.

© The Author(s), under exclusive license to Springer Nature Switzerland AG 2023
L. Karlinsky et al. (Eds.): ECCV 2022 Workshops, LNCS 13804, pp. 721–737, 2023.
https://doi.org/10.1007/978-3-031-25069-9_46

or sometimes impossible in fields like underwater image processing [27]. Alternatively employed artificial dataset based on physical model or generative adversarial network (GAN) [19,28,29] suffers from domain shift, as artificial images are less informative and may be apart from real images, resulting in limited capability of deep learning compared with other image processing tasks [3,25].

Fig. 1. Example outputs of proposed MLE. Just enhancing an image in Laplacian pyramid representation, contrast improvement (left), detail enhancement (middle), and underwater image enhancement (right) is performed in the same MLE scheme.

As requiring "zero" training data, Zero-Shot image enhancement has been getting a lot of attention and promising results are shown in denoising [24], super-resolution [11], dehazing [25], back-lit image restoration [43], and underwater image enhancement [22]. Note that Zero-Shot image enhancement recovers images only from information of an input image itself, different from typical term in general classification task [25]. As no ground truth is available, constructing task specific loss function is especially important in Zero-Shot image enhancement. Currently, loss function is mainly introduced to reflect effective prior or bias of underlying phenomena or some specific task, followed by getting feedback from parameterized latent physical model or knowledge of the target task [14,25,38]. To be specific, based on the widely used atmospheric scatter model [30,31], efficient and clear output is obtained in dehazing task [25] with loss function reflecting internal physical process composed of global atmospheric light, transmission map, and statistical property, Dark Channel Prior [18]. In back-lit image restoration, luminance is adjusted with loss of parameterized s-curve function after being mapped to YIQ color space [43]. In underwater image enhancement, Zero-Shot learning based model on the Koschmieder's physical model is first proposed in [22]. Loss function for modifying the inherent property of an image like smoothness or color balance is also effective [8,38].

On the other hand, though supervised deep learning models have improved by incorporating traditional signal processing methods such as discrete wavelet transform, whitening and coloring transform, and Laplacian pyramid, less exists in Zero-Shot learning. In order to further accelerate Zero-Shot image enhancement, this research proposes to incorporate traditional Laplacian pyramid [7] to network process. First, Multiscale Laplacian Enhancement (denoted as MLE) is formulated, which simply convolves and enhances images in multiscale Laplacian pyramid representation, depending on three hyper parameters, kernel size K and standard deviation σ of Gaussian kernel for constructing Laplacian pyramid, and pyramid level L. Compared to simple convolution, enlarged receptive field which operates to an image is naturally formulated in MLE. By employing unsharp masking filter in MLE scheme, impacts of detail enhancement, image sharpening, and contrast improvement are experimentally shown (Fig. 1), depending on its hyper parameters. Though the effectiveness of MLE is observed, hyper parameter tuning of MLE is practically inconvenience. Accordingly, Zero-Shot Attention Network with Multiscale Laplacian Enhancement (denoted as ZA-MLE) is proposed to integrate several enhanced results of MLE. ZA-MLE consists of only seven convolutional layers and process for enhancing multiscale features of an image, implemented with element wise product and addition of a contrast improved result of MLE and a sharpened result of MLE (denoted as Zero-Shot Attention). Also, we introduce "prior" specific to underwater image enhancement for selecting MLE. Namely, we experimentally found that the top component of Laplacian pyramid of a degraded underwater image (2nd column of Fig. 2) contains less original signal information, thus just removing the top component tends to extract original, high frequency signal (3rd column of Fig. 2). Despite requiring no training data and prior training, proposed ZA-MLE mainly achieves comparative performance compared to other latest supervised models in underwater image enhancement. Also compared to previous Zero-Shot learning based model [22], our ZA-MLE advantageously works fast thanks to its simple structure, as well as quantitative scores improve. Our main contributions are summarized as follows: 1. Formulation of MLE simply convolving and enhancing an image in multiscale Laplacian pyramid representation. Depending on hyper parameters, effects of detail enhancement, image sharpening, and contrast improvement are shown. 2. Propose of ZA-MLE. To the best of our knowledge, ZA-MLE is first proposed Zero-Shot learning based underwater image enhancement model combining traditional signal processing method, Laplacian pyramid and unsharp masking filter. Though working in Zero-Shot manner, proposed ZA-MLE achieves favorable performance compared to latest models. 3. Propose of elaborated loss function for Zero-Shot learning. Gradient domain loss as well as color correction loss and reconstruction loss are combined.

2   Related Work

2.1   Traditional Signal Processing Method and Deep Learning

In this section, we state the relationship between conventional signal processing methods and deep learning. Rapid advancement of deep learning is accelerated by construction of large scale dataset and sophisticated very deep network architecture [17]. More recently, deep learning architecture has further developed by incorporating traditional signal processing methods. To be specific, discrete wavelet transform is employed to preserve fine structure of an input, by passing high frequency components extracted in the encoder part to the decoder part

[13,42]. In underwater image enhancement, white balance, histogram equalization, and gamma correction are combined to mitigate domain shift between training data and test data [27]. In style transfer, content features are projected to style features with whitening and coloring transform [12]. To the best of our knowledge, these hybrid deep learning models are basically limited to supervised learning and integrating Zero-Shot underwater image enhancement especially with Laplacian pyramid is first proposed in this research.

Fig. 2. Input image (1st column), the top Laplacian pyramid component of the image (2nd column), and result of subtraction of the top component from the original input (3rd column). We experimentally found that the top Laplacian pyramid component of a degraded underwater image dominates noise signal, and just removing the component tends to extract original signal. The 3rd column is multiplied by ten for visualization.

2.2   Laplacian Pyramid and Image Restoration

Multiscale feature of an image is valuable and efficiently extracted with Laplacian pyramid [4,7]. Laplacian pyramid hierarchically presents an image as a sum of band-pass images depending on resolution or frequency, practically obtained by subtracting adjacent elements of Gaussian pyramid constructed from the original image [7,33]. An image of high frequency features as edge or contour mainly present in low pyramid levels, while low frequency features like color is decomposed in high pyramid levels [33]. Laplacian pyramid is employed in loss function [5,40] or is combined with network architecture [20] to reflect various image features in supervised learning.

3   Multiscale Laplacian Enhancement for Image Manipulation

3.1   Formulation of Multiscale Laplacian Enhancement

We formulate Multiscale Laplacian Enhancement, denoted as MLE. MLE simply, yet efficiently enhances multiscale features of an image in Laplacian pyramid domain, owing to its hierarchical image representation reflecting frequency or resolution. The detailed derivation is found in the supplementary material.

Let an input image be $I(x)$, where x is the position of the image. An input image is first divided into Laplacian pyramid representation and each pyramid elements are simply filtered followed by the reconstruction phase, formulated as:

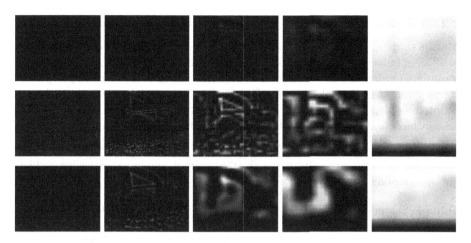

Fig. 3. Elements of Laplacian pyramid constructed from an original image (1st row), MLE-733 (2nd row), and MLE-799 (3rd row). From left to right, results of pyramid level $N = 1, 3, 5, 6, 7$ are shown. MLE-733 more enhances high frequency signal like edge or texture in lower pyramid levels, while MLE-799 more enhances low frequency signal like color in higher pyramid levels. Results are multiplied by ten for visualization.

$$MLE\left[I(x)\right] = \sum_{i=0}^{N-1} \left\{ U^i \left[L_i \left[I(x)\right] * Filter\right] \right\} + U^N \left[R_N * Filter\right]$$

where pyramid level N, a convolution $Filter$, and up-sampling operator U should be set beforehand. Here, $L_i(x)$ means the i-th level of Laplacian component obtained by subtracting adjacent elements of Gaussian pyramid constructed from the original image [7,33], defined as $L_i\left[I(x)\right] := (D \circ G)^i \left[I(x)\right] - G \circ (D \circ G)^i \left[I(x)\right]$. Above G and D mean low-pass Gaussian filter and down-sampling operator by a factor of 2, respectively. R_N is the top component (lowest resolution) of the Laplacian pyramid. Each Laplacian components are up-sampled to the original resolution of the input after filtering.

Compared with normal convolution, image size to which filters operate is different from MLE. To be specific, while MLE filters various resolution of an image of each Laplacian pyramid components, normal convolution operates only in the same size of the input, expressed as follows:

$$I(x) * Filter = \sum_{i=0}^{N-1} \left\{ U^i \left[L_i \left[I(x)\right]\right] * Filter \right\} + U^N \left[R_N\right] * Filter$$

The above equation is directly obtained using linearity of convolution and Laplacian pyramid representation of the input. Enlarged effective filter size is practically useful in extracting image features [10], which is naturally formulated in MLE. With the simple idea and implementation of MLE, multiscale features of an input image is efficiently enhanced.

3.2   Internal Results of Multiscale Laplacian Enhancement

MLE is simply defined to filter an image in Laplacian pyramid domain, resulting in efficient enhancement of overall multiscale features of an image with naturally formulated enlarged receptive field. Here, MLE depends on three hyper parameters, pyramid level N, kernel size K, standard deviation σ of Gaussian kernel, denoted as MLE-$NK\sigma$ in order. In order to examine the effectiveness of MLE, we use basic unsharp masking filter for MLE, traditionally employed in image sharpening task [36], denoted as: $Filter_{unsharp} := \left(\begin{smallmatrix} -1 & -1 & -1 \\ -1 & 9 & -1 \\ -1 & -1 & -1 \end{smallmatrix}\right)$. Selecting other sophisticated filters is our future work. Throughout numerical experiment, bicubic up-sampling is employed.

Table 1. Sharpness [15] and RMS contrast [34] of PIPAL [21] and DIV2K [1] dataset.

Dataset	Metric	MLE-133	MLE-333	MLE-599	MLE-799	UNSHARP	RAW
PIPAL [21]	RMS contrast	0.102	0.113	0.108	0.112	0.094	0.075
	Sharpness	2.961	2.799	2.692	2.615	2.768	1.063
DIV2K [1]	RMS contrast	0.085	0.094	0.091	0.098	0.079	0.069
	Sharpness	2.171	2.084	1.963	1.914	1.989	0.723

First, to comprehend the behavior of MLE scheme, the enhanced results of each Laplacian pyramid levels of an image are shown in Fig. 3. The first row shows elements of an original image, while the second and the third row show results of MLE-733 and MLE-799, respectively. Results of $N = 1, 3, 5, 6, 7$ are respectively shown from left to right. Each figures are multiplied by ten for visualization. Compared with the first row, various image features, contour captured in low pyramid levels and slightly appeared color signal in high pyramid levels, are efficiently enhanced. Also comparing the 2nd and 3rd row, each elements employing $K = 3$, $\sigma = 3$ (2nd row), preserve high frequency signal thus clearly sharpened, while results of $K = 9$, $\sigma = 9$ (3rd row), relatively cut off high frequency signal, emphasize color information more (4th column). Laplacian components of an original image (1st row) is dark and hardly be seen.

3.3   Comparison with Unsharp Masking Filter

Next, we proceed to comparison with conventional unsharp masking filter generally employed in image sharpening [36]. Qualitative and quantitative results of unsharp masking filter are respectively shown in the 4th column of Fig. 4 and Table 1. We utilize reference images from PIPAL dataset [21] and high resolution images from DIV2K dataset [1], mainly employed in image restoration task. 2nd and 3rd column of Fig. 4 respectively present results of MLE-133 and MLE-799, while 1st column presents original input images, denoted as RAW. Detail enhancement or edge emphasis is performed with MLE-133, while sharpness slightly improved with unsharp masking filter (1st row of Fig. 4). Note that filtering is performed totally twice in MLE-133. Compared with MLE-799, 3rd column, characteristically vivid, contrast enhanced results are obtained. As image

Fig. 4. Comparison with MLE and unsharp masking filter. 1st column shows input images, 2nd and 3rd column shows results of MLE-133 and MLE-799, respectively. 4th column shows results of unsharp masking filter. Image sharpening or detail enhancement is performed in MLE-133, while contrast improvement is performed in MLE-799.

features of higher pyramid levels are also enhanced, the overall sharpening effect of MLE-799 is weaker than MLE-133. For quantitative evaluation, Sharpness [15] and RMS contrast [34] are evaluated. Sharpness is the strength of vertical and horizontal gradient after Sobel filtering, and RMS contrast means the standard deviation of luminance intensities. In quantitative results from Table 1, all metrics are improved with MLE and unsharp masking filter from the original image. Sharpness is the highest in MLE-133 and decreased with higher pyramid level, as Sharpness measures image gradient and relatively higher in MLE-133 emphasizing only high frequency features. MLE-333 gets the highest score for RMS contrast in PIPAL dataset [21], while MLE-799 is the 1st in DIV2K dataset [1]. As RMS contrast measures dispersion of luminance of an image, MLE-799, also enhancing higher pyramid components, got higher scores. Image sharpening, detail enhancement, and contrast improvement are performed in the same MLE scheme depending on its hyper parameters.

3.4   Ablation Study of MLE

In this section, results of different parameter settings of MLE, namely, pyramid level N, kernel size K, and standard deviation σ of Gaussian kernel for

Fig. 5. Results of different pyramid levels of MLE. From left to right, images of MLE-133, MLE-333, MLE-733, and an input image are shown, respectively.

Fig. 6. Comparison with unsharp masking filter (3rd column). Input blurred underwater images (1st column) are enhanced with MLE-833 (2nd column).

constructing Laplacian pyramid are evaluated. Results of different pyramid levels are shown in Fig. 5. Input image (4th column) is enhanced with MLE-133 (1st column), MLE-333 (2nd column), and MLE-733 (3rd column). As we confirmed in the 2nd and 3rd columns of Fig. 4 and Table 1, high frequency signal like configuration or edge of banked up rock is sharpened with MLE-333, while contrast is improved with MLE-733. The number of filtering as well as resolution of pyramid components to which filters operate changes in accordance with pyramid levels, resulting in enhancement of various image features of an image.

As for K and σ of Gaussian kernel, the lower K and σ are, the smaller the effect of blurriness of filtering, as a consequence, wide range of frequency band tends to preserve also in higher pyramid levels, thus is strongly enhanced. Qualitative results are found in the supplementary material. Also refer to the 2nd and 3rd rows of Fig. 3. The degree of enhancement basically depends on contrast or sharpness of an original input as well as hyper parameters of MLE. As for clear land images, we experimentally observe that $K = 9$, $\sigma = 9$ are usually optimal setting for contrast improvement, and $K = 3$, $\sigma = 3$, $N = 1$ for image sharpening or detail enhancement. Practically, hyper parameters of MLE needs to be selected depending on objective task or input sharpness.

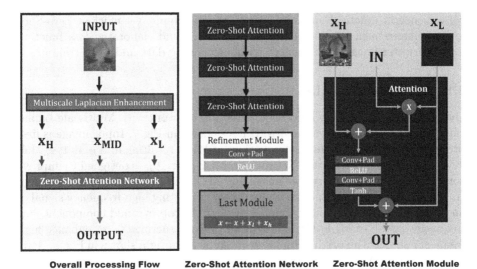

Fig. 7. Overall processing flow of ZA-MLE (left), Zero-Shot Attention Network (middle), and Zero-Shot Attention Module (right). 1: An input is first processed with different parameter settings of MLE, resulting in X_H from MLE-753 and X_L from MLE-331. Note that the top pyramid component of X_L is removed to enhance high frequency signal. X_{MID} is calculated with $X_{MID} := X_H * X_L + X_L$. 2: Input X_L, X_H, and X_{MID} into Zero-Shot Attention Network, aiming for enhancing multiscale feature of an input in Zero-Shot manner. 3: Output is obtained after training of 300 epochs. Different from elaborated learning strategy [39], we stop training in 300 epochs and fixed.

3.5 Application of MLE to Underwater Images

We proceed to application of MLE to underwater images. As many underwater images suffer from lowering of contrast or blurriness, MLE favorably sharpens severely degraded underwater images. Compared to results of unsharp masking filter shown in Fig. 6, severely degraded underwater images (1st column) are prominently recovered with MLE-833 (2nd column), though traditional unsharp masking not (3rd column). Qualitative behavior of MLE is similar both in underwater and land images, though underwater images are usually low contrast and blurred. While effectiveness of MLE is also observed in underwater images, halo sometimes appears in some MLE results like 2nd row of Fig. 4 or 1st row of Fig. 6, depending on image sharpeness and hyper parameters of MLE, caused by linearity of filter [33]. In this research, we set widely utilized linear unsharp masking filter to confirm effectiveness of MLE, and incorporating edge preserving, more sophisticated filter to MLE is our future work.

4   Zero-Shot Attention Network with Multiscale Laplacian Enhancement (ZA-MLE)

As we discuss in previous sections, results of MLE depend on input image sharpness or hyper parameters, which is practically troublesome. For the convenience in practical application, we propose simple, yet efficient Zero-Shot image enhancement scheme combined with MLE and elaborated loss function for underwater image enhancement without training data and prior training.

4.1 Process of ZA-MLE

Overall processing flow of Zero-Shot Attention Network with Multiscale Laplacian Enhancement (denoted as ZA-MLE) is shown in Fig. 7. Input image is first processed with different parameter settings of MLE, resulting in X_H and X_L, followed by Zero-Shot Attention Network. In MLE part, X_H is designed to improve contrast of an input underwater image and MLE-753 is experimentally selected. As for X_L, MLE-331 is experimentally set for extracting high frequency signal of an image like edge or configuration. Note that the top pyramid component, low frequency signal, of MLE-331 is removed in reconstructing X_L to enhance high frequency signal of an original input, inspired by the insight shown in Fig. 2. Here, $X_{MID} := X_L * X_H + X_L$ is employed as an input of ZA-MLE. After obtaining X_H and X_L, inspired by [37], element wise product and addition follow, which is designed to sharpen a contrast improved X_H with attention of sharpened X_L to integrate enhanced results of MLE, denoted as Zero-Shot Attention. After Zero-Shot Attention, X_H is added followed by layers of one convolution, Leaky ReLU activation, one convolution, and Tanh activation (right in Fig. 7). The number of channels are increased from three to six in the first convolution, and decreased from six to three in the second convolution. X_L are finally added to enhance low frequency features. This procedure is denoted as Zero-Shot Attention module. After three Zero-Shot Attention modules, Refinement Module consisting of one convolution and Leaky ReLU activation layers follows. Then, X_H and X_L are finally added to incorporate multiscale features of an image, denoted as Last Module. As each Zero-Shot Attention module contains only two convolutional layers, proposed ZA-MLE totally includes seven convolutional layers. With the power of proposed MLE, simply implemented ZA-MLE enables efficient image enhancement of challenging real underwater images.

4.2 Loss Function

The loss function for training ZA-MLE consists of three terms, reconstruction loss l_{rec}, derivation loss l_{deriv}, and color loss l_{col}, defined as follows:

$$Loss = \alpha l_{rec} + \beta l_{deriv} + \lambda l_{col}$$
$$l_{rec} := \|X_{out} - X_H\|_1$$
$$l_{deriv} := \| \left(\partial_x^2 \left(X_{out} \right) - \partial_x^2 \left(X_L \right) \right) + \left(\partial_y^2 \left(X_{out} \right) - \partial_y^2 \left(X_L \right) \right) \|_1$$
$$l_{col} := \|R_{out} - G_{out}\|_1 + \|B_{out} - G_{out}\|\|_1$$

where ∂_x^2 and ∂_y^2 respectively compute 2nd order horizontal and vertical gradients. R_{out}, G_{out}, and B_{out} respectively mean R, G, B channels of X_{out}.

The reconstruction loss l_{rec} works as regularization term and defined as l1 distance between model output X_{out} and X_H, in order to get output similar to

Table 2. Results of UIQM and UCIQE. Proposed ZA-MLE without training data and prior training achieves comparative results compared with other latest supervised and Zero-Shot learning based models. Scores of ZA-MLE show the average of three trials.

Dataset		ZA-MLE	UWCNN [26]	Water-Net [27]	U-Transformer [35]	All-in-One [29]	Koschmieder [22]
Challenging-60 [27]	UIQM	2.754	2.386	2.609	2.724	2.679	2.402
	UCIQE	4.986	3.203	4.554	4.231	4.690	4.170
Original	UIQM	2.989	2.558	3.031	2.664	2.900	1.946
	UCIQE	5.424	4.127	6.057	5.313	2.998	5.392

X_H to incorporate MLE. Note that reconstruction loss l_{rec} in this research is different from [38] which employs an original input image X_{in} instead of X_H, to get output similar to X_{in}. The derivation loss l_{deriv} is l1 distance between X_{out} and X_L in gradient domain [9] designed to reflect gradient information of X_L and inhibit noise. For correction of color distortion of underwater images, based on the gray world assumption, white balance is performed to enforce each RGB channels to have the same values as in land images [6,38]. Modified from the original gray world assumption, proposed color correction loss l_{col} is implemented to enforce R and B channels to have similar values to G channel of X_{out}, as G channel of an underwater image is less susceptible to underwater conditions [2].

Input image processed with MLE is passed through Zero-Shot Attention Network and the network is trained based on the above loss function employing X_L and X_H. Note that X_L and X_H are obtained from an input image itself thus requiring no training data. We experimentally observe that visually pleasing result is obtained around 200 to 300 epochs as in following section.

5 Experiment

5.1 Experimental Setting

We evaluate the performance of ZA-MLE for underwater image enhancement. Network parameters are initialized with [16], and used Adam optimizer [23] with the learning rate 0.001. Unlike previous elaborated learning strategy [39], we stop training at 300 epochs, experimentally selected, and fixed throughout the experiment. ZA-MLE works as Zero-Shot manner, trained per an input image without additional data or prior training. For practical application, prior training is recommended for more speed up. As discussed in previous section, hyper parameters of MLE, pyramid level, kernel size, and standard deviation, are respectively set as MLE-753 for X_H and MLE-331 for X_L in order. X_{MID}, defined as $X_{MID} = X_L * X_H + X_L$, is employed for the input. As for real underwater image dataset, Challenging-60 (4th to 6th column of Fig. 8) [27] and the Original dataset containing notably deteriorated 77 images taken in Okinawa, Japan (1st to 3rd column of Fig. 8), are employed for the evaluation. In evaluating recovered results, generally utilized non-reference metric, UIQM [32] and UCIQE [41] designed to reflect human perception, are computed as existing no ground truth

Fig. 8. Recovered results of underwater image enhancement models. 1st row: input raw images. 2nd row: results of ZA-MLE. 3rd row: results of UWCNN [26]. 4th row: results of Koschmieder [22]. 5th row: results of Water-Net [27]. 6th row: row: results of U-Transformer [35]. 7th row: results of All-in-One [29].

for real images. Coefficients of loss functions, α, β, and λ are respectively set to 1.0, 1.0, and 0.1. We implement our model with PyTorch and GeForce RTX 2080 Ti GPU. https://github.com/tkswalk/2022-ECCV-VIPriorsWorkshop.

5.2   Results and Discussions of ZA-MLE

We compare our ZA-MLE with state-of-the-art underwater image enhancement methods. Currently, as Zero-Shot learning based model is rare [22], available supervised models are also compared. First row of Fig. 8 shows input raw underwater images, and the second row shows results of proposed ZA-MLE. 4th row shows results of Zero-Shot model, denoted as Koschmieder [22]. The rest row show results of supervised models, 3rd row: UWCNN [26], 5th row: Water-Net [27], 6th row: U-Transformer [35], and 7th row: All-in-One [29].

In UIQM [32] and UCIQE [41] scores of Table 2, our ZA-MLE achieves favorable performance, getting first rank in the Challenging-60 dataset [27] and second rank in the Original dataset, compared with other supervised models which require training data or prior training. In quantitative results of Fig. 8, ZA-MLE (2nd row) basically corrects blueish (1st and 2nd column) and even yellowish (6th column) color cast, as well as visibility of blurred underwater images (3rd to 5th column) improves. Among previous methods, CNN based Water-Net combining white balance, histogram equalization, and gamma correction [27] (5th row), qualitatively and quantitatively recovers well. Owing to conducting white balance to an input image, this network recovers also a yellowish image (6th column), yet not sufficiently sharpens deteriorated images and somewhat blurred. By contrast, our ZA-MLE sharpens blurred images thanks to proposed MLE and Zero-Shot Attention. Results of U-Transformer [35] (6th row), Vision Transformer based model, are relatively well, but fail to recover yellowish image (6th column) and sometimes adds grid artifacts (5th column). UWCNN [26] and All-in-One [29] hardly improve visibility. Supervised models often suffer from domain shift between training data and test data caused by complex real underwater environment, still a challenging issue [3].

Next, comparison with results of the Koschmieder [22] (4th row of Fig. 8) are shown. Currently, to the best of our knowledge, Zero-Shot learning based underwater image enhancement model does not exist except [22]. Code and parameter setting is directly employed provided by the authors, setting epoch size to 10000. In terms of UIQM and UCIQE in Table 2, proposed ZA-MLE performs better than Koschmieder [22] in both dataset. In quantitative comparison, [22] corrects blueish and yellowish underwater images based on the Koschmieder's physical model, yet also outputs a little over enhanced results (1st, 2nd, and 3rd column of Fig. 8). Specifically, color distortion is sometimes observed as in the 2nd column. As [22] mainly corrects color cast based on the Koschmieder's model, recovered results of severely degraded underwater images are not very good, which do not likely to obey the Koschmieder's model.

Our ZA-MLE is different from [22] in that incorporating traditional signal processing method, Laplacian pyramid and unsharp masking filter. In terms of calculation time, owing to its simple structure, ZA-MLE, trained with 300 epochs, costs about 2 s, while Koschmieder, trained with 10000 epochs, costs about 5 min to process 256×256 images, 150 times faster than Koschmieder. Time analysis is conducted with NVIDIA 2080Ti GPU and Intel Core i9-9900K CPU.

Table 3. UIQM and UCIQE scores calculated from different loss functions and dataset.

Loss	Dataset	Challenging-60	Original
l_{rec}	UIQM/UCIQE	2.749/4.946	2.957/5.273
$l_{rec} + l_{deriv}$	UIQM/UCIQE	2.770/4.987	3.029/5.442
ALL	UIQM/UCIQE	2.772/4.967	3.030/5.394

Table 4. Comparison of UIQM and UCIQE scores of two color loss functions. Average scores of 10 trials are shown.

Loss	Dataset	Challenging-60	Original
Propose	UIQM/UCIQE	2.748/4.978	2.987/5.361
Gray-world [6]	UIQM/UCIQE	2.748/4.961	2.987/5.350

5.3   Ablation Study of Loss Function

Ablation study of the loss function combining three terms is shown in Table 3. UIQM and UCIQE scores of the loss function employing l_{rec}, $l_{rec} + l_{deriv}$, and all terms are compared. In terms of UIQM, weighting contrast or sharpness more, all terms contribute in all dataset. As for UCIQE, weighting chroma or saturation of an image, while l_{deriv} improves the score, color loss l_{col} a little decreases UCIQE in both datasets. l_{col} is designed to balance color channels of underwater images, making R and B channels close to G channel. Decreased UCIQE especially in the Original dataset might be caused by decreased chroma of an output, as Original dataset more includes greenish underwater images.

Proposed l_{col} is modified from the original gray world assumption [6], which enforces all channels to be the same, and UIQM and UCIQE scores are shown in Table 4. Proposed l_{col} slightly improves the scores of UCIQE. We observe that l_{col} also contributes to learning stability and adopted.

6   Conclusion

This research proposes a simple, yet efficient Zero-Shot image enhancement scheme incorporating traditional signal processing method, Laplacian pyramid. First, MLE just convolving and enhancing images in multiscale Laplacian pyramid representation is formulated. Combined with basic unsharp masking filter in MLE scheme, the effects of image sharpening, detail enhancement, and contrast improvement are shown depending on its hyper parameters. Combining MLE to network process, ZA-MLE is also proposed to enhance underwater images trained with the elaborated loss function. To the best of our knowledge, Zero-Shot learning based underwater image enhancement model incorporating Laplacian pyramid is first proposed in this research. By reflecting visual prior specific to underwater images, simply implemented ZA-MLE achieves comparative performance compared to other latest deep learning models.

References

1. Agustsson, E., Timofte, R.: Ntire 2017 challenge on single image super-resolution: Dataset and study. In: Proceedings of the IEEE conference on computer vision and pattern recognition workshops, pp. 126–135 (2017)
2. Ancuti, C.O., Ancuti, C., De Vleeschouwer, C., Bekaert, P.: Color balance and fusion for underwater image enhancement. IEEE Trans. Image Process. **27**(1), 379–393 (2017)
3. Anwar, S., Li, C.: Diving deeper into underwater image enhancement: a survey. Signal Process. Image Commun. **89**, 115978 (2020). https://doi.org/10.1016/j.image.2020.115978, www.sciencedirect.com/science/article/pii/S0923596520301478
4. Aubry, M., Paris, S., Hasinoff, S.W., Kautz, J., Durand, F.: Fast local Laplacian filters: theory and applications. ACM Trans. Graph. (TOG) **33**(5), 1–14 (2014)
5. Bojanowski, P., Joulin, A., Lopez-Pas, D., Szlam, A.: Optimizing the latent space of generative networks. In: Dy, J., Krause, A. (eds.) Proceedings of the 35th International Conference on Machine Learning. Proceedings of Machine Learning Research, vol. 80, pp. 600–609. PMLR (10–15 Jul 2018), https://proceedings.mlr.press/v80/bojanowski18a.html
6. Buchsbaum, G.: A spatial processor model for object colour perception. J. Franklin institute **310**(1), 1–26 (1980)
7. Burt, P.J., Adelson, E.H.: The laplacian pyramid as a compact image code. In: Readings in Computer Vision, pp. 671–679. Elsevier (1987)
8. Cao, X., Rong, S., Liu, Y., Li, T., Wang, Q., He, B.: Nuicnet: non-uniform illumination correction for underwater image using fully convolutional network. IEEE Access **8**, 109989–110002 (2020)
9. Chambolle, A.: An algorithm for total variation minimization and applications. J. Math. Imaging Vis. **20**(1), 89–97 (2004)
10. Chen, L.C., Zhu, Y., Papandreou, G., Schroff, F., Adam, H.: Encoder-decoder with atrous separable convolution for semantic image segmentation. In: Proceedings of the European Conference on Computer Vision (ECCV), pp. 801–818 (2018)
11. Cheng, Z., Xiong, Z., Chen, C., Liu, D., Zha, Z.J.: Light field super-resolution with zero-shot learning. In: Proceedings of the IEEE/CVF Conference on Computer Vision and Pattern Recognition, pp. 10010–10019 (2021)
12. Cho, W., Choi, S., Park, D.K., Shin, I., Choo, J.: Image-to-image translation via group-wise deep whitening-and-coloring transformation. In: Proceedings of the IEEE/CVF Conference on Computer Vision and Pattern Recognition, pp. 10639–10647 (2019)
13. Fu, M., Liu, H., Yu, Y., Chen, J., Wang, K.: DW-GAN: A discrete wavelet transform GAN for nonhomogeneous dehazing. In: Proceedings of the IEEE/CVF Conference on Computer Vision and Pattern Recognition, pp. 203–212 (2021)
14. Gandelsman, Y., Shocher, A., Irani, M.: " double-dip": unsupervised image decomposition via coupled deep-image-priors. In: Proceedings of the IEEE/CVF Conference on Computer Vision and Pattern Recognition, pp. 11026–11035 (2019)
15. Gao, W., Zhang, X., Yang, L., Liu, H.: An improved sobel edge detection. In: 2010 3rd International Conference on Computer Science and Information Technology, vol. 5, pp. 67–71. IEEE (2010)
16. Glorot, X., Bengio, Y.: Understanding the difficulty of training deep feedforward neural networks. In: Proceedings of the Thirteenth International Conference on Artificial Intelligence and Statistics, pp. 249–256. JMLR Workshop and Conference Proceedings (2010)

17. Goodfellow, I., Bengio, Y., Courville, A.: Deep learning. MIT press (2016)
18. He, K., Sun, J., Tang, X.: Single image haze removal using dark channel prior. IEEE Trans. Pattern Anal. Mach. Intell. **33**(12), 2341–2353 (2010)
19. Islam, M.J., Luo, P., Sattar, J.: Simultaneous Enhancement and Super-Resolution of Underwater Imagery for Improved Visual Perception. In: Robotics: Science and Systems (RSS). Corvalis, Oregon, USA (July 2020). https://doi.org/10.15607/RSS. 2020.XVI.018
20. Jin, C., Deng, L.J., Huang, T.Z., Vivone, G.: Laplacian pyramid networks: a new approach for multispectral pansharpening. Inf. Fusion **78**, 158–170 (2022)
21. Jinjin, G., Haoming, C., Haoyu, C., Xiaoxing, Y., Ren, J.S., Chao, D.: PIPAL: a large-scale image quality assessment dataset for perceptual image restoration. In: Vedaldi, A., Bischof, H., Brox, T., Frahm, J.-M. (eds.) ECCV 2020. LNCS, vol. 12356, pp. 633–651. Springer, Cham (2020). https://doi.org/10.1007/978-3-030-58621-8_37
22. Kar, A., Dhara, S.K., Sen, D., Biswas, P.K.: Zero-shot single image restoration through controlled perturbation of koschmieder's model. In: Proceedings of the IEEE/CVF Conference on Computer Vision and Pattern Recognition, pp. 16205–16215 (2021)
23. Kingma, D.P., Ba, J.: Adam: a method for stochastic optimization. In: ICLR (Poster) (2015)
24. Lehtinen, J., et al.: Noise2Noise: learning image restoration without clean data. In: Dy, J., Krause, A. (eds.) Proceedings of the 35th International Conference on Machine Learning. Proceedings of Machine Learning Research, vol. 80, pp. 2965–2974. PMLR (10–15 Jul 2018), https://proceedings.mlr.press/v80/lehtinen18a. html
25. Li, B., Gou, Y., Liu, J.Z., Zhu, H., Zhou, J.T., Peng, X.: Zero-shot image dehazing. IEEE Trans. Image Process. **29**, 8457–8466 (2020)
26. Li, C., Anwar, S., Porikli, F.: Underwater scene prior inspired deep underwater image and video enhancement. Pattern Recogn. **98**, 107038 (2020)
27. Li, C., Guo, C., Ren, W., Cong, R., Hou, J., Kwong, S., Tao, D.: An underwater image enhancement benchmark dataset and beyond. IEEE Trans. Image Process. **29**, 4376–4389 (2019)
28. Li, J., Skinner, K.A., Eustice, R.M., Johnson-Roberson, M.: WaterGAN: unsupervised generative network to enable real-time color correction of monocular underwater images. IEEE Robot. Autom. Lett. **3**(1), 387–394 (2017)
29. M Uplavikar, P., Wu, Z., Wang, Z.: All-in-one underwater image enhancement using domain-adversarial learning. In: The IEEE Conference on Computer Vision and Pattern Recognition (CVPR) Workshops (June 2019)
30. McCartney, E.J.: Optics of the atmosphere: scattering by molecules and particles. New York (1976)
31. Narasimhan, S.G., Nayar, S.K.: Chromatic framework for vision in bad weather. In: Proceedings IEEE Conference on Computer Vision and Pattern Recognition, CVPR 2000 (Cat. No. PR00662), vol. 1, pp. 598–605. IEEE (2000)
32. Panetta, K., Gao, C., Agaian, S.: Human-visual-system-inspired underwater image quality measures. IEEE J. Oceanic Eng. **41**(3), 541–551 (2015)
33. Paris, S., Hasinoff, S.W., Kautz, J.: Local Laplacian filters: edge-aware image processing with a Laplacian pyramid. ACM Trans. Graph. **30**(4), 68 (2011)
34. Peli, E.: Contrast in complex images. JOSA A **7**(10), 2032–2040 (1990)
35. Peng, L., Zhu, C., Bian, L.: U-shape transformer for underwater image enhancement. arXiv preprint arXiv:2111.11843 (2021)

36. Polesel, A., Ramponi, G., Mathews, V.: Image enhancement via adaptive unsharp masking. IEEE Trans. Image Process. **9**(3), 505–510 (2000). https://doi.org/10. 1109/83.826787

37. Qin, X., Wang, Z., Bai, Y., Xie, X., Jia, H.: FFA-Net: feature fusion attention network for single image dehazing. In: Proceedings of the AAAI Conference on Artificial Intelligence, vol. 34, pp. 11908–11915 (2020)

38. Sharma, A., Tan, R.T.: Nighttime visibility enhancement by increasing the dynamic range and suppression of light effects. In: Proceedings of the IEEE/CVF Conference on Computer Vision and Pattern Recognition (CVPR), pp. 11977–11986 (June 2021)

39. Ulyanov, D., Vedaldi, A., Lempitsky, V.: Deep image prior. In: Proceedings of the IEEE Conference on Computer Vision and Pattern Recognition, pp. 9446–9454 (2018)

40. Wu, H., Liu, J., Xie, Y., Qu, Y., Ma, L.: Knowledge transfer dehazing network for nonhomogeneous dehazing. In: Proceedings of the IEEE/CVF Conference on Computer Vision and Pattern Recognition (CVPR) Workshops (June 2020)

41. Yang, M., Sowmya, A.: An underwater color image quality evaluation metric. IEEE Trans. Image Process. **24**(12), 6062–6071 (2015)

42. Yoo, J., Uh, Y., Chun, S., Kang, B., Ha, J.W.: Photorealistic style transfer via wavelet transforms. In: Proceedings of the IEEE/CVF International Conference on Computer Vision, pp. 9036–9045 (2019)

43. Zhang, L., Zhang, L., Liu, X., Shen, Y., Zhang, S., Zhao, S.: Zero-shot restoration of back-lit images using deep internal learning. In: Proceedings of the 27th ACM International Conference on Multimedia, pp. 1623–1631 (2019)

Beyond a Video Frame Interpolator: A Space Decoupled Learning Approach to Continuous Image Transition

Tao Yang[1]([✉]), Peiran Ren[1], Xuansong Xie[1], Xian-Sheng Hua[1], and Lei Zhang[2]

[1] DAMO Academy, Alibaba Group, Hangzhou, China
yangtao9009@gmail.com, peiran_r@sohu.com, xingtong.xxs@taobao.com,
xiansheng.hxs@alibaba-inc.com
[2] Department of Computing, The Hong Kong Polytechnic University,
Kowloon, China
cslzhang@comp.polyu.edu.hk

Abstract. Video frame interpolation (VFI) aims to improve the temporal resolution of a video sequence. Most of the existing deep learning based VFI methods adopt off-the-shelf optical flow algorithms to estimate the bidirectional flows and interpolate the missing frames accordingly. Though having achieved a great success, these methods require much human experience to tune the bidirectional flows and often generate unpleasant results when the estimated flows are not accurate. In this work, we rethink the VFI problem and formulate it as a continuous image transition (CIT) task, whose key issue is to transition an image from one space to another space continuously. More specifically, we learn to implicitly decouple the images into a translatable flow space and a non-translatable feature space. The former depicts the translatable states between the given images, while the later aims to reconstruct the intermediate features that cannot be directly translated. In this way, we can easily perform image interpolation in the flow space and intermediate image synthesis in the feature space, obtaining a CIT model. The proposed space decoupled learning (SDL) approach is simple to implement, while it provides an effective framework to a variety of CIT problems beyond VFI, such as style transfer and image morphing. Our extensive experiments on a variety of CIT tasks demonstrate the superiority of SDL to existing methods. The source code and models can be found at https://github.com/yangxy/SDL.

Keywords: Video frame interpolation · Continuous image transition · Image synthesis · Space decoupled learning

1 Introduction

Video frame interpolation (VFI) targets at synthesizing intermediate frames between the given consecutive frames of a video to overcome the temporal limitations of camera sensors. VFI can be used in a variety of practical applications,

Supplementary Information The online version contains supplementary material available at https://doi.org/10.1007/978-3-031-25069-9_47.

© The Author(s), under exclusive license to Springer Nature Switzerland AG 2023
L. Karlinsky et al. (Eds.): ECCV 2022 Workshops, LNCS 13804, pp. 738–755, 2023.
https://doi.org/10.1007/978-3-031-25069-9_47

including slow movie generation [26], motion deblurring [53] and visual quality enhancement [68]. The conventional VFI approaches [2] usually calculate optical flows between the source and target images and gradually synthesize the intermediate images. With the great success of deep neural networks (DNNs) in computer vision tasks [16,21,51], recently researchers have been focusing on developing DNNs to address the challenging issues of VFI.

Most DNN based VFI algorithms can be categorized into flow-based [4,26, 40,67], kernel-based [32,41,53], and phase-based ones [37,38]. With the advancement of optical flow methods [5,58], flow-based VFI algorithms have gained increasing popularity and shown good quantitative results on benchmarks [4,40]. However, these methods require much human experience to tune the bidirectional flows, e.g., by using the forward [4,26] and backward [39,40] warping algorithms. In order to improve the synthesis performance, some VFI methods have been developed by resorting to the depth information [4], the acceleration information [67] and the softmax splatting [40]. These methods, however, adopt the off-the-shelf optical flow algorithms, and hence they often generate unpleasant results when the estimated flows are not accurate.

To address the above issues, we rethink the VFI problem and aim to find a solution that is free of flows. Different from previous approaches, we formulate VFI as a continuous image transition (CIT) problem. It is anticipated that we could construct a smooth transition process from the source image to the target image so that the VFI can be easily done. Actually, there are many CIT tasks in computer vision applications, such as image-to-image translation [24,69], image morphing [34,45] and style transfer [19,23]. Different DNN models have been developed for different CIT tasks. Based on the advancement of deep generative adversarial network (GAN) techniques [7,28,29], deep image morphing methods have been proposed to generate images with smooth semantic changes by walking in a latent space [25,48]. Similarly, various image-to-image translation methods have been developed by exploring intermediate domains [14,20,66], interpolating attribute [36] or feature [60] or kernel [63] vectors, using physically inspired models for guidance [47], and navigating latent spaces with discovered paths [9, 25]. Though significant progresses have been achieved for CIT, existing methods usually rely on much human knowledge of the specific domain, and employ rather different models for different applications.

In this work, we propose to learn a translatable flow space to control the continuous and smooth translation between two images, while synthesize the image features which cannot be translated. Specifically, we present a novel space decoupled learning (SDL) approach for VFI. Our SDL implicitly decouples the image spaces into a translatable flow space and a non-translatable feature space. With the decoupled image spaces, we can easily perform smooth image translation in the flow space, and synthesize intermediate image features in the non-translatable feature space. Interestingly, the proposed SDL approach can not only provide a flexible solution for VFI, but also provide a general and effective solution to other CIT tasks.

To the best of our knowledge, the proposed SDL is the first flow-free algorithm which is however able to synthesize consecutive interpolations, achieving leading performance in VFI. SDL is easy-to-implement, and it can be readily integrated into off-the-shelf DNNs for different CIT tasks beyond VFI, serving as a general-purpose solution to the CIT problem. We conduct extensive experiments on various CIT tasks, including, VFI, image-to-image translation and image morphing, to demonstrate its effectiveness. Though using the same framework, SDL shows highly competitive performance with those state-of-the-art methods that are specifically designed for different CIT problems.

2   Related Work

2.1   Video Frame Interpolation (VFI)

With the advancement of DNNs, recently significant progresses have been made on VFI. Long *et al.* [35] first attempted to generate the intermediate frames by taking a pair of frames as input to DNNs. This method yields blurry results since the motion information of videos is not well exploited. The latter works are mostly focused on how to effectively model motion and handle occlusions. Meyer *et al.* [37,38] proposed phase-based models which represent motion as per-pixel phase shift. Niklaus *et al.* [41,42] came up with the kernel-based approaches that estimate an adaptive convolutional kernel for each pixel. Lee *et al.* [32] introduced a novel warping module named Adaptive Collaboration of Flows (AdaCoF). An end-to-end trainable network with channel attention was proposed by Choi *et al.* [12], where frame interpolation is achieved without explicit estimation of motion. The kernel-based methods have achieved impressive results. However, they are not able to generate missing frames with arbitrary interpolation factors and usually fail to handle large motions due to the limitation of kernel size.

Unlike phase-based or kernel-based methods, flow-based models explicitly exploit motion information of videos [4,26,40,67]. With the advancement of optical flow methods [5,58], flow-based VFI algorithms have become popular due to their good performance. Niklaus and Liu [39] adopted forward warping to synthesize intermediate frames. This algorithm suffers from holes and overlapped pixels, and it was later improved by the softmax splatting method [40], which can seamlessly map multiple source pixels to the same target location. Since forward warping is not very intuitive to use, most flow-based works adopt backward warping. Jiang *et al.* [26] jointly trained two U-Nets [52], which respectively estimate the optical flows and perform bilateral motion approximation to generate intermediate results. Reda *et al.* [50] and Choi *et al.* [11] further improved this work by introducing cycle consistency loss and meta-learning, respectively. Bao *et al.* [4] explicitly detected the occlusion by exploring the depth information, but the VFI performance is sensitive to depth estimation accuracy. To exploit the acceleration information, Xu *et al.* [67] proposed a quadratic VFI method. Recently, Park *et al.* [44] proposed a bilateral motion network to estimate intermediate motions directly.

2.2   Continuous Image Transition (CIT)

In many image transition tasks, the key problem can be formulated as how to transform an image from one state to another state. DNN based approaches have achieved impressive results in many image transition tasks, such as image-to-image translation [24,62,69], style transfer [19,27], image morphing [9] and VFI [32,42]. However, these methods are difficult to achieve continuous and smooth transition between images. A continuous image transition (CIT) approach is desired to generate the intermediate results for a smooth transition process.

Many researches on image-to-image translation and image morphing resort to finding a latent feature space and blending image features therein [36,47,60]. However, these methods need to explicitly define the feature space based on human knowledge of the domain. Furthermore, encoding an image to a latent code often results in the loss of image details. Alternatively, methods on image morphing and VFI first establish correspondences between the input images, for example, by using a warping function or bidirectional optical flows, to perform shape deformation of image objects, and then gradually blend images for smooth appearance transition [4,33,40,65]. Unfortunately, it is not easy to accurately specify the correspondences, leading to superimposed appearance of the intermediate results. In addition to generating a continuous transition between two input images (source and target), there are also methods to synthesize intermediate results between two different outputs [22,23].

Image-to-Image Translation: Isola *et al.* [24] showed that the conditional adversarial networks (cGAN) can be a good solution to image-to-image (I2I) translation problems. Many following works, such as unsupervised learning [69], disentangled learning [31], few-shot learning [34], high resolution image synthesis [62], multi-domain translation [13], multi-modal translation [70], have been proposed to extend cGAN to different scenarios. Continuous I2I has also attracted much attention. A common practice to this problem is to find intermediate domains by weighting discriminator [20] or adjusting losses [66]. Some methods have been proposed to enable controllable I2I by interpolating attribute [36] or feature [60] or kernel [63] vectors. Pizzati *et al.* [47] proposed a model-guided framework that allows non-linear interpolations.

Image Morphing: Conventional image morphing methods mostly focus on reducing user-intervention in establishing correspondences between the two images [65]. Smythe [55] used pairs of mesh nodes for correspondences. Beier and Neely [6] developed field morphing utilizing simpler line segments other than meshes. Liao *et al.* [33] performed optimization of warping fields in a specific domain. Recently, methods [1,25,45] have been proposed to achieve efficient image morphing by manipulating the latent space of GANs [7,29]. However, these methods often result in the loss of image details and require time-consuming iterative optimization during inference. Mao *et al.* [36] and Pizzati *et al.* [47] decoupled content and style spaces using disentangled representations. They achieved continuous style interpolations by blending the style vectors. However, these methods preserve the content of source image and they are not suitable to image morphing. Park *et al.* [45] overcame this limitation by performing interpolation in both the content and style spaces.

As can be seen from the above discussions, existing works basically design rather different models for different CIT tasks. In this work, we aim to develop a state decoupled learning approach to perform different CIT tasks, including VFI, image-to-image translation and image morphing, by using the same framework.

3   Proposed Method

3.1   Problem Formulation

Given a source image I_0 and a target image I_1, the goal of VFI is to synthesize an intermediate result I_t from them:

$$I_t = \mathcal{G}(I_0, I_1, t), \tag{1}$$

where $t \in (0, 1)$ is a control parameter and $\mathcal{G}$ is a transition mapping function.

To better preserve image details, researchers [4,40,67] have resorted to using bidirectional optical flows [58,59] of I_0 and I_1, denoted by $F_{0\rightarrow1}$ and $F_{1\rightarrow0}$, to establish the motion correspondence between two consecutive frames. With the help of optical flows, I_t can be obtained as follows:

$$I_t = \mathcal{G}(I_0, I_1, \mathcal{B}(F_{0\rightarrow1}, F_{1\rightarrow0}, t)), \tag{2}$$

where $\mathcal{B}$ is a blending function. Forward [39,40] and backward [4,67] warping algorithms have been proposed to perform the blending $\mathcal{B}$ in Eq. (2).

The above idea for VFI coincides with some image morphing works [17,33, 65], where the warping function, instead of optical flow, is used to mark the object shape changes in the images. However, it is not easy to specify accurately the correspondences using warping, resulting in superimposed morphing appearance. This inspires us to model VFI as a CIT problem and seek for a more effective and common solution.

One popular solution to CIT is to embed the images into a latent space, and then blend the image feature codes therein:

$$I_t = \mathcal{G}(\mathcal{B}(L_0, L_1, t)), \tag{3}$$

where L_0, L_1 represent respectively the latent codes of I_0, I_1 in the latent space. For example, StyleGAN [28] performs *style mixing* by blending the latent codes at various scales. To gain flexible user control, disentangled learning methods [34,36,47] were later proposed to decompose the latent space into the content and style representations. The smooth style mixing can be achieved by interpolating the style vectors as follows:

$$I_t = \mathcal{G}(L_0^c, \mathcal{B}(L_0^s, L_1^s, t)), \tag{4}$$

where L_0^s, L_1^s are the style representation vectors of L_0, L_1, respectively, and L_0^c is the content vector of L_0. In this case, I_1 serves as the "style" input and the content of I_0 is preserved. However, the above formulation is hard to use in tasks such as image morphing.

Though impressive advancements have been made, the above CIT methods require much human knowledge to explicitly define the feature space, while embedding an image into a latent code needs time-consuming iterative optimization and sacrifices image details.

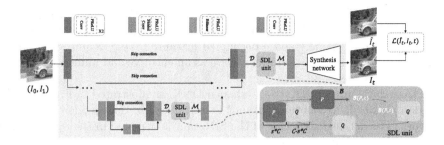

Fig. 1. The architecture of our space decoupled learning (SDL) method.

3.2 Space Decoupled Learning

As discussed in Sect. 3.1, previous works employ rather different models for different CIT applications. One interesting question is: can we find a common yet more effective framework to different CIT tasks? We make an in-depth investigation of this issue and present such a framework in this section.

The latent space aims to depict the essential image features and patterns of original data. It is expected that in the latent space, the correspondences of input images I_0 and I_1 can be well built. In other words, the latent codes L_0, L_1 in Eq. (3) play the role of optical flows $F_{0 \to 1}, F_{1 \to 0}$ in Eq. (2). Both of Eq. (3) and Eq. (2) blend the correspondence of two images to obtain the desired output. The difference lies in that the latent code representation of an image in Eq. (3) may lose certain image details, while in Eq. (2) the original inputs I_0, I_1 are involved into the reconstruction, partially addressing this problem.

From the above discussion, we can conclude that the key to CIT tasks is how to smoothly blend the image features whose correspondences can be well built, while reconstruct the image features whose correspondences are hard to obtain. We thus propose to decouple the image space into two sub-spaces accordingly: a *translatable flow space*, denoted by P, where the features can be smoothly and easily blended with t, and a *non-translatable feature space*, denoted by Q, where the features cannot be blended but should be synthesized. With P and Q, we propose a unified formulation of CIT problems as follows:

$$I_t = \mathcal{G}(Q_{0 \to 1}, \mathcal{B}(P_{0 \to 1}, t)). \tag{5}$$

The subscript "$0 \to 1$" means the transition is from I_0 to I_1. With Eq. (5), we continuously transition those translatable image components in P, and reconstruct the intermediate features that cannot be directly transitioned in Q.

Now the question turns to how to define the spaces of P and Q. Unlike many previous CIT methods [36,47] which explicitly define the feature spaces using much human knowledge, we propose to learn P and Q implicitly from training data. We learn a decoupling operator, denoted by $\mathcal{D}$, to decompose the image space of I_0 and I_1 to the translatable flow space P and the non-translatable feature space Q:

$$(P_{0 \to 1}, Q_{0 \to 1}) \leftarrow \mathcal{D}(I_0, I_1). \tag{6}$$

Specifically, we use several convolutional layers to implement the space decoupling operator $\mathcal{D}$. To gain performance, $\mathcal{D}$ is learned on multiple scales. The proposed method, namely space decoupled learning (SDL), requires no human knowledge of the domain, and it can serve as an effective and unified solution to different CIT tasks.

The architecture of SDL is a U-shaped DNN, as illustrated in Fig. 1. Unlike standard U-Net [52], a novel *SDL unit* is introduced in the decoder part of our network. The detailed structure of the SDL unit is depicted in the right-bottom corner of Fig. 1. The inputs of the SDL unit are the feature maps decomposed in previous convolution layers. Let C be the number of input feature maps and $s \in (0,1)$ be the ratio of translatable flow features to the total features. s is a hyper-parameter controlled by users (we will discuss how to set it in Sect. 4). We then split the channel number of input feature maps in P and Q as $s * C$ and $C - s * C$, and perform the blending $\mathcal{B}$ on P while keeping Q unchanged. There are multiple ways to perform the blending. For example, $\mathcal{B}$ can be achieved by scaling the features with factor t: $\mathcal{B}(P_{0 \to 1}, t) = t * P_{0 \to 1}$, which results in linear interpolation in P and is used in our experiments. Afterwards, the blended P and Q are concatenated as the output of the SDL unit. A merging operator $\mathcal{M}$ (also learned as several convolutional layers like $\mathcal{D}$) is followed to rebind the decoupled spaces on multiple scales.

A synthesis network is also adopted to improve the final transition results. We employ a GridNet architecture [18] for it with three rows and six columns. Following the work of Niklaus *et al.* [40], some modifications are utilized to address the checkerboard artifacts. The detailed architecture of the synthesis network can be found in the **supplementary materials**. In addition, it is worth mentioning that t works with the loss function during training if necessary. Details can be found in the section of experiments.

3.3   Training Strategy

To train SDL model for VFI, we adopt two loss functions: the Charbonnier loss [8] $\mathcal{L}_C$ and the perceptual loss [27] $\mathcal{L}_P$. The final loss $\mathcal{L}$ is as follows:

$$\mathcal{L} = \alpha \mathcal{L}_C + \beta \mathcal{L}_P, \tag{7}$$

where α and β are balancing parameters. The content loss $\mathcal{L}_C$ enforces the fine features and preserves the original color information. The perceptual loss $\mathcal{L}_P$ can be better balanced to recover more high-quality details. We use the *conv5_4* feature maps before activation in the pre-trained VGG19 network [54] as the perceptual loss. In our experiments, we empirically set $\alpha = 1$ and $\beta = 0.1$.

For other CIT applications including image-to-image translation and image morphing, GAN plays a key role to generate high-quality results in order to alleviate superimposed appearances. In our implementation, we use PatchGAN developed by Isola *et al.* [24] for adversarial training. The final loss is the sum of the $\mathcal{L}_1$ loss and PatchGAN loss with equal weights.

Table 1. Quantitative comparison (PSNR, SSIM, runtime) of different methods on the Middleburry, UCF101, Vimeo90K and Adobe240fps datasets. The runtime is reported as the average time to process a pair of 640×480 images. The numbers in **bold** represent the best performance. The upper part of the table presents the results of kernel-based methods, and the lower part presents the methods that can perform smooth frame interpolations. "–" means that the result is not available.

Method	Training Dataset	Runtime (ms)	Middleburry		UCF101		Vimeo90K		Adobe240fps	
			PSNR↑	SSIM↑	PSNR↑	SSIM↑	PSNR↑	SSIM↑	PSNR↑	SSIM↑
SepConv [42]	proprietary	57	35.73	0.959	34.70	0.947	33.79	0.955	–	–
CAIN [12]	proprietary	56	35.07	0.950	34.97	0.950	34.64	0.958	–	–
AdaCof [32]	Vimeo90K	77	35.71	0.958	35.16	0.950	34.35	0.956	–	–
CDFI [15]	Vimeo90K	248	37.14	0.966	35.21	0.950	35.17	0.964	–	–
SuperSloMo [26]	Adobe240fps+Youtube240fps	67	33.64	0.932	33.14	0.938	32.68	0.938	30.76	0.902
DAIN [4]	Vimeo90K	831	36.70	0.964	35.00	0.949	34.70	0.963	29.22	0.877
BMBC [44]	Vimeo90K	3008	36.78	0.965	35.15	0.950	35.01	**0.965**	29.56	0.881
EDSC [10]	Vimeo90K-Septuplet	60	36.81	**0.967**	35.06	0.946	34.57	0.956	30.28	0.900
SDL	Vimeo90K+Adobe240fps	**42**	**37.38**	**0.967**	**35.33**	**0.951**	**35.47**	0.965	**31.38**	**0.914**

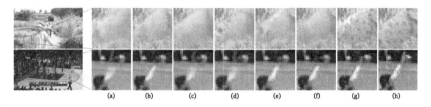

Fig. 2. Visual comparison of competing methods on the Vimeo90K test set. (a) Sep-Conv [42]; (b) SuperSloMo [26]; (c) CAIN [12]; (d) EDSC [10]; (e) DAIN [4]; (f) BMBC [44]; (g) SDL; (h) Ground truth.

4   Experiments and Applications

In this section, we first conduct extensive experiments on VFI to validate the effectiveness of our SDL method, and then apply SDL to other CIT tasks beyond VFI, such as face aging, face toonification and image morphing, to validate the generality of SDL.

4.1   Datasets and Training Settings for VFI

There are several datasets publicly available for training and evaluating VFI models, including Middlebury [3], UCF101 [56], Vimeo90K [68] and Adobe240-fps [57]. The Middlebury dataset contains two subsets, *i.e.*, *Other* and *Evaluation*. The former provides ground-truth middle frames, while the later hides the ground-truth, and the users are asked to upload their results to the benchmark website for evaluation. The UCF101 dataset [56] contains 379 triplets of human action videos, which can be used for testing VFI algorithms. The frame resolution of the above two datasets is 256×256.

We combine the training subsets in Adobe240-fps and Vimeo90K to train our SDL model. The Vimeo90K dataset [68] has 51,312 (3,782) triplets for training (testing), where each triplet contains 3 consecutive video frames of resolution

256×448. This implicitly sets the value of t to 0.5, and hence it is insufficient to train our SDL model for finer time intervals. We further resort to the Adobe240-fps dataset [57], which is composed of high frame-rate videos, for model training. We first extract the frames of all video clips, and then group the extracted frames with 12 frames per group. There is no overlap between any two groups. During training, we randomly select 3 frames I_a, I_b, I_c from a group as a triplet, where $\{a, b, c\} \in \{0, 1, ..., 11\}$ and $a < b < c$. The corresponding value of t can be calculated as $(b - a)/(c - a)$. We also randomly reverse the direction of the sequence for data augmentation (t is accordingly changed to $1 - t$). Each video frame is resized to have a shorter spatial dimension of 360 and a random crop of 256×256. Horizontal flip is performed for data augmentation. Following SuperSloMo [26], we use 112 video clips for training and the rest 6 for validation.

During model updating, we adopt the Adam [30] optimizer with a batch size of 48. The initial learning rate is set as 2×10^{-4}, and it decays by a factor of 0.8 for every 100K iterations. The model is updated for 600K iterations.

4.2   Comparisons with State-of-the-Arts

We evaluate the performance of the proposed SDL method in comparison with two categories of state-of-the-art VFI algorithms, whose source codes or pre-trained models are publicly available. The first category of methods allow frame interpolation at arbitrary time, including SuperSloMo [26], DAIN [4], BMBC [44] and EDSC [10]. The second category is kernel-based algorithms, including SepConv [42], CAIN [12], AdaCof [32] and CDFI [15], which can only perform frame interpolation iteratively at the power of 2. The PSNR and SSIM [64] indices are used for quantitative comparisons.

Table 1 provides the PSNR/SSIM and runtime results on the Middlebury *Other* [3], UCF101 [56], Vimeo90K [68] and Adobe240-fps [57] testing sets. In all experiments, the first and last frames of each group are taken as inputs. On the first three datsets, we set $t = 0.5$ to interpolate the middle frame. While on the high frame rate Adobe240-fps dataset, we vary $t \in \{\frac{1}{11}, \frac{2}{11}, ..., \frac{10}{11}\}$ to produce the intermediate 10 frames, which is beyond the capability of kernel-based methods [12,15,32,42]. All the methods are tested on a NVIDIA V100 GPU, and we calculate the average processing time for 10 runs. From Table 1, one can see that the proposed SDL approach achieves best PSNR/SSIM indices on all the datasets, while it has the fastest running speed. The kernel-based method CDFI [15] also achieves very good PSNR/SSIM results. However, it often fails to handle large motions due to the limitation of kernel size. The flow-based methods such as DAIN [4] address this issue by referring to bidirectional flows, while inevitably suffer from inaccurate estimations. The proposed SDL implicitly decouples the images into a translatable flow space and a non-translatable feature space, avoiding the side effect of inaccurate flows.

Figure 2 presents some visual comparisons of the VFI results of competing methods. It can be seen that our SDL method preserves better the image fine details and edge structures especially in scenarios with complex motions, where inaccurate flow estimations are commonly observed. SDL manages to address this difficulty by implicitly decoupling the images into a translatable flow space

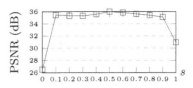

Fig. 3. Visualization of the translatable flow space and the optical flow in VFI. **Left:** the translatable flow space; **Right:** the optical flow.

Fig. 4. PSNR vs. s on the Adobe240-fps testing set. When $s = 0.5$, the PSNR reaches the peak, while the performance is very stable by varying s from 0.1 to 0.9.

Table 2. Quantitative comparison (PSNR, SSIM) between SDL and its variants on the Middleburry, UCF101, Vimeo90K and Adobe240fps datasets. The numbers in **bold** represent the best results.

Method	Training Dataset	Middleburry		UCF101		Vimeo90K		Adobe240fps	
		PSNR↑	SSIM↑	PSNR↑	SSIM↑	PSNR↑	SSIM↑	PSNR↑	SSIM↑
SDL-vimeo90k	Vimeo90K	**37.49**	**0.967**	35.27	**0.951**	**35.56**	**0.965**	26.52	0.811
SDL-w/o-sdl	Vimeo90K+Adobe240fps	36.96	0.964	35.24	0.950	35.38	0.964	26.51	0.817
SDL-w/o-syn	Vimeo90K+Adobe240fps	37.19	0.965	35.27	**0.951**	35.37	0.964	31.21	0.911
SDL	Vimeo90K+Adobe240fps	37.38	**0.967**	**35.33**	**0.951**	35.47	**0.965**	**31.38**	**0.914**

and a non-translatable feature space, and hence resulting in better visual quality with fewer interpolation artifacts. More visual comparison results can be found in the **supplementary material**. In the task of VFI, optical flow is widely used to explicitly align the adjacent frames. However, this may lead to visual artifacts on pixels where the flow estimation is not accurate. In our SDL, we decouple the image space into a translatable flow space and a non-translatable feature space, and only perform interpolation in the former one, avoiding the possible VFI artifacts caused by inaccurate flow estimation. In Fig. 3, we visualize the the translatable flow space and compare it with the optical flow obtained by SpyNet [49]. As can be seen, the translatable flow space matches the optical flow on the whole, while it focuses more on the fine details and edge structures that are import to synthesize high-quality results.

4.3 Ablation Experiments

In this section, we conduct experiments to investigate the ratio of translatable flow features, and compare SDL with several of its variants.

Translatable Flow Features. In order to find out the effect of s (*i.e.*, the ratio of translatable flow features to total features) of SDL, we set $s \in \{0, 0.1, ..., 1\}$) and perform experiments on the Adobe240-fps testing set. The curve of PSNR versus s is plotted in Fig. 4. We can see that the performance decreases significantly if all feature maps are assigned to non-translatable feature space (*i.e.*, $s = 0$) or translatable flow space (*i.e.*, $s = 1$). When $s = 0.5$, the PSNR reaches the peak, while the performance is very stable by varying s from 0.1 to 0.9. This

Fig. 5. Comparison of SDL with StyleGAN2 backpropagation on face aging. From left to right: input image, StyleGAN2 backpropagation [61] and SDL. Note that artifacts can be generated by StyleGAN2 backpropagation, while SDL can synthesize the image more robustly.

is because SDL can learn to adjust its use of translatable and non-translatable features during training.

The variants of SDL. We compare SDL with several of its variants to validate the design and training of SDL. The first variant is denoted as SDL-vimeo90k, *i.e.*, the model is trained using only the Vimeo90K dataset. The second variant is denoted as SDL-w/o-sdl, *i.e.*, SDL without space decoupling learning by setting $s = 0$. The third variant is denoted as SDL-w/o-syn, *i.e.*, the synthesis network is replaced with several convolution layers.

We evaluate SDL and its three variants on the Middlebury *Other* [3], UCF101 [56], Vimeo90K [68] and Adobe240-fps [57] testing sets, and the PSNR and SSIM results are listed in Table 2. One can see that SDL-vimeo90k achieves the best SSIM indices on all the triplet datasets, and the best PSNR indices on Middlebury *Other* and Vimeo90K by using a smaller training dataset than SDL, which uses both Vimeo90K and Adobe240-fps in training. This is because these is a domain gap between Adobe240-fps and Vimeo90k, and hence the SDL-vimeo90k can overfit the three triplet dataset. Furthermore, SDL-vimeo90k performs poorly on the Adobe240-fps dataset. This implies that training SDL using merely triplets fails to synthesize continuous frames.

Without decoupling the space, SDL-w/o-sdl performs much worse than the full SDL model, especially on the Adobe240-fps testing set. This validates that the space decoupling learning strategy boosts the VFI performance and plays a key role in continuous image transition. Without the GridNet [18], which is widely used as the synthesis network to improve VFI performance [39,40], SDL-w/o-syn maintains good VFI performance on all the datasets with only slight PSNR/SSIM decrease compared to original SDL.

4.4   Applications Beyond VFI

The proposed SDL achieves leading performance in VFI without using optical flows. It can also be used to address other CIT applications beyond VFI, such as image-to-image translation and image morphing. In this section, we take face aging and toonification and dog-to-dog image morphing as examples to demonstrate the generality of our SDL approach.

Face Aging. Unlike VFI, there is no public dataset available for training and assessing continuous I2I models. To solve this issue, we use StyleGAN [28,29], which is a cutting-edge network for creating realistic images, to generate training data. Following [61], we use StyleGAN2 distillation to synthesize datasets for face manipulation tasks such as aging. We first locate the direction vector associated with the attribute in the latent space, then randomly sample the latent codes to generate source images. For each source image, we walk along the direction vector with equal pace to synthesize a number of target images. As shown in the middle image of Fig. 5, StyleGAN2 distillation may not always generate faithful images. We thus manually check all the samples to remove unsatisfactory ones. Finally, 50,000 samples are generated, and each sample contains 11 images of 1024×1024. The dataset will be made publicly available.

The source image I_0 and a randomly selected target image I_a ($a \in 1, 2, ..., 10$) are used as the inputs to train the SDL model. The corresponding value of t is $a/10$. We also randomly replace the source image I_0 with the target image I_{10} during training, and the corresponding value of t can be set as $a/10 - 1$. In this way, the range of $t \in [0, 1]$ can be extended to $[-1, 1]$ so that our model can produce both younger (by setting $a \in [-1, 0)$) and older faces (by setting $a \in (0, 1]$). Note that SDL only needs the source image as input in inference.

Though trained on synthetic datasets, SDL can be readily used to handle real-world images. Since only a couple of works have been proposed for continuous

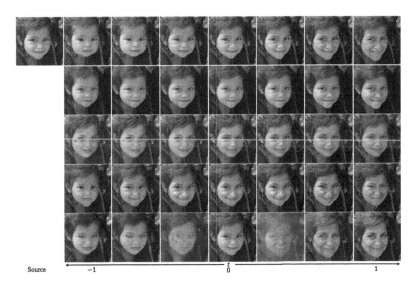

Fig. 6. Comparison of SDL with competing methods on continuous face aging. From top to bottom: SDL, StyleGAN2 backpropagation [61], SAVI2I [36], Lifespan [43] and DNI [63].

I2I translation problem, and we choose those methods [36,43,63] whose training codes are publicly available to compare, and re-train their models using our datasets. In particular, we employ the same supervised L_1 loss as ours to re-train those unsupervised methods for fair comparison. Figure 6 shows the results of competing methods on continuous face aging. One can see that SDL outperforms clearly the competitors in generating realistic images. By synthesizing the non-translatable features in reconstruction, SDL also works much better on retaining image background, for example, the mouth in the right-top corner. StyleGAN2 backpropagation [61] generates qualified aging faces; however, it fails to translate the face identity and loses the image background. SDL also produces more stable results than StyleGAN2 backpropagation, as shown in Fig. 5.

It is worth mentioning that SDL is 10^3 times faster than StyleGAN2 back-propagation which requires time-consuming iterative optimization. SAVI2I [36] fails to generate qualified intermediaries with photo-realistic details. Lifespan [43] adopts an off-the-shelf face segmentation algorithm to keep the background unchanged. However, the generated face images have low quality. To test DNI [63], we train two Pix2PixHD [62] models to generate younger and older faces, respectively, and blend their weights continuously. As can be seen, DNI [63] fails to produce reasonable transition results. Moreover, SDL can generate continuous image-to-image translations with arbitrary resolutions, while all the competing methods cannot do it. More visual comparison results can be found in the **supplementary materials**.

Face Toonification. We first build a face toonification dataset by using the method of *layer swapping* [46]. Specifically, we finetune a pretrained StyleGAN on a cartoon face dataset to obtain a new GAN, then swap different scales of layers of the two GANs (*i.e.*, the pretrained and the finetuned ones) to create a series of blended GANs, which can generate various levels of face toonification effects. Similar to face aging, we generate $50,000$ training samples, each containing 6 images of resolution 1024×1024. During training, we take the source images (*i.e.*, I_0) as input and randomly choose a target image I_a, $a \in \{1, 2, ..., 5\}$, as the ground-truth output. The corresponding value of t is $a/5$.

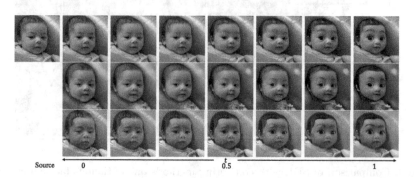

Fig. 7. Comparison of SDL with competing methods on continuous face toonification. From top to bottom: SDL, Pinkney *et al.* [46], and SAVI2I [36].

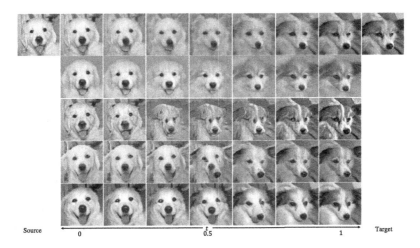

Source $\xleftarrow{\quad}$ 0 $\xrightarrow{\quad t \quad}$ 0.5 $\xrightarrow{\quad}$ 1 Target

Fig. 8. Comparison of SDL with competing methods on dog-to-dog morphing. From top to bottom: SDL, StyleGAN2 backpropagation [61], CrossBreed [45], SAVI2I [36], and FUNIT [34].

We compare SDL with Pinkney *et al.* [46] and SAVI2I [36], whose source codes are available. As shown in Fig. 7, SDL outperforms the competitors in producing visually more favourable results. Pinkney *et al.* [46] generates qualified toonification effects but it fails to retain the face identity and the image background. The generated face images of SAVI2I [36] have low quality. Furthermore, SAVI2I [36] merely synthesizes images with a resolution of 256×256, while SDL can yield results at any resolution. More visual comparison results can be found in the **supplementary materials**.

Dog-to-Dog Morphing. Similar to I2I translation, we synthesize training data for dog-to-dog morphing using StyleGAN2 [29] and BigGAN [7]. We randomly sample two latent codes as the source and target images. The intermediate images are obtained by interpolating the two codes in the latent space. We generate 50,000 training samples, each containing 11 images of resolution 512×512. During training, we take the source and target images (*i.e.*, I_0, I_{10}) as inputs and randomly choose an image I_a, $a \in \{1, 2, ..., 9\}$, as the ground-truth output.

Since few methods have been proposed for continuous image morphing, we compare SDL with I2I translation models, including CrossBreed [45], SAVI2I [36] and FUNIT [34]. (We re-train their models using our datasets and the same supervised L_1 loss for fair comparison.) As shown in Fig. 8, SDL achieves smooth morphing from one dog to another with vivid details. StyleGAN2 backpropagation [61] yields comparable results but it loses the background details. CrossBreed [45] and SAVI2I [36] fail to generate qualified intermediate results. FUNIT [34] produces smooth morphing; however, the generated dog images have low quality and it fails to retain the image content when $t = 0, 1$. Please refer to the **supplementary materials** for more visual comparisons.

5   Conclusion

We proposed a simple yet effective approach, namely space decoupled learning (SDL), for VFI problem. We implicitly decoupled the images into a translatable flow space and a non-translatable feature space, and performed image interpolation in the flow space and intermediate image synthesis in the feature space. The proposed SDL can serve as a general-purpose solution to a variety of continuous image transition (CIT) problems. As demonstrated by our extensive experiments, SDL showed highly competitive performance with the state-of-the-arts, which were however specifically designed for their given tasks. Particularly, in the application of video frame interpolation, SDL was the first flow-free algorithm that can synthesize consecutive interpolations with leading performance. In other CIT tasks such as face aging, face toonification and dog-to-dog morphing, SDL exhibited much better visual quality and efficiency with more foreground and background details.

References

1. Abdal, R., Qin, Y., Wonka, P.: Image2stylegan: how to embed images into the stylegan latent space? In: ICCV (2019)
2. Baker, S., Scharstein, D., Lewis, J.P., Roth, S., Black, M.J., Szeliski, R.: A database and evaluation methodology for optical flow. IJCV **92**(1), 1–31 (2007)
3. Baker, S., Scharstein, D., Lewis, J., Roth, S., Black, M.J., Szeliski, R.: A database and evaluation methodology for optical flow. In: ICCV (2007)
4. Bao, W., Lai, W.S., Ma, C., Zhang, X., Gao, Z., Yang, M.H.: Depth-aware video frame interpolation. In: CVPR, pp. 3698–3707 (2019)
5. Bar-Haim, A., Wolf, L.: ScopeFlow: dynamic scene scoping for optical flow. In: CVPR (2020)
6. Beier, T., Neely, S.: Feature-based image metamorphosis. ACM TOG **26**(2), 35–42 (1992)
7. Brock, A., Donahue, J., Simonyan, K.: Large scale GAN training for high fidelity natural image synthesis. In: ICLR (2019)
8. Charbonnier, P., Blanc-Feraud, L., Aubert, G., Barlaud, M.: Two deterministic half-quadratic regularization algorithms for computed imaging. In: ICIP (1994)
9. Chen, Y.C., Xu, X., Tian, Z., Jia, J.: Homomorphic latent space interpolation for unpaired image-to-image translation. In: CVPR, pp. 2403–2411 (2019)
10. Cheng, X., Chen, Z.: Multiple video frame interpolation via enhanced deformable separable convolution. IEEE TPAMI (2021)
11. Choi, M., Choi, J., Baik, S., Kim, T.H., Lee, K.M.: Scene-adaptive video frame interpolation via meta-learning. In: CVPR (2020)
12. Choi, M., Kim, H., Han, B., Xu, N., Lee, K.M.: Channel attention is all you need for video frame interpolation. In: AAAI (2020)
13. Choi, Y., Choi, M., Kim, M., Ha, J.W., Kim, S., Choo, J.: StarGAN: unified generative adversarial networks for multi-domain image-to-image translation. In: CVPR (2018)
14. Choi, Y., Uh, Y., Yoo, J., Ha, J.W.: StarGAN v2: diverse image synthesis for multiple domains. In: CVPR, pp. 8185–8194 (2020)

15. Ding, T., Liang, L., Zhu, Z., Zharkov, I.: CDFI: compression-driven network design for frame interpolation. In: CVPR, pp. 8001–8011 (2021)
16. Dong, C., Loy, C.C., He, K., Tang, X.: Image super-resolution using deep convolutional networks. IEEE TPAMI **38**(2), 295–307 (2016)
17. Fish, N., Zhang, R., Perry, L., Cohen-Or, D., Shechtman, E., Barnes, C.: Image morphing with perceptual constraints and STN alignment. In: Computer Graphics Forum (2020)
18. Fourure, D., Emonet, R., Fromont, E., Muselet, D., Trémeau, A., Wolf, C.: Residual conv-deconv grid network for semantic segmentation. In: Proceedings of the British Machine Vision Conference (2017)
19. Gatys, L.A., Ecker, A.S., Bethge, M.: Image style transfer using convolutional neural network. In: CVPR (2016)
20. Gong, R., Li, W., Chen, Y., Gool, L.: DLOW: domain flow for adaptation and generalization. In: CVPR, pp. 2472–2481 (2019)
21. He, K., Zhang, X., Ren, S., Sun, J.: Deep residual learning for image recognition. In: CVPR, pp. 770–778 (2016)
22. Hong, K., Jeon, S., Fu, J., Yang, H., Byun, H.: Domain aware universal style transfer. In: ICCV (2021)
23. Huang, X., Belongie, S.: Arbitrary style transfer in real-time with adaptive instance normalization. In: ICCV, pp. 1510–1519 (2017)
24. Isola, P., Zhu, J.Y., Zhou, T., Efros, A.A.: Image-to-image translation with conditional adversarial networks. In: CVPR, pp. 5967–5976 (2017)
25. Jahanian, A., Chai, L., Isola, P.: On the "steerability" of generative adversarial networks. In: ICLR (2020)
26. Jiang, H., Sun, D., Jampani, V., Yang, M.H., Learned-Miller, E., Kautz, J.: Super sloMo: high quality estimation of multiple intermediate frames for video interpolation. In: CVPR (2018)
27. Johnson, J., Alahi, A., Fei-Fei, L.: Perceptual losses for real-time style transfer and super-resolution. In: Leibe, B., Matas, J., Sebe, N., Welling, M. (eds.) ECCV 2016. LNCS, vol. 9906, pp. 694–711. Springer, Cham (2016). https://doi.org/10.1007/978-3-319-46475-6_43
28. Karras, T., Laine, S., Aila, T.: A style-based generator architecture for generative adversarial networks. In: CVPR, pp. 4396–4405 (2019)
29. Karras, T., Laine, S., Aittala, M., Hellsten, J., Lehtinen, J., Aila, T.: Analyzing and improving the image quality of styleGAN. In: CVPR, pp. 8107–8116 (2020)
30. Kingma, D.P., Ba, J.: Adam: a method for stochastic optimization. ArXiv (2015)
31. Lee, H.-Y., Tseng, H.-Y., Huang, J.-B., Singh, M., Yang, M.-H.: Diverse image-to-image translation via disentangled representations. In: Ferrari, V., Hebert, M., Sminchisescu, C., Weiss, Y. (eds.) ECCV 2018. LNCS, vol. 11205, pp. 36–52. Springer, Cham (2018). https://doi.org/10.1007/978-3-030-01246-5_3
32. Lee, H., Kim, T., Young Chung, T., Pak, D., Ban, Y., Lee, S.: AdaCoF: adaptive collaboration of flows for video frame interpolation. In: CVPR (2020)
33. Liao, J., Lima, R.S., Nehab, D., Hoppe, H., Sander, P.V., Yu, J.: Automating image morphing using structural similarity on a halfway domain. ACM TOG **33**(5), 168 (2014)
34. Liu, M.Y., et al.: Few-shot unsupervised image-to-image translation. Arxiv (2019)
35. Long, G., Kneip, L., Alvarez, J.M., Li, H., Zhang, X., Yu, Q.: Learning image matching by simply watching video. In: Leibe, B., Matas, J., Sebe, N., Welling, M. (eds.) ECCV 2016. LNCS, vol. 9910, pp. 434–450. Springer, Cham (2016). https://doi.org/10.1007/978-3-319-46466-4_26

36. Mao, Q., Lee, H.Y., Tseng, H.Y., Huang, J.B., Ma, S., Yang, M.H.: Continuous and diverse image-to-image translation via signed attribute vectors. ArXiv (2020)
37. Meyer, S., Djelouah, A., McWilliams, B., Sorkine-Hornung, A., Gross, M., Schroers, C.: PhaseNet for video frame interpolation. In: CVPR (2018)
38. Meyer, S., Wang, O., Zimmer, H., Grosse, M., Sorkine-Hornung, A.: Phase-based frame interpolation for video. In: CVPR (2015)
39. Niklaus, S., Liu, F.: Context-aware synthesis for video frame interpolation. In: CVPR (2018)
40. Niklaus, S., Liu, F.: Softmax splatting for video frame interpolation. In: CVPR, pp. 5436–5445 (2020)
41. Niklaus, S., Mai, L., Liu, F.: Video frame interpolation via adaptive convolution. In: CVPR (2017)
42. Niklaus, S., Mai, L., Liu, F.: Video frame interpolation via adaptive separable convolution. In: ICCV (2017)
43. Or-El, R., Sengupta, S., Fried, O., Shechtman, E., Kemelmacher-Shlizerman, I.: Lifespan age transformation synthesis. In: Vedaldi, A., Bischof, H., Brox, T., Frahm, J.-M. (eds.) ECCV 2020. LNCS, vol. 12351, pp. 739–755. Springer, Cham (2020). https://doi.org/10.1007/978-3-030-58539-6_44
44. Park, J., Ko, K., Lee, C., Kim, C.-S.: BMBC: bilateral motion estimation with bilateral cost volume for video interpolation. In: Vedaldi, A., Bischof, H., Brox, T., Frahm, J.-M. (eds.) ECCV 2020. LNCS, vol. 12359, pp. 109–125. Springer, Cham (2020). https://doi.org/10.1007/978-3-030-58568-6_7
45. Park, S., Seo, K., Noh, J.: Neural crossbreed: neural based image metamorphosis. ACM TOG 39(6), 1–15 (2020)
46. Pinkney, J.N.M., Adler, D.: Resolution dependent GAN interpolation for controllable image synthesis between domains. ArXiv (2020)
47. Pizzati, F., Cerri, P., de Charette, R.: CoMoGAN: continuous model-guided image-to-image translation. In: CVPR (2021)
48. Radford, A., Metz, L., Chintala, S.: Unsupervised representation learning with deep convolutional generative adversarial networks. In: ICLR (2016)
49. Ranjan, A., Black, M.J.: Optical flow estimation using a spatial pyramid network. In: CVPR (2017)
50. Reda, F.A., et al.: Unsupervised video interpolation using cycle consistency. In: ICCV (2019)
51. Redmon, J., Divvala, S., Girshick, R., Farhadi, A.: You only look once: unified, real-time object detection. In: CVPR, pp. 779–788 (2016)
52. Ronneberger, O., Fischer, P., Brox, T.: UNet: a convolutional network for biomedical image segmentation. Arxiv (2015)
53. Shen, W., Bao, W., Zhai, G., Chen, L., Min, X., Gao, Z.: Blurry video frame interpolation. In: CVPR (2020)
54. Simonyan, K., Zisserman, A.: Very deep convolutional networks for large-scale image recognition. Arxiv (2014)
55. Smythe, D.B.: A two-pass mesh warping algorithm for object transformation and image interpolation. Rapport Tech. 1030(31) (1990)
56. Soomro, K., Zamir, A.R., Shah, A.: UCF101: a dataset of 101 human actions classes from videos in the wild. ArXiv (2012)
57. Su, S., Delbracio, M., Wang, J., Sapiro, G., Heidrich, W., Wang, O.: Deep video deblurring for hand-held cameras. In: CVPR (2017)
58. Sun, D., Yang, X., Liu, M.Y., Kautz, J.: PWC-Net: CNNs for optical flow using pyramid, warping, and cost volume. In: CVPR (2018)

59. Teed, Z., Deng, J.: RAFT: recurrent all-pairs field transforms for optical flow. In: Vedaldi, A., Bischof, H., Brox, T., Frahm, J.-M. (eds.) ECCV 2020. LNCS, vol. 12347, pp. 402–419. Springer, Cham (2020). https://doi.org/10.1007/978-3-030-58536-5_24

60. Upchurch, P., et al.: Deep feature interpolation for image content changes. In: CVPR, pp. 6090–6099 (2017)

61. Viazovetskyi, Y., Ivashkin, V., Kashin, E.: StyleGAN2 distillation for feed-forward image manipulation. In: Vedaldi, A., Bischof, H., Brox, T., Frahm, J.-M. (eds.) ECCV 2020. LNCS, vol. 12367, pp. 170–186. Springer, Cham (2020). https://doi.org/10.1007/978-3-030-58542-6_11

62. Wang, T.C., Liu, M.Y., Zhu, J.Y., Tao, A., Kautz, J., Catanzaro, B.: High-resolution image synthesis and semantic manipulation with conditional GANs. In: CVPR (2018)

63. Wang, X., Yu, K., Dong, C., Tang, X., Loy, C.C.: Deep network interpolation for continuous imagery effect transition. In: CVPR, pp. 1692–1701 (2019)

64. Wang, Z., Bovik, A., Sheikh, H., Simoncelli, E.: Image quality assessment: from error visibility to structural similarity. IEEE TIP **13**(4), 600–612 (2004)

65. Wolberg, G.: Image morphing: a survey. Vis. Comput. **14**(8–9), 360–372 (1998)

66. Wu, P., Lin, Y.J., Chang, C.H., Chang, E.Y., Liao, S.W.: RelGAN: multi-domain image-to-image translation via relative attributes. In: ICCV, pp. 5913–5921 (2019)

67. Xu, X., Siyao, L., Sun, W., Yin, Q., Yang, M.H.: Quadratic video interpolation. In: NeurIPS (2019)

68. Xue, T., Chen, B., Wu, J., Wei, D., Freeman, W.T.: Video enhancement with task-oriented flow. IJCV **127**(8), 1106–1125 (2019)

69. Zhu, J.Y., Park, T., Isola, P., Efros, A.A.: Unpaired image-to-image translation using cycle-consistent adversarial networks. In: ICCV, pp. 2242–2251 (2017)

70. Zhu, J.Y., et al.: Toward multimodal image-to-image translation. In: NeurIPS (2017)

Diversified Dynamic Routing for Vision Tasks

Botos Csaba[1]($\boxtimes$), Adel Bibi[1], Yanwei Li[2], Philip Torr[1], and Ser-Nam Lim[3]

[1] University of Oxford, Oxford, UK
csbotos@robots.ox.ac.uk, {adel.bibi,philip.torr}@eng.ox.ac.uk
[2] The Chinese University of Hong Kong, HKSAR, Shatin, China
ywli@cse.cuhk.edu.hk
[3] Meta AI, New York City, USA
sernamlim@fb.com

Abstract. Deep learning models for vision tasks are trained on large datasets under the assumption that there exists a universal representation that can be used to make predictions for all samples. Whereas high complexity models are proven to be capable of learning such representations, a mixture of experts trained on specific subsets of the data can infer the labels more efficiently. However using mixture of experts poses two new problems, namely (**i**) assigning the correct expert at inference time when a new unseen sample is presented. (**ii**) Finding the optimal partitioning of the training data, such that the experts rely the least on common features. In Dynamic Routing (DR) [21] a novel architecture is proposed where each layer is composed of a set of experts, however without addressing the two challenges we demonstrate that the model reverts to using the same subset of experts. In our method, Diversified Dynamic Routing (DivDR) the model is explicitly trained to solve the challenge of finding relevant partitioning of the data and assigning the correct experts in an unsupervised approach. We conduct several experiments on semantic segmentation on Cityscapes and object detection and instance segmentation on MS-COCO showing improved performance over several baselines.

1 Introduction

In recent years, deep learning models have made huge strides solving complex tasks in computer vision, e.g. segmentation [4,27] and detection [10,34], and reinforcement learning, e.g. playing atari games [30]. Despite this progress, the computational complexity of such models still poses a challenge for practical deployment that requires accurate real-time performance. This has incited a rich body of work tackling the accuracy complexity trade-off from various angles. For instance, a class of methods tackle this trade-off by developing more efficient architectures [38,48], while others initially train larger models and then

---

Supplementary Information The online version contains supplementary material available at https://doi.org/10.1007/978-3-031-25069-9_48.

© The Author(s), under exclusive license to Springer Nature Switzerland AG 2023
L. Karlinsky et al. (Eds.): ECCV 2022 Workshops, LNCS 13804, pp. 756–772, 2023.
https://doi.org/10.1007/978-3-031-25069-9_48

later distill them into smaller more efficient models [12, 15, 46]. Moreover, several works rely on sparse regularization approaches [9, 36, 41] during training or by performing a post-training pruning of model weights that contribute marginally to the final prediction. While listing all categories of methods tackling this trade-off is beyond the scope of this paper, to the best of our knowledge, they all share the assumption that predicting the correct label requires a universal set of features that works best for all samples. We argue that such an assumption is often broken even in well curated datasets. For example, in the task of segmentation, object sizes can widely vary across the dataset requiring different computational effort to process. That is to say, large objects can be easily processed under lower resolutions while smaller objects require processing in high resolution to retain accuracy. This opens doors for class of methods that rely on *local experts*; efficient models trained directly on each subset separately leveraging the use of this local bias. However, prior art often ignore local biases in the training and validation datasets when tackling the accuracy-efficiency trade-off for two key reasons illustrated in Fig. 1. (**i**) Even under the assumption that such local biases in the training data are known, during inference time, new unseen samples need to be assigned to the correct local subset so as to use the corresponding *local expert* for prediction (Fig. 1 left). (**ii**) Such local biases in datasets are not known **apriori** and may require a prohibitively expensive inspection of the underlying dataset (Fig. 1 right).

In this paper, we take an orthogonal direction to prior art on the accuracy-efficiency trade-off by addressing the two challenges in an unsupervised manner. In particular, we show that training *local experts* on learnt subsets sharing local biases can jointly outperform *global experts*, i.e. models that were trained over the entire dataset. We summarize our contributions in two folds.

1. We propose Diversified Dynamic Routing (DivDR); an unsupervised learning approach that trains several local experts on learnt subsets of the training dataset. At inference time, DivDR assigns the correct local expert for prediction to newly unseen samples.
2. We extensively evaluate DivDR and compare against several existing methods on semantic segmentation, object detection and instance segmentation on various datasets, i.e. Cityscapes [8] and MS-COCO [24]. We find that DivDR compared to existing methods better trades-off accuracy and efficiency. We complement our experiments with various ablations demonstrating robustness of DivDR to choices of hyperparameters.

2   Related Work

In prior literature model architectures were predominantly hand-designed, meaning that hyper-parameters such as the number and width of layers, size and stride of convolution kernels were predefined. In contrast, Neural Architecture Search [26, 54] revealed that searching over said hyper-parameter space is feasible provided enough data and compute power resulting in substantial improvement in model accuracy. Recently, a line of research [3, 20, 25, 38, 40] also proposed

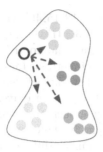

Expert assignment
during inference

Finding meaningful subsets
during training

Fig. 1. The figure depicts the two main challenges in learning local experts on subsets on subsets of the dataset with local biases. First, even when the subsets in the training dataset is presented where there is a local expert per subset, the challenge remains in assigning the local expert for new unseen samples (left Figure). The second challenge is that the local biases in the training data are not available during training time (right Figure).

to constrain the search space to cost-efficient models that jointly optimize the accuracy and the computational complexity of the models. Concurrently, cost-efficient inference has been also in the focus of works on dynamic network architectures [31,42,44,47], where the idea is to allow the model to choose different architectures based on the input through gating computational blocks during inference.

For example, Li et al. [21] proposed an end-to-end dynamic routing framework that generates routes within the architecture that vary per input sample. The search space of [21], inspired by Auto-DeepLab [25], allows exploring spatial up and down-sampling between subsequent layers which distinguishes the work from prior dynamic routing methods. One common failure mode of dynamic models is mentioned in [31], where during the initial phase of the training only a specific set of modules are selected and trained, leading to a static model with reduced capacity. This issue is addressed by Mullapudi et al. [31] through clustering the training data in advance based on latent representations of a pretrained image classifier model, whereas [40] uses the Gumbel-Softmax reparameterization [17] to improve diversity of the dynamic routes. In this work, to mitigate this problem, we adopt the metric learning Magnet Loss [35] which acts as an improvement over metric learning methods that act on the instance level, e.g. Triplet Loss [19,43], and Contrastive Learning methods [7,13]. This is since it considers the complete distribution of the underlying data resulting in a more stable clustering. To adapt Magnet Loss to resolving the Dynamic Routing drawbacks, we use it as an unsupervised approach to increase the distance between the forward paths learned by the Dynamic Routing model this is as opposed to clustering the learned representations, i.e. learning clustered dynamic routes as opposed to clustered representations.

We review the recent advances on semantic segmentation and object detection which are utilized to validate our method in this work. For semantic segmentation, numerous works have been proposed to capture the larger receptive field [4–6,49] or establish long-range pixel relation [16,37,50] based on Fully Convolutional Networks [27]. As mentioned above, with the development of neural network, Neural Architecture Search (NAS)-based approaches [3,25,32] and dynamic networks [21] are utilized to adjust network architecture according to the data while being jointly optimized to reduce the cost of inference. As for object detection, modern detectors can be roughly divided into one-stage or two-stage detectors. One-stage detectors usually make predictions based on the prior guesses, like anchors [23,33] and object centers [39,52]. Meanwhile, two-stage detectors predict boxes based on predefined proposals in a coarse-to-fine manner [10,11,34]. There are also several advances in Transformer-based approaches for image recognition tasks such as segmentation [45,51] and object detection [1,53], and while our method can be generalized to those architectures as well, it is beyond the scope of this paper.

3   DivDR: Diversified Dynamic Routing

We first start by introducing Dynamic Routing. Second, we formulate our objective of the iterative clustering of the dataset and the learning of experts per dataset cluster. At last, we propose a contrastive learning approach based on *magnet loss* [35] over the gate activation of the dynamic routing model to encourage the learning of different architectures over different dataset clusters.

3.1   Dynamic Routing Preliminaries

The Dynamic Routing (DR) [21] model for semantic segmentation consists of L sequential feed-forward layers in which dynamic *nodes* process and propagate the information. Each dynamic node has two parts: (i) the *cell* that performs a non-linear transformation to the input of the node; and (ii) the *gate* that decides which node receives the output of the cell operation in the subsequent layer. In particular, the gates in DR determine what resolution/scale of the activation to be used. That is to say, each gate determines whether the activation output of the cell is to be propagated at the same resolution, up-scaled, or down-scaled by a factor of 2 in the following layer. Observe that the gate activation determines the *architecture* for a given input since this determines a unique set of connections defining the architecture. The output of the final layer of the nodes are up-sampled and fused by 1×1 convolutions to match the original resolution of the input image. For an input-label pair (x, y) in a dataset $\mathcal{D}$ of N pairs, let the DR network parameterized by θ be given as $f_\theta : \mathcal{X} \to \mathcal{Y}$ where $x \in \mathcal{X}$ and $y \in \mathcal{Y}$. Moreover, let $\mathcal{A}_{\tilde{\theta}} : \mathcal{X} \to [0,1]^n$, where $\theta \supseteq \tilde{\theta}$, denote the gate activation map for a given input, i.e. the gates determining the architecture discussed earlier, then the training objective for DR networks under computational budget constraints have the following form:

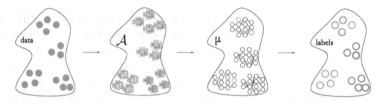

Fig. 2. Gate Activation cluster assignment. To update the local experts, DivDR performs K-means clustering on the gate activations over the $\mathcal{A}(x_i)$ $\forall i$ in the training examples with fixed model parameters θ.

Fig. 3. Gate Activation Diversification. We use the labels from the cluster assignment to reduce the *intra-cluster* variance and increase the *inter-cluster* variance by updating model parameters θ.

$$\mathcal{L}_{DR} = \frac{1}{N} \sum_{i=1}^{N} \mathcal{L}_{seg}\big(f_\theta(x_i), y_i\big) + \lambda \mathcal{L}_{cost}(\mathcal{A}_{\tilde{\theta}}(x_i)). \tag{1}$$

We will drop the subscript $\tilde{\theta}$ throughout to reduce text clutter. Note that $\mathcal{L}_{seg}$ and $\mathcal{L}_{cost}$ denote the segmentation and computational budget constraint respectively. Observe that when most of the gate activations are sparse, this incurs a more efficient network that may be at the expense of accuracy and hence the trade-off through the penalty λ.

3.2   Metric Learning in $\mathcal{A}$-space

Learning local experts can benefit performance both in terms of accuracy and computational cost. We propose an unsupervised approach to learning jointly the subset of the dataset and the soft assignment of the corresponding architectures. We use the DR framework for our approach.

We first assume that there are K clusters in the dataset for which we seek to learn an expert on each. Moreover, let $\{\mu_{\mathcal{A}_i}\}_{i=1}^{K}$, denote the cluster centers representing K different gate activations. Note that as per the previous discussion, each gate activation $\mu_{\mathcal{A}_i} \in [0, 1]^n$ corresponds to a unique architecture. The set of cluster centers representing gate activations $\{\mu_{\mathcal{A}_i}\}_{i=1}^{K}$ can be viewed as a set of prototypical architectures for K different subsets in the datasets. Next, let $\mu(x)$ denote the nearest gate activation center to the gate activation $\mathcal{A}(x)$, i.e. $\mu(x) = \arg\min_i \|\mathcal{A}(x) - \mu_{\mathcal{A}_i}\|$. Now, we seek to solve for both the gate activation centers $\{\mu_{\mathcal{A}_i}\}_{i=1}^{K}$ and the parameters θ such that the gate activation centers are pushed away from one another. To that end, we propose the alternating between clustering and the minimization of a *magnet loss* [35] variant. In particular, for a given fixed set of activating gates centers $\{\mu_{\mathcal{A}_i}\}_{i=1}^{K}$, we consider the following loss function:

$$
\mathcal{L}_{\text{clustering}}(\mathcal{A}(x_i)) = \left\{ \alpha + \frac{1}{2\sigma^2} \|\mathcal{A}(x_i) - \mu(x_i)\| \right.
$$

$$
\left. + \log \left(\sum_{k:\mu_{\mathcal{A}_k} \neq \mu(x_i)} e^{-\frac{1}{2\sigma^2}\|\mathcal{A}(x_i) - \mu_{\mathcal{A}_k}\|} \right) \right\}_{+}.
\tag{2}
$$

Note that $\{x\}_+ = \max(x,0)$, $\sigma^2 = \frac{1}{N-1}\sum_i^N \|\mathcal{A}(x_i) - \mu(x_i)\|^2$, and that $\alpha \geq 0$. Observe that unlike in *magnet loss*, we seek to cluster the set of architectures by separating the gate activations. Note that the penultimate term pulls the architecture, closer to the most similar prototypical architecture while the last term pushes it away from all other architectures. Therefore, this loss incites the learning of K different architectures where each input x_i will be assigned to be predicted with one of the K learnt architectures. To that end, our overall *Diversified* DR loss is given as follows:

$$
\mathcal{L}_{\text{DivDR}} = \frac{1}{N}\sum_{i=1}^{N}\mathcal{L}_{segm}(f_\theta(x_i), y_i) + \lambda_1 \mathcal{L}_{cost}(\mathcal{A}(x_i)) + \lambda_2 \mathcal{L}_{clustering}(\mathcal{A}(x_i)).
\tag{3}
$$

We then alternate between minimizing $\mathcal{L}_{\text{DivDR}}$ over the parameters θ and the updates of the cluster centers $\{\mu_{\mathcal{A}_i}\}_{i=1}^{K}$. In particular, given θ, we update the gate activation centers by performing K-Means clustering [29] over the gate activations. That is to say, we fix θ and perform K-means clustering with K clusters over all the gate activations from the dataset $\mathcal{D}$, i.e. we cluster $\mathcal{A}(x_i)$ $\forall i$ as shown in Fig. 2. Moreover, alternating between optimizing $\mathcal{L}_{\text{DivDR}}$ and updating the gate activation cluster centers over the dataset $\mathcal{D}$, illustrated in Fig. 3, results in a diversified set of architectures driven by the data that are more efficient, i.e. learning K local experts that are accurate and efficient.

4   Experiments

We show empirically that our proposed DivDR approach can outperform existing methods in better trading off accuracy and efficiency. We demonstrate this on several vision tasks, i.e. semantic segmentation, object detection, and instance segmentation. We start first by introducing the datasets used in all experiments along along with the implementation details. We then present the comparisons between DivDR and several other methods along with several ablations.

4.1   Datasets

We mainly prove the effectiveness of the proposed approach for semantic segmentation, object detection, and instance segmentation on two widely-adopted benchmarks, namely Cityscapes [8] and Microsoft COCO [24] dataset.

Table 1. Comparison with baselines on the Cityscapes [8] validation set. * Scores from [21] were reproduced using the official implementation. The evaluation settings are identical to [21]. We calculate the average FLOPs with 1024×2048 size input.

Method	Backbone	$mIoU_{val}(\%)$	GFLOPs
BiSenet [48]	ResNet-18	74.8	98.3
DeepLabV3 [5]	ResNet-101-ASPP	78.5	1778.7
Semantic FPN [18]	ResNet-101-FPN	77.7	500.0
DeepLabV3+ [6]	Xception-71-ASPP	79.6	1551.1
PSPNet [49]	ResNet-101-PSP	79.7	2017.6
Auto-DeepLab [25]	Searched-F20-ASPP	79.7	333.3
Auto-DeepLab [25]	Searched-F48-ASPP	80.3	695.0
DR-A [21]*	Layer16	72.7 ± 0.6	58.7 ± 3.1
DR-B [21]*	Layer16	72.6 ± 1.3	61.1 ± 3.3
DR-C [21]*	Layer16	74.2 ± 0.6	68.1 ± 2.5
DR-Raw [21]*	Layer16	75.2 ± 0.5	99.2 ± 2.5
DivDR-A	Layer16	73.5 ± 0.4	57.7 ± 3.9
DivDR-Raw	Layer16	75.4 ± 1.6	95.7 ± 0.9

Cityscapes. The Cityscapes [8] dataset contains 19 classes in urban scenes, which is widely used for semantic segmentation. It is consist of 5000 fine annotations that can be divided into 2975, 500, and 1525 images for training, validation, and testing, respectively. In the work, we use the Cityscapes dataset to validate the proposed method on semantic segmentation.

COCO. Microsoft COCO [24] dataset is a well-known for object detection benchmarking which contains 80 categories in common context. In particular, it includes 118k training images, 5k validation images, and 20k held-out testing images. To prove the performance generalization, we report the results on COCO's validation set for both object detection and instance segmentation tasks.

4.2 Implementation Details

In all training settings, we use SGD with a weight decay of 10^{-4} and momentum of 0.9 for both datasets. For semantic segmentation on Cityscapes, we use the exponential learning rate schedule with an initial rate of 0.05 and a power of 0.9. For fair comparison, we follow the setting in [21] and use a batch size 8 of random image crops of size 768×768 and train for 180 K iterations. We use random flip augmentations where input images are scaled from 0.5 to 2 before cropping. For object detection on COCO we use an initial learning rate of 0.02 and re-scale the shorter edge to 800 pixels and train for 90 K iterations. Following prior art, random flip is adopted without random scaling.

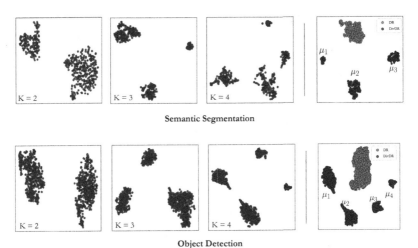

Fig. 4. Visualizing the 183-dimensional $\mathcal{A}$-space of Dynamic Routing backbones trained for semantic segmentation on Cityscapes [8] (*top*) and 198-dimensional $\mathcal{A}$-space for object detection on COCO [24] (*bottom*) using t-SNE [28]. *Left:* varying number of *local experts*, $K = 2, 3, 4$. *Right:* joint t-SNE visualization of architectures of Dynamic Routing [21] (*orange*) and our approach (*blue*). It is clear that our method not only encourages diversity of the learned routes but also reduces variance in a specific cluster. Low *intra*-cluster variance is beneficial because it facilitates feature sharing between similar tasks (Color figure online)

4.3   Semantic Segmentation

We show the benefits of our proposed DivDR of alternation between training with $\mathcal{L}_{\mathrm{DivDR}}$ and computing the gate activations clusters through K-means on Cityscapes [8] for semantic segmentation. In particular, we compare two versions of our proposed unsupervised Dynamic Routing, namely with and without the computational cost constraint ($\lambda_1 = 0$ denoted as DivDR-Raw and $\lambda_1 = 0.8$ denoted as DivDR-A) against several variants of the original dynamic routing networks both constrained and unconstrained. All experiments are averaged over 3 seeds. As observed in Table 1, while both variants perform similarly in terms of accuracy (DR-Raw: 75.2%, DivDR: 75.4%), DivDR marginally improves the computational cost by 3.5 GFLOPs. On the other hand, when introducing cost efficiency constraint DivDR-A improves both the efficiency (58.7 GFLOPs to 57.7 GFLOPs) and accuracy (72.7% to 73.5%) as compared to DR-A. At last, we observe that comparing to other state-of-the-art, our unconstrained approach, performs similarly to BiSenet [48] with 74.8% accuracy while performing better in computational efficiency (98.3 GFLOPs vs. 95.7 GFLOPs).

Table 2. Quantitative analysis of semantic segmentation on Cityscapes [8]. We report *Inter* and *Intra* cluster variance, that shows how far are the cluster centers are from each other in L_2 space and how close are the samples to the cluster centers respectively.

Method	mIoU	FLOPs	Inter	Intra
DR-A	72.7	58.7	0.4	0.3
DivDR-A	72.0	49.9	0.6	0.2
DR-Raw	75.2	99.2	1.5	1.5
DivDR-Raw	75.7	98.3	1.2	0.5

Visualizing Gate Activations. We first start by visualizing the gate activations under different choices of the number of clusters K over the gate activation for DivDR-A. As observed from Fig. 4, indeed our proposed $\mathcal{L}_{\text{DivDr}}$ results into clusters on local experts as shown by different gate activations $\mathcal{A}$ for $k \in \{2, 3, 4\}$. Moreover, we also observe that our proposed loss not only results in separated clusters of local experts, i.e. gate activations, but also with a small intra class distances. In particular, as shown in Table 2, our proposed DivDR indeed results in larger inter-cluster distances that are larger than the intra-cluster distances. The inter-cluster distances are computed as the average distance over all pair of cluster centers, i.e. $\{\mu_{\mathcal{A}_i}\}_{i=1}^K$ while the intra-cluster distances are the average distances over all pairs in every cluster. This indeed confirms that our proposed training approach results in K different architectures for a given dataset. Consequently, we can group the corresponding input images into K classes and visualize them to reveal common semantic features across the groups. For details see Fig 5. We find it interesting that despite we do not provide any direct supervision to the gates about the objects present on the images, the clustering learns to group semantically meaningful groups together.

Ablating α and λ_2. Moreover, we also ablate the performance of α which is the separation margin in the hinge loss term of our proposed loss. Observe that larger values of α correspond to more enforced regularization on the separation between gate activation clusters. As shown in Fig. 6 left, we observe that the mIOU accuracy and the FLOPs of our DivDR-A is only marginally affected by α indicating that a sufficient enough margin can be attained while maintaining accuracy and FLOPs trade-off performance.

4.4 Object Detection and Instance Segmentation

To further demonstrate the effectiveness on detection and instance segmentation, we validate the proposed method on the COCO datasets with Faster R-CNN [34] and Mask R-CNN [14] heads. As for the backbone, we extend the original dynamic routing networks with another 5-stage layer to keep consistent with that in FPN [22], bringing 17 layers in total. Similar to that in Sect. 4.3, no external supervision is provided to our proposed DivDR during training. As

Cluster #1

Cluster #2

Cluster #3

Fig. 5. Visualization of images from the validation set of MS-COCO 2017 [24] challenge. In this training $K = 3$ and we visualize the top-5 images that fall closest to their respective cluster centers μ_i. Note that the dataset does not provide subset-level annotations, however our method uses different pathways to process images containing meals (*top row*), objects with wheels and outdoor scenes (*middle row*) and electronic devices (*bottom row*).

presented in Tables 4 and 5, we conduct experiments with two different settings, namely without and with computational cost constraints. We illustrate the overall improvement over DR [21] across various hyper-parameters in Fig. 8.

Detection. Given no computational constraints, DivDR attains 38.1% mAP with 32.9 GFLOPs as opposed to 37.7% mAP for DR-R. While the average precision is similar, we observe a noticeable gain computational reduction of 5.3 GFLOPs. Compared with the ResNet-50-FPN for backbone, DivDR achieves similar performance but a small gain of 0.2% but with half of the GFLOPs (32.9 GFLOPs vs. 95.7 GFLOPs). When we introduce the computational regularization, the cost is reduced to 19.8 GFLOPs while the performance is preserved with 35.4% mAP. Compared with that in DR-A, we observe that while Div-DR constraibntconstrainted enjoys a 1.1 lower GLOPS, it enjoys improved precision of 3.3% (35.4% mAP vs. 32.1% mAP) with a lower standard deviation. We believe that this is due to the local experts learnt for separate subsets of the data.

Instance Segmentation. As for the task of instance, as observed in Table 5, unconstrainedperforms similarly to DR-R with 35.1% mAP. However, DivDR better trades-off the GLOPs with with a 32.9 GFLOPs in the unconstrained regime as opposed to 38.2 GLOPS. This is similar to the observations made in the detection experiments. Moreover, when computational constraints are introduced, DivDR enjoys a similar GLOPs as DR-A but with an improved 1.6% precision (33.4% mAP vs. 31.8% mAP).

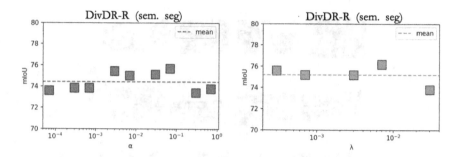

Fig. 6. Ablation on the α (*left*) and λ_2 (*right*) parameter of the diversity loss term for Semantic Segmentation. The *mean* accuracy in case of the parameter sweep for λ_2 is higher since in each case the best performing α was used for the training. We can see that the method is stable regardless the choice of the parameters over various tasks.

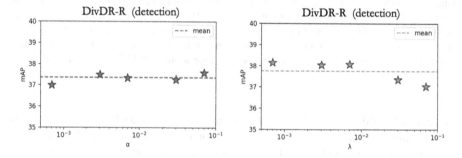

Fig. 7. Ablation on the α (*left*) and λ_2 (*right*) parameter of the diversity loss term for Object Detection. We can see that the method is stable regardless the choice of the parameters over various tasks.

Ablating K. We compare the performance of our proposed DivDR under different choices of the number of clusters K over the gate activation for both unconstrained and constrained computational constraints, i.e. DivDR-A and DivDR-R respectively. We note that our proposed $\mathcal{L}_{\text{DivDr}}$ effectively clusters the gate activation cluster centers as shown in Fig. 4. Moreover, we also observe that our proposed loss not only results in separated clusters of local experts, but also with a small intra-cluster distances as shown in Table 3. In particular, we observe that our proposed DivDR results in larger inter-cluster distances that are larger than the intra-cluster distances (in contrast with DR [21]).

Ablating α and λ_2. As shown in Fig. 7, we observe the choice of both α and λ_2 only marginally affect the performance of DivDR-A in terms of both mAP on the object detection task. However, we find that $\lambda_2 > 0.5$ starts to later affect the mAP for reduced computation.

Table 3. Quantitative comparison of Dynamic Routing [21] trained without the objective to diversify the paths and using various K for the clustering term. We omit $K = 1$ from our results as it reverts to forcing the model to use the same architecture, independent of the input image. Instead we report the baseline scores from [21] For comparison we report best Dynamic Routing [21] scores from 3 identical runs with different seeds.

(a) DivDR-A					(b) DivDR-Raw			
K	**mAP$_{val}$**	**GFLOPs**	**Inter**	**Intra**	**K**	**mAP$_{val}$**	**GFLOPs**	**Inter**	**Intra**
*	34.6	23.2	0.2	0.3	*	37.8	38.2	0.5	0.7
2	**35.1**	21.9	1.1	0.4	2	36.5	**31.0**	0.6	0.5
3	35.0	**19.2**	0.8	0.3	3	37.4	32.6	1.2	0.5
4	34.9	20.0	0.6	0.1	4	**38.1**	32.8	0.7	0.2

Table 4. Comparison with baselines on the COCO [24] **detection** validation set. * Scores from [21] were reproduced using the official implementation. The evaluation settings are identical to [34] with single scale. We calculate the average FLOPs with 800×800 size input

Method	Backbone	mAP$_{val}$	GFLOPs
Faster R-CNN [34]	ResNet-50-FPN	37.9	88.4
DR-A [21]*	Layer17	32.1 ± 5.0	20.9 ± 2.1
DR-B [21]*	Layer17	36.5 ± 0.2	24.4 ± 1.2
DR-C [21]*	Layer17	37.1 ± 0.2	26.7 ± 0.4
DR-R [21]*	Layer17	37.7 ± 0.1	38.2 ± 0.0
DivDR-A	Layer17	35.4 ± 0.2	19.8 ± 1.0
DivDR-R	Layer17	38.1 ± 0.0	32.9 ± 0.1

5   Discussion and Future Work

In this paper we demonstrate the superiority of networks trained on a subset of the training set holding similar properties, which we refer to as *local experts*. We address the two main challenges of training and employing local experts in real life scenarios, where subset labels are not available during test nor training time. Followed by that, we propose a method, called Diversified Dynamic Routing that is capable of jointly learning local experts and subset labels without supervision. In a controlled study, where the subset labels are known, we showed that we can recover the original subset labels with 98.2% accuracy while maintaining the performance of a hypothetical *Oracle* model in terms of both accuracy and efficiency.

To analyse how well this improvement translates to real life problems we conducted extensive experiments on complex computer vision tasks such as segmenting street objects on images taken from the driver's perspective, as well as

Table 5. Comparison with baselines on the COCO [24] **segmentation** validation set. * Scores from [21] were reproduced using the official implementation. The evaluation settings are identical to [34] with single scale. We calculate the average FLOPs with 800×800 size input

Method	Backbone	mAP_{val}	GFLOPs
Mask R-CNN [34]	ResNet-50-FPN	35.2	88.4
DR-A [21]*	Layer17	31.8 ± 3.1	23.7 ± 4.2
DR-B [21]*	Layer17	33.9 ± 0.4	25.2 ± 2.3
DR-C [21]*	Layer17	34.3 ± 0.2	28.9 ± 0.7
DR-R [21]*	Layer17	35.1 ± 0.2	38.2 ± 0.1
DivDR-A	Layer17	33.4 ± 0.2	24.5 ± 2.3
DivDR-R	Layer17	35.1 ± 0.1	32.9 ± 0.2

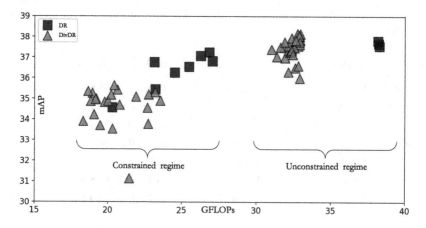

Fig. 8. Evaluations of models trained on COCO [24] across different hyper-parameters

detecting common objects in both indoor and outdoor scenes. In each scenario we demonstrate that our method outperforms Dynamic Routing [21].

Even though this approach is powerful in a sense that it could improve on a strong baseline, we are aware that the clustering method still assumes subsets of *equal* and more importantly *sufficient* size. If the dataset is significantly imbalanced w.r.t. local biases the K-means approach might fail. One further limitation is that if the subsets are too small for the *local experts* to learn generalizable representations our approach might also fail to generalize. Finally, since the search space of the architectures in this work is defined by Dynamic Routing [21] which is heavily focused on scale-variance. We believe that our work can be further generalized by analyzing and resolving the challenges mentioned above.

Acknowledgement. We thank Hengshuang Zhao for the fruitful discussions and feedback. This work is supported by the UKRI grant: Turing AI Fellowship EP/W002981/1 and EPSRC/MURI grant: EP/N019474/1. We would also like to thank the Royal Academy of Engineering. Botos Csaba was funded by Facebook Grant Number DFR05540.

References

1. Carion, N., Massa, F., Synnaeve, G., Usunier, N., Kirillov, A., Zagoruyko, S.: End-to-end object detection with transformers. In: Vedaldi, A., Bischof, H., Brox, T., Frahm, J.-M. (eds.) ECCV 2020. LNCS, vol. 12346, pp. 213–229. Springer, Cham (2020). https://doi.org/10.1007/978-3-030-58452-8_13
2. Caron, M., Bojanowski, P., Joulin, A., Douze, M.: Deep clustering for unsupervised learning of visual features. In: Ferrari, V., Hebert, M., Sminchisescu, C., Weiss, Y. (eds.) Computer Vision – ECCV 2018. LNCS, vol. 11218, pp. 139–156. Springer, Cham (2018). https://doi.org/10.1007/978-3-030-01264-9_9
3. Chen, L.C., et al.: Searching for efficient multi-scale architectures for dense image prediction. arXiv:1809.04184 (2018)
4. Chen, L.C., Papandreou, G., Kokkinos, I., Murphy, K., Yuille, A.L.: DeepLab: semantic image segmentation with deep convolutional nets, atrous convolution, and fully connected CRFs. IEEE Transactions on Pattern Analysis and Machine Intelligence (2017)
5. Chen, L.C., Papandreou, G., Schroff, F., Adam, H.: Rethinking atrous convolution for semantic image segmentation. arXiv:1706.05587 (2017)
6. Chen, L.-C., Zhu, Y., Papandreou, G., Schroff, F., Adam, H.: Encoder-decoder with atrous separable convolution for semantic image segmentation. In: Ferrari, V., Hebert, M., Sminchisescu, C., Weiss, Y. (eds.) ECCV 2018. LNCS, vol. 11211, pp. 833–851. Springer, Cham (2018). https://doi.org/10.1007/978-3-030-01234-2_49
7. Chopra, S., Hadsell, R., LeCun, Y.: Learning a similarity metric discriminatively, with application to face verification. In: IEEE Conference on Computer Vision and Pattern Recognition (2005)
8. Cordts, M., et al.: The cityscapes dataset for semantic urban scene understanding. In: IEEE Conference on Computer Vision and Pattern Recognition (2016)
9. Ding, M., et al.: HR-NAS: searching efficient high-resolution neural architectures with lightweight transformers. In: Proceedings of the IEEE/CVF Conference on Computer Vision and Pattern Recognition (2021)
10. Girshick, R.: Fast R-CNN. In: IEEE International Conference on Computer Vision (2015)
11. Girshick, R., Donahue, J., Darrell, T., Malik, J.: Rich feature hierarchies for accurate object detection and semantic segmentation. In: IEEE Conference on Computer Vision and Pattern Recognition (2014)
12. Gou, J., Yu, B., Maybank, S.J., Tao, D.: Knowledge distillation: a survey. Int. J. Comput. Vis. **129**(6), 1789–1819 (2021). https://doi.org/10.1007/s11263-021-01453-z
13. Hadsell, R., Chopra, S., LeCun, Y.: Dimensionality reduction by learning an invariant mapping. In: IEEE Conference on Computer Vision and Pattern Recognition (2006)
14. He, K., Gkioxari, G., Dollár, P., Girshick, R.: Mask R-CNN. In: Proceedings of the IEEE international conference on computer vision (2017)

15. Hinton, G., Vinyals, O., Dean, J., et al.: Distilling the knowledge in a neural network. arXiv preprint arXiv:1503.02531 (2015)
16. Huang, Z., Wang, X., Huang, L., Huang, C., Wei, Y., Liu, W.: CCNet: Criss-cross attention for semantic segmentation. arXiv:1811.11721 (2018)
17. Jang, E., Gu, S., Poole, B.: Categorical reparameterization with gumbel-softmax. In: International Conference on Learning Representations (2017)
18. Kirillov, A., Girshick, R., He, K., Dollár, P.: Panoptic feature pyramid networks. In: IEEE Conference on Computer Vision and Pattern Recognition (2019)
19. Koch, G., Zemel, R., Salakhutdinov, R.: Siamese neural networks for one-shot image recognition. In: International Conference on Machine Learning Deep Learning Workshop (2015)
20. Li, X., Zhou, Y., Pan, Z., Feng, J.: Partial order pruning: for best speed/accuracy trade-off in neural architecture search. In: IEEE Conference on Computer Vision and Pattern Recognition (2019)
21. Li, Y., et al.: Learning dynamic routing for semantic segmentation. In: IEEE Conference on Computer Vision and Pattern Recognition (2020)
22. Lin, T.Y., Dollár, P., Girshick, R., He, K., Hariharan, B., Belongie, S.: Feature pyramid networks for object detection. In: IEEE Conference on Computer Vision and Pattern Recognition (2017)
23. Lin, T.Y., Goyal, P., Girshick, R., He, K., Dollár, P.: Focal loss for dense object detection. In: IEEE International Conference on Computer Vision (2017)
24. Lin, T.-Y., et al.: Microsoft COCO: common objects in context. In: Fleet, D., Pajdla, T., Schiele, B., Tuytelaars, T. (eds.) ECCV 2014. LNCS, vol. 8693, pp. 740–755. Springer, Cham (2014). https://doi.org/10.1007/978-3-319-10602-1_48
25. Liu, C., et al.: Auto-deepLab: hierarchical neural architecture search for semantic image segmentation. In: IEEE Conference on Computer Vision and Pattern Recognition (2019)
26. Liu, H., Simonyan, K., Yang, Y.: Darts: differentiable architecture search. arXiv preprint arXiv:1806.09055 (2018)
27. Long, J., Shelhamer, E., Darrell, T.: Fully convolutional networks for semantic segmentation. In: IEEE Conference on Computer Vision and Pattern Recognition (2015)
28. Van der Maaten, L., Hinton, G.: Visualizing data using t-SNE. J. Mach. Learn. Res. 9(86), 2579–2605 (2008)
29. MacQueen, J., et al.: Some methods for classification and analysis of multivariate observations. In: The fifth Berkeley Symposium on Mathematical Statistics and Probability (1967)
30. Mnih, V., et al.: Playing atari with deep reinforcement learning. In: Neural Information Processing Systems Deep Learning Workshop (2013)
31. Mullapudi, R.T., Mark, W.R., Shazeer, N., Fatahalian, K.: Hydranets: specialized dynamic architectures for efficient inference. In: IEEE Conference on Computer Vision and Pattern Recognition (2018)
32. Nekrasov, V., Chen, H., Shen, C., Reid, I.: Fast neural architecture search of compact semantic segmentation models via auxiliary cells. In: IEEE Conference on Computer Vision and Pattern Recognition (2019)
33. Redmon, J., Divvala, S., Girshick, R., Farhadi, A.: You only look once: unified, real-time object detection. In: IEEE Conference on Computer Vision and Pattern Recognition (2016)
34. Ren, S., He, K., Girshick, R., Sun, J.: Faster R-CNN: towards real-time object detection with region proposal networks. In: Advances in Neural Information Processing Systems (2015)

35. Rippel, O., Paluri, M., Dollar, P., Bourdev, L.: Metric learning with adaptive density discrimination. arXiv:1511.05939 (2015)
36. Shaw, A., Hunter, D., Landola, F., Sidhu, S.: Squeezenas: fast neural architecture search for faster semantic segmentation. In: Proceedings of the IEEE/CVF International Conference on Computer Vision Workshops (2019)
37. Song, L., et al.: Learnable tree filter for structure-preserving feature transform. In: Advances in Neural Information Processing Systems (2019)
38. Tan, M., Le, Q.: Efficientnet: rethinking model scaling for convolutional neural networks. In: International Conference on Machine Learning (2019)
39. Tian, Z., Shen, C., Chen, H., He, T.: FCOS: fully convolutional one-stage object detection. In: IEEE Conference on Computer Vision and Pattern Recognition (2019)
40. Veit, A., Belongie, S.: Convolutional Networks with Adaptive Inference Graphs. In: Ferrari, V., Hebert, M., Sminchisescu, C., Weiss, Y. (eds.) ECCV 2018. LNCS, vol. 11205, pp. 3–18. Springer, Cham (2018). https://doi.org/10.1007/978-3-030-01246-5_1
41. Wan, L., Zeiler, M., Zhang, S., Le Cun, Y., Fergus, R.: Regularization of neural networks using dropconnect. In: International Conference on Machine Learning. PMLR (2013)
42. Wang, X., Yu, F., Dou, Z.-Y., Darrell, T., Gonzalez, J.E.: SkipNet: learning dynamic routing in convolutional networks. In: Ferrari, V., Hebert, M., Sminchisescu, C., Weiss, Y. (eds.) ECCV 2018. LNCS, vol. 11217, pp. 420–436. Springer, Cham (2018). https://doi.org/10.1007/978-3-030-01261-8_25
43. Weinberger, K.Q., Saul, L.K.: Distance metric learning for large margin nearest neighbor classification. J. Mach. Learn. Res. 10, 207–244 (2009)
44. Wu, Z., et al.: BlockDrop: dynamic inference paths in residual networks. In: IEEE Conference on Computer Vision and Pattern Recognition (2018)
45. Xie, E., Wang, W., Yu, Z., Anandkumar, A., Alvarez, J.M., Luo, P.: SegFormer: simple and efficient design for semantic segmentation with transformers. In: Advances in Neural Information Processing Systems (2021)
46. Xie, Q., Luong, M.T., Hovy, E., Le, Q.V.: Self-training with noisy student improves imagenet classification. In: Proceedings of the IEEE/CVF Conference on Computer Vision and Pattern Recognition (2020)
47. You, Z., Yan, K., Ye, J., Ma, M., Wang, P.: Gate decorator: global filter pruning method for accelerating deep convolutional neural networks. arXiv:1909.08174 (2019)
48. Yu, C., Wang, J., Peng, C., Gao, C., Yu, G., Sang, N.: BiSeNet: bilateral segmentation network for real-time semantic segmentation. In: Ferrari, V., Hebert, M., Sminchisescu, C., Weiss, Y. (eds.) ECCV 2018. LNCS, vol. 11217, pp. 334–349. Springer, Cham (2018). https://doi.org/10.1007/978-3-030-01261-8_20
49. Zhao, H., Shi, J., Qi, X., Wang, X., Jia, J.: Pyramid scene parsing network. In: IEEE Conference on Computer Vision and Pattern Recognition (2017)
50. Zhao, H., et al.: PSANet: point-wise spatial attention network for scene parsing. In: Ferrari, V., Hebert, M., Sminchisescu, C., Weiss, Y. (eds.) ECCV 2018. LNCS, vol. 11213, pp. 270–286. Springer, Cham (2018). https://doi.org/10.1007/978-3-030-01240-3_17
51. Zheng, S., et al.: Rethinking semantic segmentation from a sequence-to-sequence perspective with transformers. In: IEEE Conference on Computer Vision and Pattern Recognition (2021)
52. Zhou, X., Wang, D., Krähenbühl, P.: Objects as points. arXiv:1904.07850 (2019)

53. Zhu, X., Su, W., Lu, L., Li, B., Wang, X., Dai, J.: Deformable DETR: deformable transformers for end-to-end object detection. In: International Conference on Learning Representations (2020)
54. Zoph, B., Le, Q.V.: Neural architecture search with reinforcement learning. arXiv preprint arXiv:1611.01578 (2016)

Author Index